THIRD EDITION

MARKETING OF HIGH-TECHNOLOGY PRODUCTS AND INNOVATIONS

Jakki Mohr

University of Montana

Sanjit Sengupta

San Francisco State University

Stanley Slater

Colorado State University

Boston   Columbus   Indianapolis   New York   San Francisco   Upper Saddle River
Amsterdam   Cape Town   Dubai   London   Madrid   Milan   Munich   Paris   Montreal   Toronto
Delhi   Mexico City   Sao Paulo   Sydney   Hong Kong   Seoul   Singapore   Taipei   Tokyo

Editorial Director: Sally Yagan
Acquisitions Editor: James Heine
Product Development Manager: Ashley Santora
Editorial Project Manager: Melissa Pellerano
Editorial Assistant: Karin Williams
Director of Marketing: Patrice Lumumba Jones
Marketing Manager: Anne Fahlgren
Marketing Assistant: Susan Osterlitz
Marketing Assistant: Ian Gold
Senior Managing Editor: Judy Leale
Project Manager: Becca Richter
Senior Operations Supervisor: Arnold Vila
Operations Specialist: Benjamin Smith
Manager, Rights and Permissions: Charles Morris
Art Director: Jayne Conte

Cover Designer: Ann Marie Sole
Cover Art: Ann Marie Sole
**Manager, Cover Visual Research
  & Permissions:** Karen Sanatar
Lead Media Project Manager: Denise Vaughn
Media Manager: Lisa Rinaldi
Full-Service Project Management: Thistle Hill
  Publishing Services, LLC
Composition: Aptara Corp.
Printer/Binder: Hamilton Printing Co.
Cover Printer: Lehigh-Phoenix Color/Hagerstown
Text Font: 10/12 Times

Credits and acknowledgments borrowed from other sources and reproduced, with permission, in this textbook appear on appropriate page within text.

If you purchased this book within the United States or Canada you should be aware that it has been wrongfully imported without the approval of the Publisher or the Author.

10 9 8 7 6 5 4 3 2
ISBN 10:   0-13-136491-X
ISBN 13: 978-0-13-136491-2

To the lights in my life that make me smile: Willy and Claire. And to Michael, who also makes me smile.

  Jakki Mohr

To Amrita, Ishaan, and Ila:
For the gift of family, and the support to keep it humming.

  Sanjit Sengupta

To Paula, my bedrock, and to my mother, Anne Fant Slater, my intellectual inspiration.

  Stan Slater

To the people who see the possibilities,
  and who, sometimes with courage, sometimes with faith and hope—
    but always with effort, perseverance, and energy
      (despite self-doubt)—
        strive to make the possible become reality.
        Dreams can come true.

CONTENTS

PREFACE

What an exciting time to be involved in high-technology products and innovations! With this third edition of the *Marketing of High-Technology Products and Innovations,* we invite you to consider the enormous changes in the technology arena in just the last few years:

- Since January of 2005 the NASDAQ exchange saw a low of 1,908 in April 2005, a high of 2,810 in November of 2007, and sat at 1,532 in December of 2008.
- In 2008, purchases of computer hardware and software accounted for half of all capital spending by businesses.
- In the area of consumer electronics and technology, the most innovative products aim to change the way people purchase media and entertainment—whether in the living room, the den, or on the go.
- In the Internet arena, second-generation Web technologies (Web 2.0) such as social networking and the blogosphere have exploded. Increasingly the Web is dominated by user-generated content—increasing connectivity among communities of individuals and opening myriad possibilities for business models and marketing techniques.
- The mobile phone is revolutionizing the way people around the world gain access to information, offering the potential to mitigate the digital divide that separates technology "haves" from the "have-nots."
- Firms have greater access to the analytics necessary to optimize their online businesses; search engine optimization has moved from art to science; and websites have become much easier to navigate.
- Companies and retailers continue to harmonize distribution channels, offering customers bricks-and-clicks models for a seamless shopping experience via a multitude of channel choices.
- New technological developments in the areas of nanotechnology, biotechnology, radio-frequency identification chips, green technology, and robotics, to name just a few, are creating exciting new possibilities for a wide range of industry applications.

In addition, the past few years have seen significant challenges including soaring fuel and food costs, climate change, ongoing poverty and human needs in developing countries, and an unstable global political environment. Solving these pressing problems requires a multifaceted approach, with a key aspect being the development of new technologies. New and established high-tech companies alike are offering revolutionary solutions for transportation and efficient energy generation. Social entrepreneurs and businesses of all sizes are pioneering new products and business models, even for base-of-the-pyramid markets (comprised of individuals who have low disposable incomes). Driving companies' business strategies is an increasing focus on the triple bottom line of people, profits, and planet. While the challenges are certainly daunting, it is an exciting time to be in tech!

Yet, despite their promise, all too many high-tech inventions do not achieve commercial success. One reason is that high-tech companies often fail to develop a competency in marketing. Many of these companies—whose origins are typically found in scientific or engineering developments—mistakenly believe that marketing is superfluous, that the superiority of their technological innovations will be sufficient to convince buyers to adopt their products. Technical people have a hard time becoming customer oriented and market focused. Yet, successful commercialization of technology requires an intimate understanding of customers—their underlying needs and problems, their ways of doing business, and their environments and mind-sets. Because technological superiority alone is insufficient for ensuring the success of high-tech products, high-technology companies must develop their marketing competencies. Thriving in the high-tech marketplace requires mastery of a diverse set of marketing skills and capabilities. High-tech marketing managers must be versatile yet focused, flexible yet determined, tenacious yet open-minded. Their skills must include adroitly reading market trends; investing wisely in future technologies; leveraging the skills and capabilities of technical and marketing personnel in a dynamic, interactive fashion; understanding customers intimately; offering a compelling value proposition; developing astute marketing campaigns; pricing with an eye to customer value; and harmonizing distribution channels and supply chains.

Hence, readers of this book will learn best-practices high-technology marketing. Filled with decision frameworks and strategies that reflect cutting-edge research and practice, the material in this book is supported with a plethora of global examples and applications.

Rather than addressing marketing fundamentals, the book's primary focus is on the unique characteristics of the high-tech environment and the marketing challenges those characteristics pose. It focuses primarily on the *marketing of technology* and innovation, but also addresses the *use of technology* for marketing purposes (such as search engine positioning and new media). Related books on the *management of technology and innovation* might also be useful complements. Because of the advanced nature of this material (for a marketing novice), a book of marketing fundamentals should be used as a reference as well.

TARGET AUDIENCE FOR THIS BOOK

The book will prove useful in a variety of settings, including:

- Upper-level undergraduate and graduate courses on the marketing of high-tech products and innovations
- Technology institutes, engineering management programs, biotechnology centers, and/or telecommunications programs
- Executive education courses
- Management teams in high-tech firms
- Training programs in high-tech firms
- Technology incubators

The book has a global focus, and editions for various international markets may also be available.

APPROACH

As in the second edition, our primary aim in this third edition is to show how marketing strategies and programs must be modified and adapted for the high-tech environment. First and most basic, marketing high-technology products and innovations is not the same as marketing traditional products and services. For example, the marketing of a familiar consumer product—say, Coca-Cola—is very different from marketing products that customers may be unfamiliar with—say, the Amazon Kindle electronic book reader, the Intel Core 2 Duo computer chip, voice recognition software, or even a new computer videogame console. Customers' fear, uncertainty, and doubt about how to use and enjoy the full benefits of these high-tech products contribute to the need for different marketing considerations. In addition, there is a competitive environment in high-tech industries not found in more traditional contexts. Industry outsiders often introduce disruptive innovations that take industry incumbents by surprise. And always contributing to high-tech marketing challenges is the velocity of technological change.

Given the high degree of uncertainty in high-tech markets, the margin for error in decision making is smaller than for conventional markets. As a result, high-tech firms must execute basic marketing principles flawlessly. Knowing how to select the appropriate target market, how to clearly communicate the benefits of the innovation relative to other solutions, how to design an effective and efficient distribution channel, and how to develop solid relationships and alliances are critical marketing competencies.

In addition to basic marketing competencies, high-tech firms must also understand the nuances of high-tech environments that require adaptations to marketing basics. The complex, high-velocity high-tech environment necessitates modifications in, for example, marketing research tools, supply chain management, and strategy formulation. Grounding marketing decisions in the unique characteristics of high-tech environments helps to reduce the inherent risks and uncertainties.

In light of this complicated and challenging environment, we offer a systematic, thorough overview of the issues that high-tech marketers must address for both small and large businesses. The book provides a balance between conceptual discussions and examples, start-up and established business, products and services, and consumer and business-to-business marketing contexts. Using examples from a variety of industries and technologies to illustrate marketing tools and concepts, the book not only captures the richness of the high-tech environment, but also proves the utility of the frameworks presented. This variety also gives the reader experience in applying the frameworks to diverse situations. Some of the industries and contexts covered include cellular phones, information technology (hardware and software), the Internet and related technologies such as Web 2.0, biotechnology, and entertainment and consumer electronics including high-definition TV, MP3 players, and videogame players.

Moreover, given today's global challenges, we weave the theme of technology used to solve global problems into each chapter. For example, in Chapter 1, we introduce the important concepts of *base-of-the-pyramid markets*—markets composed of a large number of individuals who have low disposable incomes and face very different needs from the developed markets—and *corporate social responsibility,* in which companies' strategies provide not only economic profits but also social profits (through solving social problems), say, by introducing "green" or environmentally friendly technologies.

Finally, marketing does not occur in isolation in any firm, but rather is cross-functional in nature. This book brings together marketing with other business disciplines (for example, research and development, legal, and management and strategy) to offer insights on how marketing is dependent on cross-functional interactions.

ORGANIZATION

In contrast to the 12 chapters of the second edition, this revised edition of the *Marketing of High-Technology Products and Innovations* includes 13 chapters plus an end-of-book section with seven mini-cases.

Two rather long chapters in the second edition have been subdivided into distinct components. First, "Strategy and Corporate Culture in High-Tech Companies," which had been a single chapter in the second edition, now is covered in two chapters. Second, the topic of brand management (including pre-announcements), which had been included as part of the chapter on advertising and promotion in the second edition, is now covered in a chapter of its own. Also, while the second edition's Chapter 11 on "E-Business, E-Commerce, and the Internet" seems to be missing from this third edition, the content is not. More than 10 years have passed since the advent of the Web, which is now widely accepted as a key business and marketing tool. The new developments in this area have been integrated into all the chapters of the third edition.

Each chapter has been substantially revised to:

- Cover new developments in the marketing field (such as customer co-innovation and co-creation)
- Introduce new topics to the marketing field (such as biomimicry)
- Incorporate the most recent academic research
- Provide up-to-date examples
- Incorporate the themes of technology used to solve global problems, such as base-of-the-pyramid markets, energy, transportation, and health care

Table P.1 provides an overview of the key changes to content in the third edition on a chapter-by-chapter basis. Note that this table addresses changes only and does not include an exhaustive itemization of topics covered in each chapter; the Table of Contents provides a thorough overview of all the major topics covered in each chapter.

Table P.1 Chapter Updates for Third Edition*

Chapter 1: Introduction to the World of High-Technology Marketing
- Provides more thorough overview of the role of marketing in the high-tech company
- Offers a broader presentation of the types of innovation (e.g., disruptive versus sustaining innovations, architectural versus modular innovations)
- Introduces key topics such as technology life cycles and industry standards (dominant designs) earlier in the text
- Previews new topics including corporate social responsibility, base-of-the-pyramid markets, and the intersection of technology and solving global problems (energy technologies, health care, One Laptop Per Child, etc.)

Chapter 2: Strategic Market Planning in High-Tech Firms
- Covers value propositions—the foundation for competitive strategy—more explicitly, including the pros and cons of three types of value propositions (the "All Benefits," the "Favorable Points of Difference," and the "Resonating Focus" value propositions)
- Provides thorough coverage of four archetypal strategies: the Product Leader (Prospector), the Fast Follower (Analyzer), the Differentiated Defender (Customer Intimate), and the Low-Cost Defender (Operationally Excellent)
- Discusses how to develop a marketing performance measurement system, including marketing dashboards, to provide specific feedback on what is working well and what needs to be modified

Chapter 3: Culture and Climate Considerations for High-Tech Companies
- In recognition of the importance of the topic, devotes the entire chapter to culture and the organization's shared values of innovativeness
- Gives greater consideration to the characteristics of a culture of innovation, and the facilitators and obstacles to maintaining a culture of innovativeness

Chapter 4: Market Orientation and Cross-Functional (Marketing–R&D) Interaction
- Provides the latest evidence of how being market oriented leads to superior performance
- Differentiates between responsive (market-driven) and proactive (market-driving) dimensions of market orientation
- Includes questions for companies to assess their performance on each dimension of market orientation
- Integrates the section on new product development teams (which appeared in a later chapter in the second edition) into the section on cross-functional interactions
- Provides more comprehensive discussion of the benefits of, facilitating conditions for, and challenges to the effectiveness of cross-functional development teams
- Continues the focus on the marketing–R&D interaction as crucial in high-tech companies

Chapter 5: Partnerships, Alliances, and Customer Relationships
- Includes a broader presentation of the types of alliances, including, for example, alliances for new product development, industry clusters, and networks used for an open innovation model
- Provides extended coverage of the issues in managing outsourcing in high-tech companies
- Modifies the section on managing customer relationships to reflect best-practices policies in this area, including managing customers as investments to maximize customer equity

Chapter 6: Marketing Research in High-Tech Markets
- Provides benchmarking data on market research spending by a variety of industries and companies

---

*Please see the Table of Contents for the full listing of topics covered in each chapter. This table merely identifies new material and key changes appearing in the third edition.

- Provides tables with supporting "how-to" detail for each marketing research technique
- Provides thorough coverage of two new topics:
  - Customer-driven innovation and customer co-creation
  - Biomimicry (i.e., innovation inspired by nature)
- Adds coverage of an important quantitative forecasting model, the Bass diffusion model

Chapter 7: Understanding High-Tech Customers

- Highlights the important role of product design on consumers' product evaluation decisions
- Adds new information on product end-of-life (cradle-to-grave) decisions such as disposal and recycling
- In light of the latest research on topics such as feature fatigue, provides an extended discussion of complexity as a factor affecting adoption of innovation
- Offers improved structure to the section titled "Crossing the Chasm," with better discussion on identifying a beachhead, developing a whole product for adjacent markets, and other important chasm considerations (such as developing an industry standard to reduce customer risk, positioning against competition, simplifying the product by reducing features, and beefing up customer service)
- Adds a new discussion of vertical and horizontal market segments as segmentation alternatives
- Describes new tools for positioning: multiattribute models and perceptual maps
- Covers a new factor that affects customer migration decisions: the positioning and pricing of upgrades
- Concludes with the paradoxical relationship consumers have with technology and implications for high-tech marketing (covered in a later chapter in the second edition)

Chapter 8: Technology and Product Management

- Contains new material on the opportunities and challenges for companies presented by digital convergence trends
- Provides more thorough coverage of the pros and cons of choosing a *whole product* strategy (complete end-to-end solution) with a company's own proprietary platform, versus offering a more component-based/modular approach with an open platform for third-party providers to leverage
- Includes a new section on "The Role of Product Management in the High-Tech Company"
- Offers a major rewrite of the section titled "Intellectual Property Considerations"

Chapter 9: Distribution Channels and Supply Chain Management in High-Tech Markets

- Offers more extensive coverage of company-owned direct distribution channels, including:
  - The company's sales force (including an overview of the interactions between Marketing and Sales)
  - Company-owned retail outlets
  - Sales over the company's own website, and issues related to cannibalization and channel conflict
- Provides a new section on multichannel marketing
- In a new section titled "Emerging Considerations in Distribution Channels," covers new topics such as the distribution for digital goods and considerations related to "the long tail," as well as the unique distribution needs for base-of-the-pyramid markets
- Considers four types of supply chains that are matched to the different levels and combinations of demand and supply uncertainties: efficient supply chains, risk-hedging supply chains, responsive supply chains, and agile supply chains
- Explores a variety of technologies used to support best-practices supply chain management, including online platforms such as electronic marketplaces (hubs or exchanges) and e-procurement, reverse auctions, the available software for supply chain management including sales and operations planning (S&OP) software and service-oriented architectures, and RFID technologies
- Discusses the emerging trend in supply chain management called *green supply chains,* which consider environmental impacts of all decisions from sourcing to transportation to consumption and end-of-life disposal

Continued

Table P.1 (Continued)

Chapter 10: Pricing Considerations in High-Tech Markets

- Presents the concept of experience curves and how they affect pricing of high-tech durables
- Discusses cross-price elasticity of demand and how competition from other technologies affects pricing decisions
- Provides explicit examples of life-cycle costing (total cost of ownership) for a variety of types of products, including CRM software and environmentally friendly consumer durables
- Updates the section on the technology pricing paradox, in which companies must find new models for profitability given the inexorable downward pressure on price—which at the extreme means technology is free; also weaves in cutting-edge material from the popular business press
- Includes other new topics such as reference price, smart (or dynamic) pricing, and price promotions

Chapter 11: Marketing Communication Tools for High-Tech Markets

- Discusses all the traditional advertising and promotion tools with a "new media" spin
- Offers a new section titled "Internet Advertising and Promotion," grounded in permission-based marketing, that includes new media techniques for behavioral targeting, search engine optimization, Web 2.0 tools for marketing purposes (e.g., social networking, tagging, blogs, podcasting, video and virtual worlds, the role of widgets and gadgets), and mobile advertising (including location-based services)
- Covers the ways that the traditional media are dealing with the disruption posed by new media
- Covers the material on website design (previously in the e-commerce chapter in the prior edition), with an eye to the role of a company's website in its advertising and promotion mix
- Offers expanded coverage of Web analytics and ways to measure the effectiveness of a website

Chapter 12: Strategic Considerations in Marketing Communications

- Offers expanded coverage of two strategic issues in marketing communications: strategic brand management and new product pre-announcements
- Adds a new section on the benefits and risks of branding to companies and customers
- Covers new brand-building strategies—including harnessing Web 2.0 technologies and new media, thinking strategically about corporate social responsibility, and utilizing effective internal branding
- Continues the coverage on ingredient brand strategies
- Updates the section on product pre-announcements with the latest academic research and plenty of new and interesting business examples

Chapter 13: Strategic Considerations for the Triple Bottom Line in High-Tech Companies

- This new chapter incorporates the latest writings on corporate social responsibility, including the triple bottom line (profits, people, planet)
- Addresses the research findings regarding the relationship between corporate social performance and company financial performance
- Includes sections on:
  - The intersection of company strategy and environmental considerations
  - Social entrepreneurship and base-of-the-pyramid strategies
- Updates the material on the digital divide with a global focus

End-of-Book Mini-Cases

- Offers seven cases allowing more hands-on experience and in-depth analysis of key concepts from the text; also facilitates critical thinking skills

In addition, this third edition, based on feedback and reviews of the second edition, has been edited to make it more reader-friendly. It also has more application-oriented material linking chapter concepts to hands-on, real-world examples, including three special features in each chapter: detailed opening vignettes, first-person perspectives called "Technology Expert's View from the Trenches," and "Technology Solutions for Global Problems" that highlight innovations and business models to solve current pressing issues. New application material is also found in the seven mini-cases at the end of the book. These special features are detailed in the next section of this Preface. Of course, the book is also supported with a full complement of supplemental materials.

Although the material and concepts in each chapter are treated rather distinctly, much of the material, by its very nature, is interrelated and requires consideration from a multiplicity of angles. As a result, concepts are cross-referenced with treatment in other chapters. For example, the material on strategy development in high-technology organizations (Chapter 2) is part and parcel of the technology and product management chapter (Chapter 8). This approach at thematic coverage of the underlying drivers of high-tech marketing strategy offers a more comprehensive view of how these issues play out in the marketing toolkit.

SPECIAL FEATURES

In addition to the changes just noted, this new edition includes a variety of special features that capture the excitement and promise of emerging technologies. These features, designed to bring the material in each chapter to life, are summarized in the following two tables. Table P.2 provides a description of the book's special features; Table P.3 details the specific companies featured in each chapter's boxed material.

There are also discussion questions at the end of each chapter, designed to assess the reader's knowledge of the material covered and allow students to generate additional insights about the concepts. Besides the general discussion questions covering the chapter's concepts, each chapter now also includes specific application-oriented questions for readers to link the examples from the opening vignette, the Technology Expert's View from the Trenches, and the Technology Solutions for Global Problems to the chapter's concepts. Moreover, these special features can provide the opportunity for extemporaneous in class discussions.

Readers can also find useful resources at the Companion Website for the book, www .markethightech.net. The Web site provides supplementary readings (current articles from the popular business/trade press) and links to other useful Web sites.

Table P.2 **Special Features in the Third Edition**

Opening Vignettes*

Each chapter begins with a detailed opening vignette, highlighting a particular company and how it has grappled with the issues in the coming chapter. The intent is to demonstrate the relevance of the chapter material and to provide a real-world example to facilitate understanding.

Technology Expert's View from the Trenches*

Each chapter includes one or two boxed features written by technology experts—people working in the high-tech field—providing a first-person perspective on specific issues covered in the chapter. These boxes offer insights on key challenges and key success factors for high-tech marketing. The companies featured in these vignettes are shown in Table P.3.

Technology Solutions for Global Problems*

Referred to as Technology Tidbits in the prior edition, these boxes feature companies, products/technologies, and business models that solve emerging problems such as the energy crisis, global warming, or issues concerning base-of-the-pyramid markets.

Continued

Table P.2 (Continued)

End-of-Book Mini-Cases

The book now contains seven mini-cases that allow more in-depth analysis and discussion of key concepts from the text. The cases provide hands-on opportunities for students to apply the book's frameworks and concepts, and to exercise critical thinking skills. Cases include Skype, TiVo, Xerox, ESRI (GIS software), Boeing and Airbus, Goomzee Mobile Marketing, and SELCO.

Technical Appendices

Where appropriate, technical appendices provide supporting detail on chapter material, such as steps in patent procurement (Chapter 8) or Web analytics (Chapter 11).

End-of-Chapter Application Questions

In response to reviewer requests, the end-of-chapter questions now feature more application-oriented questions. Many of these tie the chapter material directly to the opening vignette, the Technology Expert's View from the Trenches, and/or the Technology Solutions for Global Problems.

*Please see Table P.3 for the specific companies featured.

Table P.3 Featured Companies

	Opening Vignette	Tech Expert's View from the Trenches	Technology Solutions for Global Problems
Chapter 1	Cars of the Future	N/A	One Laptop Per Child
Chapter 2	Medtronic	Auroras/IPTV	Dignity Toilets
Chapter 3	Google	ESRI	Star Sight
Chapter 4	Buckman Labs	Hewlett-Packard, Xilinx	Aravind Eye Hospital
Chapter 5	Apple iPhone	REI	The Sling Shot
Chapter 6	IDEO	Grupthink	bioWAVE Power
Chapter 7	RFID chips	Panasonic Mobile/Japan	Manila Water
Chapter 8	Apple iPod	Memjet Sun MicroSystems	Godisa SolarAid
Chapter 9	Cisco	Cisco	Big Boda WorldBike
Chapter 10	Apple iPhone	RightNow Tech.	Orascom Telecom
Chapter 11	Microsoft	Respond2	Lymphatic Filariasis
Chapter 12	Samsung	Foveon	IDE/Treadle Pump
Chapter 13	Eli Lilly	Intel	Nanosolar
Mini-Cases	Skype, TiVo, Xerox, ESRI, Boeing/Airbus, Goomzee Mobile Marketing, SELCO		

CONCLUSION

The world of high technology is filled with both promise and peril. Innovations across a wide spectrum of applications and industries offer tantalizing promises for the future. But, high-tech products and services are introduced in turbulent, chaotic environments where the odds of success are low. In order to successfully navigate these uncertain, turbulent markets, high-tech companies—and the students going to work in them—must know the pitfalls and mitigating factors; they must learn how to adapt and modify marketing tools and techniques for high-technology products and services. This book provides the knowledge necessary to develop a competency in best-practices high-technology marketing, providing frameworks for systematic decision making in high-tech environments that will maximize the odds of success. Welcome to the world of high-technology marketing!

ACKNOWLEDGMENTS

Although the authors' names are listed on the front of a book, it is only through the efforts of many people that a book is actually completed. First, we sincerely appreciate the time that the reviewers put in to provide us with direction for this revision:

Greg More,
  Rutgers University

Morris Pondfield,
  Towson University

Josephine Previte,
  The University of Queensland, Australia

Chatura Ranaweera,
  Wilfrid Laurier School of Business, Waterloo, Canada

James Simpson,
  University of Alabama in Huntsville

Also, one of the special features of this book—the Technology Expert's View from the Trenches—exists solely because of the generosity of the experts and their companies. We appreciate the many willing contributions of these people.

The first author, Jakki Mohr, gives special thanks to her sister, Judy Mohr, for providing expert assistance in the intellectual property sections of Chapter 8, and to Tony Ferrini for his assistance in new media marketing in Chapter 11.

We've also had much helpful assistance along the way in compiling information. Jamie Hoffman (graduate student, University of Montana) researched and wrote most of the Technology Solutions for Global Problems features. Anita Bohlert (graduate student, University of Montana) assisted in securing necessary permissions for the accompanying photographs. The support staff at the University of Montana School of Business Administration, including Heather Breckenridge, Leyli Morrow, Sarah Weatherby, and Nancy Vatoussi, provided formatting assistance with tables and figures. Sonja Grimmsmann provided capable assistance with the Instructor's Manual, and Anita Bohlert developed the PowerPoint slides.

We'd also like to thank the following Prentice Hall team: James Heine, Melissa Pellerano, Sally Yagan, Karin Williams, Becca Richter, and Judy Leale. And, our thanks are extended as well to Angela Urquhart at Thistle Hill Publishing Services.

Jakki Mohr acknowledges the supportiveness of Dean Larry Gianchetta, Department Chairs Nader Shooshtari and Jeff Shay, and her many thoughtful colleagues at the University of Montana. In addition, Jakki sincerely thanks Jeff and Martha Hamilton for their financial support. Thanks also to Sue Williams and SG Long & Company in Missoula for the use of a quiet office during the summer. I thank my family, Michael Moore and Willy and Claire Zellmer, for supporting me through the lengthy process of writing this book, during which they spent many nights and weekends enjoying life without me. Finally, I thank my students from the many schools where I have taught this course, as well as the companies with whom I have worked; without them, this process would be for naught.

Sanjit Sengupta gratefully acknowledges the participation and feedback provided by many generations of students at the different places he has taught this course: San Francisco State University, Indian School of Business, Helsinki School of Economics Executive Education, Helsinki School of Economics Mikkeli, Seoul School of Integrated Sciences and Technologies (aSSIST), Evtek Business School, and University of Alabama Huntsville. He also thanks Professor Yong Gu Ji at Yonsei University for his feedback on the previous edition of the book.

Stan Slater acknowledges his longtime friends and research colleagues, John C. Narver, professor emeritus of marketing at the University of Washington, and Eric M. Olson, professor of strategic management and marketing at the University of Colorado–Colorado Springs.

ABOUT THE AUTHORS

Jakki Mohr (Ph.D. 1989, University of Wisconsin, Madison) is the Jeff and Martha Hamilton Distinguished Faculty Fellow, Regents Professor, and professor of marketing at the University of Montana, Missoula. Prior to joining the University of Montana in the fall of 1997, Dr. Mohr was an assistant professor at the University of Colorado, Boulder (1989–1997). An award-winning teacher, Dr. Mohr teaches courses in marketing principles and the marketing of high-technology products and innovations. In addition, she has taught intensive high-tech marketing courses at many universities and executive education programs worldwide (Italy, France, Switzerland, Finland, India, Canada) and has been a keynote speaker at a number of industry trade association meetings. Dr. Mohr's research has won awards and been published in the *Journal of Marketing,* the *Strategic Management Journal,* the *Journal of Public Policy and Marketing,* the *Journal of the Academy of Marketing Science,* the *Journal of Retailing,* the *Journal of High Technology Management Research, Marketing Management,* and *Computer Reseller News.* Her interests lie primarily in the area of distribution partnerships and alliances in the marketing of high-technology products and services, focusing mainly on information and communications technologies. Before beginning her academic career, she worked in Silicon Valley in the advertising area for both Hewlett-Packard's Personal Computer Group and TeleVideo Systems. When not engaged in her academic pursuits, she pursues many outdoor activities with her family in the Rocky Mountain West, including skiing, backpacking, running, hiking, camping, fishing, and gardening.

Sanjit Sengupta (Ph.D. 1990, University of California, Berkeley) is professor of marketing at San Francisco State University. He teaches courses in strategic marketing, business-to-business marketing, and marketing of high-technology products and innovations. Prior to joining San Francisco State in the fall of 1996, Dr. Sengupta was an assistant professor at the University of Maryland, College Park, where he received two teaching awards. He has taught in many executive development programs in India, the United States, Finland, and South Korea. His research interests include new product development and technological innovation, strategic alliances, sales management, and international marketing. His research has won awards and been published in many journals including *Academy of Management Journal, Journal of Marketing, and Journal of Product Innovation Management.* Prior to his academic career, Dr. Sengupta worked in sales and marketing for Hindustan Computers Limited and CMC Limited in Bombay, India.

Stanley Slater (Ph.D. 1988, University of Washington) is the Charles and Gwen Lillis Professor of Business Administration at Colorado State University. From 1996 to 2002, he was a professor and director of the Business Administration program at the University of Washington's Bothell Campus where he was instrumental in launching an MBA program designed specifically for professionals in technology-oriented businesses. Dr. Slater's major research interests are in the areas of the role of a market orientation in organizational success and marketing's role in business strategy implementation. He has published more than 50 articles on these and other topics in the *Journal of Marketing,* the *Journal of the Academy of Marketing Science,* the *Journal of Product Innovation Management,* the *Strategic Management Journal,* and the *Academy of Management Journal,* among others. He has won "Best Paper" awards from the *International Marketing Review* and the Marketing Science Institute. Dr. Slater serves on the editorial review boards of the *Journal of Marketing,* the *Journal of the Academy of Marketing Science,* and *Industrial Marketing Management.* Prior to his academic career, he held professional and managerial positions with IBM and with the Adolph Coors Company. Dr. Slater has consulted with units of Hewlett-Packard, Johns-Manville, Monsanto, United Technologies, Cigna Insurance, Qwest, Philips Electronics, and Weyerhaeuser.

Marketing of High-Technology Products and Innovations

1

Introduction to the World of High-Technology Marketing

"Cars" of the Future

By Jamie Hoffman

In 1908, Henry Ford's famous Model T rolled off the assembly line. Within 100 years the automobile revolutionized society, putting people on wheels. Today, the average U.S. household owns 2.28 vehicles. Yet with gas prices on the rise, concerns about the environment mounting, and traffic congestion plaguing cities large and small, inventors and entrepreneurs are teaming up to transform personal automotive travel. In the very near future, popular transportation options will include personal jet-packs, flying cars, and carbon-free, stackable cars!

Photo reprinted with permission of Terrafugia, Inc., Woburn, Mass.

Futuristic Fliers

Thanks to the Terrafugia's Transition® "personal air vehicle," aggravating rush-hour traffic will become history. Brave commuters can take to the skies in a two-person light sport aircraft with automated retractable wings. Using lighter and stronger materials and more efficient engines, the vehicle aims to be classified by the FAA as the easier-to-fly light sport aircraft. Requiring only 1,500 feet to take off, the Transition will run on premium unleaded gas, fly at 120 mph, and have a range of 100–500 miles with 30 miles per gallon (mpg) in the air. On the ground, the vehicle will get 40 highway miles per gallon and 30 city mpg.[1] Other companies offer flying cars as well, such as the pioneer in this market, the Moller Skycar (www.moller.com/skycar.htm) available since 1999.

Want to fly to work, but prefer feeling the wind in your face? Try the Jet Pack T-73 created by Jet Pack International, LLC. With a range of 11 miles and maximum flying time of nine minutes, commuters can blast to work at 83 miles per hour (mph) at 250 feet above ground. The T-73 will hold 5 gallons of Jet A fuel and will retail around $200,000 (including training).[2]

(continued)

Photo reprinted with permission of Jet Pack International, LLC, Denver, Colo.

Conventional Alternatives

Now, for those who want eco-friendly transportation combined with adrenaline, but aren't quite ready to take to the skies, check out the Tesla Roadster™. The Roadster sports a base price of $109,000 and proves that a 100% electric sports car can perform just as well as the traditional models but with zero emissions. Speeding from zero to 60 mph in 3.9 seconds, the Roadster has a manual transmission, and a 248 horsepower (hp) motor with over 300 pound-feet of zero-rpm torque. The battery provides for a 220-mile range, earning the equivalent gas mileage of 256 mpg, and takes 3.5 hours to recharge.[3] Who knew being fast and being green could be achieved in one vehicle?

For those more comfortable staying grounded and keeping some change in their pockets, consider the Aptera. Designed by physicist Steve Fambro, the Aptera seats two, weighs 850 pounds, has a drag coefficient of 0.11 (compared to a typical car's drag of 0.30–0.35), maxes out at 95 mph, and impressively gets 230 mpg! Aptera is developing several versions: The classic hybrid design will sport a 12 hp diesel/19 kW electric motor combination and will be priced around $29,000; the all-electric version will be priced around $26,000 and will have a 120-mile range.[4]

For those uninterested in buying a car but still needing access to a low-cost vehicle on an as-needed basis, consider Smart Cities' foldable, electric CityCar developed at MIT's Media Lab. Designed to mitigate the negative external effects of the traditional vehicle, not only is the City-Car electric but like airport luggage carts, it is "stackable," fitting six to eight cars into a single conventional parking place. The CityCar will be available to rent at transportation hubs and can be returned when finished.[5]

Tesla Roadster™

Photo reprinted with permission of Tesla Motors, Inc., San Carlos, Calif.

Aptera

Photo reprinted with permission of Aptera Motors, Carlsbad, Calif.

CityCar

CityCar photo reprinted with permission of Franco Vairani/MIT Smart Cities, Cambridge, Mass.

India's Tata Motors recently announced an extremely low-cost automobile with a small carbon footprint: the Nano. Dubbed "India's Model T," the four-door, two-cylinder family car can fit four passengers and has lower emissions than most two-wheeled Indian vehicles—at an astounding price of only US$2,500! Touting a lean design that minimizes weight and increases fuel efficiency, as well as safety design features that protect occupants, the Nano is sure to raise a stir in India and abroad. The company has said it expects the car to revolutionize the auto industry, and analysts believe the Nano may force other manufacturers to lower their own pricing. There also is speculation that the process innovations necessary to produce a car at such a low price will threaten the operating models of market leaders. French automaker Renault SA and its Japanese partner Nissan Motor Company are trying to determine if they can sell a compact car for less than $3,000.[6]

Nano

Photo reprinted with permission of Tata Motors, Mumbai, India.

In December 2007, the world faced a constellation of powerful factors with potentially devastating consequences:

- Global warming
- Inexorable increases in the global population and the attendant pressures on food systems, health care, and education
- Environmental degradation and the depletion of natural resources
- Terrorism and its impact on global politics
- Increasing divergence between the "haves" and the "have-nots" across the globe

What are the possible solutions to these seemingly intractable problems? Radical innovations in technologies are certainly a key part of the solution, offering exciting possibilities and developments. (Of course, these technological innovations must be coupled with courageous political and regulatory decisions, and changes in individual and societal behavior as well.) Innovations in green building technologies, transportation and fuel technologies, alternative energy solutions, technologies to bring potable water and the power of information to impoverished areas of the world—all offer tantalizing promises for a different future. Yet, despite their promise, all too many inventions do not achieve commercial success.

Indeed, the world of high technology is filled with both promise and peril. Promise arises from the potential advantages that these new technologies offer: to alleviate human suffering, to enhance people's lives, to make businesses more efficient and effective, to solve social problems, to chart unexplored territories—seemingly infinite possibilities. On the other hand, the peril arises not only from the risks of the technology—say, through unintended consequences or the misuse of technology—but also from the risks faced by a new technology; the largest risk is commercial failure and the inability to deliver on its promise.

Responsible high-technology marketing provides a balanced assessment of the promises and perils of new technologies. In particular, the role of marketing is to inform the development and commercialization efforts of high-tech companies, ultimately enhancing the odds that the new technologies will deliver on their promise and avoid downside risks.

Yet, think about the origins of many high-tech companies. Genentech pioneered drugs based on recombinant DNA technology in 1976; Google's breakthrough search engine in 1998 was based on an algorithm that ranked websites based on their relevance or importance. Regardless of the specific industry—biotechnology, telecommunications, information technology, consumer electronics—the origins of many high-tech companies are frequently found in their scientific or engineering developments. As a result, they often mistakenly believe that marketing is superfluous; the superiority of their technological innovations should be sufficient to convince buyers to adopt their new products. Although they might pay lip service to the importance of marketing, they may lack the skills and competencies to develop effective marketing strategies. As a result, the role of, and need for, marketing is often misunderstood or downplayed in their organizations. They often lack marketing expertise or relegate the role of marketing to second-class status (i.e., beneath the role of engineering/R&D).

Even high-tech companies that understand the importance of marketing face uncertainty and complications in their marketing decision making. Technical people may have a hard time becoming market focused and understanding how to interact with their nontechnical customers. Marketing activities are sometimes either an afterthought to the product/technology development process or are not accorded the same importance as product/technology development. Cross-functional collaboration between engineers and marketers is a necessity but is extremely difficult to implement well. A further complication is that many people hired to do "marketing" lack an understanding of how to market in high-tech industries. Even well-known high-tech companies such as Microsoft and Intel—who are perceived as being very sophisticated marketers—have expressed publicly that they are not sufficiently "market driven."

Due to these and other complicating factors, the failure rates of more innovative products are higher than the failure rates of products in general[7] —usually well over 50%.[8] In the funding of high-tech start-ups, venture capitalists often follow a 4-3-3 rule for their technology investment portfolios. Out of 10 investments, they expect 4 to fail, 3 to break even, and 3 to generate profits exceeding all the losses on the other seven projects.[9] A robust set of empirical studies demonstrates: Technological superiority alone is insufficient for ensuring the success of high-tech products; rather, high-tech companies must complement their technological prowess with a set of marketing competencies in order to maximize their odds of success.

One important marketing competency is to think beyond the technology itself, and to gain sophisticated insights about the customers who will adopt and use the technology. As Andy

Grove, the former chairman of Intel and one of technology's big thinkers, has said: "You know that old saying that 'railroads are not in the railroad business, they are in the transportation business'? We are facing something similar to that."[10] To successfully market technology solutions, high-tech companies must have an intimate understanding of their customers' underlying needs and problems, their customers' ways of doing business, and their customers' environments and mind-sets.

This book presents marketing strategies and tools for high-technology companies to maximize their odds of success. Given the high degree of uncertainty in high-tech markets, the margin for error in decision making is likely smaller than for conventional markets. As a result, high-tech firms must execute basic marketing principles flawlessly.[11] For example, knowing how to select the appropriate target market, communicate clearly the benefits the innovation offers relative to other solutions, design an effective and efficient distribution channel, and develop solid relationships and alliances are critical marketing competencies. In addition to these marketing basics, high-tech firms must also understand how to adapt marketing principles to the nuances of high-tech environments. The complex, high-velocity high-tech environment necessitates modifications in marketing research tools, supply chain management, and strategy formulation. Readers of this book will learn strategies and tools to successfully design and implement best-practices marketing in the high-tech environment.

We make a distinction between the marketing *of* high-technology innovations and the use of technology *for* marketing purposes. When people hear the phrase *high-tech marketing,* they frequently think of things like advertising on the Internet, using social networking sites like MySpace or Facebook to market consumer products, or conducting a paid search campaign using a platform like Google AdWords. By its very nature, marketing requires the use of technologies. For example, Web 2.0 technologies (including the collaborative, user-generated websites known as wikis, RSS feeds, social networking sites, and the many other technological innovations characterized by the increased connectivity and communities of users on the Internet) are a key tool in any company's marketing arsenal—whether it is a high-tech company or not. So, naturally, the role of these tools in a high-tech company's marketing efforts is addressed. Although this book does discuss the use of technology for marketing purposes (particularly in Chapter 11), its primary focus is on the marketing of high-tech innovations (be they breakthrough or incremental, product or process, disruptive or sustaining).

A first step in developing proficiency in high-tech marketing is for companies to develop a common understanding of what is meant by the term *marketing* and the broad scope of activities and expenditures (investments) that marketing encompasses.

THE LEXICON OF MARKETING[12]

Marketing is the set of activities, processes, and decisions to create, communicate, and deliver products and services that offer value to customers, partners, and society at large.[13] At its heart, marketing is a *philosophy of doing business* that reflects shared values and beliefs about the importance of creating value for customers by solving meaningful problems. This philosophy relies on customer- and market-based information to guide internal decision making and to resolve internal conflicts.

Marketing is not something that is undertaken after engineering has developed the new innovations. Among other things, a well-developed marketing competency includes proactive consideration of the customer in the development process; it helps to guide technical specifications, determine appropriate market segments, establish cost targets to meet pricing objectives, and identify partners that will play a critical role in the value delivery process. In other words, it brings the voice of the customer into the firm.

Table 1.1 shows three levels—strategic, functional, and tactical—that encompass the scope and types of marketing activities and decisions.

TABLE 1.1 Scope of Marketing Activities

1. **Strategic:** Proactive decisions to guide the thrust of the company's efforts in the marketplace, including the best opportunities in the market and how to best develop and position the company's products.
   - In which market will we compete?
   - Which segments will we target?
   - What value will we offer customers in our target segment?
   - What will our competitive position in the marketplace be (relative to the established technologies and ways of doing things)?

2. **Functional:** Focus on marketing as a functional area of responsibility, specifically including the product/technology development function.
   - What will our decisions be regarding product, price, place, and promotion?
   - How will personnel in different departments (marketing, R&D, operations, customer service, manufacturing/production, product development teams, etc.) interact to make marketing decisions?

3. **Tactical:** Development and implementation of marketing tools, executed consistently with strategic and functional decisions.
   - How will we create and use marketing brochures, websites, and other collateral materials?
   - Which trade shows will we attend, where will we place advertisements, and how will we share information about our products?

Source: Adapted from Jakki Mohr, Stanley Slater, and Sanjit Sengupta, "Foundations for Successful High-Technology Marketing," in *Managing Technology and Innovation: An Introduction,* eds. R. Verburg, R. J. Ortt, and W. M. Dicke (London: Routledge, 2006), pp. 84–105.

Strategic Activities and Decisions

Strategic decisions chart the firm's direction in the marketplace, addressing issues such as what market segments the company competes in, what its competitive position will be, and relatedly, what its value proposition will be. In some larger high-tech companies, a strategic planning group or a formal marketing department may have primary responsibility for answering these questions. In other companies, these questions may be addressed by the top management team (such as the company's founders) or by a product development group.

Less important than the "who" in answering these questions is the need to proactively address them. Despite the need for strategic direction in allocating the firm's resources across customer market segments and across product development efforts, many high-tech companies do not proactively answer these questions. When a company lacks direction about the answers to these questions, its efforts are diffused across multiple market segments and product development projects. Although a company may think this diffused approach hedges its bets in the marketplace, it is typically a recipe for disaster. Because of the lack of focus, the company never truly understands customers' needs in any one segment; as a result, it is less likely to succeed than a company that has proactively defined a clear strategic direction.

To effectively operate at the strategic level, the company must ensure that responsibility for strategic guidance is formally vested with some group in the organization (be it a marketing department, product management group, strategic planning group, or top management team). Moreover, the company must commit to developing a competency in research and analysis for market segmentation, targeting, and positioning. Finally, the company must be willing to implement the decisions arising from the strategic planning process with focus, discipline, and requisite resources. Readers will learn more about the strategic level of marketing beginning in Chapter 2. In addition, an increasingly important aspect of a winning marketing strategy is corporate social responsibility, through which a company's strategies explicitly address social and global prob only economic profits but social benefits as well. Readers will learn more about in Chapter 13.

Functional Activities and Decisions

The second level of marketing activities encompasses what is traditionally known as the marketing mix, or the "4 P's of marketing": product, price, promotion, and place (see Table 1.2). Note that the product development arena is considered to be a subset of the marketing function. The critical management issue in the marketing mix is ensuring consistency in all decisions that support the product's position in the marketplace. For example, a product positioned as high quality must have a price that conveys that image, high-end distribution channel members that provide appropriate levels of support and service, and advertising message and media focused on the premium image. Each of the 4 P's is discussed individually, in Chapters 8–12.

The responsibility for these diverse marketing functions is scattered across many units in most organizations. Typically, a product development group—comprised of scientists, engineers, or programmers, for example—is charged with research-and-development activities. Some companies also vest the product development group with other marketing activities such as collecting market research, conducting market segmentation, or targeting and positioning activities. After product development, responsibility for product launch may be passed to product management, who coordinates with sales and "marcomm" (marketing communications); product managers may also perform

---

TABLE 1.2 The 4 P's of the Marketing Mix

1. ***Product:*** Decisions that address the new product development process (innovation management), licensing strategies with potential partners, intellectual property rights, services provided to augment the revenue stream from base-products, product name/brand decisions, development of complementary products by partners, creation of industry standards, packaging, and so forth. The critical need is to develop a stream of products with the right set of features to satisfy customer needs in a compelling yet simple fashion.

2. ***Price:*** Decisions that establish price points for the company's products, and address issues related to the cost to produce/manufacturer the goods, margins along the distribution channel, competitor's prices (pricing relative to a specific firm's market position), customer value, total cost of ownership for the customer, prices for product bundles, and profitability.

3. ***Promotion:*** Decisions that include advertising (both media and messaging decisions), sales promotion (price deals, trade incentives, etc.), personal selling (recruiting, training, compensating sales people), and public relations/publicity (garnering favorable trade press, attending trade shows, engaging in cause-related marketing, etc.). Specific issues can include developing a strong brand name, decisions about the timing and focus of new product pre-announcements, co-branding decisions with potential business partners (including cooperative advertising with channel members), leveraging the Internet and other new media to gain awareness, developing collateral materials, and so forth.

4. ***Place:*** Decisions that focus on distribution channels and supply chain management—getting the right product to the right customers at the right time. Good channel strategy is focused on effectively meeting end-user customer needs in a cost-efficient fashion. However, successful channels can be difficult to attain if channel partners have different objectives, if margins create conflicts between channel members, and if new channels (such as the Internet) cannibalize revenues from existing channels. Best-practices supply chain management is demand-driven; it harmonizes upstream logistics and manufacturing with end-user requirements.

Source: Adapted from Jakki Mohr, Stanley Slater, and Sanjit Sengupta, "Foundations for Successful High-Technology Marketing," in *Managing Technology and Innovation: An Introduction*, eds. R. Verburg, R. J. Ortt, and W. M. Dicke (London: Routledge: 2006), pp. 84–105.

other marketing activities noted above. Some high-tech companies may have a formal marketing department, whose responsibilities may vary widely depending on the level of marketing sophistication and the degree to which marketing is viewed as a strategic function. Even with a formal marketing function, the input of the marketing personnel is sometimes neither solicited nor (if solicited) respected/valued by personnel in other functional areas. Some of the disparaging comments can be found in the form of jokes, which can capture common stereotypes.[14] For example:

> Q: What is marketing?
> A: What you do when your products aren't selling themselves.

The implication of this joke is that marketing is not something companies should have to bother about unless the product is not good enough to "sell itself."

When marketing is viewed as a philosophy of doing business, the focus is less on which "function" performs a specific marketing task, and more on how every person and department in the company works together to provide value to the customer. As the famous management guru Peter Drucker stated, "There are only two functions in any organization: marketing and innovation, both of which create a relationship with the customer."[15] Given this perspective, only two types of people exist in any organization: those who serve the customer, and those who serve those who serve customers.[16]

A high-tech company that effectively coordinates its efforts across functional units to deliver superior customer value is more likely to be successful than firms that optimize decisions within each functional area in a piecemeal or independent fashion. When individual departments have independent goals and objectives, they can experience incompatibility, conflict, and, ultimately, the erosion of a customer-based focus. Regardless of the functional area involved, all personnel must understand the idea of a **moment of truth**:[17] each touch point—every interaction a customer has with a company, whether calling technical support, visiting a retail partner, asking a billing question or seeking general information—can either strengthen or undermine the relationship.

The shared values and beliefs that underlie the development of innovation and marketing decisions are addressed in Chapter 3; subsequent chapters take the discussion further, and focus on how high-tech companies use market-based information to inform and guide their decisions—in other words, how they infuse the company with a market orientation.

Tactical Activities and Decisions

At the tactical level, the actual implementation of specific marketing tools is accomplished—such as the development of marketing collateral materials and a website, decisions regarding which trade shows to attend, where to place advertisements, and so forth. Many high-tech companies equate "marketing" with only these tactical considerations, and they relegate marketing input to reactive development of communications messages. Hence, the plea, "We need to hire a marketing person," essentially means hiring someone to operate at the tactical level. Companies that operate only at this tactical level likely have not made some of the harder strategic decisions to guide the company's efforts, and in that sense, are less likely to be successful in the marketplace (regardless of how effective their advertising or trade show strategies may be).

Companies that view marketing as a strategic consideration recognize that success in high-tech markets comes from proactive consideration of where the best opportunities in the market lie, and how to best develop and position the company's products to enhance the odds of success. They work to facilitate collaborative cross-functional interaction between not only marketing and development teams, but all functional areas. And, the tactical considerations are executed in a manner consistent with the strategic foundation of the company.

With a shared understanding of marketing in general, the discussion now moves to exploring the specific realm of high-technology marketing.

DEFINING *HIGH TECHNOLOGY*

Technology is a broad concept that relates to how people use tools and knowledge—usually the product of science and engineering—to create solutions to problems.[18] **High technology** generally refers to cutting-edge or advanced technology—which means that the definition shifts over time. What was "high tech" in the 1960s—for example, a color TV—would be considered primitive technology by today's standards. This fuzzy definition has led to companies describing nearly all new products as high tech.

Innovation generally refers to introducing something new, with the intent either to increase value (either to customers or producers) or to solve some problem. These new "things" can include ideas, methods, digital content, or devices. Not all innovations are high tech in nature; later in this chapter are descriptions of the common classifications for the various types of innovations.

The traditional domains of "high tech" include areas such as information technology, computer hardware, software, telecommunications and Internet infrastructure, and consumer electronics, among others. In addition, "high tech" can encompass a broad cross section of industries including biotechnology, pharmaceuticals, medical equipment, nanotechnology, robotics, and, with the focus on using technology to solve global problems, it can also include energy and transportation technologies and green building technologies—clearly a wide range of industries and products. Hewlett-Packard's former CEO Carly Fiorina has explained, "Tech is truly becoming part of the fabric of life. Think about the big problems we have to solve now—health care, homeland security, synchronizing the world's information systems to facilitate the flow of goods and services and to prevent the flow of undesirables—all of those are technology opportunities."[19]

At least two reasons motivate the need to understand and define the domain of high technology. First, because of technology's role in the economy, economists and policy makers try to classify the economic output and purchases that are driven by high-tech industries. Second, because of the need to modify and adapt standard marketing strategies for the high-tech environment, we must specify which arenas are "high tech," and hence fall into the domain of this book.

Given their ubiquitous nature, categorizing technological innovations—say, by placing industries on a continuum ranging from low-tech industries on one end to high-tech on the other—is not as easy as one might expect. For example, although some might perceive industries such as agriculture or heavy industry to be relatively low tech, when innovations in genetically modified organisms in seed stock or innovations in nanotechnology in heavy materials are considered, those industries might be labeled high tech. Even definitions offered by experts—high-tech firms are those that are "engaged in the design, development and introduction of new products and/or innovative manufacturing processes through the systematic application of scientific and technical knowledge"[20] —are sufficiently vague that classifications are tough.

This section, therefore, presents three tools used to clarify the domain of high tech: (1) government-based classifications, (2) classifications based on shared industry characteristics, and (3) various types of innovations.

Government-Based Classifications

Motivated by the desire to classify and measure economic activity (such as employment, industry output, and so forth), most government definitions of high technology follow either an input-based or output-based approach.

Input-based definitions classify industries as high tech based on certain criteria such as the number of technical employees, the amount of R&D outlays, or the number of patents filed in a given industry. For example, the U.S. Bureau of Labor Statistics (BLS) classifies industries based on their proportion of scientists, engineers, and technicians; firms that employ a high proportion of scientific, technical, and engineering personnel (where "high" is twice the 4.9% average for all industries) comprise high-tech industries. Similarly, the Organisation for Economic Cooperation and Development (OECD) defines high tech in terms of the ratio of R&D expenditures to value added of a particular industry.[21] The National Science Foundation examines the R&D intensity, or R&D spending-to-net-sales ratios.[22] Appendix 1.A at the end of this chapter provides the list of industries classified as high tech (Level I, Level II, Level III) using the BLS approach. Technically, because these industries are categorized based on employment of technical and scientific personnel, they are "technology-oriented-occupation intensive."

Output-based approaches define an industry as high tech if its output (product) embodies new or leading-edge technologies, as determined by a panel of experts. For example, the U.S. Census Bureau identifies 10 major technology areas that produce such products: biotechnology; life sciences technology; optoelectronics; information and communications; electronics; flexible manufacturing; advanced materials (semiconductors, optical fiber cable, for example); aerospace; weapons; and nuclear technology.[23] Similarly, the American Electronic Association uses the U.S. government's NAICS (North American Industry Classification System) to classify 45 high-tech industries into three major output-based groupings: High-Tech Manufacturing, Communications Services, and Software and Computer-Related Services.[24]

Table 1.3 provides an overview of the strengths and weaknesses of defining high technology in these two ways. For the 14 industries with the highest levels of technical employment, categorized as Level I industries (with more than 5 times higher technical employment than the average of all industries), the alternative classification approaches show relatively high consistency. In other words, these industries would be classified as high tech regardless of which approach is used. For example, almost all Level I industries are also R&D intensive, and all Level I goods-producing industries have some products defined as high tech by the output-based approach used by Census Bureau.

TABLE 1.3 Strengths and Weaknesses of Government Approaches to Defining *High Technology*

Input-Based Approaches

Strengths:

- Data are generally easily obtainable.
- Classification is objective.
- Correlation between input-based classifications is reasonably high for Level I* industries.
- Data on the high-tech service sector is included.

Weaknesses:

- Because thresholds for "high" levels (say, R&D spending or technical employment) are not obvious, classifications may be deemed somewhat arbitrary.
- Classifications may include industries with products not commonly thought of as high-tech.
- Classifications may omit very new industries (e.g., biotechnology and nanotechnology are not on the list of high-tech industries).
- Different input-based measures will result in different classifications.

Output-Based Approaches

Strengths:

- Classification tends to have face value (the list of industries matches popular conceptions of high-tech).
- Relatively good correlation exists between input and output methods for Level I industries.

Weaknesses:

- Somewhat post-hoc: Judgments are somewhat subjective.
- Output-based approaches are generally not as comprehensive as input-based approaches.
- Relatively low correlation exists between input and output methods for Level II and Level III industries.

---

*Level I industries have 5 times or greater technical employment as the average of all industries; Level II industries have 3.0–4.9 times and Level III industries have 2.0–2.9 times greater technical employment as the average of all industries; industries are listed in Appendix 1.A.

However, for the 12 Level II industries (with 3.0–4.9 times the average technical employment), there is relatively low consistency among the various approaches. Some industries classified in this category have no products on the high-tech output list (e.g., oil and gas extraction, or audio and video equipment manufacturing). Moreover, R&D employment in Level II (and in Level III) industries can be high or low. And, in some cases, an industry classified as high tech using an output approach is not classified as a Level I or Level II industry based on the input approach (e.g., telecommunications). One important point is that whether or not a company uses high-tech production methods is *not* considered useful in identifying high-tech industries.[25]

Rather than focusing on objective inputs (that could lead to misclassifications) or on subjective classifications of outputs, a different approach relies on examining common characteristics that high-tech markets share. One benefit of this approach is that it is based on a manager's assessment of the uncertainty surrounding the product or service that he or she manages. Moreover, this approach is directly tied to the implications for high-tech marketing strategies.

Common Characteristics of High-Tech Environments

As shown in Figure 1.1, high-tech environments manifest a set of common characteristics[26]—most notably, (1) market uncertainty, (2) technological uncertainty, and (3) competitive volatility[27]—with specific implications for marketing.

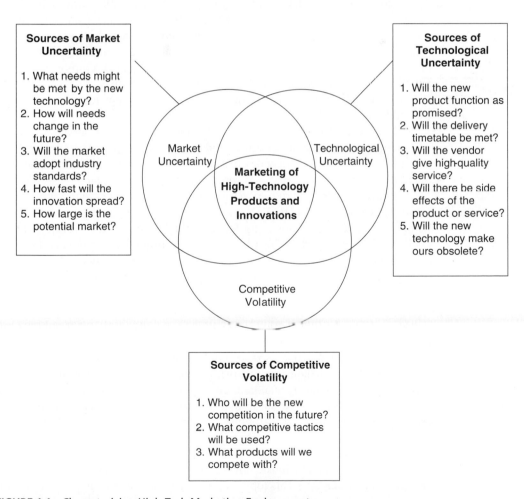

FIGURE 1.1 Characterizing High-Tech Marketing Environments

Although one or two of the three characteristics might be present in some environments, the simultaneous presence of all three factors characterizes most high-tech environments. For example, the decision about purchasing a new home might provoke customer anxiety (one of the three characteristics: market uncertainty), but if sellers aren't also simultaneously considering a radically new way of offering homes for sale (e.g., Internet channel), then it would not be characterized as high tech. Similarly, customer needs may change rapidly in some areas (such as clothing styles or music), but such purchase decisions generally do not include both a high degree of customer anxiety and totally new ways of meeting customer needs. Finally, although competitive turbulence may be present in many industries (e.g., restaurants in college towns), the new competitors don't typically offer a radically new way of meeting customer needs. These situations highlight the fact that the *intersection* of these three characteristics is what typifies a high-tech marketing environment.

In addition, a series of other characteristics, such as network externalities and unit-one costs, also pose complications for high-tech marketing; these additional characteristics are covered later in this chapter.

MARKET UNCERTAINTY   **Market uncertainty** refers to ambiguity about the type and extent of customer needs that can be satisfied by a particular technology,[28] and arises from five sources.

First, market uncertainty arises from consumer fear, uncertainty, and doubt (known as the *FUD factor*)[29] about what needs or problems the new technology will address, as well as how well it will meet those needs. Anxiety about these factors means that customers may delay adopting new innovations, require a high degree of education and information about the new innovation, and need postpurchase reassurance to assuage any lingering doubt. For example, when a business decides to automate its sales force with computers, employees are bound to have some apprehension about learning new skills, wondering if the new mode of working will be better than the old one, and so forth. Hence, marketers must take steps to allay such apprehension both before and after the sale.

Second, customer needs often change rapidly, and in an unpredictable fashion, in high-tech environments. For example, customers today may want to treat their illnesses with a particular medical regimen but next year may desire a completely different approach to the same health problems. Such uncertainties make satisfying consumer needs a moving target.

Third, customer anxiety is perpetuated by competing—and incompatible—technological standards for new products. For example, in 2007 the new high-definition DVD players were being produced in two competing formats: the Blu-ray format developed by Sony and the HD-DVD format developed by Toshiba. High-definition movies were available in either one format or the other; movies purchased in one format would not play on the incompatible high-definition players in the other format. Questions about which format would become the **dominant design**—the format that would emerge as the agreed-upon standard—hampered customer adoption, as buyers delayed purchase to minimize the odds of making a "wrong" choice. (In February 2008, Toshiba settled the issue by announcing that it would no longer produce HD-DVD products.)[30]

Therefore, coalescing disparate product development efforts around a common industry standard can help reduce the perceived risk for customers, in turn serving as a catalyst for adoptions. Having a common industry standard not only maximizes the value customers get from their investment in high-tech products, but it also stimulates the development of complementary products to create a robust industry infrastructure. A firm that pursues a unique or proprietary system in its product development faces a very different market adoption process, with very different consequences, from one that develops a system based on open standards available to multiple players in an industry. (For example, music downloads from various sites are not all compatible with all types of digital music players.) This topic of industry standards is so important that a later section of this chapter is devoted to it. Moreover, the decision about creating a product architecture using only proprietary standards (sometimes referred to as a "walled garden") is also extensively addressed in a later chapter (Chapter 8).

A fourth factor, due in large part to the prior three factors, is uncertainty among both consumers and manufacturers over how fast the innovation will spread. Figure 1.2 shows cumulative adoption figures for two different categories of products—durable goods technologies and media. In many cases, the market for high-tech innovations is slower to materialize than most would predict.[31] For example, despite its hype, WiMax (or wireless Internet access over long-range distances

Panel A: For Durable Technologies

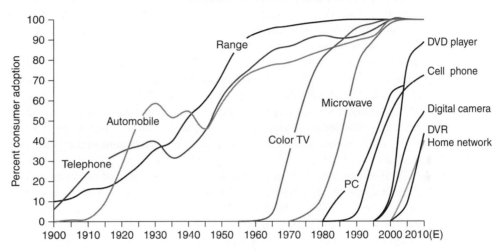

Source: Benchmark Data Overview, Forrester Research, June 2004.

Panel B: For Media Products

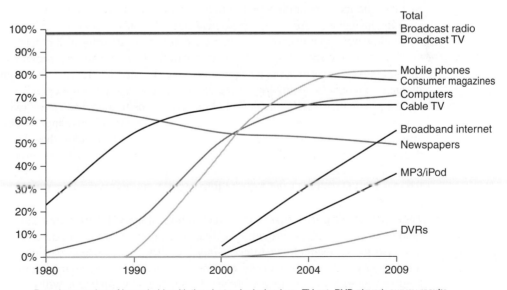

Based on number of households with the electronic device (e.g., TV set, DVD player); survey results based on share of the population who indicated using the medium in a given year (e.g., read newspapers, went to a movie) or estimates based on circulation or unit data.

Source: Communications Industry Forecast, 2005–2009.

FIGURE 1.2 Cumulative Adoption Patterns

that can maintain a connection while consumers are mobile) and muni WiFi (short for *municipal wireless:* free or cheap wireless Internet access for all citizens of a city or town) were listed as two of the biggest emerging technology disappointments of 2007.[32] Yet, for other technologies, the sales takeoff is astoundingly fast. For example, sales of the Wii video game console, made by Nintendo, took off much faster than predicted. The Wii sold more than 6 million units in the United States between its release in November 2006 and January 2008, with 981,000 units sold in November 2007 alone—more than double sales of the Sony PlayStation 3 and outpacing the Microsoft Xbox 360 by 211,000.[33]

The fifth dimension of market uncertainty, related to the uncertainty over speed of adoption, is the inability for manufacturers to estimate the size of the market. Obviously, market forecasts are crucial for cash flow planning, production planning, and staffing. Although Nintendo increased Wii production by 80% after the product launch to meet unexpectedly high market demand, chronic shortages continued to plague its sales through the early part of 2008. Nintendo of America president Reggie Fils-Aime said the company did not expect such a craze for the Wii. "We went into the launch of the Wii with very high expectations. We had expected to be in the upper range of console launches," Fils-Aime said. "But this is unheard of."[34]

Although lost sales are certainly costly to a company, just as serious is the situation of excess inventory that ties up working capital and, at the extreme, results in obsolete products that must be written off. Hence, market forecasting is an important competency that high-tech companies must develop.

Further complicating a company's ability to forecast the pace and size of adoption is what Geoffrey Moore[35] refers to as the *chasm* that high-tech products must cross in appealing to a mainstream market. When radical innovations appear in the marketplace, they appeal to "visionaries" in the market who are willing to adopt the new technology despite the often high price tag such items carry. For example, the early adopters of Apple's new iPhone were willing to pay a high initial price, despite the expected price reductions that were bound to come within a few months. Moreover, the visionaries are also typically willing to accept any inconveniences or hassles that can accompany being an early adopter. For early adopters of software, the hassle factor might come in glitches and incompatibilities with other system components. The visionaries are willing to accept such inconveniences for the psychological and substantive benefits they do receive.

However, such benefits are not sufficient for the majority of the market to adopt a new technology. "Pragmatists" comprise the majority of the market, and they require a different set of benefits and inducements to adopt new products. The **chasm**, then, represents the gulf between these two distinct segments for technology products. Visionaries are quick to appreciate the new development, but the pragmatists need more hand-holding. The transition between these two markets can be rocky at best, with many high-tech firms never crossing the chasm (crossing the chasm is covered extensively in Chapter 7). Many high-tech firms find it hard to abandon their "techie" roots and talk to this group in customer-friendly terms. The inability to predict whether, and the degree to which, the mainstream market will adopt the product—and the rate of such adoption—given the presence of the chasm, makes it extremely difficult for manufacturers to estimate the size of the market.

TECHNOLOGICAL UNCERTAINTY   **Technological uncertainty** is not knowing whether the technology—or the company providing it—can deliver on its promise to meet specific needs.[36] Five factors give rise to technological uncertainty. The first comes from questions about whether the new innovation will function as promised. For example, when new pharmaceuticals are introduced, patients might experience anxiety about whether the new treatment will be as effective for them as currently available treatments. Although early adopters are willing to be the guinea pigs on the bleeding edge of technology,[37] they might face major glitches in the functioning of the new product.

The second source of technological uncertainty relates to the timetable for development (and subsequent market availability) of the new product. In high-tech industries, product development commonly takes longer than expected, causing headaches for both buyers and sellers. For example, people have been talking about the Semantic Web since Tim Berners-Lee introduced the concept in 1999. At a very basic level, the idea of the Semantic Web is to make information actually "understandable" by computers. It would use common programming standards, based in part on XML's (**E**xtensible **M**ark-up **L**anguage) customized tagging schemes and RDF's (**R**esource **D**escription **F**ramework) flexible approach to representing data—so that through machine-to-machine interaction, people are freed of some of the tedious work involved in information search and use. Finally, in the fall of 2007, Radar Networks announced a Semantic Web application called Twine, which was built with Semantic Web technologies and was hyped as the "first mainstream Semantic Web application"[38] that would enable people to share knowledge and information. This lag of roughly eight years from vision to initial market application is not uncommon for many breakthrough technologies.

At the extreme, product delays can mean that customers get frustrated and possibly cancel orders, competitors beat the company to market, and the company faces financial disaster. For example, the Airbus 380, which made its maiden voyage in late 2007, was plagued by delays dating from 2005. Part of the reason was the sheer complexity of the A380. The cabin wiring alone amounted to more than 330 miles (531 km) of wiring with more than 40,000 connectors. Unfortunately, two incompatible versions of computer-aided design (CAD) software were used in the development process—one by the German unit in Hamburg and another by the French in Toulouse. The problem wasn't discovered until late in the process, and the solution—developing a new computer-design tool—meant yet more delays. These delays "knocked 26% off the value of the shares of the parent company, EADS, led to the resignation of three executives, and triggered investigations into the share dealings of executives and board members."[39]

A third factor in technological uncertainty arises from concerns about the supplier of the new technology: If a customer has problems, will the supplier provide prompt, effective service? When (if?) a technician arrives, will the problem even be "fixable"? For example, in the fall of 2007, Skype (the Voice-over-Internet Protocol, or VoIP, service provider that allows customers to download a free software package onto their computer and then use their broadband connection to make phone calls for free anywhere in the world) faced a disabling computer glitch in its system: A standard Microsoft Windows update led to a chain of events that overwhelmed its network and millions of users lost their service. In Skype's case, the peer-to-peer nature of its network meant that there was no quick way to recover from the failure. The result was that users lost their trust in the company, and it had to be regained.[40]

Fourth, the very real concern over unanticipated consequences or side effects also creates technological uncertainty. For example, some new companies (e.g., 23andMe and Navigenics) are offering consumers the ability to run their genetic profile for roughly $1,000 and a saliva sample. Yet, with the power of genomic information comes a burden: What do people and companies do with the information? Will a consumer who finds out that that she has a genetic predisposition for breast cancer find that information useful, or or will she live in fear of fulfilling her genetic destiny? If she decides to take a drug tailored to her genetic profile (pharmacogenomics) such as Herceptin, will it create a new set of risks and initiate a new trajectory of calculations? Will insurance companies and employers misuse the information?[41]

As another example of unintended consequences with a somewhat paradoxical outcome, some experts believe that the shift to biofuels is not accomplishing the desired outcome (minimizing oil independence), but rather, is costing an enormous sum of money and causing corn prices to skyrocket nearly 70% in the six months between late fall 2006 and early 2007. Indeed, in 2007, the United States used nearly 20% of its total corn crop for the production of ethanol—a number that was expected to rise to 25% in 2008 due to ambitious goals regarding ethanol use in the United States. The increased demand for corn in biofuels has caused a ripple effect, seen first in the price of animal feed, particularly poultry and pork. Poultry feed is about two-thirds corn; as a result, the cost to produce poultry—both meat and eggs—rose 15% due to corn prices. Corn syrup—used in soft drinks—will also get more expensive. In Mexico (which gets much of its corn from the United States), the price of corn tortillas doubled in one year, setting off large protest marches in Mexico City.[42]

Children spending more time playing with technology than playing outdoors potentially contributes to obesity; people who use cell phones while driving cause accidents; new pharmaceuticals might cause people to die: The technology world is rife with examples of technologies that are developed to solve human problems, but they also result in unforeseen and undesirable outcomes.

Finally, the fifth dimension of technological uncertainty exists because one is never certain just how long the new technology will be viable before an even newer development makes it obsolete. This uncertainty about the life span of a particular technology is directly related to competitive volatility, the third characteristic of high-tech environments, which is discussed next.

COMPETITIVE VOLATILITY **Competitive volatility** refers to both intensity in degree of change in the competitive landscape and uncertainty about competitors and their strategies. For instance, as different technologies begin to offer the same functionality—a situation known as **convergence**—new competition will be found in different product classes: Cell phones function as digital music players, or computers function as both televisions (via IPTV technology) and movie players (via either streaming downloads or an internal disk player). Because the new technologies frequently seem less sophisticated than existing technologies, incumbent companies might underestimate these new competitors—or even disregard them—until it is too late. Competitive volatility has three dimensions; firms experience uncertainty about (1) who their competitors are, (2) their competitors' market strategies, and (3) their competitors' product offerings.

First, uncertainty exists over which firms will be new competitors in the future. In some cases—though certainly not all[43]—new technologies are commercialized by companies *outside* the threatened industry.[44] Typically, because the most disruptive competitive threats use different technological platforms/underpinnings, they often don't appear on the radar screen during a company's traditional environmental scanning/competitive intelligence surveys.

This relates to the second dimension of competitive volatility: market strategies used by the competition. New competitors that come from outside existing industry boundaries often bring their own set of competitive tactics—tactics that existing industry incumbents may be unfamiliar with and that undermine their existing business models. These new players end up rewriting the rules of the game, so to speak, and changing the face of the industry for all players.[45]

For example, Skype's Voice-over-Internet Protocol (VoIP) service was a radical change in the telecommunications industry that seriously undermined existing companies' fixed investments and disrupted their business model based on paid long-distance calling. Similarly, the emergence of the dot-com e-tailers radically disrupted the retail business in the early 2000s. In the late 1990s, Amazon.com was totally unknown to retail booksellers and Expedia.com was equally unknown to airlines and travel agents. The verb "Amazoned" has since become known to mean a company that could have introduced a disruptive technology or business model but chose to stick with the status quo instead—and therefore was *Amazoned* by another, often unknown, competitor who was willing to radically change the rules of the game. As a final example, the free music download services using peer-to-peer networks over the Internet, even though illegal, radically changed the music industry. Sales of music CDs have continued to decline every year since 2000, with sales down by a whopping 19% in 2007 alone.[46] A subsequent development from these early competitive disruptions was the attractiveness of legal music download services such as iTunes and Rhapsody for customers who wanted the convenience of downloads but wanted to do so legally.

The third dimension of competitive volatility often appears as **product form competition**—new technological developments that provide a different way to satisfy the same underlying customer need or problem. For example, think about the various ways customers can connect to the Internet. They can use old technology, such as dial-up access over a modem; they can use a broadband connection in the form of either DSL or cable modem; or they can use T1 lines or Ethernet connections. Increasingly, customers are opting for wireless access, whether it's connecting through a WiFi hotspot, WiMax, or even cellular broadband (offered by wireless/mobile companies). The list of new technologies that satisfy the same underlying customer need but with a different value

TABLE 1.4 Three Sources of Marketing Myopia in High-Tech Markets

1. ***"Our technology is so new that we have no competitors."*** This type of thinking does not accurately reflect the fact that customer needs are typically already being solved, either with an older-generation technology, or in some cases, by doing nothing. Indeed, entrenched customer habits are sometimes harder to dislodge with new technology than if a competitor did exist.

2. ***"The new technology being commercialized by new competitors will not pose a large threat"*** (or some similar disparaging phrase). The high-tech field is littered with the corpses of incumbent firms who underestimated new entrants in the market. During the heyday of the Internet, this idea was captured with the saying, "You've been Amazoned"—meaning that a new competitor came in and stole an incumbent's business with a new technology or business model.

3. ***"That competitor is in a different industry, and its strategies don't/won't affect my business."*** When managers view their industry from a specific product/technology lens rather than from a customer viewpoint, they forget to acknowledge that customer needs can be solved using different underlying technology platforms. For example, in the Internet industry, broadband providers supplying a DSL connection may view their competitors only as other DSL providers. However, from the customers' perspective, cable modem, new forms of wireless and satellite, and even dated dial-up access all meet their needs, each with a different technology solution (and in some cases, a different price–performance ratio).

Source: Adapted from Jakki Mohr, Stanley Slater, and Sanjit Sengupta, "Foundations for Successful High-Technology Marketing," in *Managing Technology and Innovation: An Introduction,* eds. R. Verburg, R. J. Ortt, and W. M. Dicke (London: Routledge, 2006), pp. 84–105.

proposition (or way to do so) goes on and on: open source software, new online advertising technologies, and so forth. Innovations by both new entrants and incumbents can render older technologies obsolete, and hence the mortality rate of businesses in high-tech industries can be high, further contributing to competitive volatility.

A key implication of competitive volatility for high-tech companies is to avoid being myopic when it comes to evaluating competitive threats. **Marketing myopia** refers to the tendency of managers to be narrow-minded or shortsighted in their views about their industry contexts and their business strategies. Managers in high-tech firms suffer from three types of marketing myopia about sources of competition, as shown in Table 1.4. Indeed, the **innovator's dilemma**[47] refers to the difficulty market leaders have in developing disruptive innovations; this difficulty arises because their investments in current-generation technologies and their focus on existing customers leave them vulnerable to competitors whose new products serve different market segments that are—at least initially—less attractive. The innovator's dilemma is explored later in the chapter, in the section on types of innovations.

One way to depict the evolution in product form competition is through **technology life cycles**. As shown in Figure 1.3, these S-shaped curves track the relationship between investments in R&D in a new technology over time (on the horizontal access) against its price–performance ratio (on the vertical access).[48] As a new technology is introduced, its performance capacity improves slowly, because the fundamentals of the new technology are not yet well understood. As scientists gain a deeper understanding of the technology—including tools and processes to illuminate the underlying mechanics and science of the technology—and as additional R&D expenditures are made, the new technology's performance reaches a critical inflection point (or, in mathematical terms, the first derivative of the curve, which is its slope at the highest value) where its performance rapidly accelerates. In time, another inflection point is reached (the second derivative approaches zero), where despite further R&D investments, diminishing returns set in and performance improvements taper off.

For example, advances in semiconductor speed and processing power, relative to price, have been formalized in **Moore's Law** (formulated by Gordon Moore, cofounder of Intel; not to be confused with Geoffrey Moore, who wrote *Crossing the Chasm*), which states that semiconductor performance doubles every 18 months, with no increase in price. Stated differently, every 18 months or

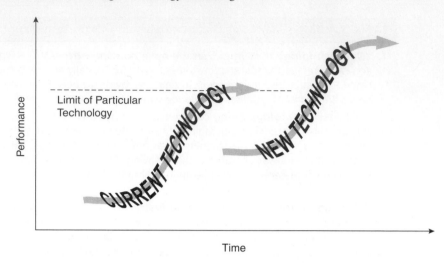

FIGURE 1.3 Technology Life Cycles

so, improvements in technology cut price in half for the same level of performance. Scientists predict that Moore's Law will hit its upper limits in about 2020.[49] Technologies that could replace conventional semiconductors are based on molecular biology and nanotechnologies. Although such research is at a preliminary stage, and although their performance is nascent, these new technologies hold the possibility of making semiconductors of today obsolete.

When a new, discontinuous technology is introduced (i.e., there is a jump to a new S-shaped curve), the performance of this new technology typically underperforms the incumbent technology. Because of this, the incumbent provider often overlooks, ignores, or downplays the potential competitive threat the new technology poses. Yet, there may be a flurry of new entrants who experiment with the new technological paradigm, each offering a slightly different version. Eventually, the market coalesces around a dominant design, and a stable architecture for the new technology emerges, providing reassurance to cautious customers and enabling firms to focus on process innovations for efficient production. When the new technology reaches its initial inflection point where its performance capability rapidly overtakes the upper limits of the incumbent technology, the new technology can make existing technologies obsolete, creating new leaders and new losers[50] in a process referred to as **creative destruction**.[51]

Empirical research employing archival data for 14 technologies in the areas of desktop memory, display monitors, desktop printers, and data transfer found evidence that technology life cycles do not always follow S-shaped curves. Rather, they can evolve through irregular step functions with long periods of no performance improvement interspersed with jumps in performance.[52]

How can a firm recognize when a current-generation technology is in danger of obsolescence? The technology life cycle curve demonstrates that one can't rely solely on economic signals: Based on incremental improvements, the revenue of the current technology can reach a peak even after the new technology is introduced. Hence, relying on economic signals may result in the firm moving too late into the new technology, and the competition will have established a stronghold. Diminishing performance returns to increasing investments in current-generation technology are the crucial indicators.[53] Other strategic lessons arising from competitive volatility, such as those found in Clayton Christensen's theory of disruption, are discussed shortly.

OTHER CHARACTERISTICS OF HIGH-TECH ENVIRONMENTS   In addition to market uncertainty, technological uncertainty, and competitive volatility, many high-tech environments share other characteristics, three of which are shown in Table 1.5: unit-one costs, tradability problems, and knowledge spillovers. A fourth and final characteristic, network externalities, has a dramatic effect in many high-tech markets and is addressed next.

TABLE 1.5 Other Characteristics of High-Tech Environments

Unit-one costs: A situation where the cost of producing the first unit of a product is very high relative to the variable costs of reproducing subsequent units; in other words, the ratio of variable costs to fixed development costs is very low. This cost structure likely exists when *know-how* (knowledge embedded in the design of the product) represents a substantial portion of the value of the product.

    A unit-one cost situation is a function of how low the variable costs of reproduction are—negligible or even zero in the case of digital distribution of information goods over the Internet.

    On the other hand, any product that is produced on a manufacturing line, despite its R&D costs, likely has significant variable costs of production (components, cables, hardware casing, etc.), and is less affected by unit-one costs.

Examples:

- *Software:* The costs of distributing software are trivial compared to the development costs of hiring programmers and specialists to write and debug the code in the first place.
- *Pharmaceuticals:* The costs of stamping (producing) a pill are trivial relative to the sizable R&D dollars to get it through clinical trials and FDA approvals.

Tradability problems: Problems that arise because high-tech companies often want to sell their underlying know-how (or intellectual property in the form of patents or licensing), but it is difficult to value that knowledge—especially when it is tacit and resides in people and organizational routines.

Knowledge spillovers: The expansion of knowledgethat occurs when synergies in the creation and distribution of know-how further enrich a related stock of knowledge. Each innovation creates the opportunity for a greater number of innovations, not only in the "home" scientific area but in a broad array of other areas as well. For example, it was once estimated that the Human Genome Project (used to map all human genes) would take at least 40 years to complete, but it took only a fraction of that time, due to knowledge building on knowledge. Know-how generated from that project is being applied not only in drugs and pharmaceuticals, but in forensics and law enforcement as well.

Sources: George John, Allen Weiss, and Shantanu Dutta, "Marketing in Technology Intensive Markets: Towards a Conceptual Framework," *Journal of Marketing* 63 (1999, Special Issue), pp. 78–91; and Raji Srinivasan, "Sources, Characteristics, and Effects of Emerging Technologies: Research Opportunities in Innovation," *Industrial Marketing Management* (August 2008), pp. 633–640.

      Network Externalities.   **Network externalities** refer to situations where the value of a product increases as more users adopt it. *Direct network externalities*—also referred to as *demand-side increasing returns* or *bandwagon effects*—exist when the value that any one user gets from the product is directly related to the number of other users; think telephones, portals on the Internet, social networking websites such as LinkedIn and Facebook, and online auctions such as eBay. As shown in Figure 1.4, the value of these technologies derives from communications and connectivity between users. For example, the first telephone was worthless, the second made the first more valuable, and so on. If a technology is driven by direct network externalities, then an individual purchaser will essentially receive no value if no other users have also adopted that technology (or limited value if only one or a handful have adopted).

    Also known as **Metcalfe's Law**, the concept of direct network externalities illustrates the power that comes from its **installed base**—the number of users who have adopted a particular technology. Metcalfe's Law, formulated by Robert Metcalfe, who coinvented Ethernet and founded 3Com, states that the value of a network scales as n^2, where n is the number of persons connected. According to Metcalfe's Law, as the number of users doubles, the value of the network quadruples. In other words, the utility received from an innovation is a function of the square of the number of users (the installed base); the rapid takeoff is where the utility or value

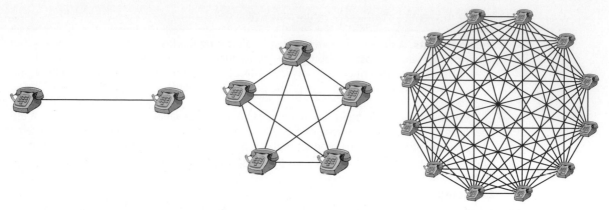

http://en.wikipedia.org/wiki/Image:Network_effect.png

FIGURE 1.4 Network Externalities

of the innovation increases exponentially because a critical mass of users has adopted it. Metcalfe's Law is derived from the fact that the number of possible pair (unique) connections in a network can be expressed as $n * (n - 1) / 2$, which follows n^2 exponentially. This explains, in part, why some firms are willing to give away their products for free in order to rapidly grow an installed base.

In a variation of Metcalfe's Law, Reed's Law acknowledges that Metcalfe's Law understates the value created by a group-forming network (GFN) as it grows. In adding up all the potential two-person groups, three-person groups, and so on, that the members of a network could form, the number of possible subgroups of network participants is $2^n - n - 1$; this grows much more rapidly than either the number of participants, n, or the number of possible pair connections, $n * (n - 1) / 2$. So the value of a network increases exponentially, in proportion to 2^n, hence, Reed's Law.[54]

Now think about a product like the Nintendo Wii game console. Although Nintendo is happy when lots of people buy its Wii platform, the fact is that any single user gets the full benefit of owning and using his or her Wii at home independently of whether any other person in the world has bought one. Rather than being affected by the number of other adopters, the value of the Wii is a function of the number of games available. The second source of network externalities—a key source of value for many high-tech products—derives from the complementary products that are used with the base product: the games in the case of products like the Nintendo Wii or Sony's Playstation, the songs in the case of MP3 players, DVDs in the case of DVD players, and so forth.

Indirect network externalities arise from the fact that as more users adopt a particular product (as the installed base grows), developers of related goods are more likely to create complementary products, which in turn creates even greater value for each customer, developing a positive feedback loop. Implicit in this characteristic is a bit of a chicken-or-egg or a Catch-22 situation for both the developers and customers: Users won't buy the underlying technology platform (say, a Sony Blu-ray DVD player) if there are no complementary products to use on it (say, movies in the Blu-ray format), but developers won't create complementary products (movies in Blu-ray format) if there isn't a sufficiently large number of users (Blu-ray DVD owners) to justify it.

Despite conventional wisdom about indirect network externalities (in which the amount and availability of complementary goods stimulates growth of the hardware/base product), recent research shows that hardware sales can take off at low levels of availability of complementary products. Indeed, in five of nine markets studied (black-and-white TV, CD, i-mode, WWW, and laser disk), hardware sales lead software availability, while in three of nine markets (color TV, DVD, and Game Boy), hardware sales and software took off at the same time. For only one of the nine markets (CD-ROM) did hardware sales take off after software availability in a fashion that would be described as an indirect network effect.[55]

Two critical success factors in industries characterized by network externalities are (1) how quickly a firm can grow its installed base of customers and (2) the establishment of industry standards.

Industry Standards.   Because new technologies in an industry are often incompatible with each other, adopters of the new technology face increased fear, uncertainty, and doubt. Therefore, the introduction of **industry standards**—agreed-upon specifications based on a common architecture or design principles that ensure technical compatibility for products offered by different firms in the market—allow customers to gain compatibility across the various components of a product (say, seamless integration across hardware and software) and across product choices in an industry (say, across different types of computers, often referred to as a *plug-and-play* capability). For example, in the cellular telecommunications industry, compatibility allows base stations, switches, and handsets to work with each other across service areas. When customers are assured that the various components of a technology solution not only will be compatible and interchangeable, but also will allow communication with other users of similar products, their fear, uncertainty, and doubt are lessened. As a result, they will be more likely to purchase the product. For products characterized by network externalities, this begins a virtuous feedback cycle: The more users who buy, the more value each user receives. Moreover, with indirect network externalities, a larger installed base leads to greater availability of complementary products, which increases the value of the product to the customer, which in turn increases demand for that product by others. As shown in Figure 1.5, each of these factors works in a self-reinforcing manner.

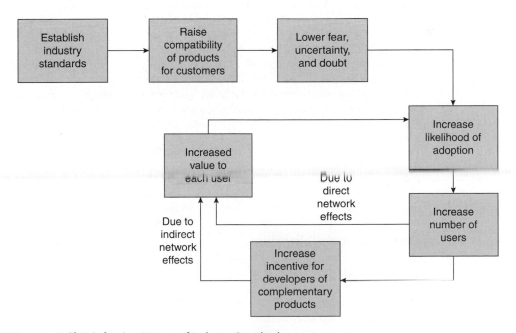

FIGURE 1.5   Self-Reinforcing Nature of Industry Standards

However, when new technologies developed by different companies operate on different—incompatible—underlying technological underpinnings, the result is frequently twofold:

1. Customers take a wait-and-see attitude to see which technology will emerge as the "winner" (de facto standard) in the marketplace. They don't want to purchase the wrong technology (that loses the battle for dominance).
2. Developers of complementary products also take a wait-and-see attitude. They don't want to spend valuable time and money developing complements for a new technology if they are not assured of a sizable revenue stream—which is a function of the size of the installed based (see #1).

Obviously, the combination of these two factors creates a double-jeopardy situation, which hopelessly stalls the sales takeoff of the new technology. See Box 1.1 for two industry examples.

Hence, an important issue is how long the industry will take to coalesce around a dominant design, when the market clearly favors one technical standard over another. Sometimes, the process for a dominant design to emerge can take many years—if one even emerges at all. For example, the camcorder category, introduced in 1984, supports multiple standards more than 20 years later, including the VHS system, super VHS, 8mm, and digital 8mm. Yet, in other cases, the process can happen quite rapidly. For example, the dominant design for a DVD player emerged just three years

---

BOX 1.1

Two Examples of a Lack of Industry Standards in U.S. High-Tech Markets:
Mobile Telecomm and HDTV

Mobile Telecomm

Why does the United States lag behind the rest of the world in mobile telecommunications? The approach taken by the Federal Communications Commission in the early 1990s was to "let the industry duke it out" to see which of the two underlying technology platforms for mobile computing (CDMA or TDMA) would win in the marketplace. The result was a bifurcated telecommunications network, with some areas of the country using one standard, and other areas using the other. In the meantime, the European Union met and established GSM (Global System for Mobile Communications) as the standard for Europe, and the rest is history: With an agreed-upon standard, the mobile phone companies could roll out next-generation cell phones; the network operators could provide services based on GSM technology; customers could rest assured that the products and services they bought in one area could be widely used. The inability to agree on industry standards—coupled with the wonderful SIM card (Subscriber Identity Module, a removable integrated circuit card that allows users to change phones by simply removing the SIM card from one mobile phone and inserting it into another mobile phone or broadband telephony device—which was *not* available in the U.S. until only recently)—ensured the superiority of the products and services available for mobile computing outside the United States at least until the time this material was written (end of 2008).

High-Definition Television

The lack of agreement on industry standards also explains, at least in part, why the adoption rate of high-definition television sets in United States has been so slow. The complementary products for HDTV include networks that must broadcast in a digital signal, as well as the producers who must make TV shows in digital format—not to mention the advertising issues as well. Yet, given that the industry couldn't agree on just what the underlying standard would be, all parts of the solution were stalled in development. It became something of an "I'll wait to see what X does" round-robin.

after product introduction in 1996. Surprisingly, one study of office products and consumer durables found that a dominant design is *less likely* to emerge in the presence of network externalities.[56] Essentially, strong network effects can create a wait-and-see attitude among consumers, leading to inertia and lowering the probability of the emergence of a dominant design. Other findings from this study showed that dominant designs are *more likely* to emerge when these conditions exist:

- The company engages in a relatively open business model, sharing know-how that allows a dominant design to emerge.
- New products are less radical. (More radical products offer initially low performance-to-price ratios; when coupled with their higher uncertainty, radical products dampen selection for dominant design.)
- There is high R&D intensity; higher investments in the category provide necessary variation and knowledge spillovers that create pressure to select a dominant design.

Moreover, dominant designs that do emerge do so *sooner* when these conditions exist:

- A large number of firms are in the "value network" (more firms have a vested interest in knowing what the dominant design will be).
- Standards emerge from a de facto process (rather than a de jure process, determined by a standards-setting board or body).

The first company to have its technology widely adopted may well become the de facto standard in the industry.[57] So, firms adopt what are known as "*get big fast*" strategies, such as low price (or even free) offerings to stimulate adoptions, licensing of underlying technology to other industry players so that more companies rely on the same platform, and other strategies to quickly grow the installed base of a particular technology. These strategies can result in what is known as **customer lock-in** (sometimes also known as *vendor lock-in*) where **switching costs,** or the additional time, money, and effort it takes to switch to a competing technology platform, make it very difficult for a customer to switch to a new vendor's offering. For example, think about the investment many people have made in their DVD collections. If they want to upgrade to a new high-definition format DVD player, lock-in arises from the lack of compatibility between their old DVD collection and the new Blu-ray players. Or, think about the ease and familiarity in using a particular software program. Switching costs are incurred from the time and frustration to learn a new program. Companies are vulnerable when introducing new product upgrades; because customers will be incurring the time and effort to learn the new version anyway, they are sometimes willing to explore competitive options.

Interestingly, the power of network externalities (arising from dominant designs, de facto standards, and lock in) can exist even when the technological standard is inferior to other designs. One of the earliest examples of a market clinging to an inferior standard is the QWERTY format for typewriter keyboards, which gets its name from the first six characters on the upper left side of the keyboard.[58] Because the type bars in the typewriters of the 1860s had a tendency to clash and jam when keys whose letters frequently appeared next to each other in words (e.g., "t" and "h") were struck in rapid succession, it was necessary to separate those letters. The result was that the QWERTY format was designed to *slow* typing speed. By the 1890s, better engineering had alleviated the problem of clashing type bars, and a new keyboard format that enabled faster typing was developed. However, these superior keyboards did not do well in the market; typists were so comfortable with the QWERTY design that attempts to use a new design seemed cumbersome. As a result, QWERTY is still the standard, even today.

Moreover, due to the power of network externalities, a very few firms—sometimes only one—control nearly all of the market share in a product category; this, in turn, can cause antitrust concerns. Some lawyers argue that dominant technology suppliers have gone beyond de facto standards and instead become *de facto monopolies* because their products control such a dominant share of the market. For example, the combined Intel/Microsoft standard, or the "Wintel (Windows/Intel)

TABLE 1.6 Reasons Companies Can't Agree on Industry Standards

- **"It's my way or the highway."** One company fervently believes that its technology is far superior to the other competing technologies that are also available. So, it is willing to adopt a go-it-alone strategy because it believes that its superior technology will prevail in the marketplace.
- **"We shall overcome!"** Even if a company doesn't believe its technology is superior, it may believe that it can overcome market forces—through sheer power, effort, and resources—to become the dominant design.
- **"But it's mine!"** Coming to terms with respect to industry standards typically requires sharing proprietary information. Yet, a company's intellectual property and underlying know-how forms the basis of its competitive advantage in the marketplace. Hence, it is unwilling to share such knowledge with other industry players, which is often required for using a common industry platform.
- **"I deserve more!"** Even if a company can accept the idea of sharing its intellectual property, the issue of how it should be compensated for doing so can rear its ugly head. Known as "tradability problems," technological innovations based on know-how are notoriously difficult to valuate in an economic/financial sense—and as a result, frequently partners cannot come to terms regarding licensing fees, royalties, or revenue sharing.

duopoly," resulted in both companies being investigated for antitrust issues. Moreover, because these two products are critical to the success of many computer companies (such as Dell, HP, and others), they are "essential facilities"[59] to which uneven access might also be a basis for unfair competition and, hence, antitrust suits. Similarly, Apple has faced antitrust investigations in the European Union for its de facto monopoly arising from its iTunes music downloads.

Although a challenger may attempt to displace an incumbent by introducing a radically improved technology, a technological advantage alone is often not enough to succeed in markets characterized by network externalities. To lure customers away from the existing standard, the new technology must somehow yield more value than the combination of value derived from the incumbent technology's functionality, its installed base, and the complementary products or services that make the incumbent more valuable.[60] This is a very difficult competitive situation.

Given the importance of establishing industry standards, why do so many industries find themselves in situations of competing technological platforms? Four reasons seem to underlie the lack of agreement on standards, as shown in Table 1.6.

Because one of the most common strategies to successfully establish industry standards relies on strategic alliances and partnerships (including industry consortia), further discussion of this topic will be delayed until Chapter 5, "Partnerships, Alliances, and Customer Relationships."

Types of Innovation

In addition to government-based definitions and identifying high-tech markets based on common characteristics, another approach to classifying high technology is based on the types of innovation represented. Technological innovations can be classified as (1) incremental versus breakthrough, (2) product versus process, (3) architectural versus component, (4) sustaining versus disruptive, or (5) organizational. Classifications of innovations—and how they relate to high-tech marketing—are discussed on the following pages. See Table 1.7 for definitions and an overview.

INCREMENTAL VERSUS BREAKTHROUGH (RADICAL) INNOVATION   Characterizing high-tech industries as only those that develop breakthrough (radical) innovations is overly simplistic. Clearly, many high-tech companies develop incremental innovations. For example, in the area of software, Windows Vista introduced many features that were already available on the Mac OS X operating system. Similarly, Dell's new colorful computers would also be considered incremental in nature.

TABLE 1.7 Common Types of Innovation

1. **Incremental versus breakthrough (radical) innovations.**

   Incremental innovations: Continuations of existing products, methods, or practices; generally minor improvements made with existing methods and technology (both process and product); evolutionary as opposed to revolutionary.

   Breakthrough innovations: Totally new products that involve considerable change in basic technologies and methods; revolutionary ideas that can create new markets.

   (Note: This is the primary classification used in this book, and the text and other figures provide substantially more detail on this classification.)

2. **Product versus process innovations.**

   Product innovations: New products that offer improvements in functional characteristics, technical abilities, ease of use, or other dimensions; embodied in the outputs of an organization–its goods and services sold on the market (which could be either incremental or breakthrough).

   Process innovations: New techniques of producing goods or services; often oriented toward improving the effectiveness or efficiency of production processes (again, could be either incremental or breakthrough) or to facilitate the discovery of underlying scientific properties of technological domains (such as lab equipment to sequence genes).

   (Note: The product innovations of one firm may be used as a process innovation by another, and vice versa. *Example:* Semitool Inc. designs and manufactures equipment used to produce semiconductors [chips]. Its key customers, including National Semiconductor and Intel, for example, run chip fabrication plants. The company's product innovations require state-of-the-art design and manufacturing processes in its own manufacturing plants. Hence, it also develops and patents many process innovations to produce its high-tech products.)

3. **Architectural (platform) versus component (modular) innovations.**

   Architectural (platform) innovations: New foundations or fundamentals of how the various components of a system work together to function; typically based on scientific principles that make the new architecture different from existing technological platforms; such architectural innovations may be considered radical.

   Component (modular) innovations: New parts or materials within the same technological platform. For example, although magnetic tape, floppy disk, and zip disk differ by components or materials, all three are based on the platform of magnetic recording.

4. **Sustaining versus disruptive innovations.**

   Sustaining innovations: New products that target demanding, high-end customers with improved performance (typically through incremental innovations, but in some cases, may be breakthrough innovations).

   Disruptive innovations: New products or services that are simpler, more convenient, less sophisticated, and/or less expensive than existing products; typically appeal to customers at the lower end of the market.

   - *Low-end disruption* attracts low-end customers initially and then moves into more upscale markets over time as the technology improves (does not necessarily create a new market, but "picks off" an established firm's least attractive customers).
   - *New-market disruption* converts previous non-customers into new customers, thereby creating a new market.

5. **Organizational innovations** that create or alter business structures, practices, and models; may include innovations in business models/strategies and marketing (either incremental or radical).

   Business strategy innovations: Changes in the way business is done in terms of capturing value; may include new methods of financial management, or innovations aimed at social needs and issues.

   Marketing innovations: Changes in the way marketing is done, for example, the use of the Internet for advertising purposes, improvements in product design or packaging, product promotion, pricing, service strategies, supply chain strategies, and so forth.

Sources: Melissa Schilling, *Strategic Management of Technological Innovations*, 2nd ed. (Boston: McGraw-Hill/Irwin, 2008); "Innovation," Wikipedia, 2007, online at http://en.wikipedia.org/wiki/Innovation; and Ashish Sood and Gerard J. Tellis, "Technological Evolution and Radical Innovation," *Journal of Marketing* 69 (July 2005), pp. 152–168.

Incremental   **Breakthrough**

Incremental	Breakthrough
• Extension of existing product or process	• New technology creates new market
• Product characteristics well defined	• R&D invention in the lab
• Competitive advantage on low-cost production	• Superior functional performance over "old" technology
• Often developed in response to specific market need	• Specific market opportunity or need of only secondary concern
• Demand-side market	• Supply-side market
• Customer pull	• Technology push

FIGURE 1.6   Continuum of Innovations

High-tech marketers must be cognizant of these two different types of innovations, because they have very different implications.

As shown in Figure 1.6, **incremental innovations** are continuations of existing products, methods, or practices already in the market; they are generally minor improvements made to existing methods and technology (both process and product), responding to short-term goals. They are *evolutionary* as opposed to *revolutionary* innovations. Both suppliers and customers have a clear conceptualization of the products and what they can do; existing products in the market are fairly close substitutes to each other.[61] Incremental innovations occur in *demand-side markets*,[62] meaning that product characteristics are well defined and customers can articulate their needs about what they would like in new innovations. Incremental innovations move forward along an existing technology trajectory (S-shaped curve), with little uncertainty about outcomes. Other examples might include innovations in internal combustion engines that improve fuel efficiency, or improvements in processor speed or battery life that enhance an existing computer's performance.

In the manufacturing context, producers of a mature product who have achieved high volume in their production processes have a strong incentive to make only incremental innovations that leverage their investments in existing technological platforms,[63] potentially making them less open to radical change and more vulnerable to obsolescence. In mature high-tech markets, competitive advantage is frequently based on low-cost production. Economies of scale may be very important, and pricing based on experience curve effects (where costs decline by a fixed and known amount every time accumulated volume doubles) may lead these producers to *process innovations*, new techniques that lower the costs of production, as well as *business strategy innovations* in which outsourcing and other arrangements are made to lower the costs of doing business.

On the other hand, **breakthrough (radical) innovations** are "so different that they cannot be compared to any existing practices or perceptions. They employ new technologies, they launch entirely novel products or services, and often, they create new markets. Breakthroughs are conceptual shifts that make history."[64] In standard marketing parlance, they are discontinuous or revolutionary innovations.[65] Generally, radical innovation involves considerable change in basic technologies and methods, often created by those working outside mainstream industry and outside existing paradigms.[66] Competitive advantage for a breakthrough technology is typically found in the superior functional performance that the new innovation offers over existing methods or products. In other words, it typically involves a leap to a new S-shaped curve. There may be considerable opposition to the proposed innovation, and questions about the ethics, practicality, or cost of the innovation may be raised. People may question if it is, or is not, a genuine advancement. Due to the considerable uncertainty of breakthrough innovations, seldom do companies achieve their breakthrough goals. But, with higher risk comes the potential for higher rewards.

Conventional wisdom generally advises listening to the market, but breakthrough innovations often arise from *supply-side markets,*[67] which are characterized by innovation-driven practice in which R&D generates new ideas, and then specific commercial applications or target markets are considered only after the innovation is developed. For these reasons, these markets are sometimes referred to as **technology-push situations:** Because the radical innovation is developed by an R&D group (in a company, university, or research laboratory) that often hasn't specifically thought about a particular commercial market application during the development process, the technology is *pushed* into the market to see where it might best be utilized. These innovations are created independently of the vision of the uses they might serve. For example, Tim Berners-Lee, a software engineer, assembled a network of interconnected computers to share and distribute information easily and cheaply in 1980, more than a decade before Marc Andreessen developed the first widely used Web browser.[68] In the United States, the Internet emerged as an outgrowth of the need for the U.S. military computer system to survive a nuclear attack.[69]

Gordon Binder, CEO of biotech firm Amgen, says, "Most pharmaceuticals companies, and quite a few biotech ones as well, are basically market-driven. They see that large numbers of people have a particular disease and decide to gather some scientists to do something about it."[70] However, rather than start with the disease and work back to the science, Amgen does the opposite. It takes a new scientific innovation and finds a unique use for it. The company developed an immune booster, for instance, that can help keep the side effects of chemotherapy from killing cancer patients. And a collaborative arrangement with a professor at Rockefeller University discovered a gene that may yield new treatments for obesity.

In other cases, breakthrough innovations are developed as a new way to meet an existing need, or in response to the identification of an emerging need. Regardless of whether the innovation originates from "pure" science or in response to a need, the new technology then creates a new market for itself.

DIFFERENCES BETWEEN BREAKTHROUGH AND DISRUPTIVE INNOVATION   The notion of sustaining versus disruptive innovation is similar to incremental versus breakthrough innovation (refer again to Table 1.7). Although the terms *breakthrough* and *disruptive* innovation are sometimes used synonymously, they refer to two different phenomena. As described earlier in this chapter, breakthrough (radical) technological innovations are based on the notion of "new to the world," or a substantially new technology relative to what already exists in the industry.

Rather than defining **disruptive innovation** in terms of technology per se, the distinguishing feature is whether the innovation appeals to a low-end or emerging customer segment rather than existing mainstream customers.[71] A key aspect of the theory of disruption is that companies continue to increase the sophistication of the feature set in their product offerings at a higher rate than customers' capabilities to use that sophisticated feature set. This gap in the marketplace then creates opportunities for new companies to enter the market with lower-end products, first selling them to lower-end customers (the least attractive to established companies), but eventually making inroads into the established company's customer base. Because disruptive innovations are (at least initially) based on a feature set that existing (mainstream) customers do not find all that attractive, and because they may underperform existing products on the attributes that mainstream customers value, the technologies they are based on may appear crude or simplistic to established companies. However, over time, not only do further developments improve the disruptive innovation's performance but mainstream customers begin to see the price/performance value as well. These two changes ultimately lead to a situation where the innovation attracts more mainstream customers—at which points it disrupts incumbents' businesses.

Established firms tend to underestimate the firms introducing new technologies, either because such firms are small or the new technology appears crude. If they believe the new technology could be a credible threat, they might attempt to acquire the start-up company that developed the technology (sometimes with the intent of "burying" it within the walls of their own corporation), or they may threaten legal action, citing patent infringement or some other infraction, with the intent either to stifle the new innovation or to burn up the start-up's cash resources.

In other scenarios, an established firm threatened by a new technology may try to make even more improvements in the current-generation technology—say, by making investments to increase the efficiency of its manufacturing facility in producing the current-generation technology—yet, being the low-cost producer of an obsolete product is not worth much.[72] Or, the established firm might try to hedge its bets by investing simultaneously in improving the current-generation technology and in developing new technology. Because of all these reasons, technology substitutes can creep up slowly on established firms and then explode in terms of market performance. Because established firms may suffer from what is called the *incumbent's curse,* they must strive to create an organizational culture that stimulates a willingness to engage in their own creative destruction—pursuing next-generation technology, even if the new technology will make their existing technologies obsolete and will cannibalize the revenue stream from their legacy-generation technologies sold to existing customers.[73]

Base-of-the-Pyramid Strategies.   Disruptive strategies are commonly used in **base-of-the-pyramid markets**—also known as **bottom-of-the-pyramid markets**—which refer to the largest but poorest socioeconomic group: the 4 billion people in the world who live on less than $2 per day, typically in developing countries. Base-of-the-pyramid business strategies and technological innovations deliberately target that group. Recent influential books address various aspects of this business strategy.[74] Although coined by U.S. president Franklin D. Roosevelt in 1932 referring to "the forgotten, the unorganized but the indispensable units of economic power . . . that build from the bottom up and not from the top down . . . of the economic pyramid,"[75] the current usage comes from C. K. Prahalad and Stuart L. Hart.[76] Rather than thinking of the poor as victims who need charitable handouts, business and government should see them as resilient and creative entrepreneurs, as value-demanding consumers, and as business partners and innovators. Multinational companies and social entrepreneurs who serve these markets in ways responsive to their needs will raise standards of living and create the middle class of tomorrow. For example, Dow Chemical is increasing its R&D in products such as roof tiles that deliver solar power to buildings and water treatment technologies for regions with shortage of clean water.[77] By developing new technologies that serve the needs of emerging markets, Dow is simultaneously pursuing new market opportunities and environmentally-friendly business practices. Similarly, Philips Electronics—citing the fact that 85% of people who live in developing nations are faced with health care shortages—has deployed medical vans with satellites for communicating with urban doctors, water-purification technologies, and smokeless wood-burning stoves.

Base-of-the-pyramid business strategies simultaneously meld a company's **corporate social responsibility (CSR)** efforts with its business initiatives. CSR efforts are designed to enhance the social welfare of some (typically impoverished or underserved) group in society. Best-practices marketing and business strategies seek the simultaneous pursuit of profit and social welfare, as described in the accompanying Technology Solutions to Global Problems box. CSR is an important topic that will be discussed more fully in Chapter 13.

IMPLICATIONS OF BREAKTHROUGH VERSUS INCREMENTAL INNOVATION   Many implications arise from the differences between breakthrough and incremental innovations. One major implication for marketing occurs in the supply chain; another involves marketing strategies.

A Supply Chain Perspective of Innovation.   Different types of innovations tend to occur at different levels in the supply chain. A **supply chain** depicts the flow of product from producer to consumer. A sample supply chain—albeit a simplistic one—for the automotive industry is shown in Figure 1.7. Customers who buy cars—both consumers for personal use as well as businesses who need fleets for salespeople, company cars, and so forth—purchase cars from car dealers (either through a physical location or on the Internet). Car dealers replenish their inventories from the car manufacturers. To produce the cars, manufacturers must either make or buy all the requisite compo-

TECHNOLOGY SOLUTIONS FOR GLOBAL PROBLEMS
One Laptop per Child
By Jamie Hoffman

Photo reprinted with permission of One Laptop per Child, Cambridge, Mass.

As developed nations embrace the reality of a technologically savvy, "flat" (i.e., connected, globalized) world, many developing nations struggle for mere access to basic necessities like clean water, health care, and education. The founders of One Laptop Per Child (OLPC) believe that alleviating poverty in emerging nations requires tapping into the potential of children's ability to learn, share, and create. To harness and develop the potential brainpower of youth, OLPC has committed itself to creating an affordable laptop for the children of the most remote and poor nations of the world. OLPC's solution is the open-source XO laptop.

Priced around $200 and with an expected five-year life span, the XO laptop features distinguishing hardware, software, interface, and design components. The ultra-durable computer is roughly the size of a textbook. It was designed for a child, with a smaller rubber-sealed keyboard and a touchpad available for pointing, drawing, or writing. The battery does not contain any hazardous materials, and every laptop can be powered by the sun or with a foot pedal. To decrease the chance of failing internal components, XO was designed without an internal hard drive and with only two internal cables.

The XO uses the open-source Red Hat's Fedora Core 6 version of the Linux operating system for its software. The computer features various open-source applications that come standard on a typical computer such as an Internet browser, word processor, e-mail client, document viewer, multimedia environment, and games, to name a few.

Production of the XO laptop began in 2007 and the first 5 million units have been ordered by Nigeria, Argentina, Brazil, Thailand, Uruguay, and Libya.

Although other low-priced laptops are available (including the Asus Eee PC, Intel's Classmate PC, and the Zonbu Notebook) and although other organizations exist to give recycled laptops to impoverished people (CareCircle; InterConnection.org), a unique feature of the OLPC initiative is its partnership with governments and non-governmental organizations (NGOs) for distribution.

Sources: Nick Bushack, "As Laptop Begins Production, OLPC Faces Competition," *The Tech* Online Edition, February 5, 2008, available at www-tech.mit.edu/V127/N66/olpc.html; and OLPC website, www.laptop.org.

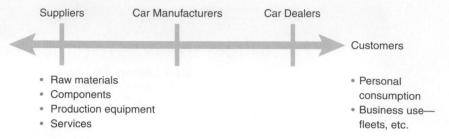

FIGURE 1.7 Supply Chain

nents: glass, metal, tires, electrical assemblies, drivetrains, and so forth, as well as production equipment (e.g., CAD-CAM systems for car design).

In many cases, breakthrough innovations occur at the supplier level (upstream) in the supply chain, rather than in the product itself (i.e., cars) at the consumer/user level (downstream), where the innovations may seem incremental in nature. For example, think of major innovations in the auto industry, such as a hybrid-powered gas/electric car, or a pure hydrogen fuel-cell-powered vehicle. At the consumer (end user) level, people will still be driving a four-wheeled vehicle using steering wheels, a gas pedal, and brakes for their transportation needs—the nature of driving hasn't really changed. Yet, these innovations have required major breakthroughs in battery design, wiring, and engine function at upstream levels in the supply chain rather distant from the end user, who is still using the same mode of transportation—a car—despite the innovation.

To be sure, innovations can occur in the nature of the product itself; for example, as this chapter's opening vignette suggests, maybe for commuting in the year 2050, we'll be using jet-packs to propel ourselves through the air or we might have flying cars! However, more often than not, breakthrough innovations occur at upstream (higher) levels in the supply chain, affecting the design, componentry, and production processes of the product, rather than revolutionizing the basic concept of the product itself.

Other examples of the prevalence and predominance of breakthrough innovations occurring at higher levels in the supply chain can be found across many industries, both high- and low-tech:

- *In the food industry:*   Breakthrough innovations have occurred at the genetic level in terms of seeds used to grow crops.
- *In the fuel industry:*   Major innovations have occurred in the transformation process of biofuels that are used to provide energy.
- *In the fashion industry:*   Computers are being used to scan a 3-dimensional image of the body in the dressing room to ensure that the sizing of clothes will be a perfect fit for each individual. Moreover, nanotechnology is being used to change the composition of fabrics to make them stain- and wrinkle-resistant—but the individual is still buying and wearing clothes (albeit at a higher price).

Many innovations occur at levels in the supply chain that are removed from end users. Hence, one cannot assume that an industry is *not* high tech simply because the products don't change much at the end user level. Has the way in which we gas up our cars changed due to the innovations in oil exploration and extraction? Not really. Do we use our computers differently because they have a new chip design? Not substantially. The supply chain can be a useful tool to help understand and define high tech in terms of the underlying technological inputs, rather than just the nature of the end products we use.

Differential Marketing Strategies.   A second implication of the differences between incremental and breakthrough innovations is that the processes firms use to manage incremental innovation are not only *not applicable* but also may be *detrimental* to the management of radical innovation.[78] The firm's challenge is to be able to manage both types of in... because both are needed for the long-term health of the organization.... 1980s and early 1990s, U.S. and European firms were competitively c... tries. U.S. firms took a beating in memory chips, office and factory a... tronics, and automobile manufacturing. These behemoth co... outmaneuvered by new competitors from other corners of the earth. K... camcorders reduced its home movie business to cinders. Xerox's lock on the ... was broken by Canon, Sharp, and others. Consumer electronics products made by Motor... Zenith, and RCA were largely displaced by better, cheaper, faster versions offered by Sony, Panasonic, and Toshiba. On the automotive front, Toyota, Honda, and Nissan expanded their inroads in the North American market, winning virtually every kudo for quality and reliability. Effective incremental innovation and dramatic improvements in operating efficiency were the two keys to success of these new leaders.

In response, firms increased their competencies in managing the development of incremental innovation in existing products and processes, with an emphasis on cost competitiveness and quality improvements. Extensive study of incremental innovation by both business managers and academic researchers led to a variety of prescriptions for improvement: Six Sigma quality in manufacturing, concurrent engineering, just-in-time inventory management, and stage-gate systems for managing the new product development process. These prescriptions were widely adopted and helped many companies gain their competitive positions in the world marketplace.

These prescriptions are based on the fundamental premise that the firm understands its market's needs and wants and is able to leverage its current technological base to fulfill those needs quickly, cheaply, and reliably. All aspects of the product development project are managed simultaneously by a team composed of representatives from every function in the business: engineering, production, marketing, cost accounting, and, often, suppliers and customers. Having all constituents present on the team, the argument goes, ensures that products are not conceived that cannot be designed, or that products are not designed that cannot be manufactured. This is the world of incremental innovation.

Managers' attention to incremental innovation, however, came at a price. It diminished the focus and capacity of their companies to engage in truly *breakthrough* innovation, which offers the promise of growth through whole new lines of business and the development of new markets. Central R&D labs, traditionally the source of radical innovation ideas, were redirected to serve the immediate needs of business units. Those business units, always under pressure to maximize short-term financial performance, were reluctant to invest in high-risk, long-term projects. The consequences can be disastrous, as dominant market leaders toppled by new, innovative competitors have found. The important message is clear: To remain successful over the long haul, firms must be adept at simultaneously managing both incremental and radical innovation.

However, managing these two types of innovation requires different tools, organizational structures, processes, evaluative criteria, and skills. We develop this notion as the "contingency theory for high-technology marketing"—the guiding framework for the remainder of this book.

THE CONTINGENCY MODEL FOR HIGH-TECH MARKETING

"Market planning that explicitly recognizes and accounts for the strategic distinction between market-driven and innovation-driven research goes a long way toward yielding better corporate performance."[79] In other words, the appropriate marketing strategy is contingent upon the type of innovation. By appropriately matching marketing strategies and tools to the different types of innovation, companies enhance their odds of success. Figure 1.8 provides a picture of how the **contingency theory** of high-tech marketing works.

Although many applications of the contingency theory of high-tech marketing are explained in subsequent chapters (e.g., outsourcing strategies and supply chain management strategies), four

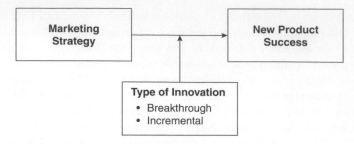

FIGURE 1.8 Contingency Theory of High-Technology Marketing

brief illustrations are provided here to give a flavor of how it works. First, the nature of the interaction between R&D groups and marketing departments depends on the *type of innovation.* Because technological prowess is key in supply-driven markets, the role of R&D is critical. Research and development is likely to give direction to marketing people in seeking commercial applications for technological advances. A critical issue is the original market that the firm chooses to pursue. The role of marketing is to identify markets.[80]

A second illustration relates to the *type of market research tools* that are appropriately used. Gathering market data to guide the development and marketing of breakthrough products can be difficult. Often the customer doesn't understand the new technology and finds it difficult to articulate performance criteria for the product.[81] For example, if a person has never used jet propulsion for transportation, how will he or she specify what is good performance? Hence, the value of customer feedback through standard marketing research can be questionable. But the voice of the customer remains vitally important; more commonly, qualitative and observational research is used to guide breakthrough product developments.

Or, in situations where the customer might understand the technology, as in the case of lead users who face a need well before the majority of other customers in a market do, the users themselves may be the innovators.[82] For example, Eric von Hippel finds that in many manufacturing processes, manufacturers who face a particular problem in the production process innovate a solution themselves. An example he cites is Lockheed Martin, which pioneered a new machining technique to speed the removal of titanium metal by up to 20 times with a new face-milling tool that shears rather than chips the metal. The tool was later introduced commercially and expanded to other applications including stainless steels and other hard-to-cut alloys.

A third implication relates to the *role of advertising.* For breakthrough products, once a viable commercial market is identified, marketers must educate customers and stimulate primary demand for the product class as a whole. This leads to the fourth marketing application of contingency theory: *pricing.* Because the breakthrough technology may offer a significant advantage over the former mode of doing things, customers may be willing to pay a premium for the new technology.

When it comes to incremental innovations (as opposed to breakthrough innovations), these four aspects—the R&D–marketing interaction, market research tools, advertising, and pricing—must be managed differently. For incremental innovations, the role of marketing is critical. In these situations, customers can play a major role in product development. They can confidently articulate their desires and preferences. In such cases, firms can use standard marketing research tools to identify customer needs, passing the information to R&D, which then develops the appropriate innovations to satisfy those needs. Marketing takes a lead role in such cases. In incremental innovation, companies tend to use more standard management controls and formal planning groups.[83] Advertising typically stimulates selective demand, building preference for the firm's specific brand or product. Pricing is more competitive.

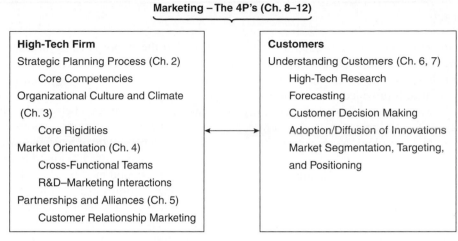

FIGURE 1.9   Framework for High-Tech Marketing Decisions

The resource allocations of a firm should match the long-term financial attractiveness of the project. Some breakthrough projects might not have a large market potential to start with. And many incremental innovations may absolutely require a major investment (e.g., if a firm's product is hopelessly out of date in a very large market). Hence, marketers should not confuse the nature of the innovation with potential payoff and wrongly assume that breakthrough innovations will have a large payoff.[84] Indeed, a 2008 *Harvard Business Review* article warns that using traditional financial tools can destroy a firm's capacity to innovate.[85] Moreover, empirical research shows that companies must balance their investments in value creation (via R&D expenditures to create new innovations) and value appropriation (advertising to reap the benefits of existing innovations) in order to maximize financial performance.[86]

Framework for High-Technology Marketing Decisions

Figure 1.9 provides the organizing framework for the remainder of this book. On the left side of the figure are the internal considerations that a firm must address and understand as the foundation to effective marketing. On the right side of the figure are the customer considerations. The link between the firm and its customers is enacted through the four elements of the marketing mix: product, place, price, and promotion. The interactions between a company and its customers occur in a regulatory and social milieu that also must be considered.

In light of the high degree of market, technological, and competitive uncertainties surrounding high-tech products and markets, the need for effective marketing in high-tech industries is paramount. High-tech products and services are introduced in turbulent, chaotic environments, in which the odds of success are often difficult to ascertain at best and stacked against success at worst. This book provides frameworks for systematic decision making about marketing in high-tech environments. It offers insights about how marketing tools and techniques must be adapted and modified for high-technology products and services. Effective high-tech marketing includes a blend of marketing fundamentals and the unique tools explored in this book. The text highlights possible pitfalls, mitigating factors, and the how-tos of successful high-tech marketing.

Because of the importance of following a basic marketing plan in conceptualizing and implementing marketing strategy for any product, be it high- or low-tech, Appendix 1.B presents an outline for developing a basic marketing plan.

Summary

This chapter provided an introduction to the world of marketing for high-technology products and innovations. In addition to providing an overview of the scope of marketing activities, it provided an in-depth examination of the various ways to define and characterize high-tech industries. Despite the prevalence of government definitions of high technology, such definitions face many drawbacks. As a result, we advocate characterizing high-tech markets based on a set of common features including market uncertainty, technological uncertainty, and competitive volatility. The intersection of these three characteristics defines the domain of high technology and offers specific implications for marketing decisions.

In addition, many high-tech markets are characterized by network externalities, a situation in which the more users that adopt a specific high-tech product, the more value that each user receives. Network externalities can give rise to de facto monopolies, and also point to the importance of industry standards in high-tech markets.

High technology can also be characterized by a wide variety of innovations. Although we rely primarily on the distinction between breakthrough and incremental innovations, disruptive innovations have played a key role in the evolution of many industries. Not only have disruptive innovations been used to serve the needs of base-of-the-pyramid markets, they also have wreaked havoc with many companies who have been relatively wedded to their existing technologies and customers. In addition, many breakthrough innovations frequently occur at levels upstream in the supply chain (versus at the end user level).

In order to be effective, marketing strategies must be tailored to the type of innovation. This notion of matching marketing to the type of innovation is known as the *contingency theory of high-tech marketing,* which will be a common theme running throughout this book.

Discussion Questions

Chapter Concepts

1. Why is the failure rate of more innovative products higher than the failures rates of new products in general? What are the implications of these failure rates for high-tech companies?

2. What is the difference between the marketing of high technology and the use of high technology in marketing?

3. What is marketing? What are the three levels that encompass the scope/domain of marketing? Provide sufficient detail for each of the three levels to communicate a sophisticated understanding.

4. What are the two approaches governments use to define high tech? What are their strengths and weaknesses?

5. What are the three characteristics common to high-tech environments? Be sure to provide detail for each dimension of each characteristic, and an example of each.

6. What are the pros and cons of the various definitions of high tech? Of these definitions, which do you think is the most useful? Why? Based on that definition, draw a continuum of low- versus high-tech industries.

7. The simultaneous presence of market uncertainty, technological uncertainty, and competitive volatility characterizes most high-tech environments. Explain.

8. What is the FUD factor? Why is it a factor in the high-tech environment?

9. What is a dominant design? Why is it important in high-tech markets?

10. Explain Geoffrey Moore's chasm and why it is important to bridge.

11. What are the five factors giving rise to technological uncertainty? Explain each factor.

12. What is convergence? How does it contribute to competitive volatility?

13. What is product form competition? How does it contribute to competitive volatility?

14. What is marketing myopia and what are its three sources? Provide an explanation for each. How does marketing myopia relate to the innovator's dilemma?

15. What are technology life cycles? Does empirical evidence support the notion of S-shaped curves in technology development? What are the implications of understanding technology life cycles?

16. What is Moore's Law?
17. What is creative destruction?
18. Define and give an example of each of the following: unit-one costs, tradability problems, and knowledge spillovers.
19. What are network externalities (both direct and indirect)?
20. How does Metcalfe's Law apply to firms who are willing to give away their products for free?
21. What are two critical success factors in markets characterized by network externalities?
22. Explain why standards are important to both marketers and customers. Explain the self-reinforcing nature of industry standards, and some of their implications.
23. What are the factors that affect how long an industry will take to coalesce around a dominant design?
24. Why do companies in some high-tech markets pursue a "get big fast" strategy?
25. Given the importance of industry standards, why do companies find it hard to establish agreed-upon standards?
26. Explain the following types of innovations: incremental versus breakthrough; product versus process; architectural (platform) versus component (modular); sustaining versus disruptive; organizational innovations.

27. What is a technology-push situation?
28. Explain disruptive innovation. How does it differ from breakthrough (radical) innovation? How do incumbents frequently respond to disruption? How should they respond?
29. What are base-of-the-pyramid markets? Why do disruptive innovations frequently occur in base-of-the-pyramid markets?
30. What is corporate social responsibility?
31. What is a supply chain? Why do many innovations arise at the upstream levels of a supply chain? Think of some examples in which low-tech industries have been transformed by high-tech innovations. Where in the supply chain have these innovations originated?
32. What are the implications for a company's strategy in terms of pursuing incremental versus breakthrough innovations?
33. What is the contingency theory of high-technology marketing? What marketing tools are appropriately used for incremental innovations? What marketing tools are appropriately used for breakthrough (radical) innovations?
34. Does high-tech marketing need to be different from marketing of traditional products? Why? How?

Application Questions

1. Based on one of the types of new cars (even flying cars and personal jet-packs) in the opening vignette, discuss the types of market, technological, and competitive uncertainty that are most important. What are the marketing implications?
2. Is the Tata Nano car (from the opening vignette) a disruptive innovation, as defined later in the chapter? If so, what are the implications for industry incumbents?
3. You've been asked to assess the market potential for mobile advertising (advertising over mobile phones, sometimes referred to as the "third screen" for advertising media, coming after TV and computer/Internet advertising). What factors will affect the adoption rate of this media platform by advertisers? To what extent do network externalities have an effect?

4. Select one car innovation that would be considered more incremental in nature, and one that would be considered more breakthrough in nature (or two innovations of your choice from another industry). Explain which of the characteristics of breakthough versus incremental innovation your examples exhibit. Draw a supply chain and show where the innovations occur. Offer specific marketing implications based on the contingency theory for high-tech marketing.
5. Select an industry that is characterized by network externalities. Discuss the implications for industry standards, get-big-fast strategies, dominant designs, and other issues.

Glossary

base-of-the-pyramid (bottom-of-the-pyramid) markets The largest but poorest socioeconomic group: the 4 billion people in the world who live on less than $2 per day, typically in developing countries.

breakthrough (radical) innovation Radical innovation that cannot be compared with any existing practices or perceptions; the technology is so new that it creates a new product class.

chasm The gulf between early adopters/visionaries and early majority/pragmatists of a high-technology product; visionaries are quick to adopt new technologies, but pragmatists take a wait-and-see attitude. Many high-tech products fail because the companies marketing them do not understand how to successfully cross the chasm.

competitive volatility Rapid and unpredictable changes in the competitive landscape: which firms are one's competitors, their product offerings, the tools they use to compete.

contingency theory Theory stating that the effects of one set of variables on another (e.g., marketing variables on new product success) depend on a third variable (e.g., type of innovation). The *contingency theory of high-technology marketing* says that in order to successfully commercialize high-tech products, the nature of the marketing strategies must be appropriately matched to the type of innovation.

convergence A situation in which different technologies begin to offer the same functionality, which means that new competition will be found in different product classes; for example, cell phones function as digital music players and digital cameras.

corporate social responsibility (CSR) Business initiatives that meld economic (profit) objectives with social welfare through such strategies as employing "green" business practices, donating to nonprofit organizations and social causes, and developing technologies and business models to solve problems in base-of-the-pyramid markets.

creative destruction The notion that in order to remain viable, a firm must be willing to destroy the basis of its current success. If a firm doesn't constantly innovate and reinvent itself, it will find its market share eroded by new competitors who are willing to do so.

customer lock-in A situation in which customers of high-tech products find it hard to switch to a competing vendor's products, either because of network externalities, industry standards/dominant designs, or product familiarity.

disruptive innovation Innovation that appeals to a low-end or emerging customer segment rather than to existing mainstream customers; initially based on a product feature that existing customers did not find attractive. May appear crude and simplistic to established companies, but further developments improve performance, attracting more mainstream customers, which disrupts incumbents' businesses.

dominant design The format that emerges as the agreed-upon standard, or industry leader, when industries have two or more competing formats for the underlying technology platform.

high technology Cutting-edge or advanced technology, the definition of which can change over time.

incremental innovation Generally minor improvements made with existing methods and technology. May involve extension of products already on the market; product features are typically well defined and understood by customers.

industry standards Agreed-upon specifications for underlying technical compatibility that are typically based on a common, underlying architecture or set of design principles for products offered by different firms in the market.

innovator's dilemma The conflict a company faces between continuing to allocate resources to serve current customers with incrementally improved products versus allocating resources to develop new products that might cannibalize a company's existing revenue stream or that its current customers might find less appealing.

installed base The users who have adopted a particular technology or platform.

knowledge spillovers The expansion of knowledge that occurs when synergies in the creation and distribution of know-how further enrich a related stock of knowledge, which creates increasing returns in the development of related technologies.

marketing The set of activities, processes, and decisions to create, communicate, and deliver products and services that offer value to customers, partners, and society at large; a *philosophy of doing business* that reflects shared values and beliefs about the importance of creating value for customers by solving meaningful problems.

marketing mix The functional level of marketing activities, composed of the "4 P's of marketing": product, price, promotion, and place. Consistency across the 4 P's is vital.

marketing myopia The tendency of managers to be narrow-minded or shortsighted in their views about their industry contexts and business strategies; they may be blindsided by new forms of competition that arise from outside their industry boundaries.

market uncertainty Ambiguity about the type and extent of customer needs that can be satisfied by a particular technology, arising from consumer fear, uncertainty, and doubt about the needs or problems the new technology will address and meet.

Metcalfe's Law The principle that the value of a network scales as n^2, where n is the number of persons connected; the rapid takeoff is where the utility or value of the innovation increases exponentially because a critical mass of users has adopted it. (See also network externalities.)

moment of truth   Each touch point a customer has with a company (technical support, at retail stores, billing, pre-sales inquiries) that can either strengthen or undermine the relationship.

Moore's Law   The principle that performance of high-tech products doubles every 18 months, with no increase in price. Stated differently, every 18 months or so, improvements in technology cut price in half for the same level of performance.

network externalities   Situations where the value of a product increases as more users adopt it. *Direct* network externalities exist when the value any one user gets from the product is directly related to the number of other users. (See also Metcalfe's Law.) *Indirect* network externalities arise from the fact that as more users adopt a particular product, developers of related goods are more likely to create complementary products, in turn creating even greater value for each customer.

product form competition   Different categories of products or technologies that satisfy the same underlying customer need but with different value propositions; for example, Skype offers phone calls over the Internet (via free software) compared to landline or cellular phone calling.

supply chain   The flow of a product from producer to consumer.

switching costs   Factors that make it hard for customers of high-tech products to switch to a new company's products; may arise from the additional time, money, and effort it takes to adopt a new technology platform.

technological uncertainty   Skepticism about whether the technology will function as promised or be available when expected by the company providing it.

technology life cycle   An S-shaped curve depicting investments made in a particular technology relative to its price–performance ratio.

technology-push   The situation that exists when radical innovations are developed by scientists without concern for commercial market applications; the technology is "pushed" into the market to see where it might best be utilized (also referred to as a *supply-side market*).

tradability problems   Problems that arise because high-tech companies often want to sell their underlying know-how (or intellectual property), but it is difficult to value that knowledge because it is tacit and resides in people and organizational routines.

unit-one costs   A situation where the cost of producing the first unit of a product is very high relative to the costs of reproducing subsequent units.

Notes

1. Gibson, Michael, "Flying Car about to Take Off?" *Technology Review,* October 10, 2007, available online at www.technologyreview.com/ Infotech/19499/?a=f; Kanellos, Michael, "Flying Car Ready for Takeoff?" *CNET News,* February 15, 2006, available online at http://news.com.com2102-11289_3-6040007.html?tag= st.util .print; and "Introducing the Transition," Terrafugia website, 2008, www.terrafugia.com/vehicle.html.

2. Jet Pack International website, 2008, www.jetpackinternational .com.

3. Reynolds, Kim, "First Drive: 2008 Tesla Roadster," Motor Trend website, 2008, www.motortrend. com/roadtests/ alternative/112_0803_ 2008_tesla_roadster; and Tesla Motors website, 2008, www.teslamotors.com.

4. Ring, Ed, "Aptera's Next Generation Car," Always On: The Insider's Network, October 5, 2007, online at www .alwayson.goingon.com/permalink/post/19936; and Aptera website, 2008, www.aptera.com.

5. "CityCar," Smart Cities, 2008, online at http://cities .media.mit.edu/projects/citycar.html; and Gibson, Michael, "A Carbon-Free, Stackable Rental Car," *Technology Review,* November 1, 2007, available online at www .technologyre-view.com/Infotech/19651/?nlid=655& a=f.

6. "Tata Motors Unveils the People's Car," Tata Motors Press Release, January 10, 2008, available online at www.tatamotors.com/our_world/press_releases.php? ID= 340&action=Pull.

7. Gourville, John T., "The Curse of Innovation: Why Innovative New Products Fail," Marketing Science Institute, Boston (2005), Working Paper #05-004.

8. Lynn, Gary, and Richard Reilly, *Blockbusters: The Five Keys to Developing GREAT New Products* (New York: Harper Business, 2002).

9. LeBaron, Dean, and Tomesh Vaitiligam, *The Ultimate Investor* (Capstone Publishing, 2001). Available

online at www.deanlebaron.com/book/ultimate/chapters/venture.html.

10. Grove, Andrew, *Only the Paranoid Survive* (New York: Currency/Doubleday, 1996).

11. Moriarty, Rowland, and Thomas Kosnik, "High-Tech vs. Low-Tech Marketing: Where's the Beef?" Harvard Business School (1987), Case #9-588-012.

12. Unless otherwise noted, this section is drawn from Mohr, Jakki, Stanley Slater, and Sanjit Sengupta, "Foundations for Successful High-Technology Marketing," in *Managing Technology and Innovation: An Introduction,* eds. R. Verburg, R. J. Ortt, and W. M. Dicke (London: Routledge, 2006), pp. 84–105.

13. American Marketing Association, "Definition of Marketing," 2007, available online at http://ama-academics.communityzero.com/elmar?go=1712138.

14. Workman, John, "Marketing's Limited Role in New Product Development in One Computer Systems Firm," *Journal of Marketing Research* 30 (November 1993), pp. 405–421.

15. Drucker, Peter, *The Practice of Management* (New York: Harper & Row, 1954), pp. 39–40.

16. Bruner, R., et al., "Marketing Management: Leveraging Customer Value," in *The Portable MBA* (3rd ed.), ed. W. D. Bygrave (New York: Wiley, 1998), pp. 103–124.

17. Ibid.

18. Capon, Noel, and Rashi Glazer, "Marketing and Technology: A Strategic Coalignment," *Journal of Marketing* 51 (July 1987), pp. 1–14; and *Technology, Innovation, and Regional Economic Development*, Washington, DC: U.S. Congress, Office of Technology Assessment, September 9, 1982.

19. From Kirkpatrick, David, "Some in Silicon Valley Have Learned to Stop Worrying and Love the Bust. Here's Why," *Fortune,* May 12, 2003. Available online at http://money.cnn.com/magazines/fortune/fortune_archive/2003/05/12/342330/index.htm.

20. Hecker, Daniel, "High-Technology Employment: A NAICS-Based Update," *Monthly Labor Review,* July 2005, pp. 57–72. Available online at www.bls.gov/opub/mlr/2005/07/art6full.pdf.

21. Hatzichronoglou, Thomas, "Revision of the High-Technology Sector and Product Classification," OECD STI working paper (1997).

22. *Science and Engineering Indicators,* National Science Foundation (1996), chapters 4 and 6, available online at http://www.nsf.gov/statistics/seind06/.

23. Hecker, "High-Technology Employment."

24. American Electronic Association, "AeA's High-Tech Industry Definition," 2002, available online at www.aeanet.org/Publications/IDMK_ definition.asp. Also see "Identifying and Defining: Life Science, Bio-Tech, High-Tech, Knowledge Industries, and Information Technology Industries," July 2007, available online at http:// lmi2.detma.org/lmi/pdf/Definitions.pdf.

25. Hecker, "High-Technology Employment."

26. Moriarty, Rowland, and Thomas Kosnik, "High-Tech Marketing: Concepts, Continuity, and Change," *Sloan Management Review* 30 (Summer 1989), pp. 7–17.

27. See also Gardner, David, "Are High Technology Products Really Different?" University of Illinois at Urbana–Champaign, Faculty Working Paper (1990), Case #90-1706.

28. Moriarty and Kosnik, "High-Tech Marketing."

29. Moore, Geoffrey, *Crossing the Chasm: Marketing and Selling Technology Products to Mainstream Customers* (New York: Harper Business, 2002).

30. Falcone, John P., and Matthew Moskovciak, "HD DVD vs. Blu-ray," *CNET's Quick Guide,* February 19, 2008, available online at http://reviews.cnet.com/hd-dvd-vs-blu-ray-guide; and "It's Official: Toshiba Announces HD DVD Surrender," *CNET News,* February 18, 2008, available online at http://news.cnet.com/8301-17938_105-9874199-1.html?tag=rb_content%3Brb_mtx.

31. Moore, *Crossing the Chasm.*

32. "The Biggest Emerging Technology Disappointments of 2007," Emerging Technologies, November 30, 2007, available online at http://etech.eweek.com//content/infrastructure/the_year_in_ emerging_technology_2007.html.

33. Enderle, Rob, "The Wii Failure: Nintendo Screws Up," *TechNews World,* January 7, 2008, available online at www.ecommercetimes.com/rsstory/61065.html?welcome=1199827491.

34. From Yi-Wyn Yen, "Nintendo Offers a Raincheck for Wii Shortage," *Fortune* Techland Blogs, December 14, 2007, available online at http://techland.blogs.fortune.cnn.com/category/nintendo.

35. Moore, *Crossing the Chasm.*

36. Moriarty and Kosnik, "High-Tech Marketing."

37. Thurm, Scott, "For Frazzled Online Brokers, Technology Is the Problem," *Wall Street Journal,* March 4, 1999, p. B6.

38. MacManus, Richard, "Twine: The First Mainstream Semantic Web App?" October 18, 2007, available online at www.readwriteweb.com/archives/twine_first_mainstream_ semantic_web _app.php.

39. "The Giant on the Runway," *The Economist,* October 13, 2007, pp. 79–82.

40. "Problems Surface at 'Net Phone Service Skype," NPR, Morning Edition, September 24, 2007, available online at www.npr.org/templates/story/story.php?storyId=13966270.

41. Goetz, Thomas, "23AndMe Will Decode Your DNA for $1,000. Welcome to the Age of Genomics," *Wired,* November 17, 2007, available online at www.wired.com/medtech/genetics/magazine/15-12/ff_genomics?currentPage=all.

42. Sauser, Brittany, "Ethanol Demand Threatens Food Prices," *Technology Review,* February 13, 2007, available online at www.technologyreview.com/Energy/18173.

43. Chandy, Rajesh K., and Gerard J. Tellis, "The Incumbent's Curse? Incumbency, Size, and Radical Product Innova-

tion," *Journal of Marketing* 64 (July 2000), pp. 1–17. Available online at www.atypon-link.com/AMA/doi/abs/ 10.1509/jmkg.64.3.1.18033.

44. Christensen, Clayton M., *The Innovator's Dilemma* (Boston: Harvard Business School Press, 1997); and Cooper, Arnold, and Dan Schendel, "Strategic Responses to Technological Threats," *Business Horizons,* February 1976, pp. 61–69.

45. Hamel, Gary, "Killer Strategies That Make Shareholders Rich," *Fortune,* June 23, 1997, pp. 70–84.

46. "Digital Music: High Volume," *The Economist,* January 30, 2008, available online at www.economist.com/ displaystory.cfm?story_id=10598460.

47. Christensen, Clayton M., *The Innovator's Dilemma* (Boston: Harvard Business School Press, 1997).

48. Schilling, Melissa, *Strategic Management of Technological Innovation* (New York: McGraw-Hill/Irwin, 2008); Foster, R., *Innovation: The Attacker's Advantage* (New York: Summit Books, 1986); and Shanklin, William, and John Ryans, *Essentials of Marketing High Technology* (Lexington, MA: DC Health 1987).

49. See, for example, Kanellos, Michael, "Intel Scientists Find Wall for Moore's Law," CNET News.com, December 1, 2003, available online at www.news.com/ 2100-1008-5112061.html.

50. Anderson, P., and M. Tushman, "Technological Discontinuities and Dominant Designs: A Cyclical Model of Technological Change," *Administrative Science Quarterly* 35 (1990), pp. 604–634.

51. Schumpeter, Joseph, *Capitalism, Socialism, and Democracy* (New York: Harper & Row, 1942).

52. Sood, Ashish, and Gerard J. Tellis, "Technological Evolution and Radical Innovation," *Journal of Marketing* 69 (July 2005), pp. 152–168.

53. Shanklin and Ryans, *Essentials of Marketing High Technology,* Chapter 7.

54. Reed, David P., "The Law of the Pack," *Harvard Business Review* 79 (February 2001), pp. 23–24.

55. Stremersch, Stefan, Gerard Tellis, Philip Hans Franses, and Jeroen L. G. Binken, "Indirect Network Effects in New Product Growth," *Journal of Marketing* 71 (July 2007), pp. 52–74.

56. Srinivasan, Raji, Gary Lilien, and Arvind Rangaswamy, "The Emergence of Dominant Designs," *Journal of Marketing* 70 (April 2006), pp. 1–17.

57. Ford, David, and Chris Ryan, "Taking Technology to Market," *Harvard Business Review* 59 (March–April 1981), pp. 117–126.

58. This example is drawn from Hill, Charles, "Establishing a Standard: Competitive Strategy and Technological Standards in Winner-Take-All Industries," *Academy of Management Executive* 11 (May 1997), pp. 7–25.

59. Gundlach, Gregory, and Paul Bloom, "The 'Essential Facility' Doctrine: Legal Limits and Antitrust Consider-

ations," *Journal of Public Policy and Marketing* 12 (Fall 1993), pp. 156–177.

60. Schilling, Melissa A., "Technological Leapfrogging: Lessons from the U.S. Video Game Console Industry," *California Management Review* 45 (Spring 2003), pp. 6–33.

61. Rangan, V. Kasturi, and Kevin Bartus, "New Product Commercialization: Common Mistakes," in *Business Marketing Strategy,* eds. V. K. Rangan et al. (Chicago: Irwin, 1995), pp. 63–75.

62. Shanklin and Ryans, "Organizing for High-Tech Marketing," *Harvard Business Review* 62 (November–December 1987), pp. 164–171.

63. Abernathy, W., and J. Utterback, "Patterns of Industrial Innovation," *Technology Review,* June–July 1978, pp. 41–47.

64. Rangan and Bartus, "New Product Commercialization," p. 66.

65. Abernathy, W., and J. Utterback, "Patterns of Industrial Innovation."

66. Sood and Tellis, "Technological Evolution and Radical Innovation."

67. Shanklin and Ryans, "Organizing for High-Tech Marketing."

68. Maney, Kevin, "The Net Effect: Evolution or Revolution?" *USA Today,* August 8, 1999, p. B2.

69. Gross, Neil, and Peter Coy with Otis Port, "The Technology Paradox," *Business Week,* March 6, 1995, pp. 76–84.

70. Lieber, Ronald B., "Smart Science Unlike Its Rivals, Biotech Leader Amgen Emphasizes Lab Research—Not Market Research," *Fortune,* June 23, 1997, available online at http://money.cnn.com/magazines/fortune/ fortune_archive/1997/06/23/228050/index.htm.

71. Govindarajan, Vijay, and Praveen Kopalle, "Disruptiveness of Innovations: Measurement and an Assessment of Reliability and Validity," *Strategic Management Journal* 27 (February 2006), pp. 189–199.

72. Shanklin and Ryans, *Essentials of Marketing High Technology.*

73. Ibid., Chapter 7.

74. See, for example, *The Fortune at the Bottom of the Pyramid* (2004) by C. K. Prahalad, *Capitalism at the Crossroads* (2005) by Stuart L. Hart, and *The 86% Solution* (2005) by Vijay Mahajan and Kamini Banga, to name just a few.

75. See Wikipedia, "Bottom-of-the-Pyramid," 2008, at http:// en.wikipedia.org/wiki/Bottom_of_the_ pyramid.

76. Prahalad, C. K., and Stuart L. Hart, "The Fortune at the Bottom of the Pyramid," *Strategy+Business* (2002), Vol. 26, pp. 54–67.

77. Engardio, Pete, "Beyond the Green Corporation," *Business Week,* January 29, 2007, pp. 50–64.

78. This section is drawn from Leifer, Richard, Christopher M. McDermott, Gina Colarelli O'Connor, Lois Peters,

Mark Rice, and Robert W. Veryzer Jr., *Radical Innovation: How Mature Companies Can Outsmart Upstarts* (Cambridge, MA: Harvard Business School Press, 2000).

79. Shanklin and Ryans, "Organizing for High-Tech Marketing," p. 167.

80. Ibid.

81. Abernathy and Utterback, "Patterns of Industrial Innovation."

82. von Hippel, Eric, "Lead Users: A Source of Novel Product Concepts," *Management Science,* July 1986, pp. 791–805.

83. Abernathy and Utterback, "Patterns of Industrial Innovation."

84. Rangan and Bartus, "New Product Commercialization."

85. Christensen, Clayton, Stephen Kaufman, and Willy Shih, "Innovation Killers: How Financial Tools Destroy Your Capacity to Do New Things," *Harvard Business Review* 86 (January 2008), pp. 1–8.

86. Mizik, Natalie, and Robert Jacobson, "Trading Off between Value Creation and Value Appropriation: The Financial Implications of Shifts in Strategic Emphasis," *Journal of Marketing* 67 (January 2003), pp. 63–76.

APPENDIX 1.A

High-Technology Industry Classifications
(Based on Employment of Scientific and Technical Personnel)

NAICS* code	Industry	Forecast 2012 employment[†]	Percent change 2002–2012	Median annual wage, US$ 2004	Per thousand employees in R&D-performing companies
	Level I Industries[‡]	6,804	15.6	—	
3254	Pharmaceutical and medicine manufacturing	361	23.2	43,930	137
3341	Computer and peripheral equipment manufacturing	182	−27.1	61,830	170
3342	Communications equipment manufacturing	201	5.4	45,520	264
3344	Semiconductor and other electronic component manufacturing	452	−14.9	39,210	180
3345	Navigational, measuring, electromedical, and control instruments manufacturing	396	−12.2	47,960	126
3364	Aerospace product and parts manufacturing	386	−17.6	51,990	43
5112	Software publishers	430	67.9	69,880	245
5161	Internet publishing and broadcasting	49	41.1	53,470	98
5179	Other telecommunications	8	−21.9	45,470	—
5181	Internet service providers and Web search portals	233	64.2	52,780	98
5182	Data processing, hosting, and related services	430	40.8	45,570	98
5413	Architectural, engineering, and related services	1,306	4.3	48,570	104
5415	Computer systems design and related services	1,798	54.6	63,350	259
5417	Scientific research-and-development services	573	6.7	57,890	302

(continued)

41

NAICS* code	Industry	Forecast 2012 employment†	Percent change 2002–2012	Median annual wage, US$ 2004	Per thousand employees in R&D-performing companies
	Level II Industries‡	**4,998**	**10.7**	—	—
1131, 1132	Forestry	10	4.0	—	—
2111	Oil and gas extraction	88	−27.8	49,290	—
2211	Electric power generation, transmission, and distribution	405	−7.1	53,330	2
3251	Basic chemical manufacturing	140	−18.0	45,970	54
3252	Resin, synthetic rubber, and artificial synthetic fibers and filaments manufacturing	89	−22.6	42,730	97
3332	Industrial machinery manufacturing	125	−4.7	39,480	72
3333	Commercial and service industry machinery manufacturing	141	6.6	35,940	72
3343	Audio and video equipment manufacturing	38	−7.7	32,460	171
3346	Manufacturing and reproducing, magnetic and optical media	63	11.1	35,720	171
4234	Professional and commercial equipment and supplies, merchant wholesalers	790	19.8	41,770	84
5416	Management, scientific, and technical consulting services	1,137	55.4	45,610	28
—	Federal Government, excluding Postal Service	1,972	2.6	—	—

* North American Industry Classification System

† Levels in thousands

‡ *Level I industries* have 5 times or greater technical employment as the average of all industries. *Level II industries* have 3.0–4.9 times greater technical employment as the average of all industries. *Level III industries* (not shown) included 20 industries whose employment was 2.0–2.9 times the average of all industries and included petroleum and coal manufacturing, pesticide, fertilizer and agricultural chemical manufacturing, engine, turbine and power transmission equipment, electrical equipment manufacturing, electronic equipment repair and maintenance.

— Not available

Source: Hecker, Daniel, "High-Technology Employment: A NAICS-Based Update," *Monthly Labor Review* (July 2005), pp. 57–72. Available online at www.bls.gov/opub/mlr/2005/07/art6full.pdf.

APPENDIX 1.B

Outline for a Marketing Plan

This section presents an outline of the steps to be systematically considered in the course of developing a marketing plan. The remainder of the book will elaborate on the content of each section. Supporting detail can also be found in any basic marketing textbook. Examples of marketing plans can be found at www.mplans.com/spm.

1.0 Executive Summary—A one- to two-page summary of the market environment and business resources, financial and nonfinancial objectives, and marketing strategy including target market(s), value proposition, and marketing mix elements.

2.0 Market Analysis
Market demographics
Market needs
Market trends
Market growth
Buyer behavior
Customer segments
Competition
Collaborators
Macroeconomic forces

3.0 Company Analysis
Tangible assets
Intangible assets
Capabilities
Areas of advantage
Key success factors and key weaknesses

4.0 Objectives
Financial
  Revenues
  Margins
  Growth rate
Nonfinancial
  Customer satisfaction
  Perceived quality
  Loyalty
  % of sales from new products

5.0 Segmentation, Targeting, and Positioning/Value Proposition
Target market
Functional, emotional, and/or self-expressive benefits
Price

6.0 Marketing Strategy
Positioning: The process of designing the company's image and value offering so that customers in the target market understand and appreciate what the company stands for in relation to competitors

Product and/or service attributes: New product development processes, decisions about what to sell and what features to include, branding strategies, packaging decisions, warranty, and ancillary services

Distribution: Locations at which product or service is made available to customer, and the channel members that offer it

Promotion:
  A. Advertising strategies, regarding both the message content in ads and media used to communicate the message; may include direct media and new media
  B. Sales promotion strategies, regarding any short-term incentives for both trade members and consumers (coupons, rebates, premiums, etc.)
  C. Public relations and publicity strategies, regarding the generation of news articles, community relations, event sponsorships, and goodwill
  D. Personal selling/trade shows

Price: What to charge for specific products, features, or services, as well as discount structures and payment plans

People: The marketing specialists who are required to execute the marketing strategy; includes systems for recruiting, motivating, and retaining them

7.0 Budgeting and Control
Financial resources required to execute the marketing strategy
System for comparing results to objectives
Processes for taking corrective action

Strategic Market Planning in High-Tech Firms

Medtronic Inc., based in Minneapolis, Minnesota, is the world's leading medical technology company. Medtronic's corporate mission, as stated on their website, is:[2]

- To contribute to human welfare by application of biomedical engineering in the research, design, manufacture, and sale of instruments or appliances that alleviate pain, restore health, and extend life.
- To direct our growth in the areas of biomedical engineering where we display maximum strength and ability; to gather people and facilities that tend to augment these areas; to continuously build on these areas through education and knowledge assimilation; to avoid participation in areas where we cannot make unique and worthy contributions.
- To strive without reserve for the greatest possible reliability and quality in our products; to be the unsurpassed standard of comparison and to be recognized as a company of dedication, honesty, integrity, and service.
- To make a fair profit on current operations to meet our obligations, sustain our growth, and reach our goals.
- To recognize the personal worth of employees by providing an employment framework that allows personal satisfaction in work accomplished, security, advancement opportunity, and means to share in the company's success.
- To maintain good citizenship as a company.

A *Fortune* magazine cover story in 1999 named Medtronic the "Microsoft of the medical device industry" and featured the company's innovative strategy and culture. Founded in 1949 as a producer of pacemakers for heart patients, Medtronic has expanded internally and through acquisitions into producing devices that do everything from controlling pain to reducing the tremors of Parkinson's disease. In 1989, Medtronic was a $1 billion company, with most of its sales coming from pacemakers. As of July 2008, Medtronic had a market value of about $56 billion; its $13.5 billion in revenues for the fiscal year ending on April 25, 2008, came from a variety of cardiovascular devices, neurological stimulators, drug-delivery systems, and spinal implants.

Medtronic's record of innovation is the result of several factors, including a rigorous strategic planning process (shown in Figure 2.1) tied to a clear mission and well-defined goals. For instance, Medtronic's financial goal is to generate a minimum of 15% revenue and profit growth over any five-year period.

Medtronic develops a new strategic plan every year, with the nature of the plan alternating between a *bottom-up strategic plan* one year and a *top-down strategic plan* the next. Both types of strategic plans are guided by and tested against the company's mission statement to ensure that

(continued)

- Articulate mission and goals
- Develop either bottom-up (driven by business units) or top-down (driven by CEO) strategic plan in alternating years; five-year planning horizon; focus on being visionary and creative
- Examine fit of the plan to the mission
- Develop annual operating plan

FIGURE 2.1   Medtronic's Planning Process

the strategy and the mission are consistent. Medtronic's management believes that when a company offers employees a clear sense of purpose—without deviating and without vacillating—then employees will adopt the company's mission and make the commitment to fulfill it. They will go the extra mile to serve customers. They may work late into the night or accelerate the timetable for a crucial new product introduction. The mission statement takes priority, and strategic plans may be reformulated to be consistent with the mission. Essentially, the mission statement and the strategy are inseparable.

In the years that Medtronic conducts a *bottom-up planning process,* all business units follow a common outline but develop their own strategy and the programs to implement that strategy. Bottom-up planning takes place in Medtronic's five key businesses: cardiac rhythm management; neurological and diabetes; spinal and ear/nose/throat surgery; cardiac surgery; and vascular products. The bottom-up planning outline varies across planning cycles, but typically includes:

- An environmental assessment
- An assessment of disruptive technologies
- The identification of key business trends such as the shift to the Internet
- A financial outlook

The bottom-up strategic planning process begins in the June–July time frame and by late fall is presented to the Medtronic Executive Committee, which challenges the plans of all of the businesses. The Executive Committee is composed of the chairman/CEO/president, the five business presidents, and key staff members including the CFO, the Chief Legal Officer, the Chief Human Resource Management Officer, the Chief Information Officer, and the Chief Medical Officer. Each business's strategic plan is laid out in 30 to 40 slides, of which about one-third are common for all businesses, with the remainder being specific to a particular business. The actual written plan for a business is likely to be 8 to 15 pages long. The individual business plans are then rolled up into a summary document by the CEO.

The rolled-up plan and the individual business plans are presented to the board of directors in a dedicated corporate strategy session in the last quarter of the fiscal year. A major goal is to keep the plan concise. The primary purpose of the written plan is to provide board members with background for their questions and discussions. Presentations and discussions are the primary way that Medtronic communicates strategy. There is substantial oversight of, and challenge to, the plan.

The strategic focus shifts in alternating years to a *top-down process.* Top-down planning is driven by the CEO. The intent of top-down strategic planning is to be visionary and creative. The objective is to identify new areas for growth, such as new disease states and new technology platforms that will allow Medtronic to expand into new markets or businesses. At this level, the CEO and the senior management team identify a small number of issues and potential new market and business opportunities that they think could be fundamental to the company's growth. The planning template is adjusted to recognize the nature of the issues that have been raised during a planning cycle. For example, in 2000, then-CEO Bill George embarked on an initiative called "Vision

2010." He instructed a number of teams composed of people from throughout the company to look at factors that he thought would impact how the company would evolve—including the role of the patient in consumer marketing and patient advocacy, the impact of the Internet on access to information, the future of biotechnology, and the role of information technology in Medtronic's medical devices. The result was a new vision and new strategies for Medtronic.

While both types of strategic plans look out over a broad five-year horizon, an *annual operating plan* is also developed for each fiscal year. The operating plan is much more detailed and contains specific objectives, milestones, a budget, and clear delineation of responsibility. Medtronic separates the strategic and operating plans in that the company's managers do not carve specific details from one year of the strategic plan and force that on top of the annual operating plan. The intent is to keep a robust strategy intact and not subtly encourage managers to manipulate numbers so they can make their targets. With this in mind, the financials in the five-year strategic plan reflect only the current, third, and fifth years and do not include the annual operating plan year specifically. In this regard, the strategic plan and the operating plan are only loosely coupled. The belief is that tightly coupling the operating plan to the strategic plan could stifle creativity and innovation. The result is that the strategic plan remains broad and visionary while the operating plan is much more detailed and focused.

As the opening Medtronic vignette demonstrates, **strategic market planning** is the process by which a company formulates its strategic plan. Serving as a road map for the company's future, a company's strategic plan guides its resource allocation decisions in specific technology development projects, specific market segments, and other projects and opportunities. As shown in Figure 2.2, this chapter overviews the critical elements and decisions that support a company in its strategy formulation process.

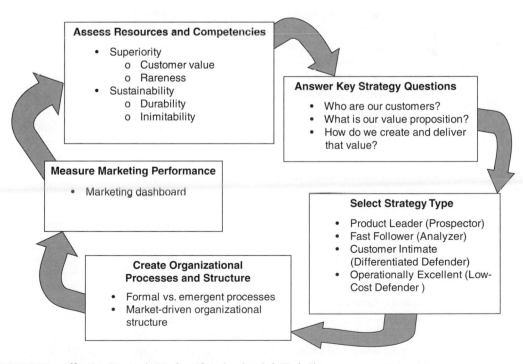

FIGURE 2.2 Effective Strategic Market Planning in High-Tech Firms

First in the process are resources and competencies—the critical ingredients that give rise to a firm's competitive advantage in the marketplace. Then, based on its assessment of its critical skills and capabilities, a company must answer three key questions in the strategy formulation process: Which customers should it serve? What value should it offer customers (the company's value proposition)? How should it create and deliver that value? The intersection of the answers to these three questions provides the "sweet spot" in strategy formulation, which allows the firm to uniquely and successfully capitalize on its core competencies.

The chapter then presents four strategy archetypes—Product Leader (Prospector), Fast Follower (Analyzer), Customer Intimate (Differentiated Defender), and Operationally Excellent (Low-Cost Defender)—that identify the strategy sweet spot in four very different ways. Next, the chapter presents important organizational characteristics that affect strategic planning: the type of process utilized (formal versus emergent) and the type of organizational structure which facilitates strategy execution. The final section in this chapter addresses the importance of measuring the performance of a firm's marketing initiatives, including the use of marketing dashboards as one tool to do so.

COMPETITIVE ADVANTAGE: THE OBJECTIVE OF MARKETING STRATEGY

The purpose of strategy is to create competitive advantage, a position where a firm is able to create more value for customers than its competitors. **Competitive advantage** exists when the firm possesses resources and competencies that are valuable, rare, durable, and difficult for competitors to imitate. Effective high-technology marketing is built on a foundation of resources and ideas that result from the strategy formulation process.[3]

Resources and Competencies

The firm's resources are the foundation for the creation of superior customer value. *Resources* can be physical assets, intangible assets, or competencies. Physical assets include such things as manufacturing plants, information systems, distribution facilities, and products. Intangible assets include brand equity, customer loyalty, distribution channels, market knowledge, and the firm's beliefs about customer needs or their responsiveness to pricing, promotional, or distribution changes. Other important assets for a high-tech company can be found in its patent portfolio and, depending upon the industry, its network effects. For example, a large installed base of existing customers and/or a well-developed partnership base that offers complementary products can provide important sources of competitive advantage.[4]

Competencies are the bundles of skills that enable a firm to achieve new resource configurations as the firm and the markets it competes in evolve. Marketing competencies in high-tech firms include processes for gathering, interpreting, and using market information; the ability to manage customer relationships and establish collaborative relationships with distributors to serve customers more effectively; service delivery; product/service development; new product commercialization; and supply chain management, among others.[5] As important as understanding what resources and competencies it possesses, the firm must also take stock of the key resources and competencies it lacks—and then must draft a plan to develop those missing skills and competencies to create a fully developed resource base.

Competencies, while important to all businesses, may be more important to high-tech businesses. A comparison of Apple and Procter & Gamble will provide some insight into this phenomenon. In January 2007, Apple was America's 7th most admired corporation; P&G was the 10th most admired. Both were ranked first in their industries on innovation and quality of products and services. While both companies had profit margins of about 13%, each dollar of Apple's equity-financed assets had a stock market value of $9.71, while the comparable figure for P&G was $3.10. This difference cannot be explained by differences in asset intensity or financial structure; rather, it can be explained by the fact that while the two companies have similar competencies, those competencies are more valuable to companies competing in high-tech markets. The primary reason is that markets for high-tech products are typically high-growth markets. The profitability that these competencies generate is magnified by growth, which is ultimately reflected in the firm's market value.

In addition, the specific type of competencies that high-technology firms need for sustainable competitive advantage likely differs from other types of firms. High-tech companies typically experience their initial success in the marketplace because of unique competencies in a technological innovation, which, in turn, is based on underlying skills and competencies in research and development. However, in order to sustain its initial technologically based competitive advantage, a high-tech company must augment its technological prowess with marketing-related competencies. For example, managing customer relationships over the long term, collecting useful market-based information, and effectively working with distribution partners are marketing-based competencies that many technology-focused firms may lack. Therefore, a critical issue for high-tech companies is to develop their marketing competencies— something that the strategies and tools presented in this book are designed to help them do.

Core competencies are distinguished from general competencies. They are the underlying skills and capabilities that give rise to a firm's competitive advantage. Core competencies exhibit three characteristics: Like other competencies, they are very difficult for competitors to imitate; they are significantly related to the benefits customers seek; and they enable the firm to access a wide variety of disparate market opportunities by applying its skills and competencies in product markets where it has not previously competed.[6]

In the high-tech arena, Hewlett-Packard serves as a good example of leveraging core competencies. One of Hewlett-Packard's core competencies is in the area of transferring digital images to paper with superior clarity, detail, and color. This core competency was exhibited in its resounding success in the laser printer business. Although other companies also made laser printers, HP's superior technology and production skills made the high quality very difficult to imitate. Moreover, the skill in transferring digital images to paper in a high-quality fashion was significantly related to the benefits customers were seeking in printing their computer images. Hewlett-Packard leveraged this core competency into a very different market: It entered the digital photography business with a digital photography package consisting of a camera, scanner, and printer. The digital photography business taps into essentially the same skills and capabilities that made HP successful in the laser printer business: transferring high-quality images to paper.[7] Moreover, HP's technology that propels tiny droplets in inkjet printing is now being used to inject medicine into the skin with a patch. HP's development of this drug-delivery technology allowed it to "repurpose its inkjet technology for use in new markets . . . allowing HP to capitalize on the booming healthcare and life sciences market."[8]

Figure 2.3 shows a diagram of Honda's core competencies, using the analogy of a tree.[9] The branches or canopy of the tree represents the widely different product markets to which the core competencies have provided access. In Honda's case, this would be the end markets in which it competes: small cars, snow blowers, motorcycles, and lawn mowers, to name a few. The trunk represents the core product, or the physical embodiment of the core competencies. The core product must be significantly related to the benefits the end user receives. The roots of the tree represent the underlying skills and capabilities that form the basis of the core competencies. In this case, Honda's superior research and development, manufacturing techniques, marketing, knowledge of customers, and its corporate culture give rise to its success in small engine technology.

A basic business tenet is that firms use criteria such as return on investment or payback period to evaluate possible investments in new projects. However, a core competencies approach to resource allocations can result in decisions that may defy the use of traditional criteria. An excellent example can be found at Amazon.com.[10] Jeff Bezos established Amazon on the concept of giving customers access to a giant selection of books without incurring the time and expense of opening stores and warehouses and dealing with inventory. However, he soon discovered that the only way to ensure a quality experience for customers and sufficient inventory at good prices for Amazon was to operate his own warehouses.

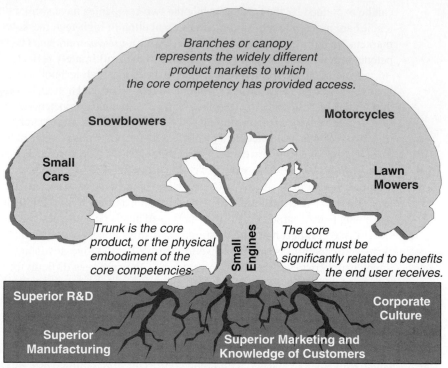

Roots are underlying skills and capabilities that represent core competencies.

FIGURE 2.3 Core Competencies for the Example of Honda

Source: Adapted from C. K. Prahalad and Gary Hamel, "The Core Competence of the Corporation," *Harvard Business Review* (May–June 1990), pp. 79–91.

Building warehouses was met with skepticism from Wall Street. At about $50 million each, warehouses are expensive to set up and expensive to operate. However, a visit to one of Amazon's six warehouses today makes it quite clear why Bezos believed he would defy financial logic. Amazon's warehouses are models of efficiency. They are so high tech that they require as much computing power to run as Amazon's website does. Computers send signals to workers' wireless receivers, telling them what items to pick off the shelves; then the computers determine everything, from which item gets picked first to whether the weight is right for sending.

Along the way the computers generate data on everything from misboxed items to backup times—and managers are expected to study the information religiously. The result is that the warehouses are extremely efficient. For example, by redesigning a bottleneck where workers transfer orders arriving in green plastic bins to a conveyor belt that automatically drops them into the appropriate chutes, Amazon increased the capacity of one warehouse by 40%. In fact, Amazon's warehouses now handle 3 times the volume they handled in 1999, and the cost of operating them fell from nearly 20% of revenues in 1999 to less than 9% in 2006. Based on Wall Street's initial skeptical reaction, it is unlikely that Amazon's investment in this logistics competency would have met traditional investment criteria.

Amazon has leveraged its investment in this competency into new market arenas by becoming a service provider to other e-tailers with a business called Fulfillment by Amazon. Amazon allows vendors who list their items anywhere on the Web—on their own sites, through Google, or even on Amazon's e-commerce rival, eBay—to use its network of more than 20 distribution centers around the world to fill orders. The program is part of a broader set of tools called Amazon Web Services, an effort by the e-commerce pioneer to rent out complicated parts of its infrastructure to smaller companies that might benefit from its hard-earned expertise, and are willing to pay for the privilege of lightening their workload. As Bezos said, "We have this beautiful, elegant, high-I.Q. part of our business that we have been working hard on for many years. We've gotten good at it. Why not make money off it another way?"[11] The most recent spinout of Amazon's underlying competencies is *cloud computing* (also known as *utility computing, computing on demand,* and even *"hardware as a service"*), in which companies can rent computing power—storage, bandwidth, processing power—without having to invest in their own expensive information technology infrastructure. While Wall Street "gnashed its teeth" about the $2 billion Amazon spent on IT, Amazon saw a new way to leverage its own necessary infrastructure investments into a new revenue model.[12]

TESTS OF COMPETITIVE ADVANTAGE Strategy, competencies, and assets form the foundation for the creation of superior customer value—but they are only the foundation. Companies must also determine whether the strategy and its supporting competencies and assets lead to a position of *sustainable superiority*. This assessment determines whether the strategy is likely to be successful or whether it needs to be adjusted. There are two tests a resource must pass to lead to a position of superiority: *customer value* and *resource rareness*. It must pass two more tests—*durability* and *inimitability*—to be sustainable.[13] These tests are explored next.

Superiority: Value and Rareness. The first test a resource must pass if it is to lead to a position of superiority is **customer value**. This is the difference between the benefits that a customer realizes from using a product and the total life-cycle costs that the customer incurs in finding, acquiring, using, maintaining, and disposing of the product. A resource is valuable if it enables the firm to develop and implement strategies that enhance its customers' effectiveness or efficiency. An effective strategy provides additional customer benefits while increasing life-cycle costs at a slower rate than the increase in benefits. An efficient strategy focuses on the cost side of the value equation. Its objective is either to reduce life-cycle costs while maintaining benefits so as to increase demand, or to reduce costs internally but not pass along the cost reductions to buyers. As long as benefits and costs are competitive, the firm will achieve "normal" market share but will see its margins and return on investment (ROI) increase.

Take green building technologies, which enhance energy efficiency by focusing on recycling, renewable energy, and water management—essentially any technologies used to build environmentally and economically sustainable buildings (e.g., see the Green Building Council's Leadership in Energy and Environmental Design—or LEED—program).[14] Studies show that the initial cost outlay for green building construction can be higher than the cost of conventional buildings—how much higher depends on who is asked. Yet, within three to four years, the savings on conservation of water and electricity pay for those higher costs, and then some.[15] For example, the Green Business Center building in Hyderabad, India, uses 55% less energy than a standard building of similar size in that country. In addition, the Toyota Motor Sales headquarters in Torrance, California, earned the LEED gold rating when it opened in 2003. Construction costs were the usual $90/square foot to design standard buildings in southern California. Both of its three-story buildings have a long, narrow footprint, and a north–south orientation to maximize interior daylighting. The perimeter is ringed with glass-enclosed private offices. More than 90% of the building's occupants enjoy natural light and outdoor views. The building's rooftop photovoltaic panels, combined with highly efficient air handling and gas-powered chillers, make it 31% more energy efficient than the company's comparable buildings. It consumes 60% less water on its 40-acre drought-tolerant landscaped sites than typical turf-planted, sprinkler-watered business sites; its use of recycled water for landscape irrigation, building cooling, and toilet flushing saves 20.7 million gallons of potable water each year. The largest benefit, according to Toyota, is its effect on employees, as seen in the very high retention rate, increases in productivity, and drops in employee absenteeism. Indeed, compared to other Toyota facilities, the Torrance building had a 14% decrease in absenteeism.

Generally, different types of buyers will have varying perspectives on the worth of different benefits or life-cycle costs. Consequently, the firm's managers must conduct careful cost–benefit analysis among members of the target market before making substantial adjustments to the customer value equation. (About the only element in the value equation on which customers generally agree is that lower prices are preferable to higher prices. However, because of the ease of imitability and the negative impact on profitability, a low-price strategy is typically an undesirable strategy and the last strategic marketing tactic that should be considered.)

The second test for a resource, if it is to provide a position of superiority, is **resource rareness**—whether the firm's resources are sufficiently rare that competitors or producers of substitutes are not able to offer the same, or similar, set of benefits and life-cycle costs. If many firms possess the same valuable resource, then each firm has the ability to deploy that resource in a similar way. In this case, because each firm can implement the same strategy, no firm achieves superiority. This docs not mean that relatively common resources, such as managerial talent, are unimportant. Indeed, such common resources may be necessary to exploit other, rare resources. However, possessing these common resources will not lead to a superior competitive position. Rareness also does not mean that just one firm can possess the valuable resource for it to be a source of superior value. As long as the number of firms in the industry that possess the resource is less than the number required for the resource to approach commodity status, the resource may be a source of advantage.

For example, within the ceramics industry there are many firms that can produce fine china capable of withstanding 5,000–15,000 pounds of pressure per square inch (352–1,055 kilograms per square centimeter). However, producing ceramic armor that can withstand 140,000 pounds of pressure per square inch (9,843 kilograms per square centimeter) is another matter. It is this type of advanced and highly specialized knowledge that has propelled Japan's Kyocera Corporation to the top spot in that industry.

The question is, how can the firm develop a valuable and rare resource base to achieve a position of superiority? One simple answer, although quite difficult to accomplish, is to *create a bundle of complementary resources*—including physical assets, intangible assets, and competencies—that produce customer value. Apple seems to possess this combination with its very user-friendly products, its highly recognizable brand, and its skill at industrial design—all of which combine to make it a darling among customers and investors.

Unfortunately, a superior position based on value and rareness does not last forever. Research shows that fewer than 5% of firms are able to generate superior profits for 10 years.[16] Changes in customer needs or in other elements of market structure can make a resource that once was a source of value no longer valuable. Sustained superiority requires continuous improvement in the resource punctuated by the regular, if infrequent, development of new resources.

Sustainability: Durability and Inimitability.   Achievement of a *sustainable* position of superiority is the Holy Grail of strategic marketing and, as mentioned earlier, is based on two more tests: durability and inimitability.[17]

Durability is concerned with how rapidly a valuable resource becomes obsolete due to innovation by current or potential competitors. The longer it takes for a resource to be rendered obsolete, the more likely it is to be a sustained source of value. Resource durability depends, in large part, on the nature of the industry. Slow-cycle industries, such as many low-tech industries, have a very slow rate of change due to low market and technological turbulence. Many consumer brands such as Coca-Cola, Ivory soap, Campbell's soup, and Kellogg's cereals have maintained strong customer loyalty, even at relatively high prices, for long periods of time. Customers have had good experiences with these brands and are reassured by the brand name. The durability of these brand names is a major reason why they are valuable as a basis for brand extensions.

Fast-cycle industries are often based on a technology or on an idea. In these industries, technology is a rapidly depreciating resource, as the list of companies that have led the video game industry illustrates: Atari→Nintendo→Sega→Sony. In 2007, Sony was challenged by Microsoft's Xbox and the Nintendo Wii. The Wii, about $250, costs considerably less than the Microsoft Xbox 360 or the Sony PlayStation 3. Each of these firms had a strong position in the industry only to see it eroded as technology advanced. Their positions were not durable. Similar evolution has occurred in the computer hardware and software industries.

In the previous edition of this book, we argued that Dell had a rare combination of resources that gave it a dominant position in the personal computer market. In 2007 Dell was knocked from its perch as the world's biggest seller of PCs by Hewlett-Packard and lost share both to HP and low-cost Asian manufacturers.[18] Although Dell is attempting to regain its leadership position in this rapidly changing industry, betting on the durability of competitively valuable resources is risky. Competitors or new entrants not only will attempt to develop a new generation of technology that renders the existing technology obsolete, but they also will develop innovations in business models that make previous business strategies obsolete.

To increase sustainability, a resource must also pass the test of **inimitability**. This concept is concerned with how easily a competitor can obtain or copy a valuable resource either through internal development or purchase in the market. Possessing a resource that is easily imitated creates only temporary advantage. Discouraging or stopping imitation of a valuable resource enhances the sustainability of the first mover's position. Some barriers limit a rival's ability to duplicate a valuable resource—including patents, brand names, corporate reputations, specialized assets, financial resources, and network effects based on an installed base of customers and/or partners who make complementary products. Even if the resource can be imitated, however, these barriers inhibit the ability to duplicate the customer's perception of the value that is created. In addition, resources that are based on complex organizational routines such as production processes, interpersonal relationships among a firm's employees, a firm's culture, or its reputation among suppliers and customers are difficult to replicate and, hence, resist imitation. The creation of values, attitudes, norms for behavior, and relationships is quite difficult and takes a long time.[19] Three factors that affect inimitability—transparency, replicability, and transferability—are highlighted in Figure 2.4.

EVALUATING COMPETITIVE ADVANTAGE   To have superior and sustainable competitive advantage, a firm must take stock of its resources and competencies. These underlying skills and competencies provide the foundation for the creation of the technology and products the firm sells in the marketplace.

With respect to competencies, firms generally face a trade-off between *competence exploitation*—investing resources to refine and extend existing knowledge, skills, and processes, aiming for greater

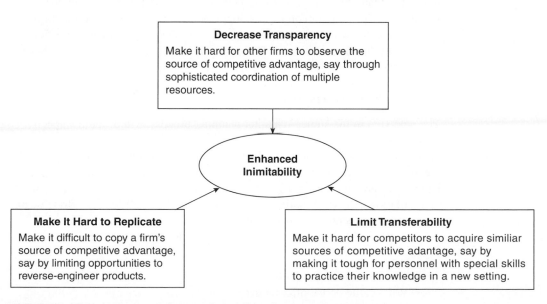

FIGURE 2.4 Three Factors to Enhance Inimitability of a Firm's Resources

efficiency and reliability, and extending its existing paradigm into new arenas—and *competence exploration*—investing resources to acquire entirely new knowledge, skills, and processes, with a goal of experimenting with novel skills. Some research suggests that competence exploitation, by focusing attention on productivity improvements in existing products, increases incremental innovation but may hinder the development of breakthrough (radical) innovations.[20] On the other hand, competence exploration involves experimentation that focuses on emerging markets and technologies to produce breakthrough innovations that offer entirely new value for new customers. Indeed, the two types of orientations (competence exploitation and competence exploration) have an interactive effect on innovation performance. To reap the advantages of competence exploration, a firm must recombine the new knowledge it gains with some level of exploitation. Existing competencies provide the necessary absorptive capacity for the firm to fully leverage the new knowledge gained in competence exploration. Conversely, a firm that is extremely competent in exploiting its current competencies will be successful with radical innovation only with a dose of exploration as well. Importantly, scoring high on both orientations has a negative effect on innovation performance.[21]

As further corroboration of this notion, a study in the pharmaceutical industry examined the variables that affect a firm's ability to successfully convert ideas into commercial products. Successful conversion ability is significantly related to firm performance, with the high-conversion-ability firms in the study exhibiting a nearly twice as high ROI compared to low-conversion-ability firms. More specifically, the study found that firms generally have a tough time successfully converting *novel* ideas for drugs into commercialized products; however, *expertise* in the particular therapeutic domain results in more successful conversion of ideas to commercial products.[22]

Finally, a third study examined the financial impact (stock returns) of a firm's relative emphasis between *value creation*—constantly innovating and investing in R&D—and *value appropriation*—investing in strategies that allow the firm to differentiate its existing offerings in order extract profits in the market (e.g., through brand-based advertising). Even for high-tech markets, the stock market reacts more favorably when a firm increases its emphasis on value appropriation relative to value creation, particularly when earnings are greater than expected.[23] One lesson from this study is that firms may overinvest in value creation (R&D) and underinvest in value appropriation (advertising and marketing to differentiate its offerings). To capture the value from their innovations, firms must also invest in marketing.

In addressing competitive advantage, small high-tech start-ups face their own set of resource challenges. One obvious challenge is securing funding. Another is securing the necessary mentoring to facilitate development. An important tool is the **technology incubator,** a facility—run by a private company, city or county development organization, or government program—that offers resources and support services on a temporary basis until the start-up is sufficiently developed to venture into the market on its own. For more about financing options and other resources available to small high-tech start-ups, see Appendix 2.A.

Key Strategy Decisions

The next step in the strategic market planning process is to develop a specific strategy: "In making strategy . . . first comes painstaking attention to the needs of customers. . . . Strategy takes shape in the determination to create value for customers."[24] Customer-focused approaches to analysis, value creation, and strategy implementation begin by assessing what customers' articulated and unarticulated needs are and how they are likely to change in the future; they focus on creating the resources that will most effectively satisfy customers' needs. This approach to strategic market planning is not focused on the technology, but rather on customers and customer value.

A firm's strategy formulation process should answer three key questions,[25] as shown in Figure 2.5 and listed here:

- Who are our target customers?
- What value do we offer them?
- How can we create and deliver that value efficiently and effectively?

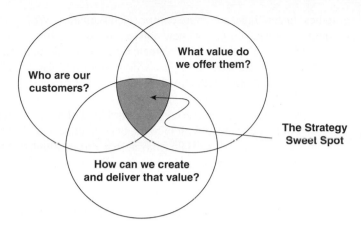

FIGURE 2.5   Key Strategy Decisions

In answering these three key questions in a complementary fashion, a firm attempts to identify the **strategy sweet spot**—the intersection of the three decisions that effectively capitalizes on the firm's resources to achieve a position of sustainable, superior competitive advantage.

WHO ARE OUR CUSTOMERS?   For new high-tech start-ups, answering the question of "who are our customers" can be very difficult. Typically, a new innovation can be used by many different types of industries and markets, and selecting the one best customer segment on which to focus the firm's initial efforts is fraught with risk. Chapter 7 presents a very specific process to address this question. The main point to remember here is that companies that do not have a clear focus on a customer segment tend to dilute their energies across too many market segments, and thus are less successful than if they had a clearly articulated customer focus.

Established companies may take the answer to the question of "who are our customers" as a given, focusing on the needs of the *served market*, the current/existing customers. This narrow definition of a firm's customers is sometimes referred to as the **tyranny of the served market,** an excessive focus on serving current customers. A narrow focus on the served market may inhibit innovation and blind the firm to the emergence of new segments in a rapidly changing market. Such a myopic focus obscures the possibility that customer needs may change over time and may be solved in radically different ways. Thus, managers must ask both "Who are our current customers?" and "Who should our customers be in five years?" Addressing the question of "which customers" with this broadened perspective requires **bifocal vision,** a simultaneous focus on both current and future customers.[26]

The success of the Nintendo Wii was due in part to how it answered the question of "which customers." The data on hard-core gamers showed them to be male, between the ages of 14 and 34, and gaming addicts (preferring to play a computer game instead of sleeping at night or watching TV). This customer market, while large and devoted—in 2005, the industry brought in $27 billion!—was stagnant. The market had sharply defined borders, due in part to the fact that many nongamers perceive hard-core gamers in a rather negative light. As stated in a *Time* magazine article, "Not everybody likes ballet, but most nonballet fans don't accuse ballet of leading to violent crime and mental backwardness."[27] Moreover, nongamers described traditional gamers as "blank-eyed joystick fondlers."

To compete in this environment, companies typically solicited feedback from their traditional customers (the hard-core gamers), asking them what they would like to see in the next generation of games and technology. And, true to form, these hard-core gamers—a very involved customer group—willingly offered insights and opinions about new features. Typically, they wanted flashier graphics, faster chips, and more complicated scenarios. When tied in with a movie license and better online service, you've got "new and improved" gaming!

However, in an insightful strategic twist, Nintendo president Satoru Iwata raised a new question: Why do people who don't play video games not play them? Most industry insiders and hard-core gamers didn't realize that to nongamers, video games are really hard:

> The standard video-game controller is a kind of Siamese-twin affair, two joysticks fused together and studded with buttons, two triggers, and a four-way toggle switch called a d-pad. In a game like Halo, players have to manipulate both joysticks simultaneously while working both triggers and pounding half a dozen buttons at the same time. The learning curve is steep.[28]

So, rather than working on incremental improvements and modifications, Nintendo threw away the controller as we know it and replaced it with a magic wand of sorts—a bit like the remote control for a TV, but one that senses a player's hand movements. With Wii, instead of passively playing the games, players physically perform them; they act them out. To make the character on the screen swing a sword, the player simply swings the controller in the desired motion. To aim a gun, the player just aims the wand and pulls the trigger.

Nintendo's answer to the question of "who are our customers" reinvented the gaming industry. Asking how to appeal to nongamers instead of hard-core gamers brought in a new franchise of customers—women, the elderly, and young girls. As Iwata said, "If you are simply listening to requests from [existing] customers, you can satisfy their needs, but you can never surprise them."

In broadening the search for an answer to "who are our customers," companies search for new market space, or **"blue ocean" strategies.**[29] The term **market space** refers to the markets or product arena in which a firm competes, and new market space represents potential—those who might be customers. Businesses in search of new market space look "across substitute industries, across strategic groups, across buyer groups, across complementary product and service offerings, across the functional-emotional orientations of an industry, and even across time."[30] New products and new market space are the foundation for organizational renewal in the customer-focused business. Rather than focusing on existing industry boundaries in strategy formulation, firms that create new wealth and play by new rules recognize that industry boundaries in today's environment are fluid rather than static.

New market space for many high-tech firms lies in *base-of-the-pyramid markets,* emerging economies that conventional wisdom says are uneconomic to serve. Because most people who can easily afford computers and cell phones already own them, companies are pushing into previously unexplored—and unappealing—markets. For example, in 2005 Motorola unveiled a no-frills cell phone priced at $40; it expected to sell 6 million of them in 6 months, in markets including China, India, and Turkey. Said Allen Burnes, Motorola vice president of high-growth markets, "You've got nearly 2 billion people who will be buying a phone—need a phone—over the next 5 to 10 years. This is the huge growth opportunity."[31] For another new product aimed at base-of-the-pyramid markets, see this chapter's **Technology Solutions for Global Problems** box.

WHAT VALUE DO WE OFFER?   The second key strategy question is to determine what value to offer the chosen customers. Products, services, and technologies should be seen as vehicles for value creation, not as something that has intrinsic value. Explains John Chambers, CEO of Cisco Systems, "I have no love of technology for technology's sake—only solutions for customers."[32] After taking a new position, the first question from a very successful divisional R&D manager in a *Fortune* 50 company, whenever an engineer brought up a new product concept, was consistently "How will this create value for our customers?" Over time, this simple question reoriented engineers to think about customers first, instead of technology, thus bringing much-needed discipline to the division's R&D efforts.

TECHNOLOGY SOLUTIONS FOR GLOBAL PROBLEMS
Ground-Breaking Dignity Toilets for Developing Countries
By Jamie Hoffman

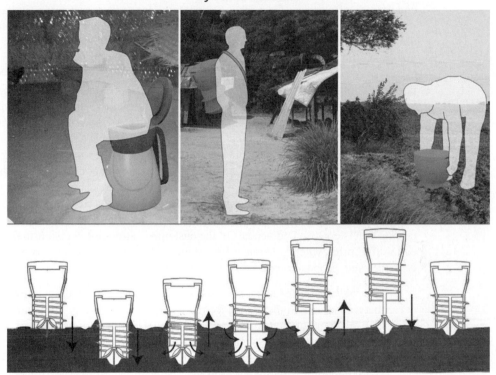

Images reprinted with permission of Cooler Solutions, Inc., Toronto, Canada.

According to the United Nations, 2.6 billion of the world's 6.6 billion people do not have access to basic toilet facilities, and 1 billion are children. The absence of proper sanitation infrastructures in developing countries and refugee camps forces residents to drink, cook, and bathe with water contaminated by human and animal feces. The lack of sanitary facilities results in widespread contraction of deadly diseases such as diarrhea, cholera, typhoid, and parasites.

The Canadian research and design company Cooler Solutions Inc. has developed an innovative solution, the Dignity Toilet, a low-cost waste disposal system providing sanitary storage of solid waste for four people up to 10 days. Once a week, the user detaches the portable toilet from its seating dock and carries it to a proper location away from residences and water sources. The user manually screws the toilet into the soil where it discharges its contents, mulches the soil and waste, and buries it, allowing for quicker decomposition.

The Dignity Toilet consists of a seating dock, toilet, and plug—and will be affordable for developing countries. All parts are made from roto-molded polypropylene, allowing for a durable, low-cost product. While the design of the dock and toilet is basic, the plug uses thread technology to safely bury waste. As the user manually turns the toilet, the plug's external auger-style threads dig the toilet 12" into the soil. As the plug is drilled into the ground a seal is broken, allowing the waste to discharge. The plug's external threads draw soil inward for mulching and to clean the female threads. The mulching agitator at the base of the plug blends and disperses the waste and soil. The user simply turns the toilet in the opposite direction to reseal the plug.

Cooler Solutions was recently recognized by the Humanitarian International Design Organization for the groundbreaking toilet and is working with HIDO to raise funds to bring the Dignity Toilet to developing nations.

Sources: Dugger, Celia, "Toilets Underused to Fight Disease, U.N. Study Finds," *New York Times,* November 10, 2006, online at www.nytimes.com/2006/11/10/world/10toilet.html; Young, Kathryn, "Canadians Create Culturally Cognizant Can," *Vancouver Sun,* November 28, 2007, online at www.coolersolutionsinc.com/articles/canadians_create_culturally_cognizant_can.html; and "The Dignity Toilet," Cooler Solutions, online at www.coolersolutionsinc.com/pdf/CoolerSolutions-HIDO-DignityToilet.pdf.

Determining how to provide customer value requires understanding not only customers, but competitors' value propositions as well. To this end, firms with innovative strategies study more than just their direct competitors. Answering the value question requires an examination of *product form competition*—competition that arises outside of an industry's existing boundaries—because competition often arises from other technologies or products that serve similar product needs.

One company that answered the value question focused on how to mitigate air pollution for customers in emerging markets. Envirofit International was established to develop well-engineered technology solutions to improve the human condition on a global scale, primarily in the developing world. It recognized that a major source of outdoor air pollution is carbureted two-stroke engines, which power 100 million "two-wheelers" (e.g., motorcycles, scooters) and "three-wheelers" (e.g., tricycles, tuk tuks) across Southeast Asia. *Each* of these carbureted two-stroke engines produces the pollution output of 50 modern automobiles, thus contributing the overall pollution equivalent of approximately *5 billion* cars. Its solution is a direct in-cylinder (DI) fuel injection retrofit kit for two-stroke engines that is less expensive, cleaner, and more fuel efficient than the replacement four-stroke engines.

The firm's answer to the value question essentially becomes its **value proposition**. The value proposition captures the essence of why a buyer in the target market should purchase a product or use a service. This statement should convince the potential customer that a company's product or service will add more value or better solve a problem than competitors' offerings. The ideal value proposition is concise and appeals to the customer's strongest decision-making drivers. Companies pay a high price when customers lose sight of the company's value proposition. The three main types of value propositions are:[33]

1. ***The "All Benefits" value proposition.*** This value proposition simply articulates the benefits that a customer would realize from purchasing the offering. For example, Joe Marengi, senior vice president of Dell Americas, said, "Dell's retail customers benefit from the increased flexibility and low costs associated with using standards-based technologies."[34] This type of value proposition is the easiest to construct, but it reflects the least knowledge of customer needs and competitor capabilities. Dell may be claiming advantages for features that actually provide little benefit to the target customer.

2. ***The "Favorable Points of Difference" value proposition.*** This value proposition contrasts the advantages of the seller's product with that of its competition. GE's value proposition for its magnetic resonance imaging (MRI) unit states, "When you demand more accuracy, more productivity and more support, GE Magnetic Resonance Imaging delivers."[35] While this approach is clearly an improvement over "All Benefits," knowing that an element of the offering is a point of difference does not convey the value of the difference to the target customers.

3. ***The "Resonating Focus" value proposition.*** This value proposition addresses the buyer's key needs by addressing the few elements that matter most to buyers in the target market and documenting the worth of the superiority. In 2002, Microsoft CEO Steve Ballmer noted, "As I talk with IT managers, I'm hearing that they must justify their technology investments more than ever before. . . . [Windows XP and Office XP] deliver more business value to our customers than any other solution available. And we'll prove it!"[36] Microsoft backs up this claim with data from recent studies, demonstrating the value to a business of Windows XP Professional. The studies showed that Windows XP Professional provided customers with an average return on investment of more than 200% and annual savings ranging from $187 per desktop to $387 for mobile users as calculated over a three-year period.

The "Resonating Focus" value proposition is the most effective of the three because it quantifies the superiority of the seller's offering. However, this is very difficult to do in most markets and requires in-depth understanding of customers' needs and competitors' product performance, regardless of the market.

HOW TO CREATE AND DELIVER VALUE?   The third of the key strategy decisions is concerned with how that value is created and delivered. As Larry Bossidy, the former CEO of Allied Signal and

Honeywell, argued, "Strategies most often fail because they aren't executed well. Things that are supposed to happen don't happen."[37] Execution requires having the right competencies, appropriate structures and systems (including compensation systems that reinforce the strategy), and good decisions in distribution, pricing, and promotion arenas. At the same time, executives must be careful not to develop an implementation program that is inflexible. Rapidly changing market conditions and strategy that emerges through learning mean that execution requirements will change as well. With the rise of contract workers, outsourcing, and supply relationships, the firm no longer has control of all its critical assets. Effective management of strategic alliances and partnerships is a key ingredient in successful implementation, which we will explore in Chapter 5. These and other factors mean that flexibility in implementation is imperative to the ability to adapt to new market forces as they emerge.

See the accompanying Technology Expert's View from the Trenches feature for insight into how one high-tech start-up wrestled with its strategic decisions.

TECHNOLOGY EXPERT'S VIEW FROM THE TRENCHES

Soliciting and Listening to the Voice of the Customer
By Diane Smith
President, Strategy and Business Affairs
Auroras Entertainment (now Avail Media Inc.), Kalispell, Montana

Auroras Entertainment was a pioneer in the MPEG-4 IPTV industry. Essentially, IPTV transmits video over any broadband infrastructure. MPEG-4 refers to a video standard that reduced bandwidth requirements by 50–75%, facilitating transmission over low bandwidth infrastructures, such as copper and wireless. Auroras's services and products were targeted at smaller telephone and fiber-to-the-home companies that wanted to compete against larger, incumbent cable providers. These smaller companies wanted to avoid the multimillion-dollar investment to build their own video transmission system. So, Auroras offers them a wholesale IPTV system, providing the technology to allow them to offer IPTV services to their own subscribers.

Background

In the beginning, we were a classic start-up: overworked and underfunded. The industry hadn't yet evolved beyond the most fundamental capabilities and we recognized that our underlying technological architecture, pricing models, and service suite would be critical to our success.

We also had to successfully position Auroras against its competitors. The competitor that caused us the most concern was a subsidiary of a multibillion-dollar multinational company. Clearly, Auroras couldn't compete against such a large company's personnel, financial resources, and vendor leverage. We had developed a good portfolio of technologies and a full suite of IPTV services, but we knew that these alone wouldn't be sufficient to compete against other players with significantly greater resources than ours.

As a result, we determined early to focus intensely on our future customers' needs. Our decision to amass deep, detailed input from our customers early, and to implement their suggestions from the first days of our business modeling, became a crucial determinant in our success.

From the outset, we had a strong knowledge of the industry we would be selling into because our CEO had worked in the telephone and wireless industries and knew in-depth the competitive and corporate profiles of many of our potential customers. We leveraged this knowledge in the process of gaining customer input.

(continued)

Gaining Customer Input

For our customer input, we selected customers somewhat opportunistically (i.e., those who were willing to talk with us), but we also used our industry knowledge to seek out customers with different cultures, sizes, technology platforms, and diversification strategies (i.e., had they expanded into other telecom businesses such as cable or wireless, nontelecom businesses such as real estate, or avoided expansion in any way).

We relied on mutual acquaintances, often from trade associations that represented our potential customers, for our initial introductions. We would ask to meet or speak by conference call with the CTO, the Sales/Marketing VP, or for smaller customers, the head of the company. *Not one customer turned us down.* Indeed, these companies, our future customers, wanted to hear from us and get our perspective on the emerging industry as much as we wanted to hear from them.

The customer interview team consisted of our CEO and CTO. If we could drive to meet with the company in person, we did; otherwise the discussions occurred via phone. The meetings typically lasted two to three hours. We were emphatically candid about our start-up position in the industry, and equally frank about the fact that information we learned in the meetings would be used to develop our initial business plan. The customers were forthcoming and generous with their time, insight, and their own companies' plans. We asked questions as broad as whether we should focus on service delivery or product development (as mentioned above, we already had a portfolio of technologies that we could commercialize in addition to our plan to offer a full suite of IPTV services), the competitive positions of the customer companies and their plans to meet emerging competition, potential pricing plans for various services and products, and market timing. In other words, together we set about creating solutions to highly complex technology and business problems. Indeed, our first $1M+ investment began with one of these first customer meetings.

For example, in order to offer the service, customers needed an array of technology equipment, along with monthly services; these services included, for example, offering 200 channels of MPEG-4 encoded and encrypted standard-definition and high-definition television. We asked the companies questions such as:

- Would you be willing to purchase this rack of equipment for $500k?
- Would you be willing to pay $6/subscriber/month for standard-definition IPTV services?
- How about a straight $15k monthly fee?
- Are you more interested in future services such as Internet access/interactivity on TV or archival programming from established programmers such as Disney?
- How important is an open standards architecture that allows you to select from multiple vendors for supporting functionalities?

With each conversation, we gained a deeper understanding of each other's objectives and needs.

Insights

Surprisingly, customers of different cultures, sizes, geographies, and technology platforms shared many of the same objectives. They wanted an IPTV service provider who understood their competitive objectives and that would *partner* with them to achieve those objectives. They wanted an IPTV vendor whose focus was channel delivery, future services, and integration of set top boxes, middleware, and encryption. They were not interested in a separate encoder or set top box vendor. They wanted to buy wholesale IPTV services from a vendor (us) who was paid, at least in part, based on how successful their companies' offerings were to their customers. They wanted access to all of the services and content already available to their competitors, including 200+ channels of standard-definition, an abundance of high-definition channels, audio, DVR and video-on-demand, and they wanted a road map for future services that would allow them to differentiate from those competitors. They wanted a service provider to do all the work of sorting through the complicated legal relationships that would govern the MPEG-4 IPTV industry. For example, content rights for MPEG-4 had to be negotiated with each programmer—Viacom, Disney, Discovery, Fox, ESPN, Turner, HBO, Hallmark, and so forth. Content rights negotiations included requirements for video quality, encryption, billing, blackout management, extended area service, and closed captioning, while each programmer required a different suite of capabilities.

As a result of these conversations, Auroras made several strategic decisions, all of which were reflected in the early drafts of the business plan we used for our Series A fund-raising. Our Series A fund-raising ultimately resulted in $2.4M and allowed us a little more than a year of operational runway.

First, we focused on becoming a service provider with superior integration capabilities. The services we proposed to offer our customers were 200+ channels of standard-definition and 10+ channels of high-definition television, all of the equipment necessary to receive this fully functional service, and all of the necessary content rights and compliance required by those rights. We limited our product development to only those technologies that (1) significantly reduced our costs or (2) significantly increased our quality.

Second, we developed a pricing model that included an upfront fee for the equipment, combined with a monthly per-subscriber fee (based on our customer's actual customer subscription base).

Third, we significantly lowered our proposed monthly minimums.

Fourth, we built a network using an open standard architecture, allowing multiple vendors of support components to sell directly to the companies.

Lastly, we focused on our future service road map, and built technologies and service capabilities that would support these future services, in order to allow our customers to more swiftly differentiate themselves from their competitors.

Summary

The insight so generously provided by our future customers allowed Auroras to quickly establish itself as an industry leader. Our competitors were forced to modify their marketing efforts to respond to our pricing, open standards architecture, and future services road map.

By spending our first months listening to our customers and revising our plan to reflect their needs, Auroras was able to establish a reputation and position in the industry that far outweighed its identity as a struggling start-up.

STRATEGY TYPES

Although there are many ways to answer the three big strategy questions, four archetypes exist: (1) Product Leader, (2) Fast Follower, (3) Customer Intimate, and (4) Operationally Excellent (see Table 2.1.) Each of these strategy types has an equal chance of being successful when properly executed.

Product Leader (Prospector)

The **Product Leader (Prospector)** strategy is based on being a market pioneer, the first with innovative new product offerings in many disparate markets.

W. L. Gore, a $2-billion-a-year company, innovates on many fronts and boldly follows its inventions into completely different businesses. W. L. Gore got its start in 1958 when Bill Gore, a DuPont chemist, envisioned some ways to use polytetrafluoroethylene (PTFE)—the smooth, slippery polymer better known as Teflon—that DuPont wasn't pursuing. In 1969 Gore's son Bob (an engineer who is now chairman of the company) found a way to stretch the polymer, creating expanded PTFE, or ePTFE. It was trademarked Gore-Tex. The material became the basis for a host of new product possibilities, including the durable outdoor fabric that was introduced in the 1970s and is still W. L. Gore's best-selling product line. Some companies might have been tempted to focus on fabrics from then on, perhaps turning into a clothing manufacturer. But not W. L. Gore. Using

TABLE 2.1 Four Strategy Archetypes				
	Who are Customers?	**What Value?**	**How Is Value Created/Delivered?**	**Pros/Cons**
Product Leader *(Prospector; Pioneer; First Mover)*	Innovators Early adopters	Innovative new products ⇒Incremental innovations have greater odds of success for pioneers than breakthrough innovations	Focus on speed, commercializing ideas quickly	Pros: • Establish barrier to entry • May gain higher profits • Define ideal product attributes Cons: • Inherently risky: market may not develop as quickly as expected • High failure rate • High development costs
Fast Follower *(Analyzer)*	Early adopters Early majority	Superior products Lower prices New business models	Focus on cost, distribution	Pros: • Innovative late entrants grow faster
Customer Intimate *(Differentiated Defender)*	Early and late majority • Narrow niches • Specific (individual) customers	Customized solutions Superior service	Relationship marketing Intimate customer knowledge	Pros: • High margins • Repeat business leads to high customer lifetime value
Operationally Excellent *(Low-Cost Defender)*	Early and late majority • Mass market • Price-sensitive customers	Superior combination of quality, price, and ease of purchase Cost leadership	Value chain efficiency	Pros: • High asset turnover rates and asset return rates

Gore-Tex as its springboard, Gore has gone on to create a variety of completely different products. Gore is so good at innovation and product development that it has become a major player in areas as diverse as guitar strings, dental floss, medical devices, and fuel cells. And it has managed to post a profit every year since its founding 45 years ago.[38]

The most successful Product Leaders target innovators and early adopters,[39] because these customers are willing to take risks and have the resources to cope with risk. Product Leaders must be creative. Being creative means recognizing and embracing ideas that may originate anywhere—inside the company or out. They habitually seek out potential users of products during development and pick their brains. For example, Gore enlisted hunters to test garments made of a new fabric called Supprescent, which had a special membrane bonded to the fabric to block human odor, allowing hunters to get closer to their prey. Managers in these Prospector companies, such as former CEO

Andy Grove of Intel, create a culture that permits discussion of half-formed embryonic possibilities. Product Leaders such as 3M use collaborative methods—networks, cross-boundary teams, supply chain partnerships, and strategic alliances—to support innovation and spread knowledge from local innovations quickly. They challenge the dogmas and the orthodoxies of entrenched competitors.

Successful Product Leaders commercialize their ideas quickly. All their business and management processes are engineered for speed. Product Leaders shape a culture of unity that derives strength from diversity; they develop common tools and measurements to put everyone "on the same page," while also encouraging everyone to "break the mold." Finally, Product Leaders relentlessly pursue ways to leapfrog their own latest product or service. If anyone is going to render their technology obsolete, they prefer to do it themselves.[40]

What are the arguments in support of being a Product Leader? First movers may have competitive advantages due to barriers established by their market entry. Such entry barriers include economies of scale, experience effects, reputational effects, technological leadership, and buyer switching costs. These barriers can lengthen the lead time between a firm's head start and the response by followers. During the time when there is no competition, the first mover is, by definition, a monopolist who can gain higher profits than in a competitive marketplace. In addition, even after competitors enter, the first mover has the established market position, which may allow it to retain a dominant market share and higher margins than later entrants. For example, Toyota's Prius first went on sale in Japan in 1997, and worldwide in 2001. Estimates are that Toyota's hybrids account for roughly three of every five hybrids sold,[41] and as a result, its manufacturing plants are reaping economies of scale that other car companies can only hope for. Based on this feat, many predict that Toyota's formidable first mover advantage is solid.

First movers are also able to "skim off" early adopters, whereas later entrants are left with potential customers who are less predisposed to purchasing new products. Again, in the case of hybrid cars, studies of initial buyers showed they were willing to pay a price premium to be first in the market. This was also true of the first purchasers of Apple's iPhone in the summer of 2007. Despite some initial customers feeling disappointed by Apple's price cut from $599 to $399 after only a few months on the market, "for many of the iPhone's early adopters, money is not and never was an issue. They were after the gratification of knowing they were among the first owners of something that was cool, even revolutionary."[42] Rather than being a drawback to initial purchase, the higher price gave these early owners a sense of specialness.

Moreover, if customers know little about the importance of product attributes or their ideal combinations, a first mover can influence how attributes are valued and define the ideal attribute combination to its advantage. The first mover becomes a prototype against which all other entrants are judged, making it harder for later entrants to make competitive inroads. First movers have a higher degree of consumer awareness, which lowers customer's perceived risk and information costs.

Despite these advantages, the pioneering aspect of the Product Leader strategy is inherently risky. A study of the personal digital assistant (PDA) industry[43] found that speeding a product to market is not always a good idea. Apple's Newton is merely a footnote in the history of that industry. Other pioneers that not only lost out to later entrants but eventually disappeared include VisiCalc, the first personal computer spreadsheet program, and Osborne, the first portable computer. And in the online grocery business, WebVan invested more than $1 billion before closing shop in July 2001. Yet in 2002, FreshDirect was thriving with a more focused, lower cost business model.[44] A historical analysis of 500 brands in 50 product categories during the period 1856–1979 found that the failure rate of market pioneers is 47% while their average market share is 10%. The eventual market leaders entered the market an average of 13 years after the pioneer. However, their failure rate was only 8% and their average market share was 28%.[45]

Somewhat counterintuitively, given their winner-take-all possibilities, the risks of pioneering are greater in markets characterized by network externalities.[46] A study of 45 product categories—including antivirus software, audiocassette players, camcorders, and DVD players—revealed that the uncertainty in the market when only a few early customers have adopted leads to a wait-and-see attitude, creating inertia in the market. In turn, this inertia leads to a delay in the pioneer's revenue relative to its development and marketing costs, decreasing the survival rate of pioneers. In such a situation, later entrants gain disproportionately because of the larger network that exists at a later point in time, and the resulting increase in product value that customers find from the network effects. In addition, larger firms have resources that allow them to have staying power as pioneers.

Another study showed that the risk of pioneering was much lower in markets started by an incremental innovation than in those that were started by a breakthrough (radical) innovation.[47] In other words, in markets pioneered by a really innovative product, the first firm to market is often the first to fail; in markets pioneered by an incremental innovation, a first mover advantage *might* protect the pioneer (unless the market is characterized by network effects). Hence, consistent with the contingency theory of high-tech marketing established in Chapter 1, matching a firm's strategy (pioneer) to the type of innovation (incremental) will enhance its odds of success. On the other hand, mismatching a pioneering strategy to a breakthrough innovation will lower the odds of success—unless the market is characterized by network effects.

The negative effect of network externalities on pioneer survival rates is moderated by breakthrough innovations and technologically intensive products: Pioneers of breakthrough innovations and technologically intensive products survive longer than pioneers of incremental, less technologically intensive products *in markets characterized by network effects.*[48] These studies highlight the very complicated intersection of product factors (type of innovation) and market conditions that affect firm strategy and performance in high-tech markets.

Clearly, firms must evaluate trade-offs among time to market, product innovativeness, and development costs. Pioneers face huge development costs against a high degree of market uncertainty. Although pioneers might enjoy sustained revenue advantages, they also suffer from persistently high costs, which may eventually overwhelm the sales gains. Over the long haul, generally, pioneers are less profitable than later entrants.[49]

Fast Follower (Analyzer)

The **Fast Follower (Analyzer)** strategy essentially imitates the Product Leader's successful product and market development efforts, attempting to improve on the Product Leader's offering in some key way. For example, in March of 2000, Bill Gates went to the annual Game Developers Conference in San Jose, California, bearing potentially worrisome news for Sony, whose hot new PlayStation 2 had rocked Japan the previous week by selling nearly a million machines in three days. After a year of top-secret development, Gates finally revealed details about Microsoft's new product, the Xbox, which would be released in the fall of 2001. Xbox, Gates promised, would have more speed, lifelike video graphics, ear-popping audio, and more memory than the PlayStation 2. "This is a huge milestone for us," Gates told an overflow crowd at the conference. "It's a new platform for the industry."[50] Similar to the situation in operating systems, applications software, and Internet browsers, Microsoft was not a pioneer in video game machines. The lack of first-mover advantage did not stop Microsoft from dominating those markets.

The most successful Fast Followers target early adopter and early majority customers.[51] They can overcome Product Leaders by identifying and fulfilling a superior but overlooked market position in the following ways:[52]

1. Innovate superior products.
2. Undercut the leader on prices.
3. Outadvertise/outdistribute the leader, thereby beating it at its own game.
4. Innovate strategies that change the rules of the game.

Innovative late entrants, relative to Product Leaders, grow faster and have higher market potential. Indeed, one study found that later entrants could replicate a pioneer's innovation at least

35% cheaper;[53] major patented innovations can be "invented around" within three years, and unpatented innovations can be imitated in one year. Hence, innovative Analyzers slow the Product Leader's growth and reduce its marketing spending effectiveness.

Customer Intimate (Differentiated Defender)

The **Customer Intimate (Differentiated Defender)** strategy focuses on delivering not what the overall market wants, but what specific customers want. Companies that follow this strategy type tend to target either more narrow niches or individual customers. They do not pursue one-time transactions; they cultivate relationships. They specialize in satisfying unique needs, which often only they recognize, through a close relationship with—and intimate knowledge of—the customer. Their proposition to the customer: "We have the best solution for you, and we provide all the support you need to achieve optimal results, or value, or both, from whatever products you buy."[54] Customer Intimates are most successful when they target the early and late majority customers in the market.[55]

Importantly, as technologies mature and become commoditized, a reputation for superior service is a valuable differentiator in high-tech markets. Delivering great customer service creates customer loyalty, which increases the likelihood of customers purchasing additional products. The Customer Intimate strategy unlocks this potential, resulting in a positive bottom-line impact on the business.

Since launching Intuit in 1983, Scott Cook's mission has been to remove the agony from necessary but odious tasks such as bookkeeping and tax preparation—Intuit calls these tasks "pain points." From the outset, the company had three cornerstones to its business strategy: (1) observational as well as traditional market research, (2) customer-driven product development, and (3) outstanding free, lifetime customer service on low-cost software.[56] At the 2006 Computer–Human Interaction Conference, Cook argued: "Innovation happens at the junction between business and customer needs, not from executive ideas or lonely geniuses within the company. . . . Creating a culture of innovation is about nurturing customer observation, incubating new ideas, celebrating failure, and staying out of the way."[57] In 2006, Intuit, a $2.6 billion company that dominates the retail software market for tax preparation and small-business accounting, had an operating profit margin of 23.9% compared to 16.6% for H&R Block, its closest competitor.

An emphasis on relationships does not mean that all customers or potential customers are worthy of the effort that building and sustaining a relationship requires. For example, when asked who its unprofitable customers are, executives at one *Fortune* 500 company responded that it had no unprofitable customers. However, this company itself was not economically profitable.[58] In a bit of perverse logic, its senior managers seemed to be arguing that the firm earned profit from every customer but destroyed economic value in the aggregate! This illustrates the importance of an active rather than a passive approach to customer relationship management. It is no longer sufficient for managers to offer the platitude "Customers are our most important asset." Building profitable customer relationships requires specific marketing activities to identify the right customers, acquire them, and retain them,[59] a topic explored in Chapter 5.

Operationally Excellent (Low-Cost Defender)

The **Operationally Excellent (Low-Cost Defender)** strategy targets early and late majority customers with a superior combination of quality, price, and ease of purchase. Once established, these firms aggressively protect their market from competitors by achieving cost leadership across their value chain. They seek technological, production, and/or distribution efficiency so that they can offer low prices.

Dell Computer is the classic example of this strategy. Dell machines are made to order and delivered directly to the customer. Until its change in 2007 to offer its products through Wal-Mart,

Dell had not used channel intermediaries. Rather, in the Dell direct model the customer can receive the exact machine he or she wants at a lower price than competitors offer. The company gets paid by the customer weeks before it pays suppliers. The company that famously started in a University of Texas dorm room has more than 65,000 full-time employees. With more than $57 billion in sales in 2006, Dell ranked 34th on the *Fortune* 500, ahead of companies like Johnson & Johnson, Microsoft, and Intel. Dell had an asset turnover ratio of 2.42 and a return on assets of 15.5% compared to HP's 1.12 and 6.4%. These high asset turnover rates and return on assets are the best financial indicators of operational excellence.

A Cautionary Note

The preceding discussion does not mean that a company should single-mindedly focus on only one strategy type. Although it is true that no company can be all things to all customers, the most successful companies have both a dominant strategy type *and* one or two supporting types. For example, Procter & Gamble, America's 7th most admired company for innovation, is also 3rd for product/service quality, indicating that the company combines the Product Leader and Operationally Excellent strategic types.

STRATEGY CREATION: APPROACHES AND STRUCTURES

Just as high-tech companies create breakthrough product innovations, they must also learn to create innovations in business strategy. Developing sound strategy requires that a company bring a unique and innovative perspective to creating customer value. Firms that are able to sustain a high rate of growth do so by radically changing the basis of competition in their industries to create new wealth. They must take risks, break the rules, be mavericks.[60]

A good strategy-creation process is a deeply embedded skill; it is a way to understand what is going on in an industry, turn it on its head, and envision new opportunities. Managers should not become enamored with a specific strategy, but should be prepared to adapt and change it based on developments in the marketplace—and to do so rapidly. Effective strategy creation is based on the paradoxical notion that one can make serendipity happen. Insights about how companies can take an innovative approach to the strategy formulation process appear in Table 2.2.

Which companies are known as being highly innovative in their strategy formulation? Intel chose not to follow the conventional wisdom that it pays to extend a product's shelf life; it keeps making its own computer chips obsolete with better designs. In the process, Intel showed that with an effective advertising campaign it is possible to "brand" a component within another product. Chevron "mined" seismic data to discover a 1.45-billion-barrel oil field. Other companies—such as Amgen, Oracle, and Iomega, to name a few—have shown their willingness to look from the outside in and, in doing so, to create new rules in established industries.

The marketing of high-technology products requires a flexible strategic posture. Because many managers in technology-intensive industries can be overwhelmed with the complexities they face, they must modify and adapt their strategy-making process to this rapidly changing environment. Sadly enough, managers in technology-intensive situations may become overwhelmed by the complexities, confronting them with strategy-making processes that are more complex and unmanageable than the situations themselves. They tie up enormous amounts of management time with extensive analysis, documentation, and a lack of cross-functional involvement. These poorly thought out processes produce plans that are quickly obsolete in the face of competitive actions and reactions. Confronted with complexity, strategic market planning processes must become simpler, faster, iterative, opportunity based, team based, and functionally integrated. Market planning and strategic action must be closely coupled.

TABLE 2.2 Insights for Innovative Strategy Creation

- **Bring in new voices.**   Says business strategy expert Gary Hamel: "Companies miss the future not because they are fat or lazy but because they are blind." Many companies are unequipped to see where the future is coming from—and lack the lens to know that. Bringing new voices into the strategy-creation dialogue, outside of the company's normal comfort zone, can provide fresh insight.
- **Make new connections**—between new voices, across boundaries of function, technology, hierarchy, business, and geography. Such new conversations can offer a rich web of insights.
- **Offer new perspectives.**   Take a new vantage point in viewing the business. Rather than based on analysis and number crunching, innovative strategies often emerge from novel experiences that yield novel insights.
- **Exude passion for discovery and novelty.**   This attitude engenders an emotional attachment of employees who are committed enough to reduce the time between conceiving and implementing an idea.
- **Be willing to experiment.**   With innovative strategies, the end target may be known, but the route to it may be unknown. The best approach is to be willing to move in the right direction and to refine strategy and process as the firm learns from its experiments. Although such a notion is an anathema to efficiency, it is a must in redefining radical business strategies.

Source: Hamel, Gary, "Strategy Innovation and the Quest for Value," *Sloan Management Review* (Winter 1998), pp. 7–14.

Two important aspects of strategy formulation in these fast-paced high-tech environments are (1) the nature of the strategy-formulation process (formal versus emergent) and (2) the type of organizational structure that facilitates adaptive strategic planning.

Emergent versus Formal Planning

Some strategy experts believe that the **formal planning process,** represented by the opening Medtronic vignette, oversimplifies strategy formation. Formal planning assumes that the firm operates in a rather predictable environment and that the organization has clear and articulated intentions that can be backed up by formal controls to ensure their achievement. Using a formal planning approach, leaders formulate their intentions in the written plan and then elaborate on this plan, with budgets, schedules and so on, to guide its implementation.

In contrast to the neat, orderly process of strategy formulation and execution implied by the formal planning approach, some argue that rapidly changing customer expectations, competitor actions, and technologies, such as those found in high-tech environments, do not allow for such a rational process. Indeed, when the environment does not provide all the details or when tactics of implementation are not clear, a "messier" approach to strategy making may be required.[61] Often referred to as an **emergent planning process,** the strategy is improvised or emerges from lower levels of organizations—whether through trial-and-error learning or incrementally with guidance from the top. In other words, strategy emerges through a series of incremental decisions over time. Hence, a process of accretion, rather than a comprehensive search, evaluation, and selection process, creates the strategy that the firm follows.[62]

As an example of an emergent planning approach, when Steve Jobs returned to Apple in 1997, there wasn't a plan for anything. "Our goal was to revitalize and get organized, and if there were opportunities we'd see them," he said. "We just had to be ready to catch the ball when it's thrown by life."[63] In the flurry, Jobs and his team initially failed to notice the revolution in digital music. Once that sea change was understood, Apple developed a slick "jukebox" application known as iTunes.

It was then that Apple's management team noticed that digital music players weren't selling. "The products stank," said Apple VP Greg Joswiak.[64] Life had handed Jobs an easy opportunity, and early in 2001 he ordered his engineers to capitalize on it. Apple's hardware chief, Jon Rubinstein, picked a team leader from outside the company to create a better music player—and have it ready for Christmas season that year. Apple's requirements: a super-fast connection to the user's computer so songs could be quickly uploaded, a close synchronization with iTunes software to make it easy to organize music, a simple-to-use interface—and it had to be gorgeous.

In April 2001 Tony Fadell was hired by Apple to assemble and run its iPod & Special Projects Group. He drew on all of Apple's talents from Jobs on down. VP Phil Schiller came up with the idea of a scroll wheel to run through the menus. Apple's industrial designer Jonathan Ive said his creative energy was focused on "a product that would seem so natural and so inevitable and so simple you almost wouldn't think of it as having been designed." This austerity extended to the whiteness of the iPod. Says Ive: "It's neutral, but it is a bold neutral, just shockingly neutral." Steve Jobs's assessment of the final product: "It's as Apple as anything Apple has ever done."[65]

This is a clear example of an emergent planning strategy, one that emerged through informal entrepreneurial processes instead of a grand strategy developed from a more formalized strategic process. The result is that the iPod dominates the personal digital music player market. (In the interest of full disclosure, the author is listening to music on his iPod as he writes this!)

Which is best: a formal strategic planning process or an emergent strategy formation process? This is a false choice. Most businesses will get nowhere without emergent strategy processes alongside formal strategic planning. The two processes complement each other. However, there are conditions that favor a tilt toward one or the other. For example, the most effective Product Leaders employ a more emergent process due to the rapidly changing conditions they face and because of their focus on innovation. On the other hand, the most effective Customer Intimate and Operationally Excellent organizations place greater weight on the formal planning process due to the importance of flawless execution in a relatively stable environment.[66]

Indeed, best-practices high-tech marketers engage in a fairly systematic strategic planning process—much like that outlined by Adrian Ryans and colleagues in *Winning Market Leadership: Strategic Market Planning for Technology-Driven Businesses.*[67] Ryans and his coauthors describe a systematic and highly integrated process for evaluating market opportunities and for developing strategies to win market leadership. They focus on the key issues and tough choices faced by executives in very demanding, technology-intensive markets. Most of their examples are drawn from the experiences of large, multinational, high-technology companies such as Intel, Hewlett-Packard, Glaxo Wellcome, and General Electric. The eight-step process they advocate, presented in Table 2.3, is a guide around which companies can structure their strategic marketing planning process. It provides a set of boundary conditions within which emergent processes can take place.

Market-Focused Organizational Structure

The second key characteristic of a company that affects its strategic marketing planning is its organizational structure. Organizations have traditionally focused on improving internal efficiency through hierarchical or purely functional structures. These organizations are governed by formal rules and procedures that provide a means for prescribing appropriate behaviors and dealing with routine aspects of a problem.[68] They also are centralized, with decision authority closely held by top managers and relatively clear lines of communication and responsibilities. The route for final approval in centralized organizations can sometimes be traveled quickly.[69] Implementation in these organizations tends to be straightforward once a decision is made.[70] This approach to organizing is adequate in relatively stable, slow-cycle markets.[71]

Because of the turbulence in high-tech markets, however, the traditional approach to organizing is not adequate. Today's innovative organizations have a multidimensional focus on customers, flexibility, and speed. What does this mean for organizational structure and design? It must adopt a

---

TABLE 2.3 Eight-Step Approach to the Strategic Market Planning Process

1. *Define the company's goals and mission.*

2. *Choose the arena.* The business must tentatively identify the markets in which it will compete. Each arena of opportunity should be defined by potential customer segments that could be served, potential benefits that could be provided to these customers, possible technologies or competencies that could deliver those benefits, and possible value-adding roles for the business in the market chain.

3. *Identify potentially attractive opportunities.* This requires understanding (a) customer needs, (b) suitability of the business's resources to satisfy those needs, (c) competitive threats, and (d) profit potential.

4. *Make tough strategic choices.* The business's management team must identify key strategic issues such as markets to target, technologies to develop, or choice of a value-creating strategy and determine how to confront each issue.

5. *Plan key relationships.* In complex and fast-moving markets, a business rarely has all of the resources to execute a strategy by itself. At this stage, key collaborators, such as suppliers, distributors, or complementors, are identified and a working relationship is defined.

6. *Complete the winning strategy.* After the broad outlines of the strategy have been determined, several additional issues must be resolved. These include articulation of a clear value proposition, development of a marketing strategy, and recruitment and placement of key personnel.

7. *Understand the profit dynamic.* Once the detailed strategy has been established, the team should assess the profit implications of each market strategy. This requires development of a financial model of the strategy and testing of the financial implications of changes to key variables.

8. *Implement the chosen strategy.* The firm must possess appropriate structure, systems, personnel, and skills to complement the strategy. Importantly, implementation is not something that follows strategy development. Rather, strategy development and strategy implementation should be tightly integrated to maximize the likelihood of strategy success.

Source: From Adrian Ryans, Roger More, Donald Barclay, and Terry Deutscher, *Winning Market Leadership: Strategic Market Planning for Technology-Driven Businesses* (Toronto: Wiley, 2000).

market-focused organizational structure, also referred to as a *customer-focused structure.* How is this accomplished? First, there must be a shift away from organizing around products to organizing around markets—that is, customers or customer groups.[72] Companies transition from being product focused to being customer centered in progressive steps. Initially, they might use informal coordination among departments or a matrix-type structure. As they progress to being customer centered, they might add integration functions, such as customer segment task forces or a customer segment manager, to better align the organization with the market.[73]

One high-tech company that has shifted from a product-centric to a customer-driven organizational structure is Intel. In 2005, the new CEO of Intel, Paul Otellini, realized that the company's organizational structure, focused on microprocessors, required a redesign based on customer's needs. The reorganization recognized that technology companies don't deliver discrete products; rather, they deliver solutions that respond to the end user's demands. As reported in *Business Week:* "So rather than relying on a structure focused on the company's discrete product lines, Intel's reorganization brought together engineers, software writers, and marketers into five market-focused units: corporate computing, the digital home, mobile computing, health care, and channel products—PCs for small manufacturers,"[74] with each unit responsible for these different market segments. For example, the mobility group builds technology platforms for notebook computers, handheld computers, and cell phones. Of course, the successful implementation of this reorganization required changes in compensation (focused on the unit's performance rather than individual product performance), changes in culture (where "gearheads have reigned supreme," the company needed experts in customer needs), and coordination between multiple units that sell across product lines and customer groups. But, as Otellini said, "Every idea should be focused on meeting customers' needs from the outset."[75]

Second, to achieve flexibility and speed, the most successful innovative firms utilize relatively few formal rules and procedures, and delegate—or decentralize—decision-making authority.[76] Less bureaucracy means improved horizontal communication, which in turn, leads to more effective information sharing and reduces the lag time between decision and action. Decentralized decision making produces more new ideas than centralized decision making.[77] Additionally, for nonroutine tasks that take place in complex environments, decentralization enhances responsiveness; it empowers managers close to the issue to make decisions and implement them rapidly.

MARKETING PERFORMANCE MEASUREMENT

How do we relate the way we run a business to results? What are results? The traditional answer—"the bottom line"—is treacherous. Under a bottom-line philosophy, we cannot relate the short run to the long term, and yet the balance between the two is a crucial test of management. The beacons of productivity and innovation must be our guideposts. If we achieve profits at the cost of downgrading productivity or not innovating, they aren't profits. We're destroying capital. On the other hand, if we continue to improve productivity of all key resources and improve our innovative standing, we are going to be profitable. Not only today, but tomorrow.[78]

Conventional wisdom says that a company can't manage what it can't (doesn't) measure. A frequent criticism of marketers is that they love to spend money but hate to measure the results of that spending. This aversion to accountability is a major reason why marketing's influence, particularly in companies with a technological orientation, is low. However, CEOs in high-tech firms that have a well-developed marketing performance measurement capability (compared to firms that don't) are more satisfied with the marketing function and realize superior performance.[79]

The issue of **marketing metrics,** or the financial accountability of marketing activities, linking marketing investments and decisions to measurable outcomes, is increasing in importance for all firms.[80] A popular framework for marketing performance measurement is the **marketing dashboard,** which is something like a balanced scorecard or multidimensional report card for a firm's performance. Performance dashboards are used to organize and report key metrics to senior management from the array of information generated by corporate information systems. Dashboards are often automated and provide (close to) real-time reporting that enables users to "drill down" to program-level details.

A marketing dashboard is more than a collection of various metrics. High-tech companies must think carefully about what types of information and metrics can best measure their marketing strategies. For example, financial performance is affected by multiple aspects of a company's decisions. A company may have a solid marketing strategy, but if it does a poor job of delivering a product and providing after-sales service and support, it may lose customers, revenues, and profits.

Moreover, the selection of the metrics should reflect leading and lagging indicators of the success of the firm's marketing strategy. The logic behind the Customer Intimate (Differentiated Defender) strategy type, for instance, is that loyal customers generate superior profits and revenue growth. In turn, customer satisfaction is a leading indicator of customer loyalty. And, customers are more likely to be satisfied when they experience high-quality service and products. Thus, a generic dashboard for a Customer Intimate company might look like the one in Table 2.4. Metrics must be linked closely to specific decisions and actions in order to draw accurate conclusions.

Tektronix, a firm that provides test and measurement equipment and services to communications and computer companies, developed an award-winning dashboard. As Martyn Etherington,

TABLE 2.4 Sample Marketing Dashboard for a Customer Intimate Company

Product/Service Quality	Customer Satisfaction	Customer Loyalty	Financial Results
Return rate	% of customers who report being "very satisfied" or "extremely satisfied"	% of sales from existing customers	Sales growth
Reject rate	Satisfaction with specific features	Duration of customer relationship	Market share
Customer-perceived quality	Satisfaction with overall experience	Share of customer's wallet	Profit growth
Time from order to delivery	Volume of customer complaints	Sales from customer referrals	Profit margin

the firm's vice president of worldwide marketing, put it: "We needed to move from activity-based marketing to outcome-based marketing."[81] To define the marketing outcomes, Etherington and his dashboard team met with executives in sales, product development, and senior management and asked them: "What would [we] need to do to obtain a grade A from you and your constituencies?"

Tektronix developed metrics for the sales department that included the number of qualified sales leads, lead-to-opportunity conversion, sales funnel (from "unqualified prospects" to those who have received delivery of the product or service and have paid for it) reporting, and lead quality. On the marketing profitability side, the team created a set of metrics including lead-to-order conversion, time from order to delivery, and order-dollar per market-dollar. The company uses these metrics throughout its marketing activities, including advertising, online marketing, direct mail, e-mail, seminars, trade shows, and sales calls.

Tektronix has had remarkable results since the dashboard was launched in 2003. It has achieved a 125% increase in responses to marketing programs and a 90% increase in qualified sales leads. In addition, the company reduced its cost per lead by 70%. Moreover, the company's marketing forecast accuracy improved from a variance of 50% before the dashboard was developed to a variance of 3%.[82]

Companies such as Salesforce.com specialize in dashboards linked to customer relationship management and other marketing activities. A sample of their dashboard appears in Figure 2.6.

In addition to metrics for measuring the various aspects of a firm's marketing activities, measuring its innovation performance is also critical.[83] Certainly, the firm's bottom line (financial performance) is an important indicator.[84] Other metrics include the innovation's usefulness and novelty, the quality, adherence to budget and schedule, speed to market, and sales or sales takeoff.[85] However, measuring innovation performance requires looking beyond a company's own performance; it should also compare the company's achieved/attained performance to that which was possible. As stated by Peter Drucker:

> Every organization needs a way to record and appraise its innovative performance. In organizations already doing that . . . the starting point is *not* the company's own performance. It is a careful record of the innovations in the entire field during a given period. Which of them were truly successful? How many of them were ours? Is our performance commensurate with our objectives? With the direction of the market? With our market standing? With our research spending? Are our successful innovations in the areas of greatest growth and opportunity? How many of the truly important innovation opportunities did we miss? Why? Because we did not see them? Or because we saw them but dismissed them? Or because we botched them? And how well do we do in converting an innovation into a commercial product? . . . It raises the right questions.[86]

These questions encourage and guide inquiry on innovation measurement not only to account for the opportunity costs, but also to consider the "why" explanations for a company's attained performance.

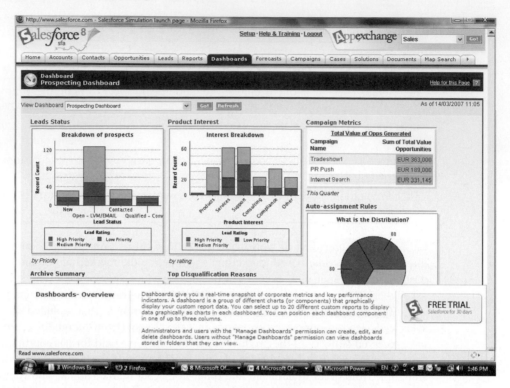

FIGURE 2.6   Sample Marketing Dashboard from Salesforce.com

Source: Salesforce.com, "Screenshots," 2007, online at www.salesforce.com/services-training.

Regardless of whether a firm chooses to use dashboards or some other tool to monitor marketing activity, the key point is that best-practices marketing strategy explicitly—and proactively—considers performance measurement. It decides *prior to* strategy implementation exactly how marketing activities will be evaluated, and the best metrics to use in monitoring marketing strategies. Furthermore, it specifically allocates funding for such measurement processes.

Summary

For a firm to achieve superior performance, it must hold a position of competitive advantage. Competitive advantage is based on having resources that are valuable, rare, durable, and difficult to imitate. However, even resources that pass these tests must be properly deployed through the firm's action plan or strategy. An effective strategy must answer three key questions: (1) Who are our customers? (2) What value do we offer them? (3) How can we create and deliver that value efficiently and effectively?

While there are countless combinations of answers to these questions, four strategy archetypes exist:

(1) Product Leader (Prospector), (2) Fast Follower (Analyzer), (3) Customer Intimate (Differentiated Defender), and (4) Operationally Excellent (Low-Cost Defender).

The means by which the strategy types answer the three key questions is the strategy formulation process. One planning approach, typified by Medtronic in the opening vignette, is a formal, rational, and comprehensive methodology. In contrast, an emergent planning approach relies more heavily on informal, market-learning processes for inspiring creative solutions. Organizational structure influences both how strategy is formed and how it is implemented. The hierarchical structure is best suited to relatively stable environments while the market/customer-focused configuration is best suited to dynamic environments.

Finally, marketers must be able to assess the effectiveness of their strategies and tactics. The marketing dashboard, a tool for reporting key metrics to senior management, is a popular and effective tool for doing this.

Discussion Questions

Chapter Concepts

1. Provide an overview of the strategic marketing planning process in high-tech firms.
2. What is competitive advantage? Provide examples of resources and competencies that high-tech firms may possess. Why are resources and competencies especially important for high-tech companies?
3. What are core competencies? Give an example of a firm's core competencies. Assess your example on each of the three characteristics of a core competency. Draw a tree diagram and explain its interpretation for the core competencies you have identified.
4. What are the tests/requirements for competitive advantage? Be sure to tie them to superiority and sustainability.
   a. Explain the idea of customer value in detail, and how valuable resources enhance customers' effectiveness or efficiency.
   b. Explain resource rareness and how it might be developed.
   c. When is a competitive advantage durable?
   d. When is a competitive advantage made inimitable? How is imitability affected by transparency, replicability, and transferability? (See Figure 2.4.)
5. What are the key questions that a company's strategy should answer? What is the strategy sweet spot?
6. How does a company approach the strategy question of "which customers"?
7. What is a "blue ocean" strategy?
8. How does a company approach the question of "what value"?

9. What is the value proposition? What are the three basic types of value propositions?
10. How does a company approach the third strategy question of how to create and deliver value?
11. What are the four strategy archetypes that address the three big strategy questions? How does each archetype answer the three questions?
12. What are the advantages of being a Product Leader (Prospector)? What are the risks?
13. How does a Fast Follower (Analyzer) slow the Product Leader's growth?
14. What is the strategy type of Customer Intimate? What customers are targeted? How does customer intimacy relate to this strategy?
15. How does the Operationally Excellent strategy type differ from the Product Leader? Why does the Operationally Excellent type have to pursue a different strategy from the Product Leader?
16. "Developing a strategy for marketing in high-tech products requires a flexible strategic posture." Explain what the implications of this statement are for innovative strategy creation.
17. How is the formal planning approach different from the emergent planning approach? How do the formal approach and the emergent process complement each other for a high-tech company?
18. How does an organization develop a market-focused organizational structure? Why is this structure conducive to high-tech markets?
19. A popular and effective tool for reporting key metrics to senior management is the marketing dashboard. What is it? How is it used? Why is it important?

Discussion Questions from the Appendix

20. Name and define the three key ways to finance high-tech start-ups.
21. What is a technology incubator? What makes an incubator successful?

Application Questions

1. Can a strategic planning system, such as Medtronic's, be a source of competitive advantage? Explain.
2. What configuration of resources (i.e., tangible assets, intangible assets, competencies) is necessary for a Fast Follower, such as Microsoft, to be successful?
3. Write a "Resonating Focus" value proposition for the Nintendo Wii.
4. *Wired* magazine named Netflix as one of the most innovative companies in the world in 2007. Clearly, Netflix's early success was largely based on first mover advantage as a Product Leader. However, Netflix launched its service in 1999. What configuration of strategies (Product Leader, Fast Follower, Customer Intimate, Operationally Excellent) has enabled Netflix to maintain its position?
5. Advances in information technology have the potential to transform how businesses organize to innovate and create customer value. Read "Virtualizing the Corporation," by Wim Elfrink, chief globalization officer of Cisco (http://www.cisco.com/en/US/solutions/collateral/ns340/ns856/ns872/Virtualizing-the-Corporation.pdf), and compare the ideas in that article to the concepts from this chapter. Where are they consistent? Inconsistent?
6. Develop a marketing dashboard composed of 10 to 15 key metrics for a Product Leader such as W. L. Gore.

Glossary

bifocal vision   A simultaneous focus on both current and future customers.

"blue ocean" strategy   A strategy that creates a new market space by serving customers whose needs are not met by current products or by serving previously unidentified customer segments.

competitive advantage   A position where a firm is able to create more value for customers than its competitors, while earning a superior return on investment. Competitive advantage requires possession of superior resources (tangible and intangible assets) and competencies that are valuable, rare, durable, and inimitable.

core competencies   The underlying skills and capabilities that give rise to a firm's source of competitive advantage; often based in hard-to-imitate embedded knowledge.

Customer Intimate (Differentiated Defender)   A company strategy focused on delivering not what the overall market wants, but what specific customers want.

customer value   The difference between the benefits that a customer realizes from using a product and the total life-cycle costs that the customer incurs in finding, acquiring, using, maintaining, and disposing of the product.

durability   How rapidly a valuable resource becomes obsolete due to innovation by current or potential competitors.

emergent planning process   A strategy-formulation approach in which strategy emerges over time through a series of incremental decisions; based on informal, market-learning processes.

Fast Follower (Analyzer)   A company strategy based on imitating the Product Leader, while attempting to improve on the Product Leader's offering and learning from its experiences.

formal planning process   A strategy-formulation approach in which a structured process is followed to develop a written plan, with budgets and schedules to guide its implementation; assumes a rather predictable environment, clear and articulated intentions, and formal controls.

inimitability   Difficulty of a competitor to obtain or copy a valuable resource (either by internal development or purchase in the market); greater inimitability enhances sustainability of a firm's competitive advantage.

market-focused organizational structure   Business structure organized around customers or customer groups (markets), rather than around products; also referred to as a *customer-focused organizational structure.*

marketing dashboard   A multidimensional report card for a firm's performance, used to organize and report key metrics to senior management from the array of information generated by corporate information systems; often automated to provide timely, detailed reporting.

marketing metrics   The financial accountability of marketing activities; the measures linking marketing investments and decisions to measurable outcomes.

market space The product/market arena in which a firm competes.

Operationally Excellent (Low-Cost Defender) A company strategy focused on a superior combination of quality, price, and ease of purchase to achieve cost leadership; often uses technological, production, and/or distribution efficiency to offer low prices.

Product Leader (Prospector) A company strategy based on being a market pioneer, the first to market with innovative product offerings.

resource rareness A situation in which competitors or producers of substitutes do not possess the same, or similar, set of resources or competencies.

strategic market planning The process by which a company formulates its strategic plan or road map for the company's future.

strategy sweet spot The intersection of the three strategy decisions—which customers to serve, what value to offer customers (the company's value proposition), and how to create and deliver that value—to achieve a position of sustainable competitive advantage.

technology incubator A building or facility that "incubates" technology start-ups by offering an array of business support resources and services for a brief period of time until they are sufficiently developed to venture into the market on their own.

tyranny of the served market A narrow focus on serving the needs of current customers at the expense of identifying possible new customers with new needs.

value proposition A description or statement of the value a company offers its customers; must be meaningful and clearly differentiate the firm from its competitors.

Notes

1. Based on an interview with Robert Guezuraga, president of Medtronic's Cardiac Surgery business, and Jan Shimanski, general manager of Biologics and Therapeutics, on June 3, 2003.
2. "Our Mission," Medtronic, 2008, online at www.medtronic.com/corporate/mission.html.
3. Cooper, Lee, "Strategic Marketing Planning for Radically New Products," *Journal of Marketing* 64 (January 2000), pp. 1–16; and Eisenhardt, Kathleen, and Shona Brown, "Patching: Restitching Business Portfolios," *Harvard Business Review* 77 (May–June 1999), pp. 72–82.
4. Morningstar Wide Moat Focus Index (2007), http://news.morningstar.com/articlenet/article.aspx?id=255465.
5. Day, George S., "The Capabilities of Market-Driven Organizations," *Journal of Marketing* 58 (October 1994), pp. 37–52; Srivastava, Rajendra K., Tasadduq A. Shervani, and Liam Fahey, "Market-Based Assets and Shareholder Value: A Framework for Analysis," *Journal of Marketing* 62 (January 1998), pp. 2–18; and Srivastava, Rajendra K., Tasadduq A. Shervani, and Liam Fahey, "Marketing, Business Processes, and Shareholder Value: An Organizationally Embedded View of Marketing Activities and the Discipline of Marketing," *Journal of Marketing* 63 (1999, Special Issue), pp. 168–179.
6. Prahalad, C. K., and Gary Hamel, "The Core Competence of the Corporation," *Harvard Business Review* 68 (May–June 1990), pp. 79–91.
7. Gomes, Lee, "H-P to Unveil Digital Camera and Peripherals," *Wall Street Journal,* February 25, 1997, p. B7.
8. Aun, Fred J., "HP Offers Shot in Arm for Aging Inkjet Tech," *TechNewsWorld* (for *E-Commerce Times*), September 11, 2007, available at www.ecommercetimes.com/story/59285.html.
9. Prahalad and Hamel, "The Core Competence of the Corporation."
10. Vogelstein, Fred, "Mighty Amazon," *Fortune,* May 26, 2003, pp. 60–67.
11. Stone, Brad, "Sold on eBay, Shipped by Amazon.com," *New York Times,* April 27, 2007, available online at www.nytimes.com/2007/04/27/technology/27amazon.html?pagewanted=print,%20downloaded.
12. Reiss, Spencer, "Planet Amazon," *Wired,* May 2008, pp. 88–95.
13. Barney, Jay, "Firm Resources and Sustained Competitive Advantage," *Journal of Management* 17 (March 1991), pp. 99–120.
14. U.S. Green Building Council, 2008, online at www.usgbc.org/DisplayPage.aspx?CategoryID=19.
15. Lockwood, Charles, "Building the Green Way," *Harvard Business Review* 84 (June 2006), pp. 129–137; and Narayan, Venkata, "Green Makes Business Sense," *SPAN* (January–February 2004), pp. 57–59.
16. Wiggins, Robert, and Timothy Ruefli, "Sustained Competitive Advantage: Temporal Dynamics and Persistence of Superior Economic Performance," *Organization Science* 13 (January–February 2002), pp. 81–105.
17. Collis, David, and Cynthia A. Montgomery, "Competing on Resources: Strategy in the 1990s," *Harvard Business Review* 73 (July–August 1995), pp. 118–128; Grant, Robert M., "The Resource-Based Theory of Competitive Advantage: Implications for Strategy Formulation,"

California Management Review 33 (Spring 1991), pp. 114–135; and Williams, Jeffrey R., "How Sustainable Is Your Competitive Advantage?" *California Management Review* 34 (Spring 1992), pp. 29–51.

18. Schenker, Jennifer, "Dell Steps Up Consumer Pursuit," *Business Week Online,* June 8, 2007, p. 8. Available at www.businessweek.com/technology/content/jun2007/tc20070607_005661.htm.

19. Barney, Jay, "Firm Resources and Sustained Competitive Advantage."

20. Danneels, Erwin, "The Dynamics of Product Innovation and Firm Competencies," *Strategic Management Journal* 23 (December 2002), pp. 1095–1121.

21. Atuahene-Gima, Kwaku, "Resolving the Capability–Rigidity Paradox in New Product Innovation," *Journal of Marketing* 69 (October 2005), pp. 61–83.

22. Chandy, Rajesh, Brigitte Hopstaken, Om Narasimhan, and Jaideep Prabhu, "From Invention to Innovation: Conversion Ability in Product Development," *Journal of Marketing Research* 43 (August 2006), pp. 494–508.

23. Mizik, Natalie, and Robert Jacobson, "Trading Off between Value Creation and Value Appropriation: The Financial Implications of Shifts in Strategic Emphasis," *Journal of Marketing* 67 (January 2003), pp. 63–76; see also Erickson, Gary, and Robert Jacosbon, "Gaining Comparative Advantage through Discretionary Expenditures: The Returns to R&D and Advertising," *Management Science* 38 (September 1992), pp. 1264–1279.

24. Ohmae, Kenichi, "Getting Back to Strategy," *Harvard Business Review* 66 (November–December 1988), pp. 149–156.

25. Markides, Constantinos, "Strategic Innovation," *Sloan Management Review* 38 (Spring 1997), pp. 9–23.

26. Hamel, Gary, and C. K. Prahalad, *Competing for the Future* (Boston: Harvard Business School Press, 1994); and Kim, W. Chan, and Renee Mauborgne, "Creating New Market Space," *Harvard Business Review* 77 (January–February 1999), pp. 83–93.

27. Grossman, Lev, "A Game for All Ages," *Time,* May 15, 2006, pp. 36–39.

28. Ibid.

29. Kim, W. Chan, and Renee Mauborgne, *Blue Ocean Strategies: How to Create Uncontested Market Space and Make Competition Irrelevant* (Boston: Harvard Business School Press, 2005); and Ulwick, Anthony, *What Customers Want: Using Outcome-Driven Innovation to Create Breakthrough Products and Services* (New York: McGraw-Hill, 2005).

30. Kim and Mauborgne, *Blue Ocean Strategies.*

31. Johnson, Kay, and Xa Nhon, "Selling to the Poor," *Time,* April 17, 2005, available online at www.time.com/time/magazine/article/0,9171, 1050276,00.html.

32. Serwer, Andy, Irene Gashurov, and Angela Key, "There's Something about Cisco," *Fortune,* May 15, 2000, pp. 114–127.

33. Anderson, James, Nirmalya Kumar, and Jim Narus, *Value Merchants* (Boston: Harvard Business School Press, 2007).

34. "Dell Joins with Leading Software Providers to Simplify Store Operations for Retailers," (2003), Business Wire, Jan 13, http://findarticles.com/p/articles/mi_m0EIN/is_2003_Jan_13/ai_96373670.

35. "MR Re-imagined," GE Healthcare Home, online at www.gehealthcare.com/usen/mr/index.html.

36. "Microsoft Windows XP Professional Offers Billions of Dollars in Savings to Enterprise Customers Worldwide," Microsoft Corporation, 2008, online at www.microsoft.com/presspass/press/2002/oct02/10-09symposiumpr.mspx.

37. Bossidy, Larry, and Ram Charan, *Execution: The Discipline of Getting Things Done* (New York: CrownBusiness, 2002), p. 15.

38. Harrington, Ann, "Who's Afraid of a New Product? Not W. L. Gore," *Fortune,* November 10, 2003, pp. 189–191. Available online at http://money.cnn.com/magazines/fortune/fortune_archive/ 2003/ 11/10/352851/index.htm.

39. Slater, Stanley, Tomas Hult, and Eric Olson, "On the Importance of Matching Strategic Behavior and Target Market Selection to Business Strategy in High-Tech Markets," *Journal of the Academy of Marketing Science* 35 (March 2007), pp. 5–17.

40. Treacy, Michael, and Fred Wiersema, *The Discipline of Market Leaders* (Reading, MA: Addison-Wesley, 1995).

41. Nunn, Peter, "Toyota Prius: Chapter Three," *Edmunds AutoObserver,* July 16, 2007, available online at www.autoobserver.com/2007/07/toyota-prius-ch.html.

42. Robertson, Jordan, and May Wong, "Apple Responds to Backlash, Offers Apology," September 6, 2007, MSNBC.com, online at www.msnbc.msn.com/id/20624042.

43. Bayus, Barry L., Sanjay Jain, and Ambar G. Rao, "Too Little, Too Early: Introduction Timing and New Product Performance in the Personal Digital Assistant Industry," *Journal of Marketing Research* 34 (February 1997), pp. 50–63.

44. Kirkpatrick, David, "The Online Grocer Version 2.0," *Fortune,* November 25, 2002, pp. 217–222.

45. Golder, Peter N., and Gerald J. Tellis, "Pioneer Advantage: Marketing Logic or Marketing Legend?" *Journal of Marketing Research* 30 (May 1993), pp. 158–171.

46. Srinivasan, Raji, Gary Lilien, and Arvind Rangaswamy, "First In, First Out? The Effects of Network Externalities on Pioneer Survival," *Journal of Marketing* 68 (January 2004), pp. 41–58.

47. Min, Sungwook, Manohar U. Kalwani, and William T. Robinson, "Market Pioneer and Early Follower Survival Risks: A Contingency Analysis of Really New versus Incrementally New Product-Markets," *Journal of Marketing* 70 (January 2006), pp. 15–33.

48. Srinivasan, Lilien, and Rangaswamy, "First In, First Out?"

49. Boulding, William, and Markus Christen, "First-Mover Disadvantage," *Harvard Business Review* 79 (October 2001), pp. 20–21.

50. Yang, Dori Jones, "Bill Gates Has His Hand on a Joystick," *U.S. News and World Report,* March 12, 2000, p. 54. Available online at www.usnews.com/usnews/biztech/articles/000320/archive_ 020966.htm.

51. Slater, Hult, and Olson, "On the Importance of Matching Strategic Behavior and Target Market Selection."

52. Shankar, Venkatesh, Gregory S. Carpenter, and Lakshman Krishnamurthi, "Late Mover Advantage: How Innovative Late Entrants Outsell Pioneers," *Journal of Marketing Research* 35 (February 1998), pp. 57–70; see also Zhang, Shi, and Arthur B. Markman, "Overcoming the Early Entrant Advantage: The Role of Alignable and Nonalignable Differences," *Journal of Marketing Research* 35 (November 1998), pp. 413–426.

53. Levin, Richard, Alvin Klevoric, Richard Nelson, and Sidney Winter, "Appropriating the Returns from Industrial Research and Development," *Brookings Papers on Economic Activity* (Winter 1987), pp. 783–820.

54. Treacy and Wiersema, *The Discipline of Market Leaders.*

55. Slater, Hult, and Olson, "On the Importance of Matching Strategic Behavior and Target Market Selection."

56. Kirkpatrick, David, "Throw It at the Wall and See If It Sticks," *Fortune,* December 12, 2005, pp. 142–150.

57. "CHI 2006: Scott Cook (Intuit)," posted April 24, 2006, at http://radio.weblogs.com/0110772/2006/04/24.html#a1754.

58. Selden, Larry, and Geoffrey Colvin, "Will This Customer Sink Your Stock?" *Fortune,* September 30, 2002, pp. 127–132. Available online at http://money.cnn.com/magazines/fortune/fortune_archive/2002/09/30/329272/index.htm.

59. Blattberg, Robert C., Gary Getz, and Jacquelyn S Thomas, "Managing Customer Acquisition," *Direct Marketing* 64 (October 1, 2001), pp. 41–55; and Reinartz, Werner, and V. Kumar, "The Mismanagement of Customer Loyalty," *Harvard Business Review* 80 (July 2002), pp. 86–95.

60. Hamel, Gary, "Strategy Innovation and the Quest for Value," *Sloan Management Review* (Winter 1998), pp. 7–14.

61. Moorman, Christine, and Anne Miner, "The Convergence of Planning and Execution: Improvisation in New Product Development," *Journal of Marketing* 62 (July 1998), pp. 1–20.

62. Mintzberg, Henry, and John Waters, "Of Strategies Deliberate and Emergent," *Strategic Management Journal* 6 (July–September 1985), pp. 257–272.

63. Levy, Steven, "iPod Nation," *Newsweek,* July 26, 2004, pp. 42–50.

64. Ibid.

65. Ibid.

66. Slater, Stanley, Eric Olson, and Tomas Hult, "The Moderating Influence of Strategic Orientation on the Strategy Formation Capability–Performance Relationship," *Strategic Management Journal* 27 (December 2006), pp. 1221–1331.

67. Ryans, Adrian, Roger More, Donald Barclay, and Terry Deutscher, *Winning Market Leadership: Strategic Market Planning for Technology-Driven Businesses* (Toronto: Wiley, 2000).

68. Ullrich, Robert, and George Wieland, *Organization Theory and Design* (Homewood, IL: Irwin, 1980).

69. Hage, Jerald, and Michael Aiken, *Social Change in Complex Organizations* (New York: Random House, 1970).

70. Ullrich and Wieland, *Organization Theory and Design.*

71. Hitt, Michael, Barbara Keats, and Samuel DeMarie, "Navigating in the New Competitive Landscape: Building Strategic Flexibility and Competitive Advantage in the 21st Century," *Academy of Management Executive* 12 (November 1998), pp. 22–42.

72. Homburg, Christian, John Workman, and Ove Jensen, "Fundamental Changes in Marketing Organization: The Movement toward a Customer-Focused Organizational Structure," *Journal of the Academy of Marketing Science* 28 (September 2000), pp. 459–478.

73. Day, George, "Aligning the Organization with the Market," *Sloan Management Review* 48 (Fall 2006), pp. 41–49.

74. Edwards, Cliff, "Shaking Up Intel's Insides: How Its Incoming CEO Intends to Make the Chipmaker More Market-Focused," *Business Week,* January 31, 2005, p. 35.

75. Ibid.

76. Olson, Eric, Stanley Slater, and Tomas Hult, "The Performance Implications of Fit among Business Strategy, Marketing Organization Structure, and Strategic Behavior," *Journal of Marketing* 69 (July 2005), pp. 49–65.

77. Ullrich and Wieland, *Organization Theory and Design.*

78. Drucker, Peter, *The Ecological Vision* (New Brunswick, NJ: Transaction, 1993), p. 99.

79. O'Sullivan, Dan, and Andrew Abela, "Marketing Performance Measurement Ability and Firm Performance," *Journal of Marketing* 71 (April 2007), pp. 79–93.

80. Frels, Judy K., Tasadduq Shervani, and Rajendra K. Srivastava, "The Integrated Networks Model: Explaining Resource Allocations in Network Markets," *Journal of Marketing* 67 (January 2003), pp. 29–46; and Srivastava, Rajendra K., Tasadduq A. Shervani, and Liam Fahey, "Market-Based Assets and Shareholder Value: A Framework for Analysis," *Journal of Marketing* 62 (January 1998), pp. 2–18.

81. Maddox, Kate, "Tektronix Wins for Best Practice," *B to B* 90 (April 2005), p. 33.

82. Ibid.

83. Hauser, John, Gerard J. Tellis, and Abbie Griffin, "Research on Innovation: A Review and Agenda for Marketing Science," *Marketing Science* 25 (November–December 2006), pp. 687–720.

84. Tellis, Gerard J., "Disruptive Technology or Visionary Leadership?" *Journal of Product Innovation Management* 23 (January 2006), pp. 34–38.

85. Sethi, Rajesh, Daniel C. Smith, and C. Whan Park, "Cross-Functional Product Development Teams, Creativity, and the Innovativeness of New Consumer Products," *Journal of Marketing Research* 38 (February 2001), pp. 73–85; Sarin, Shikhar, and Vijay Mahajan, "The Effect of Reward Structures on the Performance of Cross-Functional Prod-uct Development Teams," *Journal of Marketing* 65 (April 2001), pp. 35–53; and Foster, Joseph A., Peter N. Golder, and Gerard J. Tellis, "Predicting Sales Takeoff for Whirlpool's New Personal Valet," *Marketing Science* 23 (Spring 2004), pp. 180–191.

86. Drucker, Peter, *Management Challenges for the 21st Century* (New York: HarperCollins, 1999), p. 119.

87. National Business Incubation Association, online at www.nbia.org.

APPENDIX 2.A

Funding and Resource Considerations for Small High-Tech Start-Ups

As mentioned in the chapter, small high-tech start-ups face resource challenges that complicate their situations: access to funding and advice/mentoring. This appendix overviews the common ways that high-tech start-ups finance their budding ventures, as well as other sources for resources, advice, and mentoring.

Funding a High-Tech Start-Up

Small high-tech start-ups have three primary ways to finance their ventures:

- Friends and family
- Bootstrapping, or funding the business through early customer revenues
- Angels and venture capitalists

The first source of funds for many new companies is known as the *friends-and-family* round of financing, which is just as it sounds. An inventor or scientist asks his or her friends and family if they would like to invest in the budding venture. Some people refer to this round of financing as "friends, families, and fools," because the success rate of many new ventures is fairly low.

Absent independent wealth or wealthy family and friends, many high-tech entrepreneurs attempt to *bootstrap* their businesses. This means they attempt to fund new growth through their own cash flows. This can be a very slow way to grow a new business, as the ability to finance new initiatives depends on successfully selling products and services to at least some early customers. However, the benefit is that the company remains in the control and ownership of the initial founders.

Of course, high-tech start-ups can also rely on a variety of government-sponsored grants to fund their budding business venture. These can be useful, but can also pull the firm in directions that siphon off its energy into writing grants and reports for funding agencies, and meeting those agencies' requirements, rather than actually talking to customers and potential market partners.

Because internally generated cash flows often aren't enough to sustain growth, and banks fear the risk of issuing new loans because high-tech companies usually don't own much traditional collateral, many emerging high-tech companies consider approaching a *venture capitalist* (VC). There are two types of venture capitalists: formal and informal.

Formal venture capitalists are professional investors, such as venture capital firms and some banks. These investors often look for companies that have moved beyond the infant stage, where risks are the highest. They frequently seek a high rate of return—and some, who have felt that venture capitalists get rich off the inventors' ideas, refer to them as "vulture capitalists."

Informal venture capitalists are sometimes referred to as *angels,* probably because of their salvaging qualities and also because they are so hard to find. Angels are usually part of an informal network of investors, who hear about promising start-ups through acquaintances or friends of friends. However, angels can also act more like professional institutions. They have pooled resources, built explicit networks, and when investing they demand nothing less than a formal venture capitalist—that is, they want a business plan, clear and precise goals, a large equity stake, and very often a seat on the board.

Venture capitalists spend their days meeting with industry leaders, meeting with other venture capitalists, listening to proposals from excited entrepreneurs, and pondering the future of technology. When venture capitalists screen high-tech companies for investment opportunities, they look at four key factors, typically in this order: management, marketing, technology/product, and anticipated return on investment. Venture capitalists don't merely invest in a superior software code or specific technological innovation. Rather, they invest in the talented people who can transform that technology into a profitable product.

Utilizing Other Resources

Small high-tech start-ups are also in need of general assistance to get their budding businesses off the ground.

Technology incubators can be a useful resource for many small start-ups. A business incubator is an economic development tool designed to accelerate the growth and success of entrepreneurial companies by offering an array of business support resources and services. A business incubator's main goal is to produce successful firms that will leave the incubation program financially viable and freestanding. Critical to the concept of an incubator is on-site management, which develops and orchestrates business, marketing, and management resources tailored to the start-up company's needs. Incubators may also provide access to appropriate rental space and flexible leases, shared basic office services and equipment, technology support services, and assistance in obtaining the financing necessary for company growth.[87] The incubator concept has several benefits. For example, sharing office space allows entrepreneurs to swap ideas, and inexperienced entrepreneurs will have access to qualified consultants. Incubators can be run by private companies, city or county economic development organizations, and federal government programs. For example, Finland has a fairly sophisticated and well-developed incubator system for its high-tech start-ups.

Incubators do have mixed rates of success (hence, some people refer to them as "incinerators"). Those with the highest success rate tend to have four key elements:

- Ready access to an educated workforce/pool of talent—possibly affiliated with a university or in a university town
- Ready access to financing—possibly in partnership with the investment community
- The support of the local business community to provide advice, potential collaborative partnerships, and critical first revenues (as customers)
- A culture of innovation and risk taking (more on this in Chapter 3)

Small businesses in the United States looking to develop new ventures and markets have many other options for assistance as well, both public and private. At a basic level, start-ups may want to get in touch with the Senior Corps of Retired Executives (SCORE), a volunteer group available to counsel budding entrepreneurs. SCORE volunteers are identified by the Small Business Administration (SBA), and there are local SCORE chapters throughout the country. The nature of the expertise available may be geographic-specific, so high-tech start-ups might find this service more valuable if they are based in areas with a plethora of high-tech businesses. The SBA also offers loan guarantee programs that may be useful for a new business.

For firms that are looking to international markets, the U.S. Department of Commerce has multiple tools for identifying markets and finding partners. First, the U.S. government has commercial sections in most U.S. embassies around the world. These offices, along with a network of export assistance centers and offices throughout the United States, are available to provide counseling, market intelligence, and contacts. Indeed, the staff in these offices will help a firm develop a customized strategy to export to a particular country by identifying possible partners and trade events. Moreover, the Commerce Department conducts an annual analysis for U.S. products in export markets. These reports are made available via the Internet. Being listed on the online trade show maintained by the Department of Commerce is also possible. In addition to market intelligence, the commercial sections employ locals who are experts in targeted industries. These trade experts have the ability to identify potential partners and introduce U.S. firms. Using the introduction by the embassy can expedite the search process and add a prestige factor that can be critical in gaining access to some countries. Each of these services can be found at the website of the U.S. Commercial Service.

When a firm sells products overseas, a critical issue for high-tech products is knowing what standards and certification are necessary for different countries. The National Institutes of Standards and Technology (www.nist.gov) compiles and maintains such information for marketers and others.

Another option for small high-tech startups facing resource constraints is to partner with other firms who offer complementary skills and resources. Strategic alliances and partnerships are discussed in a separate chapter in this book.

3

Culture and Climate Considerations for High-Tech Companies

The "Googley" Culture at Google

To understand the corporate culture at Google, take a look at the toilets. Every bathroom stall has a Japanese high-tech commode with a heated seat, a bidet, and a dryer. A flier tacked inside each stall bears the title, "Testing on the Toilet," featuring a geek quiz that changes every few weeks and asks technical questions about testing programming code for bugs. The toilet reflects Google's general philosophy of work: "Generous, quirky perks keep employees happy and thinking in unconventional ways, helping Google innovate as it rapidly expands into new lines of business."[1] Some initiatives in Google's repertoire include Google Earth, Google Trends, Google Web Accelerator, Google Mars, and Google Knol, to name just a few.

Google is an example of a best practices "culture of innovation." As many previously successful businesses can attest, increasing size can be accompanied by mind-numbing bureaucratic procedures and rules. Yet somehow, despite tripling its workforce between 2003 and 2006, Google was able to launch a new product nearly every week—including some widely regarded as failures. In addition to its own internal innovation process, Google is also willing to acquire new technologies, as it did in 2006 when it acquired YouTube for $1.65 billion in stock, and in 2007 when it acquired DoubleClick. How does Google manage to continuously innovate?

- Although it places a premium on success, it shrugs off failure. This creates a culture of fearlessness that permeates the "Googleplex," as its campus of interconnected low-rise buildings is called. "If you're not failing enough, you're not trying hard enough," said Richard Holden, the product management director of Google AdWords. The stigma of failure doesn't exist: If a product doesn't work, managers encourage staff to move on.

- Employees are encouraged to propose wild, ambitious ideas often. Supervisors assign small teams to see if the ideas work. Indeed, the self-proclaimed mission of the company is equally ambitious: to organize the world's information and make it universally accessible and useful.

- All engineers are allotted 20% of their time to work on any projects they choose. Many of their personal projects yield public offerings, such as Google News and the social networking website Orkut.

- The company encourages forays into related business arenas. For example, in 2006 Google, in partnership with Earthlink, proposed a free WiFi Internet service in San Francisco. Although Earthlink pulled the plug on the project in fall 2007,[2] the foray is emblematic of Google's willingness to venture into new terrain.

- Hiring follows a rigorous procedure. New hires not only have to be very, very smart, but just as important, according to chief culture officer Stacy Sullivan, they also have to be

(continued)

"Googley"—meaning someone not too traditional or stuck in the status quo. The best candidates have a free-wheeling approach and a drive to solve big problems.[3]

• The company focuses on employee morale and productivity. Heated toilets are just one of many amenities designed to keep employees happy and on-task at work. Examples of the other conveniences the company offers: three free gourmet multicultural meals a day, free use of an outdoor wave pool, an indoor gym, a large child care facility, and free private bus service to San Francisco and other residential areas. Buildings are environmentally friendly, letting light into interiors through glass-walled workrooms shared by small groups of employees.

Google does two, seemingly contradictory, things extremely well: It has a laser-like focus on efficiency for its core assets (optimizing the performance of its search and advertising platform) coupled with a radical openness to innovation and experimentation for everything else. Google is fundamentally an engineer-driven company; engineers love to find new challenges and solve problems in novel ways. This culture is formidable to competitors: Even Google's premature ideas, launched before they are ready for prime time, may ultimately turn out to be killer apps in the industry.

Yet, Google faces increasingly serious challenges in the marketplace. One common thread in Google's business model is advertising—yet what is the risk of Google remaining focused on an advertising business model? Can it mature without losing the youthful vigor that was so important to its success?[4]

Its initial business model, built around online search by aggregating massive amounts of data and optimizing transactions with both end users and advertisers around those data, was driven by advertising revenue and search as the primary way consumers access the Internet. Yet, social networking sites and online media and entertainment are increasingly serving as entry points to the Internet. Aggregating and transacting around social data poses a similar opportunity to Google as its initial online search model. Google's release in November 2007 of OpenSocial, a set of software building blocks that allows programmers to use a single standard to create applications to perform tasks on social networking websites (share music, contacts, and other online content), is a move into this market space. Bebo, Friendster, LinkedIn, MySpace and Ning—all rivals to Facebook—plan to support Google's OpenSocial architecture, which suggests that Google is trying to position itself against Facebook. This speculation is further bolstered by the fact that Google's announcement was made just a few days before Facebook launched its own online advertising system that would allow marketing partners (Sony, Verizon, Blockbuster, etc.) to be "friends" with users. A further complicating factor is Microsoft's payment in October 2007 of $240 million for a 1.6% stake in Facebook.

Google's core competence—aggregating data to sell advertising and creating opportunities to sell more ads—might also find new market space in the mobile arena. With mobile advertising expected to grow more than 30% in 5 years, Google's release in November 2007 of its new Android software (which includes a mobile operating system and mobile software applications distributed free under an open source software license) is poised to allow it to tap into mobile advertising revenue.[5] The Open Handset Alliance—a group of wireless technology companies including Sprint, T-Mobile, NTT DoCoMo, Qualcomm, and Motorola—stated its intention to use Google's Android software to power cellular phones that became available in late 2008. The goal is to "unleash the potential of mobile technology for billions of users around the world," says Google.

Because Google already captures personal data (e.g., through Gmail) that allows it to deliver ads, some people predict that Google will move into medical data, another online market opportunity. It has invested $3.9 million in 23andMe, a company that allows people to access and understand their genetic information through Web-based software tools.

As noted previously, this expanding collection of services offers more opportunities to sell ads. What are the other revenue opportunities it could seek, and the attendant risks of diversifying its revenue base? For example, take Google Apps, a suite of open source (free) software designed to compete with Microsoft Office (calendar feature, word processor, presentation software, Web page editor, etc.). (This software is also available under a subscription model.) What if Google were to license some of its underlying software to other industry players? Or, could it become a hosted application provider and charge customers to use its computing infrastructure services (much like Amazon.com charges customers to use its Elastic Compute Cloud, also known as EC2, and its Simple Storage Service, known as S3)? Are these various initiatives too diffused, or are they sufficiently focused on a common thread—search, advertising?

Google's various initiatives reveal an entrepreneurial company that is willing to take on established players to shake up existing markets. This leads to another important question: Can Google be both a dominant player who protects its existing market position as well as an industry upstart ready to attack established companies in mature markets such as telecommunications and publishing? Moreover, even as Google strives to continue to be nimble, its dominance causes another set of issues: It must be a good partner and a fair competitor. Many other companies have found that size and dominance creates a type of backlash, from society, other industry players, and government. Despite Google's informal company motto, "Don't be evil," some people are increasingly suspicious of its motives and worried about potential abuse of the data it is amassing.

Moreover, as it grows, it must maintain its start-up-like, innovation-driven culture. To do this, it must first keep a compensation structure that motivates employees: Stock options are less motivating when the stock price is high. And second, it must keep internal teams nimble, small, and independent, with time and resources to pursue new ideas.

> Market domination tends to lull the leader to sleep. . . . Market domination produces tremendous internal resistance against any innovation and thus makes adaptation to change dangerously difficult.[6]

Paradoxically, high-tech companies founded on the development of innovative new products are as susceptible as any other companies to the complacencies and bureaucratic procedures that success breeds. This chapter turns from the concepts and processes of marketing strategy formulation to address organizational culture and climate. Because a firm's culture and climate provide the backdrop against which its strategy is developed and implemented, it can either enhance or undermine its marketing efforts.

Culture is the deeply rooted set of values and beliefs that provide norms for behavior in a company. Culture helps employees understand *why* things happen the way they do. **Climate** is the set of behaviors that are expected, supported, and rewarded. Climate explains *how* things actually happen in the company.[7] To a large extent, climate is the observable manifestation of culture. Because cultural values are so deeply ingrained, not only are they often tacit and difficult to articulate, but they are also difficult to change—which means that dysfunctional behaviors may also be difficult to change.

Successful high-tech companies must remain highly innovative; they are characterized by a **culture of innovativeness** that actively encourages, supports, and rewards breakthrough thinking and risk taking. Rather than resting on their past successes, they continually replenish their stock with ideas and knowledge to develop a steady stream of innovations, both incremental and breakthrough.

Yet, history is littered with the skeletons of many high-tech companies that became so enamored with their initial success that they were blinded to market trends and changes, unable to sustain their upward trajectory. Rather than developing a steady stream of innovative new products, their initial success bred a sense of complacency and a desire to protect the status quo. Slavish devotion to old business models plagues many companies, leading to suboptimal behaviors. Music industry executives sued individuals for illegally downloading songs (copyright infringement) rather than finding new business models early on. Verizon and Sprint sued and won judgment against Vonage over patent infringement for their Voice-over-Internet Protocol (VoIP)—which is disrupting these incumbents' telecommunications business models. Inertial forces that undermine companies' attempts to develop or embrace radical innovations are very strong, yet they must be actively resisted to ensure continued success.

Organizational size and a bureaucratic mind-set can also make it difficult for companies to remain innovative. Experts continue to debate the relationship between firm size and innovativeness. Larger firms are said to suffer from inertia; they tend to be more bureaucratic and have more to lose than other firms when they develop radical innovations that may obsolete their existing product lines. However, contrary to suffering this "incumbent's curse," one study found that since World War II, more radical innovations have been introduced by larger, established firms than by smaller firms and new entrants.[8] So, size per se does not have to be a barrier to innovativeness.

One of the major threats to the innovation process is the boom–bust nature of business cycles. When companies do well, they invest more in innovation and product development. On the other hand, when the economy deteriorates, investments in R&D tend to go down. One way to cushion innovation against cyclical funding cuts and spurts is to make R&D an indispensable part of everyday business.[9] If the output of innovation is to be reliable, it must be worked on systematically every day, even during downturns, even if blockbuster results are years away.[10] Consider the Korean conglomerate, Samsung. In 1997 Samsung was in dire straits as a result of the Asian financial crisis. Since then, it made continuous, significant investments in R&D and product development, even through the global economic downturn of the early 2000s. By 2005, Samsung had made it onto Fortune's list of the "World's Most Admired Companies" and was the global leader in memory chips and flat-screen TVs, and was the world's third-largest mobile-phone maker.[11]

For many innovative firms, a key measure of success of their innovation efforts is the percent of revenue derived from recently released products that are the outcome of new product development procedures. In many companies, the new product development process is often rational and analytical, with formal plans and procedures bordering on bureaucratic. Unfortunately, bureaucratic processes are at odds with an innovative culture and the nature of innovation, which is often tumultuous, nonlinear, and serendipitous.[12] Indeed, the development of most breakthrough innovations is nonlinear in nature, progressing in fits and starts.[13] So, rather than utilizing a stage-gate, step-by-step process through which new product ideas must pass, "managing" innovative activity as an entrepreneurial process (referred to variously as *autonomous strategic behavior,*[14] *emergent processes,*[15] or *"intrapreneuring"*[16]), the idea is to create an internal environment that fosters innovation and an entrepreneurial spirit.

To develop and maintain a culture of innovativeness, high-tech companies can actively cultivate a set of values and beliefs that are forward-looking rather than complacent and reactive. This chapter explores important facilitators of a company's culture of innovativeness, and the barriers to sustaining such a culture. Two other important aspects of a successful high-tech firm's organizational culture—its market orientation (or customer focus) and the way its marketing and R&D personnel work together in the new product development and commercialization process—are explored in Chapter 4.

FACILITATORS OF A CULTURE OF INNOVATIVENESS

Cultivating a culture and climate of innovativeness requires recognition of the nature of innovation itself. *Creative destruction* is the process, within organizations, of developing and commercializing radical or breakthrough new products, services, or business models at the risk of cannibalizing sales of existing products. *Creativity* is the cognitive foundation for the innovation.

All innovation begins with creative ideas. Successful implementation of new programs, new product introductions, or new services depends on a person or a team having a good idea—and developing that idea beyond its initial state.[17]

Creative ideas must be novel—different from what's been done before—but they must also be useful.[18] One example in Chapter 2 contrasted the success of online grocer FreshDirect with the failure of Webvan. Webvan was too optimistic about people's willingness to abandon traditional grocery stores in favor of something new and different. As Phil Terry, CEO of Creative Good, a consulting firm that has worked with online grocers, said, "One of the fundamental mistakes that everybody made is the assumption that because there are some problems with the offline experience that everyone would flock to online."[19] Ultimately, Webvan did not offer customers a sufficient advantage relative to traditional grocery stores.

As the foundation of an innovation strategy, creativity motivates the generation of new concepts for products or marketing programs that are superior to those of competitors. Furthermore, creativity is a valuable and rare competency that is inimitable and, thus, is a source of sustainable competitive advantage.[20] However, unfettered creativity, by itself, may not be good for business. It takes *disciplined creativity* to produce useful innovation for a company. Apple illustrates the principle of disciplined creativity in designing pleasing products in combination with business models that provide attractive gross margins. By contrast, Sony has been inconsistent in this respect in recent years with turf battles between its internal divisions resulting in quality problems in many products from laptop batteries to digital music players to PlayStation 3.

Companies that exhibit this culture of innovativeness share a common set of facilitating conditions (see Figure 3.1). The presence of these conditions establishes an environment that encourages, supports, and rewards breakthrough thinking and that resists the inertial forces that stymie innovation.

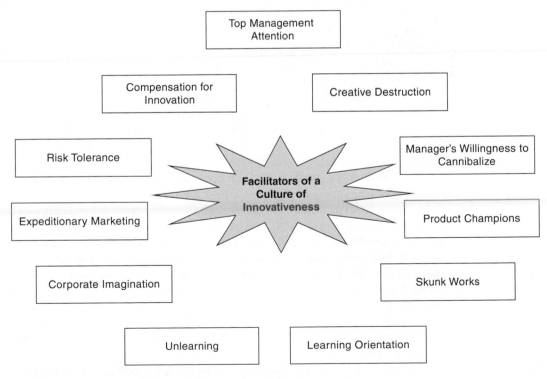

FIGURE 3.1 Facilitators of a Culture of Innovativeness

Top Management Attention

Conventional wisdom takes a skeptical view of the role of top managers in the innovation process, often characterizing them as either irrelevant or impediments; at best, they may have a positive, but indirect, influence on innovation. However, the results of a 10-year study of more than 700 new product teams show that successful new product development efforts had senior managers who either were intimately involved with virtually every aspect of the process or made it clear that they completely backed the project.[21] Thus, top management attention positively drives innovation, especially when CEOs exhibit the following:[22]

- A *future focus* that leads to greater awareness and anticipation within the firm of events in the future, rather than being preoccupied with issues of the past and present
- An *external focus* (outside the firm) that leads to greater awareness of customers' needs, market trends, and competitive actions

Both of these traits allow the firm to move quickly to detect and respond to new technologies and opportunities. Although CEOs who have an internal focus might move faster at developing new technologies, they are slower to detect new opportunities in the first place. Their key challenge, then, is to cultivate and maintain a forward-looking viewpoint; they should exhibit a generic focus on the future (rather than a focus on specific events or opportunities) and view innovation as more than product development, but opportunity detection as well.[23] Top managers must clearly support innovation through both words and actions. In this sense, top management sets the example for what values and beliefs are important and creates the organizational culture for others to follow.

Creative Destruction

Firms must recognize that products in high-tech environments typically sustain only a finite spell at the technological frontier before being made obsolete by better products. Given this reality, a firm should not be too enamored with the technology that forms the basis of its initial success, but instead strive to develop even better, next-generation technologies[24]—despite the fact that such developments may alienate some current customers, make obsolete its sunk investments in the prior technology, and render any economies of scale and experience curve advantages useless. If it doesn't, its rivals surely will. Recall that this paradoxical model of competition, known as *creative destruction*,[25] is required for ongoing leadership in innovation.

An example is Microsoft's continued desire to ride its wave of success from its operating systems software (Windows 95, 98, XP, Vista, etc.). However, rivals introduced new approaches to software business models, including the open source software movement (Linux, Mozilla/Firefox, Google Docs). Moreover, the Web services model of software delivery allows consumers to access information via a range of consumer electronic devices and "information appliances" that don't require the Windows operating system. Microsoft has struggled to compete with new competitors that base their business models on the "free" aspect of open source technology, a new approach to software deployment that is radically changing the world of desktop computing. At the extreme, it may render proprietary operating systems obsolete. So, the challenge for Microsoft is whether it will be able to successfully leverage this new business model in the software industry.

An additional example is found in companies' distribution strategies. Many companies (both manufacturers who sell through retail stores as well as many retailers themselves) have found that, in order to compete with online competitors, they must offer sales through an Internet channel. Yet, retaliation from existing stores and channel members has posed a credible threat and prevented them from leveraging the Internet effectively. However, this is the nature of creative destruction: If the company itself does not offer customers the access avenues they desire, a competitor surely will.

Because early and late majority customers resist adopting new solutions to their needs, some companies choose to set up an independent organization that is free to compete with new technologies using new business models unfettered by the company's existing strategies and routines. In 1998, Larry Ellison, largest shareholder and founder of Oracle software, launched NetSuite, a company offering software as a Web service to small and medium businesses, to better compete with newcomers like Salesforce.com. NetSuite became a public company in late 2007.

Managers' Willingness to Cannibalize

Companies that engage in creative destruction are characterized by a culture in which managers are fearful of obsolescence, and because of that are willing to cannibalize their own successful products.[26] When managers of dominant firms believe that a new technology might make their existing products obsolete, the fear of obsolescence causes them to invest aggressively in radically new technologies. This allows dominant firms to remain innovative and overcome the potential negative effects of size (arising from inertia and escalation of commitment to existing product platforms).[27] As Andrew Grove, former CEO of Intel, said:

> Business success contains the seeds of its own destruction. The more successful you are, the more people want a chunk of your business and then another chunk and then another until there is nothing left. I believe that the prime responsibility of a manager is to guard constantly against other people's attacks and to inculcate this guardian attitude in the people under his or her management. . . . I worry about competitors. I worry about other people figuring out how to do what we do better or cheaper, and displacing us with our customers. . . . When the way business is being conducted changes, it creates opportunities for players who are adept at operating in the new way. This can apply to newcomers or to incumbents.[28]

What factors in the firm might make managers more inclined to cannibalize their own successful products? Organizations have a stronger willingness to cannibalize and are more likely to introduce radically new innovations when they have:[29]

- Strong, autonomous strategic business units that compete internally for resources
- Strong product champion roles
- A focus on future markets more than on current markets

Product Champions

Product champions—integral to the successful development of many breakthrough innovations—are the people who create, define, or adopt a new idea and are willing to assume significant risk to make it happen.[30] Often referred to as *iconoclasts, mavericks,* or *crusaders,* these people break the rules, take risks, transform companies, and turn organizations upside down. They work tirelessly and lobby behind the scenes for organizational resources to help their ideas take off. The product champion is a person with drive, aggressiveness, political astuteness, technical competence, and market knowledge. Influential product champions can overcome firms' natural reluctance to cannibalize and motivate breakthrough product innovations.[31]

Although product champions are found in both innovative and non-innovative companies, those in innovating firms wield substantially more influence than in less innovative firms. The reward systems and cultures promote the influence of product champions, and top management actively supports them.[32] In contrast, product champions in non-innovative companies wield less influence and are frustrated and demoralized.

Skunk Works

To create an innovative climate conducive to thinking out of the box, companies set up **skunk works**—new venture teams that are isolated or removed from the normal corporate operations.[33] The rationale is that when large, established companies develop new innovations, they do so *despite* the corporate system, not *because* of it.[34] Hence, in order to protect imaginative individuals from corporate orthodoxies, senior managers isolate them in new venture divisions, or corporate incubators, often at a remote site from the parent organization.

The etymology of the term *skunk works* comes from the "Skonk Works," an illicit distillery in the popular comic strip from the 1940s, *Li'l Abner* by Al Capp. Because illicit distilleries were bootleg operations, typically located in an isolated area with minimal formal oversight, the term has been adopted in organizational settings to refer to a usually small and often isolated department or facility (say, for R&D) that functions with minimal supervision or impediments from the normal corporate operating procedures. Indeed, Lockheed Martin has a division, whose roots date back to the early 1940s, called "Skunk Works," which is responsible for new aircraft designs; Lockheed Martin also owns "Skunk Works" as a registered trademark with the U.S. Patent and Trademark Office.

Many firms have relied on skunk works operations to develop new lines of business. For example, IBM isolated its PC group in Boca Raton, Florida, away from corporate headquarters in New York and away from any other established IBM locations. IBM continues to rely on skunk works to develop emerging business opportunities.[35] Dow Chemical also relies on skunk works for its new venture groups. When Toys "R" Us began its online operations, it decided that being located in corporate headquarters would slow it down and hinder its innovativeness, and so it set up a separate unit in northern California.[36]

Despite the potential advantages of such isolation, some critics argue that trying to leverage corporate competencies into new businesses while at the same time protecting new ventures from the corporate culture is a contradiction in terms.[37] For an established company to become and remain innovative, it must allow individual creativity within the normal corporate operating procedures. A corporate culture that allows innovation to flourish shouldn't have to put up special protection mechanisms for it to happen. Indeed, Gary Hamel refers to such incubators as "orphanages" that isolate the creative conversations and make it hard for new ideas to emerge in the corporate hierarchy.[38]

So, the idea of isolating new product development groups from standard corporate procedures poses a dilemma. Ideally, it probably would be best if the company nourished innovation and did not need separate units to develop innovative ideas. On the other hand, if normal operating procedures stifle creativity, a skunk works may be something of a temporary or "band-aid" solution; addressing the systemic reasons for the problem may be also necessary.

Learning Orientation

"The ability to learn faster than your competitors may be the only sustainable competitive advantage."[39] Learning-oriented businesses continuously develop new knowledge or insights that have the potential to confer competitive advantage.[40] A **learning orientation** describes a firm who actively facilitates the development of new knowledge or insights that have the potential to influence its strategies and decisions.[41] A learning orientation is facilitated, in turn, by the key aspects of a firm's culture and climate that have been previously discussed: top management support, a decentralized approach to strategic planning, fostering an entrepreneurial spirit within the organization, and a market-focused organizational structure. One of the principal cultural foundations of a learning organization is a firm's *market orientation,* its ability to actively monitor and learn about customer and competitor trends in the market. The use of market-based information for organizational learning is covered in Chapter 4.

All businesses competing in dynamic and turbulent environments must pursue the processes of learning.[42] A superior ability to learn is critical due to acceleration of market and technological change, and due to the explosion of available market and technological information. It is a competency-based source of competitive advantage due to its difficulty to achieve (rareness), its usefulness in numerous activities from product development to customer service (customer value), and the difficulty that competitors have in imitating it (inimitability).[43] Learning-oriented companies that are quick to try, quick to learn, and quick to adapt will win. The faster a company learns and adapts, the more customers it wins; the more customers it wins, the faster it can learn and adapt, creating a virtuous cycle of positive feedback effects.[44]

Unlearning

Paradoxically, another cultural facilitator of innovation is a willingness to **unlearn**—the process of identifying knowledge and assumptions that are the basis for strategy, testing them for their validity, and discarding those that have become barriers to proactive change. Thus, companies must often "unlearn" traditional but detrimental practices. Peter Andrews, a consulting faculty member at IBM's Executive Development Institute, argues, "High stakes innovation requires abandoning conventional wisdom, even actively unlearning things we 'know' are true."[45] To do this, managers must surface and challenge their own assumptions and mental models about the market and the business, and encourage employees to do the same. Faced with turbulence in its established markets and an entrenched culture, Ed Artzt, former CEO of Procter & Gamble, found himself in the peculiar situation of having to make "rules that give (employees) intellectual permission to make changes."[46] As John Seely Brown,[47] the former chief scientist of the Xerox Palo Alto Research Center, explained, "Unlearning is critical in these chaotic times because so many of our hard-earned nuggets of knowledge, intuitions, and just plain opinions depend on assumptions about the world that are simply no longer true."

General Electric uses "workout" sessions to "challenge every single piece of conventional wisdom, every book, every rule." In these sessions, executives—including the CEO—take the floor of GE's management development center to respond to tough questions from managers who are taking classes at the center. This has created an environment where difficult issues can be raised without fear of retribution and where executives must respond with plans and solutions.[48] Encouraging unlearning could be the single most important task of the CEO for sustaining innovation momentum.

Corporate Imagination

Another facilitator of ongoing innovativeness is to develop **corporate imagination,**[49] a climate of creativity—even playfulness—that allows a firm to create a vision of the future that consists of markets that do not yet exist and is based on a horizon not confined by the boundaries of the current business. In that sense, it is characterized by innovations in both strategies and in products/technologies.

Engaging in corporate imagination requires challenging the status quo, including beliefs about technology performance levels and capabilities. It requires escaping the tyranny of the served market[50] and simultaneously focusing on both current customers and customers of the future. It is a capability that allows envisioning the future, generating profound insights that are not saddled with existing rules and procedures. Organizations with corporate imagination seek out new market opportunities across or between the areas of a firm's competence (see the discussion of "blue ocean" strategy in Chapter 2). Moreover, companies with corporate imagination are open to new sources of information and rely on different types of marketing research to open new doors of opportunity. These marketing research techniques, especially useful in a high-tech marketing context, include ethnographic observation ("empathic design"), customer visits, biomimicry (nature-inspired innovations), and lead users. These tools are discussed in Chapter 6.

Organizations seeking to nurture imagination might engage in a variety of creativity exercises. A classic creativity exercise is to find 100 uses for a brick, a bucket of water, a bathtub, or any other commonplace object. Many companies use one of their company's products as the focus of the challenge, assembling a group to brainstorm 100 uses for the product. This exercise not only stretches the imagination, but focuses it on a key component of the business and produces practical ideas about new usage situations or markets that the organization could readily pursue.

Expeditionary Marketing

Creating markets ahead of competitors is risky; sometimes the hoped-for market does not develop at all, or if it does, it emerges more slowly than expected. Successful new products include making the right decisions about functionality, price, and performance targeted to the correct market. Companies tend to use one of two ways to minimize the risks of mistakes and to improve their rates of success.

One way is to try to improve the "hit" rate, or the odds of success for each individual product introduction. To have a successful new product launch, a company tries to gather as much information as possible to tailor the product's functionality, price, and performance, so that when it is delivered to the market, the odds of a successful hit are as high as possible.

The other route to success is to try many "mini-introductions" in quick succession and, by learning from each foray into the marketplace, to incorporate that learning into each successive "time at bat." This objective of this strategy, known as **expeditionary marketing**,[51] is not to improve the hit rate; rather, the objective is to increase the number of "times at bat" in the market. The underlying strategy is based on learning: The firm wants to learn about the marketplace and customer needs by placing many small bets in the marketplace. These low-cost, fast-paced incursions allow the firm to learn and recalibrate its offerings each time, such that the combination of speed and learning enhances its odds of success. Over time, it allows the firm to accumulate loyal customers and higher market share. Much like an alpine ascent in mountaineering (in contrast to the traditional siege assault with massive gear hauled up the mountain on the backs of porters), the objective is to be light and fast. If the alpine ascent is unsuccessful due to unanticipated conditions, the team can withdraw, regroup, and develop a new strategy because the team has not expended all of its resources. In contrast, the failure of a siege assault often means that too many resources have been expended to safely try again.

Which of the two ways do most companies focus their attention on? Most companies focus their attention on the first way, trying to improve the hit rate. Through careful market research, competitive analysis, and the use of stage-gate procedures that specify the hurdles a new idea must overcome at each stage of development, they try to maximize the odds of success. However, such a strategy is very time-consuming, and in high-tech markets the accuracy of the information can be sketchy at best. Moreover, by the time the product is introduced to the market, the marketplace (customer needs and competitors) may have changed. Indeed, this approach might be characterized as "Ready. Aim. Aim. Aim."[52]

Hence, in high-tech markets, a series of fast-paced market incursions makes much sense, offering several advantages. First, it allows the company to learn more accurately, through successive approximations, about what customer needs are. Second, through fast-paced market incursions, it maximizes the odds that the product actually delivered to the market meets customers' needs. Fast-paced incursions imply that the time-to-market cycle is faster, and therefore, the odds of the customers' needs changing in that time period are lower. Under such a model, what counts the most is not being right the first time, but how quickly a company can learn and modify its product offerings, based on its accumulated experience in the marketplace.[53]

As an example, Storage Technology Corporation, maker of tape backup systems for large data installations (such as banks or insurance companies), uses the model depicted in Figure 3.2. The idea is that, rather than introducing the most advanced model possible based on a new technology (Model 3 in the figure), the company attempts to introduce in rapid succession a series of models based on the new technology.

Enlightened experimentation is a variation on the theme of expeditionary marketing that involves the use of new information technologies such as computer simulation, rapid prototyping,

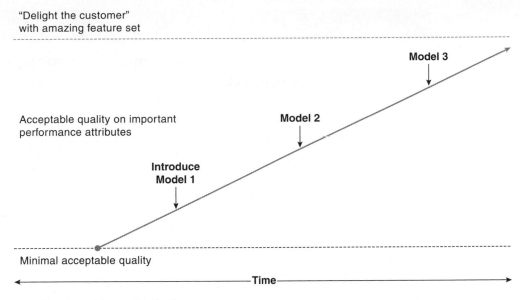

FIGURE 3.2   Expeditionary Marketing

and combinatorial chemistry to reduce the cost of testing new ideas as well as increase the opportunities for innovation.[54] It has been applied in the automotive, software, and pharmaceutical industries. For example, Millennium Pharmaceuticals uses genomics, bioinformatics, and combinatorial chemistry to test drug candidates for their toxicological profile. Unpromising candidates are eliminated before millions of dollars are spent on further clinical testing and development. This also enables Millennium to test many more potential drugs in search of the real blockbuster innovations.

Risk Tolerance

As the opening Google vignette noted, high-tech companies that have a culture of innovativeness do not rigidly penalize people for ideas that don't work out. Tolerating risks and a certain number of mistakes is part of the entrepreneurial spirit. Indeed, learning from such ostensible "mistakes" may be the basis of the company's next new success. Thomas Watson, chairman and CEO of IBM from 1956 to 1971, once called an engineer who had managed a failed $25 million project into his office. The young engineer said, "I suppose you've called me in to fire me." Watson replied, "I can't fire you, I've just spent $25 million educating you!"[55]

Indeed, both employees who innovate successfully and those who take calculated risks that do not pay off should be celebrated and rewarded in a tangible fashion.

Compensation for Innovation

People who perform complex jobs such as R&D are most creative when they are offered the opportunity to do challenging work and given autonomy, regardless of the amount of incentive compensation.[56] However, an appropriate reward system is necessary to attract and retain key employees. Thus, compensation systems in more innovative firms are quite different from those in less innovative firms. Particularly, the compensation package for members of the R&D group relative to other groups in firms with a strong commitment to innovation is higher than in less innovative firms. Also, firms with a strong commitment to innovation weight their compensation packages more

TECHNOLOGY SOLUTIONS FOR GLOBAL PROBLEMS
StarSight: Solar-Powered Street Lighting + WiFi
BY JAMIE HOFFMAN

Imagine a city where every lamppost provides wireless Internet access, solar-powered street lighting, and power to charge your mobile phone. Although such technological inspirations might be expected in a developed city, the first installations of StarSight (www.starsightproject.com) are in a city in Cameroon, in west-central Africa. Developed in conjunction with the Kolam Partnership, countries like Turkey, Cameroon, and the Republic of Congo are using StarSight to decrease crime, create a safer environment, establish connectivity for emerging economies, and provide emergency communication with law enforcement. StarSight provides a drastically cheaper means of providing a lighting infrastructure with extremely low carbon emissions; its excess solar energy can be resold to a national grid if one is available.

Costing merely $30–50 per unit, StarSight's core product is made up of two components: the lighting system and the communication system. Stationed at the top of a pylon, the lighting system houses a solar power panel, wireless monitoring and control, an antitheft system, and a modern LED lighting-head. The communication system employs a cutting-edge combination of wireless communication via integrated WiFi access point as well as StarSight AP, which can service more than 100 nodes of StarSight radios. StarSight's designers created a product capable of evolving alongside rapidly changing technologies including disaster warning systems, pollution monitors, and cell phone chargers.

StarSight's developers also encourage customers to carry out the final stages of manufacturing, primarily assembly, in the country of installation as a means to provide additional socio-economic benefits for that area.

Photo reprinted with permission of StarSight International Ltd., London, UK.

Sources: "New Sustainable Solar Street Lighting: Strategic Partnership Makes Scotland the European Showcase for StarSight Project," press release, December 2005, University of Albertay Dundee, available online at www.abertay.ac.uk/News/NewsPopup.cfm?NewsID=937; "Award Winners 2005," Mobile Data Association, November 2005, available online at www.mdaawards.org/index.php?btnid=winners; and Kolam-StarSight brochure, personal correspondence with Yannick Gaillac and Chris Huchon, founding partners, Kolam Partnership Ltd., July 28, 2008.

Table 3.1 Additional Steps to Create a Culture of Innovativeness
1. Ensure that each employee understands the organization's strategy and how innovation efforts are aligned with it.
2. Teach each employee the critical importance of diversity of thinking styles, experiences, perspectives, and expertise.
3. Maintain close relationships with the company's most innovative customers to jointly determine what those customers will want next.
4. Create metrics and a procedure for frequent evaluation of project progress.
5. Make innovation part of everyone's job description and build it into the performance review process.
6. Provide considerable freedom of action.
7. Allocate substantial resources to educate employees about emerging technologies.
8. Use teams of employees who possess many skills among them.
9. Create a process for rapidly communicating new ideas across the company.

heavily to long-term components such as stock options and stock grants than to short-term components such as base pay, profit-sharing, and bonuses because innovative firms must have a longer term perspective.[57]

Other Facilitators of Innovativeness

Table 3.1 provides a summary of other characteristics that contribute to an innovative climate.[58] For an example of introducing innovative technology to a growing and largely untapped market, see the accompanying **Technology Solutions for Global Problems** feature.

OBSTACLES TO MAINTAINING A CULTURE OF INNOVATIVENESS

High-tech companies face two major roadblocks in their efforts to maintain a culture of innovativeness—core rigidities and the innovator's dilemma, both related to inertial forces arising from existing competencies and business models.

Core Rigidities

Although core competencies are an essential ingredient for success, they might also become **core rigidities** and possibly hinder new product development into new domains. Core rigidities arise from well-established skills and competencies that are so entrenched that they prevent a firm from seeing new ways of doing things. For example, new product ideas built on familiar skills and capabilities are more likely to be embraced by a firm than those built on unfamiliar technologies. However, when market conditions are changing—for example, when new technologies are being developed by other firms in the industry—it may be important for a firm to examine closely the viability of a new technology. But ingrained routines, procedures, preferences for information sources, and existing views of the market, all of which can be related to underlying core competencies, can become barriers to a forward-looking assessment of new market opportunities. In such a situation, core competencies can become core rigidities, which strangle a firm's ability to act on novel information.[59]

More broadly, core rigidities are straitjackets that inhibit a firm from being innovative and can include:

• Cultural norms in the firm

- Preferences for existing technology, products, business models, and routines
- Status hierarchies that give preference to, say, technical engineers over marketers

Because they are the basis for the success many companies enjoy, cultural norms, technologies, routines, and company leaders' beliefs are valued. At some point, however, such skills, values, and routines may not be as well suited to the changing business environment as in the past, and they warrant reexamination themselves. Firms that are able to reevaluate their skills and capabilities on a regular basis—and update and modify them as needed—are not as burdened with rigidities as other firms may be. In part, this is due to a culture focused on organizational learning, and open to unlearning, as discussed earlier.

For example, Sony, long the leader in the portable audio device market with its Walkman tape and compact disc players, was overtaken by MP3 players. In 2004, Sony did not rank among the top five companies in market share for digital audio players. Even when MP3 became the de facto industry standard for digital music, Sony devices continued to play only files encoded with its own proprietary ATRAC music file format. Sony's core rigidity was its commitment to its own proprietary technological standards, with sometimes unhappy results. Famously, Sony's Betamax video recording format lost out in the market to VHS in the 1980s. Sony's proprietary Memory Stick flash memory technology has struggled to gain support from device makers, keeping it boxed in a niche.[60] Sony's most notable victory came in February 2008, when the standards war between Sony's Blu-ray and Toshiba's HD-DVD high definition DVD players ended with Toshiba's announcement that it would no longer produce HD-DVD products.[61]

The Innovator's Dilemma

As noted in Chapter 1, a frequent pattern in business is the failure of leading companies to stay at the top of their industries when technologies or markets change. Why do established companies face an *innovator's dilemma,*[62] experiencing difficulty both innovating and responding to disruptive innovations? Market leaders may encounter obstacles to introducing breakthrough innovations because they divert resources away from incremental innovations in established markets, toward new markets and customers that seem insignificant or do not yet exist. The difficulty is not due solely to failures in product development; it is also due to sunk costs that create a bias in managerial decision making, as well as an excessive reliance on a certain class of customers:[63] "Focusing on their current customers, managers literally do not see other opportunities."[64] Hence, successful companies may be overly dedicated to their existing business infrastructure as well as their existing customer base serving those existing customers' needs.[65]

A radical innovation provides customer value in a very different way from the traditional solution. Therefore, if a new innovation arises that serves different customers in different ways, established companies may tend to overlook or to minimize its potential impact on their businesses. One radical innovation that disrupted an entire industry is the digital camera. Kodak, an innovator in photographic film, had a very difficult time adjusting to digital photography because its entire business model was focused on camera film. It lost 50% of its market value from the beginning of 2000 to the beginning of 2007.

It is common for leaders who stay on top of wave after wave of incremental innovations to fail when a breakthrough innovation disrupts their way of doing business. Rather than responding to them, they tend to be "disrupted" by them. Companies must guard against the inertial forces arising from core rigidities and the innovator's dilemma by attending carefully to maintaining a culture of innovativeness.

To read how one company successfully managed to remain competitive in light of disruptive technologies introduced by industry outsiders, see the accompanying Technology Expert's View from the Trenches feature.

TECHNOLOGY EXPERT'S VIEW FROM THE TRENCHES

Coping with Disruption

BY BRYANT RALSTON
Strategic Account Executive for Idaho and Montana
Environmental Systems Research Institute (ESRI), Clinton, Montana

ESRI is a privately held company that leads the geographic information systems (GIS) software market in terms of reputation, innovation, revenue, market share, and customer base. GIS is a horizontally applied information technology (IT) for managing all kinds of diverse geographic data. Although not necessarily immediately apparent, GIS is fundamental to an extremely diverse set of organizations throughout the world, both in the public and private sectors. The ESRI "ArcGIS" family of products consists of solutions tailored for four types of GIS deployment environments: desktops, servers, embedded GIS, and mobile solutions.

Over its 35-year history, ESRI has successfully applied the practices of listening to its customers, aligning products to real customer needs, bringing innovations to the marketplace, working with its worldwide channel to enhance the "whole product," and watching the competitive landscape. The traditional GIS systems in the 1980s were primarily the domain of specialists utilizing stand-alone workstations running specialized GIS software to visualize, create and update, analyze and integrate, and plot geographic information. ESRI developed and priced its products to meet the needs of this type of specialized user. As the GIS market grew, so did revenues, the user base, and the company.

Web-based virtual globes—notably Google Earth and Microsoft Virtual Earth—have "disrupted" traditional GIS by providing free, easy-to-use, but powerful and immersive GIS-like visualization and exploration capabilities. These virtual globes are coupled with Internet search capabilities, providing users a well-known search paradigm tied to geographic location. They come from the leading technology companies in the world (Google and Microsoft) and appeal to a larger, mass-market, consumer-oriented audience and primarily "compete against nonconsumption." Virtual globes have proven to be extremely popular, experiencing hundreds of millions of downloads, and have attracted a large online following where users often post their own personalized "geo-tagged" content including vacation and travel photos, restaurant and lodging reviews, and even the locations of known Bigfoot sightings.

These new applications created several key challenges for ESRI. There initially was fear that the popularity of these virtual globes would lead the new competitors to "move upward" into our traditional GIS market segments. In addition, due to their popularity, virtual globes began to confuse the GIS marketplace, especially with the many newer customers. Proper product positioning suddenly became critical and a frequent topic of "water cooler" discussions and company conference calls. User expectations were raised in terms of ease-of-use, performance, user interface design, data collection and integration, and application interoperability. The mass-market consumer exposure of virtual globes has also been challenging: Google and Microsoft are household names and virtual globes have been utilized on TV news to provide a locational context to the stories. While this exposure raises geographic awareness to "the mainstream," the business models of the companies new to this space made it difficult to determine whether virtual globes were really a threat to traditional GIS. We had to determine how to explain these new options to existing GIS customers inquiring as to "what they mean for us."

Insights for Coping with Disruption

1. ***Find comfort in the realization that we have been here before.*** Disruptive technology is not new to GIS, and in fact, over the years GIS has proven to be fast paced and hungry for innovation. All of the technologies that were originally considered "disruptive" were eventually

(continued)

incorporated into the GIS value chain. This included innovations that have affected all the major parts of GIS: software, hardware, data networks, analytical methods, best practices, and even people.

2. ***Move from "threat frame" to "opportunity frame."*** Take advantage of the energy created by a threat perception to turn the corner—to recognizing it as an opportunity to be leveraged. The prevailing attitude during the threat frame was "it's not *real* GIS so it's not really our core business." But after much healthy dialogue coupled with undeniable exploding popularity, it became clear that virtual globes were, in fact, not a threat per se but were enhancing the traditional GIS industry by raising awareness. The opportunity frame attitude then became "these help us grow mainstream awareness, which helps grow the potential user base for our GIS powered solutions."

3. ***Differentiate: the "don't bring a switchblade to a gunfight" approach.*** While these virtual globes provide GIS-like visualization capabilities, they lack the powerful data management, analytical, and modeling richness of traditional GIS. Certain customers need the functionality of traditional GIS, while others may need only a virtual globe solution. Virtual globes are not being positioned as direct competitors of traditional GIS products, and many industry analysts are quick to point that out. Many in the traditional GIS industry now note how complementary virtual globes can be to existing solutions.

4. ***Interoperate.*** Traditional market segmentation techniques separate consumer solutions from traditional GIS solutions, but the investments in each can be leveraged through proper interoperability. Being able to leverage the popularity of virtual globes and have them become clients in a "presentation layer" harnesses the best of both worlds. Connecting the virtual globes to powerful, functionally rich, IT-compliant "back office" GIS servers—notably ESRI's ArcGIS server—makes business sense, especially in the later phases of disruption.

5. ***Never let them see you sweat.*** Do not get defensive about the technology doing the disrupting, but rather figure out how you can be part of a future solution. Do not contribute to entrenching your problem by being rigid in your thinking. Rather, envision new possibilities for the future.

Note: The information presented here represents the opinions of the author and not necessarily those of Environmental Systems Research Institute Inc.

Sources: Moore, Geoffery, *Crossing the Chasm: Marketing and Selling Technology Products to Mainstream Customers* (New York: HarperCollins, 1991); Christensen, Clayton M., *The Innovator's Dilemma* (Cambridge, MA: Harvard Business School Press, 1997); Gilbert, Clark, and Joseph L. Bower, "Disruptive Change: When Trying Harder Is Part of the Problem," *Harvard Business Review* 80 (May 2002), pp. 94–101, 134; Ball, Matt, "Google Raises the Profile of Geospatial Information," interview with John Hanke, July 6, 2005, available online from GeoPlace.com at www.geoplace.com/ME2/dirmod.asp?sid=119CFE3ACE2A48319AA7DE6A39B80D66&nm= News&type=news&mod=News&mid=9A02E3B96F2A415ABC72CB5F516B4C10&tier=3&nid=88D55CF122BD4254AD2D2CCA617 4D4B8; and Byrd, Jennifer, "Q&A with John Hanke, president, Google Earth," November 11, 2005. Available online from CrediblePress.org, www.crediblepress.org/2005/11/11/qa-with-john-hanke-president-google-earth.

Summary

As Linda Sanford, IBM's senior vice president of Enterprise on Demand Transformation & Information Technology, said, "The companies that can create a culture of innovation are the companies that will succeed in the next era of business."[66] Her views are supported by results from a study on firms competing in 17 major world economies whose authors concluded that "we find firm culture is the single most important driver of radical innovation."[67]

What are the values and beliefs that characterize a culture of innovation?

- Top management supports innovative activities.
- Managers are willing to cannibalize sales of their own products through product or process innovation rather than lose those sales to competitors; they engage in creative destruction.
- Product champions break the rules, take risks, transform companies, and turn organizations upside down; their efforts are encouraged and celebrated.
- The company has a structure for new product development projects that encourages breakthrough thinking; such a structure might be a skunk works.

- The company exhibits a learning orientation, creating the knowledge that enables it to stay ahead in a rapidly changing market.
- Simultaneously, because knowledge has a relatively short half-life in high-tech conditions—in other words, what once was valuable may soon be obsolete—barriers to progress are abandoned and disregarded in the process of unlearning.
- Imagination and disciplined experimentation are essential to create the products, markets, and technologies of the future.
- Failure, when it is not the result of faulty analysis or planning, is not penalized.
- In addition to being tangibly compensated, employees who develop innovations are given challenging work and substantial autonomy to complete that work.

Creating and maintaining such a culture is challenging because it requires constant organizational change. Core competencies—those skills that made the organization successful in the past—may become core rigidities as organizational members are reluctant to let them go.

In the next chapter, we pay close attention to two specific elements of organizational culture that are critically important in high tech firms: market orientation and the way in which personnel from different departments and units, particularly those in marketing and R&D, interact in cross-functional teams.

Discussion Questions

Chapter Concepts

1. What is organizational culture? What is organizational climate?
2. What is a culture of innovativeness? What factors make it hard for firms to remain innovative over time?
3. What is the incumbent's curse?
4. How is creativity related to innovation?
5. What is the role of top managers in maintaining a culture of innovativeness?
6. What is creative destruction? How does it help a company maintain a culture of innovativeness?
7. What factors affect a manager's willingness to cannibalize?
8. Who are product champions? What are their characteristics? How does their role differ in innovative companies compared to non-innovative companies?
9. What are skunk works? What are the arguments in favor of and against their role in innovation?
10. What is a learning orientation and what are its benefits?
11. What is unlearning and why is it important?
12. What is corporate imagination and how can it be stimulated?
13. What is expeditionary marketing and what are its advantages? How does expeditionary marketing differ from a more traditional approach of bringing new products to market successfully?
14. What is enlightened experimentation?
15. What is the relationship between risk tolerance and a company's culture of innovativeness?
16. What are the characteristics of a compensation system that facilitate innovativeness?
17. What other steps can a company take to create a culture of innovativeness?
18. What are the two major barriers to a culture of innovativeness?
19. What are core rigidities? How do they inhibit innovativeness?
20. What is the innovator's dilemma? How does it inhibit innovativeness?

Application Questions

1. Of the characteristics of a culture of innovation, which does the opening Google vignette specifically address? Which of these appear to be most important for Google? Elaborate sufficiently to show your logic and your detailed knowledge of the concepts.

2. Of the factors that inhibit a culture of innovativeness, which might Google need to be particularly wary of? How should Google ensure that those factors don't become concerns in the future?

3. Pick three other concepts from this chapter and offer insights about how Google might adopt them in order to continue creating a culture of innovativeness. What other insights do you have to offer?

4. In 1999, Dell was on *Wired*'s list of most innovative companies but had dropped completely off by 2007. What aspects of Dell's culture might explain this?

5. Google's famous catchphrase, "Don't be evil," has become a shorthand mission statement for Silicon Valley, encompassing a variety of ideals that proponents say are good for business and good for the world: *Embrace open platforms. Trust decisions to the wisdom of crowds. Ensure that your customers have a positive experience. Treat your employees like gods.*

It's ironic, then, that one of Silicon Valley's most successful companies ignored all these tenets. Apple's rules:

- Design software to work on your own hardware— and not on anyone else's.
- Never talk to the press. Never leak product news until you're ready to announce it. Threaten to sue children who send you their ideas.
- Please yourself, not your customers. Release iMacs without floppy drives. Release MacBook Airs without optical drives.
- Motivate through fear. Don't be afraid to scream. Threaten to fire employees. Withhold praise until it's truly deserved.[68]

How can Apple continue to be an innovation and market leader with values such as these?

Glossary

climate   The observable manifestation of culture shown in the set of behaviors that are expected, supported, and rewarded in a company.

core rigidity   Well-established skills and competencies that are so entrenched that they prevent a firm from seeing new ways of doing things.

corporate imagination   An aspect of a firm's culture characterized by creativity and playfulness that allows it to create breakthrough products and strategies by challenging the status quo, envisioning new markets, and using novel sources of information.

culture   The deeply rooted set of values and beliefs that provide norms for behavior in a company or organization.

culture of innovativeness   A set of forward-looking values and beliefs that actively encourages, supports, and rewards breakthrough thinking and risk taking, allowing the company to replenish its stock of ideas to develop a steady stream of both incremental and breakthrough innovations.

expeditionary marketing   A strategy for new product success based on trying many mini-introductions in quick succession and incorporating what was learned into each successive attempt, such that over time, the firm has accumulated loyal customers and higher market share than firms with fewer incursions into the market.

learning orientation   A characteristic of a firm's organizational culture that actively facilitates the development of new knowledge or insights that have the potential to influence its strategies and decisions.

product champion   A person who is so committed to a particular idea that he or she is willing to work tirelessly advocating the idea, to work outside normal channels to pursue it, and to bet future successes on the idea.

skunk works   New venture teams that are isolated or removed from normal corporate operations in order to foster an innovative culture that allows them to think out of the box.

unlearning   The process of identifying knowledge and assumptions that are the basis for strategic action, testing them for their validity, and discarding those that have become barriers to proactive change.

Notes

1. Goo, Sara Kehaulani, "Building a 'Googley' Workforce; Corporate Culture Breeds Innovation," *Washington Post*, October 21, 2006, p. D01.

2. Reardon, Marguerite, "Earthlink's Citywide Wi-Fi Biz for Sale," NewsBlog, February 8, 2008, http://news.cnet.com/8301-10784_3-9867634-7.html.

3. Goo, Sara Kehaulani, "Building a 'Googley' Workforce; Corporate Culture Breeds Innovation."

4. "Google: In Search of Itself," Knowledge@Wharton, November 14, 2007, online at http://knowledge.wharton.upenn.edu/article.cfm?articleid=1839

5. Wu, Tim, "Yes, Google Is Trying to Take Over the World; Next Step: Take Out Ma Bell," *Slate,* Nov. 16, 2007, online at www.slate.com/id/2178158.

6. Drucker, Peter, *Management: Tasks, Responsibilities, Practices* (New York: Harper & Row, 1973), pp. 105–107.

7. Deshpande, Rohit, and Frederick E. Webster Jr., "Organizational Culture and Marketing: Defining the Research Agenda," *Journal of Marketing* 53 (January 1989), pp. 3–15.

8. Chandy, Rajesh K., and Gerard J. Tellis, "The Incumbent's Curse? Incumbency, Size, and Radical Product Innovation," *Journal of Marketing* 64 (July 2000), pp. 1–17.

9. Prahalad, C. K., and Gary Hamel, "The Core Competence of the Corporation," *Harvard Business Review* 68 (May–June 1990), pp. 79–91.

10. Leonard-Barton, Dorothy, "Core Capabilities and Core Rigidities: A Paradox in Managing New Product Development," *Strategic Management Journal* 13 (Summer 1992; Special Issue), pp. 111–125.

11. Hjelt, Paola, "The World's Most Admired Companies," *Fortune International (Europe),* March 7, 2005, pp. 40-45.

12. Quinn, James, "Managing Innovation: Controlled Chaos," *Harvard Business Review* 63 (May–June 1985), pp. 73–85.

13. Leifer, Richard, Christopher M. McDermott, Gina C. O'Connor, Lois S. Peters, Mark P. Rice, and Robert W. Veryzer, *Radical Innovations: How Mature Companies Can Outsmart Upstarts* (Cambridge, MA: Harvard Business School Press, 2000).

14. Burgelman, Robert, "Corporate Entrepreneurship and Strategic Management: Insights from a Process Study," *Management Science* 29 (December 1983), pp. 1349–1364.

15. Hutt, Michael, Peter Reingen, and John Ronchetto Jr., "Tracing Emergent Processes in Marketing Strategy Formulation," *Journal of Marketing* 52 (January 1988), pp. 4–19.

16. Pinchot, Gifford, *Intrapreneuring: Why You Don't Have to Leave the Corporation to Become an Entrepreneur* (San Francisco: Berrett-Koehler, 2000).

17. Amabile, T. M., R. Conti, H. Coon, J. Lazenby, and M. Herron, "Assessing the Work Environment for Creativity," *Academy of Management Journal* 39 (October 1996), pp. 1154–1184.

18. Amabile, Teresa, "Motivating Creativity in Organizations," *California Management Review* 40 (Fall 1997), pp. 39–58.

19. Glasner, Joanna, "Why Webvan Drove off a Cliff," *Wired,* July 10, 2001, online at www.wired.com/techbiz/media/news/2001/07/45098.

20. Hermann, Andreas, Oliver Gassman, and Ulrich Eisert, "An Empirical Study of the Antecedents for Radical Product Innovations and Capabilities for Transformation," *Journal of Engineering and Technology Management* 24 (March 2007), pp. 92–120; Im, S., and J. P. Workman Jr., "Market Orientation, Creativity, and New Product Performance in High-Technology Firms," *Journal of Marketing* 68 (April 2004), pp. 114–132; and Narver, John, Stanley Slater, and Douglas MacLachlan, "Responsive and Proactive Market Orientation and New-Product Success," *Journal of Product Innovation Management* 21 (September 2004), pp. 334–347.

21. Lynn, Gary, and Richard Reilly, *Blockbusters: The Five Keys to Developing* GREAT *New Products* (New York: Harper Business, 2002).

22. Yadav, Manjit, Jaideep Prabhu, and Rajesh Chandy, "Managing the Future: CEO Attention and Innovation Outcomes," *Journal of Marketing* 71 (October 2007), pp. 84–101.

23. Ibid.

24. Shanklin, William, and John Ryans, *Essentials of Marketing High Technology* (Lexington, MA: Heath, 1987).

25. Schumpeter, Joseph, *Capitalism, Socialism, and Democracy* (New York: Harper & Row, 1942).

26. Chandy, Rajesh K., and Gerard J. Tellis, "Organizing for Radical Product Innovation: The Overlooked Role of Willingness to Cannibalize," *Journal of Marketing Research* 35 (November 1998), pp. 474–487.

27. Chandy, Rajesh K., Jaideep C. Prabhu, and Kersi D. Antia, "What Will the Future Bring? Dominance, Technology Expectations, and Radical Innovation," *Journal of Marketing* 67 (July 2003), pp. 1–18; and Sorescu, Alina B., Rajesh K. Chandy, and Jaideep C. Prabhu, "Sources and Financial Consequences of Radical Innovation: Insights from Pharmaceuticals," *Journal of Marketing* 67 (October 2003), pp. 82–102.

28. Grove, Andrew, *Only the Paranoid Survive* (New York: Currency, 1996).

29. Chandy and Tellis, "Organizing for Radical Product Innovation."

30. Maidique, Modesto, "Entrepreneurs, Champions, and Technological Innovations," *Sloan Management Review* 21 (Spring 1980), pp. 59–70; see also Howell, Jane, "Champions of Technological Innovation," *Administrative Science Quarterly* 35 (June 1990), pp. 317–341.

31. Chandy, and Tellis, "Organizing for Radical Product Innovation."

32. Ibid.

33. Tabrizi, Behnam, and Rick Walleigh, "Defining Next-Generation Products: An Inside Look," *Harvard Business Review* 75 (November–December 1997), pp. 116–124.

34. Hamel, Gary, and C. K. Prahalad, "Corporate Imagination and Expeditionary Marketing," *Harvard Business Review* 69 (July–August 1991), pp. 81–92.

35. Deutschman, Alan, "Building a Better Skunk Works," *Fast Company,* March 2005, pp. 68–73.

36. Byrnes, Nanette, and Paul Judge, "Internet Anxiety," *Business Week,* June 28, 1999, p. 84.

37. Hamel, Gary, and C. K. Prahalad, "Corporate Imagination and Expeditionary Marketing," *Harvard Business Review* 69 (July–August 1991), pp. 81–92.

38. Hamel, Gary, "Killer Strategies That Make Shareholders Rich," *Fortune,* June 23, 1997, pp. 70–84.

39. de Geus, Arie P., "Planning as Learning," *Harvard Business Review* 66 (March–April 1988), pp. 70–74.

40. Slater, Stanley, and John Narver, "Market Orientation and the Learning Organization," *Journal of Marketing* 59 (July 1995), pp. 63–74.

41. Sinkula, James M., William Baker, and Thomas Noordewier, "A Framework for Market-Based Organizational Learning: Linking Values, Knowledge, and Behavior," *Journal of the Academy of Marketing Science* 25 (September 1997), pp. 305–318.

42. Schilling, Melissa A., "Technology Success and Failure in Winner-Take-All Markets: The Impact of Learning Orientation, Timing, and Network Externalities," *Academy of Management Journal* 45 (April 2002), pp. 387–398; Hanvanich, Sangphet, K. Sivakumar, and G. Tomas Hult, "The Relationship of Learning and Memory with Organizational Performance: The Moderating Role of Turbulence," *Journal of the Academy of Marketing Science* 34 (September 2006), pp. 600–612; Calantone, Roger J., Tamer Cavusgil, and Yushan Zhao, "Learning Orientation, Firm Innovation Capability, and Firm Performance," *Industrial Marketing Management* 31 (September 2002), pp. 515–524; and Baker, William E., and James Sinkula, "The Synergistic Effect of Market Orientation and Learning Orientation on Organizational Performance," *Journal of the Academy of Marketing Science* 27 (September 1999), pp. 411–428.

43. Day, George S., "Continuous Learning about Markets," *California Management Review* 36 (Summer 1994), pp. 9–31.

44. Hamel, Gary, and Jeff Sampler, "The e-Corporation," *Fortune*, December 7, 1998, pp. 8–92.

45. Andrews, Peter, "Unlearn to Innovate: Breaking Patterns and Getting Rid of Assumptions," IBM Global Business Services Executive Technology Report, August 2006, available online at www-935.ibm.com/services/us/gbs/bus/pdf/g510-6313-etr-unlearn-to-innovate.pdf.

46. Saporito, Bill, and Ani Hadjian, "Behind the Tumult at P&G," *Fortune,* March 7, 1994, pp. 74–80.

47. Brown, John Seely, "Research That Reinvents the Corporation," *Harvard Business Review* 69 (January–February 1991), pp. 102–111.

48. Potts, Mark, "Toward a Boundary-less Firm at General Electric," in *The Challenge of Organizational Change,* eds. Rosabeth Moss Kanter, Barry A. Stein, and Todd D. Jick (New York: Free Press, 1992), pp. 450–455.

49. Hamel and Prahalad, "Corporate Imagination and Expeditionary Marketing."

50. Leonard-Barton, Dorothy, Edith Wilson, and John Doyle, "Commercializing Technology: Understanding User Needs," in *Business Marketing Strategy,* eds. V. K. Rangan et al. (Chicago: Irwin, 1995), pp. 281–305.

51. Hamel and Prahalad, "Corporate Imagination and Expeditionary Marketing."

52. MacDonald, Elizabeth, and Joann Lublin, "In the Debris of a Failed Merger: Trade Secrets," *Wall Street Journal,* March 10, 1998, p. B1.

53. See also the notion of waste in Gross, Neil, and Peter Coy with Otis Port, "The Technology Paradox," *Business Week,* March 6, 1995, pp. 76–84.

54. Thomke, Stefan, "Enlightened Experimentation: The New Imperative for Innovation," *Harvard Business Review* 79 (February 2001), pp. 67–75.

55. Slater, Stanley, "The Challenge of Sustaining Competitive Advantage," *Industrial Marketing Management* 25 (January 1996), pp. 79–86.

56. Baer, M., G. Oldham, and A. Cummings, "Rewarding Creativity: When Does It Really Matter?" *Leadership Quarterly* 14 (August–October 2003), pp. 569–586; and Shalley, C., J. Zhou, and G. Oldham, "The Effects of Personal and Contextual Characteristics on Creativity: Where Should We Go from Here?" *Journal of Management* 30 (December 2004), pp. 933–958.

57. Yanadori, Yoshio, and Janet H. Marler, "Compensation Strategy: Does Business Strategy Influence Compensation in High-Technology Firms?" *Strategic Management Journal* 27 (June 2006), pp. 559–570.

58. McGosh, Andrew, Alison Smart, Peter Barrar, and Ashley Lloyd, "Proven Methods for Innovation Management: An Executive Wish List," *Creativity and Innovation Management* 7 (December 1998), pp. 175–193; and Wycoff, Joyce, "The 'Big Ten' Innovation Killers: How to Keep Your Innovation System Alive and Well," *Journal for Quality Progress,* Summer 2003, pp. 17–21. Available online from ThinkSmart Innovation Network (www.thinksmart.com/library/BigTenInnovationKillers.htm).

59. Leonard-Barton, "Core Capabilities and Core Rigidities."

60. Guillemin, Christophe, Pierre Labousset, and Richard Shim, "Sony to Support MP3," *CNET News,* September 22, 2004, online at www.news.com/Sony-to-support-MP3/2100-1027_3-5377625.html.

61. Falcone, John P., and Matthew Moskovciak, "HD DVD vs. Blu-ray," *CNET's Quick Guide,* February 19, 2008, available online at http://reviews.cnet.com/hd-dvd-vs-blu-ray-guide; and "It's Official: Toshiba Announces HD DVD Surrender," *CNET News,* February 18, 2008, available online at http://news.cnet.com/8301-17938_105-9874199-1.html?tag=rb_content%3brb_mtx.

62. Christensen, Clayton M., *The Innovator's Dilemma* (Boston: Harvard Business School Press, 1997).

63. Danneels, Erwin, "The Dynamics of Product Innovation and Firm Competencies," *Strategic Management Journal*

23 (December 2002), pp. 1095–1121; and Danneels, Erwin, "Disruptive Technology Reconsidered: A Critique and Research Agenda," *Journal of Product Innovation Management* 21 (July 2004), pp. 246–258.

64. Henderson, Rebecca, "The Innovator's Dilemma as a Problem of Organizational Competence," *Journal of Product Innovation Management* 23 (January 2006), pp. 5–11, quote on p. 7.

65. Atuahene-Gima, Kwaku, Stanley F. Slater, and Eric M. Olson, "The Contingent Value of Responsive and Proactive Market Orientations for New Product Program Performance," *Journal of Product Innovation Management* 22 (November 2005), pp. 464–482; Narver, John C., Stanley F. Slater, and Douglas L. MacLachlan, "Responsive and Proactive Market Orientations and New Product Success," *Journal of Product Innovation Management* 21 (September 2004), pp. 334–347; and Greenley, Gordon E., Graham J. Hooley, and John M. Rudd, "Market Orientation in Multiple Stakeholder Orientation Context: Implications for Marketing Capabilities and Assets," *Journal of Business Research* 58 (November 2005), pp. 1483–1494.

66. Sanford, Linda, "Corporate Culture Is the Key to Unlocking Innovation and Growth," IBM Corporate Responsibility Report, 2005, pp. 1–2, available online at www.ibm .com/ibm/environment/annual/ibm_crr_061505.pdf.

67. Tellis, G., J. Prabhu, and R. Chandy, "Innovation in Firms across Nations: New Metrics and Drivers of Innovation," working paper, University of Southern California, 2007.

68. Kahney, Leander, "Breaking the Rules: Apple Succeeds by Defying 5 Core Valley Principles," *Wired,* March 18, 2008. Available online at www.wired.com/techbiz/it/magazine/ 16-04/bz_apple_rules.

4

Market Orientation and Cross-Functional (Marketing–R&D) Interaction

Market Orientation at Buckman Labs[1]

In 1988, while lying in bed after injuring his back, Bob Buckman thought about what he really needed to run his business. What he wanted was information, not just for himself but for all his people—a steady stream of information about products, markets, customers. As he said, "The customer is most important. . . . We need to be effectively engaged on the front line, actively involved in satisfying the needs of our customers." According to Buckman, the real questions were these: "How do we stay connected? How do we share knowledge? How do we function anytime, anywhere—no matter what?"

Established in 1945, Buckman Laboratories is a privately held chemical company that provides target industries with a range of specialty chemicals including microbiocides, scale inhibitors, corrosion inhibitors, polymers, dispersants, and defoamers. This is a highly technical industry with many new treatments and technologies developed each year. With 2006 revenues of $82 million, the company serves customers in three industries: pulp and paper, water treatment, and leather treatment.

In 1992, Buckman realized that the company's previous strategy of product leadership—being the first to market with innovative product offerings in many disparate markets—by itself, was no longer viable. Much larger competitors were, as he put it, "eating our lunch." To survive and prosper, the company had to change. Bob Buckman described the challenges: Close the gap with the customer. Stay in touch with each other. Bring all of the company's brainpower to bear in serving each customer. "We have to be so tuned into our customers that we anticipate what they need," says Buckman. "If an employee is not effectively engaged with the customer, why are they employed?" Consequently, the company adopted the business strategy of customer intimacy, focusing on delivering a customized version of the company's product mix to provide the best total value to the customer. This way of doing business places a premium on knowledge and the capacity to act on it.

As Buckman noted, "We recognized early on that the greatest knowledge base in the company did not reside in a computer database somewhere, but was in the heads of the individual associates (employees) of the company. Over 72% of our associates worldwide have degrees from a college or university. Then it became a question of how do we get each individual to share what they know, freely and openly with others, and how do we get each individual to assume responsibility for their actions."

Buckman needed to create an organization that would support this strategy by getting employees to move from hoarding information to sharing it. "Over the years, people have taught themselves to hoard knowledge to achieve power," says Buckman. "We have to reverse that: The most powerful people are those who become a source of knowledge by sharing what they know."

Toward this end, the company developed a global knowledge sharing platform that included access to the worldwide network of 1,200 employees, a library of critical knowledge, and forums for sharing knowledge—open places where anyone in the company can post a message, question, or request help or technical support. Buckman Labs has numerous forums, each with a particular focus. The company coupled these knowledge-sharing tools with a type of carrot-and-stick approach, mixing visible incentives with invisible pressure to stimulate teamwork and knowledge sharing. For example, as one "carrot," Buckman organized a one-time celebration at a fashionable resort in Scottsdale, Arizona, to recognize the 150 best knowledge sharers. Those selected received a new IBM ThinkPad 755 and listened to a presentation by management guru Tom Peters. Within the company, the high-profile event sparked a good deal of discussion, particularly among those who failed to make the cut.

Buckman also used a powerful "stick" component. Early in the life of the knowledge-sharing program, Buckman laid out his expectations in a speech to his people. "Those of you who have something intelligent to say now have a platform or forum in which to say it," he told them. "Those of you who will not or cannot contribute also become obvious. If you are not willing to contribute or participate, then you should understand that the many opportunities offered to you in the past will no longer be available."

Melissie Rumizen, knowledge strategist at Buckman Labs, stated, "As we applied our strategy of customer intimacy, we realized that we needed a new mission statement to reflect it. While working together in our forums, our global associates discussed (at times hotly) what the mission statement should be." The results, according to Rumizen, were this statement.

> We, the associates of Buckman Laboratories, will excel in providing measurable, cost-effective improvements in output and quality for our customers by delivering customer-specific services and products, and the creative application of knowledge.

"We also refined our purpose," said Rumizen, "stating that it is to foster customer loyalty, attaining and maintaining long-term, loyal customers for mutual profitability and growth. Doing this requires us to link our success with our customers, demanding deeper relationships. It also reaffirms the importance of having skilled, knowledgeable associates."

In 1997, the company started the Buckman Learning Center with an emphasis on training. However, over time, as more companies adopted their own knowledge management and distance learning initiatives, Buckman's customers asked for assistance in developing their own knowledge-management and learning programs. Such cooperative development is a natural extension of the overall strategy of customer intimacy; it deepens and strengthens the company's partnership with its customers. It is also a valuable avenue for learning what really matters to customers, both strategically and tactically. In 2001 Buckman formed a consulting arm within the learning center where customer representatives work to better support its clients. Executives from AT&T, 3M, Champion International, International Paper Company, and Samsung have made the pilgrimage to Buckman to look and learn. What they've seen is a company that is fast, global, and interactive, built on a philosophy of intimate understanding of customers, harnessing the power of all employees in the service of those customers, and knowledge sharing to benefit the whole.

For companies thinking about embarking on such a program, Bob Buckman emphasizes one lesson before all others: "What's happened here is 90% culture change. You need to change the way you relate to one another. If you can't do that, you won't succeed." Looking back, Buckman says that incorporating the knowledge transfer system into a corporate culture is at least a three-year process.

The prior chapter focused on the internal culture of high-tech companies, characterized by the shared values and beliefs that affect decisions and behaviors. This chapter continues the focus on organizational culture, examining two additional topics: market orientation and cross-functional teamwork—particularly teams comprised of marketing and R&D personnel. Both market orientation and cross-functional teamwork reflect the company's shared values and beliefs about how decisions are made and how people work together in making them.

This chapter's first section on market orientation highlights the role of market-based information—about customers, competitors, and other important stakeholders and trends—in strategic decisions. Simply put, information about customers and their needs becomes the beacon to guide the company through the messy, often political, process of decision making. Companies operating in high-tech environments characterized by the triple threat of market, competitive, and technological uncertainty stand to benefit disproportionately from effectively gathering and using market-based information. With a shared understanding about the value and use of market intelligence, a company can create a powerful knowledge-based competency that allows it to gain superior advantage in the marketplace.

The second section in this chapter focuses on how such information is used and shared by cross-functional teams—particularly those that involve marketing and R&D personnel. Effectively harnessing the power of cross-functional teams is one of the key dimensions of a firm's market orientation.

WHAT IT MEANS TO BE MARKET ORIENTED[2]

To compete successfully in high-tech markets, companies minimize the inherent risks by proactively gathering, sharing, and acting on market-based information. Paradoxically, technological innovations are more market dependent than other types of innovations.[3] Superior technology alone is insufficient for achieving marketplace success for high-tech firms. Rather, careful analysis of the needs and the capabilities of the intended user is essential to successful development of new high-tech products. Studies show that a market orientation leads to greater creativity and improved new product performance in high-tech firms.[4]

High-tech firms must excel at three activities: opportunity identification, product and process innovation, and product commercialization. Because one of marketing's tasks is to listen to the customer and define a broad set of opportunities, a strong marketing capability implies that marketing is able to identify a wide range of markets and customer applications for the innovative technology. The voice that marketing brings to the innovation process must be joined with the knowledge that R&D brings in order to develop an offering that effectively addresses customer needs.

Businesses that value and rely on market information to guide strategic decision making are commonly described as *market oriented*. Distilled to its essence, a **market orientation** simply means gathering, sharing, and using information about the market (customers, competitors, collaborators, technology, trends, etc.) to make decisions that lead to the creation of superior customer value. As noted in this chapter's opening vignette, some companies use the terms *market oriented* and *customer focused* interchangeably, referring to an organizational culture that bases all decisions solidly in customer analysis and market-based trends; a customer-focused culture puts customers' interests first.[5] This philosophy of decision making requires effective management of the knowledge that resides in different places within the firm; it also recognizes that all people, departments, and divisions must work collaboratively to create and deliver superior value to customers.

As shown in Figure 4.1, market-oriented businesses are characterized by four key dimensions. First, they *generate intelligence* about the expressed and latent needs of their customers, the capabilities and strategies of their competitors, and a realistic assessment about the potential opportunities

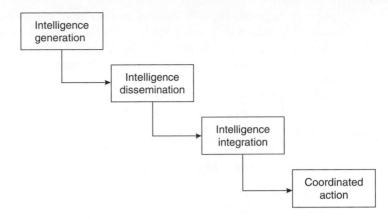

FIGURE 4.1   Dimensions of a Market Orientation

and threats of emerging technologies. Second, they *disseminate* this information broadly throughout the organization and, third, they *integrate* the disparate pieces of information—creating a repository of market knowledge assets to guide decisions. Fourth, the company's decisions arising from that knowledge are based on *coordinated action across departments and functions* to create superior customer value. Prior to describing these steps in more detail, the next section examines the effects of a market orientation on a company's performance.

The Effect of Market Orientation on Company Performance

A significant body of research has established a positive relationship between a firm's market orientation and its performance.[6] As shown in Figure 4.2, by continuously improving both product quality and service quality, and by developing innovative products that meet evolving customer needs, a market orientation delivers superior sales growth and profitability.

Some research has found that the positive relationship between market orientation and firm performance holds true regardless of the environmental conditions in which a firm competes[7] (e.g., high or low levels of market turbulence, technological turbulence, competitive intensity, market growth rates). However, other studies report that the relationship between market orientation and firm performance is *stronger* in highly dynamic markets,[8] which are characteristic of technology-oriented industries.

Therefore, the positive relationship between a firm's market orientation and performance outcomes is especially important for high-tech firms. Firms in high-tech markets need to excel

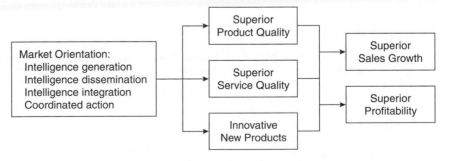

FIGURE 4.2   The Impact of Market Orientation on Key Outcomes and Firm Performance

not only at generating new innovations, but also at commercializing these innovations. Studies show that firms with a strong R&D and technological base stand to gain the most from a strong marketing capability.[9] Moreover, a strong market orientation is one of the most fertile sources of ideas for innovation. Conversely, a strong market orientation without commensurate development of a strong innovation/technological capability can have a negative effect on new product and market performance.[10] Hence, superior technology and innovation capabilities must be combined with an effective market orientation to achieve the highest levels of success in high-tech markets.

Dimensions of a Market Orientation

INTELLIGENCE GENERATION   Market-oriented firms generate a wide array of intelligence about market forces. **Market intelligence** includes useful information about trends and stakeholders in the market, including (1) current and future customer needs, (2) competitors' capabilities and strategies, and (3) emerging technologies both inside and outside the industry.

The acquisition of information can be done via customer hotlines, trade shows, customer visits, working with lead users, alliances, cooperative relationships with universities, or some of the more high-tech-oriented research tools discussed in Chapter 6. In addition, competitive intelligence can be gleaned through a variety of online and offline resources. The Society of Competitive Intelligence Professionals (www.scip.org) and the Competitive Intelligence Resource Index (www.bidigital.com/ci) are two useful starting places. Competitive intelligence can include information on competitors' customer lists, product and pricing information, new product plans and R&D efforts, job postings that provide insights into new business arenas, or information about partnerships, alliances, distributors, and managers.

Although both customer and competitor orientations are important, a competitor orientation is more strongly associated with a firm's ability to invest resources into existing products, and hence is more likely to result in incremental innovations. On the other hand, a customer orientation, particularly a proactive orientation as discussed below, is more strongly associated with a firm's tendency to invest in new knowledge and skills, which in turn is more likely to result in breakthrough (radical) innovations.[11]

Regardless of the specific tools used to gather market-based information, it is imperative that the firm allocates resources to the information-gathering process. However, this is much easier said than done. Many high-tech companies do not give adequate resources to initiatives that seemingly are not directly tied to underlying technology development or commercialization. Indeed, one $2 billion U.S.-based high-tech company was unwilling to invest any meaningful money in purchasing secondary market data and analyst reports, or in collecting its own primary market research. Yet this company wanted to make a "cultural transformation" to being market driven. Clearly, a company that says it wants to become more customer focused must put its money where its mouth is!

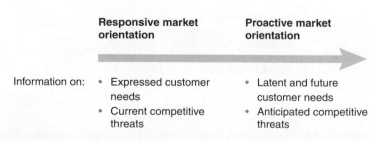

FIGURE 4.3   The Intelligence Continuum: Responsive to Proactive

As shown in Figure 4.3, intelligence can be viewed along a continuum from current to antici-pated.[12] Businesses that predominantly respond to current intelligence are well positioned to dis-cover, understand, and satisfy the expressed needs of existing customers, and to deal with existing competitive threats.

Responding to customers' expressed needs or existing threats is usually inadequate to create superior, sustainable competitive advantage for at least three reasons. First, customers of high-technology products may find it very difficult to articulate their needs clearly; they may be unaware of the capabilities that new technologies offer or may lack insight into how technology can address their needs. Second, even if customers are able to articulate their needs, what they articulate to one seller can and will be articulated to competitive sellers as well. Third, responding to existing threats means that the business is frequently one step behind the competition.

Firms that primarily respond to current intelligence are said to have a **responsive market ori-entation**. Although current intelligence provides important information, some argue that listening to current customers too carefully can inhibit innovation, constraining it to ideas that customers can envision and articulate—which may lead to safe, but bland, offerings. Firms with a responsive mar-ket orientation may suffer from *marketing myopia,* a tendency to focus very specifically on solving existing customers' needs with a current technology. Firms that focus too narrowly on their estab-lished customers (the *tyranny of the served market*) may be constrained in the strategies and tech-nologies they choose to pursue. Such a myopic focus obscures the possibility that customer needs may change over time and may be solved in radically different ways, allowing new, disruptive inno-vations to creep up like a stealth attack.[13]

For example, a firm's best customers may be the last to embrace a disruptive technology and the company may believe that the people who embrace a new technology first, say a simple or inex-pensive new technology, may be an unattractive market segment—and as a result, the company pays it little attention. Take the 5.25-inch disk drive, a technology introduced in the early 1980s that was embraced by the emerging desktop personal computer marketplace—yet was inconsistent with mainframe and minicomputer customer demands.[14] Established disk drive firms failed not because they were unable to develop innovative technologies. Rather, because established customers were uninterested in new technologies that didn't address their immediate needs, industry leaders did not allocate resources to the new technologies. This decision allowed new entrants to gain leadership in the new market.

At the other end of the intelligence continuum are businesses that gather anticipatory intel-ligence; these businesses have a **proactive market orientation**. Anticipatory customer intelli-gence is concerned with customers' latent and future needs, and it enables the firm to proactively pursue market opportunities that are not evident to competitors. Latent needs are real needs that are not yet in the customers' awareness. If these needs are not satisfied by a provider, there is no customer demand or response. Customers are not necessarily dissatisfied, because the need is unknown to them. However, if a company understands such a need and fulfills it, the customer is "wowed," and rapidly delighted. Offering products and services that address these latent needs in a compelling fashion delights and excites customers and inspires loyalty. A simple example is 3M Post-it Notes, for which a need had long been present but was not articulated until the product existed. The product met a latent need, generated great enthusiasm, and became wildly successful. Another example is found in the accompanying Technology Solutions for Global Problems box.

Similarly, think about the 1 billion people in the world who do not have access to safe drinking water—which results in 1.8 million children dying each year from diseases attributed to contami-nated water. In a radical innovation that wowed people, the Vestergaard Frandsen Group (a Swiss company with Danish roots) developed a drinking straw—aptly named LifeStraw—that filters bacte-ria from drinking water in base-of-the-pyramid markets. As shown in the accompanying photo, indi-viduals can literally drink straight out of a contaminated water source using LifeStraw, and each LifeStraw can filter approximately 700 liters of water—about enough for one individual per year.[15]

Photo and illustration reprinted with permission from Vestergaard Frandsen, New York.

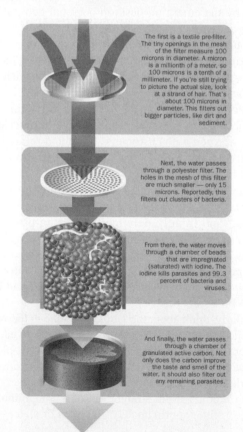

The first is a textile pre-filter. The tiny openings in the mesh of the filter measure 100 microns in diameter. A micron is a millionth of a meter, so 100 microns is a tenth of a millimeter. If you're still trying to picture the actual size, look at a strand of hair. That's about 100 microns in diameter. This filters out bigger particles, like dirt and sediment.

Next, the water passes through a polyester filter. The holes in the mesh of this filter are much smaller — only 15 microns. Reportedly, this filters out clusters of bacteria.

From there, the water moves through a chamber of beads that are impregnated (saturated) with iodine. The iodine kills parasites and 99.3 percent of bacteria and viruses.

And finally, the water passes through a chamber of granulated active carbon. Not only does the carbon improve the taste and smell of the water, it should also filter out any remaining parasites.

TECHNOLOGY SOLUTIONS FOR GLOBAL PROBLEMS
Curing Blindness among India's Poor: Aravind Eye Hospital

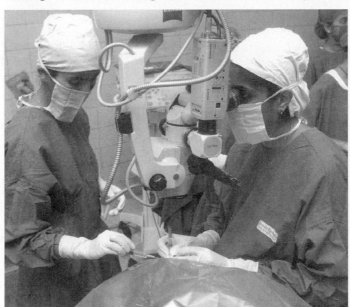

Photo reprinted with permission of Aravind Eye Hospital/Aurolab, Madurai, India.

Cataracts are a leading cause of blindness among India's poor. To address this problem, which could also be seen as a market opportunity, Dr. Govindappa Venkataswamy had to create an innovative product, business model, and marketing strategy. Aravind's main product is a low-cost cataract operation. It costs Aravind about $10 to perform a cataract operation, compared to about $1,650 in the United States. Aravind keeps costs low by using ophthalmic paramedical staff to do all the preparatory and postoperative work on each patient, allowing ophthalmologists to perform an increased number of surgeries. Each ophthalmic surgeon has two tables, allowing a surgeon to perform one 10- to 20-minute operation, and then swivel around to do the next. Aravind produces its own high-quality, low-cost ($4) intraocular lenses in its Aurolab division.

To implement his vision of giving eyesight to the blind regardless of ability to pay, Dr. Venkataswamy set up hospitals in South India that serve both the rich, who pay for state-of-the-art cataract surgery, and the poor, who receive almost identical services for free. This economically self-sustaining model is based on generating enough revenue from the paying patients to cover the costs of the providing free or low-cost eye care to the majority.

The sales, advertising, and promotion activities of Aravind Eye Hospital focus on attracting free rather than paying patients. A representative from Aravind visits a village and meets with its leaders. Together they do the planning necessary to organize a weekend camp. Then Aravind doctors and technicians set out for the village, sometimes driving for days. Once there, they work around the clock, examining people and working to identify those who will need to be taken to the hospital for surgery.

Aravind Eye Hospital developed an innovative business model coupled with innovations in the surgical process and its own proprietary technology for high-quality, low-cost intraocular lenses to successfully compete in a base-of-the-pyramid market. Aravind is so productive that this nonprofit organization has a gross margin of 50%—despite the fact that more than 65% of its patients do not pay. In a true market-driving spirit, in one of Aravind's classrooms, a sign says "If You Are Looking for a Big Opportunity, Find a Big Problem."

Sources: Kumar, N., Lisa Scheer, and Philip Kotler, "From Market Driven to Market Driving," *European Management Journal* 18 (April 2000), pp. 129–142; and Rubin, Harriet, "The Perfect Vision of Dr. V.," *Fast Company*, January 2001, available online at www.fastcompany.com/online/43/drv.html.

An extreme type of proactive market orientation, potentially very useful in the high-tech arena, is known as **market driving** in which a firm actively seeks to (1) redefine the structure of the market, and/or (2) introduce an innovative value proposition that enables the firm to reduce, or even avoid, competition.[16] Market-driving activities may be focused on many different stakeholders, including customers and competitors as well as vendors, potential partners and allies, and regulators.[17] A market-driving strategy necessitates a multidimensional, resource intensive, and long-range planning approach.

A market-driving strategy can be quite risky since managers are attempting to change the structure of a market and/or the behavior of players in the market by introducing a discontinuous leap in the customer value system or an unproven technology. For example, Apple anticipated the emergence of the PDA (personal digital assistant) market when it introduced the Newton in 1993. The Newton Message Pad featured a variety of personal-organization applications, such as an address book, a calendar, and notes, along with communications capabilities such as faxing and e-mail. It featured a pen-based interface that used a word-based, trainable handwriting-recognition engine. Unfortunately, this engine was notoriously difficult to use which, along with competition from the PalmPilot, led to the Newton's withdrawal in 1998.[18]

How do companies gather anticipatory intelligence? Significantly, it cannot be gathered through standard market research approaches. Users may be unable to envision new solutions that technology may have to offer.[19] So, anticipatory intelligence is gleaned from observing customers' use of products or services in normal routines, conducting a structured customer visit program, or working closely with lead users who recognize a need in advance of the majority of the market, among other techniques. A future focus leverages *bifocal vision,* considering both the future needs of current customers as well as customers that may comprise the firm's future market. Businesses with a future focus also attempt to forecast the future moves of current competitors and the emergence of new competitors or of substitute products.

Importantly, both responsive and proactive market orientation are consistent with the contingency theory of high-tech marketing. Recall that contingency theory matches different marketing strategies to the different types of innovations (incremental versus breakthrough). Research shows that a responsive market orientation (market driven) may generate incremental innovations, while a proactive market orientation (market driving) is most likely to lead to breakthrough innovation.[20] Therefore, firms must be ambidextrous, combining both responsive and proactive market orientations.

INTELLIGENCE DISSEMINATION   As useful as intelligence generation is, the information generated is of limited value until it is shared across the organization and potentially combined with other information. A firm's knowledge-based competitive advantage—one that increasingly resides in its know-how—is only as strong as its ability to share and use knowledge within and across the organization's boundaries. Effective knowledge management requires a *boundaryless organization,* which takes good ideas from disparate functions and even outside organizations and uses them in many areas. Indeed, as stated by Lew Platt, former CEO of Hewlett-Packard, "If only we knew what we know."[21] These words reflect a tough truth: The knowledge and know-how of most organizations is too often underused—isolated in departments and functional units.

Buckman Labs recognized the vital role of knowledge dissemination in its transformation to becoming more market oriented; it created a variety of incentives and technology platforms to stimulate the sharing of information. Firms committed to a market orientation disseminate information throughout the company—across people, departments, and divisions. Information may be shared formally in meetings, conferences, newsletters, and databases, or informally through "hall talk" and informal mentoring. People in the organization must be able to ask questions and augment or modify the information to provide new insights.

However, freely sharing and using information is easier said than done. An organization is a coalition of individuals or groups, each with its own goals that may conflict with organizational goals. One way that individuals or departments promote their own self-interests is by creating and protecting their "proprietary" knowledge base. They believe that hoarding or withholding information protects their turf and generates power and status. However, this dysfunctional belief does not allow the organization to fully leverage its knowledge. Effective dissemination increases the value of information, allowing each piece of information to be seen in its broader context by all organizational players who might be affected by it, utilize it, or possess complementary information about it.

Thus, the challenge is to create an organizational environment in which group success does not come at the expense of individual success. One way to ensure this is to cultivate a *team orientation* as the foundation for effective information sharing, a topic addressed in the second half of this chapter.

INTELLIGENCE INTEGRATION   Third, market-oriented companies integrate intelligence to create knowledge assets. Arno Penzias, the 1978 Nobel laurcate in physics and vice president of research at AT&T Bell Laboratories, stated that "knowledge is not chunks of information, it's a belief about how the world works."[22] Reaching this state of a shared belief system requires achieving a shared interpretation of the information and its implications for the business. Shared interpretation goes far beyond simply sharing the information. Consider this variation on an old joke:

> *Marketing: The glass is half full.*
> *R&D: The glass is half empty.*
> *Operations: That glass is twice as large as it needs to be.*

Each of these perspectives carries substantially different views of what a particular piece of information means, and, by extension, implications for what the firm should do. Integrating intelligence in order to achieve a shared understanding is vital.

Indeed, firms in dynamic and complex high-tech markets must strive for consensus internally in order to successfully develop and execute strategy.[23] However, prior to achieving consensus, companies in volatile industries may benefit from a relatively high level of disagreement among managers in interpreting the information they have gathered.[24] Such disagreement allows a closer inspection of the validity of different assumptions and alternatives, as well as an assessment of the relative importance of the company's objectives and competitive methods. In his 1999 book, *Only the Paranoid Survive,* Intel's chairman, Andy Grove, described his view of spirited debate to his middle managers: "Your criterion for involvement should be that you're heard and understood. . . . All sides cannot prevail in the debate, but all opinions have value in shaping the right answers."[25] A company must reach a shared interpretation of the information it has gathered, but it should not do so prematurely. It should actively facilitate debate, discussion, disagreement, and dialogue in order to fully tap the value of its knowledge.

Due to the boom and bust nature of high-tech industries, and to the mobility of highly skilled employees, knowledge workers frequently leave the company and take their knowledge with them. Compounding this traditional problem is the looming retirement of the 76 million members of the baby boom generation. The question is, how can organizations create an organizational memory to retain that knowledge?

Starting in 1999 the Tennessee Valley Authority (TVA), the nation's largest public power company, began asking line managers three questions:

- What knowledge is likely to be lost when particular employees leave? ("What?")
- What will be the business consequences of losing that knowledge? ("So what?")
- What can be done to prevent or minimize the damage? ("Now what?")

TVA gave each employee a score of 1 to 5, with 1 denoting a person who didn't plan to leave for six years or longer, and 5 earmarking someone who would be gone within a year. At the same time, managers assigned the people reporting to them a second score of 1 to 5, reflecting how essential their knowledge was to the company's operations, with a score of 5 being the most critical. When they multiplied everyone's attrition score by their critical-knowledge score, they had a clear picture of where their most valuable knowledge was.[26] Once a business has identified where its most valuable knowledge is, it must develop a strategy for preserving that knowledge in its organizational memory. Companies that take steps to preserve their organizational memory and disperse it broadly throughout the organization exhibit greater creativity and improved new product performance.[27]

The company's strategy for storing knowledge in organizational memory depends on the nature of the knowledge. *Explicit knowledge* is the obvious knowledge that can be stored in manuals, documentation, patents, blueprints, reports, databases, and other accessible sources.[28] A more vague concept is *tacit knowledge*. As described by Scott Schaffar, Northrop Grumman's director of knowledge management, tacit knowledge represents "what is held in our heads and includes facts, stories, biases, misconceptions, insights, and networks of friends and acquaintances, as well as the ability to invent creative solutions to problems."[29] Since tacit knowledge is not easily recorded and shared, it must be transferred person-to-person. Mentoring, communities of practice (e.g., companywide groups that meet, in person and online, to share information), and action learning teams that put people together from several disciplines (e.g., manufacturing, sales, marketing, legal, finance) to solve particular problems are valuable ways to disseminate knowledge throughout the organization.

More broadly, **knowledge management** comprises the range of practices companies use to identify, create, represent, and distribute knowledge, often under the auspices of the Information Technology or Human Resource Management departments; some organizations also have a chief knowledge officer reporting directly to the head of the company. Many companies rely on technology and software—a multibillion-dollar worldwide market—for their knowledge management practices; these include, for example, knowledge bases, expert systems, knowledge repositories, corporate intranets and extranets, wikis (collaborative websites, often password protected, that allow people to generate and share knowledge on specific topics) and document management/content management software programs that facilitate collaboration and enhance the knowledge transfer process. Although knowledge management is frequently linked to the idea of a learning orientation (recall this concept from Chapter 3), knowledge management focuses fairly specifically on knowledge assets and the mechanisms used to share and transfer it.

IMPLEMENT THE DECISION WITH COORDINATED ACTION   Finally, the market-oriented firm implements its decisions through coordinated action. An organization can generate and disseminate intelligence—but unless it acts on that intelligence, nothing will be accomplished.[30] Mark Hurd, CEO of Hewlett-Packard, quoting Einstein, once said, "Vision without execution is hallucination."[31] Acting on market intelligence involves making decisions including the selection of target markets; the development of products/services that address customers' current and anticipated needs; and production, distribution, and promotion of the products to engender both customer satisfaction and customer loyalty.[32]

Moreover, the idea of coordinated action means that all functions in a market-oriented company—not just marketing—participate in responding to market needs. To drive new products from concept to launch more rapidly and with fewer mistakes, "all functional interfaces (contacts) must jointly share in discussions and information exchange."[33] Coordinated action requires a greater emphasis on multifunctional activities—activities that are the joint responsibility of multiple functions in the business.[34] When decisions are made interfunctionally and interdivisionally, a more accurate representation of the information and a closer connection to the market issues will occur. Moreover, interfunctional decision making implies that the people who will be involved in *implementing* the decisions are the ones actually involved in *making* the decisions—the rationale being that if someone is involved in making a decision, he or she will be more committed to implementing it. Indeed, two firms may have the same information, but what allows one firm to leverage that knowledge more successfully is not the knowledge itself, but rather, how that knowledge is coordinated and integrated among different functions and departments.[35]

Yet, organizational politics can make such multidisciplinary interaction tough. Engineers and other technical personnel are often given higher status in the organization relative to marketing or customer-focused organizational members.[36] Either implicitly or explicitly, the preference for engineering-related knowledge and skills can become a type of core rigidity that often results in the disregard of information about users—unless it comes from someone with status in the organization.

For example, during the design of the Deskjet printer at Hewlett-Packard, marketers tested early prototypes in shopping malls to determine user response. They returned from their studies with a list of 21 changes they believed essential to the success of the product; however, the engineers accepted only five. Unwilling to give up, the marketers persuaded the engineers to join them in the mall tests. After hearing the same feedback from the lips of users—feedback they had previously rejected—the product designers returned to their benches and incorporated the other 16 requested changes.[37] Unfortunately, because marketing personnel are often not a part of the distinctive competencies of technology-oriented firms, the information they bring to the design process goes unheeded—which can have negative consequences.

In addition, interfunctional rivalry can impede internal knowledge transfer. Functions must compete for scarce organizational resources and are thus often reluctant to share information. This sets up a situation sometimes referred to as *"co-opetition,"* or the simultaneous need for cooperation and competition by various departments within a company. True, functional areas compete with one another for resources, but they also cooperate to work toward the firm's common interests. A study of these complicated intraorganizational dynamics found that fostering cooperation but squelching internal competition can actually limit a firm's performance potential.[38] Constructive conflict can promote learning and knowledge sharing, and can improve performance. Importantly, market learning is the mechanism by which the simultaneous effects of interdepartmental cooperation and competition are transferred to higher performance. Hence, problems in a firm's market orientation may be due not only to low cross-functional cooperation, but also to an inability to assimilate and deploy market knowledge.

Because the notion of interfunctional coordination is so important, the second half of this chapter is devoted to the topic. In particular, it explores one of the most important interdepartmental dynamics in high-tech companies: the relationship between R&D and marketing.

In sum, a firm's market orientation is determined by its ability to generate and disseminate market intelligence, to come to shared understanding of what that information means, to create an organizational memory for it, and to make and execute decisions based on the information with cross-functional representation, including multiple departments that are affected by the decisions. How does a company know how market oriented it is? See Box 4.1 for guidelines.

BOX 4.1

Assessing a Firm's Degree of Market Orientation

A manager should rate his or her business on each of the questions below using the following scale:

Strongly Disagree	Disagree Moderately	Disagree Slightly	Agree Slightly	Agree Moderately	Strongly Agree
−3	−2	−1	1	2	3

After rating each item, sum the scores for the four items in each of the categories. Negative scores indicate much room for improvement, low positive scores indicate the need for selective or incremental improvement, and high scores indicate a strong capability. High scores are relatively rare.

Responsive Customer Intelligence Generation:

- We continuously work to better understand our customers' needs.
- We pay close attention to after-sales service.
- We measure customer satisfaction systematically and frequently.
- We want customers to think of us as allies.

(continued)

Responsive Competitor Intelligence Generation:

- Employees throughout the organization share information concerning competitors' activities.
- Top management regularly discusses competitor's strengths and weaknesses.
- We track the performance of key competitors.
- We evaluate the strengths and weaknesses of key competitors.

Proactive Customer Intelligence Generation:

- We continuously try to discover additional needs of our customers that they might be unaware of.
- We incorporate solutions to unarticulated customer needs in our new products and services.
- We brainstorm about how customers' needs will evolve.
- We work with lead users—customers who face needs that eventually will be in the market—but face them months or years before the majority of the market.

Proactive Competitor Intelligence Generation:

- We try to anticipate the future moves of our competitors.
- We monitor firms competing in related products/markets.
- We monitor firms using related technologies.
- We monitor firms already targeting our prime market segment but with unrelated products.

Intelligence Dissemination:

- We have interdepartmental meetings to discuss market trends and developments.
- Marketing personnel spend time discussing customers' needs with other functional departments.
- We share information about major market developments.
- When one function acquires important information about customers or competitors, it shares that information with other functions.

Intelligence Integration:

- We have cross-functional meetings for the purpose of intelligence integration.
- We reach organizational consensus regarding the holistic meaning of related pieces of information before taking action.
- We utilize cross-functional teams or task forces for important initiatives to ensure that all points of view are considered before decisions are made.
- We value collaboration in this business.

Coordinated Action:*

- We are quick to take advantage of market opportunities.
- The activities of different functions in this business are well-coordinated.
- We make sure that all critical functions understand our objectives and strategy before we take action.
- There is a high level of cooperation and coordination among functional units in setting the goals and priorities for the organization to ensure effective response to market conditions.

---

* The second half of this chapter provides additional ways to assess the degree of cross-functional interaction.

Becoming Market Oriented: Facilitating Conditions

The logic for and evidence of the value of a market orientation is clear and compelling. Why, then, aren't more firms able to become market oriented? As noted in the opening vignette, becoming market oriented requires a cultural change in the organization. It requires shared values and beliefs about the need to gather, share, and compile market-based information; it requires a common desire to actively discuss the meaning of that information, and to ensure that the company's decisions are soundly grounded in the relevant information. It also requires a resource commitment to the

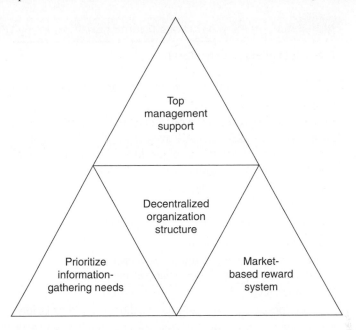

FIGURE 4.4   Facilitating Conditions for a Market Orientation

processes involved in a market orientation: data collection expenditures and staffing, time to share and discuss the meaning of the information, and so forth.

The fact is, many firms do not achieve a market orientation because they do not have the proper facilitating conditions in place. As shown in Figure 4.4, these conditions include, at a minimum: (1) prioritizing a firm's information-gathering focus to match its strategy, (2) top management advocacy, (3) a flexible, decentralized structure, and (4) a market-based reward system.

PRIORITIZE INFORMATION-GATHERING NEEDS   Resource constraints coupled with highly complex markets make it impossible to comprehensively scan the high-tech environment. To make sense of complex environments, managers must focus their scanning efforts on the market forces that are most salient to their strategies.[39] This focus is especially critical in high-tech markets; due to their turbulence and dynamism, without the ability to simplify, structure, and focus their learning efforts, managers would suffer from "paralysis by analysis." One approach to prioritizing their market-scanning efforts is to identify the critical issues faced by each of the four different strategy types.[40] Table 4.1 shows that, depending on whether a company's strategy is Product Leader (Prospector), Fast Follower (Analyzer), Operationally Excellent (Low-Cost Defender), or Customer Intimate (Differentiated Defender), it will collect and use market-based information differently.

TOP MANAGEMENT COMMITMENT   If a firm's *top managers* are not unequivocally and visibly committed to its customers and the collection of market-based information, then the firm will not bring its resources to bear on developing solutions to its customers' needs. As Michael Dell, CEO of Dell Computers, said, "We have a relentless focus on our customers. There are no superfluous activities here. Once we learn directly from our customers what they need, we work closely with partners and vendors to build and ship relevant technologies at a low cost. Our employees feel a sense of ownership when they work directly with customers."[41] The opening vignette also clearly demonstrated the commitment of Buckman's CEO to the cultural change involved in become market oriented.

TABLE 4.1 Scanning Focus and Information Priorities by Different Types of Firm Strategy

Product Leaders (Prospectors)

Supply new technology solutions to address their customers' expressed and latent needs; most successful when targeting innovator and early adopter market segments.

Information focus:

- Must place highest priority on understanding customers' unarticulated needs through techniques such as:
  - Observing customers' use of products or services in normal routines
  - Conducting a structured customer visit program
  - Working closely with lead users (customers who face new needs in advance of the majority of the market)
- Must stay ahead or abreast of technological developments

Fast Followers (Analyzers)

Bring out improved or less expensive versions of products introduced by Product Leaders while simultaneously defending core markets and products.

Information focus:

- Should closely monitor customer reactions to Product Leaders' offerings
- Must learn about customers' preferences from Product Leaders' successful and unsuccessful efforts
- Must also monitor competitors' activities, successes, and failures
- Must limit new product introductions to categories that have already shown promise in the marketplace

Operationally Excellent (Low-Cost Defenders)

Provide quality products or services at the lowest overall cost; will generally have less technologically sophisticated product lines than firms pursuing other strategies; role of technology is in process innovations that are critical to their operations.

Information focus:

- Must keep an external focus, with competitors serving as a benchmark against which prices, costs, and performance can be compared

Customer Intimate (Differentiated Defenders)

Focus on maintaining position with established—early and late majority—markets; value proposition based on a nuanced understanding of customers.

Information focus:

- Must be skilled at segmenting the early and late majority markets to identify customer segments that value superior quality and service
- Must closely monitor customer satisfaction
- Must identify opportunities to increase share of customer's wallet
- Must analyze reasons for customer defections
- Must assess customer profitability

Source: Slater, Stanley, Tomas Hult, and Eric Olson, "On the Importance of Matching Strategic Behavior and Target Market Selection to Business Strategy in High-Tech Markets," *Journal of the Academy of Marketing Science* 35 (March 2007), pp. 5–17.

DECENTRALIZED ORGANIZATIONAL STRUCTURE Market-oriented behaviors thrive in an organization that is *decentralized,* with fluid job responsibilities and extensive lateral communication processes. Members of these organizations recognize their interdependence and are willing to cooperate and share market intelligence to sustain the effectiveness of the organization. Effective information sharing in a market-oriented organization demands that bureaucratic constraints on behavior and information flow be dismantled. The uncertainty in high-tech markets requires frequency and informality in communication patterns among organizational units for effective intelligence dissemination. This decentralization is consistent with the market-focused organizational structure discussed previously in Chapter 2.

MARKET-BASED COMPENSATION SYSTEM   By rewarding employees for generating and sharing market intelligence and for achieving high levels of customer satisfaction and customer loyalty, a market-based reward system is the single organizational factor that has the greatest impact on a market orientation.[42] Market-oriented firms place less emphasis on short-term sales and profit goals than their more financially or internally focused competitors.

At its heart, creating a market-oriented organizational culture within a firm requires a major transformation, characterized by four stages: initiation, reconstitution, institutionalization, and maintenance.[43] Essentially, members of the company recognize a threat. To address the threat, a group of empowered managers creates a coalition to mobilize the larger organization; they create a process that reconnects the company's personnel with its customers. Through this process, personnel build common experiences and perspectives and build a consensus for more formal organizational changes to solidify the cultural shift. These formal changes then sustain the new orientation.

Effectively harnessing the power of cross-functional teams is one of the key dimensions of a firm's market orientation. Not only do cross-functional teams reflect the company's shared values about how decisions are made, but they also play a key role in how the process of innovation unfolds in the company. In the high-tech arena, close collaboration between marketing personnel and R&D personnel is especially vital. These are the topics this chapter will now address.

CROSS-FUNCTIONAL INTERACTION: NEW PRODUCT DEVELOPMENT TEAMS AND MARKETING–R&D INTERACTION

A firm's degree of market orientation is determined, in part, by the level of cross-functional integration and quality of its interdepartmental relations.[44] For example, in addition to the marketing–R&D relationship, also important are the integration between marketing and manufacturing,[45] logistics,[46] and even finance.[47] Marketing is considered a **boundary-spanning activity**—meaning that for the firm to make effective decisions, marketing personnel must interact cross-functionally with their counterparts *internally* in other departments (R&D, manufacturing, operations, etc.) as well with partners, customers, retailers, and other stakeholders *external* to the firm. The process of innovation itself, or the new product development process, is a cross-functional activity.[48] Therefore, this section of the chapter explores cross-functional teamwork in the new product development process in high-tech firms, with a particular focus later in this section on interactions between marketing and R&D personnel. (Relationships with constituents outside the firm's organizational boundaries are the topic of Chapter 5.)

Cross-Functional Teamwork in Product Development

Most technology companies organize new product development through **cross-functional product development teams**, which include individuals from different functional areas (e.g., marketing, R&D, manufacturing, engineering, purchasing). In order to maximize new product success—creating products that deliver superior customer value and organizational performance—all functional areas in the organization must be closely integrated. For example, Eric Kim, executive vice president of Global Marketing at Samsung Electronics from 1999 to 2004, achieved significant results by building bridges between his Global Marketing Operations office and the various business units designing and manufacturing products in different geographic areas (see Chapter 12). In addition, Apple Computer's many innovative products in recent years—such as the iMac in 1998, the PowerBook G4 Titanium in 2001, the iPod in 2003, and the iPhone in 2007—have been developed by many different teams that work on new products. For Apple, one of the most critical of these teams, early on,

---

BOX 4.2

Assessing the Degree of Cross-Functional Integration

Below is a set of questions to assess the level of cross-functional integration. A manager should rate his or her business on each item using the following scale:

Strongly Disagree	Disagree Moderately	Disagree Slightly	Agree Slightly	Agree Moderately	Strongly Agree
−3	−2	−1	1	2	3

After rating each item, sum the scores. Negative scores indicate much room for improvement, low positive scores indicate the need for selective or incremental improvement, while high scores indicate a strong capability. High scores are relatively rare.

- The activities of functional units are tightly coordinated to ensure better use of our market knowledge.
- Functions such as R&D, marketing, and manufacturing are tightly integrated in cross-functional teams in the product development process.
- R&D and marketing and other functions regularly share market information about customers, technologies, and competitors.
- There is a high level of cooperation and coordination among functional units in setting the goals and priorities for the organization to ensure effective response to market conditions.
- Top management promotes communication and cooperation among R&D, marketing, and manufacturing in marketing information acquisition and use.
- People from marketing, R&D, and other functions play important roles in major strategic market decisions.

Source: Adapted from Narver, John C., and Stanley F. Slater, "The Effect of Market Orientation on Business Profitability," *Journal of Marketing* 54 (October 1990), pp. 20–35.

---

is the hardware engineering and design team.[49] These teams follow a five-phase development process including a rigorous conceptual stage where the goals of a product are hashed out. Steve Jobs, Apple cofounder and CEO, is a frequent attendee at these meetings, providing a clear product strategy and motivation. Apple's cross-functional process delivers a time-to-market in about 12 to 18 months.

Box 4.2 provides a way for managers to assess the level of cross-functional integration in their companies.

Based on a 10-year study of more than 700 new-product teams, 95% of cross-functional teams that exhibit all five of the following characteristics produced either successful or "blockbuster" products—defined as those that not only met or exceeded the company's goals for sales and profit, but that also met or exceeded customer expectations *and* received awards from either the company, the industry, or prestigious national publications such as *BusinessWeek* for their companies:[50]

1. ***Commitment of senior management:*** Blockbuster teams had the full cooperation of the highest level of management. Senior managers were either intimately involved with virtually every aspect of the process, or they made it clear that they completely backed the project, and then gave the team the authority it needed to proceed.
2. ***Clear and stable vision:*** Blockbuster teams stayed on course by establishing "project pillars" early on—specific, immutable goals for the product that the team must deliver.
3. ***Improvisation:*** Blockbuster teams did not follow a structured, linear path to market. Instead they were flexible, trying all kinds of ideas and iterations in rapid succession until they developed a prototype that clicked with their customers.

4. *Information exchange:* Blockbuster teams did not limit their information exchange to formal meetings. They shared knowledge in dozens of small ways—from coffee klatches to video conferencing to streaming in and out of a room covered in Post-it Notes to hundreds of e-mails.

5. *Collaboration under pressure:* Blockbuster teams focused on goals and objectives as opposed to interpersonal differences. They built coherent teams, but they were not especially concerned about building friendships or even insisting that everyone like each other.

Importantly, consistent with the contingency theory of high-tech marketing, greater interfunctional coordination facilitates the development of radical innovations.[51] It allows reinterpretation of one another's perspectives in linking information to the firm's competencies, developing new solutions and recombining existing competencies to generate breakthrough ideas.

WHAT DETERMINES THE EFFECTIVENESS OF CROSS-FUNCTIONAL TEAMS? As Figure 4.5 shows, many factors affect the performance of these complicated team interactions.[52] Product innovativeness is negatively related to product quality, reinforcing the common observation that some breakthrough products are "buggy" (have flaws or operating problems) in the beginning. Customer influence results in higher product quality, as does the developer firm's quality orientation. Encouraging risk taking and the monitoring of team progress by senior management both result in greater product innovativeness. The remainder of this discussion, however, will focus on three key factors: communication, team orientation, and reward systems.

Communication, Use of Information, and Conflict Management. The effectiveness of cross-functional teams is based on the degree to which members interact, communicate, and coordinate with one another to collect and use market information, to synthesize, integrate, and apply

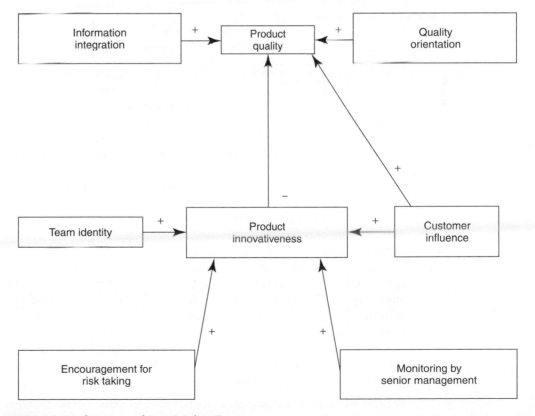

FIGURE 4.5 Performance of New Product Teams

knowledge to make decisions.[53] Lateral communication deepens knowledge flows across functional boundaries, builds trust, and harnesses the diverse functional perspectives in the use of market information, which in turn allows cross-fertilization of ideas, enhancing product quality.

Moreover, effective cross-functional teams that address threats and opportunities reduce the degree of confrontational (dysfunctional) conflict between different functions. Members of such teams learn the language of both technologists and marketers, and later act as "translators" for others in their home functions. This reduces the language barriers across functions in the organization and the perceived conflict.[54] As noted previously, organizational functions must simultaneously cooperate and compete.[55] Constructive conflict can promote learning and knowledge sharing, and improve performance.

Team Orientation. A **team-oriented organization** relies on team effort to get work done; it places value on working cooperatively toward common goals—goals for which all employees feel mutually accountable. The development of an overall team identity and teams' encouragement of risk taking results in greater product innovativeness. A team-oriented organization exhibits the following characteristics:[56]

1. Leaders who have a clear set of values and demonstrate them in every action—challenging employees with high standards, communicating optimism about future goals, and providing meaning for the task at hand
2. Confidence that other team members will do quality work, meet commitments, and treat others with respect
3. A reward system that weights organizational performance more than individual performance, so that all functions share the rewards when the organization is successful

Reward Systems. Because of its impact on both individual and group behavior, the nature of a reward system has a large impact on team performance. Two important managerial issues with regard to rewarding teams are how to distribute rewards among team members and what criteria to use as the basis for team rewards. There are two approaches:

1. *Reward the team as a group.* Here the reward could be distributed to all team members *equally,* or individuals could *get different amounts based on their position* or status in the organization. When an individual's contribution to team performance is easy to evaluate, position-based rewards are positively related to team member satisfaction; however, when individual performance is difficult to evaluate, equal rewards are negatively related to team member satisfaction.[57] Taken together, if individual performance is difficult to evaluate, neither equal nor position-based rewards work satisfactorily. Thus, technology companies should invest in monitoring systems and procedures to better measure individual performance within teams.
2. *Reward individual performance of team members.* Individual team members could be given process-based or outcome-based rewards. *Process-based rewards* are tied to procedures or behaviors (e.g., completion of certain phases of the product development project), whereas *outcome-based rewards* are tied to bottom-line profitability of the project. Team performance can be measured by *external* criteria (e.g., speed to market, adherence to budget and schedule, innovation, product quality, market performance) or by *internal* criteria (e.g., self-rated team performance and team member satisfaction). For long and complex product development projects, outcome-based rewards are positively related to team performance, while process-based rewards are negatively related to team performance.[58] Employee stock option plans are used by technology companies as a specific way of providing outcome-based rewards for long-term individual and team performance.

Because of its vital role in high-tech marketing success, the next section provides a closer look at one particularly important cross-functional dynamic: R&D–marketing interaction.

R&D–Marketing Interaction

Although the origins of many high-tech firms come from engineering breakthroughs, a successful transition to being market driven requires that input from marketing also be heard and responded to. Managing **R&D–marketing interactions**, or the cross-functional collaboration between personnel from research and development and from marketing, is vital for the firm to succeed as the market evolves from a supply-side, innovation-driven market to a demand-side, market-driven market.[59] Although R&D tends to play a stronger, more influential role in breakthrough products and marketing tends to play a greater role in more incremental products, for either R&D or marketing to overlook the vital perspective that the other brings contributes to failure. Therefore, successful high-tech firms overcome the internal bias that elevates engineering personnel. Firms with tightly integrated teams of marketing and engineering personnel that link R&D and marketing efforts[60] experience higher rates of new product success.[61]

Moreover, the need for effective marketing–R&D integration increases as the complexity of the innovation and environmental uncertainty increase.[62] The lack of such integration, as evidenced by rivalry between the two functions, severely reduces the use of relevant marketing information by R&D. Such neglect of useful information can be based on misperceptions about the quality of information supplied by marketing.[63]

So, high-tech firms have much to gain by understanding how to manage the dynamic between developers and marketers. The following pages present an overview of best practices and strategies for collaborative marketing–R&D interaction. First, as shown in Figure 4.6, the nature of the marketing–R&D interaction must be effectively matched to the nature of the innovation (i.e., breakthrough versus incremental). Second, firms must understand the barriers to interaction. Finally, strategies to overcome barriers to interaction and, more specifically, to enhance communication between marketing and engineering personnel are addressed.

NATURE OF R&D–MARKETING INTERACTION: BREAKTHROUGH VERSUS INCREMENTAL INNOVATIONS R&D–marketing interaction is most important during the early stages of a product

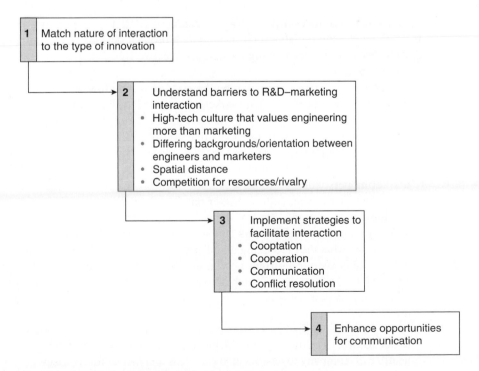

FIGURE 4.6 Steps to Effective R&D–Marketing Interaction

development project,[64] often referred to as the "fuzzy front end" of product development.[65] Marketing should have the knowledge about customer preferences and competitive offerings that is crucial for resolving design and positioning issues. And, because R&D has primary responsibility for translating technology into a design that addresses customers' needs, its knowledge is also crucial for resolving design issues. As a project moves into the production and commercialization stages, product engineering and operations supercede R&D in importance, while the focus for marketing shifts from product definition to development of a marketing program.[66]

Although effective R&D–marketing interaction is critical for both breakthrough and incremental products, consistent with the contingency model of high-technology marketing, the nature of the interaction must be tailored to the type of innovation. Greater R&D–marketing integration in technical activities is required for breakthrough products than for incremental products.[67] Because the possibilities for the application of the new technology tend to be either nonobvious or very numerous, engineering should not proceed in isolation from market-related feedback. Hence, for breakthrough products, the early interface efforts between R&D and marketing should address what industry the company should compete in, what the conceivable market opportunities are, and what the market development priorities are.[68] The cross-functional interaction helps to determine desired product features and to assess engineering feasibility.

For incremental innovations, R&D again should actively participate in the market-planning process, especially in setting objectives. R&D can ensure that marketing does not lose sight of R&D's vision for the product. Marketers can offer parameters for the engineers' efforts. Through give and take, the team members can agree on the target market, priorities, expectations, and timing. Moreover, R&D efforts don't end once selling begins; engineers should continue to help with brochures, research, pricing, sales promotion, trade shows, and customer visits. R&D–marketing interaction for incremental innovations establishes the direction for commercialization, helps to design marketing plans, and implements the launch of the new product.[69] Similarly, marketing should participate during the precommercialization period, bringing the voice of customer and marketplace into the development process.

BARRIERS TO R&D–MARKETING COLLABORATION   Despite the crucial need for effective R&D–marketing collaboration in high-tech product development, the reality of the interaction is very different. Understanding barriers to effective R&D–marketing collaboration is crucial to removing them.

One important barrier can be the culture of a high-tech firm that respects and values engineering knowledge more than marketing knowledge. This dominant engineering culture can create a core rigidity in many high-tech firms[70] and manifests itself in several ways in the organizational culture—in the jokes that are told (see Table 4.2 for a few examples), the seating arrangements that are made, the job titles (and pay) that are given, and how marketing decisions get made. This type of technology-driven culture translates into a lack of regard and respect for marketing personnel.

The engineering culture further reinforces the dominant role of engineering and justifies not listening to marketing.[71] One of the most frequent reasons engineers give for not using market research or input from consumers is that "customers don't know what they want" or that "marketers don't know what they're talking about" (because they lack technical expertise). For example: "Marketing wants everything right now at no cost—they have no concept of feasibility—they want a $5,000 Cadillac tomorrow."[72] Obviously, it is very difficult for marketing personnel to be effective when the prevalent view is that "engineering does its thing and then marketing helps get it out the door."

As a case in point, a health care company adopted a new information system to manage customer records. This new system included customized software from the vendor, which allowed the health care company to customize the screens and menus for its caseworkers, based on certain treatment criteria and standards. Unfortunately, in tailoring the software to its needs, the health care

TABLE 4.2 Jokes Engineers Make about Marketing

Marketing research: When Marketing goes down to Engineering to see what they're working on.

Marketing: What you do when your products aren't selling.

A software manager, a hardware manager, and a marketing manager are driving to a meeting when a tire blows. They get out of the car and look at the problem.

The software manager says, "I can't do anything about this—it's a hardware problem."

The hardware manager says, "Maybe if we turned the car off and on again, it would fix itself."

The marketing manager says, "Hey, 75% of it is working—let's ship it!"

What high-tech salespeople say and what they mean by it:

> *All new:* Parts not interchangeable with previous design.
> *Field-tested:* Manufacturer lacks test equipment.
> *Revolutionary:* It's different from our competitors' products.
> *Breakthrough:* We finally figured out a way to sell it.
> *Futuristic:* No other reason why it looks the way it does.
> *Distinctive:* A different shape and color from the others.
> *Redesigned:* Previous faults corrected (we hope).
> *Customer service across the country:* You can return it from most airports.
> *Unprecedented performance:* Nothing we ever had before worked *this* way.

company encountered a series of bugs and other issues (i.e., security concerns) that required extensive rework on the part of the software vendor. In order to communicate their needs and concerns, the health care company personnel had talked to either the salesperson or a customer support person, who, in turn, expressed the customer's concerns to the software engineers. The filtering and recommunicating of the customer's needs had resulted in several misunderstandings between the engineers and the customer. Despite this, the customer was not allowed to communicate directly with the software engineers. And why not? The reason given by the software vendor was that "its engineers' time was too costly to be spent in talking to customers!"

High-level executives from firms with a dominant engineering culture typically come out of engineering; they are expected to develop a business orientation and understanding of customers as they advance. Indeed, the disdain for marketing can manifest itself in engineering taking on many tasks traditionally thought of as marketing, such as competitor analysis or product management. This can also contribute to a bias reflected in the semantics of the company, with some high-tech companies saying they don't have a "marketing department." Although this may be technically true, someone has to perform necessary marketing activities (e.g., attending trade shows, developing and managing a company's websites, preparing the brochures and spec sheets); more importantly, someone has to make the important strategic marketing decisions, including its choices of market segments, product development projects, and acquisitions and alliances. Although these decisions may not be called "marketing decisions" and they may not be made by "marketing personnel," they are, nonetheless, in the marketing domain. So, the subtle bias can be perpetuated by not recognizing marketing for what it is—a philosophy of decision making that prioritizes organizational initiatives based on customers, competitors, and value creation in the marketplace.

A second barrier to effective R&D–marketing interaction is that marketing and R&D personnel tend to differ on a number of dimensions such as education, goals, needs, and motivation. More importantly, they tend to differ with regard to their values, as illustrated in Table 4.3.[73] Of course, these differences are generalities and do not apply in all situations. However, they do illustrate why marketing and R&D may have a difficult time understanding each others' goals and decision processes.

Third, spatial dynamics contribute to the problem: Locating R&D and marketing in different parts of the building—or even worse, different geographical locations—further contributes to the

TABLE 4.3 Different Orientations between R&D and Marketing Personnel

	R&D	Marketing
Time Orientation	Long	Short
Projects Preferred	Breakthrough	Incremental
Ambiguity Tolerance	Low	High
Department Structure	Informal	Moderately formal
Bureaucratic Orientation	Less	More
Orientation to Others	Permissive	Permissive
Professional Loyalty	Profession	Firm
Professional Orientation	Science	Market

lack of collaboration. Yet, studies have found that increased R&D–marketing interaction increases the likelihood of a new product development project's success[74]—and such interaction is greatly enhanced by co-location.

Fourth, rivalry may exist between R&D and marketing if they perceive each other as competitors for scarce resources. If rivalry between the two functions is high, trust and perception of information quality is likely to be low. In this situation, the use of information that marketing provides to engineering is low.[75]

In light of these barriers, how can companies structure R&D–marketing interaction to enhance effectiveness?

ACHIEVING R&D–MARKETING INTEGRATION Many firms specify a number of formal systems and processes for marketing groups to provide information to engineering groups.[76] For example, during specific review phases of the new product development project, marketing offers input into the product requirements document and an understanding of the trade-offs involved in the myriad attributes being considered. In addition, during the annual planning process, marketing groups forecast revenue and profit for their market segments and indicate products and programs needed to achieve their goals. Finally, marketing communicates with engineering via the sales forecasting system.

Although these formal mechanisms exist for R&D–marketing interaction, they tend not to be the primary means by which such interaction occurs; indeed, they offer little in the way of marketing influence in the process. Such formal measures are often a chimera, erecting a façade of R&D–marketing interaction but doing little to make it productive. The question is, how can R&D–marketing interaction be made productive for the firm? Useful techniques fall into the categories of cooptation, cooperation, communication, and constructive conflict resolution.

Cooptation. The objective of cooptation is to merge the interests of R&D and marketing. In particular, effective high-tech marketers use the following methods to gain the support of R&D in the product development process:[77]

- Effective marketers use informal networks and build bridges to engineering. They know the right people; they are in close physical and organizational proximity to engineering.
- Effective marketers understand products and technology, which gives them credibility with engineers. Engineers "don't mind talking to marketers" if they know what they're talking about.
- Effective marketers don't tell others what to do; rather, they ask questions, tell stories, and build consensus across groups.
- Effective marketers form strategic coalitions that include high-level managers who push through changes that engineering resisted. However, there is a cost to this strategy, because it can alienate peers in engineering, so this option is the least preferred and should be saved for the most important issues.

- Effective marketers recognize that minor improvements to the new innovations can be particularly important. "It's the 5% that's uninteresting to the engineering folks that can produce five times the levels of sales you would otherwise have."[78] In such a situation, effective marketers either undertake development themselves or turn to external partners to complete the work.

Cooperation. Based on their extensive review of the research in this area, Griffin and Hauser[79] found that the following cooperative strategies enhance marketing–R&D interaction:

- Co-locate marketing and R&D to overcome the barrier of physical separation and encourage information transfer. To realize the benefits of co-location, it must be complemented with techniques that foster communication and collaboration.
- Move personnel across functions to give them insight into the challenges their counterparts face. Personnel movement blurs distinctions between groups, may decrease market or technological uncertainty, and reduces barriers that stem from differences in values and language.
- Develop informal cross-functional networks to encourage open communication and provide contact across the functions within the team.
- Create a structure and systems that foster cooperation. This would include clear responsibilities and performance standards, decentralized decision making, tolerance of failure that is not the result of poor planning or execution, and joint reward systems.

Even with these strategies, the reality still remains that marketing and R&D have very different worldviews, which can result in misunderstanding and conflicts in goals and solutions.[80] Experts say the best way to overcome these differences is enhanced communication.

Communication. One commonly studied factor in enhancing R&D–marketing interaction is communication. Many argue that simply increasing the frequency of interaction between marketing and R&D will help improve understanding and harmony between the two functions, will increase their ability to cope with complex, dynamic environments, and will lead to greater product success. Indeed, the use of market information provided by a marketing manager to nonmarketing managers (including not only R&D managers but also manufacturing and finance managers) requires a minimum threshold of interactions. For those who are more quantitative in orientation and appreciate the specificity of the research on this topic, one study defines this threshold at approximately 125 interactions in a three-month period.[81] However, too frequent communications (beyond 525 interactions in a three-month period) can hurt perceptions of information quality. The key lesson is that increasing frequency of cross-functional communication may be warranted if the minimum threshold has not been reached, but once it is reached, increased frequency may not always prove beneficial.

The same study[82] also found that when disseminated through *formal* means (those that are planned and verifiable), market information is used to a greater extent than when disseminated through *informal* channels. Although informal channels—which are spontaneous and unplanned communication interactions—may provide greater openness and clarification opportunities, formal interactions are more credible.

Two other factors that affect the nature of communication between marketing and R&D are (1) the presence of information-sharing norms within the organization and (2) the degree to which engineering goals are integrated with marketing goals (in which case the marketing manager's goal attainment depends on actions of the engineering counterpart, and vice versa).[83]

Information-sharing norms indicate organizational expectations about the exchange of information between functions and can promote increased communication behaviors. However, the degree to which such norms really influence marketers' communication with engineering depends on how strongly marketing managers identify with the marketing function. Marketing managers who identify more strongly with the organization as a whole (than with the marketing function) communicate more bidirectionally when information-sharing norms are stressed.[84]

Similarly, *integrated goals* suggest that the organization's needs are superordinate to the goals of the individual functional units; such goals can promote increased collaboration and cooperation. Marketing managers who identify very strongly with the marketing function are more likely to communicate more frequently and more bidirectionally when integrated goals are stressed. However, marketing managers who identify very strongly with their functional area of marketing also resort to coercion to ensure that the engineering contacts comply with marketing's perspective on organizational issues. In this case, integrated goals increase the coerciveness of influence attempts by high-functional-identification managers.[85]

So, how should managers in high-tech firms use this information to facilitate R&D–marketing interaction? First, managers need to establish policies to encourage information-sharing norms when marketing managers identify more strongly with the organization as a whole. Second, managers should set integrated goals for marketing and R&D when marketing managers identify more strongly with the marketing function specifically. Note, however, the increased risk of coercive influence attempts in this latter case.[86]

Constructive Conflict Resolution.   When marketers perceive a greater frequency of interaction, they also perceive more conflict with R&D personnel; however, that conflict doesn't necessarily result in a less effective relationship.[87] In fact, although many studies have corroborated the need for close R&D–marketing interaction for new product development success, when relationships are too close, the desire to retain harmony precludes alternative viewpoints from emerging.[88] When such groupthink takes over, adverse opinions are not expressed and potential problems are not addressed, resulting in lower product performance. A real question, then, is how to structure the R&D–marketing collaboration for frequent interaction with the simultaneous ability to challenge others' viewpoints. Having formalized roles within the group, with certain members assigned the role of devil's advocate, may be helpful.

In addition, different conflict-handling strategies have differential effects. For example, two conflict-handling strategies—*integrating* (high concern for both self and others by bringing all issues into the open and fully exploring all parties' concerns) and *accommodating* (low concern for self and high concern for others by trying to satisfy others' expectations)—are both associated with constructive conflict and higher levels of innovation performance.[89] On the other hand, two other conflict-handling strategies—*forcing* (high concern for self and low concern for others by treating conflict as a win-lose situation) and *avoiding* (low concern for both self and others by smoothing over conflicts or ignoring them)—are associated with destructive conflict and lower levels of innovation performance. *Compromise* (moderate concern for self and others) is associated with lower levels of destructive conflict.

Managers can assess how effectively personnel from R&D and marketing interact by employing the questions in Box 4.3.

A CAVEAT   Although cooptation, cooperation, communication, and constructive conflict resolution can enhance the quality of marketing and R&D interactions, such interactions must be solidly grounded in an *understanding of customer needs and wants*. As Figure 4.7 shows, marketing and engineering may think they are doing a good job of interacting, but in reality they have not accurately captured or conveyed the customer's needs in the process.

Because of the inherent differences between marketing and R&D, this chapter offers two Technology Expert's View from the Trenches features: one from a worldwide product marketing manager at Hewlett-Packard, who provides insights about effectively advocating for customer needs and working collaboratively with cross-functional teams, and the other from an engineer who speaks to the respective roles of marketing and engineering. In addition, because many high-tech companies struggle with the cultural transformation from being engineering-driven to becoming market-oriented, Appendix 4.A delivers a senior executive's perspective on what it takes to make a successful transition.

BOX 4.3

Assessing the Degree of Marketing–R&D Integration

Below is a set of questions to assess the degree of integration between the marketing and R&D functions. A manager should rate his or her business on each item using the following scale:

Strongly Disagree	Disagree Moderately	Disagree Slightly	Agree Slightly	Agree Moderately	Strongly Agree
−3	−2	−1	1	2	3

After rating each item, sum the scores. Negative scores indicate much room for improvement, low positive scores indicate the need for selective or incremental improvement, while high scores indicate a strong capability. High scores are relatively rare.

Marketing and R&D:

- Coordinate work activities smoothly
- Have senior managers who share values and perspectives
- Enhance each other's performance
- Cooperate with each other
- Have compatible goals and objectives
- Agree on the priorities for each function
- Respect each other's capabilities

Source: Maltz, E., and A. Kohli, "Reducing Marketing's Conflict with Other Functions: The Differential Effects of Integrating Mechanisms," *Journal of the Academy of Marketing Science* 28 (September 2000), pp. 479–492. Copyright © 2000. With kind permission of Springer Science and Business Media.

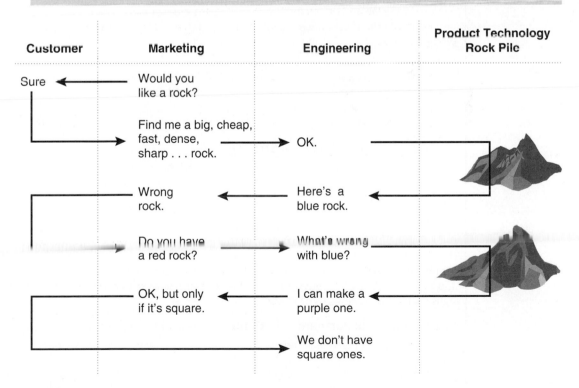

FIGURE 4.7   The Rock Game

Source: Reprinted with the permission of Storage Technology Corporation, Louisville, Colorado. Copyright © StoreTek. Reprinted with permission.

TECHNOLOGY EXPERT'S VIEW FROM THE TRENCHES

Product Management: Serving as a Customer Advocate and Collaborating with Cross-Functional Teams

BY TRINA FINLEY
Business Development Manager, Hewlett-Packard, Boise, Idaho

As a worldwide product manager for five years at Hewlett-Packard, I have had the privilege of working on numerous exciting products, including the HP LaserJet 4100 MFP, HP LaserJet 9000 MFP, HP LaserJet 4345 MFP, HP Color LaserJet 4730 MFP, and other commercial multifunction printers. The functionality and pricing of multifunction printers (MFPs) varies depending on the segment (e.g., consumer vs. small to medium-sized business vs. enterprise). Commercial/enterprise products typically include printing, copying, scanning, and faxing functions and range in price between approximately $1,000 and $20,000.

Worldwide product managers at Hewlett-Packard are responsible for product and pricing decisions, including managing products from definition of objectives through obsolescence. I was fortunate to be well regarded as highly effective in the worldwide product management role, and new worldwide product managers frequently ask me the "secret of my success." Rather than approaching my role as an adversarial position to R&D, product engineering, manufacturing, finance, and even the regional product management teams, I've viewed these relationships as partnerships. In order to get the right product at the right price, I first needed the regional marketing and sales teams to equip me with field-based information to become the customer expert. Then, empowered by their knowledge and serving as their advocates, I can work with the lab and other cross-functional teams more effectively.

Indeed, the most critical components to effective product management that maximize the product manager's contribution to the organization are:

- Understanding the overall business strategy and using it as a guideline to drive decision making
- Being an advocate or champion for customer and market needs and expectations
- Coming to the table as the customer expert with lab and cross-functional partners
- Building credibility and effective working relationships across functions
- Proactively communicating with partners and being responsive to their requests

Understanding the Business Strategy

The most important starting point for the product manager is to have a clear understanding of the overall strategy, which is typically set by business management. For example, if the strategy for the business is to gain market share, then that strategy can drive one set of decisions by the product manager, while a strategy to increase margins and revenue might drive a different set of decisions. For example, I was product manager for the HP LaserJet 4100 MFP, one of our first products to push Hewlett-Packard into the commercial MFP business. The strategy at that time was to gain market share to get a foothold in the market. As a result, we made the decision to lower our price significantly, trimming margins but increasing demand, shipments, and ultimately market share. Knowing and understanding the strategy enables the product manager to make day-to-day marketing decisions that support the goals of the business.

Cross-Functional Partnering with R&D Labs and Other Teams

A worldwide product manager needs to come to the table with engineering teams from the lab and other cross-functional partners as the customer expert. Bringing customer insights to the table allows the product manager to gain credibility and be effective in his or her role—to more effectively negotiate product design and/or product enhancements, as well as changes in features, accessories, pricing, and so forth. To gain customer knowledge, the product manager can participate in customer visits, focus groups, and customer research, and—in my opinion, most importantly—should actively engage and listen to the sales teams interfacing with customers daily. An effective product manager communicates insights and customer expectations with lab partners and cross-functional teams (e.g., technical marketing, quality assurance, manufacturing, planning, finance, information engineering, product completion) and quickly responds when they submit requests. When the product manager makes a request of the lab or cross-functional team, he or she needs to have thorough customer knowledge and a solid business case behind it.

For example, one of the most challenging times in my role was when we first entered the MFP market with the HP LaserJet 4100 MFP and HP LaserJet 9000 MFP. We learned a great deal about where our products were falling short after launch, and we were making every effort to be responsive to customer feedback to meet customer expectations, preserve HP's brand identity, and drive more sales. As a result, I worked closely with the regional marketing and sales teams. We learned that we were missing key features that customers expected. I put together a business case including revenue and share impacts due to lost sales because of missing features, a list of evaluations in potential jeopardy, and the competitive landscape. Providing this solid business case backed with specific customer knowledge allowed the lab and current product engineering teams to make easy cost–benefit comparisons and quickly move to add the features via firmware revisions. The result was that we began to see increased shipments and grow market share in a space where we had not been able to get a foothold in the past.

The Product Manager's Role as a Customer Champion

Worldwide product managers need to position themselves as champions for customer needs to gain credibility and mutual respect with the regional product management and sales teams. The regional product management teams at Hewlett-Packard typically manage the go-to-market and distribution plans in their respective regions (North America, Latin America, Europe/Middle East/Africa, and Asia Pacific) and interface with and support the sales teams directly on a daily basis. By listening to, filtering, and prioritizing feedback and requests from these teams, worldwide product managers can actively and effectively advocate for product design and development decisions that are based on customer needs and expectations. For example, when we launched the HP LaserJet 9040 and 9050 MFPs, we were able to make significant improvements in the products because we listened to the regional product management and sales teams, understood and prioritized their requests based on business impact and customer satisfaction, and more effectively negotiated product design and development decisions where it mattered most to our channel partners and customers.

Proactively Communicating with Partners and Being Responsive to Their Requests

Effective worldwide product managers maintain open communication with the regional product management and sales teams—proactively providing product updates and changes, reporting and properly positioning the issues, actively providing meaningful updates on region requests, and maintaining regular contact. A good worldwide product manager ensures that these partners are equipped with the necessary information and content to be proactive and successful in their roles and partnerships with sales representatives, technical sales consultants, channel partners, and customers.

Likewise, being responsive to requests by these partners creates a trusting and credible relationship. If the regional product management teams and sales teams view the worldwide product manager as an advocate, they will openly share information with this person that helps the worldwide product manager gain more customer knowledge.

Building Credibility and Effective Working Relationships across Functions

By building credible relationships with the labs and cross-functional teams in the global business unit and the product management and sales teams in the regions, the worldwide product manager is a critical conduit for the flow of information needed for product success. The product manager understands and advocates customer and market needs and expectations with the lab and cross-functional teams designing new products or updating current products. He or she also ensures that product management and sales teams in the regions are adequately prepared and equipped to launch new products or support existing products. Successful product managers are able to launch or revise products that meet and exceed customer and market expectations.

TECHNOLOGY EXPERT'S VIEW FROM THE TRENCHES

Engineering/Marketing Collaboration
JENNIFER LONGSTAFF, BS, MS, MBA
Technical Marketing Engineer 1997–2008, Xilinx Inc., Longmont, Colorado

The relationship between engineering and marketing is often a source of conflict, even in the most cooperative, consensus-driven, proactive companies.

In a market-driven, product-based company, *marketing's role* is typically to define a product based on input from customers and its knowledge of market needs. Marketing delivers the finished product, and then solicits customer feedback on how well the product has met needed solutions and has fared against competition. This feedback gets rolled into the next cycle of product definition as the product is improved. In this way, marketing can assure a consistent long-term evolution of an improving product.

Engineering's role is to use the marketing definition and create an efficient implementation of the product within a development time frame. Part of engineering's product design must include extensibility for inclusion of future enhancements within the long-term vision. Engineering must make it known whether marketing's product definitions are feasible, and then work with marketing to revise the product definitions if not.

In the high-tech environment, new products are often new-to-the-world innovations with no existing models to compare with customer experiences. Marketing receives general requirements from customers, often expressed vaguely as "I need the ability to do something, but I'm not sure how I want that ability to look." Marketing may need to envision a new paradigm to deliver compelling new benefits that satisfy the customer. Engineering can assist marketing with development of the new paradigm by creating a prototype to examine and assess if it will meet customer need. However, engineering may view the prototyping stage as a time-consuming step and a rationale for marketing indecision. Therefore, it is very important for marketing to do sufficient research and study feasibility before requesting prototype work from engineering. Ideally, the prototype will be close to a final product and may need only minor changes. Marketing (as the driver of new products) must be responsible and must use engineering's time expeditiously.

In the most common scenario, engineers serve both as *developers* (R&D for new products and feature enhancements) as well as *maintainers* of the existing installed product in the field. Marketing, in contact with customers, is also aware of the issues in the field with existing products; however, engineering takes the brunt of the "resource hit" because maintenance is difficult to schedule, and it is a distraction from new development. Marketing may not always understand the disruptive nature of the maintenance distraction and the tension it creates with new development.

Engineering and marketing personnel would probably all agree with the foregoing concepts and roles. It would seem that both groups, in following their role definitions, should be able to work together productively. Yet, there seems to be an abundance of conflicts and problems between these two groups during the development process. Inherent problems seem to be:

- *Each group assumes it can do the other's job better.* Engineering questions marketing's ability to define products, especially when marketing can't immediately finalize a definition and requests prototyping. Engineers don't acknowledge marketing's experience with customers or marketing's ability with long-term "crystal ball" prediction of future product needs. Marketing doesn't understand why engineering takes so long to develop products, and why they react against late definition changes. Each group often sees only a superficial view of the other's tasks.

- *Communication between the groups is often poor.* Engineers, even those who have moved into marketing, may be introverted and reluctant to take the initiative to contact others, even when definition is needed in order for work to continue. This communication is especially difficult when negative feedback must be given. But the marketing–engineering dynamic is based on a foundation of open and frequent communication. Without this communication, the interaction degenerates quickly, usually due to simple misunderstandings.

The solutions are obvious! Each group must acknowledge the other's skills. This is easier to do if each group consists of competent and hardworking individuals. If either group fails to deliver its commitments in a competent manner, it will cause unnecessary work for the other and will increase feelings of resentment. And, again, the key solution to all misunderstandings and conflicts is *frequent, open communication.*

Summary

This chapter focused on two aspects of a company's values and beliefs about the role of market-based information and cross-functional teamwork in decision making. A firm's market orientation and its degree of cross-functional interaction are part and parcel of its strategy development process and its culture of innovativeness.

At the most basic level, market-oriented firms excel at generating, sharing, and using information about the market (customers, competitors, collaborators, technology, trends, etc.) to make decisions that lead to the creation of superior product and service quality, and, in turn, superior performance. However, some firms are more proactive than others, both in the type of intelligence they generate and in the action they take. Proactive businesses seek intelligence not only about current customer needs and competitor threats, but also about latent and future customer needs and competitive threats. Proactive businesses are more likely to develop market-driving strategies that actively seek to redefine the structure of the market and/or introduce an innovative value proposition that enables the firm to reduce, or even avoid, competition. A market-driving strategy, while riskier than a responsive strategy, is more likely to lead to a breakthrough position. To successfully transition to becoming a market-oriented firm, there must be strong leadership from top management, decision-making that takes place as close to the customer as possible, and a reward system that recognizes the importance of information sharing and coordinated action.

Cross-functional interactions are a key driver in diffusing market and customer knowledge among all members of a project team in high-tech firms. These interactions ensure that an understanding of market needs, desires, and behavior underlies the early stages of product development—thus generating technological applications that are valued by customers.

In addition, integration between marketing and R&D is especially important. The diffusion of customer and market knowledge to the engineering team is enhanced by allowing them to have direct and repeated contact with customers and other outside sources of information. High-tech companies must attend carefully to the ways in which their internal practices either facilitate or stymie R&D–marketing interaction.

Discussion Questions

Chapter Concepts

1. What is a market orientation? What is its role in a high-tech company?
2. What is the effect of a firm's market orientation on its performance outcomes? Is this effect stronger or weaker for high-tech environments? What are the implications for a high-tech company?
3. What is meant by intelligence generation? What types of information are gathered? What techniques

might be used? What is the corollary (budget implications) of gathering market-based information?

4. Compare and contrast a responsive with a proactive market orientation. Be sure to discuss the pros and cons of each, as well as the types of approaches used to gather information for each orientation.

5. What is market driving? What are its pros and cons for high-tech companies?

6. Explain how the contingency theory of high-technology marketing relates to a firm's responsive versus proactive market orientation.

7. Describe how a firm accomplishes the second dimension of a market orientation, intelligence dissemination. What are the hurdles in freely sharing information in the organization and how might they be overcome?

8. What is meant by intelligence integration? How does a firm accomplish this?

9. How does a firm create an organizational memory to retain its knowledge? Why is this important?

10. Compare and contrast the tools a company might use to preserve its explicit versus its tacit knowledge.

11. What is a knowledge management system? What various technologies might be used for knowledge management purposes?

12. What is the role of cross-functional teams in implementing market-oriented decisions? Why is this role important? What makes it difficult to achieve?

13. What are the facilitating conditions for a firm to successfully create a market-oriented organizational culture? Explain each one with sufficient detail to communicate its meaning and importance.

14. For each of the four strategy types (Product Leader, Fast Follower, Operationally Excellent, and Cus-

tomer Intimate), describe what the information/scanning focus should be.

15. What is the relationship between a firm's market orientation and its use of cross-functional teams?

16. Why is marketing considered a boundary-spanning activity?

17. What is the purpose of cross-functional product development teams?

18. What are the five characteristics of cross-functional teams that produce blockbuster products?

19. How does the contingency theory of high-tech marketing apply to the use of cross-functional teams?

20. Elaborate on the many factors that affect the effectiveness of cross-functional teams.

21. What is a team orientation? What are its characteristics?

22. Which of the reward systems are best for technology companies? Why?

23. What is R&D–marketing interaction? Why is it important in high-tech companies?

24. How does the contingency theory of high-tech marketing apply to R&D–marketing interaction?

25. What are the barriers to effective R&D–marketing interaction? Describe the many ways a high-tech company's culture elevates technical personnel over marketing personnel. Also, explain the different backgrounds and values that the two groups have.

26. Why are formal mechanisms for R&D–marketing interaction generally less effective than other mechanisms?

27. What are the four ways to facilitate effective interaction between R&D and marketing personnel? Elaborate on each of the four ways with supporting detail.

Application Questions

1. Which of the characteristics of a market-oriented company does the Buckman Labs vignette best describe? Are any major pieces missing?

2. Does any information in the vignette suggest that Buckman Labs has achieved a proactive level of market orientation?

3. Explain how a market orientation enables a business to achieve superior quality and develop innovative new products.

4. Is cross-functional teamwork necessary to achieving an optimal level of market orientation? Why or why not?

5. In 2007, Michael Dell returned as CEO of his company after leaving that post to serve as chairman. But after he left the day-to-day operations at Dell in 2004, the company faced increasing pressure from rival Hewlett-Packard. Since returning as CEO, Michael Dell said a major focus was to "bring back the cus-

tomer-centricity" into the company, adding that he thought "we lost some of that customer focus." Based on the opening vignette, the Technology Expert's View from the Trenches features, and the material in the chapter, what five to seven pieces of advice would you give him?

Glossary

boundary-spanning activity   A corporate activity that spans both internal boundaries (across various departments and divisions) and external boundaries (across various firms and partners—suppliers, customers, channel members, etc.).

cross-functional product development team   A group of individuals from different functional areas (e.g., marketing, R&D, manufacturing, engineering, purchasing) who work together to develop new products.

knowledge management   The range of practices a company uses to identify, create, represent, and distribute knowledge—often relying on technology and software such as knowledge bases, expert systems, knowledge repositories, corporate intranets and extranets, wikis, and document management/content management software programs.

market driving   An extreme type of proactive market orientation in which a firm actively redefines the structure of the market, and/or introduces an innovative value proposition, allowing it to reduce, or even avoid, competition.

marketing–R&D interaction   *See* R&D– marketing interaction.

market intelligence   Useful information about trends and stakeholders in the market including current and future customer needs, competitors' capabilities and strate-

gies, and emerging technologies both inside and outside the industry.

market orientation   A business philosophy that emphasizes the use of market-based information ("intelligence") to guide decision making.

proactive market orientation   An organizational culture whereby the firm actively gathers forward-looking market information that seeks out customers' latent and future needs, identifies market opportunities that are not evident to competitors, and anticipates (future) threats; strongly associated with the development of radical innovations.

R&D–marketing interaction   Cross-functional teamwork between personnel from research and development and from marketing, to share information and collaborate on decision making.

responsive market orientation   An organizational culture that primarily responds to current intelligence, focusing on expressed customers' needs and existing competitive threats and often suffering from myopia and running the risk of being disrupted; tends to be associated with the development of incremental innovations.

team-oriented organization   A company that relies on team effort and places value on working cooperatively toward common goals that all employees feel mutually accountable for.

Notes

1. Based on Buckman, Robert, "Knowledge Transfer," *Montague Institute Review,* January 1997, available online at www.montague.com/review/buckman.html; Rifkin, Glenn, "Buckman Labs Is Nothing but Net," *Fast Company,* June 1996, available online at www.fastcompany.com/magazine/03/buckman.html; Rumizen, Melissie, "How Buckman Laboratories' Shared Knowledge Sparked a Chain Reaction," *Journal for Quality and Participation* 21 (July–August 1998), pp. 45–53; and Rumizen, Melissie, "Leader of the Pack," *Inside Knowledge,* October 1, 2003, available online at www.ikmagazine.com/xq/asp/sid.0/articleid.C5B9FF64-2EDF-4DA6-87D8-15E0B3516D57/ qx/display.htm.

2. Shapiro, Benson, "What the Hell Is 'Market Oriented'?" *Harvard Business Review* 66 (November–December 1988), pp. 119–125.

3. Drucker, Peter, "The Discipline of Innovation," *Harvard Business Review* 63 (May–June 1985), pp. 67–73.

4. Im, Subin, and John P. Workman Jr., "Market Orientation, Creativity, and New Product Performance in High-Technology Firms," *Journal of Marketing* 68 (April 2004), pp. 114–132.

5. Deshpande, Rohit, John Farley, and Frederick E. Webster Jr., "Corporate Culture, Customer Orientation, and Innovativeness in Japanese Firms: A Quadrad Analysis," *Journal of Marketing* 57 (January 1993), pp. 23–37.

6. For excellent reviews and syntheses of this work, see Ellis, Paul, "Market Orientation and Performance: A Meta-Analysis and Cross-National Comparisons," *Journal of Management Studies* 43 (July 2006), pp. 1089–1107; and Kirca, A. H., S. Jayachandran, and W. O. Bearden, "Market Orientation: A Meta-Analytic Review and Assessment of Its Antecedents and Impact on Performance," *Journal of Marketing* 69 (April 2005), pp. 24–41.

7. Jaworski, B., and A. Kohli, "Market Orientation: Antecedents and Consequences," *Journal of Marketing* 57 (July 1993), pp. 53–70; and Slater, Stanley, and John Narver, "Does Competitive Environment Moderate the Market Orientation–Performance Relationship?" *Journal of Marketing* 58 (January 1994), pp. 46–55.

8. Homburg, Christian, and Christian Pflesser, "A Multiple-Layer Model of Market-Oriented Organizational Culture: Measurement Issues and Performance Outcomes," *Journal of Marketing Research* 37 (November 2000), pp. 449–462.

9. Dutta, Shantanu, Om Narasimhan, and Surendra Rajiv, "Success in High-Technology Markets: Is Marketing Capability Critical?" *Marketing Science* 18 (April 1999), pp. 547–568.

10. Baker, William E., and James M. Sinkula, "Market Orientation and the New Product Paradox," *Journal of the Academy of Marketing Science* 33 (Fall 2005), pp. 461–475.

11. Atuahene-Gima, Kwaku, "Resolving the Capability–Rigidity Paradox in New Product Innovation," *Journal of Marketing* 69 (October 2005), pp. 61–83.

12. Slater, Stanley, and John Narver, "Customer-Led and Market-Oriented: Let's Not Confuse the Two," *Strategic Management Journal* 19 (October 1998), pp. 1001–1006; and Narver, John, Stanley Slater, and Douglas MacLachlan, "Responsive and Proactive Market Orientation and New Product Success," *Journal of Product Innovation Management* 21 (September 2004), pp. 334–347.

13. Byrnes, Nanette, and Paul Judge, "Internet Anxiety," *Business Week,* June 28, 1999, p. 84.

14. Christensen, Clayton, and Joseph Bower, "Customer Power, Strategic Investment, and the Failure of Leading Firms," *Strategic Management Journal* 17 (March 1996), pp. 197–218.

15. See LifeStraw at a Glance, online at www.vestergaard-frandsen.com/lifestraw.htm. Our thanks to Ashley Grant for helping with this example.

16. Jaworski, Bernard, Ajay Kohil, and Arvind Sahay, "Market-Driven versus Driving Markets," *Journal of the* *Academy of Marketing Science* 28 (December 2000), pp. 45–54; and Kumar, N., Lisa Scheer, and Philip Kotler, "From Market Driven to Market Driving," *European Management Journal* 18 (April 2000), pp. 129–142.

17. Hills, Stacey Barlow, and Shikhar Sarin, "From Market-Driven to Market Driving: An Alternative Paradigm for Marketing in High-Technology Industries," *Journal of Marketing Theory and Practice* 11 (Summer 2003), pp. 13–24.

18. Linzmayer, O., *Apple Confidential 2.0* (San Francisco: No Starch Press, 2004).

19. Leonard-Barton, Dorothy, Edith Wilson, and John Doyle, "Commercializing Technology: Understanding User Needs," in *Business Marketing Strategy,* eds. V. K. Rangan et al. (Chicago: Irwin, 1995), pp. 281–305.

20. Atuahene-Gima, Kwaku, Stanley Slater, and Eric Olson, "The Contingent Value of Responsive and Proactive Market Orientations for Product Innovation," *Journal of Product Innovation Management* 22 (November 2005), pp. 464–482; Chandy, Rajesh, and Gerard Tellis, "Organizing for Radical Product Innovation: The Overlooked Role of Willingness to Cannibalize," *Journal of Marketing Research* 35 (November 1998), pp. 474–487; and Narver, Slater, and MacLachlan, "Market Orientation, Innovativeness, and New Product Success."

21. See "What Lew Platt Said . . . and Why," www.spiralpartners .com/ know_prac.html.

22. Cited in Tjaden, Gary, "Measuring the Information Age Business," Technology Analysis and Strategic Management, 8 (September 1996), pp. 233–246.

23. Dess, Gregory G., and Nancy K. Origer, "Environment, Structure, and Consensus in Strategy Formulation: A Conceptual Integration," *Academy of Management Review* 12 (April 1987), pp. 313–330.

24. Dess, Gregory G., "Consensus on Strategy Formulation and Organizational Performance: Competitors in a Fragmented Industry," *Strategic Management Journal* 8 (May–June 1987), pp. 259–277.

25. Grove, Andrew S., *Only the Paranoid Survive* (New York: Doubleday Business, 1999), p. 105.

26. Fisher, A., "Retain Your Brains," *Fortune,* July 24, 2006, pp. 49–50.

27. Moorman, Christine, and Anne Miner, "The Impact of Organizational Memory on New Product Performance and Creativity," *Journal of Marketing Research* 34 (February 1997), pp. 91–107.

28. Inkpen, A. C., and A. Dinur, "Knowledge Management Processes and International Joint Ventures," *Organization Science* 9 (July–August 1998), pp. 454–468.

29. Cited in Fisher, A., "How to Battle the Coming Brain Drain," *Fortune,* March 21, 2005, pp. 121–128.

30. Noble, Charles, Rajiv Sinha, and Ajith Kumar, "Market Orientation and Alternative Strategic Orientations: A Longitudinal Assessment of Performance Implica-

tions," *Journal of Marketing* 66 (October 2002), pp. 25–39.

31. Cited in Murray, Alan, "After the Revolt, Creating a New CEO," *Wall Street Journal,* May 5, 2007, p. A1.

32. Kohli, A., and B. Jaworski, "Market Orientation: The Construct, Research Propositions, and Managerial Implications," *Journal of Marketing* 54 (April 1990), pp. 1–18.

33. Gupta, Ashok K., S. P. Raj, and David L. Wilemon, "A Model for Studying R&D–Marketing Interface in the Product Innovation Process," *Journal of Marketing* 50 (April 1986), pp. 7–17.

34. Cooper, Robert G., and Elko J. Kleinschmidt, "New Product Processes at Leading Industrial Firms," *Industrial Marketing Management* 20 (May 1991), pp. 137–148.

35. Atuahene-Gima, "Resolving the Capability-Rigidity Paradox in New Product Innovation."

36. Leonard-Barton, Dorothy, "Core Capabilities and Core Rigidities: A Paradox in Managing New Product Development," *Strategic Management Journal* 13 (Summer 1992, Special Issue), pp. 111–125.

37. This example is cited in Leonard-Barton, "Core Capabilities and Core Rigidities."

38. Luo, Xueming, Rebecca Slotegraaf, and Xing Pan, "Cross-Functional 'Coopetition': The Simultaneous Role of Cooperation and Competition within Firms," *Journal of Marketing* 70 (April 2006), pp. 67–80.

39. Day, George, and Prakash Nedungadi, "Managerial Representations of Competitive Advantage," *Journal of Marketing* 58 (April 1994), pp. 31–44.

40. Slater, Stanley, Tomas Hult, and Eric Olson, "On the Importance of Matching Strategic Behavior and Target Market Selection to Business Strategy in High-Tech Markets," *Journal of the Academy of Marketing Science* 35 (March 2007), pp. 5–17.

41. Mears, Jennifer, "Customer Focus Keeps Dell Productive," *Network World,* April 21, 2003, p. 51.

42. Jaworski, B., and A. Kohli, "Market Orientation: Antecedents and Consequences," *Journal of Marketing* 57 (July 1993), pp. 53–70.

43. Gebhardt, Gary, Gregory Carpenter, and John Sherry, "Creating a Market Orientation: A Longitudinal, Multifirm, Grounded Analysis of Cultural Transformation," *Journal of Marketing* 70 (October 2006), pp. 37–55.

44. Narver, John C., and Stanley F. Slater, "The Effect of Market Orientation on Business Profitability," *Journal of Marketing* 54 (October 1990), pp. 20–35; Jaworski, B., and A. Kohli, "Market Orientation: Antecedents and Consequences," *Journal of Marketing* 57 (July 1993), pp. 53–70; and Kohli, A., and B. Jaworski, "Market Orientation: The Construct, Research Propositions, and Managerial Implications," *Journal of Marketing* 54 (April 1990), pp. 1–18.

45. Maltz, Elliot, and Ajay K. Kohli, "Market Intelligence Dissemination across Functional Boundaries," *Journal of Marketing Research* 33 (February 1996), pp. 47–62; and Ruekert, Robert W., and Orville Walker, "Marketing's Interaction with Other Functional Units: A Conceptual Framework and Empirical Evidence," *Journal of Marketing* 51 (January 1987), pp. 1–19.

46. Rinehart, Lloyd M., M. Bixby Cooper, and George D. Wagenheim, "Furthering the Integration of Marketing and Logistics through Customer Service in the Channel," *Journal of the Academy of Marketing Science* 17 (December 1989), pp. 63–71.

47. Srivastava, Rajendra K, Tasadduq A. Shervani, and Liam Fahey, "Market-Based Assets and Shareholder Value: A Framework for Analysis," *Journal of Marketing* 62 (January 1998), pp. 2–18.

48. De Luca, Luigi M., and Kwaku Atuahene-Gima, "Market Knowledge Dimensions and Cross-Functional Collaboration: Examining the Different Routes to Product Innovation Performance," *Journal of Marketing* 71 (January 2007), pp. 95–113.

49. Tam, Pui-Wing, "Designing Duo Helps Shape Apple's Fortunes," *Wall Street Journal,* July 18, 2001, page B1.

50. Lynn, Gary, and Richard Reilly, *Blockbusters: The Five Keys to Developing GREAT New Products* (New York: Harper Business, 2002).

51. Atuahene-Gima, "Resolving the Capability-Rigidity Paradox in New Product Innovation."

52. Sethi, Rajesh, "New Product Quality and Product Development Teams," *Journal of Marketing* 64 (April 2000), pp. 1–14; and Sethi, Rajesh, "Cross-Functional Product Development Teams, Creativity, and the Innovativeness of New Consumer Products," *Journal of Marketing Research* 38 (February 2001), pp. 73–86.

53. Menon, Ajay, Bernard Jaworski, and Ajay Kohli, "Product Quality: Impact of Interdepartmental Interactions." *Journal of the Academy of Marketing Science* 25 (June 1997), pp. 187–200; Olson, Eric, Orville Walker, and Robert Ruekert, "Organizing for Effective New Product Development: The Moderating Role of Product Innovativeness," *Journal of Marketing* 59 (January 1995), pp. 48–62.

54. Maltz, Elliot, and Ajay Kohli, "Reducing Marketing's Conflict with Other Functions: The Differential Effect of Integrating Mechanisms," *Journal of the Academy of Marketing Science* 28 (September 2000), pp. 479–492.

55. Luo, Xueming, Rebecca Slotegraaf, and Xing Pan, "Cross-Functional 'Coopetition.'"

56. Slater, Stanley, "Learning How to Be Innovative," *Business Strategy Review,* 19 (December 2008), pp. 46–51.

57. Sarin, Shikhar, and Vijay Mahajan, "The Effect of Reward Structures on Cross-Functional Product Development Teams," *Journal of Marketing* 65 (April 2001), pp. 35–54.

58. Ibid.

59. Shanklin, William, and John Ryans, "Organizing for High-Tech Marketing," *Harvard Business Review* 62 (November–December 1984), pp. 164–171.

60. Song, X. Michael, and Mark E. Parry, "A Cross-National Comparative Study of New Product Development Processes: Japan and the United States," *Journal of Marketing* 61 (April 1997), pp. 1–18.

61. Gatignon, Hubert, and Jean-Marc Xuereb, "Strategic Orientation of the Firm and New Product Performance," *Journal of Marketing Research* 34 (February 1997), pp. 77–90; and Olson, Walker, and Ruekert, "Organizing for Effective New Product Development."

62. Gupta, Ashok K., S. P. Raj, and David Wilemon, "A Model for Studying R&D–Marketing Interface in the Product Innovation Process," *Journal of Marketing* 50 (April 1986), pp. 7–17.

63. Maltz, Elliot, William E. Souder, and Ajith Kumar, "Influencing R&D/Marketing Integration and the Use of Marketing Information by R&D Managers: Intended and Unintended Effects of Managerial Actions," *Journal of Business Research* 53 (July 2001), pp. 69–82.

64. Olson, E. M., O. C. Walker, R. W. Ruekert, and J. M. Bonner, "Patterns of Cooperation during New Product Development among Marketing, Operations, and R&D: Implications for Project Performance," *Journal of Product Innovation Management* 18 (July 2001), pp. 258–271.

65. Reid, Susan E., and Ulrike de Brentani, "The Fuzzy Front End of New Product Development for Discontinuous Innovations: A Theoretical Model," *Journal of Product Innovation Management* 21 (May 2004), pp. 170–185; and Kim, Jongbai, and David Wilson, "Focusing the Fuzzy Front-End in New Product Development," *R & D Management* 32 (September 2002), pp. 269–280.

66. Workman, John, "Marketing's Limited Role in New Product Development in One Computer Systems Firm," *Journal of Marketing Research* 30 (November 1993), pp. 405–421.

67. Song, X. Michael, and JinHong Xie, "The Effect of R&D–Manufacturing–Marketing Integration on New Product Performance in Japanese and U.S. Firms: A Contingency Perspective," Report Summary #96-117 (Cambridge, MA: Marketing Science Institute, 1996).

68. Shanklin and Ryans, "Organizing for High-Tech Marketing."

69. Song and Xie, The Effect of R&D–Manufacturing–Marketing Integration."

70. Leonard-Barton, "Core Capabilities and Core Rigidities."

71. Kunda, Gideon, *Engineering Culture: Culture and Control in a High-Tech Organization* (Philadelphia: Temple University Press, 1992).

72. Workman, "Marketing's Limited Role in New Product Development in One Computer Systems Firm."

73. Griffin, Abbie, and John Hauser, "Integrating R&D and Marketing: A Review and Analysis of the Literature," *Journal of Product Innovation Management* 13 (October 1996), pp. 191–215.

74. Ayers, Doug, Robert Dahlstrom, and Steven J. Skinner, "An Exploratory Investigation of Organizational Antecedents to New Product Success," *Journal of Marketing Research* 34 (February 1997), pp. 107–116.

75. Maltz, Souder, and Kumar, "Influencing R&D/Marketing Integration and the Use of Market Information by R&D Managers."

76. Workman, "Marketing's Limited Role in New Product Development in One Computer Systems Firm."

77. Ibid.

78. Ibid.

79. Griffin and Hauser, "Integrating R&D and Marketing."

80. Gupta, Raj, and Wilemon, "A Model for Studying R&D–Marketing Interface in the Product Innovation Process"; and Griffin, Abbie, and John R. Hauser, "Patterns of Communication among Marketing, Engineering, and Manufacturing—A Comparison between Two New Product Teams," *Management Science* 38 (March 1992), pp. 360–373.

81. Maltz, Elliot, and Ajay K. Kohli, "Market Intelligence Dissemination across Functional Boundaries," *Journal of Marketing Research* 33 (February 1996), pp. 47–61.

82. Ibid.

83. Fisher, Robert J., Elliot Maltz, and Bernard J. Jaworski, "Enhancing Communication between Marketing and Engineering: The Moderating Role of Relative Functional Identification," *Journal of Marketing* 61 (July 1997), pp. 54–70.

84. Ibid.

85. Ibid.

86. Ibid.

87. Ruekert, Robert, and Orville Walker, "Interactions between Marketing and R&D Departments in Implementing Different Strategies," *Strategic Management Journal* 8 (May–June 1987), pp. 233–248.

88. Ayers, Dahlstrom, and Skinner, "An Exploratory Investigation of Organizational Antecedents to New Product Success."

89. Song, Michael, Barbara Dyer, and R. Jeffrey Thieme, "Conflict Management and Innovation Performance: An Integrated Contingency Perspective," *Journal of the Academy of Marketing Science* 34 (Summer 2006), pp. 341–356.

APPENDIX 4.A

What Does It Take to Become Customer Focused and Market Oriented?

By Jack Trautman, Former Senior Vice President and General Manager,
Automated Test Group, Agilent Technologies
Loveland, Colorado

The challenge in any high-tech organization is creating a customer-centered, or market-oriented, culture. On the surface, it sounds easy—listen to the customer and invent what the customer wants. In practice, it's a complex task to choose a target customer, accurately anticipate the target's exact needs, translate those needs into product features, and generate not only the right product, but also the total customer experience that captures the customer's money today and their loyalty tomorrow—and do it better than competitors.

For a lot of high-tech start-ups, the founders are the customers. Their bright new ideas are spawned from their own user experiences. Through their engineering expertise, they solve their own problems with the products they invent, and in so doing solve the same problem for all of their customers.

Such was the case in the early days of Hewlett-Packard. Engineers developed test and measurement equipment for engineers. When they wanted to know what the customer wanted, they asked their fellow engineers. This "Next Bench Syndrome" served HP well for many years.

But then the business got more complex. HP diversified into computers, printers, and a host of other product types, and even spun off Agilent Technologies to give the test and measurement parts of its business more focus. Suddenly the customer was not sitting at the next bench. In the absence of real customers, engineers will still do their best to anticipate needs. But their instincts are more likely grounded in technical prowess than in customer requirements. "I'll give them the most technically elegant solution, and surely they'll love it!" Maybe.

Enter marketing. Through market research, customer segmentation, competitive analysis, focus groups, a customer-relationship management process, and a host of other techniques, a surrogate for the customer can be constructed. These are well-understood processes that can yield great results.

But an effective product marketing team is not enough to ensure success. It takes teamwork between marketing and engineering to truly bring the customer into the center of the product generation process. This teamwork comes through mutual respect, working together side by side, and an open sharing of ideas. Enlightened companies are co-locating product marketing and engineering teams beside each other to maximize interaction and a free-flowing exchange of ideas. Insights into ways to best satisfy customers will not occur in a "throw it over the wall" environment between marketing and engineering.

Finally, and most importantly, it takes management leadership to commit the entire organization to a customer-centered culture. Managers must budget the time and money to get product marketing and product development engineers out with customers so they can experience the customer's environment firsthand. They must identify early adopter customers who help complete the final product feature set through hands-on testing of early prototypes. Capturing the customer's insights early in the prototyping phase can often make the difference between the ultimate success or failure of a product or service. It is vital to involve the customer in each of the key phases of product generation: design, develop, test, and launch.

Beyond marketing and engineering, every department and every person must embrace the total customer experience. Managers must be willing to invest in all employees to give them the tools and skills to get close to the customer. And then those managers must insist that every decision in the company be made with the customer in mind.

It's hard work to keep a focus on the customer. But the payoffs are big: increased customer satisfaction and loyalty, high employee motivation and morale, and increased shareholder value.

CHAPTER

5

Partnerships, Alliances, and Customer Relationships

Partnerships at Apple[1]

The Apple iPhone launched in the United States in June 2007 through an exclusive partnership with AT&T. Lauded as a breakthrough innovation in the mobile phone industry, it relied on a constellation of partners during development. Moreover, it radically altered the dynamics of the mobile telecom industry, with a major rethinking of how to work with partners.

The wireless carriers, or "network operators," in the $11 billion mobile phone industry in the United States typically treated phone manufacturers "like serfs." They used the power they derived from owning the networks—which controlled access to the end customers—to dictate what phones would get made and what their pricing and feature sets would be. Indeed, the phones themselves were viewed as promotional freebies, to snare new customers and lock them in to the carrier's network.

The ingenuity of the Apple iPhone, however, upset that power imbalance—highlighting that creative innovations, even expensive ones, could attract new customers. Indeed, in the initial months after its release, AT&T found that 40% of iPhone buyers were new customers to its network. It forced America's big phone companies, notoriously conservative, to reexamine their orientation toward the status quo. Moreover, other phone manufacturers realized that reinventing their innovation processes—creating products to delight and wow customers rather than to meet with the carriers' approval—could change the balance of power between the handset manufacturers and the carriers. This would ultimately affect how the revenue in the industry was divided between the various players in the market.

The history of the development of the iPhone began as early as 2002.

The Background

In 2002, Steve Jobs, Apple's CEO, was beginning to realize that rather than carrying multiple devices (phones, MP3 players, BlackBerrys, etc.), people would prefer just one device to perform multiple functions. And, as cell phones and mobile e-mail devices delivered more features, they would challenge the Apple iPod's dominance as a music player.

In 2004, Apple's iPod business had become important—but at the same time, more vulnerable than ever. Even though the iPod represented 16% of Apple's revenue, its long-term position as the dominant music device seemed at risk: 3G phones were gaining popularity, WiFi phones would be coming soon, the price of storage was dropping, and music stores to rival iTunes were proliferating. Further, trends were indicating that mobile phones would increasingly be the device of choice for customers to access the Internet—and for marketers to reach customers.[2]

So, even while denying that he was doing so (Apple never preannounced products), Jobs began work on a mobile phone product in the summer of 2004. His initial partner (in an effort to

(continued)

bypass the carriers) was Motorola, which had shown its ability to create breakthrough innovations with the RAZR. Apple would concentrate on developing the music software while Motorola and Cingular, working in the established pattern of manufacturer–carrier relationship, would work on hardware details. Yet, the three companies did not see eye-to-eye on much of anything: How would songs get into the phone? How much music could be stored? How would each company's name be displayed? And the first prototype? "Ugly" was the verdict.[3] Other drawbacks were that it couldn't download music directly and it held only 100 songs. Despite these problems, and a sense that consumers were an afterthought in the development process, this phone, the ROKR, was introduced in the fall of 2005.

Yet, Jobs realized even as early as February 2005 that the ROKR was a dud, and began separate talks with Cingular (which AT&T acquired in December 2006) for a revolutionary "phone of the future." Jobs indicated that Apple was so committed to the project that it would consider an exclusive partnership with the carrier—but, if that did not work, it would buy wireless minutes wholesale and become a de facto carrier itself.

Cingular's views on this project were mixed. On the one hand, the company realized that the future of wireless telecom would need to be based on a data (rather than voice) business model. As consumers used their phones to download music and video and to surf the Web at WiFi speeds, margins would be higher. In addition, Cingular realized that with a saturated wireless market, new customers would be found, not by bribing them with a cheap handset, but by enticing them with must-have devices with unique functionality and design. Who better to partner with than Apple? And, if Cingular didn't partner with Apple, it was clear that someone else would. On the other hand, the terms and control that Jobs was asking for were unprecedented.

While Cingular hemmed and hawed, Jobs met with executives from Verizon—who promptly turned him down! They didn't want to turn the Verizon network into a "mere conduit for content" rather than the source of that content.

Ultimately, Cingular bet that a product like the iPhone would result in more data traffic that would make up for lost revenue on content deals. (Their prediction was borne out: Within the first several months of release, the iPhone tripled the volume of data traffic on the AT&T network in major cities such as New York and San Francisco.) So, they made a deal: Cingular (AT&T) received five years of exclusivity on its network, 10% of iPhone sales in AT&T stores, and a slice of Apple's iTunes revenues. Apple received AT&T's commitment to create a new feature for its network (visual voice mail), a major overhaul of the time-consuming in-store sign-up process, $10 per month from every iPhone customer's AT&T bill, and—importantly—unprecedented power with complete control over all design, manufacturing, and marketing of the iPhone. Jobs, a notorious control freak,[4] wasn't about to let a group of "suits" from the networks (which he actually referred to as "orifices" through which content is sent to end users) tell him how to design his phone.

The Product Development Effort

Apple spent at least $150 million to build the initial iPhone. (Because the focus of this vignette is on partnerships, the stress at Apple surrounding the product development process, including screaming matches in hallways, engineers frazzled from all-night work sessions, slammed doors that broke the handles, is not mentioned. See the Wired.com source below for details.) In order to make the vision of the iPhone a reality, many new technologies were required:[5]

- The data network that could fully support a handheld Internet device (delivering Web pages and images on the smaller mobile device's display) came from AT&T. Components that connected local WiFi networks to the main data network came from Infineon, Skyworks, RF Micro Devices, and Marvell Technology.
- A new operating system to manage complicated networking and graphics had to be designed (Apple's OS X would be too much for a cell phone chip to handle). Jobs refused to use

someone else's software, such as Linux, which had already been rewritten for use on mobile phones. Ultimately, Apple decided to revise the Mac OS X software for the iPhone.

- The main microprocessor chip with technology licensed from British firm ARM Holdings was manufactured by Samsung, as was the NAND flash memory for storage.
- Power management chips came from NXP Semiconductor, Texas Instruments, and Linear Technology. Display chips came from National Semiconductor, Broadcom, and NXP. The imaging chip for the camera came from Idaho-based Micron Technology, and the accelerometer (chip that senses physical motion and changes the orientation of pictures accordingly) was made by STMicroelectronics.
- New hardware considerations such as antenna design, radio-frequency radiation, and network compatibility had to be addressed. The antenna design alone required millions of dollars of investments in special robot-equipped testing rooms. The radiation problem required models of human heads, complete with goo to simulate brain density, to measure effects. To simulate network performance, Apple invested in nearly a dozen simulators for millions of dollars each.
- The touch screen, which required glass rather than hard plastic like on the iPod, was from a little-known German company, Balda.
- It is speculated (without confirmation) that the entire iPhone was outsourced for manufacture to Foxconn, a division of the Hon Hai Precision Industry in Taiwan.[6]

Thus, the Apple iPhone had an international network of supplier partners all sworn to secrecy. Other procedures were followed as well, to keep details of the project confidential. Product development teams were split up and scattered across the company's campus in Cupertino, California. Hardware and software teams were kept in the dark about the product they were actually working on. When Apple executives visited Cingular, they registered as employees of Infineon, the company making the phone's transmitter. By the time the product was announced in January 2007, only about 30 or so of the most senior people in the company new about it.

The Outcomes

Following its U.S. launch through AT&T in June 2007, the iPhone was launched in November 2007 in the UK through O2 and in France through Orange, with a plan to distribute the iPhone globally over time. The financial analysis on the iPhone suggests that Apple nets roughly $80 for every $399 iPhone it sells, plus $240 for the $10 per month (for 24 months) from each AT&T contract. Some 50% of iPhone purchasers replaced another phone: 24% replaced a Motorola RAZR, 14% replaced a Windows mobile device, and 20% replaced a BlackBerry or Palm.[7]

Although not a perfect product (e.g., the initial price of $599 was too high; the first version ran only on AT&T's rather slow EDGE network; users couldn't perform e-mail searches or record video; the browser wouldn't run programs written in Java or Flash), consumers got an amazing device—the functionality of which will become even more powerful since Apple released a developer's kit so anyone can write programs for the iPhone. Moreover, based on the success of the iPhone, industry experts predict that consumers will have more influence over what gets built in the future.

Carriers are realizing that they need to find innovative devices that offer functionality to attract users. They must partner with device makers to do so. Indeed, Verizon, sometimes called "one of the most intransigent carriers," said that it will open up its network for any compatible handset, as did AT&T. This will allow applications to work on any device over any network, and will add flexibility and functionality to the mobile Internet experience.

Application developers gain more opportunities as wireless carriers begin to show signs of abandoning their proprietary networks approach. For example, T-Mobile and Sprint have signed on as partners with Google's Android mobile operating system for mobile application development.

(continued)

The net result is that the iPhone "cracked open the carrier-centric structure of the wireless industry and unlocked a host of benefits for consumers, developers, manufacturers—and potentially the carriers themselves."[8] Rather than being a carrier's nightmare—that is, giving all the power to consumers, developers, and manufacturers while turning wireless networks into "dumb pipes"—the iPhone fostered more innovation and in turn made carriers' networks more valuable, not less. Consumers will spend more time on devices (e.g., doing mobile banking), and thus on networks, generating larger bills and more revenue for companies.

Even though Apple locks the iPhone by default so it works exclusively with its authorized carrier in each country, hackers wasted little time in unlocking the iPhone to make it work on other carriers' networks. The exception is in France, where the law requires Apple to sell unlocked iPhones. The availability of unlocked iPhones and the pent-up demand in countries like China, where there is no official distributor yet, have created an international gray market problem for Apple and its carrier partners; they lose service revenue when consumers buy unlocked iPhones and use them with unauthorized carriers. Estimates in February 2008 indicated that roughly 1.7 million more iPhones were sold than had signed service contracts with approved carriers—resulting in a possible loss of revenue to Apple of $240–400 million![9] This is a major problem that Apple needs to address.

The business landscape in which high-tech companies operate is frequently referred to as a **knowledge-based economy**,[10] in which competitive advantage is more likely to be found in intangible resources (such as know-how, expertise, and intellectual property) rather than in more traditional means of production (tangible economic resources). Table 5.1 lists some key characteristics of a knowledge-based economy.

TABLE 5.1 Some Characteristics of a Knowledge-Based Economy

- Unlike most resources that deplete when used, information and knowledge can be shared, and actually grow through application; commonly stated as "the economics of abundance rather than scarcity."
- The effect of geographic location is diminished in some economic activities; using appropriate technology and methods, virtual marketplaces and virtual organizations that offer benefits of speed, agility, round-the-clock operation, and global reach can be created.
- The effect of geographic location is reinforced in other cases, say, by the creation of industry clusters and universities and research centers of worldwide excellence.
- Laws, barriers, and taxes are difficult to apply on solely a national basis. Knowledge and information "leak" to where demand is highest and the barriers are lowest.
- Knowledge when locked into systems or processes has higher inherent value than when it can "walk out the door" in people's heads.
- Knowledge-enhanced products or services can command price premiums over comparable products with low embedded knowledge or knowledge intensity.
- The same information or knowledge can have vastly different value to different people, or even to the same person at different times. Therefore, pricing and value depends heavily on context.
- Human capital is a key component of value in a knowledge-based company, yet few companies report competency levels in annual reports. In contrast, downsizing is often seen as a positive "cost-cutting" measure.
- Communication is increasingly seen as fundamental to knowledge flows. Social structures, cultural context, and other factors influencing social relations are therefore of fundamental importance to knowledge economies.

Source: http://en.wikipedia.org/wiki/Knowledge_economy, downloaded 12/28/07.

Moreover, globalization and connectivity arising from computer networking, communications technologies, the Internet, enhanced transportation, and so forth have resulted in what journalist Thomas Friedman calls a "flattening" of the world:

> . . . a global, Web-enabled playing field that allows for multiple forms of collaboration—the sharing of knowledge and work—in real time, without regard to geography, distance, or even language.[11]

One key implication of competing effectively in this globalized, knowledge-based economy is that more and more people will have access to the tools of collaboration, which Friedman says will "ensure that the next generation of innovations will come from all over 'Planet Flat.' The scale of the global community that is soon going to be able to participate in all sorts of discovery and innovation is something that world has simply never seen before."[12]

This transition is occurring at an increased velocity in recent years, according to Friedman, due to ten forces that have "flattened" the world.[13] At least five of these forces have very specific implications for partnering and strategic alliances, the focus of this chapter: outsourcing, offshoring, global technology-enabled supply chains (more on this in Chapter 9 as well), "insourcing" (meaning a company takes on additional capabilities within its own boundaries), and open source innovation. (Four of the other forces are due directly to the impacts of technology, including the development of Netscape, the development of software that enabled computer-to-computer interoperability, the ease and availability of online information, and the new forms of digital, mobile, and virtual technology. The tenth force was the tearing down of the Berlin Wall.)

High-tech markets necessitate the use of partnerships and alliances. Because the time-to-market cycle is short and development costs and risks are high, firms often find it faster and more cost-efficient to develop products jointly than alone. For example, BP teamed up with Ford and DaimlerChrysler (among others) in the development of alternative-fuel (hydrogen) vehicles, seeing the race to produce a viable alternative-technology auto as key to their futures. The battle is less between individual automakers, and more between networks—often transcontinental—of partners. A desired outcome arising from synergistic, win-win partnerships is to strengthen both companies against other competitors.

Moreover, most companies lack the requisite technical expertise to develop new high-tech products on their own. Indeed, a complete, end-to-end solution typically entails many different products, some of which may be outside a company's own skill set and found in partnering relationships. And, as noted in Chapter 1, the importance of setting industry standards necessitates collaborating with other companies.

This chapter explores important issues in the formation and management of the partnerships and alliances—referred to as *tie-ups* in many areas of the world—that are vital to success in high-tech markets. The chapter explores a variety of types of partnerships, including those with both upstream suppliers and downstream distribution channel members, partnerships with customers, and even partnerships with competitors. Moreover, many companies outsource some of their firm's internal processes, such as manufacturing, customer service, or even design and innovation. Working carefully with outsource providers brings its own set of opportunities and challenges. Another increasingly important alliance in high-tech contexts is partnerships formed for new product development purposes. Given the increased prevalence and complexity of these alliances, a special section of this chapter is devoted to their unique issues, including learning from partners. Finally, one of the most important types of partnerships a company can have is with its customers. Hence, the final section of this chapter is devoted to customer relationships.

PARTNERSHIPS AND STRATEGIC ALLIANCES

Types of Partnerships

A variety of partnerships can be formed at all levels of the supply chain, as shown in Figure 5.1.

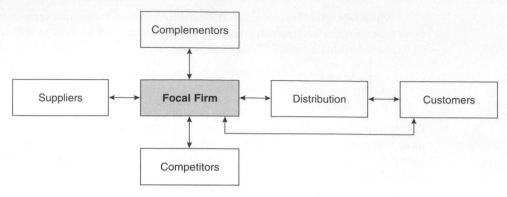

FIGURE 5.1 Partnerships along the Supply Chain

VERTICAL PARTNERSHIPS   Alliances between companies that operate at different levels in the supply chain are called **vertical partnerships**. These are typically referred to as buyer–supplier relationships, meaning that one partner is the customer who is buying goods and services from the upstream partner. For example, vertical partnerships can be found between:

- Suppliers and original equipment manufacturer (OEM) customers
- Outsource (service) providers and business customers
- Manufacturers and distribution channel members
- Companies and their most important end-user customers

Relationships with *suppliers* are often formed to gain efficiencies in accessing parts and materials. For example, a firm may choose to form partnerships with key suppliers whose skills and experiences complement its strengths to develop next-generation technology. Chip manufacturers and computer manufacturers work together to develop next-generation computers. Because chips are a supply used in computers, this is an example of a supplier–OEM vertical relationship. An **original equipment manufacturer (OEM)** is a company that buys components, such as disk drives, from suppliers to integrate via a manufacturing process into a finished product, such as a computer. Early supplier involvement (ESI) is useful in developing innovations in supplies that can help differentiate the customer's product in downstream markets. Collaborative relationships built around common procedures and intensive information sharing mean that the supplier's operations can be more closely fitted to the customer's needs.

Often, vertical partnerships with suppliers are formed when a company chooses to outsource some aspect of its business. The unique considerations involved in outsourcing relationships are addressed later in this chapter.

Another type of vertical partnership is found in relationships with *channel members*. These alliances can be a source of competitive advantage to gain efficiency and effectiveness in accessing downstream markets. Channel partners are key to effectively accessing end-user customers, to implementing marketing programs, and to funneling market information back to the manufacturer. Relationships with channel intermediaries are discussed more fully in Chapter 9.

Other important vertical partnerships are relationships with *customers*—either business customers who use the product in their businesses or consumers who purchase the technology product for their own personal use. Close, long-term relationships with customers are vital in many types of markets, and particularly in high-tech markets. Because of the need to rely on customers for beta test sites and ideas for innovations, firms that have close relationships with customers have a strong source of market-based information. Moreover, long-term relationships with customers generate a long-term revenue stream, allowing a firm to capture the lifetime value of a customer's purchases in a particular product category. Customer relationships are discussed in detail later in this chapter.

HORIZONTAL PARTNERSHIPS   Alliances between firms that operate at the same level of the supply chain are called **horizontal partnerships**—relationships between firms that provide jointly used, complementary products or firms that are competitors.

Complementary alliances are formed between companies that offer different components of the end-to-end solution (product) that is sold to other customers; the members are referred to as *complementors*.[14] For example, Apple's partnerships with car manufacturers to ensure that the car stereo system is compatible with its iPod music devices, and with clothing and apparel companies such as Nike (whose shoes are embedded with chips to send workout data to the iPod and from there to social networking sites to share with a community of like-minded athletes) would be considered complementor relationships. Similarly, when Facebook partners with third-party application providers (say, to build a FunWall or other applications), and when Intel and Microsoft work together to ensure compatibility in chips and operating systems for consumer electronics, they have established complementary alliances. Because both firms are producers of goods that, in turn, are resold to customers or channel intermediaries, this form of horizontal partnering enables each firm to maintain flexibility and focus on its own core competencies; the partnership stimulates demand through synergistic innovation.

Competitive alliances (also referred to as *competitive collaboration*) are formed between firms that typically compete at the same level of the supply chain in some market domain, but choose to join forces to collaborate in some other domain. For example, competitors may join forces to develop next-generation technology, to define standards for new technologies, to provide market access in an area that one firm lacks, or to be a stronger force against a larger competitor. Another name given for this type of collaboration is **co-opetition**,[15] a blend of the words *cooperation* and *competition,* meaning that the firms collaborate in some arenas and compete in others.

Recent research has found that a company's level of cooperation with competitors is related to its financial performance.[16] Companies that have either a very low or a very high competitor alliance intensity (meaning they cooperate very little or very much with competitors in alliances) experience a *lower* return on equity than companies with an intermediate level of competitor alliance intensity. In other words, a company can benefit from forming alliances with competitors; a company's sophistication in its competitor-oriented strategy (e.g., the ability to make sense of and respond to competitor moves) enhances the relationship between competitor alliance intensity and return on equity. However, if a company's objective in the competitive alliance is to "beat" the alliance partner, then competitor alliance intensity has a negative effect on return on equity. In practical terms, companies must be cautious when they cooperate with competitors, but it can be a win-win situation if the partners don't think of it as a zero-sum game.

Companies often establish an *industry consortium,* an industry-wide coalition composed of companies—typically competitors—that have a shared interest in some arena, such as setting industry standards, influencing government regulations, or pursuing international markets. For example, the DVD Forum (founded in 1995 as the DVD Consortium) is an international association of hardware manufacturers, software firms, content providers, and other users of DVDs. Its purpose is to exchange and disseminate ideas and information about the DVD format, its technical capabilities, and innovations/improvements. Founding companies included those traditionally identified as competitors: Hitachi, Pioneer, Sony, and Toshiba, to name a few.[17]

Increasingly, companies in many industries are forming industry-wide coalitions to develop metrics for *sustainability* (the degree to which their products are environmentally friendly with minimal degradation to the environment). For example, the Green Electronics Council's mission is to "inspire and support effective design, manufacture, use and recovery of electronic products to contribute to a healthy, fair and prosperous world."[18] One of its initiatives was the Electronic Product Environmental Assessment Tool (EPEAT) to allow customers in the public and private sectors to evaluate, compare, and select desktop computers, notebooks, and monitors that have been certified to conform to the environmental performance standard for electronic products.[19] Participating companies include 25 companies commonly viewed as competitors: Dell, HP, Lenovo, and others.

As this discussion on vertical and horizontal partnerships highlights, the nexus of relationships may be quite complex, with the same players serving as suppliers in some arenas, complementors in others, and competitors in still others.

Reasons for Partnering

Companies engage in partnerships and alliances for a variety of reasons. In general, partnerships are formed to provide one firm access to resources and skills that, if it had to develop in isolation, would be costly in terms of either money or time. So, by partnering, firms are able to gain access to such resources and skills in a timely, more cost-efficient manner. For example, although Google is a huge company and seems to have all the talent and resources it needs to innovate and grow, it needs partnerships with large websites that can provide the eyeballs needed to view its ads. In 2006, Google signed a partnership with MySpace for the privilege of providing ads on that website in return for shared ad revenue of about $900 million to MySpace over three years.[20]

The product life cycle provides a useful way to look at the reasons for partnering at different points in the life cycle for technology-based products,[21] as illustrated in Figure 5.2 and discussed below. Recall from Chapter 1 that *product innovations* are new technologies that will be sold to customers in the marketplace, while *process innovations* are new technologies used in manufacturing and business processes.

AT THE EMERGENCE STAGE   During the emergence stage of the life cycle, substantial uncertainty surrounds the product. As discussed subsequently in Chapter 7, purchasers at this stage are innovators or technology enthusiasts who are willing to take substantial risks. As a buyer group, innovators have few requirements but they are very demanding when it comes to those requirements. First, they want an accurate portrayal of both the benefits and the liabilities of the innovation. In their role as gatekeepers in the system, innovators have the ability to undermine the prospects of any product that holds a surprise for them. Second, when they have a problem (and there will be problems due to the early stage of development), they require access to the most technically knowledgeable person in the organization to help them work through it. Finally, they want the new technology early and they want it at a low cost. They often see themselves as providing a service to the seller, acting as a beta test site by testing early versions of the new technology prior to its full-scale launch. They will refuse to pay a high price to be guinea pigs.[22] Furthermore, barriers to entry may be low during the

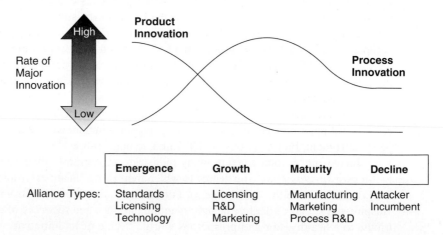

FIGURE 5.2   Reasons for Partnering across the Stages of the Product Life Cycle   *Sources:* Adapted from Utterback, James M., *Mastering the Dynamics of Innovation* (Boston: Harvard Business School Press, 1994); and Roberts, Edward B., and Wenyun Kathy Liu, "Ally or Acquire? How Technology Leaders Decide," *Sloan Management Review* 43 (Fall 2001), pp. 26–34.

emergence stage because brand loyalty has not been established, little cumulative learning has occurred, and major investments in production facilities have not been made.

From a competitive perspective, although little direct competition may exist at the emergence stage, often competing standards do exist. For example, at the emergence of the high-definition DVD technology, Toshiba proposed the HD-DVD format while Sony proposed the Blu-ray format. Toshiba's allies included Microsoft, Viacom's Paramount, and NBC Universal, while Sony's allies included Matsushita, Twentieth Century Fox, and Walt Disney. When the major movie studio, Warner Brothers, decided to back Blu-ray exclusively, the scales tipped in Sony's favor. Thereafter, retailers Blockbuster, Best Buy, Wal-Mart, and Netflix all endorsed Blu-ray. Finally, in February 2008, Toshiba withdrew its support of HD-DVD, conceding the format standard war to Sony, in a reversal of fortune from Sony's Betamax debacle 30 years earlier. Now Sony is well positioned to earn large revenues and profits by licensing Blu-ray technology to manufacturers of Blu-ray-compatible disks and players.

Thus, one of the key reasons for partnering at this stage is that alliances among potential competitors are valuable to establish standards.

Setting Industry Standards.   Firms have four different strategies available to them in pursuit of setting the industry standard:[23]

1. Licensing and OEM agreements
2. Strategic alliances
3. Diversification into complementary products
4. Aggressive product positioning

The first two of these strategies require partnering relationships and are explored here. The second two strategies are "go-it-alone" strategies and are described in Table 5.2. Note that this discussion does not address the use of government intervention, standards-setting bodies such as the International Standards Organization, or trade associations' attempts to establish an industry standard.

Managers of companies at the emergence stage pursue *licensing agreements* in which the company licenses its technology to other manufacturers and competitors in order to establish its technology as the standard. For example, in a major departure from its prior strategy (and in order to more effectively compete with Microsoft's entry into cellular phone technology), Nokia began to license its software to other cell phone makers in 2002; its objective was to try to make the Finnish company's software the industry standard.[24] Similarly, Palm spun off its operating system division as a separate subsidiary, PalmOne, in part to allow it the opportunity to license its software to other companies. As discussed subsequently, a firm can choose only to license its technology (and not enter the market with its own branded product), or can both license to others and also enter the market with its own brand. For example, Dolby licensed its high-fidelity sound technology to all players in the audio player market with relatively low licensing fees, and has prevented competitors from developing a superior technology. Despite the low licensing fee, the large market volume provides Dolby a good revenue base.

By licensing its technological design to others, a firm can help to grow the market quickly using that design. A spin-off of this strategy is known as an OEM strategy, in which a firm sells subcomponents to other companies (OEMs) that compete in the same market. For example, Matsushita licensed its VHS format to Hitachi, Sharp, Mitsubishi, and Philips NV, who produced their own VHS-format videocassette recorders and tapes. In addition, Matsushita also provided to GE, RCA, and Zenith, on an OEM basis, the components needed to assemble their companies' equipment.

The licensing strategy ensures a wide supply base for the technology, which helps to build the installed base of customers. In addition, this strategy co-opts competitors that might have had the capability to produce their own competing technology. It also limits the number of technologically incompatible product choices that customers face, reducing their confusion and doubt and hastening market acceptance. Licensing also signals to suppliers of complementary products the possibility of a larger installed base, providing an incentive for them to pursue development.

TABLE 5.2 Go-It-Alone Strategies for Setting an Industry Standard

Product Diversification into Complementary Products:

A company offers multiple elements of the whole product solution to become the de facto industry standard.

Rationale: Customers are hesitant to adopt a new technology unless complementary products are available. For example, customers won't adopt DVD players if there are no movies available to rent or buy in DVD format. Yet, independent suppliers of complementary products have no incentive to develop them if no installed base of customers for the new technology exists. Given this chicken-or-egg situation, a firm may have to diversify into producing the complementary products whose wide availability is crucial to the success of a new technology.

Examples: Matsushita had its in-house record label, MCA, develop an extended offering of digital compact cassette (DCC) recordable audiotapes in order to jump-start the adoption of its DCC technology. Similarly, Philips issued a wide selection of prerecorded DCC tapes under its in-house PolyGram Records label when it introduced its DCC digital audio recording equipment in 1992. In addition to providing the impetus in the market to generate indirect network externalities, this strategy allowed the company to realize revenue from not only sales of the base product but also sales of the complementary products. Other companies that have followed a product diversification strategy in setting industry standards are Apple, with its combined iPod/iTunes solution, and Sony, with its Blu-ray DVD players and Sony Films.

Risks: Developing complementary products from scratch can entail a significant capital commitment, as well as possibly straying from a company's core competencies. If the technology fails to become the standard, the costs of failure are that much greater.

Bottom line: If no potential suppliers of complementary products will respond rapidly to the firm's introduction of a new technology, then the firm may have no choice but to diversify into complementary products.

Aggressive Product Positioning:

Attempts to maximize the installed base of customers by relying on a combination of penetration (low) pricing, product proliferation (many models/versions), and wide distribution.

Rationale/Examples: Penetration pricing, including pricing below current costs, makes sense in high-tech markets with network externalities. This is why, for example, cellular phone service companies are willing to give away handsets for free. In each case, the company plans to grow the installed base in order to establish its product as the standard in the market. Moreover, in these cases, the companies try to recoup the base product's costs with sales of the complementary products (phone services). HD-DVD, the competing standard to Sony's Blu-ray DVD player, took this approach. Toshiba's HD-DVD players were priced below the Sony Blu-ray players, and when coupled with its licensing strategy, attempted to become the industry standard.

Product proliferation attempts to serve as many customers as is feasible by developing various product variations that appeal to different segments in the market. *Wide distribution* can also help to build a large customer base.

Risks: Aggressive positioning requires considerable investments in production capacity, product development, and market share building, all of which are sunk costs if the strategy fails.

Poor implementation of this strategy may have been why DCC technology failed. Although DCC technology would replace cassette tapes (based on analog technology) in much the same way that CDs and CD players replaced analog record players and albums, customers were confused over the benefits of the digital recording technology. Philips did not mention the fact that the DCC tape decks would play both existing analog and new digital tapes and did not highlight the benefits of the new recording technology. Moreover, the initial market price of $900 to $1,200 per DCC tape player was very high. Consumers were also worried about the presence of another, incompatible digital recording standard, Sony's minidisc system, and so adopted a wait-and-see attitude. Retailers were unable to move their initial inventory of DCC players and tapes and were wary about trying again. Philips's initial offer was limited to home entertainment centers and did not include portable players or car players. This lack of proliferation further limited the potential installed base.

Source: Hill, Charles, "Establishing a Standard: Competitive Strategy and Technological Standards in Winner-Take-All Industries," *Academy of Management Executive* 11 (May 1997), pp. 7–25.

The main drawbacks of a licensing strategy include the following:

- Licensees may attempt to alter the technology to avoid paying licensing fees or royalties.
- By increasing the number of suppliers in the market, the original developer loses a possible monopoly position, having to share revenues with its licensees.
- Competition may result in lower prices in the market, thus lowering profits for the original developer as well.

Alternatively to licensing, a firm can enter into a *strategic alliance,* a cooperative agreement with one or more actual or potential competitors to jointly sponsor development of a particular technological standard. For example, the WiMAX Forum has more than 522 members comprising the majority of operators and component and equipment companies in the communications ecosystem. Established in 2001, its purpose is to "certify and promote the compatibility and interoperability of broadband wireless products" in order to "accelerate the introduction of these systems into the marketplace." By establishing standards-based, interoperable solutions, the industry can enable economies of scale that, in turn, drive price and performance levels unachievable by proprietary approaches. The goal is to make WiMAX Forum certified products the most competitive at delivering broadband services on a wide scale.[25]

Strategic alliances that establish industry standards can also help to ensure a wide supply base for the technology, can co-opt competitors, and can help to build positive expectations for market demand, inducing other companies to develop complementary products. Alliances can help reduce confusion in the marketplace and build momentum behind the jointly sponsored standard, persuading potential competitors to commit to it. A particularly compelling advantage of this strategy for standards development is that alliances may be able to produce a superior technology by combining the best aspects of two companies' know-how, which can also increase the probability of the joint development becoming the industry standard. For example, during the development of the digital audio technology used in compact disc players, at least four companies were pursuing incompatible designs. In a turnabout from their VCR experience, Philips NV and Sony partnered to develop the first compact disc system. Philips NV contributed a superior basic design, and Sony provided the error correction system. This alliance threw momentum behind the Philips-Sony standard, and 18 months prior to product introduction, more than 30 firms had signed agreements to license the Philips-Sony technology.[26] At the emergence stage, because the product technology is likely to be immature, potential competitors should collaborate to improve the technology so that the next buyer group in the diffusion process can be addressed.

Of course, as discussed in the next section, the risk exists that a partner might appropriate the firm's know-how in an opportunistic fashion. This risk might be mitigated by careful structure and management of the alliance. For example, structuring the joint venture so that each company has a significant financial stake in the shared outcome serves as a "credible commitment" that prevents undermining the alliance. Alternatively, the alliance might be structured in such a way that, after the standard has been developed, each party is free to go its own way with regard to future extensions of the technology.

When does it make sense for a company to pursue strategic alliances to establish industry standards? When should it use a licensing strategy? Or, when should it attempt to "go it alone"? As detailed in Table 5.3, consideration of three factors helps a firm decide which strategy to pursue:[27]

1. **Barriers to imitation**, such as those found in patents or copyright protections, for example.
2. **A firm's skills and resources**, including manufacturing and marketing capabilities, financial resources, and company reputation. Technological skills alone are insufficient for establishing a technology as a standard.
3. **The existence of capable competitors**, which pushes the firm to build an installed base of customers rapidly despite the risks. The absence of capable competitors who might develop their own, possibly superior technology means a firm can move on its own.

TABLE 5.3 Which Strategy to Set an Industry Standard?

	Barriers to Imitation*	Firm Has Requisite Skills†	Existence of Capable Competitors‡
Aggressive Sole Provider (avoid licensing agreements and alliances, develop key complementary products, adopt aggressive positioning)	High	Yes	No
Passive Multiple Licensing (license to as many players as possible and don't enter the market on its own)	Low	No	Yes
Aggressive Positioning + Licensing (license to many companies and enter market with own brand as well)	Low	Yes	Yes
Selective Partnering (partner with one or a few companies)	High	No	Yes

*Patents or copyright protections, for example.

†Skills in manufacturing and marketing capabilities, financial resources, company reputation, etc.; technological skills alone are insufficient for establishing a technology as a standard.

‡The absence of capable competitors (who might develop their own, possibly superior, technology) means a firm can move on its own.

Source: Hill, Charles, "Establishing a Standard: Competitive Strategy and Technological Standards in Winner-Take-All Industries," *Academy of Management Executive* 11 (May 1997), pp. 7–25.

AT THE GROWTH STAGE   One of the signals that the life cycle has shifted from emergence to growth is that a dominant design becomes the industry standard. Customer needs become increasingly clear. Purchasers in this stage are early adopters to whom technology allows breakthrough performance in their own businesses. Because these early adopters can envision the potential that technology has to offer, they tend to be the least price sensitive of the innovation adopter groups. Their lack of price sensitivity is offset by the fact that they are in a hurry to reap the rewards of the innovation before the opportunity window closes. A critical conclusion is that the seller must manage early adopters' expectations by understanding their goals and clearly communicating how the innovation will help attain those goals.[28] The early adopter category is important because it is large enough to generate meaningful revenue and is often quite profitable due to low price sensitivity. While innovation in product technology declines during the growth stage, process technology becomes more important due to the increased emphasis on quality and efficiency (recall Figure 5.2).

In the growth stage, the winners of the race to establish the dominant design license their technology to the losers. In addition, R&D alliances are formed to enable companies to improve the dominant design and develop product extensions and new features. If the dominant design is controlled by a small or a technology-oriented company, the company's managers will want to establish a marketing alliance with a larger, well-established firm that has well-developed distribution and other marketing competencies. This is a common strategy in the biotech/pharmaceutical industry.

AT THE MATURITY STAGE   In the maturity stage, sales volume and revenue are high but their growth rate slows dramatically. Purchasers are members of the mass market (early and late majority markets). Mass market purchasers utilize a rational decision-making style to evaluate the relative benefits of the innovation. They typically seek a measurable and predictable improvement in performance or productivity compared to the previous solution to this particular need. They put strong pressure on price. To maintain their profits in the face of price pressure, sellers emphasize cost control from purchasing through distribution and post-sale service. Process innovation dominates product innovation, and process R&D alliances take center stage. In the area of improving innovation processes, many companies facing maturity, such as Samsung, have partnered with design consulting firms, such as IDEO, to learn how to be creative and quickly get new products out the door.[29]

Also at this stage, many firms form alliances to outsource production to contract manufacturers with competency in efficient production. Outsourcing capitalizes on each firm's core competencies, enabling the developer to focus its resources on developing a next generation of the technology and the contract manufacturer to focus on efficient manufacturing. As in the growth stage, marketing alliances are formed to address new markets.

AT THE DECLINE STAGE A major reason that products start to decline is that new technologies are developed and introduced. As the transition is made from the old to the new technology, the industry newcomers form technology and marketing alliances to take on incumbents. Because incumbents often have a difficult time developing and introducing a technology that will obsolete or cannibalize their cash cow, they are often in the position of having to license the disruptive technology from a new competitor. And then, the cycle starts again. (Note that this stage of the cycle is consistent with the concepts of technology life cycle, core rigidities, and creative destruction from Chapters 1 and 3.)

Table 5.4 lists the reasons for companies to establish partnerships and alliances. Because of their ostensible benefits, partnerships and alliances are frequently prescribed as the panacea for success, the modus operandi for success in today's business environment. However, despite the many reasons for partnering, the prescription to partner overlooks the reality that the overwhelming majority of partnerships fail to achieve the objectives set by at least one of the partners.[30] In addition to outright failure, many risks are inherent to partnering efforts. These inescapable realities of strategic alliances highlight the need to understand fully the risks as well as the factors that contribute to the potential success and viability of the partnership.

Risks of Partnering

As shown in Table 5.5, firms face serious impediments to realizing the benefits of their partnerships. For starters, working in tandem with another organization increases the project's complexity. More serious is the potential loss of autonomy and control that accompanies a joint effort. In teaming with another firm, decisions must be made jointly, and the success of the project becomes, to some extent, dependent on the efforts of another. Sharing decision-making control is very difficult for many firms—and because some are never able to give up their autonomy, their partnerships fail (at worst) or do not function effectively (at best).

TABLE 5.4 Reasons for (Benefits of) Partnering Relationships

To access resources and skills

To gain cost efficiencies

To speed time to market

To access new markets (marketing and distribution)

To define industry standards

To develop innovations and new products

To develop complementary products

To gain market clout

To maintain focus on core competencies

To learn from partners

TABLE 5.5 Risks of Partnering
Increased project complexity
Loss of autonomy and control
Loss of trade secrets
Dilution of competitive advantage ("de-skilling")
Legal issues and antitrust concerns
Failure to achieve objectives

Some identify the loss of trade secrets as the least noticed, but potentially riskiest, aspect of alliances.[31] Indeed, product managers have cited the leakage of information as the greatest risk in joint product development.[32] Although companies typically sign confidentiality agreements that ostensibly prevent partners from exploiting what they learn about each other, shared secrets are a hazard of partnerships. Many experts recommend that a firm never forget that one partner might be out to "disarm" the other. As one business manager put it: "If they were really our partners, they wouldn't try to suck us dry of technology ideas they could use in their own products. Whatever they learn from us they are going to use against us worldwide."[33]

The potential leakage of information can lead to one partner learning valuable skills and knowledge from the other. As the software company Oracle notes, "The partners we catch up to in our core applications will cease to be useful partners." Although partners want to learn as much as possible from each other in order to maximize the effectiveness and efficiency of the alliance, each partner must also limit transparency and leakage of information so as not to dilute its sources of competitive advantage.[34]

Another risk that strategic alliances face is that of legal issues and antitrust problems.[35] Cooperative ventures may run afoul of U.S. and European Union antitrust laws, especially when they involve large firms. Public policy officials wrestle with finding an appropriate balance between maintaining national competitiveness in an increasingly global market and protecting consumer welfare. On one hand, collaborative ventures are necessary to compete globally. On the other hand, collaborative relationships can result in less competition in domestic markets, potentially harming consumer welfare.

The historically strong antitrust environment in the United States necessitated two laws to facilitate the ability of firms to collaborate. The National Cooperative Research Act (1984) was passed to ease the prohibition against competitive collaboration and to promote R&D, encourage innovation, and stimulate trade. The act applies to R&D activities up to and including the testing of prototypes. The National Cooperative Production Amendment (1993) expanded the 1984 act to allow for joint production as well. The 1993 amendment excludes marketing and distribution agreements, except for the products manufactured by the venture, and also excludes the use of existing facilities by the joint venture. The Federal Trade Commission's *Antitrust Guidelines for Collaborations among Competitors* is a useful document that outlines the concerns the FTC and the Justice Department have about competitive collaboration.[36] Any partnership that has the potential to directly or indirectly affect pricing is of great concern. Hence, partnerships that affect allocation of customers or supply in a market are likely to be carefully scrutinized because of the indirect impact on pricing. On the other hand, partnerships that arise because firms are unable to pursue projects alone, and that create some common good in the market, are likely to be encouraged. For example, partnerships focused on costly R&D efforts are typically easily justified. Another factor involves the impact of the collaboration on market share. If, together, the partners will not control more than 20% of the market being affected, their partnership is unlikely to raise concerns.

Possibly the largest risk of any partnership, the unfortunate reality in the overwhelming majority of cases, is the failure to achieve the partners' objectives. Reasons can be attributed to a host of factors: incompatible cultures between the two firms, lack of attention and resources allocated to the ongoing

TABLE 5.6 Critical Ingredients for Alliance Success
Interdependence
Appropriate governance structure (bilateral, relational norms)
Mutual commitment
Trust
Communication
Perceived fairness in the relationship
Compatible corporate cultures
Integrative conflict resolution
Judicious use of legal contracts

management of the relationship, lack of trust in the other party's motives, or the inability to deliver its part of the agreement, and so forth. The risks highlight the need to understand the factors that contribute to the potential success and viability of the partnership.

Factors Contributing to Partnership Success

Certain traits, such as trust and communication, are important in most—if not all—business relationships. However, because strategic alliances and close partnering relationships tend to have higher levels of risk, such relationships have a concomitant need to exhibit a greater amount or intensity of these characteristics.[37] As shown in Table 5.6, effective strategic alliances have been shown to exhibit several characteristics, explored next.

INTERDEPENDENCE   To enhance the odds of partnership success, both parties must be dependent on the other for some important resource that is valued and hard to obtain elsewhere. Shared mutual dependencies form the basis for a give-and-take relationship in which both parties are equally motivated to ensure the success of the alliance. Uneven, or asymmetrical, dependence undermines the dual nature of the relationship, can lead to exploitation, and may leave one party more vulnerable than the other. Alliances with low levels of interdependence suffer from a lack of commitment and tend not to fare well.

A special case of interdependence arises with partners of very disparate sizes. Past research has shown that partnerships between relatively equal-sized partners are more likely to be successful than those between partners of unequal size.[38] However, in the technology arena, small start-ups commonly partner with industry behemoths. In a typical case, the small start-up has an exciting new technology, while the large company brings needed resources, access to markets, and management and marketing expertise. When small companies partner with large companies, the risks can loom large and special attention should be paid to the governance structure of the relationship.

APPROPRIATE GOVERNANCE STRUCTURE   Governance structures are the terms, conditions, systems, and processes used to manage the ongoing interactions between two companies. At a simple level, governance structures can be *unilateral* in nature, granting one party the authority to make decisions, or *bilateral,* based on mutual expectations regarding behaviors and activities.

Generally, the governance structure should be matched to the level of risk in the partnership. When one partner has assets that are at risk (say, if the other party behaves opportunistically and takes advantage of the other) or when uncertainty is high, crafting a governance structure to reduce that risk is crucial. Governance structures based on "credible commitments" (mutual investments that make both parties vulnerable should the partnership fail) are one way to achieve interdependence. Alternatively, a vulnerable partner could adopt an interfirm agreement that only loosely couples the organizations,[39] relying on a narrow marketing agreement or licensing agreement rather than a joint venture. However, these looser, arm's-length relationships can compromise the ability of the partnership to function effectively. As a last resort, the more vulnerable party might either trust or rely on its partner's reputation as a governance strategy. However, such a strategy is a bit of a gamble.

A governance structure based on bilateral, relational norms, including expectations for future interactions (commitment), acting in the best interests of the partnership (trust), and intensive information sharing (communication), can also provide a reasonable solution, as discussed in the next three factors. These factors are so important that at least one theory of collaborative partnering has been called "the commitment–trust theory."[40]

COMMITMENT Commitment, or the desire to continue the relationship into the future, is an important element for strategic alliances to succeed. Partners who are committed to the relationship are less likely to take advantage of the other partner or to make decisions that might sabotage the long-run viability of the alliance. Commitment can be demonstrated by making investments in the partnership that are dedicated solely to that relationship. Importantly, commitment should arise from a positive feeling and regard for each other's contributions, rather than from feelings of desperation or economic necessity.[41] When the nature of commitment between partners is of the "have to be committed" variety (rather than because of a positive, voluntary desire), the impact on the alliance is negative.

TRUST Trust refers to the sense that the other partner will make decisions that serve the best interests of the partnership when one party is vulnerable and will act honestly and benevolently. Trust leads to more effective information sharing, a willingness to allocate scarce and sensitive resources to a shared effort, and the sense that both parties will benefit in the long run.

COMMUNICATION Effective communication in strategic alliances is absolutely critical to success. Effective communication is characterized by frequent sharing of information, even information that may be considered proprietary. Such communication flows bidirectionally, with both partners participating in the flow of information about their needs and potential problems. The quality of the communication, in terms of its credibility and reliability, is vital. Communication needs to be somewhat structured, with someone accountable for maintaining open lines of information. Informal, unplanned, and ad hoc interactions are also an important component of communication.

PERCEIVED FAIRNESS Especially in relationships between unequal partners (asymmetric dependence), the more dependent party is vulnerable to potential exploitation by the more powerful party. Moreover, concerns or fears about a partner's opportunism or lack of integrity also may surface, particularly in some cross-cultural alliances. In addition to the governance components mentioned previously, perceptions of fairness also affect relationship performance and success. Three types of fairness, or justice, operate in relationships: *distributive fairness,* or fairness or equity in the distribution of rewards and outcomes; *procedural fairness,* or fairness of the process used to determine the distribution of rewards; and *interactional fairness,* which refers to fairness of the nuances of interpersonal treatment. Studies indicate that procedural fairness is relatively more important than distributive fairness in developing effective long-term relationships.[42] Despite managers' attention to "dividing up the pie" between partners, the process used to allocate returns from the partnership warrants more effort.

Indeed, fairness in one aspect tends to reinforce justice in another. With respect to interactional fairness in particular, the careful attention to interpersonal interactions creates an environment where procedural justice can flourish.[43]

COMPATIBLE CORPORATE CULTURES Although two firms may have synergistic skills that could usefully be shared in a partnership, such synergies are difficult to realize if corporate cultures clash. Companies like Microsoft and Apple historically have not had "partner-friendly" cultures. In the world of online retailing, there is an industry perception that Amazon is difficult to partner with.

Toys "R" Us won a lawsuit against Amazon for violation of its 10-year exclusive partnership to manage the former's website and provide fulfillment services.[44] The Microsoft–IBM partnership for OS/2 ruptured in 1991 because Microsoft wanted to pursue Windows as an alternative. Even though Microsoft and Apple have partnered over the years, Apple sued Microsoft for copyright infringement of its Macintosh graphical user interface in 1988. At the All Things Digital conference in 2007, Steve Jobs and Bill Gates were being interviewed together by the *Wall Street Journal*. When Jobs was asked what he had learned most from Bill Gates and Microsoft, he said he learned how to be more open to partnering with other companies and how to do it well.[45] So it seems like both companies are attempting to change their partnering culture.

INTEGRATIVE CONFLICT RESOLUTION AND NEGOTIATION TECHNIQUES   Conflict arises in any relationship, and strategic alliances are no different. Some level of conflict is likely functional, because it indicates that problems are being identified and addressed rather than being ignored. In determining the alliance's future success, the ways in which such conflicts are addressed are more important than the sheer level of disagreement. Parties must be willing to resolve conflict in a way that allows for both partners to have a stake in the outcome, that addresses both partners' needs simultaneously, and that is mutually beneficial. Although this may seem idealistic, many creative problem-solving techniques can identify solutions that do not result in a win-loss outcome, but rather win-win solutions.

Often, conflicts arise at operational levels in executing alliances. Companies that have a successful track record of partnering have senior executive champions responsible for the alliance, in addition to operational alliance managers. The mechanism for conflict resolution at operational levels is to escalate the problems to the level of executive champions. Because these champions are senior executives who understand the strategic importance of the relationship and have authority over resources, they can usually resolve differences and get things back on track. It is far better and cheaper to resolve differences through negotiation rather than legal recourse. Despite past differences between Apple and Microsoft, after Steve Jobs returned to Apple as its CEO in 1996 he was able to negotiate a $100 million equity investment from Microsoft and a commitment to continue developing Office for the Macintosh.

JUDICIOUS USE OF LEGAL CONTRACTS   Certainly, alliances also tend to be governed by a contract that specifies each party's roles and responsibilities, as well as legal remedies in the case of failure. The degree to which legal governance is compatible with positive relationship functioning (or whether it undermines the relationship) has been studied extensively. Findings indicate that contracts must be used judiciously. For example, some studies find that the use of explicit contractual terms is at odds with a relational approach to governance.[46] Moreover, when the relationship operates in uncertain environments, increasing contractual specificity can lower performance of the partnership. However, it is also possible that when the parties have a foundation of trust, solidarity, and mutual flexibility, contractual agreements can help clarify obligations and expectations and provide the formal structure within which adaptations can be made to unforeseen contingencies.[47] Given the conflicting evidence in this regard, partners probably ought to have some type of contract at a minimum, but the contract should not be used as a substitute for establishing more relational norms that guide the ongoing management of the alliance over time.

SUMMARIZING PARTNERSHIP SUCCESS   In sum, the spirit of cooperation, identified in the traits mentioned, signals the likelihood of future success of partnerships and alliances. In fact research scholars have conceptualized, measured, and tested ideas such as "cooperative competency,"[48] "alliance competence,"[49] or partnering orientation at the level of an organization and found positive relationships with performance.

Recreational Equipment Inc. (REI) is a member-owned co-op, a multichannel retailer of outdoor goods (camping and backpacking gear, technical climbing gear, etc.), footwear, and apparel. In addition to its own private label, REI carries such well-known brands as North Face, Patagonia,

Prana, and Mountain Hardware, to name just a few. What's high-tech about REI's business and products? The answer is multidimensional, but one high-tech aspect is found in the very sophisticated "ingredients" used in fabrics and textiles of the company's products. (Recall from Chapter 1 that many radical innovations occur in upstream supply innovations, rather than at the end user level.) In addition to investing in innovative technologies to enhance the performance and functionality of their products, apparel and gear manufacturers also invest in innovative technologies to make their products more environmentally friendly. This includes, for example, sourcing goods from non-petroleum-based plastics, such as bioplastics. In addition, it includes researching new technological processes to recycle fabrics for next-generation uses. Although other companies in the industry strive to make their business models more sustainable, REI has taken a unique leadership role to form an industry-wide coalition to create sustainability standards for vendors/suppliers. The coalition seeks to address the issue of what it means when a company puts a "green" or "eco-sensitive" label on its products. REI is working cooperatively with other companies to develop and implement such standards— particularly from an R&D/technology perspective—to bring positive changes to upstream providers. Yet, predictably, the functioning of such a coalition is fraught with challenges. Hence, the accompanying Technology Expert's View from the Trenches has an important message for all types of industry coalitions and partnerships.

TECHNOLOGY EXPERT'S VIEW FROM THE TRENCHES

Forming a Coalition to Collaborate on Industry-Wide Product Eco-Metrics

By Kevin Myette

Director of Product Integrity and R&D
Recreational Equipment Inc. (REI), Seattle, Washington

Being "green"—or environmentally responsible—is an expression that has been used with growing frequency throughout the past decade. In the last few years the use of the term has accelerated and has taken on a new urgency in our increasingly climate-change-aware society. Yet, simply claiming a company's products are environmentally considerate does not make them so. Indeed, one of the big challenges in defining a company's "greenness" is that there is no common framework to measure the claim—either absolutely or in degrees. Given the lack of an agreed-upon standard, it's easy for companies to "greenwash," claiming their products and business practices are environmentally responsible when in fact they are not—or more commonly, not as good as they could be. Not only is this situation confusing for consumers, it's bad for industry—particularly for companies whose products truly are superior from an environmental perspective.

To level the playing field, a number of companies in the outdoor industry agreed to collaborate on the creation of a framework that defines an eco-index that companies can effectively implement and customers can ultimately trust. To that end, in May 2007, REI and the Timberland Company convened the Outdoor Industry Association's Eco Working Group, whose formal mission became:

> . . . to take a leadership role to develop environmental impact evaluation tools, programs, education and communication to stakeholders and consumers that will direct product life cycle and informed purchasing decisions.

The goal is to establish the metrics that measure the environmental footprint of a company's products in a truthful fashion and—just as importantly—to convey a product's eco-impact to the consumer in a consistent, meaningful and understandable manner.

For REI, our motives, although pure, are not entirely altruistic. Three good reasons explain REI's dedication to this work. First, REI has been hard at work making our company's global footprint smaller. Making products more sustainable is a very significant way to have a very large impact. To that end, we want to make smarter choices about suppliers and their products—choosing the ones that offer not only excellent quality and performance, but also allow us to be more environmentally considerate. Second, as a retailer of many brands, we found it confusing to sort through our various vendors' multiple measures, marketing claims, and in some cases, pseudoscience about the "greenness" of their products. We wanted to create a common measure across our industry. Third, we wanted to make information accurate, easier to understand, and more consistent so that we can communicate to and educate our customers about the sustainability of our company's products.

Solving a noble problem doesn't make forming an effective industry-wide coalition any easier. Some issues are technical, others are complicated by corporate values, and others are rooted in the marketing and competitive positioning of our work group members. Paradoxically, the junction of these issues creates the raison d'etre for the group. Below are the critical challenges and success factors that we have identified.

Challenge #1: Who Should Be Involved?

When the working group was formed, the industry appetite for addressing the problem—creating a platform to lessen the environmental impact of outdoor products—could not have been stronger. In fact, Timberland had just introduced its Green Index, which was an eco-metric applied to footwear. The recruitment of members was easy but it was difficult to keep the initial group intimate enough to be effective.

The group cofounder, Betsy Blaisdell from the Timberland Company, and I determined early that it would be best to convene the work group under an industry trade association rather than as an effort of just REI and/or Timberland. This step was taken to promote the best atmosphere for full collaboration.

Although no organization has ever been *excluded,* initially companies were *included* for previously demonstrated leadership in the area of sustainable product design, general stature in the outdoor industry, and general interest. We took extra effort to reach out to companies both large and small and with diverse product offerings so that, rather than solving the challenge for just a small set of companies, we could solve it for the industry as a whole. We saw participation quickly balloon to 40-plus major product brands.

Challenge #2: Divergent Goals and Objectives

Members of the Eco Working Group fiercely compete for customer's heads, hearts, and wallets. When something as significant as the "green" movement emerges as a means to build brand loyalty, it becomes another logical point of differentiation and, therefore competition.

The challenge within the work group was to agree that we should not compete by creating better metrics, but—using a codeveloped and shared framework—by making our own respective products less environmentally harmful; this is good competition, all for the right reasons.

Another very interesting phenomenon is the "invented here, but willing to share" behavior exhibited by leaders in environmental responsibility in our industry. As innovation occurs on the product sustainability front, it is often shared with many companies—including direct competitors. Water-based instead of solvent-based glues, recycled polyester, lower impact dyes and finishes—companies are regularly creating new technologies, processes and methodologies that help move product sustainability forward. Because the premise of these efforts is to "do good," they have the most impact when they are utilized by multiple companies and not kept exclusive.

What has emerged as most important to the eco-innovator companies is not the exclusive right to use their sustainable inventions, but rather, recognition by peers and consumers that they invented and commercialized these great ideas.

Challenge #3: The Science behind the Metrics

The information necessary to understand the full life-cycle impact of the materials and substances that make up our products is hard to obtain. The devil is in the details, which are both more extensive and more specific than supply chains are currently set up to deliver. However, this information is critical in order to make the eco-metrics credible.

(continued)

Further, not all the information utilized for determining environmental impact is quantitative. Much of it remains qualitative, subjective, and based on individual, personal, or corporate values. For example, are corn-based bioplastics more or less environmentally responsible than plastics derived from petrochemicals? Science can show the material consumption, energy, and carbon impact of either option—but if the corn came from genetically modified organisms, does that make the product less clean? How should issues like these be factored into the framework that are meaningful, yet don't trump or compromise the individual values of companies?

Challenge #4: Communication Complexity

Initiatives such as ours gain the greatest momentum through regular dialogue and debate in face-to-face meetings. However, with participants scattered across the continent in different time zones, keeping our group's communication lines open has been challenging. We've improved our effectiveness between in-person meetings by using a collaborative website so all information about our efforts is easy to access, easy to find, and easy to contribute to. Users can participate in virtual and real-time meetings, discussions, wikis, voting, sharing of documents, and so forth. The site requires users to be identified to participate and add content, so everyone knows who is participating and how—just like in a face-to-face meeting.

One of the stated goals for the way we conduct ourselves during the creation of our eco-index is to be as transparent as we can possibly be. A positive side benefit of the website is that it serves as a permanent record of our activities to be shared with anyone who has interest in learning about our process.

The website does not completely eliminate the need for in-person meetings, but it does help us make progress between meetings. Plus, the carbon footprint of the bits and bytes necessary to make an online community work pales in comparison to the emissions necessary to get 30-plus people to a single location—a good thing for a group that is trying to promote environmental responsibility.

Success Factors to Successfully Navigate the Challenges:

- ***Identify and engage stakeholders very early in the process.*** These people—key influencers and subject-matter experts—must be included and consulted constantly so that they remain engaged and committed.
- ***Openness and a spirit of collaboration trump all else in this process.*** If companies don't check their self-interests and competitive positioning at the door, then the atmosphere for true collaboration will never exist and the effort will fail.
- ***Develop a clear shared mission and vision, a well-defined structure and governance, and an agreed-upon set of operating rules and procedures.*** Without them, we have no common ground as we navigate difficult tactical tasks and competing interests.
- ***Constantly seek out examples, models, and other means that illustrate the business case for proceeding.*** We must show that making products more environmentally responsible for moral reasons also adds value to our businesses; without this business case, our efforts will lose steam and die.
- ***Don't be afraid to shake up the model of how collaboration takes place*** especially in this busy and connected world.

Desired Outcome

Our goal is to establish one—and only one—framework for eco-comparison and selection among vendors and products. It is our vision that some day consumers will be able to make quick and meaningful comparisons among similar products about the true environmental impact of each choice. Our belief is that this will positively influence consumer purchasing behavior as well as drive business innovation toward more sustainable options. Ultimately this not only will make tangible, measurable differences in the waste, emissions, and energy consumption of our products—making our planet healthier—but also will drive business value for those who embrace the environmental practices, because consumers will expect and demand environmental responsibility just as they have grown to expect quality and performance.

OUTSOURCING: HIGH RISK/HIGH OPPORTUNITY VERTICAL PARTNERSHIPS

In the mid-1990s, Hewlett-Packard decided to outsource 100% of its manufacturing. Indeed, HP today does not manufacture any of its own products. With its outsourcing decision, HP's outsource provider (Jabil in many U.S. locations) took ownership of HP's manufacturing facilities, and HP's manufacturing employees were "rebadged" as Jabil employees. As one might imagine, this change was quite stressful. HP had been revered for its excellent benefits and organizational culture, and employees were fearful that a different company wouldn't be as generous. In another major change, when HP decided to offer its products for sale via an Internet distribution channel, it originally performed many of those functions, including order fulfillment and shipping, from its own facilities. However, HP quickly realized that the business of performing e-commerce was not one in which it had a core competency; moreover, it distracted HP from its true core competencies (product innovation and marketing)—and so it outsourced the order fulfillment and distribution of all HP.com orders to FedEx.

Because of its increasing prevalence in high-tech industries and its increased complexity as a type of partnership, this section focuses specifically on outsourcing as a unique form of vertical partnership.

Outsourcing is an arrangement in which one company (the customer or the client) hires another company (the supplier or outsource provider) to perform a particular function on its behalf. It involves the transfer of the management and/or day-to-day execution of an entire business function to an external service provider. Based on a contractual agreement that defines the transferred services, the client agrees to procure the services from the supplier for the term of the contract. In addition, the outsource provider may acquire the means of production in the form of a transfer of some combination of people, assets, and other resources from the client.

The trend toward outsourcing began in the mid-1980s, when many companies outsourced their manufacturing to providers in emerging economies, such as China, in order to reduce labor and manufacturing costs. Businesses whose manufacturing operations were located in places with relatively high labor and materials costs found that to remain competitive, they had to lower the costs of doing business. Hence, contract manufacturing was a key solution, not only in high-tech industries with electronics manufacturing companies such as Jabil, Solectron, and Flextronics, but also across a wide swath of industries.

Next to outsourced manufacturing (often referred to as *contract manufacturing*), the growth in outsourcing has occurred mostly in two categories: One of these is **business process outsourcing (BPO)**, which outsources business processes such as customer service, finance and accounting, human resources, analytics, or perhaps even new product development. The other is **information technology outsourcing (ITO)**, which outsources IT functions such as website design or IT infrastructure. Another type of outsourcing receiving quite a bit of attention from high-tech companies in recent years is outsourced innovation (including R&D, product development, and design work), sometimes known as **original design manufacturing (ODM)**. These outsourced product designs, in which companies use a predesigned platform, can shave 70% off development costs for a mobile phone, for instance.[50] In 2005, the percentage of product development/product design efforts that were handled by outsource providers included:[51]

- 70% of PDAs (handheld devices such as Palm Treos and RIM BlackBerrys)
- 75% of notebook PCs
- 65% of MP3 players
- 30% of digital cameras
- 20% of mobile phones

In theory, outsourcing allows a customer to gain access to best-in-class services performed by experts. Outsource providers become highly skilled in performing the outsourced function. When coupled with the aggregation of business from many customers, outsource providers develop

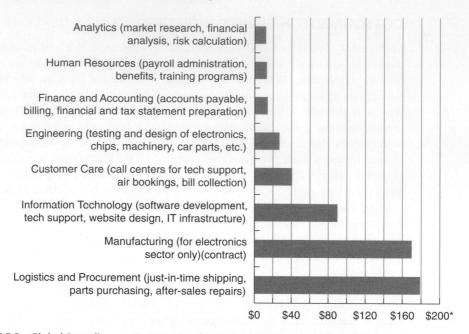

FIGURE 5.3 Global Spending on Outsourcing for Various Business Functions, 2005
* in billions
Source: Engardio, Pete, "The Future of Outsourcing," *Business Week*, January 30, 2006, pp. 50–64.

economies of scale that further refine their knowledge base about the function. The combination of specialized knowledge, skill proficiency, and scale economies drives the cost efficiencies that underlie a key reason for outsourcing. A second reason for outsourcing is that companies can then better focus on their own business model, allowing them to increase revenues.

Despite its controversies (discussed subsequently), outsourcing has grown steadily over the years in both the variety of services that are outsourced as well as the volume of business. As Figure 5.3 shows, business functions outsourced (outside of a company's domestic borders) include manufacturing, logistics and distribution, IT management, customer service, human resource management, accounting and finance, website development and management, product development (including software engineering) and product testing, and sales and marketing (including search engine positioning), to name just a few. Despite the volume of business represented in Figure 5.3, the offshore component (relative to outsourcing to domestic partners) varies widely by business function. For example, with respect to business process outsourcing, the offshore component was estimated to be roughly just 10% in 2005.[52]

Figure 5.4 shows the geographic regions involved in this major business trend. India accounts for nearly 60% of the volume of business process outsourcing due in part to English as a primary language, but also to the talented workforce, solid education system, and ruthless efficiency in their service operations. Emerging areas of the global economy—including the BRIC countries (Brazil, Russia, India, and China) as well as eastern Europe, Egypt, and the Philippines—have developed proficiencies not only in manufacturing, but also in software engineering and other skill sets that are highly desired by global multinational corporations. Table 5.7 shows the major outsource providers in key industries.

More Outsourcing Terminology

The jargon surrounding outsourcing can be formidable. In addition to the terms presented above, following are definitions for other common outsourcing terms.

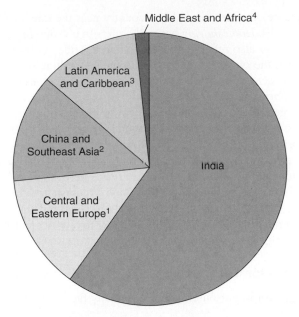

FIGURE 5.4 Business Process Outsourcing, by Region
Sources: Engardio, Pete, "The Future of Outsourcing," *Business Week,* January 30, 2006, pp. 50–64; and A. T. Kearney Global Services Location Index, 2005.

1. Czech Republic, Bulgaria, Slovakia, Poland, Hungary *Up-and-comers:* Romania, Russia, Ukraine
2. China, Malaysia, Philippines, Singapore, Thailand *Up-and-comers:* Indonesia, Vietnam, Sri Lanka
3. Chile, Brazil, Mexico, Costa Rica, Argentina *Up-and-comers:* Jamaica, Panama, Nicaragua, Colombia
4. Egypt, Jordan, United Arab Emirates, Ghana, Tunisia *Up-and-comers:* South Africa, Israel, Turkey, Morocco

Offshoring generally refers to performing functions outside the client's home country. *Captive offshoring* refers to a company performing the function in its own company-owned facilities in another country. For example, Microsoft has an office in Hyderabad, India, that does software development in collaboration with its headquarters based in Redmond, Washington. *Reverse outsourcing* is a situation whereby an outsource provider, say an Indian-based company performing work for a U.S.-based company, opens an office in the client's home country (e.g., the United States) in order to fulfill the outsourcing contract. *Near-shore outsourcing* is provided by an

TABLE 5.7 Major Outsource Providers, by Function

Business Services	Software Development	Call Centers
Hewitt Associates (U.S.)	Tata Consultancy Services (TCS) (India)	Convergys (U.S.)
ACS (U.S.)	Infosys Technologies (India)	Wipro (India)
Accenture (U.S.)	Wipro (India)	ICICI OneSource (India)
IBM (U.S.)	Accenture (U.S.)	ClientLogic (U.S.)
EDS (U.S.)	IBM (U.S.)	24/7 Customer (India)
Hewlett-Packard (U.S.)	Cognizant Technology Solutions (U.S.)	SR. Teleperformance (France)
Wipro (India)	Satyam (India)	eTelecare International (U.S.)
HCL Technologies (India)	Patni Computer Systems (India)	SITEL (U.S.)
Tata Consultancy Services (India)	EDS (U.S.)	Teletech (U.S.)
WNS Global Services (India)	CSC (U.S.)	CustomerCorp (U.S.)

Source: Engardio, Pete, "The Future of Outsourcing," *BusinessWeek,* January 30, 2006, pp. 50–64; and Gartner Inc.

outsource provider located in a country near the client's own home boundaries and in proximate time zones. *Homeshoring* generally refers to outsourcing the function to a domestic provider. However, it can also refer to (and be confused with) the hiring of independent domestic workers who perform the outsourced function in their own homes. For example, Dell hires customer service workers across the United States to whom customer service inquiries are routed.

Farmshoring refers to outsourcing to rural areas, typically in a company's own country. Given the large volume of business from government contracts, many believe a government has an obligation to outsource contract work within its own boundaries prior to seeking an offshore provider. Farmshoring often requires a build-out of broadband infrastructure and other technology centers in rural areas. Indeed, calls for "universal service" of broadband technology are often tied to economic development opportunities for rural areas. One example of farmshoring is found in the hiring of Native Americans who face dire economic circumstances on their reservations, where unemployment rates can be as high as 80%. Yet, these American Indians have found outsourcing to be an economic opportunity that allows them to stay close to family and tribal customs that are an important part of their heritage.[53]

Reasons for Outsourcing

The growth in outsourcing is due to many factors, as illustrated in Figure 5.5 and discussed next.

COST SAVINGS By taking advantage of lower costs in other markets, economies of scale, and specialization, outsourcing offers the potential of huge cost savings. With respect to contract manufacturing in particular, outsourcing's lower costs of production are driven primarily by economies of scale. By accumulating volume over many OEM customers' product needs, the outsource provider gains economies of scale in the learning process, secures volume discounts from the upstream suppliers of raw goods and materials, and gains efficiencies in manufacturing and supply chain processes.

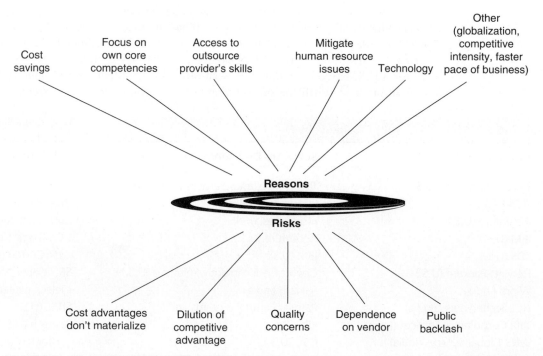

FIGURE 5.5 Reasons for and Risks of Outsourcing

Pharmaceutical companies such as GlaxoSmithKline and Eli Lilly partner with Asian biotech companies to lower the costs of developing new drugs, in part because their internal R&D spending isn't yielding enough "bang for the buck." Eli Lilly estimates that it invests $1.1 billion for each new drug it releases, with that cost expected to grow to $1.5 billion by 2010.[54] As a result, it relies on outsourcing to lower that cost to $800 million. For example, 20% of its chemistry work is done in China—at one-quarter the U.S. cost. The pharmaceutical industry also outsources clinical trials on human patients (costs range from $50 million to $300 million per drug). Of the roughly $1 billion needed to bring a drug to market, more than half that cost is tied to the four trial phases in which new medicines are tested on humans for safety and efficacy. Recruiting patients is the most expensive part of drug development, accounting for 40% of a trial's budget.[55] Therefore, pharmaceutical companies hire contract research organizations, who in turn subcontract the work to hospitals and physicians in emerging markets. Roughly 10% of clinical trials were outsourced in 1999; by 2005 that number had grown to 40%.[56] Running a trial cost about $30,000 per patient in the U.S. in 2005, but only $3,000 in Romania. The benefits include not only cost saving and faster time-to-market, but also more ethnically diverse testing and access to new markets. The downside, however, comes from the risk of exploiting vulnerable foreign patients and improper assurances of human subjects' protections. Moreover, meeting procedural requirements for the U.S. Food and Drug Administration can be tough in places where electricity is unreliable and other infrastructure problems exist.

HONE AND FOCUS ON CORE COMPETENCIES   Outsourcing allows a company to pare away many functions necessary for effective business management but which distract the company from its primary focus. Whether a small high-tech start-up is learning to manage its books and employee payroll or a large multinational company is trying to stay abreast of cutting-edge R&D, outsourcing nonessential tasks allows a company to focus on its own core competencies. For many mature technology companies, these competencies are in customer relationships, marketing, breakthrough innovation, and other more tacit skills. Atul Vashistha, coauthor of the book *The Offshore Nation,* is quoted by *BusinessWeek* as saying that the issue for companies is: "Don't tell me how much I can save. Show me how we can grow by 40% without increasing our capacity [fixed investments of our own]."[57]

CAPABILITIES OF OUTSOURCE PROVIDERS   Companies hire outsource providers because of their deeply embedded skills in specific domains. For example, IBM's legendary prowess in R&D (with its "legions of Ph.D.'s and closets of patents") is now available for hire for all sorts of companies.[58] IBM and Boeing have a 10-year R&D alliance to create technologies to link all branches of the military and intelligence services via an information-sharing network. IBM has teamed with the Mayo Clinic to search its digital database of 4.5 million patients for doctors to give instant treatment guidance. Although this latter project seems a stretch for IBM, scientists at both organizations had genomics projects that, when combined, yielded new insights and efficiencies. In addition, IBM has taken over BP's finance and accounting, allowing it to focus on ocean drilling and energy markets.

Sophisticated skill sets are as likely to be found in emerging countries as in developed countries. Boeing worked with HCL Technologies in India to codevelop software for its 787 Dreamliner jet, including navigation systems, landing gear, and cockpit controls. The solid education and knowledge base in emerging market countries, coupled with the lower wage rates and costs of doing business, make them well suited to perform many of today's business functions. Access to this skilled, low-cost talent pool, as well as the efficiency of these outsource providers' businesses, are key benefits.

TECHNOLOGY   Technology developments such as the Internet, Web conferencing, and other mobile technologies make outsourcing a practical undertaking. These technologies facilitate outsourcing by making it easier for companies to communicate with their outsource providers on a regular basis and to share the necessary information to facilitate the outsourcing process.

MITIGATE HUMAN RESOURCE MANAGEMENT ISSUES   A key driver in outsourcing is mitigating the overhead associated with labor costs in mature economies, including pension plans, medical insurance, and other employee expenses. Outsourcing often involves a transfer of personnel to the outsource provider's business ("rebadging"), and the outsource provider takes on these employee-related expenditures.

OTHER GENERAL BUSINESS TRENDS   Other good reasons for outsourcing include general business trends such as globalization, competitive intensity, and the faster pace of business in today's world economy. Each of these has driven companies to perform more capably under time and cost pressures, making a go-it-alone strategy particularly difficult and risky.

But Does It Work? Problems and Risks in Outsourcing

Despite the surge in outsourcing, studies question whether the promised benefits materialize. One study of 420 IT professionals found that only half of them rated their IT outsourcing efforts a success; one-third were neutral and 17% called them disasters.[59] Among the reasons cited (multiple responses allowed) for the dissatisfaction were:

- Poor customer service or vendor responsiveness (cited by 45% of respondents)
- Hidden vendor costs (39%)
- Insufficient up-front planning by "our" company (the customer) (37%)
- Insufficient vendor technical expertise (33%)
- Not enough contract/vendor management by "our" organization (the customer) (28%)
- Insufficient vendor business or industry expertise (25%)
- Inherent weaknesses in outsourcing model (24%)

One-quarter of companies surveyed in 2005 had brought some business functions back in-house after realizing poor results from outsourcing. In another study, only 42% of C-level and other executives at 650 companies in 32 countries said outsourcing contracts "definitely improved their financial performance."[60] Other findings from this study included these facts:

- 53% thought their service providers did not bring new business experience to the table that the customer didn't already have
- 38% believed that up to half of outsourcing arrangements fail
- Only 27% said that outsourcing helped their company improve its competitiveness

Clearly, outsourcing comes with a host of risks that make its benefits elusive.[61] "Lawsuits, recriminations, millions squandered on failed projects" make outsourcing a high-risk business practice.[62] Outsourcing can introduce added complexity and costs, and it requires more attention and management than anticipated. Moreover, the trend toward divvying up outsourcing among a team of vendors further increases complexity. Following are the key risks, also shown in Figure 5.5.

COST SAVINGS DON'T MATERIALIZE   The true costs of the function being outsourced are very hard to calculate in advance of the contract. With respect to IT outsourcing in particular, many companies underestimate what their in-house IT departments do and the flexibility they offer.[63] Outsource providers quote a set rate, but then charge for add-ons and extras.

QUALITY CONCERNS   Outsource contracts are signed with the intent that the outsource provider will deliver the same level of quality that the company's in-house function delivered. Yet, quality concerns have been a key reason cited for bringing outsourced functions back in-house: "Suppliers, for all their technical smarts, just don't understand the businesses of their customers."[64] Sprint Nextel canceled its outsourcing arrangement with IBM for system analysis and design. In other circumstances, a company's own customers experience the quality problems. Customers are finding that calls to toll-free numbers end with a customer service staffer who can handle basic inquiries—but complex problems seem to cause befuddlement at best and insensitive handling, causing customer outrage, at worst. The caller experiences an endless cycle of transfers as the customer service staff pass the call from person to person, trying to find someone who can answer the question.

DEPENDENCE ON THE VENDOR   Companies come to rely on their outsource provider to perform a key business function, but if problems arise, they find it difficult to pull the function back in-house. Referred to as "switching costs," outsourcing can make it costly for customers to find alternative sources for the function that the provider is performing, leaving them vulnerable to quality problems and price increases.

DILUTION OF COMPETITIVE ADVANTAGE AND RISK OF FOSTERING NEW COMPETITORS   Companies who outsource run the risk of diluting their competitive advantage.[65] This dilution can come from sharing important trade secrets and intellectual property with outsource providers—who may then use the information to compete on their own ("de-skilling" the partner), first in their own (host) country, and then in the customer's geographic locale. This risk of forward integration by the outsource provider into the customer's own markets is very real. Motorola hired Taiwan's BenQ Corporation to design new mobile phones, but then BenQ began selling phones in China under its own brand. This prompted Motorola to pull its contract.[66]

In addition, when a company outsources its R&D activities in particular, the products developed by an outsource provider may be less differentiated from competitors' products—particularly if the outsource provider is also developing products for other players in the same industry. When technology is standardized on a common underlying platform/architecture (e.g., on Microsoft's operating system and Intel's chip), designing and assembling products on an outsource basis is cheaper, but harder to differentiate. Dell, Motorola, and Philips buy completed designs of some of their products from Asian developers with minor tweaking and then merely add their own brand names. Says one outsource provider: "Many just take our product. Because of price competition, they have to."[67] At the extreme, companies who outsource more and more of their business functions—particularly the innovation function—may find themselves with less and less to call their own. Sometimes referred to as being *hollowed out*, companies "incrementalize themselves to the point where there isn't much left."[68] Especially in an area like technology innovation, where knowledge spillovers and serendipitous insights occur, companies can find themselves on a slippery slope and lose the incentive altogether to keep investing in new technology.

PUBLIC BACKLASH   Society in general has been critical of businesses who outsource, accusing them of undermining employment levels in their own country by giving jobs to people in other countries. In addition, government work that has been outsourced has become a focus of critics, saying that government projects should be performed within the country, rather than taking revenues (a source of taxes and income) elsewhere. This has made outsourcing a political issue. Outsourcing is also accused of taking unfair advantage of poor working conditions in developing countries, exploiting workers with low wage rates, poor working conditions, child labor, and other unsavory business practices.

As in so many partnerships, companies come to an outsource relationship with differing goals and objectives. OEM customers want to gain access to lower cost structures while retaining their proprietary advantages in product design and customer knowledge. Outsource providers want to aggregate volume across customers and to spread risk across a larger customer base. These are conflicting goals, with one party pressing for more commonalities across the customer set, and the other pressing for "cheap uniqueness" (an oxymoron at best).

Yet, despite these risks, 89% of companies surveyed in 2007 by KPMG said they planned to maintain or increase their current level of outsourcing.[69] So, the question most high-tech managers ask is not "Should we outsource?" but rather "How can we outsource effectively in order to reap the benefits and avoid the drawbacks?"

A Contingency Approach to Managing Outsourcing for Success

Managers are faced with a daunting "damned if they do and damned if they don't" choice: If they don't outsource, their companies won't be able to survive the brutally competitive global business environment; if they do outsource, they run a gamut of risks, pitfalls, and criticisms. When outsourcing goes bad, disgruntled employees leave, customers are frustrated, brand names are damaged, and outsourcing is given a bad name. At the extreme, outsourcing disasters provide ready fodder for critics who say that not only is offshored work of low quality, but it does not even reduce costs. The key for effective management of the outsourcing process is careful consideration of a variety of contingent factors,[70] as shown in Figure 5.6.

CRITICALITY OF THE BUSINESS FUNCTION   A company must consider very carefully what should be outsourced in the first place. It must carefully define its "mission-critical" business processes—those forming the foundation for its business mission and presence in the market. All activities that are core to these mission-critical activities must be kept in-house. With respect to the outsourcing of innovation in particular, a company's newest, breakthrough innovations—developed via right brain tasks that entail artistry, creativity, and empathy with the customer—should remain in-house as physically close to the customer as possible. Indeed, as reported in *BusinessWeek:* "Ownership of design strikes close to the heart of a company's intrinsic value. . . . The key is to guard some sustainable competitive advantage, whether it's control over the latest technologies, the look and feel of new products, or the customer relationships themselves. . . . A company should draw the line where core intellectual property is above it (meaning it shouldn't be outsourced) and commodity technology is below it."[71] In line with this approach, Apple designs its hit products in-house. Close partnership relationships that follow an open innovation model (R&D partnerships, joint ventures, strategic alliances) would be appropriate to tap into the benefits of knowledge outside the firm's boundaries. On the other hand, for incremental innovations, outsourcing the design work probably makes sense. Given competitive and market pressures to bring costs down over the product life cycle, outsourcing minor product modifications and "versioning" of established technologies into multiple product versions probably makes sense.

NATURE OF BUSINESS PROCESS AND DEGREE OF CUSTOMIZATION   Uday Karmarker recommends that companies match the nature of the business process (simple or more complex) and the degree of customization of the service being provided to the type of outsourcing that is adopted:[72]

- For *simple processes* that are relatively *standardized across customers* (e.g., retail banking, billing, telemarketing), outsourcing can be used to save costs.

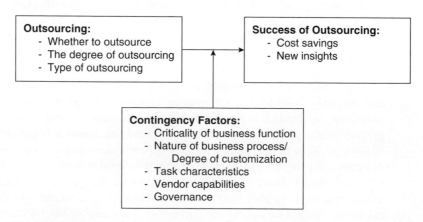

FIGURE 5.6   Contingency Approach to Outsourcing

- For *simple processes* that must be *customized for specific customers* (e.g., retail sales, website design, technical support), companies can save costs not only by selective outsourcing, but also by investments in self-service technologies and automation.
- For more *complex processes* that are relatively *standardized across customers* (e.g., credit analysis, technical research, software development tax preparation, medical diagnosis), captive offshoring and selective outsourcing can save costs.
- For *complex processes* that are *customized for each customer* (e.g., personal financial planning, relationship marketing, engineering, design work), companies should keep such activities in house, or decouple their processes and selectively outsource some components.

TASK CHARACTERISTICS Cost savings from outsourcing are premised on the following:

- *Economies of scale:* Is the business function that is being outsourced one that can be aggregated across customer/OEM businesses? If not, how will the outsource provider generate economies of scale?
- *Transfer of explicit, codified knowledge so the outsource provider can perform the desired task accurately:* Can the business function that is being outsourced be clearly mapped and communicated to the outsource provider? One estimate is that 80% of engineering tasks can be outsourced: Prototypes can be translated into product design specs; mature products can be upgraded; quality can be tested; user manuals can be written. These left brain tasks, which can be "boiled down to a spec sheet," will be shipped to cheaper areas.
- *Clearly specified ownership of intellectual property rights and risks:* Is the business function or process one where intellectual rights and responsibilities can be clearly identified and articulated? If not, then outsourcing risks increase. Moreover, if an outsource provider performs similar functions for competitors, are there security procedures in place to guard one company's proprietary information from seeping to other personnel who work on competitors' tasks?

VENDOR CAPABILITIES A company should evaluate and select its potential vendor very carefully. As an outsource provider "moves up the value chain" (meaning that it takes on more and more functions related to the core process, at higher and higher margins), a key question becomes whether the provider has the requisite capabilities to take on more challenging tasks. Manufacturing requires a set of capabilities in logistics, operations, and supply chain management that are very different from the capabilities required for software engineering or for product design. For example, the latter requires capabilities in market analysis, customer needs requirement, and trade-offs in functionality, design, and costs. Clear articulation of the underlying capabilities that are required for the outsource provider to be successful in performing the specific functions is necessary. Moreover, clear articulation of the customer's business and process is key.

GOVERNANCE A company should consider carefully the specific types of governance that it uses to manage and control the interorganizational relationship with its provider. Both parties must manage dependence, risks of sharing proprietary information, and trust issues. Governance is particularly important in managing outsourced creative work such as new product development and innovation (areas where those task characteristics mentioned above can be particularly weak). In general, a client's attempts to control a supplier's creative work place limits on the supplier's creativity and performance, clearly detrimental to the success of the outsourcing effort. Yet, a client must have some assurance that the supplier's efforts will meet its needs, despite the governance challenge. To walk this fine line, companies should distinguish between *ex ante* control (meaning that governance is designed into the outsourcing contract before the creative work is performed) and *ex post* control (meaning that the client exercises actual control during the performance of the creative work).[73] Research shows that to get the best supplier performance for outsourced new product

TABLE 5.8 Keys to Best-Practices Outsourcing

- **Have clear reasons.** Don't outsource just because competitors are. Rather than outsourcing at all, a company may find that it can boost its own efficiency by re-engineering its own operations or improving its own technology.
- **Don't outsource a mess.** Merely shifting a poorly performed business function to someone else won't fix the problem. Dissect workflow process meticulously; map out everything; build a tight outsource process.
- **Set up the right relationship.** A company may decide to run its own offshore subsidiary via a "captive outsourcing" model. For example, to keep control of its proprietary technology and processes, Boeing used captive offshoring in Moscow, Russia, where it hired 1,100 skilled aerospace engineers for a range of projects including the design of new titanium parts for the 787 Dreamliner jet.
- **Be ready for possible backlash** from customers and employees.
- **Invest time and effort to make it work.** Offshore workers need training. Management must be involved. Quality control must be carefully defined and monitored.
- **Treat partners as equals.** Invite the partner to be involved.

Source: Kripalani, Manjeet, "Five Offshore Practices That Pay Off," *BusinessWeek,* January 30, 2006, pp. 60–61.

development, a client should carefully design controls into a contract *before* the creative work is performed; exercising less control during the actual performance of the work allows the supplier to perform to its highest capability. This seems to make sense as creative work is best done with agreed-upon review meetings and deadlines set before the work begins, and minimal interference beyond these while the work is being undertaken.

Other key success factors for outsourcing are shown in Table 5.8.

The Future of Outsourcing

Outsourcing will continue to evolve as a unique type of partnership. Its evolution will encompass global, political, and managerial aspects.

Globally, the performance of specific business functions will continue to migrate to those low-cost areas of the world that have the requisite skills to perform the tasks. This evolution is inescapable, but lessons from the past will tell us that costs even in those areas will rise. For example, while India has experienced an economic boom as companies have sought out its low-cost, talented labor pool to perform a wide variety of business functions, its costs are also rising. Turnover of talented employees, rising pay scales, hiring and annual bonuses, and the need for a competitive benefits package—which can even include a dating allowance for employees—will inevitably erode the cost differences over time.

Politically, countries and governments engage in the rhetoric of lost jobs (wages, taxes, earnings)—particularly in election years. Although providing symptomatic relief is a necessary safety net, long-term viability is found in an educated workforce with cutting-edge skills for jobs of the future.

Finally, from a business/managerial perspective, the pendulum will continue to swing as companies strike the right balance between in-house work, strategic alliances, and outsourced relationships.

OPEN INNOVATION NETWORKS AND ALLIANCES FOR NEW PRODUCT DEVELOPMENT

The prior section addressed R&D as one of the business functions that companies have been outsourcing. Although cost efficiencies often drive the outsourcing of innovation, the costliness of developing new products is only one factor affecting companies' approaches to innovation. The growing complexity and uncertainty of R&D, the globalization of industries, the convergence of technologies, and resource constraints all make the innovation process challenging in any environment, but particularly so in high-tech environments.

Open Innovation Networks

To deal with these challenges, companies are examining different models to pursue new product development. One such model, referred to as **open innovation**,[74] develops an innovation ecosystem or network, tapping into partners' expertise to develop the next breakthrough innovation. Companies are also using other new techniques to bring new perspectives to the innovation process, such as customer co-creation; we cover this technique in Chapter 6.

IBM is reinventing the way it innovates based on the open innovation model,[75] recruiting R&D partners aggressively. It does chip R&D with nine partners including Advanced Micro Devices, Sony, Toshiba, Freescale Semiconductor, and Albany Nanotech, a university research center. IBM partners have contributed more than $1 billion in equipment purchases as well as skilled brainpower from scientists and engineers. In addition, IBM has made 500 software patents freely available to outside programmers; its proprietary technology is envied by competitors and it remains the world's top patent holder with more than 40,000 per year. It has helped fund the Open Innovation Network, a company formed to acquire patents and make them available, royalty free, to support the open source software movement. According to John E. Kelly III, IBM's senior vice president of technology and intellectual property, "We want to do the things that make markets grow . . . [because] we know we'll get at least our fair share."[76] "We are the most innovative when we collaborate," adds IBM CEO Sam Palmisano.[77]

BusinessWeek magazine, in collaboration with the Boston Consulting Group, conducts an annual survey of senior executives in global companies to determine its list of 50 most innovative companies. It asks them several questions with a view to summarizing best practices in innovation. One of the most important practices that came out strongly from the surveys conducted in 2007 and 2006 was companies' realization that they can't do innovation alone. Increasingly, they have to draw in suppliers, customers, and even competitors to get creative ideas, speed up innovation, and share costs and risks of development. Boeing, for its 787 Dreamliner, departed from its tradition of designing all components in-house, relying instead on a global network of suppliers: Mitsubishi Motors of Japan created the wing, for instance, while Italy's Alenia Aeronautica produced the rear fuselage and horizontal stabilizer.[78]

Companies that follow an open innovation model develop innovation processes that transcend local industry clusters and national boundaries; they harness diverse knowledge sources that include both technical know-how and market knowledge to generate new innovations. This "new imperative" in innovation attempts to match the complexity of market and technological knowledge to a company's innovation process to maximize success.[79] In addition, companies who understand the importance of their web, or network, in stimulating innovation realize that the same inertial forces that can stymie internal innovation can also operate at the network level as well: "The ties that bind may become that ties that blind."[80] Toward this end, companies who want to harness their networks for new insights, competencies, and relationships must actively manage them. They can create new relationships with network partners in either local or distant areas, they can build relationships with unusual partners; or they can even move into totally uncharted territory. The idea is that while innovation can be the result of serendipity, just as likely it is the result of careful cultivation of relationships that create new opportunities—and then the company must be smart enough to seize them. As the saying goes, "Good luck doesn't necessarily just happen, but is created!"

New Product Alliances

New product alliances are a unique form of strategic partnership that represent another approach to meeting the challenges of developing innovations. New product alliances are formalized, collaborative arrangements among two or more organizations to jointly acquire and utilize information and know-how related to the research and development (R&D) of new product (or process) innovations. As with open innovation, a key motivation of new product alliances is to access new sources of knowledge. However, the "logic of innovation" clashes fundamentally with the "logic of alliances."[81] Alliances succeed when the goals and responsibilities of partners are detailed clearly— but at the same time, innovation requires flexibility and spontaneity. Moreover, although companies enter into alliances to tap into partners' knowledge and expertise, many alliances are characterized by a lack of trust that restricts the free flow of information critical for new product success.

To maximize the effectiveness of new product alliances, firms must pay attention to the nature of the ties between alliance partners, in addition to the governance factors previously covered in this chapter. Studies of new product alliances suggest that cooperation and appropriate relationship governance are critical to their success.[82] A study of 106 new product alliances between U.S. firms found that those composed primarily of horizontal partners have "weak ties." They display lower levels of organizational closeness but higher levels of redundancy (similarity) in knowledge than vertical alliances—and that this combination is associated with lower levels of knowledge sharing.[83] Thus, managers who want to develop new products through horizontal alliances with competing firms face the challenge of cooperating with firms that (1) can provide relatively little complementary knowledge and (2) are reluctant to share what knowledge they do possess.[84]

Despite the difficulty of sharing knowledge, horizontal new product alliances exhibit higher levels of new product creativity and faster speed of development, possibly due to the synergies created by the overlap in product development related knowledge, which allows each firm to fully harness the knowledge because it has the requisite skills and capabilities to do so.

An interesting issue is the role of geographic proximity of partners in new product alliances. Does geographic proximity help improve face-to-face communication between partners, which, in turn, leads to better knowledge transfer and new product success? One study found that geographic proximity is probably less important than the relational ties that firms establish (regardless of geographic proximity).[85] Even firms in close geographical proximity must not take closeness for granted; whether partners are located near or far, high-tech firms must nurture strong relational ties with their partners via face-to-face communication. After firms establish close ties, then e-mail can be an efficient and effective medium for product and process knowledge that can be codified; however, face-to-face communication is still better for sharing tacit or uncodified knowledge.

Another important aspect of geographic proximity tied to innovation is the development of **industry clusters**.[86] Industry clusters are concentrations of economically and socially linked companies and institutions; the embedded relationships facilitate knowledge sharing across boundaries, leading to knowledge spillovers and other desirable characteristics (e.g., economies of scale and scope in the employment of capital, transportation and communications infrastructure, demand, and the creation of social capital). Such traits result in a high level of innovativeness, as seen in places like Silicon Valley and Route 128 (the Boston Corridor), to name just two. Governments work to stimulate the positive returns that clusters provide, and the virtuous cycle of innovativeness seems to be fairly stable over time. Even in base-of-the-pyramid markets, clusters can develop that allow local market economies to flourish and their distinctively indigenous products to be marketed in a global arena.[87]

A key purpose of alliances for new product development is to harness new sources of knowledge and to learn from partners.[88] However, learning from partners in strategic alliances has both an upside and a downside. On the one hand, interfirm learning can lead to a win-win situation, in which both parties improve their skills and performance in the marketplace. Further, learning about each other can make the partnership more successful and contribute to positive relationship dynamics. On the other hand, as mentioned in the discussion of outsourcing, interfirm learning can involve the "de-skilling" of a partner, whereby one party absorbs the proprietary information of its partner, ultimately leading to a situation in which the de-skilled partner is no longer needed. These two views of interfirm learning are not mutually exclusive, but instead are complementary sides to the same coin. Hence, the real challenge is to manage the paradox by using tools that allow firms to maximize the upside potential of interfirm learning while limiting its downside risks.[89]

TECHNOLOGY SOLUTIONS FOR GLOBAL PROBLEMS
Empowering the People: Access to Safe Drinking Water
By Jamie Hoffman

"Water, water, everywhere, nor not a drop to drink," lamented Coleridge's Ancient Mariner upon drifting, parched on an ocean of undrinkable water. Some 20% of the planet's human population suffers from the same plight as the Mariner, as they do not have access to clean drinking water; the 1.1 billion people who drink or bathe in tainted water often acquire deadly and physically debilitating diseases. Dean Kamen, world-famous inventor of the Segway mobile transportation device, has delivered a groundbreaking technology to address this problem. The machine, dubbed "the Slingshot" by Kamen (using the term *slingshot* to reference the simple technology that David used to bring down Goliath) puts safe water back into the hands of the common people. The Slingshot is a portable, efficient, and economical alternative to large water purification plants.

The size of a dorm fridge, the Slingshot employs vapor compression distillation in an incredibly efficient way to transform any water source (urine, ocean, puddles, or even arsenic) into drinking water. With a Slingshot, villages could produce 1,000 liters of drinking water per day. It runs on less than 1 kW of power, weighs 100 kg (220 pounds) and will cost around US$1,000–2,000 to manufacture.

This technology provides the means to produce clean water without large water purification plants. Developers envision pairing the Slingshot with an entrepreneurial spirit. For example, through microlending, village entrepreneurs could bring the Slingshot into their villages and provide clean water to their village, for a fee.

Sources: Pearson, Ryan, "Segway Inventor Drinks His Own Pee (Really)," *Orange County Register,* December 15, 2005, online at www.ocregister.com/ocregister/healthscience/atoz/article_890581.php; and Whitesides, Loretta, "Colbert and Kamen Solve the World's Water Problems," *Wired,* March 25, 2008, online at http://blog.wired.com/wiredscience/2008/03/colbert-and-kam.html.

To the extent that the most valuable knowledge is tacit—which is also the most difficult to transfer—firms have an incentive to structure the closest type of partnership agreement possible. Indeed, the only way to learn skills and competencies that are highly embedded in organizational routines is to partner closely. Firms seeking to learn and internalize tacit knowledge must absorb it through an apprenticeship model of learning, which often entails collaborative agreements such as joint ventures and R&D consortia. Close collaborative agreements allow organizational routines to be examined and understood, in terms of what is done, why it is done, and how it is done. Because more tightly linked relationships face greater risks from interfirm learning, the higher the firm's technological innovativeness, the less likely it is to use more transparent types of partnering arrangements such as joint ventures and research and development agreements, compared to marketing and licensing agreements.[90]

The conditions under which interfirm learning is optimized and its impact on the performance of the alliance are similar to the other factors covered previously in Table 5.6. Interested readers are referred to the resources cited in the chapter endnotes.[91]

CUSTOMER RELATIONSHIP MARKETING/MANAGEMENT

Since its founding in 1984, Cisco has been a leader in the development of Internet Protocol (IP) networking technologies. In 2007, Cisco earned $8 billion on revenues of $38 billion, yielding a profit margin of 21.4%—far exceeding that of any of its competitors. Cisco appeared on the 2007 list of the world's 50 most innovative companies compiled by *Fast Company* magazine. One of the reasons for Cisco's success is its marketing philosophy, as explained by Robert Lloyd, senior vice president: "Looking for a sustainable competitive advantage? Make each customer feel like a market of one."[92]

Consider Cisco's relationship with PictureVision.[93] PictureVision, an independent subsidiary of Eastman Kodak, possesses a core technology that provides the foundation for digital photography services. PictureVision was rolling out Cisco PhotoNet packages to Wolf Camera's locations across the country and needed a cost-effective and efficient way to meet the specific networking needs of each site. Cisco's Customer Service and Internet Commerce teams worked with PictureVision to ensure that the routers were configured correctly and efficiently. Prior to the rollout, Doug Lavanchy, a Cisco customer service specialist, trained the PictureVision staff on the Cisco online tools including Configuration Express. With Configuration Express, PictureVision was able to order and configure routers that were shipped directly to customers, reducing costs for warehousing, shipping, and manually configuring the devices. Effective communication was essential to meet the required project delivery time frames. To ensure easy tracking of the shipments, Lavanchy maintained a spreadsheet for each location that was forwarded to PictureVision, Cisco Systems Capital (the leasing company used by PictureVision), and Cisco's manufacturing facilities. By relying on the Cisco Customer Service department and Configuration Express, PictureVision achieved a 20% reduction in the costs of delivering a state-of-the-art networking solution to Wolf Camera.

Customer relationship marketing (CRM), also sometimes called *customer relationship management,* refers to the development of close, long-term relationships with customers that provide mutual benefits, or win-win solutions. Companies that follow this philosophy recognize that marketing techniques can be used to identify desirable customers, and then keep them satisfied and keep them coming back. Indeed, the cost of retaining customers is only about one-fifth the cost of finding and acquiring new customers. Moreover, as important as acquiring new customers is, retaining the most valuable customers provides a greater lift to a company's profits. A 1% increase in customer retention increases profitability by 3–7%.[94] As a result, many companies have shifted their philosophy from managing the enterprise as a portfolio of products and services to managing it as a portfolio of segments and customers, with the goal of building customer equity.[95]

A company with a relationship marketing philosophy changes its thinking about marketing activities. More specifically, a decision to acquire customers should be treated like other investment decisions: Investments in customer acquisition costs (cash outflows) must be evaluated relative to the resulting profit stream from retained customers (cash inflows).[96] *Acquisition costs* include such things as price discounts or rebates, advertising campaigns, direct marketing, and personal selling. For example, in 2005 the Internet telephone service provider Vonage incurred an acquisition cost of roughly $400 per new customer.[97] By 2006, this cost had come down to $239 per new customer.[98] Customer acquisition costs are higher for newer brands that customers are not familiar with, especially when these new brands are introduced with large national advertising campaigns.

As the Vonage example suggests, to compute customer acquisition investment, the individual who performs the analysis, whether a marketer or financial analyst, should determine the *acquisition cost per new customer:*

(Total cost of the marketing campaign) / (Number of prospects who become customers)

Then, subtracting the acquisition cost per customer from the revenue generated by a new customer's first set of purchases yields the net acquisition investment that a company can spend on a per customer basis:[99]

Revenue from customer's first purchase – Acquisition cost per customer =
Acquisition investment (spending) per customer

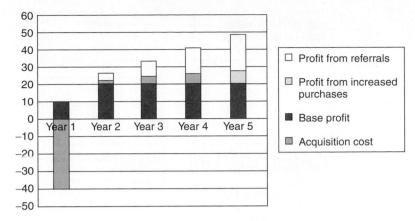

FIGURE 5.7 Computing Customer Equity

Customer equity refers to the net present value of the cash flows associated with a particular customer or segment. Calculation of customer equity—sometimes also called **lifetime value (LTV)**—requires estimating the future profit stream generated by the customer, discounting it to its present value at the firm's cost of capital, and subtracting the customer acquisition cost and associated retention costs over time. The profit stream stems from the base profit generated by the customer, increased business from the customer over time, and referrals of new customers from the loyal customer. See Figure 5.7 for an illustration of this concept. For positive customer equity to be created, the net present value (NPV) of the cash inflows must exceed the present value of the cash outflows. For example, the customer equity of American Airlines was estimated in 1999 at $7.3 billion, pretty close to the $9.7 billion market capitalization of the company.[100]

An excellent resource for simplifying the quantitative aspects of this approach to managing customers as investments, including many extremely useful and interesting examples, can be found in Sunil Gupta and Don Lehmann's book by the same name.[101] A brief overview of the types of data a company needs in order to quantify its CRM approach is shown in Table 5.9.

The development of relationships that build positive customer equity is a challenging process for most businesses. The process follows three steps, illustrated in Figure 5.8 and discussed next. Any company interested in developing a relatively sophisticated customer equity management system should also invest in CRM software to create a database of customers and reliably measure the return on marketing investments; this can also lead to opportunities to cross-sell or up-sell customers.[102] CRM software is covered later in this chapter.

TABLE 5.9 Quantitative Numbers (Financials) Needed for Customer Equity Management

1. Total marketing cost to acquire new customers
2. Number of prospects reached during the campaign
3. Number of prospects who became customers
4. Revenue from a new customer's initial purchase
5. Expected retention duration for a customer
6. Annual revenues expected over the customer's relationship duration with the company
7. Costs to serve a customer (retention costs)
8. Firm's cost of capital (discount rate)
9. Present value chart (time value of money)

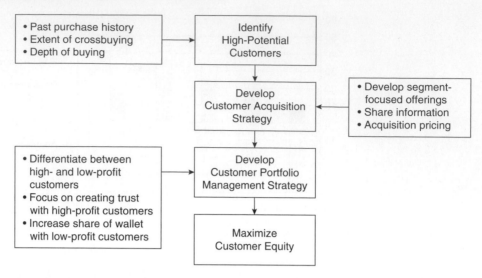

FIGURE 5.8 The Customer Equity Management Process

Step 1: Identify High-Potential Customers

High-potential customers are those who will generate a profitable revenue stream over time for the company. Identifying these customers requires estimating the lifetime value (LTV) of a customer or customer group. As discussed previously, LTV, or customer equity, is computed by estimating the revenue stream over a likely time horizon that the customer will stay with the company, discounting that stream to its present value, and subtracting the acquisition and retention costs. If the net present value (NPV) is positive, the company should pursue a relationship. If it is negative, the marketing team should decrease its costs to acquire or serve the customer, or should increase prices to the customer. If these actions cause the seller to lose the customer, that is acceptable because the customer would not be profitable.

This customer evaluation ideally should be computed at the customer level. While this is realistic for producers of big-ticket items such as MRI scanners (used in the diagnosis of many medical conditions) or ERP software systems, and by companies such as Dell and Charles Schwab that have very sophisticated customer databases, it may not be feasible for many sellers of high-volume, low-price goods. In the latter case, the analysis should be done at the segment level and an average acquisition investment per customer computed.

Alternatively, when such an analysis is not feasible, the analyst may instead examine criteria associated with positive customer equity. Specifically, "customer share of wallet" and "customer cross-buying" are significant predictors of relationship potential.[103] *Customer share of wallet* refers to the percentage of business in a specific category that a customer does with a particular vendor or company. Customers who have a large proportion of purchases in a category are likely prospects for relationship expansion and enhancement. More satisfied customers spend a greater share of wallet and have longer relationships with their service providers. *Cross-buying* refers to the degree to which customers purchase other products or services from the seller. Customers who engage in cross-buying are typically motivated by the convenience of one-stop shopping with a specific vendor, as they, too, are more likely to continue a relationship.

A final approach that may be used to identify high-relationship-potential targets is first to identify the characteristics of loyal, profitable customers, and then to target new customers who share the same characteristics.[104] For example, in 2002 Best Buy examined sales records and demographic data in its computer database to identify good customers—those who boosted profits by purchasing new consumer electronics without waiting for markdowns and rebates—and bad customers—those who, say, bought products, applied for rebates, returned the purchases, and then bought them back at returned-merchandise discounts; they loaded up on "loss leaders" (severely discounted merchandise used to boost store traffic) and then flipped the goods on eBay; they bargain-hunted online to bring in price quotes from websites to get Best Buy to meet the lowest price. Best Buy estimated that as many as 100 million of its 500 million customer visits each year were from these less desirable customers![105] From this analysis, it concluded that its most desirable customers fell into five distinct groups: (1) upper-income men ("Barrys," who tend to be enthusiasts of action movies and cameras), (2) suburban mothers ("Jills," who are busy but like to talk about helping their families), (3) technology enthusiasts ("Buzzes," who are interested in showing off the latest tech gadgets), (4) price-conscious family guys ("Rays"), and (5) small business owners ("Mr. Storefronts"). Each store was asked to analyze the demographics of its local market, and then focus on two of these groups that best matched the local market characteristics and stock merchandise accordingly. Store clerks received training on how to identify desirable customers according to their shopping preferences and behaviors based on quick interviews to categorize shoppers. "Barrys" and "Jills" were lured with time-saving services like a personal shopper to help them find desirable items or to coordinate service calls. Best Buy also modified its interior layout as well, to appeal more to these shoppers. Best Buy also worked to deter customers who drove profits down by enforcing a 15% restocking fee on returned merchandise, culling their names from mailing lists, and cutting links between online discounters and BestBuy.com.

Step 2: Develop a Customer Acquisition Strategy

After determining which customers to pursue, the next step focuses on how much to spend on pursuing them. The aggressiveness of a company's acquisition strategy (the amount of its investment) in acquiring a customer or customer group depends on the likelihood of realizing adequate future cash flows. Companies that fail to link their customer acquisition strategies to customer retention metrics will potentially have unprofitable customers in their portfolio. As shown in Figure 5.9, four generic acquisition strategies help a company to balance its investment in acquisition with the returns from retention:[106]

- *Full Throttle*   (low risk/high return): When retention probability is high and the time horizon to recoup the acquisition investment is low, customer acquisition represents a major opportunity; therefore, the firm should invest as much as possible in acquisition until the NPV of the marginal customer is negative.

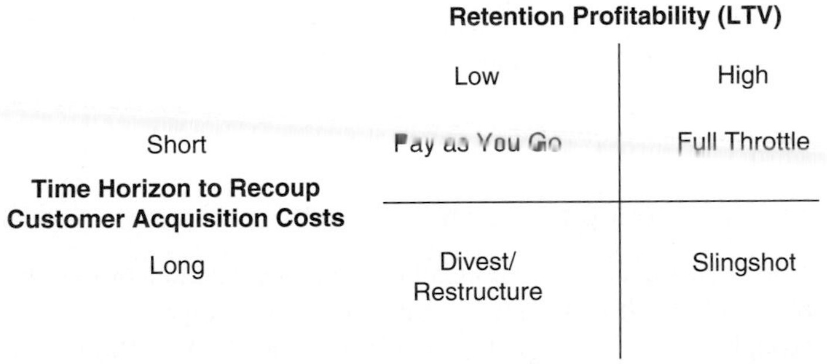

	Retention Profitability (LTV)	
	Low	High
Short	Pay as You Go	Full Throttle
Time Horizon to Recoup Customer Acquisition Costs		
Long	Divest/ Restructure	Slingshot

FIGURE 5.9   Four Customer Acquisition Strategies (Balance time horizon to recoup customer acquisition costs against lifetime value)
Source: Blattberg, Robert C., Gary Getz, and Jacquelyn S. Thomas, "Managing Customer Acquisition," *Direct Marketing* 64 (October 2001), pp. 41–55.

- *Slingshot* (high risk/high return): When retention probability is high but the time to recoup the acquisition investment is long, the long time until payout makes the investment in customer acquisition risky. However, the more invested in acquisition, the greater the future payout. In this situation, the firm must "bet" on a high NPV driven by high retention profits. A slingshot strategy is typical of the customer-investment environment among Internet companies. Many e-tailers invested heavily in acquisition spending without recognizing the risk associated with lower-than-expected retention profits.
- *Pay as You Go* (low risk/low return): With a low customer acquisition cost and short payback period (most of the investment recovery occurs during the first period), the firm's acquisition strategy should be to invest as though all profits will accrue in the current period. In other words, short-term profit goals drive the firm's investment.
- *Divest/Restructure* (high risk/low return): In this situation, because customer acquisition will not pay off (the initial payback is low and retention and add-on sales are low), a firm must restructure its marketing system. The firm will not be profitable in the long run, and it will have very low customer equity.

Customer acquisition strategies are also affected by differentiation and pricing decisions.

DIFFERENTIATION To fully benefit from investments in customer acquisition, a company must provide a differentiated product that is adapted to needs of the target customers or market segments.[107] Service support and personal interaction are core differentiators, as are the seller's expertise in and its ability to improve the customer's operational effectiveness and efficiency. Interestingly, in business-to-business settings, service support, personal interaction, and the ability to improve the customer's operations are more important than product quality, delivery performance, and acquisition spending in helping a firm maintain its key supplier status with customers over the long term.[108]

PRICE Using pricing tactics (i.e., discounting) as a customer acquisition strategy is a double-edged sword. Some discounting may be necessary to induce switching from one seller to another, but customers who buy only when deep discounts are offered tend to buy in much smaller quantities than average. In other cases, they defect to another company before their acquisition and other up-front costs are recovered. These customers are "price butterflies," bargain hunters chasing deep discounts.[109] One study found that lost customers who are reacquired at higher prices have longer second tenures.[110] That study also found that after the relationship is reestablished, price increases have no effect on second tenure duration.

The specific recommendation for high-tech companies is to clearly articulate the superiority of non-price elements in the value proposition while offering a modest price inducement to encourage trial and switching. The signal that a price focus may send is that the seller has to compensate for an inferior product/service offering by undercutting the competition on price. The new customers who will be attracted by a low-price strategy are the price-sensitive customers who are neither loyal nor profitable.

Step 3: Develop the Customer Portfolio Management Strategy

Not all customers in the portfolio are equally valuable or worthy of efforts to retain them. For example, when Skelton-Tomkinson, an Australian firm, raised prices to some unprofitable customers, although revenues dropped from $20 million to $8 million, losing those customers paid off as net profits increased by 98%.[111]

Rather than assuming long-term loyal customers are profitable, Reinartz and Kumar[112] show that some loyal customers may be more costly to serve and may demand lower prices than customers who are not as loyal. At a minimum, the decision to maintain customer relationships must include an assessment of both customer profitability and the projected duration of the customer's relationship with the company. With a more fine-grained analysis based on these two dimensions, different relationship strategies are appropriate for different market segments that the firm serves. In particular, Reinartz and Kumar identify four categories of customers and strategies for managing customers in those categories:

1. *True Friends:* The most valuable customer group, true friends are highly profitable, loyal customers who find a good fit between the company's offerings and their needs. At least until 2006, PC manufacturers like Dell might have been considered true friends to Intel and Microsoft, because they offered PCs based only on Intel microprocessors and the Windows operating system.

These relationship-oriented customers seek to build strong and extensive social, economic, service, and technical ties with the intent of increasing mutual benefit.[113] True friends often have a business philosophy that is similar to the seller's philosophy, have roughly the same power in the relationship as the seller, anticipate conflicts of interest and develop resolution processes, and pay substantial attention to measuring benefits and costs for both parties.

The greatest risk in marketing to true friends is that of *overkill,* or intensifying the level of contact and potentially drowning them in marketing messages. Because they are already loyal, the overkill strategy can needlessly increase costs or even alienate customers without changing their purchasing habits. Rather, a company should attempt to stimulate these customers to engage in word-of-mouth communications with others, and to reward them for their loyalty by giving them preferential treatment. These are the customers that the company should strive to retain.

Retention is accomplished by keeping the relationship fresh. All relationships have the potential to go stale or sour. It is essential for both parties to be honest with each other. Although there may be a contractual obligation, a relationship is doomed if the parties feel that they can't trust one another. The larger or stronger partner should be careful about the way it wields power; the abuse of power may produce an imbalance in benefits, jeopardizing the future of the relationship. Rather, the relationship should be mutually beneficial. The seller should continuously ask questions that encourage customers to relate problems and sources of dissatisfaction and regularly update the offering to better address the customer's needs. Open and frequent communication can go a long way toward identifying opportunities and avoiding conflict.

2. *Butterflies:* The second most valuable customer group are the butterflies, short-term (transient) customers who also are highly profitable. Butterflies are shoppers, constantly searching for the best value at a particular point in time. Korean cell phone manufacturers LG and Samsung got their initial entry into the U.S. market by capitalizing on butterfly customers like Sprint and Verizon who wanted the latest features at an affordable price within very tight deadlines.

The goal with butterflies is to capture as much of their business as possible in the short time they buy from a particular company. Although it is tempting to try and convert butterflies into true friends, these attempts are rarely successful. A common mistake in marketing to this group is to continue to invest in retaining or reacquiring them after they have switched their business to a new company. Absent the ability to increase these customers' switching costs, the company should market intensively during the time the butterfly customer buys from them, follow up with marketing efforts after their most recent purchase, and if these follow-up efforts are unsuccessful, stop marketing to these customers altogether.

3. *Barnacles:* Barnacles are very loyal customers in that they desire long-term relationships. However, they are not very profitable because the size and volume of their transactions are too low or the cost to serve them is too high. In this sense, these customers can be the most problematic: Like barnacles on the hull of a cargo ship, they create drag. In July 2007 cell phone service provider Sprint sent letters to about 1,000 subscribers terminating their contracts because they were making more than 25 customer service calls per week, a rate 40 times higher than average customers.[114] Properly managed, however, barnacles can sometimes become profitable. The selling company should attempt to determine whether it can renegotiate the value proposition or capture a greater share of their business.

Renegotiation begins by engaging customers in a dialogue about the value proposition. The aim is to educate them about the value of the secondary and tertiary benefits the seller provides. This is the foundation for negotiating a new pricing structure that is fair both to the seller and the customer.[115]

If a company decides to attempt to capture a greater share of the customer's business, it first must estimate the share that it currently has. Estimates by the sales force and market research, whether conducted by the company or a third party, are the most common routes to developing a customer share database. With a picture of its share of wallet, the company can better understand the growth potential with that customer. If there is substantial potential, the seller must develop sufficient knowledge about the customer's business to develop a compelling value proposition that clearly articulates the advantage of the seller's offering relative to the next best alternative. However, since this customer may already have a preferred supplier, the seller must be prepared to document or even guarantee productivity improvements or cost savings.[116]

If the seller is not able to have the buyer accept a new value proposition that reflects increased prices or reduced benefits, or if it cannot capture a greater share of the customer's business, it may lose many barnacles—but if these are unprofitable customers, that is acceptable.

4. *Strangers:* Transaction-oriented customers, or strangers, tend to focus on price rather than value. They focus on the timely exchange of products for highly competitive prices. Each transaction is treated as a discrete event. These transactions are typically characterized by limited communication between the buyer and seller. Because of this attitude toward exchange, they will have the lowest profit potential. The company should identify these customers early on and not invest in marketing to them. Just as these customers treat every interaction as a transaction to be exploited for maximum benefit, the seller must profit on every transaction.

A company that has learned how to segment its customer base for relationship-specific efforts is Tech Data, a computer distributor. During the industry downturn in 2001, it was able to remain profitable by applying sophisticated CRM techniques. Tech Data's customers include corporations, smaller distributors, and computer retailers. For each customer, Tech Data computed 150 customer service costs, ranging from average order size to freight charges. Based on these data, it calculated gross expenses and margin on every account. CEO Steven Raymund was able to persuade Tech Data's barnacles to order more efficiently—for example, by placing one order of $1 million rather than 10 orders of $100,000 each—or their business was terminated. It was also willing to link its true friends directly to its inventory system as a way to bond relationship-oriented customers through high-value services.[117]

In addition to carefully analyzing its customer portfolio to determine which customers are worthy of retention efforts, a company must also carefully attend to how it creates value for customers over the life of their relationship. Evidence suggests that a 1% increase in customer satisfaction can lead to a 3% increase in the market capitalization of a company.[118] Careful attention to measuring and improving value for customers is a key to retaining them over time. Many useful resources exist to assist a company with this aspect of its customer relationship management strategy.[119]

Customer Relationship Management Software

The sophisticated strategies used in customer relationship management require investments in CRM software. Research by the Gartner Group projected worldwide CRM software revenue to surpass $7.8 billion in 2008, a 14.2% increase from 2007.[120] While a custom CRM implementation can be quite expensive for a company, a new trend is to use an application service provider like Salesforce.com to provide similar benefits through a simple Internet browser at a low annual per-user fee.

At a broad level, these software systems are used to capture data about customers from any contact within the enterprise. These one-to-one marketing applications provide the ability to track profitability per customer; to detect customers' dissatisfaction before they leave; and to improve product selling, retention, loyalty, and revenue. Independent surveys have shown that business customers of Siebel, one of the largest vendors of CRM software, report a 13–24% increase in contentment of their customers after installing Siebel CRM software.[121]

More specifically, customer relationship management software includes the following:

- *Sales force automation*—to allow sales reps to track accounts and prospects
- *Call-center automation*—to create customer profiles, to provide scripts to help service reps answer customer questions or suggest new purchases, and to coordinate phone calls and messages on websites
- *Marketing automation*—to help marketers analyze customer purchasing histories and demographics, design targeted marketing campaigns, and measure results
- *Web sales*—to manage product catalogs, shopping carts, and credit card purchases
- *Web configurators*—to walk a customer through the process of ordering custom-assembled products
- *Web analysis and marketing*—to track online activities of individual shoppers (offering merchandise they're likely to buy based on past behavior and offering special on-the-spot prices for specific customers) and to target marketing via e-mails to individual customers

When a customer calls a company that has installed CRM software, the representative at the other end of the line can immediately access that customer's information on historical interactions culled from a variety of databases,[122] quickly solve customer problems, and make new offers to sell additional products and services. For example, salespeople can use CRM software to figure out how much a prospect is probably authorized to spend, profile a product against competition, or find an advocate within the customer's organization. Customer service reps can use the data in these programs; the knowledge enables a meaningful, intelligent dialogue with the individual customer. These programs provide a strong value proposition not only for businesses but also for their customers.

Parata Systems—a young, fast-growing company that provides automated prescription dispensing systems that count, cap, label, and sort prescriptions in virtually any pharmacy setting—wanted to improve the quality and responsiveness of its service and support functions to ensure the level of quality and commitment it had already established on the front end of its customer relationships. Parata decided to combine call center and field service support with improved efficiency and effectiveness in its sales and marketing efforts. This approach focused the company's search on a CRM solution that provided a range of tools in a single package. Parata wanted all customer contacts to come through the call center, so its service representatives have the first opportunity to answer questions and resolve issues over the telephone before field service is dispatched. E-mail is tied into the call center as well, although the company currently receives very few service requests via e-mail. Web inquiries are primarily sales focused. Service representatives quickly and efficiently sort information requests and forward them to the appropriate sales director.[123]

The CRM solution that Parata Systems adopted also improved the way service engineers and sales directors access information. When a dispatch is waiting, the field service technician is automatically paged by the system. "Our people log in and update information for work performed on previous dispatches," says Tom Bowen, Parata's director of customer satisfaction. "Then, they can review any new dispatches waiting, accept them, and head to the appointment. When the call is complete, pertinent information is recorded and the ticket is closed—all online."[124]

"From our founding, one of our mantras has been to provide world-class service. 'Good enough' was not good enough for us in our larger quest to become a world-class company." To accomplish this, Parata had to closely monitor and ensure high customer satisfaction, and provide rapid issue resolution and high reliability. "We didn't have a mastery of our customers' specific pain points or a clear profile of what was happening at the store level following implementation," says Bowen. "In building our own service and support organization, we wanted a solution that would keep our fingers on the pulse of our customers over the life of our relationship—from the initial sale through service."[125] A CRM solution was key to its success.

Despite some notable successes, according to AMR Research nearly one-third of CRM deployments end in failure, and Forrester Research found that fewer than 50% of IT executives were fully satisfied with their CRM deployments. AMR senior research analyst Robert Bois says that sales professionals reject CRM because it is not easy to use and because they don't see how it helps them do their jobs better. Bois believes that CIOs need to get sales executives more involved in the selection and deployment of the CRM system; the sales executive understands the capabilities, responsibilities, and needs of the sales team better than the CIO does.[126] Indeed, companies that align CRM goals with the idea of creating successful employees through competence-enhancing technology have more successful CRM installations.[127]

As with any technology adoption decision, implementation of a CRM system requires both adept customer marketing on the part of the vendor and a receptive customer who understands that technology is not a silver bullet for systemic organizational problems. Indeed, companies that attempt to adopt and implement a CRM system prior to developing a relationship marketing strategy are likely to find the CRM benefits elusive.[128] Cultivating a relationship marketing philosophy requires a customer-focused organization, including job descriptions, performance measures, compensation systems, and training programs that are organized around customer needs.

Moreover, as discussed in Chapter 4, top management plays a critical role in the success of a CRM installation. When senior management views CRM as a critical or strategic tool, the probability of success is much greater than when senior management is merely supportive (viewing CRM as useful, but not view a strategic initiative).[129]

Summary

This chapter opened with the notion of a knowledge-based economy, in which know-how and collaboration are increasingly vital to a company's success in a globalized, interconnected world. Based on this foundation, the chapter covered the array of strategic alliances that high-tech companies use at all levels of the supply chain. Certainly, strategic alliances are fraught with risks; proper governance and careful attention to key success factors can help to mitigate those risks.

Moreover, outsourcing relationships pose their own challenges and opportunities. Because of their increasing prevalence in high-tech industries, particular attention was paid to developing a comprehensive understanding of outsourcing relationships.

The desire for high-tech companies to leverage new sources of knowledge and to save time and money in their own internal R&D efforts have led to new models of innovation. One such model, the open innovation model,

forms innovation networks to harness knowledge from multiple sources. Industry clusters can facilitate innovativeness across business networks. In addition, new product alliances, formed for the express purpose of new product development, require careful attention to inter-firm dynamics.

Finally, relationships with customers also warrant careful attention. The philosophy of customer relationship marketing views customers as investments, and has a corresponding "finance" language that accompanies the creation of customer equity. The costs to acquire a new customer must be carefully computed relative to the present value of the customer's revenue stream over time. The three-step customer equity management process helps a company to carefully segment the customers in its portfolio and to refine its customer relationship approach based on each type of customer and its potential to the firm.

Discussion Questions

Chapter Concepts

1. What is a knowledge-based economy? What are some of its key characteristics?
2. What is meant by the phrase "The World Is Flat"? What are some examples of "flatteners"?
3. What are the various types of partnerships a firm might form? Provide an example of each. Draw a supply chain to demonstrate the nature of the relationships.
4. What are the various reasons for a firm to form partnerships? How do these change over the course of the product life cycle?
5. What are the four strategies a firm has to set industry standards? What are the pros and cons of each?
6. What are the three factors a company should consider in choosing its strategy to set industry standards? Under what conditions should a company choose the following strategies: Aggressive sole provider? Passive multiple licensing? Aggressive positioning + licensing? Selective partnering?
7. What are the risks of partnering arrangements? How can each of these risks be mitigated?
8. What factors are associated with the success of strategic alliances? Provide sufficient detail to show a sophisticated understanding.
9. Why is outsourcing considered a vertical partnership? What are the drivers of the increasing prevalence of outsourcing?
10. What is information technology outsourcing? Business process outsourcing? Original design manufacturing?
11. Define the following terms: offshoring; captive offshoring; reverse outsourcing; near-shore outsourcing; homeshoring; farmshoring.
12. What are the reasons for outsourcing? How do these reasons mirror the risks of outsourcing?
13. Elaborate on the nature of the tension (conflicting goals) between the parties in an outsource relationship.
14. What is the contingency approach to managing outsourcing for success? Describe each of the following contingent factors and how it affects a company's outsourcing model:
    - Criticality of the business function
    - Nature of the business process and degree of customization
    - Task characteristics
    - Vendor capabilities
    - Governance
15. Overview the other success factors for outsourcing described in Table 5.8.
16. What are the three factors that will continue to affect outsourcing trends in the future?
17. What is open innovation? Why are companies pursuing open innovation in contrast to internally developed new products?
18. Why are new product alliances more difficult to manage than other types of alliances? How should companies manage these alliances to generate successful new product innovations?
19. How are industry clusters used to generate innovation?
20. What are the upside benefits and downside risks of learning from partners? How can this tension be navigated?
21. What is customer relationship marketing?
22. How does a firm compute actual acquisition costs per customer? How does a firm decide how much to spend on customer acquisition?
23. What is customer equity? How is it computed?
24. What are the three steps in implementing a relationship marketing process (to manage it for customer equity) in a company?
25. What are the three ways that high-potential customers (those a company would like to acquire and cultivate for a long-term relationship) are identified?
26. What is customer share of wallet? What is cross-buying? How do these affect identification of high-potential customers?
27. What are the four generic customer acquisition strategies?
    a. What else affects a company's acquisition strategy?
    b. To what extent should a company use price discounting as an acquisition strategy?
28. How does a company determine which customers are worthy of retention efforts? What are the four categories of customers, and what are the appropriate retention strategies for each category? What is the role of customer satisfaction in customer loyalty and profitability?

29. What are the functions of CRM software? What is the success record of CRM installations? What are the reasons for failure? What can a company to do enhance the probability of a successful installation?

30. "Thought" question: How do the unique characteristics of the high-tech environment affect the desirability of and techniques used to build customer relationships?

Application Questions

1. Of the partners mentioned in the opening Apple vignette, which are vertical and which are horizontal? Of the horizontal partnerships, are they complementary or competitive?

2. Compare and contrast the challenges and success factors mentioned by this chapter's technology expert (Kevin Myette, R&D director, REI) in his efforts to form an industry-wide coalition to develop a metric for sustainable products.

   a. To what extent do the challenges he notes overlap with the risk factors identified in the chapter?

   b. To what extent do his success factors overlap with those identified in the chapter?

   c. Based on your assessment, do you predict the Eco Working Group will be successful in accomplishing its objectives?

   d. What advice would you offer to maximize the odds of their success?

3. What kind of partnerships do you think Dean Kamen will need in order to maximize the odds of success for his revolutionary water purifying machine, the Slingshot? (See this chapter's Technology Solutions for Global Problems feature.) What other insights do you have for Kamen's business and marketing strategy?

Glossary

business process outsourcing (BPO) A type of outsourcing in which various business processes (such as customer service, finance and accounting, human resources, analytics, and even new product development) are performed by a third-party provider.

competitive alliances Partnering relationships between firms that typically compete in the marketplace (usually at the same level of the supply chain) in order to collaborate on some aspect of their business; also referred to as *competitive collaboration.*

complementary alliances Collaborative relationships between firms that make products that are jointly used in a downstream customer's end-to-end solution (e.g., the partnership between Intel and Microsoft); partners are referred to as *complementors.*

co-opetition A combination of the words "cooperation" and "competition," implying that firms both collaborate in some arenas while they compete in others.

customer equity The net present value (NPV) of the cash flows associated with a particular customer or segment; calculated by estimating the future profit stream generated by the customer, discounted to its present value at the firm's cost of capital, less the customer acquisition cost.

customer relationship marketing (CRM) The development of close, long-term, win-win relationships with customers; views marketing as a process of identifying desirable customers and then implementing retention strategies to keep them satisfied and loyal; also referred to as *customer relationship management.*

horizontal partnership An alliance between firms at the same level of the supply chain—either competitors or "complementors" (providing complementary products).

industry clusters Concentrations of economically and socially linked companies and institutions whose embedded relationships facilitate knowledge spillovers and other desirable characteristics that stimulate innovation.

information technology outsourcing (ITO) A type of outsourcing in which IT functions (such as IT infrastructure or website design) are performed by a third-party supplier.

knowledge-based economy An economy where the source of competitive advantage is more likely to be found in intangible resources (such as know-how, expertise, and intellectual property) than in more traditional means of production (tangible economic resources).

lifetime value (LTV) *See* customer equity.

open innovation A model of innovation that develops an innovation network, tapping into partners' expertise to develop breakthrough innovations.

original design manufacturing (ODM) A type of outsourcing in which innovation (including R&D, product development, and design work) is performed by a third-party supplier.

original equipment manufacturer (OEM) A company that buys components from a supplier to use in manufacturing its company's products.

outsourcing An arrangement in which one company hires another to perform a particular function (say, manufacturing or customer service) on its behalf; the outsource provider then manages and performs that business function on a contractual basis.

vertical partnership An alliance between firms at different levels of the supply chain—for example, between upstream suppliers of components and OEM customers, between manufacturers and distribution channel members, between outsource providers and their customers, or even between companies and their end-user customers.

Notes

1. Based on Vogelstein, Fred, "The Untold Story: How the iPhone Blew Up the Wireless Industry," *Wired,* January 9, 2008, online at www.wired .com/gadgets/wireless/magazine/16-02/ff_ iphone?currentPage=all.
2. Ibid.
3. Ibid.
4. Ibid.
5. Hesseldahl, Arik, "Taking the iPhone Apart," *Business Week,* July 2, 2007, online at www.businessweek.com/technology/content/jul2007/tc2007072_957316.htm.
6. Chandler, Clay, "Meet the Other Mr. iPhone," *Fortune,* May 12, 2008, p. 32.
7. "The Apple iPhone: Successes and Challenges for the Mobile Industry," Rubicon Consulting, April 2008, online at http://rubiconconsulting.com/insight/whitepapers/2008/04/the-apple-iphone-is-easily.html.
8. Vogelstein, "The Untold Story."
9. "On the Trail of the Missing iPhones," *BusinessWeek,* February 11, 2008, pp. 25–29; and Barboza, David, "Turning a Blind iPhone," *National Post,* February 19, 2008, p. FP7.
10. Drucker, Peter, *The Age of Discontinuity: Guidelines to Our Changing Society* (New York: Harper & Row, 1969).
11. Friedman, Thomas, *The World Is Flat* (New York: Farrar, Straus & Giroux, 2005), p. 176.
12. Ibid., p. 182.
13. Ibid.
14. Brandenburger, Adam, and Barry Nalebuff, *Co-opetition* (New York: Currency/Doubleday, 1996).
15. Ibid.
16. Luo, Xueming, Aric Rindfleisch, and David K. Tse, "Working with Rivals: The Impact of Competitor Alliances on Financial Performance," *Journal of Marketing Research* 44 (February 2007), pp. 73–83.
17. "DVD Forum's Mission," 2004, online at www.DVDForum.org/about-mission.htm.
18. See "GEC Mission," Green Electronics Council, 2005–2006, online at www.greenelectronicscouncil.org/index.htm.
19. See "Welcome to EPEAT," Electronic Product Environmental Assessment Tool, online at www.epeat.net.
20. "Google Signs $900m News Corp Deal," BBC News, August 7, 2006, online at http://news.bbc.co.uk/2/hi/business/5254642.stm.
21. Utterback, James M., *Mastering the Dynamics of Innovation* (Boston: Harvard Business School Press, 1994); and Roberts, Edward B., and Wenyun Kathy Liu, "Ally or Acquire? How Technology Leaders Decide," *Sloan Management Review* 43 (Fall 2001), pp. 26–34.
22. Moore, Geoffrey A., *Inside the Tornado* (New York: Harper Business, 1999).
23. Hill, Charles, "Establishing a Standard: Competitive Strategy and Technological Standards in Winner-Take-All Industries," *Academy of Management Executive* 11 (May 1997), pp. 7–25.
24. Pringle, David, "Nokia Signs Accord with Samsung," *Wall Street Journal,* September 3, 2002, p. B6.
25. See WiMAX Forum Overview, online at www.wimaxforum.org/about.
26. Hill, Charles, "Establishing a Standard: Competitive Strategy and Technological Standards in Winner-Take-All Industries," pp. 7–25.
27. Ibid.
28. Moore, Geoffrey A., *Inside the Tornado.*
29. Nussbaum, Bruce, "The Power of Design," *BusinessWeek,* May 17, 2004, online at www.businessweek.com/magazine/content/04_20/b3883001_mz001.htm?chan=search.
30. Mohr, Jakki, and Robert Spekman, "Characteristics of Partnership Success: Partnership Attributes, Communication Behavior, and Conflict Resolution Techniques," *Strategic Management Journal* 15 (February 1994), pp. 135–152.
31. MacDonald, Elizabeth, and Joann Lublin, "In the Debris of a Failed Merger: Trade Secrets," *Wall Street Journal,* March 10, 1998, p. B1.
32. Littler, Dale, Fiona Leverick, and Margaret Bruce, "Factors Affecting the Process of Collaborative Product Development," *Journal of Product Innovation Management* 12 (January 1995), pp. 16–32.

33. Hamel, Gary, "Competition for Competence and Inter-Partner Learning with International Strategic Alliances," *Strategic Management Journal* 12 (Summer 1991, Special Issue), pp. 83–103, quote on p. 87.

34. For more detail on the exact types of governance strategies recommended to navigate this tension, see Mohr, Jakki, and Sanjit Sengupta, "Managing the Paradox of Inter-firm Learning: The Role of Governance Mechanisms," *Journal of Business and Industrial Marketing* 17 (2002, Special Issue), pp. 282–301.

35. Mohr, Jakki, Gregory T. Gundlach, and Robert Spekman, "Legal Ramifications of Strategic Alliances," *Marketing Management* 3 (May 1994), pp. 38–46; and Gundlach, Greg, and Jakki Mohr, "Collaborative Relationships: Legal Limits and Antitrust Considerations," *Journal of Public Policy and Marketing* 11 (November 1992), pp. 101–114.

36. Federal Trade Commission and U.S. Department of Justice, *Antitrust Guidelines for Collaborations among Competitors,* April 2000, online at www.ftc.gov/os/2000/04/ftcdojguidelines.pdf.

37. Mohr and Spekman, "Characteristics of Partnership Success"; Mohr, Jakki, and Robert Spekman, "Perfecting Partnerships," *Marketing Management* 4 (Winter–Spring 1996), pp. 34–43; and Sarkar, M. B., Raj Echambadi, Tamer Cavusgil, and Preet Aukakh, "The Influence of Complementarity, Compatibility, and Relationship Capital on Alliance Performance," *Journal of the Academy of Marketing Science* 29 (September 2001), pp. 358–373.

38. Bucklin, Louis P., and Sanjit Sengupta, "Organizing Successful Co-Marketing Alliances," *Journal of Marketing* 57 (April 1993), pp. 32–46.

39. Dutta, Shantanu, and Allen M. Weiss, "The Relationship between a Firm's Level of Technological Innovativeness and Its Pattern of Partnership Agreements," *Management Science* 43 (March 1997), pp. 343–356.

40. Morgan and Hunt, "The Commitment–Trust Theory of Relationship Marketing," *Journal of Marketing* 58 (July 1994), pp. 20–38.

41. Kumar, Nirmalya, Jonathon Hibbard, and Louis Stern, "The Nature and Consequences of Marketing Channel Intermediary Commitment," working paper #94-115 (Cambridge, MA: Marketing Science Institute, 1994).

42. Kumar, Nirmalya, Lisa Scheer, and J. B. Steenkamp, "The Effects of Supplier Fairness on Vulnerable Resellers," *Journal of Marketing Research* 32 (February 1995), pp. 54–65.

43. Luo, Yadong, "The Independent and Interactive Roles of Procedural, Distributive, and Interactional Justice in Strategic Alliances," *Academy of Management Journal* 50 (June 2007), pp. 644–664.

44. Mangalindan, Mylene, "Court Rules against Amazon in Toy Dispute," *Wall Street Journal,* March 3, 2006, online at http://online.wsj.com/article/SB114130712302487542-search.html?KEYWORDS=amazon&COLLECTION=wsjie/6month.

45. "D5 Highlights," All Things Digital, interview of Steve Jobs and Bill Gates by Kara Swisher and Walter Mossberg, May 31, 2007, online at http://d5.allthingsd.com/20070531/d5-gates-jobs-transcript.

46. Lusch, Robert, and James Brown, "Interdependency, Contracting, and Relational Behavior in Marketing Channels," *Journal of Marketing* 60 (October 1996), pp. 19–38; Jap, Sandy, and Shankar Ganeson, "Control Mechanisms and the Relationship Life Cycle: Implications for Safeguarding Specific Investments and Developing Commitment," *Journal of Marketing Research* 37 (May 2000), pp. 215–226; and Poppo, Laura, and Todd Zenger, "Do Formal Contracts and Relational Governance Function as Substitutes or Complements," *Strategic Management Journal* 23 (August 2002), pp. 707–725.

47. Cannon, Joseph, "Contracts, Norms, and Plural Form Governance," *Journal of the Academy of Marketing Science* 28 (Spring 2000), pp. 180–194.

48. Sivadas, Eugene, and F. Robert Dwyer, "An Examination of Organizational Factors Influencing New Product Success in Internal and Alliance-Based Processes," *Journal of Marketing* 64 (January 2000), pp. 31–49.

49. Lambe, C. Jay, Robert E. Spekman, and Shelby Hunt, "Alliance Competence, Resources, and Alliance Success: Conceptualization, Measurement, and Initial Test," *Journal of the Academy of Marketing Science* 30 (Spring 2002), pp. 141–158.

50. Engardio, Pete, and Bruce Einhorn, "Outsourcing Innovation," *BusinessWeek,* March 21, 2005, pp. 85–94.

51. Ibid.

52. Engardio, Pete, "The Future of Outsourcing," *Business Week,* January 30, 2006, pp. 50–64.

53. Walker, Carson, "Outsourcing Seen as Boon to Reservations," *Missoulian,* July 24, 2005, pp. D1, D3.

54. Engardio, "The Future of Outsourcing."

55. Lustgarten, Abrahm, "Drug Testing Goes Offshore," *Fortune,* August 8, 2005, pp. 67–72.

56. Ibid.

57. Engardio, "The Future of Outsourcing."

58. Hamm, Steve, "Beyond Blue," *Business Week,* April 18, 2005, pp. 68–76.

59. McDougall, Paul, "In Depth: When Outsourcing Goes Bad," *Information Week,* June 19, 2006, online at www.informationweek.com/story/showArticle.jhtml?articleID=189500043.

60. Preston, Rob, "Down to Business: Outsourcing Customers Send Mixed Signals," *Information Week,* February 17, 2007, online at www.informationweek.com/story/showArticle.jhtml?articleID=197006870.

61. Shi, Yuwei, "Today's Solution and Tomorrow's Problem: The Business Process Outsourcing Risk Management Puzzle," *California Management Review* 49 (Spring 2007), pp. 27–44.

62. McDougall, "In Depth: When Outsourcing Goes Bad."

63. Ibid.; see also Alter, Allan, "I.T. Outsourcing: Expect the Unexpected," *CIO Insight,* March 7, 2007, online at www .CIOinsight.com.

64. Preston, "Down to Business."

65. Arrunada, Benito, and Xose Vazquez, "When Your Contract Manufacturer Becomes Your Competitor," *Harvard Business Review* 84 (September 2006), pp. 135–145.

66. Engardio and Einhorn, "Outsourcing Innovation."

67. Ibid.

68. Ibid.

69. Preston, "Down to Business."

70. Aron, Ravi, and Jitendra V. Singh (2005), "Getting Offshoring Right," *Harvard Business Review,* 83 (December 2005), p. 154.

71. Engardio and Einhorn, "Outsourcing Innovation."

72. Karmarkar, Uday, "Will You Survive the Services Revolution?" *Harvard Business Review* 82 (June 2004), pp. 100–107.

73. Carson, Stephen J., "When to Give Up Control of Outsourced Product Development," *Journal of Marketing* 71 (January 2007), pp. 49–66.

74. Chesbrough, H. W., "The Era of Open Innovation," *MIT Sloan Management Review* 44 (Spring 2003), pp. 35–41.

75. Chesbrough, Henry William, *Open Innovation: The New Imperative for Creating and Profiting from Technology* (Cambridge, MA: Harvard Business School Press, 2003).

76. "The World's Most Innovative Companies," *BusinessWeek,* April 24, 2006, online at www.businessweek.com/magazine/content/06_17/b3981401.htm.

77. Hamm, Steve, "Radical Collaboration," *BusinessWeek,* August 30, 2007, online at www.businessweek.com/innovate/ content/aug2007/id20070830_258824.htm? chan=innovation_special+report+–+inside+innovation_ inside+innovation.

78. McGregor, Jena, "The World's Most Innovative Companies," *Business Week,* May 4, 2007, online at www.businessweek. com/innovate/content/may2007/id20070504_051674.htm? chan=search.

79. Santos, Jose, Yves Doz, and Peter Williamson, "Is Your Innovation Process Global?" *MIT Sloan Management Review* 45 (Summer 2004), pp. 31–37.

80. Birkinshaw, Julian, John Bessant, and Rick Delbridge, "Finding, Forming, and Performing: Creating Networks for Discontinuous Innovation," *California Management Review* 49 (Spring 2007), pp. 67–84.

81. Sivadas, Eugene, and F. Robert Dwyer, "An Examination of Organizational Factors Influencing New Product Success in Internal and Alliance-Based Processes," *Journal of Marketing* 64 (January 2000), pp. 31–49.

82. Ibid.

83. Rindfleisch, Aric, and Christine Moorman, "The Acquisition and Utilization of Information in New Product Alliances: A Strength-of-Ties Perspective," *Journal of Marketing* 65 (April 2001), pp. 1–18.

84. Rindfleisch, Aric, "Organizational Trust and Interfirm Cooperation: An Examination of Horizontal versus Vertical Alliances," *Marketing Letters* 11 (February 2000), pp. 81–95.

85. Ganesan, Shankar, Alan J. Malter, and Aric Rindfleisch, "Does Distance Still Matter? Geographic Proximity and New Product Development," *Journal of Marketing* 69 (October 2005), pp. 44–60.

86. Porter, Michael, "Clusters and the New Economics of Competition," *Harvard Business Review* 76 (November–December 1998), pp. 77–90.

87. Arnould, Eric, and Jakki Mohr, "Dynamic Transformations for Base-of-the-Pyramid Market Clusters," *Journal of the Academy of Marketing Science* 33 (Summer 2005), pp. 254–274.

88. Lei, David T., "Competence-Building, Technology Fusion, and Competitive Advantage: The Key Roles of Organizational Learning and Strategic Alliances," *International Journal of Technology Management* 14 (2/3/4, 1997), pp. 208–237; Hamel, "Competition for Competence and Inter-Partner Learning"; Inkpen, Andrew C., and Paul W. Beamish, "Knowledge, Bargaining Power, and the Instability of International Joint Ventures," *Academy of Management Review* 22 (January 1997), pp. 177–202; and Hamel, Gary, Yves Doz, and C. K. Prahalad, "Collaborate with Your Competitors—And Win," *Harvard Business Review* 67 (January–February 1989), pp. 133–139.

89. Mohr and Sengupta, "Managing the Paradox of Inter-firm Learning."

90. Dutta, Shantanu, and Allen M. Weiss, "The Relationship between a Firm's Level of Technological Innovativeness and Its Pattern of Partnership Agreements," *Management Science* 43 (March 1997), pp. 343–356.

91. Dyer, Jeffrey H., and Nile W. Hatch, "Using Supplier Networks to Learn Faster," *MIT Sloan Management Review* 45 (Spring 2004), pp. 57–63; Hult, G. Tomas M., Robert F. Hurley, Larry C. Giunipero, and Ernest L. Nichols Jr., "Organizational Learning in Global Purchasing: A Model and Test of Internal Users and Corporate Buyers," *Decision Sciences* 31 (Spring 2000), pp. 293–325; Hult, G. Tomas M., David J. Ketchen Jr., and Ernest L. Nichols Jr., "An Examination of Cultural Competitiveness and Order Fulfillment Cycle Time within Supply Chains," *Academy of Management Journal* 45 (June 2002), pp. 577–586; Johnson, Jean L., Ravipreet S. Sohi, and Rajdeep Grewal, "The Role of Relational Knowledge Stores in Interfirm Partnering," *Journal of Marketing* 68 (July 2004), pp. 21–36; Kotabe, Masaaki, Xavier Martin, and Hiroshi Domoto, "Gaining from Vertical Partnerships: Knowledge Transfer, Relationship Duration, and Supplier Performance Improvement in the U.S. and Japanese Automotive Industries," *Strategic Management Journal* 24 (April 2003), pp. 293–316; and Selnes, Fred, and James Sallis, "Promoting Relationship Learning," *Journal of Marketing* 67 (July 2003), pp. 80–95.

92. Lloyd, Robert "Creating the Total Customer Experience," *Executive Perspective* (3rd Quarter 2007), online at http://www.cisco.com/en/US/solutions/collateral/ns340/ns857/ns861/rLloyd_total_cust_experi_20070627.pdf.

93. "PictureVision," Customer Service Information, Cisco Systems, 2008, online at www.cisco.com/web/about/ac49/ac162/about_cisco_success_story09186a00800a8936.html.

94. Gupta, Sunil, Donald Lehman, and Jennifer Stuart, "Valuing Customers," *Journal of Marketing Research* 41 (February 2004), pp. 7–18; and Reichheld, Fred, and Christine Detrick, "Loyalty: A Prescription for Cutting Costs," *Marketing Management* 12 (September–October 2003), pp. 24–25.

95. Rust, Roland, Katherine Lemon, and Das Narayandas, *Customer Equity Management* Upper Saddle River, NJ: Prentice Hall, 2005).

96. Gupta, Sunil, and Donald Lehmann, *Managing Customers as Investments* (Upper Saddle River, NJ: Wharton School Publishing, 2005).

97. Malik, Om, "Vonage Scary Big Spending Ways," GigaOM, May 28, 2005, online at http://gigaom.com/2005/05/28/vonage-scary-big-spending-ways.

98. "The Short Case on Vonage: Why No Price Is Cheap Enough," Seeking Alpha, September 11, 2006, online at http://seekingalpha.com/article/16608-the-short-case-onvonage-why-no-price-is-cheap-enough.

99. Blattberg, Robert C., Gary Getz, and Jacquelyn S. Thomas, "Managing Customer Acquisition," *Direct Marketing* 64 (October 2001), pp. 41–55.

100. Rust, Roland T., Katherine N. Lemon, and Valerie A. Zeithamel (2004), "Return on Marketing: Using Customer Equity to Focus Marketing Strategy," *Journal of Marketing* 68 (January 2004), pp. 109–127.

101. Gupta and Lehmann, *Managing Customers as Investments.*

102. Solomon, Michael, Greg Marshall, and Elnora Stuart, *Marketing: Real People, Real Choices,* 5th ed. (Upper Saddle River, NJ: Prentice Hall, 2007).

103. Reinartz, Werner, and V. Kumar, "The Mismanagement of Customer Loyalty," *Harvard Business Review* 80 (July 2002), pp. 86–95.

104. Blattberg, Getz, and Thomas, "Managing Customer Acquisition."

105. McWilliams, Gary, "Minding the Store: Analyzing Customers, Best Buy Decides Not All Are Welcome," *Wall Street Journal,* November 8, 2004, pp. A1, A17.

106. Blattberg, Getz, and Thomas, "Managing Customer Acquisition."

107. Johnson, Michael, and Fred Selnes, "Customer Portfolio Management: Toward a Dynamic Theory of Exchange Relationships," *Journal of Marketing* 68 (April 2004), pp. 1–17.

108. Johnson, Michael, Andreas Herrmann, and Frank Huber, "The Evolution of Loyalty Intentions," *Journal of Marketing* 70 (April 2006), pp. 122–132; Palmatier, Robert, Rajiv Dant, Dhruv Grewal, and Kenneth Evans, "Factors Influencing the Effectiveness of Relationship Marketing: A Meta-Analysis," *Journal of Marketing* 70 (October 2006), pp. 136–153; and Ulaga, Wolfgang, and Andreas Eggert, "Value-Based Differentiation in Business Relationships: Gaining and Sustaining Key Supplier Status," *Journal of Marketing* 70 (January 2006), pp. 119–136.

109. Cao, Yong, and Thomas Gruca, "Reducing Adverse Selection through Customer Relationship Management, *Journal of Marketing* 69 (October 2006), pp. 219–22; and Lewis, Michael, "Customer Acquisition Promotions and Customer Asset Value," *Journal of Marketing Research* 43 (May 2006), pp. 195–203.

110. Thomas, Jacquelyn, Robert Blattberg, and Edward Fox, "Recapturing Lost Customers," *Journal of Marketing Research* 41 (February 2004), pp. 31–45.

111. "What To Do about Unprofitable Customers," Fusion-Brand blogs, posted July 2, 2004, at http://fusionbrand.blogs.com/fusionbrand/2004/07/what_to_do_abou.html.

112. Reinartz and Kumar, "The Mismanagement of Customer Loyalty."

113. Anderson, James C., and James A Narus, "Partnering as a Focused Market Strategy," *California Management Review* 33 (Spring 1991), pp. 95–113; and Blattberg, Getz, and Thomas, "Managing Customer Acquisition."

114. Srivastava, Samar, "Sprint Drops Clients over Excessive Inquiries," *Wall Street Journal,* July 7, 2007, online at http://online.wsj.com/article/SB118376389957059668.html?mod=rss_whats_news_technology.

115. Mittal, Vikas, Matthew Sarkees, and Feisal Murshed, "The Right Way to Manage Unprofitable Customers," *Harvard Business Review* 86 (April 2008), pp. 94–102.

116. Anderson, James, and James Narus, "Selectively Pursuing More of Your Customer's Business," *MIT Sloan Management Review* 44 (Spring 2003), pp. 41–49; Anderson, James, James Narus, and Wouter van Rossum, "Customer Value Propositions in Business Markets," *Harvard Business Review* 84 (March 2006), pp. 90–99.

117. Tech Data website, www.techdata.com.

118. Fryer, Bronwyn, "High Tech the Old-Fashioned Way: An Interview with Tom Siebel of Siebel Systems," *Harvard Business Review* 79 (March 2001), pp. 119–125.

119. Anderson, James, Nirmalya Kumar, and James Narus, *Value Merchants* (Boston: Harvard Business School Press, 2007); Anderson, James, and James Narus, *Business Market Management: Understanding, Creating, and Delivering Value,* 2nd edition (Upper Saddle River, NJ: Prentice Hall, 2004); and Kordupleski, Ray, *Mastering Customer Value Management* (Cincinnati: Pinnaflex Educational Resources, 2003).

120. Beal, Barney, "Gartner: CRM Spending Looking Up," SearchCRM.com, April 29, 2008, online at http://searchcrm.techtarget.com/news/article/0,289142,sid11_gci1311658,00.html.

121. Fryer, "High Tech the Old-Fashioned Way."

122. Ibid.

123. "Parata Systems Improves Customer Satisfaction and Reduces Costs," online at http://www.iconarabia.com/downloads/case/parata-siebel-casestudy.pdf, 2006.

124. Ibid.

125. Ibid.

126. McGillicuddy, Shamus, "CRM Projects Fail Because Users Say 'No Thanks,'" SearchCIO.com, October 18, 2007, online at http://searchcio.techtarget.com/news/article/0,289142,sid182_gci1277542,00.html.

127. IBM Business Consulting Services, "Doing CRM Right: What It Takes to Be Successful with CRM," 2004, online at www-935.ibm.com/services/us/gbs/bus/pdf/ge510-3601-01f_crm_study_exec_summ_jun04.pdf.

128. Rigby, Darrell, Frederick Reichheld, and Phil Schefter, "Avoid the Four Perils of CRM," *Harvard Business Review* 80 (February 2002), pp. 101–109.

129. IBM Business Consulting Services, "Doing CRM Right.

6

Marketing Research in High-Tech Markets

IDEO: The Innovation Factory[1]

In 2008, *Fast Company* magazine designated IDEO, the Palo Alto–based design firm, as one of the world's five most innovative companies. From its inception, IDEO has been a force in the world of design. It has designed hundreds of products and won more design awards over the past decade than any other firm. IDEO designed such high-tech products as the Palm V, the Treo communicator/organizer/cell phone, Hewlett-Packard's OmniBook 4100 subnotebook computer, Polaroid's i-Zone cameras, Steelcase's Leap chair, and Zinio interactive magazine software.

IDEO's approach is to identify opportunities for growth by revealing people's latent needs, behaviors, and desires—and visualizing new ways to serve and support them. IDEO's unique approach to marketing research is the means through which this is accomplished. All IDEO-designed products are inspired by watching real people. As Tom Kelley, IDEO general manager, says, "We are not fans of focus groups. We don't much care for traditional market research either." Alan South, director of service design, elaborates: "The main reason why market research and focus groups are not design tools is that they are only able to address explicit user needs."

Interdisciplinary design team members employ a range of observational ("empathic") techniques to understand the issues people face. For example, the TEX group at IDEO (*TEX* stands for "technology-enabled experiences") aims to take new high-tech products that first appeal only to early adopters and remake them for a mass consumer audience. IDEO's success with the Palm V led AT&T Wireless to call for help on its mMode consumer wireless platform. The company launched mMode in 2002 to allow AT&T Wireless mobile phone customers to access e-mail and instant messaging, play games, and find nearby stores, restaurants, or other mMode users. Technology enthusiasts liked mMode, but average consumers were not signing up. "We asked [IDEO] to redesign the interface so someone like my mother who isn't Web savvy can use the phone to navigate how to get the weather or where to shop," explained mMode's vice president, Sam Hall.

IDEO formed a team of design, human factors, and business factors specialists who conducted immersive workshops, interviewed content providers, observed users in different markets, and created working interaction prototypes for testing on phones. For example, IDEO sent AT&T Wireless managers on a scavenger hunt in San Francisco to see the world from their customers' perspective. They were told to find a CD by a certain Latin singer that was available at only one small music store, find a Walgreens that sold its own brand of ibuprofen, and get a Pottery Barn catalog. They discovered that it was simply too difficult to find these kinds of things with their mMode service and wound up using the newspaper or the phone book instead. IDEO and AT&T Wireless teams also went to AT&T Wireless stores and videotaped people using mMode. They saw that consumers couldn't find the sites they wanted. It took too many steps and clicks. "Even teenagers didn't get it," said Duane Bray, leader of the TEX practice at IDEO.

The next stage in the IDEO process is brainstorming. IDEO has created an intense, quick-turnaround, brainstorm-and-build process dubbed "the Deep Dive." IDEO mixes designers, engineers, and social scientists with its clients in a room where they intensely scrutinize a specific problem, encourage wild ideas, defer judgment, combine and build on the ideas of others, and allow individuals to "tune out" for 5 or 10 minutes to sketch a solution inspired by the conversation, and then jump back into the conversation to show the others their idea.

IDEO designers then mock up working models of the best concepts that emerge. Rapid prototyping has always been a core tool of the company. Tom Kelley calls prototyping "the shorthand of innovation." IDEO believes that just about anything can be prototyped—a new product or service, a website, or a new space. Ranging from simple proof-of-concept models to looks-like/works-like prototypes that are practically finished products, prototyping—paradoxically—allows failure early in order to ensure success sooner.

After dozens of brainstorming sessions and many prototypes, IDEO and AT&T Wireless came up with a new mMode wireless service platform. The opening page starts with "My mMode," which is organized like a Web browser's favorites list and can be managed on a website. A consumer can make up an individualized selection of sites, such as ESPN or Sony Pictures Entertainment, and ring tones. Nothing is more than two clicks away. The whole design process took only 17 weeks. "We are thrilled with the results," said Hall. "We talked to Frog Design, Razorfish, and other design firms, and they thought this was a Web project that needed flashy graphics. IDEO knew it was about making the cell phone experience better." (*Note:* mMode is no longer available to new subscribers, following Cingular's takeover of AT&T wireless and the emergence of superior smartphone platforms such as Symbian, Linux, and Palm OS, as well as Apple's iPhone.)

IDEO's approach to design highlights the importance of accurate customer-based information. An intimate understanding of customers—particularly with respect to how they approach high-technology products and services—makes the difference between success and failure.

As presented in the prior chapters, technology marketers face a paradox. On the one hand, customers can't always articulate what their specific needs are; on the other hand, high-tech firms must keep a finger on the pulse of the market in order to enhance their odds of success. Winning high-tech firms should not first develop new products and then later worry about how to market them. They ought to take calculated risks without ignoring customers. High-tech firms must incorporate customers into the product development process, despite the inherent difficulties and the all-too-common tendency to overlook them. After all, the innovation that matters isn't what the innovator/company offers; it's what the customer adopts![2]

For example, in the online arena, Amazon built its website with the visitor experience in mind. However, because site features are easily copied, Amazon's ability to innovate is based on identifying novel ways to deal with customers and leapfrog its competitors. "We ask customers what they want," says Jeff Bezos, CEO.[3] Amazon encourages feedback, sorts through purchase histories to identify customer preferences, conducts focus groups, and collects information in ways that don't impose on customers. Because of this superior customer knowledge, even if other sites offer better prices, customers tend not to switch. As one customer said, in a useful analogy: "I'm happily married, so it doesn't matter how cute the guys are I meet."[4]

In another situation, enthused by the success of hybrid cars that combine a gasoline engine with battery power (e.g., the Toyota Prius and Ford Escape), automakers are testing concept cars for the future. The Chevy Volt is an electric car that may not be commercially launched until 2010, but it will be designed to operate on a chargeable lithium ion battery, which can be charged by plugging the car into a standard household electric socket for six hours to provide a driving range of 40 miles, good enough for local commuting. Another concept car further away in time is based on hydrogen fuel cells (e.g., the Honda FCX and Chevrolet Equinox); hydrogen atoms will fuse together with oxygen atoms to release huge amounts of energy, producing only water vapor as a by-product; the released energy will be stored in a lithium ion battery that will run the motor. Because these concepts are so radical, marketers need a way to monitor potential market response to them.

Ultimately, what separates successful companies from the competition is the kind of information they collect and whom they collect it from. Such information can be used in a variety of ways, such as helping to determine the specifications and/or pricing of new products, determining the best target market for a new product, gauging the most compelling value proposition, and even generating ideas for innovations. Hence, this chapter focuses on gathering information in high-tech markets. It explores marketing research techniques, customer-driven innovation, and biomimicry, a relatively new tool that is used to develop breakthrough innovations. In addition, because effective forecasting in high-tech markets is paramount to making good decisions, the last section in the chapter addresses various forecasting methods and tools.

GATHERING INFORMATION: HIGH-TECH MARKETING RESEARCH TOOLS

Collecting useful market-based information in high-tech environments is difficult. Customers aren't aware of what new technologies are available or how those technologies might be used to solve current problems. They might not even be aware of the needs they have. Moreover, in this environment, firms must accelerate the product development process, closing the time between idea to market introduction. Despite these difficulties, successful firms in high-tech markets are indeed able to collect useful information to guide decisions.

Data in Table 6.1 from a survey[5] of 80 marketing executives provides useful benchmarking information on typical market research expenditures by companies in various industry sectors. For example, technology companies in the business-to-business sector spent 0.25% of their revenue on market research (including general and administrative expenses of the research unit) and had a research staff of 15 people, on average. (The sample sizes are a bit low across some sectors.) By company size, companies with more than $5 million in revenue spent from 0.50% to 0.69% on market research, with a market research staff size ranging from 13 for companies earning $5 million to $9 million in revenue, to 41 for companies earning more than $20 million in revenue. The bulk of the

TABLE 6.1 Market Research Expenditures and Market Research Staffing

	% of Revenue	Number of Market Research Personnel
By Industry Sector		
Pharmaceutical	0.78%	52 people
Media companies	0.68%	22
Consumer goods	0.51%	18
Technology in the B2B sector	0.25%	15
Telecommunications	0.07%	15
By Size of Company ($ Revenue)		
Less than $1 million	0.07%	5
Greater than $5 million	0.50–0.69%	13–41
($5–9 million)		(13)
(> $20 million)		(41)

Source: Corporate Executive Board, Market Research Executive Board, Member Benchmarking Survey Analysis, "2003–2004 Benchmarking the Research Function," March 2004.

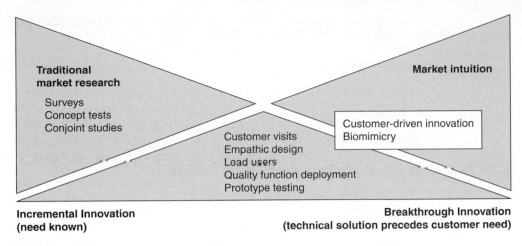

Traditional market research

Surveys
Concept tests
Conjoint studies

Market intuition

Customer visits
Empathic design
Load users
Quality function deployment
Prototype testing

Customer-driven innovation
Biomimicry

**Incremental Innovation
(need known)**

**Breakthrough Innovation
(technical solution precedes customer need)**

FIGURE 6.1 Aligning Market Research with Type of Innovation

research—39%—was spent on vendor-executed research (conducted by a market research agency), followed by purchases of secondary and syndicated research (roughly 30% of research expenditures) and company-executed research (about 9%). About 22% of expenditures were spent jointly between an outside research agency and the company. Wide variance in industries existed across these average expenditures by research category, with technology companies spending a large amount of their research monies on syndicated and secondary research (47%).

These data show that research expenditures are not a trivial matter for high-tech companies. As noted in Chapter 4, companies that are market oriented must gather information to ensure a solid understanding of the markets in which they compete and the customers to whom they sell. Such an understanding can be accomplished only by allocating resources toward gathering customer- and market-based information.

As Figure 6.1 shows, consistent with the contingency theory of high-technology marketing, research methods must be aligned with the type of innovation being developed.[6] For incremental innovations, new product developments are in alignment with the current market. Customer needs are generally known, and traditional marketing research can help companies understand such needs. Indeed, traditional marketing research techniques are most effective when a product or service is well understood by customers, or when the customer is familiar with possible solutions because of related experience in other contexts. Traditional marketing research techniques such as focus groups, surveys, concept tests, conjoint studies, and test markets can be useful to match new product characteristics with customer needs.

Standard marketing research tools typically don't address new uses or new benefits and are less effective when customers are unfamiliar with the product being researched. Moreover, standard research tools cannot help firms develop truly breakthrough (radical) innovations. Hence, for breakthrough products or for rapidly changing markets, standard marketing research techniques might not provide useful information. In the extreme, where technical solutions precede customer needs, market research might consist largely of guided intuition. Industry experts may be helpful, and the creation of different future scenarios can be used to guide decision making based on intuition.[7]

In the midrange (between incremental and radical innovation), useful market research techniques include, for example, customer visits, empathic design, the lead user process, quality function deployment, and prototype testing.

Figure 6.1 also depicts the flow of the chapter. Covered first are traditional tools of marketing research that are most appropriate for identifying opportunities for incremental innovation or for managing existing products. This first section touches on some of the traditional market research methods, and more information about them can be found in the many excellent resources available.[8]

The chapter then continues with the tools that are particularly useful for providing insight into opportunities for breakthrough innovations.

Concept Testing

Typically, the new product development (NPD) process starts with the generation of ideas for products that will address identified customer needs. As many concepts as possible should be generated, since only a small percentage of new product ideas ultimately prove to be profitable.[9] Also, keeping multiple options open and freezing the concept late in the development process affords the flexibility to respond to market and technology shifts and may actually shorten total product development time.[10] Common approaches to idea generation include:

- The various observational techniques that will be discussed subsequently in the chapter
- Brainstorming, where employees from engineering, marketing, sales, and manufacturing are guided through a series of creativity exercises to generate new product ideas
- Focus groups, where members of the target market are asked to think about the ways different product or service ideas could satisfy their needs
- Depth interviews, where target customers participate in lengthy, nondirective, one-to-one interviews regarding their needs and potential solutions to those needs

Concept testing then evaluates these early-stage ideas in order to determine which of them are good enough to be developed further. The firm's goal is to focus limited R&D and marketing resources on the one or two concepts that have the greatest probability of market success. In the initial concept testing, concepts are described in one or two paragraphs, sometimes with a name and a price, and potential customers are asked to rate them on dimensions such as interest in trying the product, purchase intent, uniqueness, and perceived value. The results can give a better idea of customer interest, eliminating concepts that generate little customer interest, and allowing the firm to refine other concepts to improve the chances of success before going to a full-blown, predictive concept test.

The process continues through a series of stages, narrowing the initial large pool of product concepts to a smaller and smaller number. Ultimately, the number of concepts is reduced to a manageable set that can be thoroughly assessed. In this last stage, a representative sample of potential customers is asked to view a small number of new product concept finalists and complete a battery of questions and diagnostic ratings. This is commonly done on a secure Internet site due to the speed, power, convenience, and flexibility of the Internet.[11] Conjoint analysis, to be discussed next, is often used to evaluate the final concepts at this stage.

Concept testing is employed quite a bit by the video game industry. Publishers and developers use concept testing as a simple initial "reality check" on their game idea. Using a one-sheet description that summarizes the game and describes game-play mechanics, the developer can compare new game concepts to a baseline or norm. Traditional concept testing is done qualitatively in focus groups and one-on-one interviews. The latest method for testing game concepts involves the integration of one-sheet text descriptions and raw game-play footage from prototype models. Most gamers can watch a video of game-play footage and understand what it's like to play a game without directly experiencing it. Because of the nature of gaming, watching videos of games is not only entertaining but incredibly insightful into what the game experience will eventually be for a consumer. Testing demos over the Internet using video footage is a rapid and cost-effective way to get an early read on consumer perceptions. Using these data helps to predict the demand for new and existing game software with good probability.[12]

Conjoint Analysis[13]

Many high-tech products are made up of a dizzying array of features (computers, cell phone calling programs, manufacturing equipment). During product development, how does a company decide what product characteristics, branding strategy, or price point will hit the "sweet spot" in the market, generating maximum sales and profits? And how does the consumer evaluate the offering vis-à-vis other alternatives in the marketplace?

Conjoint analysis is a survey research tool that can statistically predict which combination of product attributes across various brands and prices customers will prefer to buy. In conjoint analysis, respondents are asked to make judgments about their preferences for combinations of product attributes (e.g., price, brand, speed, warranties, technical services) that involve various levels such as high or low price, premium or value brand, and so forth. The objective is to determine the trade-offs that people make for each of the attributes and to derive the importance they place on each attribute. For example, conjoint studies might assess whether consumers in the market would be willing to pay a higher amount for a more sophisticated feature set in a product, or whether they would rather trade off the sophisticated feature set for a lower price (and how much lower). In other words, conjoint analysis examines the specifics of how various product features fit together to deliver a complete offering.

One of the first steps in designing a conjoint study is to develop a set of attributes and levels of those attributes that adequately characterize the range of product options. Focus groups, customer interviews, and internal corporate expertise are some of the sources used to structure the sets of attributes and levels that guide the rest of the study. Suppose a manager must find the right combination of features for a new brand of Global Positioning System (GPS) device to provide maps and directions based on the location of a person's car or mobile phone.[14] Because each feature is costly to provide, the company should develop a specific feature only if consumers' willingness to pay (WTP) for the feature exceeds the cost of providing it. Let's say focus groups or internal expertise suggest that consumers are most concerned with accuracy, color of the display, battery life, and price—and that there are two levels for each of these four features. Let's also say that engineers can design a GPS device with the ability to indicate a person's location within either 10 feet or 50 feet (accuracy), with either a color display or a black and white display, with a battery that could last either 12 hours or 32 hours between recharges, and with a price of either $250 or $350. Thus, 16 product profiles (four features with two levels each, or $2 \times 2 \times 2 \times 2$) are possible; Table 6.2 shows what one of the 16 product profiles would look like. Each of the 16 profiles is presented, one at a time, to a respondent who evaluates each one on a rating scale of 1 (low preference) to 100 (high preference).

An ordinary least squares (OLS) regression on the data yields a consumer utility function (regression equation). Let's say that it showed the following importance weights (also called *part-worths*) for each feature:

Accuracy	9.6
Display	14.9
Battery life	30.4
Price	40.6

Hence, accuracy is the least important feature, while price is the most important. This analysis suggests that if the price is reduced from $350 to $250 (the two levels that were evaluated in the study), the consumer's utility would increase by 40.6 units (*utils*). The value per util to the consumer is therefore $2.46 ($100 / 40.6). The willingness to pay (WTP) for each feature is then $23.62 for accuracy (9.6 × 2.46), $36.65 for color display (14.9 × 2.46), and $74.78 for battery life (30.4 × 2.46). The marketer, in consultation with the engineers, can now decide if the company can develop these specific features at a lower cost than the WTP. The results from this analysis are then used to make critical product development, positioning, and pricing decisions.

TABLE 6.2 Illustrative Product Profile for GPS Conjoint Study

Product Concept	Accuracy	Battery Life	Display	Price
#1	10 feet	32 hours	Color	$250

Repeating the conjoint task for many potential consumers provides a more robust estimate of the consumer utility function for the whole market. With a calibrated utility function, a company can plug in the feature values of existing competing brands and figure out how its own brand could trump them with a distinctive set of feature levels. In addition, these data can be used to make market share predictions for each brand in the market. Conjoint simulators can be used to estimate demand curves, substitution effects (e.g., from which competitors does a company take the most share if it increases the processor speed?), or cannibalization effects (e.g., what happens to overall share if the company comes out with another product with lesser performance at a lower price?).

Our simplified GPS example involved just four features with two levels each. When products involve many more features and levels per feature, the product profiles that a consumer must consider proliferate. More sophisticated conjoint techniques, such as choice-based conjoint (CBC) analysis, as well as better statistical estimation techniques, such as hierarchical Bayes estimation, can be used to minimize the number of alternatives each respondent will evaluate.

Today, thousands of conjoint studies are conducted each year—over the Internet, through person-to-person interviews, or via mail surveys. Companies save a great deal of money on R&D costs, successfully using the results to design new products or line extensions, reposition existing products, and make more profitable pricing decisions.

Customer Visit Programs

The idea of using customer visits for market research has developed in response to the challenges faced by managers in many industries. A **customer visit program** is a systematic program of visiting customers with a cross-functional team to understand customer needs. Customer visit programs are useful not only for new product development ideas but also for customer satisfaction studies, identification of new market segments, and other myriad issues. Teams should include, at a minimum, an engineer, a product-marketing representative, and the account manager or salesperson. For cross-functional teams to work smoothly in customer visits, good teamwork must exist between engineering and marketing. Effective customer visits leverage the following characteristics:

- *Face-to-face communication:* Development of new-to-the-world products benefit from the unique capacity of personal communication to facilitate the transfer of complex, ambiguous, and novel information.
- *Field research:* Doing research at the customer's place of business allows personnel to see the product in use, talk to actual users of the product, and gain a better understanding of the product's role in the customer's total operation.
- *Firsthand knowledge:* Everyone believes his or her own eyes and ears first. When key players hear about problems and needs from the most credible source—the customers—responsiveness is enhanced.
- *Interactive conversation:* The ability to clarify, follow up, switch gears, and address surprising and unexpected insights provides depth to interactions. Interesting questions to ask include:
  - If you could change any one thing about this product, what would it be?
  - What aspects of your business are keeping you awake at night?
  - What things do we do particularly well or poorly, relative to our competition?
  - What things do we do particularly well or poorly, relative to your expectations?
- *Inclusion of multiple decision makers at the customer location:* Many technology products are purchased by groups of people, and customer visits allow all of the players' various needs and desires to be addressed.

How should customer visits be structured to maximize the benefits? Table 6.3 provides the key elements of such a program.

TABLE 6.3 Elements of Effective Customer Visit Programs

1. Get engineers in front of customers.

Relying solely on marketing personnel to conduct customer visits makes cross-functional collaboration unlikely, and marketing may lack credibility with key technical people. The people who participate in the visits must be the ones who will use the information.

2. Ensure that the corporate culture embraces the value of the customer visit program.

For a customer visit program to be successful, it must be enthusiastically embraced by the technical team. R&D managers who say "Go see the customers yourself" or "Take the project team out to visit customers" are vital to communicating the appropriate attitude. Having only marketers go out to visit customers does not substitute for a commitment on the part of the entire organization to understand customers.

Having only high-level executives on customer visits makes other company personnel question the degree to which a customer focus is real or just window dressing.

3. Visit different kinds of customers.

The common tendency in customer visit programs is to visit only national accounts. Although visiting national accounts may result in increasing satisfaction with these accounts, market share may shrink if the firm falls into the trap of developing products that exactly suit an ever-smaller number of customers. Ideally, teams should visit multiple customers to get more than just an idiosyncratic reading on customer needs.

Often the freshest perspectives and greatest surprises come from atypical sources, such as competitors' customers, global customers, lost leads, lead users, distribution channel members, or "internal" customers of the firm's own field staff.

Customer councils are another important source of information. Typically designed to get feedback, share perspectives, and build stronger customer relationships, they offer the potential of synergy through group action.

4. Visit customers in their own settings: Get out of the conference room!

All too often, companies think that a "customer visit program" means inviting customers to their own premises—say, in the company's visit center. Such a policy might cut costs and save time, but it puts the customers in a passive role; the hidden agenda in such an approach is that the company is typically showcasing its products and giving VIP treatment to customers.

Effective customer visit programs do not take place in the conference room (neither the company's nor the customer's). Because customers often don't realize and cannot vocalize specific needs, it is important to listen and observe what they do. This can only happen when the visit takes place in the customer's usage setting and captures realistic elements of the customer's work environment.

5. Conduct programmatic visits.

A systematic approach including between 15 and 40 visits will yield a depth of understanding and illumination that goes well beyond what a few scattered visits can offer. The company must coordinate the visits so that customers are not confused or irritated by a series of haphazard visits from different divisions and levels in the firm. Promptly log and review customer visits in a central database. Reviewing all profiles that are kept in a central database allows the firm to spot trends, define segments, identify problems, and glimpse opportunities.

Source: McQuarrie, Edward, "Taking a Road Trip," *Marketing Management* 3 (Spring 1995), pp. 9–21; and McQuarrie, Edward, *Customer Visits: Building a Better Market Focus,* 3rd ed. (Armonk, NY: Sharpe, 2008).

Customer visits are more than a tool to groom customer relationships; when implemented correctly, they offer significant insights and benefits for high-tech product development. For example, a manager for a manufacturer of factory test equipment recounted that as part of each customer visit, the visiting team would examine where and how each customer's test equipment was set up.[15] The team soon noticed that their customers' testing areas tended to be crowded with a variety of instruments, often more than the space could comfortably hold. In particular, customers for this vendor's test equipment resorted to a number of jury-rigged attempts to locate the machine at eye level. After these visits, the vendor came up with a new model of the instrument that they dubbed "Skyhook." This was the same basic instrument but with a hook welded to the top of the casing, making it easy to suspend from the ceiling at any desired height, in any orientation, anywhere in the testing area. The new modification was quite successful, highlighting the fact that sometimes less sophisticated solutions create value for customers of high-tech products.

Empathic Design

Observation of customers (what they do) of high-tech products is often more useful in developing novel insights than asking them direct questions (what they say). For example, users may have developed "workarounds"—modifications to usage situations that are inconvenient yet so habitual that users are not even conscious of them. Or customers may not be able to envision the ways new technology could be used. Indeed, a key component of IDEO's research approach is based on the notion that users may not be able to articulate their needs clearly. **Empathic design** is an approach that focuses on understanding user needs through empathy with the user world, rather than from users' direct articulation of their needs.[16] Based in anthropology and ethnography, empathic design research allows the marketer to develop an appreciation of the current user environment, to extrapolate the evolution of that environment into the future, and to imagine the future needs that technology can satisfy.[17]

The design team of the Power Tool Division at Ingersoll-Rand combined customer visits with empathic design to improve its products. Upon visiting factories where the tools were used, the design team found that half the people using wrenches on an auto assembly line were women, who typically have smaller hands than men. The women found it difficult to grasp the tools properly. The team developed a two-size variable-grip wrench that was made even easier to hold by using rubberized plastic. An unexpected bonus was the wrench's success in Japan, where both men's and women's hands are generally smaller than in the United States.[18]

In another example, Marriott wanted to improve the experience of young, tech-savvy road warriors at its Marriott and Renaissance Hotels. It hired IDEO to do ethnographic research in hotels, cafes, and bars in 12 U.S. cities. IDEO found that hotel lobbies are usually dark and good for killing time but not for casual business meetings in small groups. Business travelers want comfortable spaces to work outside their rooms. In response, Marriott built a social zone in its lobbies with smaller tables, brighter lights, and WiFi access.

INSIGHTS FROM EMPATHIC DESIGN Observation of customers can provide illuminating insights when customers find it difficult to articulate their needs. For example, observations of customers using the product allow marketers to identify the following:[19]

- *Triggers of use:* These are the circumstances that prompt people to use a product or service.
- *How users cope with imperfect work environments and surfacing of unarticulated user needs:* For example, when engineers from a manufacturer of lab equipment visited a customer, they noticed that the equipment emitted an unpleasant odor when used for certain tasks. The customers were so accustomed to the smell that they had never mentioned it. The company added a venting hood to its product line, which actually became a compelling sales point when comparing the product to those of competitors.

  With respect to unarticulated user needs, Intel ethnographer Tony Salvador spent two years visiting dozens of villages in rural India. He noticed that thousands of Internet kiosks had sprouted all over. But one problem the kiosks faced were long power outages lasting several hours per day. From this observation, Intel rolled out the India Community PC, a $500 computer with a dust filter that could withstand temperatures of 113°F and could run off a truck battery.
- *Different usage situations:* These are situations where consumers are observed using a product, but in a different context/manner than the company initially intended. When Apple

realized how popular the iPod was with runners, they actively targeted them with the Nike + iPod Sport Kit, which gives runners real-time updates about the speed and length of their workouts and includes TrailRunner software that enables runners to plan workout routes on a geographical map, then export the route directions as small NanoMaps onto an iPod.

- *Customization of products that marketers are unaware of:*   These are ways that customers have modified products they've purchased to get more value or to make them easier to use. Purchasers of hybrid cars have been known to modify their engines to yield even more fuel efficiency. Automakers can see how customers have customized their products and then offer similar features in subsequent offerings of their own.
- *The unarticulated importance of intangible attributes:*   These intangible attributes include sound, feel, and so forth, which frequently aren't addressed in traditional surveys. Customers may not say how important the look or feel of a particular product is, but the aesthetics may be a significant source of value.

Empathic design techniques exploit a company's existing technological capabilities in the widest sense of the term. Company observers carry the knowledge of what is possible for the company to do. When that knowledge is combined with what customers need, existing organizational capabilities can be redirected toward new product ideas or markets. A note of caution: Empathic design techniques do not replace traditional market research; rather, they contribute to the flow of ideas that warrant additional testing before committing to the project.[20]

Table 6.4 overviews the five-step process used to conduct empathic design.

At the extreme, companies use empathic design to develop detailed "maps" of how their customers operate and the sequence of steps involved in, say, beginning a project, getting it financed, developing a strategy, evaluating vendor options, managing the implementation process, and follow-on use and maintenance. In order to truly understand their customers' environments and get inside their lives, companies must focus on more than just the point at which customers come into contact with them. Rather, companies must do what Patricia Seybold refers to as "customer scenario planning"—generating an intimate understanding of customers—which in turn allows them to deliver more value and reap greater loyalty.[21]

Increasingly, high-tech firms such as Hewlett-Packard, IBM, Motorola, Nokia, and Intel are using empathic design to augment their traditional marketing research practices. They are hiring social scientists, anthropologists, and psychologists to help them figure out how people use products. Ethnographers tend to study relatively few subjects, chosen with great care, looking for big insights rather than amassing large volume statistical data.

For example, how does Intel learn about the ways customers work and use electronic equipment?[22] How does the knowledge it learns help Intel design more effective products in the future? Intel has an eight-person team of "design ethnographers" who go to customer sites and observe customers in their natural work settings. Their goal is to learn how customers navigate their daily environments, and then use this information to help Intel design more effective products in the future.

At first, the corporate culture within Intel—particularly the R&D folks—did not take the ethnographers seriously. Indeed, their presence was an acknowledgment that the personal computer had shortcomings, and in the Intel culture that acknowledgment put them at odds with most other employees. But the success rate is only 20% when engineers design what they *think* other people want, according to former Intel chairman Andy Grove. For example, Grove believes that Intel wouldn't have sunk millions of dollars into its ProShare videophone, introduced by Intel in the early 1990s, if it had done more ethnographic research. The quality of the video was slow, jerky, and not synchronized with the sound. However, Intel loved the phone because it required significant computing power, based on the underlying microprocessors. Yet consumers hated it, because the out-of-sync video resulted in miscues when people nodded or shook their heads.

So, Intel has used empathic design in a variety of industries and with different customer groups to gain new insights. For example, its design team has spent time observing people working

---

TABLE 6.4 Process to Conduct Empathic Design

1. **Observation.** Researchers should clarify:

   - **Who should be observed?** Although "customers" is a logical answer, often noncustomers, customers of customers, or a group of individuals who collectively perform a task may provide useful information.

   - **Who should do the observing?** Differences in perception and background lead different people to notice very different details when observing the same situation. Hence, it is best to use a small cross-disciplinary team to conduct observational studies. Members should be open-minded and curious, and they should understand the value of observation. For this reason, hiring trained ethnographers to assist in the study is useful. Moreover, cross-functional collaboration among sales, marketing, and R&D allows knowledge of the customer to be combined with the capabilities of a particular technology.

   - **What behavior should be observed?** Observe the "subjects" in as normal an environment as possible. Although some believe that observation changes people's behavior (which is probably unavoidable), alternatives to observation would be experiments in highly artificial lab settings or focus groups, both of which also have limitations. The goal is to gather new kinds of insights that other research techniques cannot.

2. **Capture the data.** Researchers need to establish how to record the information. Most data from empathic design projects are gathered from visual, auditory, and sensory cues. Hence, photographs and videographs can be useful tools that capture information, such as spatial arrangements, lost in verbal descriptions.

   Whereas standard research techniques might rely on a sequence of questioning, empathic design asks very few questions other than to explore, in a very open-ended fashion, why people are doing things. Researchers may want to know what problems the user is encountering in the course of the observed activity.

3. **Reflection and analysis.** The team members and other colleagues review the team's observations contained in the captured data. The purpose is to identify all of the customers' possible problems and needs.

4. **Brainstorm for solutions.** Brainstorming is used to transform observations into ideas for solutions.

5. **Develop prototypes of possible solutions.** Researchers consider very concretely how possible solutions might be implemented. The more radical an innovation, the harder it is to understand how it should look and function. Researchers can stimulate useful communication by creating some prototype of the idea. Such prototypes, because of their concreteness, can clarify the concept for the development team, allow insights from others who weren't on the team, and stimulate reaction and discussion with potential customers. Simulations and role-playing can be useful prototypes when a tangible representation of the product cannot be made.

Source: Rayport, Jeffrey F., and Dorothy Leonard-Barton, "Spark Innovation through Empathic Design," *Harvard Business Review* 75 (November–December 1997), pp. 102–113.

in the salmon industry off the coast of Alaska. The team was trying to understand how technology, such as satellite-guided locators instead of helicopters to monitor fishing boats, would help. (Who said "fish and chips" wasn't a good combination!) Other insights have come from observing business owners. In observing their often harried schedules, the design team learned that businesspeople needed a tool to capture all the messages and phone numbers they write down on self-stick notes, such as an electronic organizer that recognizes handwriting.

Does the nature of being observed change people's behavior? The Intel team found that most people love to be observed and eventually lower their guard when they are being studied by the researchers. And ethnographic researchers are masters at getting people to feel at ease under observation. For example, one of the Intel team members spent hundreds of hours with teenagers in their bedrooms, using videotapes to catalog their behaviors and belongings, from dirty laundry to posters. His goal was to find out more about how they live and what technology they might find useful. Some of his insights were: Teens should be able to send pictures to each other instantaneously, over phone lines to computers and into flat-display bedside picture frames. They also need handheld computers that allow them to communicate schedule changes to their parents when they're out and about. The bottom line is that what a user does with a product, rather than what the product can do, ultimately drives its success.

Lead Users

Another research technique helpful in high-tech environments is the lead user process.[23] Customers who are well ahead of market trends, who have needs that go far beyond those of the average user, and—critically—who innovate solutions to their own problems are known as **lead users**.[24] Because lead users—say, Olympic athletes or aerospace companies—may face more extreme needs months or years before more typical customers in a marketplace, they benefit significantly by inventing solutions to their own needs. As a result, marketers can look to lead users for useful insights for innovations and possibly even commercialize lead users' innovations for other customers. For example, Lockheed Martin pioneered a new machining technique in the development of titanium aircraft; the innovation was later commercialized by a machine tool operation that refined the Lockheed tool.[25] Other lead users may not have developed a solution but are simply aware of the need. Examples of companies using the lead user process to improve their ability to match product development with customers' needs include Bose (maker of consumer electronics) and Cabletron (designer of fiber-optic networks).[26]

Research on lead users shows that many products are initially thought of and even prototyped by users rather than manufacturers.[27] For example, Table 6.5 shows that across a variety of industries, the number of innovations conceived of by users is quite high.

The lead user process systematically collects information about both needs and solutions from the leading edge of a market, which faces more extreme forms of problems, and uses that information to generate breakthrough innovations. It transforms the difficult job of creating breakthrough products into a systematic task of identifying lead users and learning from them. The development team actively attempts to track down promising lead users and adapt their ideas to its

TABLE 6.5 Innovations Developed by Lead Users

	Percent of Products Developed By		
	User	*Manufacturer*	*Other*
Computer industry	33%	67%	
Chemical industry	70%	30%	
Poltrusion-process machinery	85%	15%	
Scientific instrument with major functional improvements	82%	18%	
Semiconductor-electronic process equipment with major functional improvements	63%	21%	16%[*]
Electronic assembly	11%	33%	56%[†]
Surface chemistry instruments with new functional capability	82%	18%	

[*] Joint user–manufacturer innovation
[†] Supplier innovations
Sources: Adapted from von Hippel, Eric, "Lead Users: A Source of Novel Product Concepts," *Management Science* 32 (July 1986), pp. 791–805; and von Hippel, Eric, Stefan Thomke, and Mary Sonnack, "Creating Breakthroughs at 3M," *Harvard Business Review* 77 (September–October 1999), pp. 47–57.

customers' needs. General Electric's health care division refers to these customers as "luminaries," and they might include well-published doctors and scientists from leading medical institutions. GE developed its LightSpeed VCT, a scanner that creates a 3-dimensional image of a beating heart, in conjunction with these luminaries.[28] Lead users might not be within the firm's usual customer base; they may be customers of a competitor or outside the industry. Moreover, if lead users have already solved a problem, they may no longer articulate the solving of that need as an issue; hence, using a survey to identify them may be unproductive. In such a situation, the use of empathic design to observe lead users' use of the product is particularly wise.

As shown in Table 6.6, Eric von Hippel advocates the use of a four-step process to incorporate lead users into marketing research. The process is conducted by a cross-disciplinary team that includes marketing and technical departments. The lead user process begins with identification of key trends in the market, as one cannot specify what the leading edge of a target market might be without first understanding the major trends in the heart of the market.[29] For example, at 3M, the firm determined that a critical trend in the medical industry, particularly in developing countries, was the need to find inexpensive methods of infection control during surgery. The entire process can be time-consuming, with each step taking about four to six weeks, and the entire process four to six months.[30]

TABLE 6.6 Steps in the Lead User Process

1. *Identify important market/technical trends.*

Before a company can identify lead users—defined as being in advance of the market with respect to an important dimension, which is changing over time—it must identify the underlying trend on which these users have a leading position. In the strategic planning process, firms assess the external environment in which they operate for key trends, identifying competitive, economic, regulatory, physical (natural), global, sociocultural, demographic, and technological opportunities and threats.

2. *Identify and question lead users.*

To track down lead users efficiently, product development or market research teams may use telephone interviews to network their way into contact with experts on the leading edge of the target market's trends; they seek to identify those users who are actively innovating to solve problems. In business-to-business markets, manufacturers typically have a better understanding of their key customers than may be possible in consumer markets. Hence, personal knowledge of customers may identify lead users in B2B markets, whereas surveys may be used to identify lead users in consumer goods industries.

In addition, people with a serious interest in any topic, such as research professionals who have written articles on the topic or have presented research at conferences, tend to know of others even more knowledgeable than themselves. This networking can lead the team to the users at the front of the target market or in very different markets.

3. *Develop the breakthroughs.*

The team may begin this phase by hosting a workshop that includes several lead users who have a range of expertise, as well as a number of representatives from different areas of the company (e.g., marketing, engineering, manufacturing). During the workshop, the group combines insights and experiences to provide ideas for the sponsoring company's needs. The company may ask respondents to sign a document specifying intellectual property rights for ideas that arise from the workshop.

4. *Project the lead user data onto the larger market.*

Firms must assess how well lead user data apply to more typical users (rather than simply assuming such data transfer in a straightforward fashion). Prototyping the solution and asking a sample of typical users to use it is one way to gather data to make the projection.

Source: von Hippel, Eric, Stefan Thomke, and Mary Sonnack, "Creating Breakthroughs at 3M," *Harvard Business Review* 77 (September–October 1999), pp. 47–57.

Experience with the problem is what makes the lead user's insights so valuable. Let's say an automobile company wanted to design an innovative braking system.[31] It might try to identify some users who had a strong need for better brakes—such as auto racing teams or perhaps even some Air Force pilots. In fact, because military aircraft must stop before reaching the end of the runway (say, on an aircraft carrier), antilock braking systems were first developed in the aerospace industry. The data from lead users are examined for the solutions they have found to their problems. Companies rarely adopt a lead user innovation "as is." Rather, they adapt and modify the ideas based on their own insights and other information. Based on a determination of how the new concept fits the needs of a larger target market, the team will present its recommendations to senior managers. The team may assess the business potential of ideas that emerge from this process—including evidence about why customers would be willing to pay for the new products—and the ideas' fit with the company's interests.

What are some of the benefits of the lead user process?[32] Ultimately, the process allows a firm to gather and use information in a different way, which leads to new insights. The process brings cross-functional teams into close working relationships with leading-edge customers and other sources of expertise. In addition, because the process involves a cross-functional team from the organization, no one person feels like the lone ranger, working in isolation on new product ideas. But the lead user process is not a panacea for the difficulties in gathering research to develop breakthrough products. Without adequate corporate support, skilled teams, and needed time, the process may not succeed.

Moreover, a key issue in the lead user process relates to intellectual property concerns. The lead user project team should be up front about its company's commercial interest in the ideas being discussed. In a lead user study devoted to improving credit reporting services, a team found that at least two major users of such services had developed advanced online credit reporting processes. One of the users was unwilling to discuss details because the service was viewed as a significant source of competitive advantage. The other said, "We only developed this in the first place because we desperately needed it—we would be happy if you developed a similar service we could buy."[33] Many users will participate simply for the intellectual challenge. Because their focus tends to be on their own products and industries, sharing their ideas for these innovations is something they are generally willing to do; these ideas are generally not directly related to their sources of competitive advantage within their fields. Moreover, by transferring their knowledge to a willing supplier (either voluntarily or on a licensing basis), they can continue to focus on their own core competencies and have an improved source of supply through transferring the innovation. However, if a customer hesitates to talk, it is better to not pursue that interview due to the intellectual property concerns.

A detailed example of how the lead user process works at 3M Corporation is addressed in Box 6.1.

Quality Function Deployment

Quality function deployment (QFD) is an engineering tool that incorporates a customer orientation into design decisions. It first identifies what the customer's requirements are (through customer visits, empathic design, working with lead users, etc.) and then maps those requirements onto the product design process.[34] QFD uses the voice of the customer in the new product development process to ensure a tight correlation between customer needs and product specifications.[35] The process ensures that all design decisions take into account the importance of that design requirement from the customer's perspective. The ultimate outcome is a new product that provides superior value to the marketplace via a customer-informed design team. QFD requires close collaboration between marketing, engineers, and customers.

QFD emerged from the total quality management (TQM) movement in manufacturing and has become closely linked with the notion of market orientation. The TQM paradigm is originally based on the concept of using the process to create value for customers, which requires that engineering and marketing are closely integrated to focus on customer value.

BOX 6.1

The Lead User Process at Work: 3M Corporation

Example 1: Medical Imaging

Step 1: Identify important market/technical trends. A team focused on medical imaging knew that a major trend was the development of capabilities to detect smaller and smaller features in medical images—very early stage tumors, for example. Its initial goal was to develop new ways to create better high-resolution images.

Step 2: Identify and question lead users. Through networking with research experts in the field, the team identified a few radiologists who were working on the most challenging medical problems. They discovered some lead users in radiology who had developed imaging innovations that were ahead of commercially available products. Networking to other fields that were even further ahead in *any* important aspect of imaging led the team to specialists in pattern recognition and people working with images that show the fine detail in semiconductor chips. The lead users in pattern recognition were very valuable to the team. Military specialists relied on computerized pattern recognition in reconnaissance. These users had actually developed ways to enhance the resolution of the best images by adapting pattern recognition software. This discovery of the use of pattern recognition helped the development team refine its initial goal (developing new ways to create better high-resolution images) to finding enhanced methods for recognizing medically significant patterns in images, whether by better imaging or by other means.

Step 3: Develop the breakthroughs. In the course of a two-day workshop, lead users with a variety of experiences were brought together: people on the leading edge of medical imaging, those who were ahead of the trend with ultra-high-resolution images, and experts on pattern recognition. Together, they created a solution that best suited the needs of the medical imaging marketplace and represented a breakthrough for 3M.

Example 2: Infection Control

Another 3M team was charged with developing a breakthrough product for the division's surgical drapes unit, which designs the material that prevents infections from spreading during surgery. Surgical drapes are thin, adhesive-backed plastic films that adhere to a patient's skin at the site of the surgical incision prior to surgery. Surgeons cut directly through these films during an operation. Drapes isolate the area being operated on from the most potential sources of infection: the rest of the patient's body, the operating table, and the members of the surgical team. But drapes don't cover catheters or tubes being inserted into the patient. The drapes' cost prohibited market entry into less developed countries.

Step 1: Identify important market/technical trends. In looking for a better type of disposable surgical draping, the 3M team first had to learn about the causes and prevention of infections by reading research articles and interviewing experts in the field. Then, the team gathered information about important trends in infection control. During this process, the team realized it didn't know about the needs of surgeons in developing countries where infectious diseases are still major killers, so the team traveled to more hostile surgical environments to learn how people keep infections from spreading in those operating rooms. Some surgeons combated infections by using cheap antibiotics as a substitute for more expensive measures. The team saw a coming crisis with the doctors' reliance on antibiotics: Bacteria would become resistant to the drugs.

Step 2: Identify and communicate with lead users. The team networked to find the innovators at the leading edge of the trend toward cheaper, more effective infection control. As is usually the case, some of the most valuable lead users turned up in surprising places. For example, specialists at leading veterinary hospitals were able to keep infection rates very low, despite facing difficult conditions and time constraints. As one vet said, "Our patients are covered with hair, they don't bathe, and they don't have medical insurance, so the infection controls we use can't cost much." Another surprising source of ideas was from Hollywood: Makeup artists in Hollywood are experts in applying materials that don't irritate the skin and are easy to remove. Because infection control materials can be applied to the skin, those attributes were very important.

Step 3: Develop the breakthroughs. During the lead user workshop, the participants were invited to brainstorm about revolutionary ideas for low-cost infection control. The outcome of this session were the following ideas:

- An economy line of surgical drapes, made with existing 3M technology, targeted to the increasingly cost-conscious medical world.
- A "skin doctor" line of handheld devices to layer antimicrobial substances onto a patient's skin during an operation and to vacuum up blood and other liquids during surgery; this line again would be developed from existing 3M technology and would offer a new infection prevention tool.
- An "armor" line to coat catheters and tubes with antimicrobial protection, created with existing 3M technology; this line could open up major new markets outside surface infections, including blood-borne, urinary tract, and respiratory infections.
- A revolutionary approach to infection control based on the idea that some people enter the hospital with a greater risk of contracting infections—for example, those suffering from malnutrition or diabetes. Rather than providing every patient the same degree of infection prevention from the same basic drapes, this approach worked on a different philosophy. Through "upstream" containment of infection (treatment before people went to surgery), doctors could reduce the likelihood of these higher-risk patients contracting disease during an operation. This approach, however, would require 3M to radically change its strategy in the market and require new competencies, products, and services. After much debate, 3M decided to fund a "discovery center" to develop and diffuse the new approach to infection control, and the product lines needed to deploy it are currently being developed.

Source: von Hippel, Eric, Stefan Thomke, and Mary Sonnack, "Creating Breakthroughs at 3M," *Harvard Business Review* 77 (September–October 1999), pp. 47–57.

The implementation of QFD is a multistage process, including the following:[36]

- *Collect the voice of the customer.* Through customer visit programs or empathic design, identify customer needs, in the customers' own words, regarding the benefits they want the product to deliver. Roughly 10 to 12 customers will yield close to 80% or more of the customers' needs (assuming a relatively homogeneous market segment). These desired benefits and attributes can be weighted or prioritized to help the product development team in design trade-offs later (e.g., to trade off processing speed versus price, in the case of a computer chip).[37]
- *Collect customer perceptions of competitive products.* Surveys of customers can be used to assess how well current products fulfill customer needs. These data are an important component in identifying any gaps or opportunities in the market.
- *Transform customer insights into specific design requirements.* Sometimes called *customer requirements deployment,* the idea in this step is to identify the product attributes that will meet the customers' needs. It is important here to understand the interrelated nature of various attributes. For example, although customers may want more speed in processing, they may also want a lower price. This step is sometimes also referred to as the *house of quality,* or the planning approach that links customer requirements, competitive data, and design parameters.

At the heart of the process lies one of the key tools in QFD, the Kano concept (or Kano dimensions/diagram). As shown in Figure 6.2, the **Kano concept** provides a graphical representation of the relationship between three types of product attributes and customer satisfaction or dissatisfaction. The first type, deemed *one-dimensional quality* attributes, are linearly related to customer satisfaction. Increasing the performance (or level) of these attributes leads to a linear increase in satisfaction. These attributes are typically known and voiced by the customer. For example, in a laptop computer, lengthening the life of the battery would probably lead to a predictable increase in satisfaction.

The two other types of attributes have nonlinear relationships with satisfaction. *Must-be quality* attributes must be present in order for the customer to be satisfied at all. The absence of a must-be attribute is exponentially related to *dissatisfaction,* but increasing the level of that attribute does not

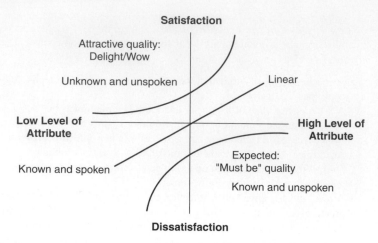

FIGURE 6.2 The Kano Concept
Source: Adapted from Kano, Noriaki, Shinichi Tsuji, Nobuhiko Seraku, and Fumio Takerhashi, "Miryokuteki Hinshitsu to Atarimae Hinshitsu (Attractive Quality to Must-Be Quality)," Japanese Society for Quality Control, 1984.

increase customer satisfaction with the product. Moreover, these attributes are so essential to product functionality that they may not be explicitly voiced by the customer. For example, in the laptop industry, the computer must be fairly immune to bumps and roughness in handling. If the computer failed upon booting every time the laptop received rough handling, customers would be horribly dissatisfied. On the other hand, allocating significant resources to improve the degree of roughness the laptop can handle likely would not appreciably increase the satisfaction of most laptop users.

Attractive quality attributes exhibit an exponential relationship with satisfaction. When the attribute is lacking, the customer is not dissatisfied—but the presence of the attribute leads to an extremely favorable reaction, delighting the customer and providing the "wow" factor in the product usage experience. Often, customers cannot articulate these attributes, and hence they must be discovered through some of the techniques mentioned earlier (empathic design and lead users). In the laptop example, the "wow" factor might be found in a laptop that is decompressable into pocket-size for carrying but expands upon opening. Many experts in product innovation believe that firms that have the capabilities and skills to identify these attributes that delight the customer are destined for success.

Rather than culminating in a specific design solution, the QFD process reveals friction points in the design process. It allows the product development team to develop a common understanding of the design issues and trade-offs, bases resolution of those trade-offs in customer needs, and enhances the collaborative processes among marketing, manufacturing, and engineering.[38] One study found that using QFD reduces design time by 40% and design costs by 60%, while enhancing design quality.[39]

Although being fast to market is important, the firm must have the ability to hit customer requirements accurately. Because customer needs and expectations are a rapidly moving target in high-tech markets, faster cycle time also ensures a higher correlation with quality—as defined by the customer. Indeed, in striving to get to the market quickly, firms may initially develop a product with a relatively basic combination of attributes, such that the attributes exceed the "must-be" quality level and are in the desired one-dimensional quality space. As the firm brings additional versions of the product to market, it adds additional features, approaching the attractive quality threshold in a process consistent with the notion of expeditionary marketing, introduced in Chapter 3. Guy Kawasaki refers to this willingness to commercialize an initial version of the product with only basic features as Rule #2 for revolutionaries: "Don't worry, be crappy."[40] Although he uses fairly inflammatory rhetoric, he means that companies can strive, not for perfection, but for the minimum level of market acceptability with the first generation of a radical new product. Of course, in pursuit of a fast time-to-market cycle, companies should not introduce products that overlook key attributes in customer choice (which would fall below acceptable, or "must be" quality). Rather, they should use QFD to be quick to market with an acceptable level of quality.

This prescription, although counterintuitive, makes good business sense for at least two reasons. First, many high-tech firms have failed because, in their striving to attain a complicated combination of product attributes—many of which require design time, testing, and debugging well beyond what was initially projected—either customer needs have changed or competitors have beat them to the market with similar products or products that serve the customers' needs in a different way. Storage Technology encountered this situation in the lengthy development of one of its radical innovations in data storage devices. Delays in technological development due to the high level of complexity resulted in a competitor beating it to market. Second, many high-tech customers are faced with switching costs; getting to market quickly captures an installed base of customers for later upgrades and versions. Hence, a firm should work incrementally on features and functionality, guided closely by marketing input. Firms must ensure that their first foray in the market is quick and at least "must-be quality"; over time, additional product extensions can strive for attractive quality.

Many seminars are offered on the QFD process and many research firms can provide expertise in implementing it. Interested readers are referred to additional reading.[41]

Prototype Testing

A *prototype* is a model of the ultimate product or service. As a model, the prototype provides only the essential elements of the planned final product while ignoring minor or purely supporting elements. The first step in **prototype testing** is to ensure it meets technical design specifications. If the prototype does not meet technical specifications, appropriate adjustments are made. After it meets specifications, the prototype is then evaluated by potential customers. The earlier example of product testing in the video game industry combined concept and prototype testing in a useful manner. The IDEO vignette also incorporated prototype testing.

The *information acceleration (IA)* technique relies on a virtual representation of a new product to assist in product development and forecasting.[42] Such representations are more vivid and realistic than traditional concept descriptions but less expensive than actual prototypes. Hence, they provide a useful middle ground between traditional concept descriptions and actual physical prototypes. Feedback from customers is obtained through the use of the virtual representation of the new product idea.

In addition to providing customer feedback about the product, virtual prototyping has other benefits. First, to simulate a future environment, the design team must agree on the implications of that future environment. This forces the team to carefully define the target group of customers and the core product benefits early in the process. Other issues that are brought to the fore are:

- The requisite infrastructure required for product usage (e.g., recharging stations for electric vehicles)
- Technology requirements for future generations of the innovation (e.g., new battery technology for electric vehicles)
- Competitive forecasting of new market entrants
- Available alternatives to the new technology (e.g., hybrid electric vehicles that combine gas power with electric)

To simulate one product, the team must plan for the entire product line (including vans, two-seaters, sedans, etc.) and cannibalization of existing products.

Beta Version Testing

Beta versions of new products are prerelease versions that the company provides to existing or potential customers to try. In a **beta test**, a customer agrees to provide feedback on this early version of the new product to the producer so that the product can be improved prior to its commercial release. For example, Symantec provides a broad range of content and network security software and appliance solutions to individuals, enterprises, and service providers. The Symantec External Test Program is designed to expose prerelease software to a wide range of equipment and real-world usage. Participants in the External Test Program receive prerelease software, test scripts, and documentation to review and test. Participants are expected to remain active throughout a project's life cycle and communicate issues effectively to Symantec team members.[43]

CUSTOMER-DRIVEN INNOVATION

The first research tool discussed in this chapter, concept testing, noted that the goal of much research is to get customer input to refine the company's ideas for new product development. This model of innovation—one driven and guided primarily by the company—although common, has very high failure rates. Indeed, when researchers are sent out to "discover" unmet customer needs, to test the ideas against a sample of customers, to share the information and ideas internally, and then to decide which ideas to pursue, three-fourths of such new product development projects fail.[44]

In contrast, **customer-driven innovation** taps the collective wisdom of a community—customers or otherwise—for product improvements and innovations. Customer-driven innovation requires a radical rethinking of how the innovation process works, moving away from R&D-driven innovation in the lab (with input from marketing and customers used as a guide), to active co-creation of innovation with customers themselves. In contrast to the traditional high-tech innovation process—brilliant scientists working in labs where "backroom boys and boffins come up with bright ideas that they pass down a pipeline to waiting consumers"[45]—user-driven innovation moves from "R&D to R&We."[46] As stated in *The Economist:* "The customer not only is king, now the customer is market-research head, R&D chief, and product development manager, too."[47] Indeed, the new dominant logic for business in general, and marketing in particular, emphasizes the role that customers play in co-creating value with the firm.[48]

Collectives and cooperatives of various kinds, from city-sponsored farmer's markets to the Israeli kibbutz, are examples of customer-driven innovation. Barter, or exchange of goods of one kind for those of another kind, prevalent since ancient times, continues today because it involves co-creation of value. The entire system of microcredit pioneered by Nobel Laureate Muhammad Yunus's Grameen Bank in Bangladesh is based on trust rather than collateral. Grameen Bank provides microcredit to impoverished women in rural Bangladesh and leaves it up to them to invest in business activities that will be beneficial to their local community. In a highly successful program called GrameenPhone, 950 rural women were each provided with a cell phone and service that they could resell to 65,000 village folk to make a profit.[49] Impoverished sections of the population in any country can use the local wisdom of crowds to decide on the appropriate products and services needed by the local community and co-create value with appropriate technology, as discussed further in Chapter 13.

Many different terms are used to refer to customer-driven innovation, including *customer co-creation, customer co-production, "do it yourself" (DIY) innovation, feedback-influenced design, peer production,* and *mass production,* to name a few. To be sure, the notion of user-based innovation is not new; Eric von Hippel has been writing about this concept for more than 20 years in his lead user innovation research. What is new is the degree and intensity of user participation in the innovative process (due in large part to new technologies that facilitate collaboration and harnessing the power of communities of customers) as well as companies' willingness to allow this form of innovation to flourish. The increasing prevalence of this model of innovation is underscored by the many books on this topic that have emerged in the last few years, of which the following are a sample:

The Wisdom of Crowds, by James Surowiecki (2005)

Democratizing Innovation, by Eric Von Hippel (2006)

Wikinomics: How Mass Collaboration Changes Everything, by Don Tapscott and Anthony Williams (2006)

We Are Smarter Than Me: How to Unleash the Power of Crowds in Your Business, by Barry Libert and Jon Spector (2007)

TABLE 6.7 Examples of Companies Using Customer Co-Creation and the Wisdom of Crowds

Open source software development (Apache, Linux, Firefox) is designed by a community of developers, very few of whom actually "work" for the "company" behind it.

Electronic Arts (EA), a computer game developer, ships programming tools to customers, posts their modifications online, and works their creations into new games.

Boeing 787 Dreamliner airplane has a team of 120,000 individuals around the world serving as voluntary advisers.

BMW posted a toolkit on its website to let customers develop ideas showing how the firm could take advantage of advances in telematics and in-car online services. From the 1,000 customers who used the toolkit, BMW chose 15 and invited them to meet its engineers in Munich. Some of their ideas are now in prototype stage.

Staples, the office-supplies retailer, held a competition among customers to come up with new product ideas and received 8,300 submissions. The product called Wordlock, a padlock that uses words instead of numbers, came from this process.

Dell's IdeaStorm, launched in February 2007 using vendor Salesforce.com's IdeaExchange platform, received 7,200 ideas that garnered 510,000 votes and 43,000 comments. The information gathered allowed Dell to deliver new features such as the Ubuntu 7.04 Linux operating system on two of its consumer PCs in Europe.

HP runs "prediction markets" to extract collective wisdom from scientists to help gauge whether the government will approve a new drug or how well a product will sell.

Eli Lilly's "prediction markets" lets groups of employees buy or sell virtual stocks to predict the outcome of drug trials before the data are complete.

A variety of **company-sponsored user communities**—such as SAPfans.com, Cisco Systems' Networking Professionals Connection, Hewlett-Packard's IT Resource Center forums, Intel's Developer Services Forums, Microsoft's Developers Network, SAP IT Community, Sun Microsystems' Java Center Community, and Xerox's Eureka Community for service technicians—allow interactions and information sharing among a variety of users.

Sources: "The Rise of the Creative Consumer," *The Economist,* March 12, 2005, pp. 59–60; Hof, Robert, "The Power of Us," *BusinessWeek,* June 20, 2005, pp. 74–82; Wells, Melanie, "Have It Your Way," *Forbes.com,* February 14, 2005, online at www.forbes.com/forbes/2005/0214/078_print.html; Stix, Gary, "When Markets Beat the Polls," *Scientific American,* March 2008, pp. 38–45; and Dell IdeaStorm, online at http://en.wikipedia.org/wiki/Dell_IdeaStorm.

Open Innovation: Researching a New Paradigm, by Henry Chesbrough, Wim Vanhaverbeke, and Joel West (2008)

The Future of Competition: Co-Creating Unique Value with Customers, by C. K. Prahalad and Venkat Ramaswamy (2004)

The *McKinsey Quarterly* reports that three-fourths of business executives planned to maintain or increase their technology investments to encourage user collaboration.[50] Their diverse purposes include broader sharing of ideas, solving problems, acquiring customer feedback, and improving productivity. Table 6.7 highlights some examples of this new model of innovation.

At least three factors explain the interest in customer-driven innovation: (1) technology tools to facilitate it, (2) the economics of product development costs and high failure rates, and (3) customer's expectations (society's beliefs) about the role of customers in business strategy. With respect to the latter, customers are increasingly concerned about the environmental impact of the products they buy and the safety of the products they use. By participating in generating ideas for sustainable, safe products, customers receive psychological and social benefits, including a heightened sense of social responsibility. With user-led innovation, customers seem to be willing to "donate" their ideas freely. Their motives may derive from the enhanced reputation and network effects to be gained. Or, they may realize that their odds of successfully commercializing their ideas on their own are low. (Regardless, careful attention to intellectual property rights is crucial to make this model work.)

Moreover, the development and growth of the Internet in general, and of Web 2.0 technologies in particular, have facilitated the use of customer co-creation. User-generated "content" (content on the Internet, product development, marketing campaigns, etc.) is one of the largest transformations that Web 2.0 technologies have afforded to both online and traditional companies. The significance of this trend led *Time* magazine to name "You" as its Person of the Year for 2006—meaning the collective power of all of us is being harnessed in new and creative ways never imagined prior to this generation of technology. The Internet connects like-minded individuals in communities that can build on each individual's knowledge and insights, leading to the notion of the "wisdom of crowds." More than a billion people online worldwide—sharing knowledge and social contacts—represent a collective force of unprecedented power and mass cooperation that is being used to disrupt activities and harness new insights.[51] In addition, the Internet can be used to share development tools with customers (such as programming tools, etc., in the case of software). Moreover, a variety of companies are developing online platforms that harness the power of user-generated content.

Companies are flocking to the "virtual commons" for the business benefits. By opening themselves up to contributions from enthusiastic customers, they can create products and services faster with fewer failures, at lower cost, with less risk. Indeed, findings indicate that companies who open their internal innovation procedures to share with customers typically acquire greater influence with them, and grow the number of possible ideas that customers co-create.[52] The Cooperative Corporation reports that by making better use of resources from outside the company, P&G boosted sales per R&D person by 40%.[53]

Although this model of innovation sounds deceptively smart and simple, the fact remains that many companies actively resist customer-driven innovation. One example cited by *The Economist* is that American farmers in the early 1900s lobbied car manufacturers for detachable backseats. But the car companies didn't "invent" the pickup for more than a decade. Similarly, many companies today respond to customer modifications of their equipment by voiding the warranty. In contrast, when Lego launched its Mindstorms build-it-yourself robot kit in 1997, some 1,000 "hackers" downloaded its operating system, improved it, and posted their work freely online. Lego at first was uncertain about how to handle this situation, but ultimately realized the merits of this community's work, and now actively encourages it by allowing programs written by hackers to be uploaded to the Mindstorms' website.

Company resistance to user-driven innovation comes from many places. Especially in high-tech markets, some companies mistakenly think that customers lack the sophisticated knowledge and skills that new product development requires. At the extreme, engineers may view customer innovators as rivals who undermine their creativity. Mistakenly, companies may worry that it gives customers too much power. And, as we note in Chapter 8, many high-tech companies suffer from the mistaken belief that the only good technology is one that they invent themselves. Sometimes referred to as the *NIH syndrome* ("not invented here"), engineers' beliefs in their own brilliance sometimes blind them to the value of innovations developed by others.

Yet, paradoxically, technology companies are leading the way in harnessing the collective knowledge of their customers.[54] In contrast to the NIH syndrome, Don Tapscott, author of *Wikinomics: How Mass Collaboration Changes Everything,* refers to customer-driven innovation as the *PFE approach:* "proudly found elsewhere."[55] The new paradigm of best-practices marketing actively cultivates relationships that involve customers in developing customized, compelling value propositions.[56] Effective core competencies harness "communication, involvement, and a deep commitment to working across organizational boundaries"; they involve collective learning about how to coordinate diverse product skills.[57] This new paradigm emphasizes that value, rather than being embedded in the output of a firm, is defined by and co-created with customers. The notion of customer-oriented goes beyond simply collecting information from customers; it means collaborating with and learning from customers.

The accompanying Technology Expert's View from the Trenches is about a small high-tech start-up that harnessed the power of customers to design its software solution in the customer feedback/market research industry.

TECHNOLOGY EXPERT'S VIEW FROM THE TRENCHES

Trials and Tribulations of a High-Tech Start-Up
STEVEN SUNDHEIM
Cofounder, Grupthink, Missoula, Montana

In 2006, I saw chaos in online discussions. Video game discussion forums, for example, were plagued with the same questions and topics being posed over and over, scattered across multiple pages and times. Even within a single topic or discussion thread, the same answers appeared and reappeared. Often, good topics and answers would get lost in this "conversational soup"—to the detriment of both writers and the readers. Ultimately, these problems fed off each other to create chaotic discussion that was hard for gamers and the game companies to harness effectively.

This, then, was the genesis of Grupthink—an online feedback community that combined aspects of wikis, photo sharing, and social networking to create a community-powered survey system. Our technology prevented duplicate posts; allowed users to post useful, thoughtful responses that weren't lost in a noisy, chaotic discussion format; and let users rate and rank other people's ideas. Removing the chaos and noise would ultimately give consumers a clearer voice and, hence, a more valuable role in companies' product development activities.

Despite our sense that there was real potential in the feedback community concept, we had no idea who our target market be, nor the value customers would find in the service. Grupthink was an entirely new way for people to have discussions and we didn't know how it would be used, or if people would find it useful. In addition to addressing these targeting and positioning questions, we had other challenges to address.

Challenges

First, we already owned a profitable business (Modwest) and managed 12 employees. It was through Modwest's profits that we would fund Grupthink, so we had make sure Modwest stayed healthy, while giving Grupthink the attention it needed.

Second, we had no direct experience with feedback management or market research—likely areas where Grupthink could become useful. How would we build a valuable solution for clients in that market without knowing exactly what they needed?

Playing to Our Strengths: Technology and Willingness to Experiment

We did have one important strength to work with. We had expert technical know-how in building and hosting Web applications. We decided on a "fast and small" approach that capitalized on that strength. We dedicated three people, including myself, to build and launch a free, public version of Grupthink within three months and get as many people using it as possible.

We figured that if people could have fun using Grupthink.com, which had no real focus other than social networking, the Grupthink format could be valuable to a business trying to establish a community and connection with consumers. We would gain insights into how people reacted to the new format so we could improve upon the idea. There was also the small chance that Grupthink.com would go viral and become the next MySpace or Facebook, so we had little to lose and everything to gain by getting it into the hands of real users.

Refining the Business Model: Gathering Feedback and Information

By the end of 2006, the Grupthink.com community had grown to 5,000 users and was still expanding without any effort on our part. We knew from the Grupthink.com community that people enjoyed our novel approach to online discussion, and that Grupthink could provide businesses with instant, continuously updated views of consumer opinion.

(continued)

Launching the public site and listening to members helped us overcome the initial challenges of getting started and allowed us to continually revise our ideas of what Grupthink could possibly become. We even used Grupthink as it was intended—for feedback management on our business ideas for the platform. We tossed ideas around, like:

- Shopping prediction tool
- Matchmaking site
- Product review database
- Feedback management for the gaming community

My cofounder, John Masterson, and I attended the Web 2.0 summit in 2006 to engage people with these ideas and see how they would be received. We continued to read about industry trends to see where the Grupthink concept would be most valuable. One noteworthy trend was in the feedback management and online market research industry. Spending in these areas was growing by more than 35% a year and there was an incredible amount of buzz related to corporate transparency, collective wisdom, and user-centered innovation—things that a Grupthink feedback community could provide.

So, in 2007 we decided—with some trepidation—that the biggest opportunity for Grupthink was in feedback management and the online market research industry.

Given our unfamiliarity with the world of market research, we knew we had a steep learning curve ahead of us if we were going to sell feedback communities. To get some objective criticism, we talked to our two longtime advisers. After that, I traveled to Europe and to Seattle to meet with potential customers in the gaming industry, to learn more about their challenges with acquiring user feedback and to see if Grupthink could fill a need in their organizations.

We were encouraged and enlightened by these conversations and felt like we were ready to start building Grupthink "Pro." Unfortunately, the demands of Modwest were prohibiting us from dedicating developers to Grupthink. Things felt like they had stalled, and the only way out was to find either investment or an early client who could help fund initial development of Grupthink Pro.

The Serendipitous Turning Point

I was contacted by the business manager for one of the game development companies I had visited. He connected me with a renowned professor at MIT who specialized in user-centered innovation. I was honored when he invited me to present my experiences with online communities at the MIT innovation lab.

Through that meeting I was introduced to leaders in a large corporation who were struggling to leverage online communities and harness consumer insights. They were intrigued by the idea of community-powered feedback, and thought Grupthink could help keep their customers engaged, overcome "survey fatigue," and allow them to pose qualitative questions that weren't possible in any other online format. Six months later, that company became our first customer.

The importance of this first client went far beyond the revenue stream. First, we had demonstrated that an established, well-known multibillion-dollar company saw value in the concept of Grupthink Pro, even before it was built. Second, we could confidently tell potential investors that our product would sell, because it already *did*. Third, and probably most importantly, we got to collaborate with market research professionals in a large company who would help guide and refine the development of our product to best suit their needs. This was user-centered innovation at work.

Looking to the Future

By the time this book is published, we will have installed feedback communities for a handful of other companies, including our own. We'll continue to use these early adopters to refine the product, determine its real value, and build case studies. We have interested investors who will be instrumental in helping us promote the service, and plan for growth. Although we still have challenges, such as refining our positioning and pricing for the market, I'm finally confident that we have the momentum needed to bring Grupthink Pro all the way to market.

We've overcome our challenges through a combination of our strengths as well as our weaknesses, with a bit of good fortune thrown in. A start-up with a large seed investment may have hired experts, consultants, salespeople with contacts in the industry, plus a team of developers. We instead relied on our technical strengths and existing resources, stayed light on our feet, and constantly listened to our customers and advisers to help guide development. This approach, although arguably slower than other tactics, has allowed us to work at our own pace and keep the business owned 100% by the founders and employees.

The great thing about our approach is that we were able to work on something we cared about and could communicate effectively. That kept us motivated and led to important industry contacts. Ultimately, it's impossible to know for sure if we chose the best approach. I just think it's best to make the most informed decisions along the way, work hard at building something you are passionate about, and enjoy the ride.

BIOMIMICRY[58]

Certainly, customers are an important source of ideas for new innovations. However, ideas can be found in many places. The term *biomimicry* means, quite literally, mimicking or emulating ideas from the natural world: *bios* means "life" and *mimesis* means "to imitate." So, **biomimicry** is the conscious seeking of inspiration, the search for finding new and better ways to do things, through understanding nature and the principles of biology; it is a process of looking for nature's advice to solve human challenges. In addition to generating novel insights, biomimicry tends to generate environmentally friendly insights. Disenchanted with the negative impacts and unintended consequences of existing technologies, biomimics recognize that nature achieves its purposes without harsh chemicals or excess energy. Indeed, nature's production methods don't guzzle fossil fuels, require toxic chemicals, or pollute the planet. So, biomimics harness and borrow from nature's successes in their quest for new innovations that offer a sustainable future. In the process, biologists work hand in hand with engineers, architects, and product designers to find sustainable solutions to an array of problems.

With respect to high-technology innovations, biomimicry generates novel insights for disruptive/breakthrough innovations in many areas. For example, if a company wants to develop a fabric to keep athletes cool in very hot, dry weather, it might look to the sand grouse from Africa for insights. Sand grouse typically live 10 to 15 kilometers from any water source. To provide this necessity for their young, the males fly to water, bury their bellies and soak up water in their breast feathers (which have a very unique coil shape), and then return to their nests with water for their babies. A company, then, might develop a fabric with fibers that mimic the coil shape and the weight–volume ratio found in the birds' feathers. Table 6.8 provides other examples of biomimicry in action.

The Biomimicry Process

The biomimetic path was originally outlined by Janine Benyus in her 1997 book, *Biomimicry: Innovation Inspired by Nature*. Her colleague, Dayna Baumeister, works with companies to implement this path through a seven-step process using a spiral model, shown in Table 6.9. The spiral emphasizes the iterative nature of the process; after one set of challenges is addressed, another set arises and the process begins anew. The specifics of the methodology were developed by the Biomimicry Guild and are used by permission here.

To aid in the use of the biomimicry process, the Biomimicry Guild will match biologists with product designers and engineers. The guild has worked with Procter & Gamble, Nike, and

TABLE 6.8 Examples of Biomimicry

The ***Namibian beetle*** has learned how to find water in the middle of a desert, catching water from the fog that periodically rolls across. On the underside of its wings, a series of bumps attract and capture the water. The tips of each bump are fashioned from a substance that attracts the moisture, while the sides shed it. So, when the fog rolls in, the beetle unfurls its wings; water droplets form on the tips and run down the sides of the wings into the beetle's mouth. Now, at refugee camps in the Namibian desert, people hang sheets made of a checkerboard of water-loving and water-hating materials to collect water from the air. Indeed, Inventa Partners and QinetiQ received the "Bright Spark" BSRIA/ BSJ building services industry award for their development and application of a revolutionary new material that copies the properties of the wing surface of the Namibian desert beetle for collecting precious drinking water from an invisible mist.

 Prototypes are now under test for industrial applications such as recycling the water lost by the evaporating cooling towers in air-conditioning systems. The new material could save many billions of litres currently lost every day—a major cost and environmental problem for developed regions. There are more than 50,000 new cooling towers installed globally each year, and one large system will evaporate more than 500 million litres (over 132 million gallons) each year.

The leading edge of ***humpback whale flippers*** are scalloped with prominent knobs called *tubercles.* In wind tunnel experiments, the scalloped flipper proved a more efficient wing design than the smooth edges used on airplanes. In tests of a scalloped versus a sleek flipper, the scalloped flippers have 32% lower drag, 8% better lift properties, and withstood stall at a 40% steeper wind angle. This discovery has the potential to optimize not only airplane wings but also the tips of helicopter rotors, propellers, and ship rudders. The improved stall angle would add a margin of safety while making planes more maneuverable, and the drag reduction would improve fuel efficiency. The design has also been used in wind turbine blades.

Mussels (the little sea creatures that live inside purplish-blue shells) make a glue that lets them anchor themselves firmly to a rock and remain there, drenched by water, buffeted by the ocean's waves. The mussel foot produces an epoxy with adhesive properties that rival any super-glue on the market. What other adhesives are capable of such strength over long periods of time under such conditions? Molecular biologists at Oregon State have mimicked the adhesive properties of a mussel's secretions and added soy protein to make an environmentally friendly adhesive used in products such as Columbia Forest's PureBond hardwood plywood. Although biomimetic epoxy is still in the development stage, researchers project that it will prove useful to the private marine industry and medical and dental fields—where scientists believe that the glue could one day repair shattered bones.

Locusts, those annoying insects of biblical renown, have a special ability to fly in dense swarms without colliding. Possible commercial applications? Volvo is using insights from studying locusts to provide anticollision devices in cars.

Source: Biomimicry Institute, Case Studies, 2008, online at www.biomimicryinstitute.org/case-studies/case-studies.

General Mills. A separate organization, the Biomimicry Institute, promotes the new field and generates sustainable innovations. The biomimicry design portal, www.AskNature.org is a digital library of nature's solutions and an information exchange between biologists and innovators. Consistent with the biomimicry process as well as current trends toward best-practices corporate social responsibility, the Innovation for Conservation program at the Biomimicry Institute encourages companies that use the biomimicry process to donate a percentage of their sales of biologically inspired products, or savings due to biologically inspired processes, to restore the habitat of the organism that inspired their breakthrough design.

TABLE 6.9 Steps in the Biomimicry Process

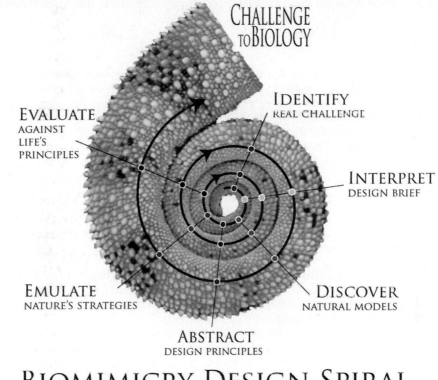

CHALLENGE to BIOLOGY

IDENTIFY
REAL CHALLENGE

EVALUATE
AGAINST
LIFE'S
PRINCIPLES

INTERPRET
DESIGN BRIEF

EMULATE
NATURE'S STRATEGIES

DISCOVER
NATURAL MODELS

ABSTRACT
DESIGN PRINCIPLES

BIOMIMICRY DESIGN SPIRAL
©2008 Biomimicry Guild. Used with permission.

Step 1: Identify the problem to be resolved. The goal in this step is to identify the heart of the problem the company is trying to solve. This step asks a series of "why" questions to open the process for creativity in innovation. For example, say a company's water filters in its plants clog and are expensive to replace. Rather than asking "What do I want to design?" (a better water filter), a company's interests are better served by asking "What problem do I want to solve?" (access to clean water). Then ask:

- Why do filters clog? (Do they filter all particulate from the water, and algae grows on the filters?)
- Why do the filters filter everything? (Because the toxic molecules are the smallest, and to remove them, everything bigger is removed too.)
- Why do algae grow on the filters? (Because it's a wet medium and the tank is in the sunlight.)

This "why" approach in getting to the heart, or the root, of what the company wants to accomplish opens up a broad set of possibilities for what the answer might be. For example, after this series of "why" questions, one solution might be a better filter. But, another solution might be a totally new approach to purifying water that removes the small toxic molecules.

Step 2: Interpret the problem in nature's terms. The goal in this step is to "biologize" the issue, by asking "How does nature achieve this function in the environment?" To discover answers in the biological world, one has to identify functions that must be performed (rather than the products that could perform those functions) to solve the problem.

In addition to requiring a very specific identification of the exact functions involved, this step also requires identification of the habitat/context ("operating parameters") where it occurs. For example, typical considerations in this step include:

- How does nature perform this function? And, conversely, how does nature *not* perform this function?

(continued)

TABLE 6.9 (Continued)

- What are the climate and nutrient conditions where this function is performed (e.g., temperature, pressure, moisture, pH)?
- Is there a social aspect for the organism in nature (e.g., cooperation, competition, predation)? Is there a temporal aspect (e.g., aging, growing)?

The question for the example above is "How does nature clean water?" Many different ideas and insights could be generated for a variety of organisms and contexts.

In order to ensure close correspondence between the biomimicry process and the outcome, a *design brief* is developed. This document helps to cross the bridge between business people and "naturalists" in the service of innovation, allowing engineers and biologists to come together with a common focus and to use a common language. Especially in situations when innovation and design are outsourced, the design brief carries a critical role.

The design brief captures the essentials of the problem, deconstructing it in specific terms, such as:

- What does the design need to do? (Not "What do you want to design?")
- Who is involved with the problem and who will be involved with the solution?
- Location: Where is the problem, and where will the solution be applied? (How does nature do that function in these conditions?)

Step 3: Discover the best natural models that answer/resolve the challenges. The goal in this step is to identify as many solutions as possible from organisms in nature that offer either literal or metaphorical solutions to the problem. The identification can be done by combing through literature or brainstorming with biologists.

Creative insights are found by turning the problem inside out and on its head. For example, in another illustration, say the problem was to find a way to dry out humid air. Creative insights wouldn't come only from looking in the tropics; one might also look in the desert where, surprisingly, cockroaches drink water from air.

Often, special insights can be gained by identifying *champion adapters,* or those organisms that are most challenged by the problem—possibly those whose survival depends on it—but are unfazed by the situation. These champion organisms can sometimes be found in the extremes of habitat, and can be identified by talking with biologists and field specialists who are experts on the organism and its habitat.

Step 4: Abstract from the examples in prior step to identify patterns and to create a taxonomy. The goal in this step is build a *taxonomy*—a clustering structure for the ideas generated in the prior step. Based on all the ideas discovered in step 3, what does each organism's strategy have in common? What are the repeating patterns and processes within nature that accomplish the desired function? How are they different? Different clustering techniques such as Venn diagrams, graphs, and visual representations can be helpful. Careful examination of outliers can also be useful. What strategies appear most promising, given the habitat conditions and design parameters? This can be a very difficult step to accomplish and often requires the assistance of biologists and other scientists trained in the biomimicry method.

Step 5: Emulate nature and apply the ideas and solutions to the challenge at hand. This step, the heart of the design phase, develops concepts and ideas that apply the lessons from nature to solve the problem at hand. Standard R&D approaches can be used—as long as they adhere to the design brief that was previously developed. The best applications will apply these lessons as deeply as possible into the designs that are inspired by nature. Again, consultations with biologists and other scientists can provide valuable insight into how the organisms function.

There are three ways biomimicry solutions can be applied to solve human problems:

1. Solutions can mimic nature's *forms;* these focus on "what" nature does in terms of the shape of the organism, for example.
2. Solutions can mimic nature's *processes;* these focus on how nature created that shape, or the "manufacturing" process, for example.

3. Solutions can mimic nature's *ecosystems;* these focus on the cycle and interrelationships of nature's organisms and include, for example, considerations about what happens at the end of life, how to reuse/recycle different pieces, and how the various components relate to other parts of system.

Step 6: Evaluate how well the proposed ideas/solutions compare to successful principles of nature and continue to improve the design by asking another layer of questions. The goal in this step is to consider how consistent the innovation is with nature and to further improve the design for sustainability. Again, this step may lead to new questions, possibly related to the packaging, manufacturing, marketing, or transporting of the innovation.

Step 7: Begin the process anew, with a new identification step. Nature works with small feedback loops, constantly learning, adapting, and evolving. High-tech innovators can also benefit from this thinking—evolving new innovation designs in repeated steps of observation and development, unearthing new lessons, and applying these constantly throughout the design exploration process.

Source: Dayna Baumeister telephone interview, February 23, 2008.

Biomimicry Benefits

Biomimicry develops innovations that are conducive to life. In particular, biomimicry designs have the potential to:[59]

- *Be sustainable:*   For example, they use life-friendly materials and processes, engage in symbiotic relationships, and enhance the biosphere.
- *Perform well:*   Nature has been vetting strategies for 3.8 billion years; in nature, if a design strategy is not effective, its carrier dies. Similarly, innovations based on biomimicry allow companies to thrive in the marketplace.
- *Save energy:*   Energy in the natural world is even more expensive than in the human world. Plants have to trap and convert it from sunlight and predators have to use it to hunt and catch prey. As a result of the scarcity of energy, life tends to organize extremely energy-efficient designs and systems, optimizing energy use at every turn. Emulating these efficiency strategies can dramatically reduce the energy use for companies.
- *Cut material costs:*   Nature builds to shape, because shape is cheap and material is expensive. By studying the shapes of nature's strategies and how they are built, biomimicry can minimize the amount a company spends on materials while maximizing the effectiveness of its products' patterns and forms to achieve desired functions.
- *Redefine and eliminate "waste":*   By mimicking how nature transitions materials and nutrients within a habitat, companies can set up various units and systems to optimally use resources and eliminate unnecessary redundancies. Organizing based on nature's designs will drive profitability through cost savings and/or the creation of new profit centers focused on selling waste to companies who desire it, say, as biomass.
- *Define new product categories and industries:*   Biomimicry reveals stale product categories in a radically different light, creating an opportunity for disruptive technologies that transform industries or build entirely new ones.
- *Build a company's brand:*   Creating biomimetic products and processes allows a company to develop a reputation as both innovative and proactive about the environment.

The innovations that arise based on the biomimicry process will most certainly be a source of products that will radically alter the technology landscape and redefine innovation as we currently know

TECHNOLOGY SOLUTIONS FOR GLOBAL PROBLEMS
Learning from Kelp to Harness the Power of the Ocean
By Jamie Hoffman

Photo reprinted with permission of BioPower Systems Pty. Ltd., Eveleigh, Australia.

Earth's predictable and powerful moon-induced tidal forces have long remained an elusive potential energy source for humans to unlock. For decades, engineers and scientists have struggled to develop a system that would efficiently and economically convert the power of the ocean into electricity. However, the traditional systems implemented were large, stationary, expensive, and typically did not provide enough revenue to warrant the cost. By taking a revolutionary, biologically based approach to the issue, researchers at the University of Sydney in Australia have developed bioWAVE,™ an innovative tidal energy conversion system that has nominal capacity of about 1 MW (megawatt) per unit, and will be installed in "farms." The bioWAVE technology is being commercialized by BioPower Systems to locate strategic partners and raise private equity capital in hopes of taking the technology to market.

The bioWAVE's engineering mimics the biological design of ocean plants such as kelp, which anchor into the seabed and move in synch with the ocean's ebb and flow. The result is a cost-effective renewable energy resource that is environmentally benign and performs efficiently. Each unit has hydro-dynamically optimized blades, an O-DRIVE generator module, a self-orienting base that continually aligns itself with the varying wave movement, and a bioBASE anchor system. Easily installed, bioWAVE has a small footprint, barely disturbs the seabed, operates silently, and does not harm sea life. Additionally, no pollutants or foreign substances are dispersed into the ocean and the system does not affect the salinity or turbidity of the water. The product continuously adjusts itself to align with the wave direction, has direct-drive power conversion, and functions autonomously and unmanned. Lastly, in hurricanes or other extreme conditions, the blades will lie flat against the seabed.

Source: Technology Brief: bioWAVE, BioPower Systems, 2006, online at www.biopowersystems.com/briefs/bioWAVE_Technology_Brief.pdf; and "Biomimetics to Harness Ocean Power," Warren Centre for Advanced Engineering, *E-Bulletin,* November 2006, online at www.warren.usyd.edu.au/bulletin/NO48/ed48art3.htm.

it.[60] This chapter's Technology Solutions for Global Problems focuses on the development of a new energy source based on biomimicry.

FORECASTING IN HIGH-TECH MARKETS

In addition to gathering customer feedback and market research to guide new product development and to generate ideas for breakthrough innovations, high-tech firms are faced with the daunting prospect of forecasting demand for new products. Forecasting future sales of high-tech products is difficult for many reasons. Quantitative methods typically rely on historical data, but for radically new products there are no historical data. Moreover, data obtained through traditional techniques are of dubious value, because customers are unable to articulate their preferences and expectations when they have no basis for understanding the new technology. In addition, commitment biases to existing generations of technology can cast a shadow on managers' ability to accurately see the potential of newer technologies. For example:[61]

- In reacting to the addition of audio technology to silent movies (circa 1927), Harry M. Warner said, "Who the hell wants to hear actors talk?"
- His later colleague, Darryl F. Zanuck, head of 20th Century Fox Films in 1946, predicted that "Television won't be able to hold onto any market it captures after the first six months. People will soon get tired of staring at a plywood box every night."
- Ken Olsen, president and founder of the DEC Corporation, said in 1977, "There is little reason for any individual to have a computer in their home."
- Lou Gerstner, CEO of IBM, said in 1999, Internet companies are "fireflies before the storm. They shine now but will eventually dim out."[62]

Although gathering information regarding customers in high-tech markets is difficult, companies should not resort to the "crystal ball" technique to develop forecasts for high-tech products; such an approach is imprecise at best and flat-out wrong at worst. Rather than being overwhelmed by the challenge, high-tech marketers must seek out the available tools to help them build a forecast. And because the task is fraught with uncertainty and many sources of error, using a systematic process to develop the forecast is more important for high-tech products than other types of products.

Forecasting Methods

In selecting a forecasting method, the manager should address the following questions:[63]

- What do we want to forecast (e.g., market demand, company sales, technology trends)?
- Why do we need the forecast (e.g., for new product development investments, to determine manufacturing capacity)?
- How important is the past in predicting the future?
- What influence do we have in constructing the future?
- What factors could change the forecast?

Forecasting tools can be categorized into quantitative and qualitative tools; some techniques use a combination of qualitative and quantitative. Following the contingency theory of high-technology marketing, Figure 6.3 presents a framework to select a forecasting method. As the figure shows, for relatively incremental innovations with which the company has previous experience, standard quantitative tools that rely on historical data—moving averages, exponential smoothing, and time series regressions—are appropriate. Collecting primary research and using conjoint analysis (previously discussed) can help a company to predict market share forecasts for incremental innovations with which it does not have a prior sales history. Readers interested in these forecasting tools should consult one of the many excellent resources available.[64]

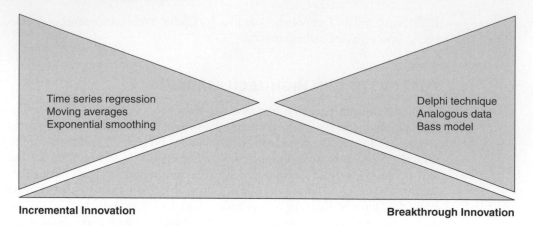

Incremental Innovation   **Breakthrough Innovation**

FIGURE 6.3   Aligning Forecasting Type with the Type of Innovation

For breakthrough products, a different set of forecasting tools is appropriate. Following is an overview of some of the more applicable tools for high-tech companies.

QUALITATIVE TOOLS   The **Delphi method** is probably the most common qualitative method. In this technique, a panel of experts is convened and asked to address specific questions, such as when a new product will gain widespread acceptance. These experts are purposefully kept separate, so that their judgments will not be influenced by social pressures or group dynamics. The answers to initial questions are sent back to the participants, who are asked to refine their own judgments and to comment on the predictions of the others, in an attempt to find a consensus. Anonymity among the panel members allows for open debate.[65]

Although this method has limitations, including lack of reliability assessment and potential sensitivity to the experts selected, the same limitations also apply—possibly even more so—to other subjective estimates. Selection of the experts also warrants careful attention. Experts from the industry in general, including lead users, can offer their knowledge as a useful benchmark against the estimates generated internally by a firm.

Another useful forecasting tool in high-tech markets relies on **analogous data**—information about a related, similar product—to make inferences about the new technology.[66] Recent research shows that when a structured approach is used to guide the development of an analogous forecast, the forecast is substantially more accurate than if obtained from experts alone.[67] This forecasting method uses data about another product currently on the market, or one that existed at an earlier time, to forecast a new product's expected growth pattern. For example, in predicting sales of high-definition TV (HDTV) equipment, forecasts can be based on the history of similar consumer products, such as color TVs or videocassette recorders. In selecting the analogous product, it is critical to establish a logical connection between the two. For example, do the two products serve a similar need or share other important characteristics? One also must take into consideration environmental factors and market conditions that may uniquely affect the new product's growth pattern. Then, based on the sales pattern for the analogous product, the use of intuitive judgment traces the expected pattern of sales for the new product.

This technique is valid only to the extent that the analogy holds true. The degree to which the analogy is appropriate depends on the logical connection between the products involved. For example, in 1999 to forecast the demand for PDAs (personal digital assistants, or handheld computers), possible analogous products might include personal computers and cell phones.[68]

The degree to which these analogous products are logically connected to handheld computers depends on similarities in the attributes of importance to the consumer in making purchase decisions and in the business factors that contribute to product success. Important attributes include technical support, ease of use, and product form/design considerations. Critical business factors include distribution considerations, brand name considerations, and model options. Based on consideration of these factors, Handspring Inc. of Mountain View, California, concluded that both products served as useful benchmarks but neither alone was entirely appropriately analogous.

QUANTITATIVE TOOLS   For breakthrough innovations, one of the most widely used quantitative tools for forecasting is the **Bass model**. Named after Frank Bass, the model can be used as a prelaunch forecasting technique, meaning that is estimated prior to the introduction of a new product, before preliminary sales figures are available. This model is most appropriate for forecasting sales of a new technology for which there is no closely competing alternative. A long history of research has validated the viability of this approach for forecasting a variety of innovations.[69] Formulations of the Bass model have been used by corporations such as Kodak, IBM, RCA, Sears, and AT&T.[70]

Historical analysis of new product sales curves indicates that one of the most common patterns of new product cumulative sales is an S-shaped curve. An S-curve implies that new product sales may start slowly, then grow at a rapid rate, then the rate of growth tapers off, and finally declines with time, as illustrated in Figure 6.4.[71]

The Bass model explains this S-shaped curve based on diffusion theory.[72] Diffusion theory examines the reasons why innovations spread through markets. The consumer product adoption process, detailed in Chapter 7, categorizes individuals as innovators, early adopters, early majority, late majority, and laggards, based on relative adoption time. When a new product is introduced, uncertainty exists in the minds of potential adopters regarding the relative superiority of the new product. Potential customers attempt to reduce this uncertainty by acquiring information about the new product. Earlier adopters tend to rely on mass media and other external sources for product information. Later adopters are more likely to acquire such information from interpersonal channels such as word-of-mouth communication and observation.[73]

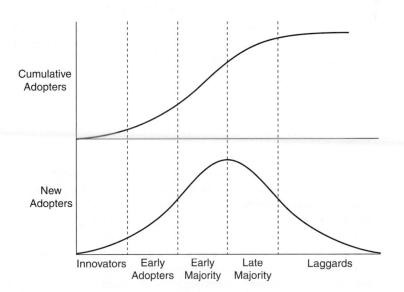

FIGURE 6.4 Diffusion of Innovation

The Bass model thus assumes that new product adopters are influenced by two types of communication: mass media and interpersonal communications. The mass media effects, which have a greater impact on earlier adopters, will be greater at the outset of the product launch, whereas the interpersonal communication effects, which have a greater impact on the much larger number of later adopters, will be greater during the later periods of the diffusion process.[74]

The model requires that a marketing manager develop estimates for first-year sales and for total product lifetime sales (i.e., year one adopters and total adopters). The manager must also estimate the *coefficient of innovation* and the *coefficient of imitation*. These coefficients are simply referred to as *p* (mass media) and *q* (word of mouth). These can be estimated from data on a similar product (analogous data) or from industry values. The formula for the Bass model is:

$$n_t = (p \times \text{Remaining potential adopters}) + (q \times \text{Adopter Proportion} \times \text{Remaining potential adopters})$$

where n_t = the number of adopters at time *t* (Sales), and *p* and *q* are as above

Despite its seeming complexity, the model is easy to implement in a standard spreadsheet program. A recent Webinar (an online seminar delivered via the Web) provides illustrative applications.[75]

Numerous assumptions underlie the Bass model that should be recognized prior to and during its application; these include:[76]

- The size of the potential market of total number of adopters remains constant over time.
- There is only one product bought per new adopter.
- The coefficients of innovation and imitation remain constant over time.
- The new product innovation itself does not change over its life cycle.
- The innovation's sales are confined to a single geographic area.
- The impact of marketing strategies is adequately captured by the model's parameters.

Thus, despite the usefulness and proven viability of the Bass method, it may not be reliable in some situations.

Other Considerations in Forecasting

Whichever method or combination of methods is used, the forecaster must ensure that bias does not enter into the forecast due to personal or organizational desire for the success of the new technology. Stakeholders in a new technology often inflate predictions of its future success, and "since their bullish statements of technical potential are often misleadingly packaged as precise market forecasts, unwary businesses and investors often suffer."[77] Marketing researchers can avoid bias by studying a new technology's potential buyers, who have less of a stake in its success. However, this primary research may not be done because the group of potential customers can be difficult to reach, making accurate market research expensive and time-consuming.[78]

Another problem with forecasting new technologies is the "cross-competition of current technologies with new technologies serving the same market."[79] Although Chapter 1 addressed the "incumbent's curse," in which existing firms downplay the competitive threat posed by new technologies, new start-ups also suffer from their own curse: that of overenthusiasm. In developing a forecast for a new technology, managers must consider the entrenched market position of the incumbents. The main advantage of incumbent technologies is that they already have developed markets with established distribution channels and loyal consumers. Also, they have proven production processes and higher production volumes. All of these factors allow established technologies to be marketed with a cost leadership strategy, pushing prices downward and helping them maintain or even increase market share.[80] To avoid inaccuracies due to this problem, forecasters must fully consider the advantages of the established technologies and, at a minimum, temper their enthusiasm for the success of the innovation while adjusting their predictions of how quickly the new technology will overtake the existing technology.

Many times, decision makers are less than confident in the prepared forecast for a certain technology, and this lack of confidence can sometimes lead to indecisiveness or bad decisions. Although forecasting demand for new technologies is difficult, it provides critical information to decision makers. Forecasters should keep in mind that the success of the forecast is based not on whether it comes true, but on the quality of information provided to the decision makers who use the forecast.

Summary

This chapter has covered a range of tools used to collect market-based information in high-tech markets. Such information can be used for a variety of purposes, including generating ideas for novel innovations, evaluating ideas for innovation that companies have, determining desirable features and pricing in new products, and forecasting sales and technology trends.

Based on the contingency theory of high-tech marketing, companies must match the type of research techniques they use to the type of innovation. Tools that are best suited to the complexities inherent in the high-tech environment include:

- Customer visit programs
- Empathic design
- Lead user innovation processes
- Quality function deployment
- Prototype and beta testing

In a break from the paradigm of company-driven innovation (in which a company seeks customer feedback and input for the innovation ideas a company's personnel develop), customer-driven innovation allows customers themselves to generate and refine new product ideas. This model of innovation is transforming the way companies and industries think about the innovation process.

Biomimicry is another new technique used to generate ideas for innovation. Seeking innovation ideas that are inspired by nature, innovations derived from the biomimicry process have built-in benefits: environmental friendliness and reputation effects.

Finally, a range of forecasting tools lend assistance to companies seeking to make solid business decisions. Delphi techniques, analogous forecasting, and the Bass model were the tools emphasized in this section.

A key point to take away from this chapter is the overwhelming need to diligently and assiduously gather information from the marketplace. In addition to collecting information on competitors, high-tech marketers must work with customers—to understand them, to have an ongoing dialogue with them, to study them, and to incorporate their needs into the product development and marketing process. High-tech firms sometimes end up with baffled, frustrated, and unhappy customers, which is itself a threat to the health of the high-tech economy. Technical people find that, much of the time, users don't know what they want; and when they do know, they all want something different. Despite that, users are the customers; developers should delight the customers and be responsive to their needs and anticipate them in designs.[81]

The next chapter takes another step toward understanding customers and explores issues related to customer adoption decisions for high-technology products.

Discussion Questions

Chapter Concepts

1. Using contingency theory from Chapter 1, identify how marketing research techniques must be matched to the type of innovation to ensure greater success and insight.

2. What is concept testing and how is it used by high-tech marketers?

3. What is conjoint analysis, and how can high-tech marketers use it to refine the product development process?

4. What is a customer visit program, and what benefits does it offer? What are the elements that make a customer visit program successful?
5. What is empathic design? What insights can it generate? What are the steps in the process?
6. How are customer visits similar to and different from empathic design?
7. Who are lead users? What are the four steps in the process of using lead users in marketing research?
8. What is QFD? What are its benefits? Identify and describe the three steps in the QFD process.
9. What is the Kano concept? What are the three levels of attributes it depicts? How does a Kano concept add insight into customer needs?
10. What are the trade-offs in being fast to market versus having the right set of attributes in the product? How should a firm manage this tension?
11. What are prototype testing and beta testing? How are they different from each other?

12. What are the benefits of virtual prototype testing?
13. Explain customer-driven innovation and how it differs philosophically from the other research techniques covered previously in the chapter.
14. What are some other terms used for customer-driven innovation? Why is it increasing in prevalence today?
15. Compare and contrast the NIH syndrome with the PFE approach.
16. What is biomimicry? Give an overview of the seven steps in the process. What are its key benefits?
17. Why is it so difficult to develop forecasts in high-tech markets?
18. Describe the Delphi method of forecasting.
19. How does forecasting based on analogous data work?
20. What is the Bass model? How does this approach to forecasting work?
21. What are the caveats for high-tech marketers in building their market forecasts?

Application Questions

1. Which of the market research tools does IDEO incorporate into its design process for its clients? What are the pros/cons of its approach?
2. Find examples of three of the following and comment on their insights for high-tech marketing:
   • biomimicry
   • customer-driven innovation
   • lead users
   • empathic design
   • beta testing.
3. Provide examples of the three types of attributes from the Kano Concept for a product category of your choice. What are the implications for product design?

Glossary

analogous data   Information based on a related product that is similar in some way to a new innovation; used to develop a market forecast for the new product.

Bass model   A quantitative forecasting technique that incorporates both mass media influences and interpersonal influences to predict the timing of adoption for breakthrough technologies.

beta test   A market research tool by which a customer is given an early (beta) version of a new technology in order to provide feedback to the producer to make needed improvements prior to its commercial release.

biomimicry   Seeking ideas for innovations from nature; biologically inspired design.

concept testing   The process of asking potential customers to evaluate early-stage ideas, to determine which concepts have the greatest probably of market success and warrant further development.

conjoint analysis   A market research tool in which respondents are asked to make judgments about their preferences for various combinations of product attributes; statistical tools are then used to estimate how much each attribute is valued and to use this information in making trade-offs in the product design process.

customer-driven innovation   A user-led model of innovation that uses technology to tap into the collective insights of a group of customers; a radical departure from a company-centered model, in which research is used primarily to evaluate company-generated ideas.

customer visit program   A systematic program of visiting customers with a cross-functional team to understand customers' needs, how they use products, and their environment, with the objective to develop new products/features.

Delphi method   A qualitative forecasting tool by which a panel of experts makes judgments based on the likelihood of future scenarios.

empathic design   A research technique for product innovation based on understanding user needs through observation of the customer, rather than through traditional questioning methods (focus groups, surveys); aimed at generating insights for new products and features that customers may not be conscious of wanting or able to articulate.

Kano concept   A graphical representation of the relationship between certain product attributes and customer satisfaction or dissatisfaction; depicts impact on customer satisfaction/dissatisfaction as the level of the attribute increases or decreases.

lead users   Customers who are well ahead of market trends and who innovate their own solutions to their own problems, providing market researchers with useful insights for innovations.

prototype testing   The evaluation of a model of a desired product that replicates the product's critical features while ignoring minor or noncritical features.

quality function deployment (QFD)   An engineering tool that incorporates customer requirements into product design decisions.

Notes

1. Based on "iPhone Greedily Eats North American Market Share," Gizmodo, December 16, 2007, online at http://gizmodo.com/gadgets/smartphones/iphone-greedily-eats-north-american-market-share-334516.php; IDEO website, at www.ideo.com; Kelley, Tom, with Jonathan Littman, *The Art of Innovation* (New York: Currency, 2001); Nussbaum, Bruce, "The Power of Design," *Business Week,* May 17, 2004, p. 96; Borden, Mark, et al., "The World's Most Innovative Companies," *Fast Company,* March 2008, pp. 73–117, online at www.fastcompany.com/magazine/123/the-worlds-most-innovative-companies.html; and South, Alan, "Abstract Truth," *Aircraft Interiors International,* March 2004, pp. 116–122.

2. Schrage, Michael, "My Customer, My Co-Innovator," *Strategy+Business,* August 31, 2006, online at www.strategy-business.com/press/enewsarticle/enews083106.

3. Quoted in Brown, Eryn, "9 Ways to Win on the Web," *Fortune,* May 24, 1999, pp. 112–124.

4. Ibid.

5. Corporate Executive Board, Market Research Executive Board, Member Benchmarking Survey Analysis, "2003–2004 Benchmarking the Research Function," March 2004. (Survey was updated for 2006, but detailed breakouts by category were unavailable to us.)

6. Leonard-Barton, Dorothy, Edith Wilson, and John Doyle, "Commercializing Technology: Understanding User Needs," in *Business Marketing Strategy,* eds. V. K. Rangan et al. (Chicago: Irwin, 1995), pp. 281–305.

7. Leonard-Barton, Wilson, and Doyle, "Commercializing Technology."

8. Churchill, Gilbert, and Dawn Iacobucci, *Marketing Research: Methodological Foundations,* 9th ed. (Fort Worth: SouthWestern, 2005); McQuarrie, Edward, *The Market Research Toolbox* (Thousand Oaks, CA: Sage, 2005); and Ereaut, Gill, Mike Imms, and Martin Callingham, *Qualitative Research: Principles and Practice* (Thousand Oaks, CA: Sage, 2002).

9. Stevens, Greg A., and James Burley, "3,000 Raw Ideas = 1 Commercial Success!" *Research-Technology Management,* May–June 1997, pp. 16–27.

10. Iansiti, Marco, "Shooting the Rapids: Managing Product Development in Turbulent Environments," *California Management Review* 38 (Fall 1995), pp. 37–58.

11. For an example, go to Confirmit Case Studies at http://www.confirmit.com/customers/case-studies.aspx.

12. Davis, Owen, "Using Xbox for Concept Testing," *Edge,* August 14, 2007, online at www.next-gen.biz/index.php?option=com_content&task=view&id=6831&Itemid=2.

13. Our thanks to Bryan Orme of Sawtooth Software (Sequim, Washington) for preparing a Technical Appendix that appeared in the second edition of our book. Parts of this section are pulled from that prior version.

14. This example is from Hauser, John, "Note on Conjoint Analysis," MIT Sloan Courseware, online at http://web.mit.edu/hauser/www/Papers/NoteonConjointAnalysis.pdf.

15. McQuarrie, Edward, *Customer Visits: Building a Better Market Focus,* 3rd ed. (Armonk, NY: Sharpe, 2008).

16. Leonard-Barton, Wilson, and Doyle, "Commercializing Technology."

17. Rayport, Jeffrey F., and Dorothy Leonard-Barton, "Spark Innovation through Empathic Design," *Harvard Business Review* 75 (November– December 1997), pp. 102–113.

18. Nussbaum, Bruce, "Hot Products," *Business Week,* June 7, 1993, pp. 54–57.

19. Rayport and Leonard-Barton, "Spark Innovation through Empathic Design."

20. Ibid.

21. Seybold, Patricia, "Get inside the Lives of Your Customers," *Harvard Business Review* 79 (May 2001), pp. 81–89; and Guillart, Francis, and Frederick Sturdivant, "Spend a Day in the Life of Your Customer," *Harvard Business Review* 72 (January–February 1994), pp. 116–125.

22. Takahashi, Dean, "Doing Fieldwork in the High-Tech Jungle," *Wall Street Journal,* October 27, 1998, pp. B1, B22.

23. von Hippel, Eric, "Lead Users: A Source of Novel Product Concepts," *Management Science* 32 (July 1986), pp. 791–805; von Hippel, Eric, Stefan Thomke, and Mary Sonnack, "Creating Breakthroughs at 3M," *Harvard Business Review* 77 (September–October 1999), pp. 47–57; and Urban, Glen L., and Eric von Hippel, "Lead User Analyses for the Development of New Industrial Products," *Management Science* 34 (May 1988), pp. 569–582.

24. von Hippel, Thomke, and Sonnack, "Creating Breakthroughs at 3M."

25. von Hippel, "Lead Users."

26. DeYoung, Garrett, "Listen, Then Design," *Industry Week,* February 17, 1997, pp. 76–80.

27. von Hippel, Eric, "Users as Innovators," *Technology Review* 80 (January 1978), pp. 3–11.

28. "The Rise of the Creative Consumer," *The Economist,* March 12, 2005, pp. 59–60.

29. von Hippel, Thomke, and Sonnack, "Creating Breakthroughs at 3M."

30. Ibid.

31. Ibid.

32. Ibid.

33. Ibid.

34. Burchill, G., and D. Shen, *Concept Engineering* (Cambridge, MA: Center for Quality Management, 1995), Document ML0080.

35. Griffin, Abbie, and John R. Hauser, "The Voice of the Customer," *Marketing Science* 12 (Winter 1993), pp. 1–27; and Hauser, John R., and Don Clausing, "The House of Quality," *Harvard Business Review* 66 (May–June 1988), pp. 63–73.

36. Burchill and Shen, *Concept Engineering.*

37. Another way to determine the value customers place on various features, attributes, or benefits, to better understand possible trade-offs, is conjoint analysis.

38. Griffin, Abbie, and John Hauser, "Patterns of Communication among Marketing, Engineering, and Manufacturing: A Comparison between Two New Product Teams," *Management Science* 38 (March 1992), pp. 360–373.

39. Hauser and Clausing, "The House of Quality."

40. Kawasaki, Guy, and Michele Moreno, *Rules for Revolutionaries* (New York: Harper Business, 1999).

41. Clark, Kim, and Steven Wheelwright, *Managing New Product and Process Development* (New York: Free Press, 1992).

42. Urban, Glen, John Hauser, William Qualls, Bruce Weinberg, Jonathan Bohlmann, and Roberta Chicos, "Information Acceleration: Validation and Lessons from the Field," *Journal of Marketing Research* 34 (February 1997), pp. 143–153.

43. See "Corporate Responsibility," About Symantec, online at www.symantec.com/corporate.

44. "The Rise of the Creative Consumer," pp. 59–60.

45. Leadbeater, Charles, and Paul Miller, *The Pro-Am Revolution* (London: Demos, 2004), p. 64.

46. Libert, Barry, and Jon Spector, *We Are Smarter Than Me: How to Unleash the Power of Crowds in Your Business* (Upper Saddle River, NJ: Wharton School Publishing, 2007), p. 19.

47. "The Rise of the Creative Consumer," p. 59.

48. Vargo, Stephen, and Robert Lusch, "Evolving to a New Dominant Logic for Marketing," *Journal of Marketing* 68 (January 2004), pp. 1–17.

49. "How Businesses Are Using Web 2.0: A McKinsey Global Survey," *McKinsey Quarterly,* March 2007; Manyika, James, Roger Roberts, and Kara Sprague, "Eight Business Technology Trends to Watch," *McKinsey Quarterly,* December 2007; and Bughin, Jacques, "How Companies Can Make the Most of User Generated Content," *McKinsey Quarterly,* August 2007.

50. Richardson, Don, Ricardo Ramirez, and Moinul Huq, "Grameen Telecom's Village Phone Programme: A Multi-Media Case Study," Canadian International Development Agency, March 2000, online at www.telecommons.com/villagephone/credits.html.

51. Hof, Robert, "The Power of Us," *Business Week,* June 20, 2005, pp. 74–82.

52. Schrage, "My Customer, My Co-Innovator."

53. Hof, "The Power of Us."

54. Hof, "The Power of Us."

55. Tedeschi, Bob, "Putting Innovation in the Hands of a Crowd," *New York Times,* March 3, 2008, online at www.nytimes.com/2008/03/03/technology/03ecom. html?_r=1&scp=1&sq=Putting%20Innovation%20in%20the%20Hands%20of%20a%20Crowd&st=cse&oref=slogin.

56. Vargo, Stephen, and Robert Lusch, "Evolving to a New Dominant Logic for Marketing," *Journal of Marketing* 68 (January 2004), pp. 1–17; Bendapudi, Neeli, and Robert Leone, "Psychological Implications of Customer Participation in Co-Production," *Journal of Marketing* 67 (January 2003), pp. 14–28; and O'Hern, Matthew, and Aric Rindfleisch, "The Emerging Logic of Customer Co-Creation," working paper, University of Wisconsin–Madison, 2006.

57. Prahalad, C. K., and Gary Hamel, "The Core Competence of the Corporation," *Harvard Business Review* 68 (May–June 1990), p. 82.

58. Benyus, Janine, *Biomimicry: Innovation Inspired by Nature* (New York: McGraw-Hill, 1997); Backus, Perry, "Stevensville Scientist Launches the Biomimicry Institute," *Missoulian,* February 12, 2007, pp. A1, A6; Underwood, Anne, "Nature's Design Workshop: Engineers Turn to Biology for Inspiration," *Newsweek,* September 26, 2005, online at www.newsweek.com/id/104676; and Vella, Matt, "Using Nature as a Design Guide," *Business Week,* February 11, 2008, online at www.businessweek.com/innovate/content/feb2008/id20080211_074559.htm.

59. "What Is Biomimicry?" Biomimicry Guild, 2006, online at www.biomimicryguild.com/guild_biomimicry.html#.

60. Benyus, Janine and Pauli Gunter, *Nature's 100 Best: World-Changing Innovations Inspired by Nature* (White River Junction, Vermont: Chelsea Green Publishing, 2009).

61. These first three examples came from a presentation by Rajesh Chandy, professor of marketing, University of Minnesota.

62. Byrnes, Nanette and Paul Judge (1999), "Internet Anxiety," *Business Week,* June 28, pp. 79–88, quote p. 84.

63. Rangaswamy, Arvind, "New Technology Forecasting with the Bass Model," Institute for the Study of Business Markets, Webinar, November 14, 2007, online at www.bostonconferencing.net/index. php?option=com_content&task=view&id=116&Itemid=148.

64. Makridakis, S., S. C. Wheelwright, and V. E. McGee, *Forecasting: Methods and Applications* (New York: Wiley, 1997); Kress, G., and John Snyder, *Forecasting and Market Analysis Techniques* (Westport, CT: Greenwood, 1994); Armstrong, J. Scott, *Long Range Forecasting,* 2nd ed. (New York: Wiley, 1985); Armstrong, J. Scott, *Principles of Forecasting: A Handbook for Researchers and Practitioners* (New York: Springer, 2001); and Lilien, Gary, and Arvind Rangaswamy, *Marketing Engineering* (Upper Saddle River, NJ: Prentice Hall, 2003).

65. Krajewski, Lee, and Larry P. Ritzman, *Operations Management, Strategy, and Analysis,* 6th ed. (Upper Saddle River, NJ: Prentice Hall, 2002).

66. Weiss, Allen, *Hitchhiker's Guide to Forecasting,* February 20, 2000, no longer available (originally from www.marketingprofs.com/Tutorials/Forecast).

67. Green, K. C., and J. Scott Armstrong, "Structured Analogies in Forecasting," *International Journal of Forecasting* 23 (July–September 2007), pp. 365–376.

68. This example is based on a presentation by Donna Dubinsky, CEO of Handspring Technologies, to the American Marketing Association Summer Educators' Conference, San Francisco, August 1999.

69. Bass, Frank M., "The Future of Research in Marketing: Marketing Science," *Journal of Marketing Research* 30 (February 1993), pp. 1–6; Bass, Frank M., "Empirical Generalizations and Marketing Science: A Personal View," *Marketing Science* 14 (Summer 1995), pp. G6–G19; Mahajan, Vijay, Eitan Muller, and Frank M. Bass, "New Product Diffusion Models in Marketing: A Review and Directions for Research," *Journal of Marketing* 54 (January 1990), pp. 1–26; Mahajan, Vijay, Eitan Muller, and Frank M. Bass, "Diffusion of New Products: Empirical Generalizations and Managerial Uses," *Marketing Science* 14 (Summer 1995), pp. G79–G88; Mahajan, Vijay, and Yoram Wind, *Innovation Diffusion Models of New Product Acceptance* (Cambridge, MA: Ballinger, 1986); Norton, John A., and Frank Bass, "A Diffusion Theory Model of Adoption and Substitution for Successive Generations of High-Technology Products," *Management Science* 33 (September 1987), pp. 1069–1086; Pae, Jae H., and Donald R. Lehmann, "Multigeneration Innovation Diffusion: The Impact of Intergeneration Time," *Journal of the Academy of Marketing Science* 31 (December 2003), pp. 36–45; Sultan, Fareena, J. U. Farley, and D. Lehmann, "A Meta-Analysis of Applications of Diffusion Models," *Journal of Marketing Research* 27 (February 1990), pp. 70–77; and Van den Bulte, Christophe, "Want to Know How Diffusion Speed Varies across Countries and Products? Try Using a Bass Model," *PDMA Visions* 26 (October 2002), pp. 12–15.

70. Rogers, Everett M., *Diffusion of Innovations,* 5th ed. (New York: Free Press, 2003), p. 208.

71. Michelfelder, Richard, and Maureen Morrin, "Overview of New Product Diffusion Sales Forecasting Models," The Intellectual Property Management Institute, 2000, online at www.ipinstitute.org/onpdsfm.html.

72. Bass, Frank M., "A New Product Growth Model for Consumer Durables," *Management Science* 15 (January 1969), pp. 215–227.

73. Rogers, *Diffusion of Innovations.*

74. Ibid.

75. Rangaswamy, "New Technology Forecasting with the Bass Model"; see also Bass, Frank M., Kent Gordon, Teresa L. Ferguson, and Mary Lou Githens, "DIRECTV: Forecasting Diffusion of a New Technology Prior to

Product Launch," *Interfaces* 31 (May–June 2001), pp. S82–S93.

76. Mahajan and Wind, *Innovation Diffusion Models of New Product Acceptance*.

77. Brody, Herb, "Great Expectations: Why Technology Predictions Go Awry," *Technology Review* 94 (July 1991), p. 38.

78. Ibid.

79. Stevenson, Mirek J., "Advantages of Incumbent Technologies," *Electronic News,* July 20, 1998, p. 8.

80. Ibid.

81. Wildstrom, Stephen, "They're Mad as Hell Out There," *Business Week,* October 19, 1998, p. 32.

7

Understanding High-Tech Customers

RFID Technology: How Far Can It Go?

RFID technology—radio-frequency identification—consists of a small chip, or tag, that can be embedded or placed on items ranging from pets to grocery items to clothing to hospital supplies. Either individual items (say, pill bottles) or cases/pallets can be tagged. The chip or tag sends out information, over its radio transmitter, about what and where the tagged item is. The technology tracks the tagged items wirelessly, transmitting to a central database, which in turn provides a wide range of information ranging from inventory quantity, location of tagged items, freshness dating, handling instructions, and so forth.

Tags are one of two types: passive or active. Passive tags are about the size of a postage stamp, and, as of early 2008, cost approximately 15–30 cents each. Active tags are about the size of a credit card and cost $15–20 each; they have enhanced functionality (e.g., they can record cash balances, temperature and exact location data) and require a power supply to function.

A complete system consists of the tags (also referred to as *chips* or *labels*); software that encodes the tags about what the item is, and readers that read the tag via wireless technology. The RFID tags pass by an RFID reader—say, at a port, warehouse, or loading dock; the radio in the tag sends out an identifying packet of data; the reader then passes the data to the appropriate software/database (say, a company's supply chain software). Because the system doesn't require human scanning like barcodes do, it increases productivity and accuracy. Additionally, unlike barcodes, tags can monitor any unauthorized movement of a tagged item and generate an automatic text message that is sent to a manager.

Many companies have jumped into supplying parts of the RFID solution. Suppliers include NXP Semiconductors (formerly Philips Semiconductor), Alien Technology, and Motorola (formerly Symbol Technologies), to name a few. Avery Dennison makes every part of the tag except the radio chip. As they state: "We're chip agnostic; we will work with the six or so major RFID chip designs on the market."

As is true for so many new technologies, the hype surrounding RFID chips in the earlier years, around 2002—such as the promise to streamline global supply chain management—was overly optimistic. Technology optimists are fond of making statements such as "If Napoleon had had RFID technology, he might have taken Moscow in 1812 instead of running out of supplies"[1] (with obvious reference to military tagging of supplies). For example, the early forecast for RFID predicted an imminent effect on retailers and manufacturers, allowing them to follow products from the factory to store shelves, helping to manage inventory and reduce costs. In addition, in 2003 consulting firms predicted that by 2006, more than a billion RFID tags would be used on cases and pallets in the complex network of suppliers, storage facilities, transporters, distributors, and retailers. In 2006, that forecast was radically revised to only about 350 million tags. Despite the less-than-predicted

(continued)

performance in some areas, item-level tagging is growing faster than expected, from $200 million in 2006 (200 million items) to $13.2 billion (550 billion items) in 2016, according to IDTechEx.

In looking back at the evolution and takeoff of RFID technology, some are asking why the "business case [has] not materialized in line with what was expected."[2]

What Went Wrong?

Not only did demand not meet expectations, but technical glitches and vendor instability also made customers a bit wary. Early RFID devices were hard to read. If they got wet or were affixed to metal or liquid, they did not work well (metals reflect radio waves; liquids absorb them). In addition, the placement of tags affected their readability, and depending on the application, group tag interference could occur. Costs were high. Moreover, to work well in a supply chain context, the technology has to be adopted by most, if not all, the companies in a given network (manufacturers, shippers, retailers, etc.).

Customer Applications

Surprisingly, RFID technology has experienced greater success in unexpected places. Indeed, "20% of RFID applications are created out of thin air."[3] (Astute readers will recognize this quote as a classic illustration of breakthrough technologies in which the technology was not developed in response to a known need.) Some of these unexpected, novel applications include:

- Automated toll collection on roads, allowing cars with RFID tags to go through tollbooths while the fee is automatically deducted
- Car entry and security
- Contact-less payment options such as ExxonMobil's Speedpass and other credit card uses in which customers can wave tagged cards in front of a reader
- Airline tracking of baggage
- In Japan, cigarette vending machines that sell only to people with chip-equipped ID cards
- In Asia, RFID tags that pay for purchases ranging from train rides to snacks (China has spent $6 billion to issue 1.3 billion RFID cards to people)
- For parts on Boeing's 787 Dreamliner; tags that contain identification information and maintenance/inspection data, helping to reduce maintenance and inventory costs
- Passports, parking lot access passes, and tracking of such diverse objects as pets, library books, hospital equipment, and even prisoners

Health Care.   Drilling down into one specific application of RFID technology—hospital applications—offers additional insight into customer behavior issues surrounding this new technology.

In hospitals, nurses are sometimes known to hide or "squirrel away" essential equipment so it is available when it is needed. For example, at the Beth Israel Deaconness Medical Center in Boston, nurses sometimes put patient-controlled anesthesia pumps up in the ceiling tiles. The hospital didn't lack sufficient equipment; rather, the nurses couldn't find the equipment at the right time. They felt that by hiding the equipment, they would be able to find it when they needed it.

The hospital undertook a pilot project in the emergency department, putting RFID tags on equipment and using an asset-tracking application to tell health care workers where to find intravenous pumps, ventilators, and other devices. The $50 tags contained a battery and a transmitter, using the hospital's high-speed wireless network to broadcast the location of the equipment. The results from the trial showed that health care workers spent an average of 20 fewer minutes per day looking for equipment. The trial also showed a reduction in equipment losses ($600,000 annually), some of which was attributable to theft, but more often to poor handling—for example, when a $1,000 portable monitor was rolled up in a sheet and thrown down a laundry chute. The RIFD tag signal can be read and within five minutes, a clerk can find the item and retrieve it.

Another benefit from the trial showed that not stashing equipment at a particular worker's station resulted in maximum use and minimized the need for overstocking expensive devices. For example, the hospital estimated savings of $1–2 million by not buying extra devices: "It's the kind of investment that pays for itself within one year."[4]

Moreover, the RFID tags allowed the hospital to prevent errors. For example, RFID tags were placed on breast milk stored for infants. The tag on the milk would be read and matched to the infant's wristband prior to feeding. RFID tags are also used for accuracy in dosages of medication. AstraZeneca began tagging syringes of an anesthetic called Diprivan, used during surgery, for delivery to Europe and Japan. They've tagged 40 million infuser syringes and have totally eliminated errors.

Another health care application is tracking pharmaceuticals. The World Health Organization estimated in 2006 that 10% of global pharmaceutical sales were counterfeit. In December 2006, the Food and Drug Administration began requiring wholesale distributors of prescription drugs to provide a statement of origin that identified each drug prior to sale, purchase, or trade of the drug, essentially ensuring its "pedigree." RFID tags were used for such popular drugs as Viagra and OxyContin, as well as the HIV drug Trizivir. The use of the tags also allowed companies to flag shipments that went missing, battling theft of the drugs from the manufacturer to the pharmacy by people who would try to divert them to "street use" or to people who adulterated them for other purposes.

Retail.   Of course, the initial vision of RFID was to smooth supply chain management and retail operations. Wal-Mart, one of the largest users, expects 600 of its 20,000 suppliers to use RFID tags on cases and pallets. Their goal is to reduce the number of out-of-stock items. Industry-wide, 8% of products are out of stock on store shelves at any given time—a whopping 1 of 12 items! In some cases (22% of the time in a sample of Wal-Mart stores), the out-of-stock products were in the store, but not on the shelf. RFID helps reduce such stock-outs by about 30%, and for faster moving items (those that sell 7 to 15 products per day), RFID can reduce stock-outs by 62%. The result is about a 1% uplift in sales for both retailers and suppliers. Other retailers adopting RFID include Best Buy and Target.

Other retail uses include tracking in-store promotional displays and the timing and availability of advertised products. For example, P&G used RFID to monitor the movement of promotional displays from Wal-Mart's distribution center to the back room of a specific store, and ultimately to the store floor. This tracking yielded a 19% increase in sales for Gillette products. Moreover, in 2003 a British supermarket chain, Tesco, tested selling Gillette razor blades packaged with RFID tags that triggered a camera when a package was removed from the shelf, and then a second camera at checkout. Security staff then compared the two images, and in at least one case presented photos of a shoplifter to police. (Tesco stopped this practice amid a public outcry over privacy concerns.)

Marks & Spencer (a British retailer that sells its own branded products), the world's largest adopter of item-level tags, began its RFID initiative in 2002. The company tagged returnable food-produce delivery trays to track delays in delivery that would indicate perishability. The tagging helped to boost sales via fresher produce. Then, in 2006, it began tagging men's suits at 52 stores. For a company such as this, the RFID technology is easier to deploy because it runs a "closed loop" supply chain composed of its own suppliers. Although all retailers require coordination with suppliers, when the retailer "owns" those suppliers, it can roll out the technology seamlessly rather than waiting for independent players to also adopt the technology.

Continued Adoption Concerns

As is true for many new technologies, for RFID to gain greater market penetration, the industry needs to establish global standards around a compatible set of technical specifications and architecture. In addition, customers' pricing concerns must be addressed. RFID is in a classic catch-22

(*continued*)

situation: It needs price to come down to a penny per tag before its usage will become more wide-spread, but it needs increased demand to drive costs down. Privacy concerns also loom large. RFID can be used in surveillance or to track an individual's buying behavior. As technology experts say, however, it is not the technology that is inherently evil; rather, it's the way the technology is used.

Moreover, as with any technology, new breakthroughs are on the horizon. Hitachi has developed a chip only 0.05 mm on a side and 0.005 mm thick—nearly invisible. In a vial of liquid, the tiny particles look like gold dust. These chips can be put on items such as securities, concert tickets, gift certificates, and cash to prevent counterfeiting. They have a 128-bit architecture with an almost infinite number of digit combinations: 10^{38}.

Along with a host of possible new uses, like embedding sidewalk tiles or crosswalks with RFID tags to help auto-navigation systems in wheelchairs, the future potential for this technology is wide open.

Sources:

"RFID Success Signals," *ComputerWorld,* August 14, 2006; "Best Buy Sees a Growth Future in RFID," *InformationWeek,* March 27, 2007; "Pharma RFID: FDA Products, IBM Cheers," *eWeek,* August 8, 2006; King, Rachael, "Radio Shipment Tracking: A Revolution Delayed," *Business Week,* October 9, 2006; "Tag—You're IDed," *Fortune,* Spring 2006, p. 14; "Tag, You're It!" *Fortune,* special advertising section, May 29, 2006; "Wal-Mart Rethinks RFID," *InformationWeek,* March 26, 2006; "Gillette's Fusion Launch Makes a Good Business Case for RFID," *eWeek,* August 11, 2006; "Motorola Hatching Big Plans for Symbol: RFID, Wi-Fi, and WiMax," *InfoWorld,* September 25, 2006; "RFID Vendors Raise the Stakes with New Products," *eWeek,* October 12, 2006; Godinez, Victor, "Firms Find New Uses for Radio Identification," *Dallas Morning News,* March 20, 2006; Hornyak, Tim, "RFID Powder," *Scientific American,* February 2008, pp. 68–71; and Jha, Alok, "Tesco Tests Spy Chip Technology: Tags in Packs of Razor Blades Used to Track Buyers," *The Guardian,* July 19, 2003, online at www.guardian.co.uk/business/2003/jul/19/supermarkets .uknews.

In order to develop effective marketing strategies, firms must have a solid understanding of how and why customers make purchase decisions for high-technology products and innovations.[5] For example, take the case of a large company deciding to purchase enterprise resource planning (ERP) software to help manage and integrate a variety of business functions and applications.[6] Some of the key vendors in the ERP software market include Oracle, SAP, i2 Technologies, and Trilogy Software. *Front-office products* are designed both to help companies find and sell to customers and to automate and track data related to sales force management, marketing, and customers. *Back-office solutions* are designed to handle functions that do not interface with customers: functions such as supply chain management, accounting and finance, and so forth.

Effectively implementing enterprise resource planning requires a company to eliminate its stovepipe (or silo) mentality, which keeps each functional area isolated from the others. Where different functional areas operate autonomously, goals may not be well integrated. For example, salespeople might be rewarded on volume, operations personnel on cost of products and conformance to specifications, and so forth. The success of ERP planning hinges on integrating and collaborating across functions, such that each is working toward common goals.

Moreover, these programs are not cheap. To install an ERP system in a *Fortune* 500 company may cost tens of millions of dollars in license fees for global rollouts, with an additional expenditure for consulting, typically in the ratio of one to three times the license fees, plus investments in computers and networks. Issues such as application interfaces, compatibility ("plug and play"), interoperability between disparate systems, scalability across the enterprise, and linking new and legacy applications in an enterprise-wide system must be addressed. The time line can take one to three years, or more. Thus, the adoption of an ERP system by a large enterprise is likely to take time. On the other hand, subscription-based Web services (also known as *Software as a Service,* or *SaaS*) such as Salesforce.com—which have some of the limited functionality of ERP software—have had rapid adoption by medium and small enterprises.

Firms must examine at least six critical issues in assessing the motivations of customers to buy their products, as shown in the organizing framework in Figure 7.1. Marketing must be tailored to address these issues:

1. What steps do customers go through in making technology purchase (adoption) decisions? How do these steps affect marketer's strategies?
2. What is the process by which a technological innovation is adopted and diffused throughout the market? What factors affect customers' purchase decisions? Who is likely to buy? Are there categories of customers who are predisposed to adopt an innovation earlier than others?
3. What happens when the sales trajectory of a new technology stalls, or falls into a "chasm"? What strategies can marketers follow to cross the chasm?
4. How can technology markets be segmented?
5. What affects the timing of customers' technology migration decisions and upgrades? Are they likely to postpone purchases or bypass new generations of technology in anticipation of better options coming in the near future?
6. What are consumers' paradoxical relationships with technology?

To understand the behavior of technology buyers, basic models of consumer behavior (business to consumer, or B-to-C) and organizational buyer behavior (business to business, or B-to-B), as depicted in Figure 7.2, can be a useful first step. Indeed, some people argue that there is no such thing as a "high-tech consumer" and that the buying decision for technology products is the same as for any product.[7] However, others have concluded that when it comes to high-tech products and services, conventional consumer behavior models "don't go far enough."[8]

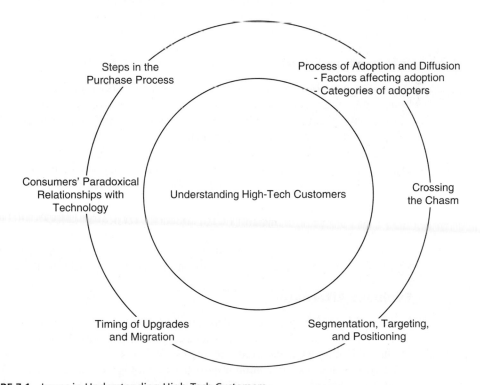

FIGURE 7.1   Issues in Understanding High-Tech Customers

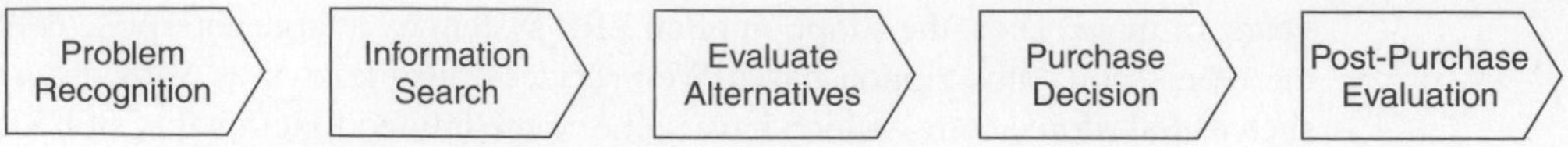

| Problem Recognition | Information Search | Evaluate Alternatives | Purchase Decision | Post-Purchase Evaluation |

FIGURE 7.2   Stages in the Purchase Process

Consistent with the basic foundation of this book, our position is that basic models of buyer behavior offer a useful starting perspective, but they must be augmented and modified to account for the unique factors of the high-tech environment. Hence, we begin this chapter with an overview of the standard model of the customer purchase process. This is followed by a discussion of the traditional model of adoption and diffusion of innovation, which examines different categories of adopters and the factors that affect their adoption decisions.

The chapter continues with adaptations to understanding customer behavior in high-tech markets, based on the notion of a "chasm" between early market adopters and later market customers. Then, a process for segmenting, targeting, and positioning in high-tech markets is covered. The next topic covers the complications in high-tech customer decision making that arise from customers' desire to avoid obsolescence (the timing issue in adoption). The final topic discusses the different paradoxical relationships that consumers have with technology.

CUSTOMER PURCHASE DECISIONS

Problem Recognition

As shown in Figure 7.2, the purchase process begins when the buyer recognizes a need, be it a problem or an opportunity. Need recognition for high-tech customers can be stimulated by factors internal to the customer or by external factors in the environment. An example of an internal factor is when a business customer recognizes that a bottleneck exists in its order fulfillment process. An example of an external factor is an advertisement that creates awareness of a new type of technology offering, or observation of other customers' purchase behaviors.

Information Search

At this stage the buyer actively seeks out information about how to solve the problem. Information search often helps the customer identify alternative solutions for solving the problem. The amount of information customers require varies by product category and customer type. The buyer may utilize personal sources for information such as friends or colleagues, commercial sources such as advertising or a vendor, public sources such as the Internet or reviews in trade publications, or experiential sources such as examining the product during a demonstration. Particularly for distributors or retailers of high-tech products, trade shows (such as the International Consumer Electronics Show) are an important source of information about which new products and cutting-edge technologies to carry in their stores.

Additional insights regarding the information search behaviors of organizational buyers are found in Box 7.1. Rapid and uncertain changes in the environment make information time sensitive, which leads organizational buyers to use information differently than in more traditional contexts.

Evaluate Alternatives

For many customers, whether to adopt a new technology is a high-risk, anxiety-provoking decision. Market and technological uncertainty lead customers to worry about making a bad decision, including the switching costs involved, training needs, and so forth. Therefore, understanding the factors that affect customers' evaluation of possible technology solutions is vital. The research techniques discussed in Chapter 6 can be very helpful in this regard.

BOX 7.1

Organizational Buyer Behavior in High-Tech Markets

By Allen Weiss, Marshall School of Business, University of Southern California, Los Angeles, California

One defining feature of a high-technology market is the fast pace of technological change. Such change creates high levels of uncertainties for buyers. The uncertainties that buyers face in high-tech markets is really about the *information* in these markets.

Unlike in slower-paced markets, the information in high-technology markets is time sensitive. Time-sensitive information quickly loses its value. For example, computers and the microprocessors on which they are based have been rapidly improving in terms of speed, capabilities, and so forth. Consequently, knowledge associated with a given generation of computer quickly diminishes, and both customers and engineers are finding it difficult to maintain up-to-date knowledge. There are two broad implications for buyers in these markets.

First, as it relates to their purchase behavior, customers who perceive a rapid pace of technological improvements in computer workstations do tend to recognize the short shelf life of the received information. As a result, they tend to search for shorter periods of time because the information they receive is time bound. In low-technology markets, uncertainties are typically resolved by longer periods of information search.

The short shelf life of information also affects buyers who have an existing relationship with a current (incumbent) vendor, but who decide to consider other vendors for a subsequent purchase. Paradoxically, this causes buyers both to expand their information collection effort and at the same time to restrict the tendency to switch vendors.

For incumbent vendors, these information search characteristics of buyers pose an apparent dilemma. On the one hand, it would appear important for incumbent vendors to convince buyers that they are remaining technologically active. On the other hand, to the extent that such efforts contribute to increasing buyer perception of technological change, it may create an incentive for buyers to consider the products of competitors. Interestingly, it appears that buyers end up staying with existing vendors. As such, rapid technological change actually buffers incumbent vendors from competition.

Rapid technological change also generates expectations in prospective customers that they may purchase a soon-to-be-obsolete technology. These expectations have been shown to induce prospective customers to "leapfrog" current generations of high-technology products. Presumably, these expectations reduce the perceived benefits of owning a current product generation.

When information is time sensitive and customers anticipate rapid improvements, marketers of high-technology products face several other challenging product decisions. In particular, they must decide when to introduce new generations of a product and whether older generations should be sold concurrently. Although the pressures of producing leading-edge products are high, managers who quickly introduce new generations may both cannibalize their existing products and increase buyers' perceptions that their technology is changing rapidly. This may reduce the benefits of owning a current generation and ultimately encourage customers to leapfrog.

Sources: Glazer, R., and A. M. Weiss, "Marketing in Turbulent Environments: Decision Processes and the Time-Sensitivity of Information," *Journal of Marketing Research* 30 (November 1993), pp. 509–521; Weiss, A. M., and J. Heide, "The Nature of Organizational Search in High-Technology Markets," *Journal of Marketing Research* 30 (May 1993), pp. 220–233; Heide, J., and A. M. Weiss, "Vendor Consideration and Switching Behavior for Buyers in High-Technology Markets," *Journal of Marketing* 59 (July 1995), pp. 30–43; Grenadier, S., and A. M. Weiss, "Investments in Technological Innovations: An Options Pricing Approach," *Journal of Financial Economics* 44 (1997), pp. 397–416; and Weiss, A. M., "The Effects of Expectations on Technology Adoption: Some Empirical Evidence," *Journal of Industrial Economics* 42 (December 1994), pp. 1–19.

One factor playing an increasingly important role is the impact of design on a customer's product evaluation.

DESIGN   An attractive form factor, or "cool" design (equivalent to a "wow" factor in the Kano concept language of Chapter 6), can play a big part in a customer's evaluation of a product or service. Smart design can also help to reduce perceived complexity or increase ease of use, further facilitating adoption of complex technology products and services. Many examples highlight the role of design in the commercial success of technology innovations. Consider the Sony Walkman (1979), PalmPilot (1996), Nokia Communicator 9210 (2000), Apple iPod (2001), Motorola Razr V3 (2004), and Apple iPhone (2007), among many others. Certainly, many high-tech companies understand the important role that spectacular design can play in crafting a winning product.

Solid empirical evidence has established the business case for design. In one study, the UK Design Council selected 166 public companies that had won numerous design awards; it tracked the shareholder returns of these companies from 1995 to 2003. After assigning points to the companies based on their design track record during that period, it divided these 166 companies into two groups: 63 were classified as belonging to a Design Portfolio because they had won more design awards, while 103 were classified as part of the Emerging Portfolio that had won fewer design awards. Both portfolios consistently outperformed the FTSE 100 index over this eight-year period. The Design Portfolio peaked in February 2000 with a 296% increase over the starting value. The Emerging Portfolio peaked in March 2000 with a 244% increase over the starting value.[9]

What exactly is "design" and how can technology companies leverage design factors both in product development and organizational processes? Simply put, **design** is a conscious effort to produce a product, service, or experience that combines both functionality and aesthetics (i.e., it is both useful and aesthetically pleasing). Design is a multidisciplinary field based on applied art (e.g., architecture, industrial design), technology, ethnography, and business. The Design Management Institute in Boston (www.dmi.org) seeks to increase awareness of design as an essential part of business strategy by providing tools and training through its programs and publications.

Many companies have different philosophies and operationalizations for practicing design. Design adds value to an organization by serving as (1) a differentiator of product, brand, or image by integrating aesthetics into product, marketing, and communications; (2) an integrator of internal business processes to deliver a better customer experience and improve organizational performance; and (3) a change agent in the culture of a company to make it more responsive to opportunities and challenges.[10] In addition, designers play a key role in the "greening" of technology products. Incorporating sustainability issues at the outset of the design process can be not only a source of competitive advantage, but socially responsible as well.[11]

IDEO (the design consulting firm in the opening vignette to Chapter 6) emphasizes what is known as *design thinking*.[12] As shown in Figure 7.3, design thinking emphasizes three multidisciplinary considerations in decision making: *human factors* (that require organizational personnel who are skilled in ethnography or usability issues), *technology factors* (that require individuals skilled in industrial design or engineering) and *business factors* (that require individuals skilled in

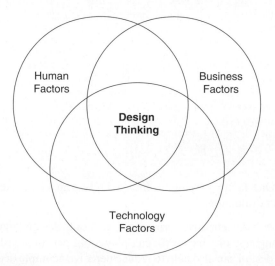

FIGURE 7.3 Design Thinking Factors

understanding cost and revenue implications of various strategic decisions). The focus of design thinking is on how human beings interact with products and their environment, and the challenge is to uncover their unarticulated needs or problems.

Good members of design-oriented teams have specialized knowledge and training and are also quick to learn about new domains. Such individuals are metaphorically called "T-people" because they demonstrate both intellectual breadth and depth. Design thinking is very different from analytical thinking. The analytical problem-solving process goes from problem specification to analysis of data to synthesis of the solution. In contrast, the design process (sometimes referred to as "managed chaos") goes from analysis (based on observational data) to brainstorming to specification of alternative solutions (rapid prototyping) to synthesis of the final solution. While analytical thinking is well suited for linear, incremental change, design thinking is better for breakthrough insights and transformative change. Companies like Samsung have benefited a great deal from IDEO's design thinking and process. Other design consulting companies like Design Continuum, Frog Design, and Ziba Design have variations of this process that they offer their clients. Designers' use of emotion and user-centered methods give them incredible power to influence people by imagining things in new and innovative ways and by communicating ideas in visually appealing ways.

Purchase Decision

During the prior evaluation stage, the buyer forms opinions about the desirability of different alternatives. At the purchase stage, the buyer reaches agreement with the selected seller on the terms of purchase, including scope of the offering, price, terms of payment, and delivery. Retailers and other channel members can play a key role in at this stage of the customer's decision.

Postpurchase Evaluation

At this stage, the buyer assesses how well the product has lived up to its potential. Issues such as the following arise for the customer:

- Was I able to successfully learn how to use the new technology?
- Did the technology deliver the promised benefits?
- Were there hidden costs to using the new product?

These postpurchase issues (and potential buyer's remorse) loom large for tech buyers. For example, in the case of the ERP installations mentioned earlier, many expensive implementations have been scrapped—because after years of attempting to configure the organizational processes to reap the benefits, companies are frustrated. They see many hidden costs, such as training, customization, data integration between legacy and new systems, and so forth. In such situations the vendor must be diligent in follow-up—in ensuring that the company and product deliver on its promises and that this phase of the process be a positive one, particularly if the vendor hopes to rely on customer testimonials or word-of-mouth referrals.

Another important issue in the postpurchase phase includes a company's efforts at customer relationship management for the long term. Managing upgrades and customers' migration to subsequent generations of a technology product are topics deferred to a later section in this chapter. Moreover, these *post-adoption usage patterns* can be extremely useful to marketers in targeting particular customers for future opportunities.[13] A study of personal home computers developed the following taxonomy, based on the amount and variety of usage of a technological innovation:

- *Intense users* have a high usage of the product in a given time period and use it for many different applications.
- *Specialized users* also have a high usage of the product within a given time period, but use it for fewer applications.
- *Nonspecialized users* have a higher variety of usage but amount of usage is low.
- *Limited users* are low on both variety and amount of usage.

User satisfaction with the product and intent to use future innovations decreases from intense users to specialized users to nonspecialized users to limited users. Thus, the first three segments (intense users, specialized users, nonspecialized users) can be combined and targeted as a "repeater" segment ready to repurchase due to their experience and satisfaction.[14]

What variables affect a consumer's variety and amount of usage? Communication with others and the presence of complementary technologies both have positive effects on increasing the variety of use. Thus, marketers should encourage user groups, offline and online, to help users learn from each other. They can use the presence of complementary technologies in the household to further target users most likely to use and enjoy new technological innovations.

Finally, at this postpurchase stage of decision making, customers and companies must also consider "end-of-life" issues for high-tech products and appropriate disposal/recycling. Products such as cars, refrigerators, televisions, and computers contain complex materials, some of them toxic, which have to be safely disposed of. Nontoxic waste has to be disposed of in landfills or incinerators, or be recycled. Landfills take up space and release harmful methane gas to the environment. Incinerators use energy and also release harmful greenhouse gases to the environment. Environmental impact studies have demonstrated that recycling consumes less energy and releases fewer harmful emissions than the creation of new materials. Europe and Japan have laws that make electronics manufacturers responsible for taking back end-of-life products from consumers and recycling content as much as possible. In the U.S., only a few states have passed such legislation. Under pressure from environmental groups, Hewlett-Packard has voluntarily set up recycling factories in California and Tennessee, while Dell takes back its old computers at no charge. These are all positive trends, which should increase in future.[15]

However, one of the biggest barriers to more recycling—both now and in the future—is that products have not been designed with recycling in mind. Industrial processes need to be redesigned for cradle-to-cradle closed-loop product life cycles where there is minimal waste.[16] Innovation opportunities abound in this area for creative, concerned technology firms.

ADOPTION AND DIFFUSION OF INNOVATIONS

One of the most influential models to illustrate customers' adoption of innovations is based on Everett Rogers's framework for the evaluation and adoption of innovations.[17] His model describes the critical characteristics that influence a customer's potential adoption of a new innovation as well as the various categories of adopters that adopt over time as a new innovation diffuses through a market. An important adaptation to the model by Geoffrey Moore introduces the notion of a "chasm" in market diffusion; the chasm is discussed later in this chapter.

Factors Affecting Adoption of Innovation

As shown in Table 7.1 and discussed here, six factors affect customer's likelihood of adoption of new technology. High-tech marketers must be able to articulate their vision of how their product

---

TABLE 7.1 Six Factors Affecting Customer Purchase Decisions

1. Relative advantage	The benefits of adopting the new technology compared to the costs and in relation to other alternatives
2. Compatibility	The extent to which adopting and using the innovation is based on existing ways of doing things and standard cultural norms
3. Complexity	The difficulty involved in using the new product
4. Trialability	The extent to which a new product can be tried on a limited basis
5. Ability to communicate product benefits	The ease and clarity with which benefits of owning and using the new product can be communicated to prospective users
6. Observability	The extent that benefits of the new product are observable to everyone

fares on each of these factors. To the extent that a new innovation fares well on each of these factors, the more likely it will be quickly adopted and diffused throughout the market.

1. *Relative advantage.* Relative advantage refers to the benefits of adopting the new technology compared to the costs. In addition to the dollar price, the ambiguity of high-tech products can lead to emotional worry, a type of psychic cost. The customer will have fear, uncertainty, and doubt about whether (1) the technology will deliver the promised benefits, and (2) the customer will have the skills and capabilities to realize those benefits.

Many high-tech entrepreneurs believe that their invention is the Holy Grail, a better mousetrap, and the next best thing to sliced bread—all rolled into one. However, the factor of relative advantage suggests that it is not sufficient for the inventor to believe that he or she truly has a better product; the improvements must be readily perceived by the customer *and* be worth the monetary and other costs of adoption.

For example, with respect to the adoption of hybrid-engine (gas plus electric) cars, despite the fact that many car owners are concerned about not only high fuel prices but also environmental pollution, the payback on the higher-priced hybrid cars remains elusive to many mainstream-market customers. The incremental $2,000–4,000 in the car's price relative to the fuel savings over time simply doesn't make the relative advantage (compared to a cheaper, highly fuel-efficient compact, gasoline-powered car) compelling enough for many consumers—yet.

2. *Compatibility.* Compatibility refers to the extent to which customers will have to learn new behaviors to adopt and use the innovation. Compatibility with existing ways of doing things, and with cultural norms, can hasten adoption and diffusion. Products that are incompatible with standard ways of doing things require more time in getting up to speed and require more education from the marketer. Especially in high-tech markets, issues of compatibility arise in terms of interfaces to legacy systems (e.g., between new desktop computers and older mainframes in which data are stored) and in terms of compatibility with complementary products (e.g., between charging stations and purely electric cars).

3. *Complexity.* Complexity refers to how difficult the new product is to use. Very complex products have slower adoption and diffusion rates compared to those that are less complex. Obviously, many new high-tech products are complex. Marketers must proactively consider the level of complexity of their products during the development process, as well as how they communicate (set expectations) for complexity for customers in their marketing communications strategies.

With respect to feature complexity, a study using high-tech products (DVD player, PDA/handheld computer, and an online product-rating database) showed that customers experience a tension between their desire to adopt a product with many features and their ability to learn to use the more complex, feature-laden product.[18] Prior to purchase (product use), customers tend to weight their desire for more features more heavily than their concerns about difficulty of use. However, after they actually use the product, customers experience a backlash from overly complex products that they initially desired; after use, they express a preference for more simple products—even after they have invested the time to master the feature-laden version. This finding was true for both novices and more experienced product users!

A different study found that learning to use a complex technology product evokes an emotional response that affects customer's product evaluations *independent of the features and benefits of the product.* The emotional effect arises from whether or not the customer's expectations about ease (or difficulty) of use are confirmed during actual experience with the product. Expectations can be set by company or retail salespeople who possibly "demo" the product prior to purchase, or from a customer's own expertise with the product category. When customers mistakenly believe a product will be easy to use, they experience a backlash from negative emotions, which in turn, negatively affects their product evaluation, satisfaction level, and future use (return) of the product.[19] For example, if consumers expect that Apple products are easy to use, and then they experience that the new Apple TV (which streams video from the computer to a TV screen) is hard to use, they inevitably experience negative emotions that inhibit product evaluation, adoption, and use of such a product.

A third study showed gender-based differences in handling complex, new product features.[20] Even when both men and women have been trained in using new product features (such as a new software feature that assists in debugging formula errors in a spreadsheet program), women used the advanced feature less frequently than men. Modifications to the software feature that "softened" the presentation style of the feature (from a black-or-white, right-or-wrong type of presentation to "possibly," "perhaps," and other gentler interactions) resulted in women using the advanced, complex features as often as men. Because marketers are interested in ensuring that all users of their high-tech gadgets both use and are satisfied with the functionality their products offer, a key implication of this study is to ensure that women's preferences are explicitly considered during the design phase when adding in complex new features. This does not imply, however, that a company ought to offer a "pink" version and a "blue" version—both male and female users have more in common than such a strategy would suggest.

These studies highlight the need to carefully manage the level of complexity in technological innovations. Companies should ask themselves how they can simplify their products and whether the level of complexity is absolutely necessary, in terms of customer requirements. For example, data showed that customers of TiVo had to call customer service an average of six times to get their systems properly installed. Such installation complexities provide a large barrier to a positive evaluation of technology. Even mere perceptions of complexity are enough to stymie customer adoption.

Fully loaded (feature-laden) products can create "feature fatigue" in which customers mistakenly desire the fully loaded version but then are overwhelmed by its complexity during use. Unfortunately, market research during product development may not reveal this problem, as consumers would experience the negative effects only after product use/purchase. In addition, because the marginal cost of adding features—particularly for digital products, such as software—approaches zero, firms are tugged in the direction of adding additional product features. However, adding even costless features can damage a firm's profitability via a negative effect on customer satisfaction, repurchase, and lifetime customer value. The solution?

- Offer a wider assortment of simpler products, rather than all-purpose, feature-rich products.
- Conduct extended product trials.
- Influence expectations on ease-of-use prior to consumer trial. Accurate explanations by salespeople that set reasonable expectations for complexity (don't oversimplify) are key.
- Product demonstrations in person or through video can help build realistic consumer expectations for ease-of-use or complexity.

A final caveat about complexity: A different aspect of complexity—that arising from product proliferation itself—is a different beast. Product proliferation can overwhelm customers with too many choices, and it greatly complicates all aspects of operations (supply chain, manufacturing, packaging, marketing, customer service) for companies. A variety of resources are available to help companies manage this type of complexity.[21]

4. _Trialability._ Trialability is the extent to which a new product can be tried on a limited basis prior to the actual purchase. When customers perceive a new product as risky (whether due to perceived complexity or incompatibility with older technologies), trialability can help reduce perceived risk. New products that can be tried for a limited time without a commitment or can be tried on a modular basis are generally adopted more rapidly than products that require irrevocable purchase or that are not divisible.

5. _Ability to communicate product benefits._ The likelihood of customer purchase is influenced by the ease with which the product benefits can be communicated to prospective customers. Two issues are pertinent to high-tech marketers with respect to this factor. First, for many high-tech products, the benefits are difficult to convey to customers. Second, many high-tech marketers tend to talk in technical terms when communicating about the product. Such communication typically focuses on product features and specifications, rather than the real benefit the customer will receive, which becomes an obstacle to adoption. This lack of relevance is exemplified by the opening paragraph of a company press release:

Oracle today introduced Oracle® Communications IP Service and Network Management, a new market offering designed to simplify the lifecycle management of complex IP-based services. This offering enables communications service providers to manage growing IP service complexity, scale operations efficiently and facilitate ongoing network change by providing one integrated solution for IP service management and network change control. Included in the offering are updated versions of two key Oracle Service Fulfillment Suite applications, Oracle Communications Configuration Management and Oracle Communications IP Service Activator.[22]

6. *Observability.* Observability refers to, first, how observable the benefits are to the consumer using the new product and, second, how easily other customers can observe the benefits the product provides to a customer who has already adopted the product. For example, how does a person who drinks one of the new nutrient-enhanced waters really know that the drink has made him or her healthier? Moreover, can this person's friends see that the nutrient-enhanced water has made their friend healthier? For products that are used in a *public manner* and for which the *benefits are clearly observable,* the likelihood of purchase is greater.

Inventors must assess the foregoing six factors in order to understand just how quickly their new product might take off in the marketplace. Although the factors sound deceptively simple, they can pose crucial barriers. High-tech marketers must educate buyers to overcome the "FUD" factor (fear, uncertainty, and doubt) and highlight benefits. Because breakthrough products don't connect easily with buyers' existing expectations, traditional approaches to marketing—which assume that customers understand the usefulness of the product and have the know-how to evaluate its features—are often insufficient.

In assessing the rate of adoption and diffusion of innovation, inventors must take the perspective of the majority of the possible users of the product and not be biased by their own familiarity and ease with using technology. Insight about the factors can be gained by involving customers in the new product development process and by involving innovative customers who might be early adopters in evaluating new product ideas. If a new idea does not fly well with innovators, it should raise a red flag. Even if the new idea does fly with early adopters, inventors should still be cautious. Excited early adopters are important, but not a guarantee of success. However, without excited innovators, a new product rarely survives.

Categories of Adopters[23]

Understanding which customers might be the first to embrace a new technology is key to astute high-tech marketing. A company's best customers might be the last to embrace an innovation if it poses a disruption in their current routines and procedures or if it does not provide the service and performance they prefer. Moreover, although an early market for a product might exist, the first customers to adopt a new technology (known as "innovators") are usually not representative of "typical" customers. Therefore, understanding the different categories of customers in terms of their likelihood of early adoption is key. In addition, understanding the differing motivations (or reluctance, as the case may be) between the early adopters and more typical customers (the mainstream market) is also critical.

As shown in Figure 7.4, the categories of adopters discussed in Everett Rogers's traditional adoption and diffusion model include innovators, early adopters, early majority, late majority, and laggards. Innovators, early adopters, and the early majority adopt an innovation *prior* to the average time of adoption, whereas the late majority and laggards adopt *after* the average time of adoption. Based on his extensive review of the research in this area, Rogers found that earlier adopters (innovators, early adopters, and early majority) tend to be younger and better educated, have greater upward social mobility, have a greater capacity to cope with uncertainty and change, and have greater exposure to mass media and interpersonal communications than later adopters (late majority and laggards).[24]

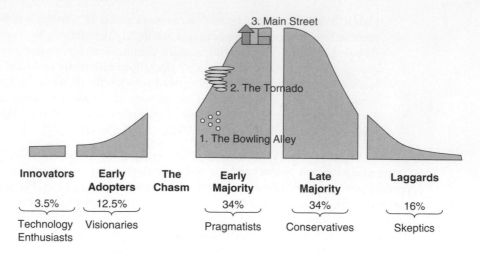

Innovators	Early Adopters	The Chasm	Early Majority	Late Majority	Laggards
3.5%	12.5%		34%	34%	16%
Technology Enthusiasts	Visionaries		Pragmatists	Conservatives	Skeptics

Descriptions of Customer Categories*

Innovators	People who are fundamentally committed to new technology on the grounds that, sooner or later, it is bound to improve our lives. Moreover, they take pleasure in mastering its intricacies, just fiddling with it, and they love to get their hands on the latest and greatest innovations. Thus they are typically the first customers for anything that is truly brand-new.
Early Adopters	The first constituency who can and will bring real money to the table. They help to publicize the new innovations, which helps give them a necessary boost to succeed in the early market.
Early Majority	These people make the bulk of all technology infrastructure purchases. They do not love technology for its own sake, but rather, are looking for productivity enhancements. They believe in evolutionary not revolutionary, products and innovations.
Late Majority	These customers are pessimistic about their ability to gain any value from technology investments and undertake them only under duress, typically because the remaining alternative is to let the rest of the world pass them by. They are price sensitive, highly skeptical, and very demanding.
Laggards	Not so much potential customers as ever-present critics. As such, the goal of high-tech marketing is not to sell to them but, rather, to sell around them.

FIGURE 7.4   The Categories of Adopters

Sources: Diagram from *Crossing the Chasm* by Geoffrey A. Moore. Copyright © 1991 by Geoffrey A. Moore. Reprinted by permission of HarperCollins Publishers.
*Table prepared by Jacob Hachmeister, Missoula, Montana.

Geoffrey Moore has adapted the theory of adoption and diffusion of innovations for the purchase of high-technology products. His key insight about how to modify the traditional adoption and diffusion theory to account for the unique environment of high-tech products is the large gap, or *chasm,* between the early market (composed of innovators and early adopters) and the mainstream market (early majority, late majority, and laggards). The reason for this chasm and the specific marketing strategies to cross the chasm are the focus of his book, *Crossing the Chasm.* This discussion integrates his work with traditional adoption and diffusion theory, beginning with a description of the various categories of adopters.

INNOVATORS   The early market for high-tech products is comprised of *technology enthusiasts—* customers who appreciate technology for its own sake and are motivated by the idea of being change agents in their reference groups. Their interest in new ideas leads them out of narrow circles of peers into broader circles of innovators. They are willing to tolerate initial glitches and problems that may accompany any innovation just coming to market and are willing to develop makeshift solutions to such problems. In the realm of consumer electronics, such individuals are also called *gadget lovers,* who not only adopt early and enjoy playing with their new toys but are also consulted by others for their opinions on technology.[25] In the computer industry, often these technology enthusiasts work closely with the company's technical people to troubleshoot problems. Although not much revenue may come from this group, these technology enthusiasts are key to accessing the next group.

EARLY ADOPTERS   The next category, the early adopters, are *visionaries* in their market. They are looking to adopt and use new technology to achieve a *revolutionary* breakthrough to gain dramatic competitive advantage in their industries. These customers are attracted by high-risk, high-reward projects, and because they envision great gains in competitive advantage from adopting new technology, they are not very price sensitive. Customers in this early market typically demand personalized solutions and quick-response, highly qualified sales and support. Competition is typically between product categories (e.g., between videoconferencing and air travel) at the primary demand level (referred to as *product form competition* in Chapter 1). Communication between these early customer adopters cuts across industry and professional boundaries.

As an example, the early adopters of Apple's first iPhone paid $200 more than did people who bought it 10 weeks after launch. These visionaries were willing to accept the higher price and hassles that accompanied being an early adopter. For early adopters of iPhones, the "hassle factors" came in the form of long waits at retail locations, the fact that they had to get cell phone service exclusively through AT&T, and somewhat limited functionality in the early version of the phone. The visionaries were willing to accept such inconveniences for the psychological and substantive benefits they received.

EARLY MAJORITY   The next group, moving into the mainstream market, are the *pragmatists,* or the early majority. Rather than looking for revolutionary changes, this group is motivated by *evolutionary* changes to gain productivity enhancements in their firms. They are averse to disruptions in their operations and, as such, want proven applications, reliable service, and results.

Pragmatists generally want to reduce risk in the adoption of the new technology and therefore follow three principles:[26]

1. "When it is time to move, let us all move together." This principle is why adoption increases so rapidly at this point in the diffusion process, causing the tornado of demand.
2. "When we pick the vendor to lead us to the new paradigm, let us all pick the same one." This obviously determines which firm will become the market leader.
3. "Once the transition starts, the sooner we get it over with, the better." This is why this stage occurs very rapidly.

From a marketing perspective, these customers are not likely to buy a new high-tech solution without a reference from a trusted colleague, typically in the same industry or line of business. A trusted colleague to a pragmatist is—who else?—another pragmatist, *not* a visionary or enthusiast who has a different view of technology. Obviously, this need for a reference from a pragmatist from within the same industry poses a real catch-22 to selling to this group: how to get just one pragmatist to buy, when the first won't buy without another pragmatist's reference. Yet, pragmatists are the bulwark of the mainstream market.

LATE MAJORITY   The late majority *conservatives* are risk averse and technology shy; they are very price sensitive and need completely preassembled, "bulletproof" (fail safe) solutions. They are motivated to buy technology just to stay even with the competition and often rely on a single, trusted adviser to help them make sense of technology.

LAGGARDS   Finally, laggards are technology *skeptics* who want only to maintain the status quo. They tend not to believe that technology can enhance productivity and are likely to block new technology purchases. The only way they might buy is if they believe that all their other alternatives are worse and that the cost justification is absolutely solid. Laggards hang on to older technologies out of loyalty, satisfaction, fear of time-consuming upgrades and even inertia.[27] At the extreme, some laggards may characterize themselves (or be characterized by others) as *Luddites,* people opposed to technology and changes brought about by technology.[28]

These categories of adopters fall into a normal, bell-shaped curve. Although under most circumstances, a firm would likely target the innovators in a new-product launch, in some cases it might be more worthwhile to target the majority directly instead of the innovators. Firms will find it worthwhile to target the majority in these circumstances:[29]

- When word-of-mouth effects are low
- In consumer products industries (versus business-to-business situations)
- When there is a low ratio of innovators to majority users
- When profit margins decline slowly with time
- The longer the time period for market acceptance of a new products

Each category of adopters has unique characteristics. Moore characterizes the degree of these differences as gaps between each group in the marketplace.[30] Each gap represents the potential difficulty that any one group will have in accepting a new product if it is presented in the same way as it was to the group to its immediate left. Each of the gaps represents a possibility for marketing to lose momentum, to miss the transition to the next segment, and never to gain market leadership, which comes from selling to a mainstream market. The differences between the early market and the mainstream market are more pronounced than differences between the other categories, and hence warrant special attention.

CROSSING THE CHASM[31]

The *chasm* is the gulf between the visionaries (early adopters) and the pragmatists (early majority, mainstream market) and derives from critical differences between the two. Visionaries see pragmatists as pedestrian, whereas pragmatists think visionaries are dangerous. Visionaries will think and spend big, whereas pragmatists are prudent and want to stay within the confines of reasonable expectations and budgets. Visionaries want to be first in bringing new ideas to the market, but pragmatists want to go slow and steady. The chasm arises because the early market is saturated but the mainstream market is not yet ready to adopt. Hence, there is no one to sell to.

What contributes to this chasm, and how can it be overcome? The nature of a firm's marketing strategy in selling to visionaries is very different from the marketing that is required to be successful with pragmatists. Many firms do not understand this difference and are unable to make the necessary shift in strategies to be successful.

Early-Market Strategies: Marketing to the Visionaries

As mentioned previously, visionaries require customized products and technical support. Because such customization for several visionaries can pull a firm in multiple market directions, they can be a costly group of customers to support. However, for a high-tech start-up, sales to these visionaries represent the initial cash flows to the firm. Hence, given the demand from visionaries and need for cash flows, high-tech companies experience pressure both to support these customers' customization needs and to release products early to them. Just as customization can pull the firm in multiple market directions at a steep cost, early release of a product can backfire if it has not been adequately tested.

The goal of the firm at this point is to establish its reputation. In high-tech start-ups, this time of selling to the early market is exciting and energizing. The product is often the focus; engineering and R&D folks play a critical role, and brilliance and vision are embraced. Firms are eager to develop the *best possible* technology for these early market customers.

The Chasm

The bloom falls off the rose, however, when the firm takes on more visionaries than it can handle, given the high degree of customization and support they expect. No pragmatists are yet willing to buy, in part because visionaries are not a credible reference for them. Hence, revenue growth tapers off or even declines. The goal of the high-tech marketer should be to minimize the time in the chasm. The longer the time a firm spends in the chasm, the less likely it is to get out. In 2008 examples of products that may be stuck in the chasm are the Segway Personal Transporter, Apple TV, and Second Life virtual world.

One implication of the chasm relates to relationships with venture capitalists and investors. Lack of knowledge of the existence of the chasm can create a crisis. Key personnel become disillusioned and management becomes discredited. Investors may pull out at the very time that more financing is necessary to get the product to the mainstream market. The ultimate demise of early-market success stories might be explained by a lack of financial support right at the time a company is posed to cross the chasm.

Anecdotal evidence suggests that the chasm has a large—negative—impact on the sales trajectory of many high-tech products, and one empirical study documents its presence. An analysis of historical sales data of 32 consumer electronics innovations (ranging from audiocassettes to projection TVs) found that sales in one-third of these product categories exhibited an initial peak after takeoff, followed by a trough, followed by another rise in sales that exceeded the initial peak. The explanation for what this study called the "saddle" is the difference between the early market and the late market in terms of word-of-mouth communications. The people within each adopter category communicate well with each other but not with people across the categories, leading to the dip in sales.[32] This empirical evidence supports Moore's chasm theory: The lack of communication between categories of adopters means that the proverbial "baton is dropped" between the visionaries and the pragmatist adopters.

How does a firm overcome this problem? As mentioned previously, the solution requires recognition that *the marketing strategies that led the firm to its initial success in the early market will not be effective with pragmatist adopters.* To successfully speak to the mainstream market, high-tech companies must explicitly recognize the underlying motivations that drive pragmatists to adopt technology in the first place, and to craft a marketing strategy that specifically addresses their buying motivations.

Crossing the Chasm: A Beachhead and a Whole Product Solution

Crafting a marketing strategy that specifically addresses the buying motivations of pragmatist customers will allow a firm to successfully cross the chasm. Moore recommends a two-pronged approach, as follows:

1. Identify a beachhead, a single target market from which to pursue the mainstream market.
2. Partner to develop a whole product solution, an integrated, end-to-end solution that allows the customer a seamless experience in buying the company's product.

In addition to these two major points, other unique aspects about pragmatists and how they require adaptations to a company's marketing strategy are also addressed in the discussion that follows.

TABLE 7.2 Compelling Reasons Customers Adopt Technology

1. Purchase of the new technology provides the customer a *dramatic competitive advantage in a previously unavailable domain in a critical market*.
  • This reason to buy is difficult to quantify in terms of costs/benefits.
  • Although appealing to the visionary, this reason to buy is unpalatable to pragmatists and conservatives.

2. Purchase of the new technology *radically improves productivity on an already well-understood critical success factor, and no other alternative to achieving a comparable result is available*.
  • This reason has the greatest appeal to a pragmatist, because the cost savings (in terms of a better return on resource expenditures) can be quantified, typically in terms of incremental dollars.

3. Purchase of the new technology *visibly, verifiably, and significantly reduces current total overall operating costs*.
  • This reason will have the greatest appeal to a conservative because of the hard dollar savings. However:
    ◇ The risk factors surrounding the new technology are still too high for conservatives to take a chance.
    ◇ The surrounding infrastructure for the whole product may not be sufficiently developed to convince these conservatives to take the purchase risk.

IDENTIFY A BEACHHEAD   First, to successfully cross the chasm, a company must identify a **beachhead**—a single target market from which to pursue the mainstream market.[33] A good beachhead exhibits two critical characteristics: (1) its customers have a single, compelling, "must have" reason to buy that maps fairly closely onto the capabilities of the firm, and (2) the selected beachhead provides "adjacencies" to enter related, contiguous segments.

A variety of compelling reasons to buy and their relative attractiveness to different categories of adopters are shown in Table 7.2. According to Moore, because it speaks directly to the pragmatist's concerns to generate incremental revenues from technology investments, *only the second reason* represents a good choice for crossing the chasm: Purchase of the new technology will drastically improve the company's productivity on a well-understood, critical success factor. The issue is whether the firm's capabilities offer this compelling reason for the customer.

Second, a good beachhead should provide clear opportunities to enter adjacent segments. To use a bowling pin analogy, as shown in Figure 7.5, the beachhead is the lead pin; adjacent market

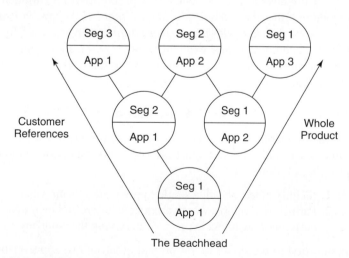

FIGURE 7.5   Bowling Alley Market Development

Source: Moore, Geoffrey, *Inside the Tornado* (New York: Harper Business, 1995). Copyright © 1995 by Geoffrey A. Moore Consulting, Inc. Reprinted by permission of HarperCollins Publishers, Inc.

opportunities are the pins immediately behind the beachhead. The objective of the bowling pin model is to proactively identify new market opportunities by leveraging knowledge about technologies or segments. As Figure 7.5 illustrates, adjacent "pins" can be identified based on either (1) new market segments where the firm can sell its existing technological solution or (2) current market segments where the firm can sell new technological solutions.

For example, the business networking website LinkedIn targeted recruiters as its beachhead to achieve critical mass and cross into the mainstream. The presence of recruiters naturally attracted business executives who were seeking a job or career change, an adjacent segment. Later these executives spread positive word of mouth and invited their business associates to join the network even if they were not interested in changing their job, yet another adjacent segment.

Alternatively, some companies may penetrate more deeply into a current market segment by developing products or services that are related to the original application. The package of products and services comprises a total solution that addresses more of a customer's need. The classic example is Microsoft's development of its Office Suite of application programs that complements its Windows operating system. One risk of a market penetration strategy is that, by keeping a company focused within one segment, competitors may outflank it.

Regardless of whether a company identifies subsequent opportunities in terms of new markets or new solutions, adjacent market segments should exhibit two key attributes:

1. Word-of-mouth relationships
2. Similarities in whole product needs

Moreover, the segment should be capturable in a "short" period of time. *Short* is a relative term and, in most marketing contexts, references the length of the buying cycle for a specific customer for a particular technology solution. For example, "short" for a federal government agency considering purchase of a military GIS software solution would be different from "short" for an individual real estate agent considering whether to use a mobile marketing solution to complement her existing marketing strategies in the next quarter.

With respect to the first attribute for adjacencies, segments adjacent to the selected beachhead can be "knocked over" more easily when word of mouth relationships exist between customers in the two segments. Communication between pragmatist customers in the mainstream market tends to be vertical, or within industry and professional boundaries (rather than horizontal, across industry boundaries, as in the early market). Recall that the higher level of communication between the early market adopters and the mainstream market adopters is a key factor in mitigating the sales slump of the chasm.[34] Yet, when such communication doesn't occur naturally, the company's marketing strategy must explicitly attempt to build word-of-mouth effects by careful selection of its beachhead.

For example, in targeting a mobile text message marketing solution to real estate agents, Goomzee Mobile Marketing explicitly considered the fact that real estate agents are embedded in a larger industry network that involves construction companies, bankers and mortgage providers, title and homeowner's insurance providers, home inspectors, and others. Given these existing word-of-mouth relationships and Goomzee's knowledge of how the real estate industry worked, its subsequent marketing efforts could offer similar text-alert-based systems to other players in the industry.

Many high-tech firms make the mistake of attempting to pursue too many market segments at the outset. They are enthusiastic about the potential their innovations can offer to many different types of segments and are unwilling to limit their potential market opportunities. They also want to hedge their bets against selecting the wrong segment, and so pursue several. However, most firms simply do not have the resources to be effective in multiple segments. They cannot learn the industry- (customer-) specific knowledge and language they need to be credible to customers in a particular segment, nor can their fledgling marketing initiatives rise to a sufficiently influential level. They end up spreading their resources too thinly to be effective in any of the segments and, as a result, fail. A firm must be ruthless in paring its opportunities. Indeed, success in just one segment can be the catalyst required to succeed in other segments.

DEVELOP THE WHOLE PRODUCT   In addition to being based, in part, on word-of-mouth relationships, adjacencies are also determined by similarities in whole product needs. In contrast to marketing to the visionaries who are willing to tolerate some incompleteness in the product and will fill in the missing pieces, marketing to the mainstream, pragmatist market requires that the vendor assume total responsibility for system integration. This need demands the development of a *complete, end-to-end solution* for the customer's needs, or the **whole product**. Identifying the whole product requires an exhaustive analysis of what it takes to fulfill the reasons the customer is buying. Asking what else the customer will need, from a systems perspective, makes apparent possible switching costs and exposure. Indeed, the definition of the whole product should be done within the confines of a single target market because needs will be relatively common within that target market but may differ significantly across segments. To develop a whole product solution requires that a firm either develop or partner with firms to provide the whole solution to the initial customers in the mainstream market.

For example, in the computer industry, the whole product includes hardware, software, peripherals, interfaces and connectivity, installation and training, and service and support. In conducting electronic business over the Internet, the whole product includes website design, hosting of the site on a server, connection to the Internet, purchase of a domain name, security, financial transaction ability, search engine optimization and building site traffic, order fulfillment and logistics, and ongoing customer relationship management.

Hence, so that pragmatist customers are not overwhelmed by the complexity of cobbling together a complete solution, and to ensure compatibility across various aspects of the solution, establishing partnerships with other industry players to provide a whole product solution is critical.

Firms that succeed in the mainstream market have complemented their initially strong competencies in technological development with equally strong competencies in partnering and collaborative skills. Partners often drive further expansion, and so the firm's ability to interact with partners becomes a critical success factor. For example, Cisco has partnered with AT&T to offer its telepresence services more broadly to enterprise customers.[35] Partnering requires critical skills that were discussed earlier in Chapter 5.

SAP, maker of enterprise resource planning (ERP) software, did an excellent job of identifying what a whole product solution would look like and creating a series of partnerships to develop the whole product. SAP developed a "solution map" for each of 17 vertical markets, including automotive, media, oil and gas, and utilities. The maps identified every function in each industry, specifying where SAP's software already offered a solution, where products from partner software developers were required, and where SAP would fill in gaps later. In essence, its solution map functioned as a technology map for SAP, corresponding closely to the steps and needs of high-tech product development (see Chapter 8). Moreover, in terms of partnerships, SAP readily recognized that it needed the smaller developers, but the smaller developers were somewhat hesitant to partner; they wondered how long their investment in developing their own company's products based around a SAP product would be viable, or whether SAP would ultimately enter that market too. So, to entice small developers to partner, to some extent SAP had to guarantee that it would not enter the partner's space in a two- to three-year time frame. (Despite this, SAP and its partners sometimes experienced conflict. For example, in working with i2, SAP announced it would enter i2's lucrative business of supply chain software, and the partnership died.) The challenge for the smaller developers was to partner with key market players (such as SAP or Oracle,) as a seal of approval in order to even be considered a viable option by large *Fortune* 500 customers (the pragmatists in the mainstream market). The smaller developers had to show *Fortune* 500 companies how well their products worked with the companies' existing ERP installations.

Partnerships generally revolve around the issues of power.[36] In the early market (enthusiasts and visionaries), power belongs to the technology providers and the systems integrators (firms that bring different suppliers' products together to create one integrated solution for a customer's needs). The technology providers and systems integrators make the decisions about whom to bring into the process as partners. In crossing the chasm and approaching the early mainstream pragmatists, the power is centralized in the hands of the company that has effectively picked the target (beachhead) customers, understands why they buy, and designed the whole product. In the pragmatist market, the market leader and its partners have the power. In later markets (mainstream conservatives), power is vested in the distribution channels or companies that provide superior distribution of product.

OTHER IMPORTANT STRATEGIES TO EFFECTIVELY MARKET TO THE PRAGMATISTS AND CROSS THE CHASM Recall that pragmatist customers experience fear, uncertainty, and doubt about adopting high-tech products. This manifests itself not only in uncertainty over the value of complex feature sets, but also uncertainty over which particular vendor platform may ultimately become the de facto industry standard. Because the company's reputation was established with its early market successes, the goal at this point is to focus on growing revenue. This requires a very different skill set than marketing to the early market.

First, customers typically demand some sort of industry standard to minimize their perceived risk. For example, now that the high-definition DVD standard war is over, the adoption rate of Blu-ray DVD players is likely to take off because pragmatist customers can safely abandon their wait-and-see attitude.

Second, pragmatists want to see a competitor's proposal and product offering before making a decision. In contrast to the product form competition between the new technology and older technology platforms that existed in the early market, competition now is between vendors within the new technology category. Competition, to the pragmatist, is actually a sign of legitimacy for the new technology. So, rather than saying, "no company makes a product like ours" (or, "we have no competition"), high-tech companies can legitimize their new technology by differentiating on benefits, service, and more traditional positioning tactics.

Third, to successfully cross the chasm and speak to the needs of the mainstream market, a company must simplify, rather than add additional features to its product. Indeed, gadget makers tend to make new models bigger and more complicated—but not necessarily better—right at the time the mainstream market would buy if they were more user-friendly. For example, Cisco's telepresence (high-end videoconferencing) service was originally available only to business customers internally between specially equipped rooms prepared at a cost of $200,000 per room. Now, to reach a broader market, the service is available on an intercompany basis between any rooms that are equipped by Cisco.[37] The negative emotional effects and feature fatigue that customers experience with complex, featured-loaded products was previously covered.

Rather than developing the "best possible solution," the company now must develop the *best solution possible*. Now, the R&D team, rather than basing development on engineering solutions, must work closely with partners and allies on a project-oriented approach. For many, this is much less exciting than the pursuit of technological brilliance and requires painstaking work on compatibility, standards, and so forth.

Fourth, this period may require that engineers go to customer sites to observe them in action. Customer service is a critical component of crossing the chasm.

Inside the Tornado

Until high-tech firms have established themselves in the mainstream market, they have not established staying power. How to succeed when the mainstream market takes off is addressed in Geoffrey

Moore's second book, *Inside the Tornado,*[38] in which there are three distinct phases: the bowling alley, the tornado, and Main Street.

1. *The bowling alley:* The bowling alley is the period during which the new product gains acceptance in niche markets within the mainstream market, but has yet to achieve general, widespread adoption. During the bowling alley stage of market development, the market is typically not large enough to support multiple industry players. The successful firm will establish itself as the dominant market leader. One of the best ways to become the market leader is to follow a whole product strategy and partner extensively to create the de facto standard in the market.

2. *The tornado:* The tornado is a period when the general marketplace switches over to the new technology. From 2001 to 2006, Apple's iPod rode out the tornado to a commanding 70% market share of the U.S. digital music player market.

The tornado typically is driven by the development of a "killer app," or an application of the technology that is based on a universal infrastructure, is appealing to a mass market, and is "commoditizable"—meaning that it can be manufactured, distributed, and marketed at increasingly attractive price points.

The massive number of new customers entering the market in a rapid time period can swamp a company's existing system of supply, as happened with the Nintendo Wii during its first year in the market. During this stage, companies have a huge opportunity to develop their distribution channels. In fact, when companies hit this stage of the cycle, they need to focus on operational excellence: getting their products out to the consumers through streamlining the creation, distribution, installation, and adoption of their whole product. This is typically best done by beefing up internal systems to handle the high-volume workload.

An important caveat during the tornado: Do not bet on preventing a tornado. If a market leader begins to develop, even if it is not your company, it is important to switch efforts to follow the emerging leader.

As noted previously, competing effectively in the tornado requires effective partnering skills. Moore looks at a number of the issues that must be addressed, one of the most important of which is: How do you dance with the market leader, "a gorilla," in a tornado and come away in one piece?[39] The answer is to sustain only enough ongoing innovation to stay out of the market leader's reach. In other words, don't try to pass the market leader by creating the next big thing. Do enough to remain stable, but not so much that the gorilla feels threatened—or it will crush you!

3. *Main Street:* *Main Street* refers to the period when the tremendous growth in the early majority/pragmatist market stabilizes. This period of after-market development is when the base infrastructure for the product's underlying technology has been deployed and the goal now is to flesh out its potential. Rather than focusing on generating sales from new customers, companies must sell extensions of their products to their current customer base to be competitive. Overall, it is important to emphasize operational excellence and customer intimacy, rather than product leadership.

Finally, for continued success in the mainstream market, the high-tech firm will also need to reach out to the conservative market. This requires making the product even simpler, cheaper, and more reliable and convenient, and possibly splitting the product line into simpler components. From an engineering perspective, this is the anathema of good engineering work. Rather than adding more interesting features and cool "wow" factors, engineers should actually do the opposite. Because this is so foreign to product development in high-tech firms, many high-tech companies leave the conservatives' money on the table.

In summary, crossing the chasm requires a different type of marketing from the strategies used in the early market. Selecting a beachhead and developing the whole product are critical success factors in crossing the chasm and reaching out to the mainstream market. Bringing the whole product together is expensive and time-consuming. Moreover, targeting only one market segment feels quite risky. However, a high-tech firm *must* make a decision about what one key market segment to put its resources behind and focus its efforts there. Going after too many market segments at once results in spreading its resources too thin and not building a strong reputation within that segment.

RightNow Technologies offers a final example to highlight the type of thinking and insight that can be gained from the strategies used to cross the chasm. RightNow's software automates customer service inquiries via a Web-based interface. Using an algorithm that is similar in flavor to artificial intelligence, responses to a specific customer inquiry are based on a continually updated knowledge base dependent on the accuracy (or lack thereof) of prior responses.

The market that RightNow initially targeted was *Fortune* 500 companies. The company's logic was that the greater volume of inquiries that large companies received—and therefore could automate for a significant cost savings—would be a "compelling benefit" (increase productivity on a well-understood success factor). Initial results showed, however, that the renewal rate on the software varied widely, depending on the number of inquiries a company received. So, even though all companies were large enterprise accounts, some of those companies received many more inquiries than others, and hence, the accuracy of the software was better for these companies than for others.

A reexamination of RightNow's initial beachhead selection allows the benefit of 20-20 hindsight. A key question that might have been asked was this: Rather than targeting large companies in general, to what extent does a particular company's success depend on a large volume of calls to a customer call center—that, if automated, would have a disproportionately larger impact on productivity improvements than other companies? The answer to this question might include companies that have a large volume of calls that require technical support or direct-response marketing companies (whose orders go directly to the company). Although all large companies have a large volume of calls, some types of companies rely heavily on call volume as a critical foundation of their business. This approach to RightNow's market segmentation might have identified a vertical market segment in the direct-response industry (catalog, mail order, and Web-based companies).

Proactive, explicit consideration of adjacencies (subsequent market segments) would also be considered at the time this initial market segment is identified. Possible adjacent segments would be other vertical market segments that also have a heavy reliance on call centers, such as the banking or financial services industries. The question is, do companies in these vertical markets have word-of-mouth relationships with the initial market segment? Certainly, trade associations may exist for call center operations where personnel from different industries share best practices for such customer service operations. However, absent this information, it's hard to make a case for word-of-mouth adjacencies between these industries.

Another possibility would be to pursue the same market segment (industry) but with another technology solution that leveraged the same underlying approach to inquiry automation. *Fortune* 500 companies (regardless of industry vertical) also experience a large volume of calls in other areas of their company: human resources (e.g., benefits management and planning, retirement, recruitment), media/analyst relations, and so forth. So, targeting the same initial market segment with a new technology application would leverage word-of-mouth relationships among company personnel.

THE CHOICE OF CUSTOMER TARGET MARKET: SEGMENTATION, TARGETING, AND POSITIONING

One of the most important issues that high-tech firms wrestle with is the choice of an initial target market to pursue with their promising new technologies. Yet, the choices that seem obvious in hindsight are rarely clear at the time of the decision. For example, Intel cofounder Gordon Moore rejected a proposal in the 1970s for a home computer built around an early microprocessor. He didn't see anything useful in it, so he never gave it another thought. In a list of possible uses for its 386 chip (written before the IBM PC), Intel omitted the personal computer, thinking instead of industrial automation, transaction processing, and telecommunications.[40]

The rationale behind segmenting markets and selecting a target is to identify groups of customers who share similar needs and buyer behavior characteristics and who are responsive to the

---

TABLE 7.3 Steps in the Segmentation Process

1. Divide the market into groups, based on variables that meaningfully distinguish between customers' needs, choices, and buying habits.
2. Describe the customer's profile within each segment.
3. Evaluate the attractiveness of the various segments and select a target market.
4. Position the product within the segment selected.

---

firm's offering. Directing marketing efforts toward a specific target is both more effective and more efficient than loosely attempting to reach as many customers as possible in the hope that some of them might be interested and respond.

As shown in Table 7.3, market segmentation includes four steps.

Step 1: Divide Possible Customers into Groups

This step is based on important characteristics that distinguish between customer groups in terms of the choices they make and the reasons they buy.

For consumer marketing, the traditional bases of segmentation include:

- Demographic variables, such as age, income, gender, occupation, and so forth.
- Geographic variables, such as geographic location, rural versus urban, etc.
- Psychographic variables, or consumers' values and beliefs that affect their lifestyles and, hence, purchasing behavior, such as an orientation toward a healthy lifestyle or being technologically current or environmentally friendly.
- Behavioral variables related to the customer's behavior with respect to the specific product category, such as:
  - Frequency/volume of usage of a product (i.e., heavy versus light users, and the 80/20 rule)
  - The benefits desired in a product (i.e., ease of use)
  - The usage occasion (such as use for work versus home)

High-tech companies have only recently begun to tailor their marketing strategies to account for the role of women in technology purchases.[41] Consider these facts:

- Women now head 33 million households, up from 21 million in 1980.
- Women's median income has soared 63% in the past three decades, compared to just 0.6% for men.
- Women are starting new businesses at twice the rate of men.

According to *Business Week,* women are far less likely than men to feign understanding of a new technology, and they expect a high level of customer service and user-friendly designs.[42] However, until about 2003, companies such as Dell, Samsung, and Best Buy acted as if men were the primary market segment. As *Business Week* stated, "Blame the male geek culture at digital hardware companies for ignoring women in the past. ... Perhaps consumer electronics marketers wouldn't have taken so long to appreciate women if only they had a few more in their personnel ranks." For example, women comprised only one-third of Dell's management ranks in 2005. The goal for most high-tech companies is to ensure that their employee base reflects the demographic diversity of their customer base.

Many companies have found clear differences in how the two genders shop for and use technology products, and are altering their marketing strategies accordingly. Best Buy launched its "Jill Initiative" to focus on time-pressured suburban moms who are interested in style rather than price. As part of this project, Best Buy changed lighting at the stores, made them less cluttered, trained salespeople to avoid "geek speak" (tech jargon), and generally tried to create an improved shopping experience. Dell altered its marketing communications strategy to include more women's-oriented media as well as improved customer service. Samsung no longer tests its products exclusively with all-male focus groups and attributes its recent spate of design awards to the feedback that women have offered.

For B-to-B marketing, traditional segmentation variables such as SIC (Standard Industrial Classification) or NAICS (North American Industry Classification System) codes, firm size, and corporate culture can be useful. One common segmentation strategy in B-to-B high-tech markets is based on vertical or horizontal market segments. **Vertical market segments** are industry-specific while **horizontal market segments** cut across industry boundaries. Horizontal segments often use a specific technology application to automate a specific business process or function. For example, CRM software targets a variety of different industries, and in that sense, uses a horizontal market segmentation approach. Although GIS software is also a horizontal technology, many GIS providers have typically focused on a vertical market segmentation approach, identifying particular industries, or "verticals," that would receive a disproportionate benefit from adoption of software that links geographic data with business decisions. For example, forestry and vegetation management was an initial vertical market segment targeted by GIS companies.

When should a company adopt a horizontal or vertical market segmentation approach? Horizontal segmentation approaches can make sense when customers across industries share a common approach to the use or benefits they find in adopting technology. On the other hand, vertical segmentation is appropriate when customer use or value of technology products varies by industry, and they require different value propositions. Moreover, a "solutions" (whole product) orientation often varies by industry. Some criticize horizontal market approaches as an excuse for not doing the hard work of identifying a clear segmentation strategy. The risks of a diffused (horizontal) market approach (dilution of efforts across multiple market segments, which doesn't allow a clear edge in any market, leading to global mediocrity) must be weighed against the larger market volume that such a strategy seemingly offers. While vertical segments may appear to have smaller potential, the fact is a high-tech company can garner a larger share of the market within the segment, commanding a leadership position that can be exploited elsewhere.

Step 2: Profile the Customers in Each Segment

This step describes a "typical" customer within each segment. Research that focuses specifically on high-tech buyers can provide insights that more traditional segmentation variables may not. Some firms that specialize in technology-related buyer issues are Odyssey Research, Yankelovich partners' Cyber Citizen, and SRI Consulting.

For example, a 2007 study by the Pew Internet & American Life Project provides an example of steps 1 and 2. Based on a survey of 4,001 U.S. adults aged 18 and over, this study divided the U.S. population into three tiers encompassing 10 key consumer segments, which have been profiled in terms of attitude toward and use of different information and communication technologies.[43] The 10 segments are presented in Table 7.4.

The 10 segments have demographic profiles associated with them as well. For example, the Omnivores, who are the most active consumers of content and services on the Web, are young (median age 28), ethnically diverse, and mostly male (70%). The attitudinal and behavioral profiles together with the demographic information make it easy for technology marketers to target certain segments. For example, if RIM wanted to target the U.S. consumer market with the BlackBerry Pearl, it would probably go after the Omnivores, the Connectors, and the Mobile-Centrics. In other cases, and with a broad market lens, other companies would find opportunity with the less experienced part of the market.

Step 3: Evaluate and Select a Target Market

After identifying meaningful segments in the market and understanding the customers within each of those segments, the third step in the segmentation process requires that the firm evaluate

TABLE 7.4 Segmentation of U.S. Information and Communication Technology (ICT) Users

Tier	Segment	% of U.S. Adults	Attitude and Behavior
Elite tech users (31% of U.S. adults)	Omnivores	8	They have the most information gadgets and services, used voraciously to participate in cyberspace and express themselves online and do a range of Web 2.0 activities such as blogging or managing their own Web pages.
	Connectors	7	Between feature-packed cell phones and frequent online use, they connect to people and manage digital content using ICTs—all with high levels of satisfaction about how ICTs let them work with community groups and pursue hobbies.
	Lackluster Veterans	8	Frequent users of the Internet and less avid about cell phones, they are not thrilled with ICT-enabled connectivity.
	Productivity Enhancers	8	They hold strongly positive views about how technology lets them keep up with others, do their jobs, and learn new things.
Middle-of-the-road tech users (20% of U.S. adults)	Mobile Centrics	10	They fully embrace the functionality of their cell phones, use the Internet (but not often), and like how ICTs connect them to others.
	Connected but Hassled	10	They invested in a lot of technology, but find the connectivity intrusive and information something of a burden.
Few tech assets (49% of U.S. adults)	Inexperienced Experimenters	8	They occasionally take advantage of interactivity—but with more experience, might do more with ICTs.
	Light but Satisfied	15	They have some technology, but it does not play a central role in their daily lives; they're satisfied with what ICTs do for them.
	Indifferents	11	Despite having either cell phones or online access, they use ICTs only intermittently and find connectivity annoying.
	Off the Network	15	They have neither cell phones nor Internet connectivity; tend to be older adults who are content with old media.

Source: Horrigan, John B., "A Typology of Information and Communication Technology Users," Washington, DC: Pew Internet & American Life Project, May 7, 2007, online at www.pewinternet.org/pdfs/PIP_ICT_Typology.pdf.

the attractiveness of the various segments in order to narrow its choice of which to pursue. Four important criteria on which to evaluate each segment are useful:

1. *Size:* Estimates of the potential sales volume within each segment are needed in order to identify large segments. Historically, marketers based their size estimates on purchase volume. Segments with fewer people may have larger dollar purchases. For example, the **80/20 rule** says that 80% of the sales in any one category are typically purchased by only 20% of the

customers. More recently, however, with base-of-the-pyramid market strategies, companies have realized that rather than selecting a target market based on purchasing volume, the number of sheer customers in a market can also be an important metric.

2. *Growth:* Estimates of the growth rates of various segments are also needed to help evaluate possible attractiveness. Segments that are growing in size are attractive for at least two reasons. First, the growth means that the firm will be able to capitalize on customers' needs and grow with the market. Second, the growth means that, rather than stealing customers away from other firms, firms can capture new customers coming into the market.

3. *Level of competition:* Estimates of the level of competitive intensity within each segment help a firm to identify how costly it may be to pursue that segment. High numbers of competitors or even a few powerfully entrenched competitors can pose formidable risks to a new firm. The issue is not so much whether the new firm's technology is better—but rather, how hard these competitors will work to defend their existing base of customers.

4. *Capabilities to serve the needs of that segment:* Finally, firms must take a good hard look at their core competencies and strengths to determine if they have the capabilities to serve the needs of a particular segment. Although partnering can augment some deficiencies, the reality is that customers will look to firms that offer the right set of skills to address their needs.

One company used these four criteria to help select a market segment in the emerging area of blue-green algae (spirulina) protein supplements. After growing, harvesting, and drying algae, the resulting product can be sold in powder form (as a supplement), either to actual consumers (e.g., through online channels, in select retail outlets) or to business customers who would use the supplement in their own product line (e.g., Odwalla makes a protein bar with this supplement and other companies add it to their nutrient-enhanced beverages). In addition, it can be used in cosmetics and animal feed, both pet and farm (such as poultry feed and in aquaculture). The company collected sales data as well as sales growth rates for the various product categories and market applications. It combined the insights from these data with insight about its start-up position in the industry, the increased regulation in food products, and the trends toward organic farming and non-hormonal-based supplements for animals, and targeted the poultry feed market as a viable prospect.

Step 4: Positioning the Product within the Segment

The fourth and final step of the segmentation process is to create a meaningful positioning strategy for the new technology. A company's **positioning** is the image of the product in the eyes of the customer, relative to competitors, on critical attributes of importance. In other words, the company's positioning is the way customers perceive its value proposition. Some important aspects of this definition warrant mention.

First, positioning is based on *customers' perceptions.* After all, customers will be making the decision to purchase (or not purchase) the new technology, and what matters is what they believe about the new technology. Whether the firm thinks the customers' perceptions are wrong is totally irrelevant. The firm must effectively communicate its value proposition to customers to create the positioning it desires.

Second, a company's positioning is always *relative to competitors.* Many new high-tech firms mistakenly believe that they have no competition, that their innovation is so radical that no other firm provides anything remotely similar. Although this may be technically true, the customer's reality is that there are always other options (i.e., competition). The customer can choose to do things the old way or even do nothing. This is why positioning of the new product is generally achieved by focusing on how the new technology fits within existing market categories by referencing the older technology that is being displaced (i.e., product form competition).

Marketing a product that violates the categorical scheme of the marketplace creates confusion both among consumers and in the retail channel.[44] If positioned as something that is totally new, retailers will not know (1) if it is something that their store should carry or (2) which department to put the product in. As a result, consumers will not know where to find the product. Moreover, if positioned as something totally new, pragmatist consumers will not be able to compare products because they don't know what to compare it with. Firms may need to collaborate with their rivals who also offer the new technology to overcome the prior technology successfully.

As the new technology is adopted by customers and begins to diffuse through the marketplace, competition based on the new technology will develop. Relative positioning strategies must now implicitly reference this brand competition (i.e., selective demand). Finally, in late stages of the adoption process, the market leader needs to create new products that will cannibalize its old products.[45] As a result, positioning may then reference the company's prior version of the product as the competition.

Two tools that can assist a company with its positioning strategies are multi-attribute models and perceptual maps. A **multi-attribute model** collects data from a sample of customers about the relative importance of the attributes (product benefits and features) in their technology purchase decision, and then assesses how well a company's and its key competitors' products stack up on those features. If a company cannot afford primary market research as input to this process, its frontline sales staff, customer service personnel, and installation/maintenance people are a rich source of customer perspectives on what matters.

The data in Table 7.5 show the relative importance of three attributes in an emergency communication software solution as well as the company's and key competitors' scores on each attribute. The total score is computed by weighting each company's score on each attribute by its importance. Certainly, the attributes themselves, as well as the number of attributes and competitors listed, will vary widely by product category.

The focal company in this example was consistently losing orders to Competitor #2. Its seemingly outstanding score on the most important attribute (scalability) was emphasized consistently in its marketing literature and data sheets. The source of this advantage was its software-as-a-service business model, while its two competitors still operated on a client/server model, making it more difficult for customers to scale the solution as their emergency communication networks widened. The company's salespeople consistently cited the barrier that the low perception of ease-of-use posed to customers. Although the company's software developers insisted that the Web-based interface was "intuitive" to use—and why wouldn't they? after all, they had designed it!—customers lacked familiarity and comfort with the new operating approach, where users themselves could sign up for the emergency notification network rather than relying on a network administrator to do so. Moreover, although Web-based software services are highly secure, customers' lack of familiarity with this software delivery model resulted in low scores on perceived security as well. The company worried about explicitly addressing

TABLE 7.5 Illustration of a Multi-Attribute Model

Attribute (importance)*	My Company	Competitor #1	Competitor #2
Scalability (8)	10[†]	5	5
Security (6)	3	5	7
Ease of use (7)	3	6	8
Total Score	**119**	**112**	**138**

* Scaled from 1–10, with 10 being rated as most important.
† Scaled from 1–10, with 10 = customer perception that that company has the highest level of that attribute possible.

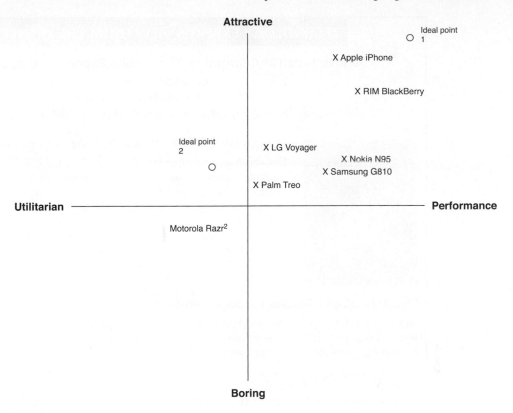

FIGURE 7.6 Perceptual Map for Smart Phones

ease-of-use and security in its marketing messages, because after all, it did not score well on these factors, relative to competitors.

However, best-practices marketing must explicitly address customer perceptions. Certainly, a company wants to position on its strengths, but to ignore the barriers customers perceive is naïve. The solution for this company was twofold: First, it created easy (realistic) online tutorials and demonstrations that walked prospective customers through the actual process of using the software from sign-on to emergency notification. It helped devise pilot tests so that customers could see the system in use. Second, it offered customer testimonials about the security of Web-based software relative to client/server architecture. These efforts resulted in improvements in its scores on both these attributes.

A **perceptual map** uses similar data from consumers as in Table 7.5. It consolidates multiple attributes into two higher-level attributes using statistical techniques (either factor analysis or multi-dimensional scaling). Perceptual mapping produces a simple visual graphic of a company's offering (positioned on these two key attributes) versus the competition. A perceptual map can also display what consumers' ideal points or preferences are. An example of a perceptual map for smart phones is presented in Figure 7.6.

Figure 7.6 can be interpreted easily to draw competitive implications. For example, both Palm and Motorola need to improve product performance and attractiveness to compete better with the other brands for the segment of consumers that prefer ideal point 1. As an alternative, Motorola could launch a new product to serve the needs of ideal point 2, which is currently being ignored by the existing players.

To provide a holistic picture of the segmentation, targeting, and positioning process, the accompanying Technology Expert's View from the Trenches discusses the mobile phone market in Japan.

TECHNOLOGY EXPERT'S VIEW FROM THE TRENCHES

Segmentation and Targeting: The Mobile Phone Industry in Japan

CHISATO OKAZAKI
Market Trend Analysis Team
Panasonic Mobile Communications Company, Yokohama, Japan

Background: Mobile Phone Industry in Japan

The population in Japan is about 1.2 billion people; the number of mobile phone subscribers is about 1 billion: four out of five people in Japan have a mobile phone. The mobile phone market in Japan is in the mature phase, and is getting similar to a commodity-type market.

The number of different mobile phones on the market in any given year is more than 90. The technology of mobile phones in Japan is amazingly well developed. People can take a picture, send e-mail, watch TV, play games, and create business documents with their phones.

The Role of and Process for Segmentation

To be successful in this environment, it is critical to accurately identify the target market of a product and its value proposition. The different ways to approach segmentation vary across companies. The basic process we follow at Panasonic is:

State purpose of segmentation → Set hypothesis → Research/Analyze → Report

1. State the purpose of segmentation. The first and most important step is to establish the purpose of segmentation. Also, it is critical to decide how segmentation will be done and who will use the segmentation analysis *before* starting the research. One key purpose for segmentation might be to identify the key product features that will be attractive to various segments in the market, and to increase the effectiveness of the promotion of those products.

2. Set hypothesis. After the purpose of segmentation is carefully stated, the company has to decide how to divide the users in the market. This step is very critical to the fourth step (analyzing the data to identify the factors that underlie the segments in the market). One way to develop hypotheses is to gather people from different departments in the company and, based on their daily experiences, to brainstorm about what kinds of customer types might exist in the market. These hypotheses then become the basis for specific items that are included in questionnaire during the research stage.

When I came to Panasonic six years ago, the company segmented the market based on gender and age. Although this was useful when the market was in the growth stage, it was less suitable for segmenting a mature market; segmentation based on gender and age does not provide insights about what users value in mobile phones. Especially in Japan, because income and racial differences are small, psychological segmentation works better than demographic approaches. Indeed, customers' needs for mobile phones are very diverse in Japan. For example, some customers want cheap phones, while others care more about the latest technology. Instead of using gender and age, we moved to segmentation based on users' motivation level or the psychological value they receive from their phones. In today's market, it seems that customers are experiencing "feature fatigue," meaning they are getting tired of the rapid pace of technological changes in their phones' features; as a result more and more segments seem to care more about the phone's design and its size (rather than the latest features and functionality). Understanding what users value is key to effective segmentation.

3. Research and analyze. At this stage, the company must develop a questionnaire to collect data to use in the market segmentation process. Questionnaires must be designed to provide insight for segmentation and information to understand each segment's characteristics. The researcher should take care not to overwhelm the respondents with too many items in the questionnaire, which could compromise the quality of the data.

Because of cost and time considerations, online surveys are commonly used to gather data for segmentation. In the case that the respondents' ages are under 18 or over 60, it is better to use mail surveys because of lower Internet usage in these age groups.

When I develop the research protocol for segmentation, finding the right wording to ask customers what they would like to see in new mobile phones is difficult. This problem is made more complicated by the fact that sometimes the engineering and planning group members in our company have their own specific ideas about what kinds of new technology each segment would want to use. However, even though it is difficult for customers to imagine how they might use new mobile phone features, it is not meaningless to ask. For example, sometimes after I have identified various market segments, I can use depth or group interviews to show new technology ideas to selected users to get their reactions.

After the research is completed, the data are analyzed to identify market segments. Usually, factor-cluster analysis is the statistical method that we use. Five things that the researcher should assess before reporting the results are:

- Is the size of each cluster suitable for segmentation purposes?
- Are some segments similar to others? If so, the data should be reanalyzed so that each segment is different from the others.
- Are the results presented in such a way that those in other departments who have no background in segmentation can understand them? (A complicated segmentation analysis might be accurate, but not useful.)
- Does the segmentation analysis fit its purpose and solve the company's problem?
- Is each segment's name easy to imagine and understand its characteristics?

4. *Report findings.* The segmentation report should be easy to read and should summarize the critical information needed to guide new product development. Using pictures and graphs makes it easy for other people in the company to understand each cluster (or market segment).

I usually create a catchy, descriptive name for each segment, so other people in the company can easily understand the characteristics of the segment just by reading the name. For example, a market segment (cluster) that cares only about the latest technology might be named "Digital Maniac." Furthermore, it is very useful to use a picture that symbolizes each segment. These devices (easy to understand, descriptive names and images) are helpful in getting other people in the company to use the segmentation results. Often I hear that engineers and designers are more motivated and find their work more easy when they understand who the product is targeted for and how the product satisfies the customers.

Summary

It is very challenging to create segmentation and decide on a target strategy in the mobile phone industry; there are so many complicating factors that must be considered. Manufacturers such as Panasonic cannot control price and distribution channels because of the nature of the industry (we rely on the wireless carriers). However, segmenting the end user market helps the company to develop products based on a customer view, not on carrier's demand. Moreover, ongoing segmentation helps the company to stay abreast of trends in the industry.

CUSTOMER STRATEGIES TO AVOID OBSOLESCENCE: IMPLICATIONS FOR UPGRADES AND MIGRATION PATHS

High-tech markets are blessed (cursed?) with fast and significant (revolutionary) improvements, which result in "inflection points" and technological discontinuities in the marketplace. The steady stream of improved and overlapping product generations typically makes the customers' investments in prior generations obsolete, even while those investments are still perfectly functional in use.[46] For example, computer microchips show a rapid pace of constantly improved generations available to the marketplace, even while customers are using prior generations. Indeed, as noted in

Chapter 1's discussion of technology life cycles, successive generations tend to arrive when the current generation's sales curve is still rising and may continue to rise for some time.[47]

In high-tech markets, customers must make important decisions about if and when to adopt a new generation of technology. Customer investments in prior-generation, outdated **legacy systems** create a "core rigidity," acting as an inhibitor to purchase of newer generations. Business customers often face a gap between a product's useful technological life (shorter than three years, in many cases) and its "accounting" life (for depreciation purposes, usually five years for durable assets). As a result, they are interested in upgrades that allow them to protect their investments in technology by providing benefits of "new and improved" without scrapping the old version entirely.

Upgrades for businesses are particularly important in driving growth in technology spending. Spending on new technology and technology upgrades by businesses fell 6.2% between 2001 and 2003;[48] however, technology spending by businesses grew 4.7% in 2004, 6.9% in 2005, 6.1% in 2006, and 6.9% in 2007. In 2008 the economic slowdown was expected to keep technology spending flat.[49] Tech spending by enterprise/business customers is vital for PC hardware and software makers, because corporate buyers allow hardware and software makers to survive (subsidize) the razor-thin margins in the high-volume home PC market. But some corporate buyers have decided that the price/performance calculations don't warrant buying next-generation computers just for fractions of increases in speed. Because most firms can make the case that the technology in place is pretty workable, the motivation for another buying round for new technology would have to be a real technological breakthrough or a "killer app" that compellingly convinces buyers their dollars are worth it. The emphasis in a slow growth economic environment will be on improving worker productivity and asset efficiency.

For example, as the price of oil escalates and business travel becomes more expensive, more companies will embrace high-end videoconferencing (telepresence) from companies like HP and Cisco. In another area, virtualization software from companies like VMWare will see increasing revenues as most companies try to increase the utilization of existing servers instead of replacing them. Finally, as entertainment companies send more rich media content to consumers' laptops, cell phones, and flat panel displays, the demand for speed and capacity of broadband networks is likely to increase. Companies that offer WiFi networks may within a few years upgrade to a WiMax network that can provide high-speed wireless broadband access over large distances (several kilometers as opposed to WiFi's hundreds of meters range). Sprint Nextel, Comcast, Time Warner Cable, Intel, and Google have all invested in a company called Clearwire to roll out a national WiMax network by 2010.[50]

In addition to making investments in technology that will improve worker productivity and asset efficiency, companies are increasingly willing to invest in information technologies that will save on energy-related operating expenses. A Forrester report in December 2007, "Green Progress in Enterprise IT," surveyed 130 executives in charge of IT operations and procurement in U.S., Canadian, and European companies. Some 38% of those surveyed said they were using "environmental criteria" to evaluate and purchase IT equipment, and 55% reported pursuing "more sustainable IT operations" in order to reduce their energy-related operating expenses.[51]

Customers' expectations and fears of obsolescence require that the firm carefully balance introducing new, state-of-the-art technology and maintaining interfaces to legacy systems. Proactively charting a migration path for customers from older generations of technology to newer generations is one way to negotiate this situation.

Customer Migration Decisions

What affects a customer's decision to migrate from the older to newer generation of technology? Customers' expectations about the *pace* and *magnitude* of performance improvements and price declines play a big role in their adoption decisions. The customer must balance the value of the existing products against the value of new offerings and even future arrivals. When products improve rapidly and significantly, the adage about starting with a high price (to "skim" those who buy early) and then lowering the price to entice later purchasers may not hold.[52] In general, the greater the anticipated product improvement or expected price decline, the greater the customer's propensity to delay purchase. In the extreme, customers may engage in "**leapfrogging**"—passing entirely on purchasing a current generation of technology in anticipation of a new, better innovation coming down the pike in the near future. This behavior, based on customer expectations of imminent improvement, can have a chilling effect on sales of current products.[53] The customer says, "If I wait to buy tomorrow, the product will not only be cheaper, it will also be better." Moreover, if the customer has already purchased, then the greater the product improvements and price decrements of successive generations, the greater will be the customer's regret factor—especially for early adopters.

Marketers' Migration Options

To help customers to manage the transitions between generations, marketers can offer a **migration path**, or a series of upgrades to help the customer's transition between generations.[54] The various options along the migration path are based on the degree to which the customer's options in the transition are *more constrained* versus *enlarged:*

1. *Withdraw the older generation as soon as the new one is launched, with no assistance to the installed base* (the set of existing customers who adopted the firm's prior-generation technology). Lack of parts and service for these customers forces them into a decision about migration sooner than they might like.
2. *Withdraw the older generation when the new one is launched, but offer migration assistance,* which can be in the form of technical help, trade-ins, backward compatibility with gateways, and the like. Customers can upgrade, maintain the old version and move to the new version later.
3. *Sell old and new generations together for a period of time,* after which the old generation is withdrawn. Customers can continue with installed version A, migrate to next generation B, or skip B entirely (leapfrog) and go to C.
4. *Sell both generations as long as the market desires them* and provide migration assistance to the installed base.

Based on an options model developed by Grenadier and Weiss,[55] the firm's choice of which migration path to offer is affected by the factors shown in Table 7.6 and as discussed next.

EXPECTATIONS OF PACE OF ADVANCEMENTS When customers expect rapid advances, albeit small ones, it pays for a firm to increase options. Customers who anticipate a rapid pace of change tend to wait for price declines or bug fixes in the newly launched version. These customers can also bypass completely, waiting for some yet-to-be-launched future version.[56] Both stalling and

TABLE 7.6 Migration Considerations and Options

Customer Perceptions	Implication for Customer Behavior	Implication for Migration Path
Customers expect rapid pace in technology advancements	Willing to wait for price declines	Marketer should provide migration assistance
Customers expect large magnitude of change in technology advancements	Recognize smooth upgrading is unlikely; therefore, waiting to purchase an older model at a lower price may result in obsolescence	Migration path less crucial, because the latest technology effectively obsoletes any path that was available
Customers have anxiety about making a decision	Need to feel that their decisions are safe	Marketer should provide migration path, possibly selling old and new models together for a period of time

leapfrogging are mitigated by migration assistance. Without such assistance, firms will find their revenues swinging wildly.

EXPECTATIONS OF MAGNITUDE OF ADVANCEMENTS In contrast to the pace of advancements, customers who expect significant advances in technology recognize that smooth upgrades are simply not possible. In such a situation, few customers are willing to wait to purchase an older generation of the product at a reduced price, which will be made obsolete. In essence, there is less to be gained by keeping the customers' options open. As a result, where customers anticipate large discontinuities between generations, the firm may choose not to offer migration assistance. Even if the firm did, the reality is that the customers' existing investments have been destroyed.

CUSTOMER UNCERTAINTY When customers have fear, uncertainty, and doubt about their expectations, such a situation warrants migration assistance. A firm can choose to sell both the old and new versions to encourage customers with even older installed versions to migrate to the next step.

POSITIONING AND PRICING OF UPGRADES Upgrades can attract both new customers and existing customers who have previously purchased an earlier generation of the product. In mature product categories (e.g., PC operating systems), upgrades generally attract more existing customers than new customers. Research shows that existing customers prefer upgrades that are *dissimilar from the previous generation product* and that have *new features* (versus when they merely enhance existing features of the previous-generation product). Also, if the upgrade does involve enhancement of existing features, existing customers prefer enhancements of a few features rather than enhancement of all the features of a previous-generation product.[57] These findings offer an explanation for the lukewarm reception that Windows Vista received in the marketplace. Perhaps existing customers of Windows XP did not see Vista as dissimilar; rather than adding new features, they saw Vista as enhancing existing features of Windows XP. Thus, they couldn't justify scrapping their sunk costs in XP for Vista's enhanced benefits. Indeed, despite Microsoft's attempts to discontinue XP in January 2008, customers clamored to keep it because of a lack of willingness to migrate to Vista. Hence, Microsoft is making XP available to PC manufacturers, at least through June 2009.

Hewlett-Packard uses pricing strategies to manage customer migration decisions. Because demand drops for older Unix servers when new machines are nearly ready for the market, HP uses sophisticated dynamic pricing software to analyze market trends. Rather than trying to slash prices on old machines too late in the process, HP calculates when to start discounting the old machines and by how much. So far, the markdowns made possible by this pricing strategy to manage customer migration decisions helped increase sales 1–1.5% in 2002.[58]

SUPPLY CHAIN COMPLEXITIES Any decision to offer upgrades must also account for complexities in managing relationships across the supply chain:

$$\text{Supplier} \rightarrow \text{Manufacturer/OEM} \rightarrow \text{Channel Members} \rightarrow \text{Customers}$$

Revenue from upgrades often flows to other members of the channel (e.g., the supplier in the case of chip upgrades, or manufacturers in the case of add-on components) and is smaller in size than that from selling entirely new units. Upgrades can also cannibalize sales of new products. So, the revenue implications and possible conflicts with other members of the supply chain must be monitored carefully with any decision.

CONSUMERS' PARADOXICAL RELATIONSHIPS WITH TECHNOLOGY AND UNINTENDED CONSEQUENCES

The final section in this chapter on understanding high-tech customers serves as a reminder to high-tech companies: Do not be blinded by internal enthusiasm for technology. Technology evokes complicated, often opposing, reactions in customers, and remaining aware of these reactions helps companies to stay in touch with customers' concerns.

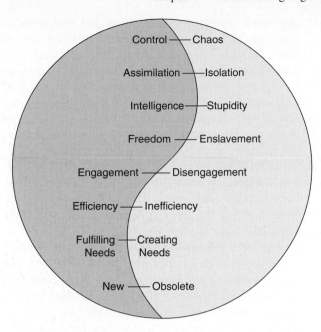

FIGURE 7.7 Consumers' Paradoxical Relationships with Technology

For example, in the 1930s, the fierce, biting South American fire ant entered the United States as a ship stowaway. Although it was targeted for eradication with DDT and super-pesticides, after three decades the pesticides had done more damage to the ants' predators than to the invaders. The chemicals actually ended up helping to *increase* the ant population rather than to eradicate it.[59]

This example shows that the best-laid technological plans can, and often do, go awry. Many people are susceptible to overinflated enthusiasm about the possible benefits of high-tech solutions. Yet technology is inherently paradoxical, with every positive quality potentially countered by an opposing negative quality. Although technology enthusiasts would have us believe in their promises that technology can transform our lives in positive ways, the pace, complexity, and unintended consequences of technological development are all too apparent. Yes, technology has provided people freedom, control, and efficiency; but it also has degraded the environment, undermined the social fabric of our lives, and, at the extreme, put us on the brink of obliteration.

As shown in Figure 7.7, David Mick and Susan Fournier identified eight paradoxes that characterize the relationship between consumers and technology.[60] These paradoxes highlight the fact that technological developments do have unintended consequences. Of course, the unintended consequences are not always negative.

1. ***Control–Chaos:*** Technology is supposed to help bring order to our lives and businesses. However, once purchased, technology can disrupt our everyday routines. For example, as a society, we have become so dependent on computers that if they fail, we can no longer do our work. Indeed, in this wired world, we are more susceptible than ever to viruses, worms, and cyberattacks that bring work to a standstill.

2. ***Assimilation–Isolation:*** Many technologies help us connect with other people, either by facilitating communication or by providing shared experiences. But these technologies have also become a substitute for face-to-face communication, personal contact, and other social activities. Ultimately, when we log on to virtual communities, we sit alone at our computers. Research shows the more hours people spend on the Internet, the more

depressed, stressed, and alienated they feel, even if most of their time is spent sending messages or "interacting" in chat rooms: "We've become so removed from reality that we don't even know how to feed our increasing hunger for intimacy."[61] Moreover, 6% of Internet users suffer from some form of addiction to it, such as compulsive Internet use, online gambling, accessing porn/sex sites, or stock trading.[62] False feelings of intimacy, timelessness, and lack of inhibition contribute to the addictive force of the Internet.

3. *Intelligence–Stupidity:* Sophisticated products are supposed to help us be smarter by allowing us to perform complex tasks. But if the products are difficult to master or cause us to lose old skills, we actually feel inadequate or inferior in mastering new technologies.

4. *Freedom–Enslavement:* Products that offer freedom can end up creating new restrictions. For example, voice mail gets us away from our offices, but makes us feel obligated to check our messages constantly. Cell phones give us freedom, but now we are always available. Many fear that wireless access to the Internet (via WiFi technology) means there will be no safe haven from the ubiquitous reach of technology.

5. *Engagement–Disengagement:* Although some technologies are designed to make participating in activities easier, they may detract from the quality of the experience. For example, instead of smelling and squeezing lemons for ripeness, we can now order them from a website. We watch food heat under plastic wrap behind glass microwave doors instead of carefully stirring pots and inhaling aromas that have filled the house with comforting smells.[63] The very things that are supposed to simplify our lives and allow us to stay in touch with each other are putting us further out of reach, with sense-dulling "conveniences."

6. *Efficiency–Inefficiency:* New technology can help us perform tasks faster, but also creates time-consuming new chores. In 2005, U.S. adults received, on average, 18.5 spam e-mail messages per day that took, on average, 2.8 minutes per day to delete. The time employees spend on deleting junk e-mail is estimated to cost companies nearly $22 billion per year.[64]

   Additionally, many companies invested in information technology with the expectation that such investments would make their businesses more productive. Although most statistics do suggest that the increases in economic productivity are, in large part, due to IT investments, at the personal level, inefficiencies and frustration arise from such things as continued efforts to keep current with software upgrades, the use of computers for personal activities (such as e-mail and Internet surfing), and so forth.[65]

7. *Fulfilling needs–Creating needs:* Many technology products make us aware of new needs even as they fulfill others. For example, although we might be able to use software to perform more sophisticated tasks, we also now need to attend training programs to learn how to use the software to do those tasks. Similarly, instant messaging has become a popular tool not only among consumers, but among corporate/business users as well. Yet, network security managers report that instant messaging poses a dangerous security risk to their networks, and as a result, network security managers are having to develop and implement new technologies to handle the threat.[66]

8. *New–Obsolete:* Although people are excited by owning cutting-edge products, their excitement is chronically undermined by a fear of falling behind. This paradox creates the simultaneous desires both to purchase new technologies and to delay purchase in order to wait for the latest/best technology.

In a similar vein to the paradoxes in the Mick and Fournier study, Robert Kozinets offered four ideological elements of technology, each offering somewhat opposing views and implications:[67]

- *Techtopian:* Technology is a critical underpinning to social progress.
- *Green/Luddite:* Technology is a destructive force in society.
- *Work Machine:* Technology is an important component of the global economic engine.
- *"Techspressive":* Technology is a key ingredient in delivering pleasure.

These four ideological elements again offer a useful perspective that customers' views of technology are complicated; their views are not unidimensional, nor mere love/hate relationships.

Marketing Implications of Consumers' Paradoxical Relationship with Technology

How do these paradoxes affect the way people view technology and, ultimately, their purchase strategies and behaviors? Potential purchasers of technology create coping strategies to lessen the presence of the paradoxes. Understanding such coping strategies can give marketers insight into segments of high-tech consumers and the differing types of marketing messages that may be useful, given consumers' concerns. For example, buyers of computers may place limits on how often they use them so that they don't take over their lives. Indeed, some Silicon Valley executives decidedly reject the use of technology during their personal time, refusing to own computers and other electronic gadgets.[68] For others, yoga, massage, or Tai Chi can reduce the stress caused by the technology paradox. At its extreme, the backlash against technology can be seen in echoes of the Luddite movement that continue even today in sporadic antitechnology protests that equate technology with corporate power.[69]

Unfortunately, technology marketers, with their enthusiasm for the products they market, are sometimes blind to the dual impacts of technology from a customer's perspective. At a minimum, marketers must be alert to the presence of the paradoxes, try to actively consider what the unintended consequences of a new technology might be, and develop contingency plans for their marketing efforts.

An interesting case is that of nanotechnology, which has been in development for the past 20 years. *Nanotechnology* is the study and application of material properties at very small sizes, typically less than 100 nanometers.[70] A nanometer is one billionth of a meter, the size of a marble compared to the size of planet Earth. At this tiny scale, materials behave very differently than at larger sizes. For example, pieces of aluminum at sizes of 20 to 30 nanometers can explode, so nano-alumnium could be added to rocket fuel to improve combustion. Carbon nanotubes, when sprinkled into epoxy, strengthen the glue by 30%. Silica nanoparticles layered over fire-resistant glass can help panes withstand temperatures up to 1,800°F for up to two hours.

The two ways to create nanoparticles are (1) "top-down," where a bulk material is chopped up into smaller and smaller pieces, or (2) "bottom-up," where molecules are grown under controlled conditions (as in crystals) and then put together into specific configurations. Bottom-up nanotechnology holds greater promise because it can create longer nanoparticles. For example, fibers spun from carbon nanotubes should be stronger than Kevlar, the material in bullet-proof vests. The governments of the European Union, Japan, and the United States each spent about $1 billion for nanotechnology research in 2005, and corporations spent about $4 billion on nanotechnology research in 2006. Silicon Valley start-up Nanosys is working on non-carbon nanomaterials like flexible plastics, silicon germanium, and gallium arsenide for applications in solar power generation and semiconductors. University researchers are testing processes with gold nanoparticles to kill cancer tumors in mice without any side effects.

However, not much research has investigated the problem of nanoparticle pollution. The toxicity of materials at these sizes is hard to predict, especially the impact on human cells if nanoparticles get into the water and food chains. Hence, nanotechnology, like other new technologies, illustrates the paradox of promise and pitfall.

Ultimately, firms must be aware that blind pursuit of the technological imperative can potentially be viewed as threatening and interfering with sociological needs for safety and human dignity. Because customers will ultimately make the decision whether to buy, firms must pay attention to the tension these paradoxes pose. The controversy surrounding genetically modified food provides an example.

TECHNOLOGY SOLUTIONS FOR GLOBAL PROBLEMS
Creative Solutions for Water Services in Base-of-the-Pyramid Markets

Companies competing in base-of-the-pyramid markets must understand and solve what seemingly are intractable problems. Consider the following challenges:

- Low disposable incomes limit the amount consumers can buy at any one time.
- Consumers in low-income segments have trouble saving money, which makes it hard for companies to collect money, especially for products when consumers receive bills after consumption (such as for utility services).
- Low-quality roads, postal services, electricity, and other basic forms of infrastructure make it harder and costlier to support production and distribution in many areas.
- Sellers are vulnerable to theft and extortion.

In such situations, community leaders such as school principals, teachers, religious leaders, as well as residents themselves, are in the best position to help companies deal with these challenges of doing business in low-income areas. These community leaders have the information that sellers need, with the ability to monitor and influence what happens in the local community. If a company can show that its own interests are aligned with the community's interests in employment and commerce, it can then enlist community support for security, collection, and system monitoring.

Working with community leaders, Manila Water devised a market-oriented business model: letting communities themselves decide if they want individual or collective installation, metering, and billing. The company offers three options: (1) one meter per household, (2) one meter for three or four households, or (3) a bulk meter for 40 to 50 households. Where households band together, the connection fee (ordinarily 7,000 pesos a household) can fall by as much as 60%, depending on the number of customers who shoulder the cost of pipes, the meter, and installation. Submeters measure water usage in each household, and everyone on a group meter takes responsibility for paying the total bill, an arrangement that in effect gives consumers (and Manila Water) group insurance coverage on payment.

Source: Beshouri, Christopher, "A Grassroots Approach to Emerging-Market Consumers," *McKinsey Quarterly* 26 (2006), available at www.mckinseyquarterly.com/article_page.aspx?ar=1866&L2=21&L3=34&srid=253&gp=0.

Summary

This chapter has provided an in-depth look at issues related to customer purchase decisions: steps in customer technology purchase decisions; the factors that affect their purchase decisions; different categories of adopters; the chasm that exists in high-tech markets; segmenting, targeting, and positioning in high-tech markets; customer strategies to avoid obsolescence; and the technology paradoxes that customers face. Marketers must understand these issues in order to develop successful marketing strategies.

Discussion Questions

Chapter Concepts

1. What are the stages in the purchase process? What are the implications for marketing strategy in each stage?

2. What is meant by "design" and "design thinking"? What impact does design have on a company's financial performance? What are the three ways that design

can add value to an organization? How does the composition and process of a design-oriented team differ from a more traditional, analytical team?

3. What are some of the issues buyers face in the postpurchase evaluation stage? How does a vendor help buyers manage these concerns?
   a. How can marketers target customers based on post-adoption usage patterns?
   b. How can high-tech companies find opportunity in end-of-life issues for their products?

4. What factors influence a customer's potential adoption of a new innovation? What are the implications of each of the factors for high-tech marketers?

5. Studies show that customers desire complex products prior to purchase, but then experience negative emotions after using them. What are the implications of these studies for the ways marketers ought to manage complexity in high-tech products?

6. What are the categories of adopters and their characteristics? What are the appropriate marketing strategies for each of the categories?

7. What is the chasm? Compare and contrast the marketing strategies that are necessary in the early market versus the mainstream market.

8. What are the two key decisions a company must make to cross the chasm?

9. What are the two critical characteristics of a good beachhead? Why?

10. Explain a compelling reason to buy for pragmatist customers.

11. Explain the bowling pin model used to identify adjacencies. What are the key attributes for adjacent market segments?

12. How many market segments can a high-tech company realistically pursue in crossing the chasm? Why?

13. What is a whole product solution? How does a high-tech firm typically develop it?

14. What are four other issues that companies must address in crossing the chasm?

15. What are the three phases of the pragmatist market (from *Inside the Tornado*)?

16. What are the unique marketing issues in reaching the conservatives in the market?

17. What are the four steps in segmenting markets?

18. What is the difference between vertical market segmentation and horizontal market segmentation? When is one approach versus the other more likely to be used?

19. What are the four criteria commonly used to evaluate and select a target market? What is the 80/20 rule and how is it adapted for base-of-the-pyramid market strategies?

20. What is positioning and how should a new innovation's position change over the diffusion process?

21. Explain how multi-attribute models and perceptual maps can be used to help diagnose a company's positioning strategy.

22. What are the issues customers face in attempting to avoid obsolescence of their technology purchases? What are the impacts of customers' expectations about the pace and magnitude of performance improvements and price declines on their technology adoption decisions? Of the four types of migration paths companies can offer, what path should be offered under which conditions?

23. Of the eight paradoxes of customers' relationships with technology, identify and describe three that you find most interesting. Give your own example of each. What are the marketing implications?

24. Of the four ideological views of technology, which combination/juxtaposition most accurately captures your own complicated relationship with technology?

25. Identify and describe one unintended consequence arising from technology. If you were the manager marketing products based on this technology, how would you handle this unintended consequence?

Application Questions

Based on the opening vignette of this chapter, answer the following questions:

1. What are the sources of market, technology, and competitive uncertainty for this technology (review from Chapter 1)?

2. Collect some recent data regarding RFID sales and adoption. At what stage of the adoption-and-diffusion process is this technology?

3. Research a vendor of RFID technology. What would make a compelling beachhead for RFID technology? What are the elements of the whole product solution?

4. Evaluate RFID technology on the factors that affect customer adoption decisions.

5. Come up with a radical but useful new application of RFID.

Glossary

beachhead   A single target market from which to pursue the mainstream market.

design   The combination of both functionality and aesthetics in creating a product, service, or experience.

80/20 rule   The principle that 80% of the revenues/profits in a market (or for a particular company) are typically generated by roughly 20% of the customers.

horizontal market segments   Markets that span industry boundaries, addressing a specific function (e.g., human resources or customer management) that serves several different industries.

leapfrogging   Passing entirely on purchasing a current generation of technology in anticipation of a new, better innovation coming down the pike in the near future.

legacy systems   Outdated technology that customers still use despite newer versions being available, usually because of a considerable investment in time and money.

migration path   A series of marketing tools (upgrades, pricing strategies, etc.) to help customers move from a prior generation of technology to a new generation.

multi-attribute model   A tool used to assist a company in understanding both its market position as well as strengths and weaknesses that need to be addressed in its marketing strategy; collects data from a sample of customers on both the importance of various attributes in customer purchase decisions for a specific technology, as well as customers' perceptions of various companies' ratings on how well each company's product scores on those various attributes.

perceptual map   A two-dimensional graphic of a company's offering (positioned on two key attributes) versus the competition.

positioning   The image of the product in the eyes of the customer, relative to competitors on critical attributes of importance; the customer's perception of a company's value proposition.

vertical market segments   Industry-specific markets (e.g., forestry, health care, retail businesses).

whole product   The complete end-to-end solution of what it takes to fulfill the customer's needs. For example, in the computer industry, the whole product includes hardware, software, peripherals, interfaces and connectivity, installation and training, and service and support.

Notes

1. "Tag, You're It!" *Fortune,* May 29, 2006, advertising supplement, p. S2.
2. King, Rachael, "Radio Shipment-Tracking: A Revolution Delayed," *BusinessWeek,* October 9, 2006, p. 2.
3. Ibid., p. 3.
4. Ibid.
5. Interested readers may also want to review the book by Allan Reddy, *The Emerging High-Tech Consumer: A Market Profile and Marketing Strategy Implications* (Westport, CT: Quorum Books, 1997).
6. Kirkpatrick, David, "The E-Ware War," *Fortune,* December 7, 1998, pp. 102–112.
7. Cahill, Dennis, and Robert Warshawsky, "The Marketing Concept: A Forgotten Aid for Marketing High-Technology Products," *Journal of Consumer Marketing* 10 (Winter 1993), pp. 17–22; and Cahill, Dennis, Sharon Thach, and Robert Warshawsky, "The Marketing Concept and New High-Tech Products: Is There a Fit?" *Journal of Product Innovation Management* 11 (September 1994), pp. 336–343.
8. Judge, Paul, "Are Tech Buyers Different?" *BusinessWeek,* January 26, 1998, pp. 64–66.
9. Rich, Harry, "Proving the Practical Power of Design," *Design Management Review* 15 (Fall 2004), pp. 29–34.
10. Borja de Mozota, Brigette, "The Four Powers of Design: A Value Model in Design Management," *Design Management Review* 17 (Spring 2006), pp. 44–53.
11. Chochinov, Allan, "'Greener Gadgets' Isn't an Oxymoron," *BusinessWeek* (Green Design Issue), February 11, 2008.
12. Based on personal discussions of Sanjit Sengupta with David M. Kelley at IDEO, Palo Alto, California, on August 4, 2004.
13. Shih, Chuan-Fong, and Alladi Venkatesh, "Beyond Adoption: Development and Application of a Use-Diffusion Model," *Journal of Marketing* 68 (January 2004), pp. 59–72.
14. Ibid.
15. "The Truth about Recycling," *The Economist,* June 7, 2007, available at http://economist.co.uk/PrinterFriendly.cfm?story_id=9249262.
16. McDonough, William, and Michael Braungart, *Cradle to Cradle: Remaking the Way We Make Things* (New York: North Point Press, 2002).

17. Rogers, Everett, *Diffusion of Innovations,* 5th ed. (New York: Free Press, 2003).

18. Thompson, Debora Viana, Rebecca W. Hamilton, and Roland T. Rust, "Feature Fatigue: When Product Capabilities Become Too Much of a Good Thing," *Journal of Marketing Research* 42 (November 2005), pp. 431–442.

19. Wood, Stacy L., and C. Page Moreau, "From Fear to Loathing? How Emotion Influences the Evaluation and Early Use of Innovations," *Journal of Marketing* 70 (July 2006), pp. 44–57.

20. Mintz, Jessica, "High-Tech Gap," Associated Press, September 30, 2007.

21. Gottfredson, Mark, and Keith Aspinall, "Innovation versus Complexity: What Is Too Much of a Good Thing?" *Harvard Business Review* 83 (November 2005); Schwartz, Barry, *The Paradox of Choice: Why More Is Less* (New York: Harper Perennial, 2004); and Mariotti, John, *The Complexity Crisis: Why Too Many Products, Markets, and Customers Are Crippling Your Company—And What to Do about It* (Avon, MA: Adams Media, 2008).

22. "Oracle Introduces Oracle Communications IP Service and Network Management to Simplify Lifecycle Management of Complex IP-Based Services," press release, February 6, 2008, online at www.oracle.com/corporate/press/2008_feb/ipsnm.html.

23. Much of the material in this section is derived from Moore, Geoffrey, *Crossing the Chasm: Marketing and Selling Technology Products to Mainstream Customers* (New York: HarperCollins, 1991); and Rogers, Everett, *Diffusion of Innovations,* 5th ed. (New York: Free Press, 2003).

24. Ibid.

25. Bruner, Gordon C. II, and Anand Kumar, "Gadget Lovers," *Journal of the Academy of Marketing Science* 35 (September 2007), pp. 329–339.

26. Moore, Geoffrey, *Inside the Tornado* (New York: Harper Business, 1995).

27. Helft, Miguel, "Tech's Late Adopters Prefer the Tried and True," *New York Times,* March 12, 2008, available at http://www.nytimes.com/2008/03/12/technology/12inertia.html.

28. To protest unemployment caused by the Industrial Revolution in the early 19th century, English workers known as Luddites, who either were led by Ned Ludd or were from Ludlam, resorted to a campaign of breaking machinery, especially knitting machines.

29. Mahajan, Vijay, and Eitan Muller, "When Is It Worthwhile Targeting the Majority Instead of the Innovators in a New Product Launch?" *Journal of Marketing Research* 34 (February 1998), pp. 488–495.

30. Moore, *Crossing the Chasm.*

31. The majority of the information in this section is derived from Moore, *Crossing the Chasm;* revised edition (2002) by Harper Business.

32. Goldenberg, Jacob, Barak Libai, and Eitan Muller, "Riding the Saddle: How Cross-Market Communications Can Create a Major Slump in Sales," *Journal of Marketing* 66 (April 2002), pp. 1–16.

33. Moore, *Crossing the Chasm.*

34. Goldenberg, Libai, and Muller, "Riding the Saddle."

35. Cheng, Roger, "Cisco, AT&T Look to Expand High-End Video-Conference System," *Wall Street Journal,* April 21, 2008, online at http://www.humanproductivitylab.com/archive_blogs/2008/04/29/att_rolls_out_intercompany_tel.php.

36. Moore, *Inside the Tornado.*

37. Cheng, Roger, "Cisco, AT&T Look to Expand High-End Video-Conference System."

38. Ibid.

39. Readers interested in the idea of market gorillas might reference Moore, Geoffrey A., *The Gorilla Game: Picking Winners in High Technology* (New York: Harper Business, 1999).

40. Gross, Neil, and Peter Coy with Otis Port, "The Technology Paradox," *Business Week,* March 6, 1995, pp. 76–84.

41. Gogoi, Pallavi, "Meet Jane Geek," *Business Week,* November 28, 2005, pp. 94–95.

42. Ibid.

43. Horrigan, John B., "A Typology of Information and Communication Technology Users," Washington, DC: Pew Internet & American Life Project, May 7, 2007, online at www.pewinternet.org/pdfs/PIP_ICT_Typology.pdf.

44. Moore, *Inside the Tornado.*

45. Ibid.

46. John, George, Allen Weiss, and Shantanu Dutta, "Marketing in Technology Intensive Markets: Towards a Conceptual Framework," *Journal of Marketing* 63 (Special Issue 1999), pp. 78–91.

47. Norton, John A., and Frank Bass, "Diffusion Theory Model of Adoption and Substitution for Successive Generations of Technology Intensive Products," *Management Science* 33 (September 1987), pp. 1069–1086.

48. Mullaney, Timothy, et al., "The E-Biz Surprise," *BusinessWeek,* May 12, 2003, pp. 60–66.

49. Tam, Pui-Wing, and Ben Worthen, "Tech Spending May Veer off Fast Track," *Wall Street Journal,* December 11, 2007, online at http://online.wsj.com/article/SB119731174726119648.html.

50. Sharma, Amol, and Vishesh Kumar, "Tech Firms to Build WiMax Network in U.S.," *Wall Street Journal,* May 8, 2008, online at http://online.wsj.com/article/SB121015567027273579.html?mod=djempersonal.

51. Jana, Reena, "Green IT: Corporate Strategies," *BusinessWeek* (Green Design Issue), February 11, 2008.

52. Dhebar, Anirudh, "Speeding High-Tech Producer, Meet the Balking Consumer," *Sloan Management Review* 37 (Winter 1996), pp. 37–49.

53. Weiss, Allen, "The Effects of Expectations on Technology Adoption: Some Empirical Evidence," *Journal of Industrial Economics* 42 (December 1994), pp. 1–19.

54. John, Weiss, and Dutta, "Marketing in Technology Intensive Markets."

55. Grenadier, S., and Allen Weiss, "Investments in Technological Innovations: An Options Pricing Approach," *Journal of Financial Economics* 44 (1997), pp. 397–416.

56. Weiss, "The Effects of Expectations on Technology Adoption."

57. Okada, Erica Mina, "Upgrades and New Purchases." *Journal of Marketing* 70 (October 2006), pp. 92–102.

58. Keenan, Faith, "The Price Is Really Right," *BusinessWeek,* March 31, 2003, pp. 62–67.

59. Dibbell, Julian, "Everything That Could Go Wrong . . . ," *Time,* May 20, 1996, p. 56.

60. Mick, David Glen, and Susan Fournier, "Paradoxes of Technology: Consumer Cognizance, Emotions, and Coping Strategies," *Journal of Consumer Research* 25 (September 1998), pp. 123–143.

61. Donato, Marla, "Sense-Dulling Conveniences Creating Alienated World," *Missoulian,* January 12, 1999, p. A8, from *Chicago Tribune.*

62. Down, Jeff, "Plugged In to Excess," *Missoulian,* August 23, 1999, p. A1.

63. Donato, "Sense-Dulling Conveniences Creating Alienated World."

64. Swartz, Nikki, "Deleting Spam Costs Businesses Billions," *Information Management Journal* 39 (May–June 2005), available at http://findarticles.com/p/articles/mi_qa3937/is_ 200505/ ai_n13638937/pg_1.

65. Siegel, Matt, "Do Computers Slow Us Down?" *Fortune,* March 30, 1998, pp. 34, 38; Landauer, Thomas, *The Trouble with Computers: Usefulness, Usability, and Productivity* (Cambridge, MA: MIT Press, 1995); and Brynjolfsson, Erik, "The Productivity Paradox of Information Technology," *Communications of the ACM* 36 (December 1993), pp. 67–77.

66. Woods, Bob, "IM Use a Big Security Threat," Instant Messaging Planet, June 5, 2002, online at www.instantmessagingplanet.com/security/article.php/1268921.

67. Kozinets, Robert, "Technology/Ideology: How Ideological Fields Influence Consumers' Technology Narratives," *Journal of Consumer Research* 34 (April 2007), pp. 865–881.

68. Tam, Pui-Wing, "Taking High Tech Home Is a Bit Much for an Internet Exec," *Wall Street Journal,* June 16, 2000, p. A1.

69. Bailey, Ronald, "Rebels against the Future: Witnessing the Birth of the Global Anti-Technology Movement," ReasonOnline, February 28, 2001, online at www.reason.com/news/show/34773.html.

70. This section is summarized from Kahn, Jennifer, "Welcome to the World of Nanotechnology," *National Geographic,* June 2006, pp. 100–119.

8

Technology and Product Management

"Walled Garden" or "Open Plain": Which Business Model Is Music to Your Ears?[1]

In 1999, Napster was the first company to demonstrate that consumers liked listening to music downloaded via the Web. Unfortunately the courts ruled that peer-to-peer file sharing encouraged copyright infringement and ordered Napster to shut down its illegal operations in 2001 even though it had about 60 million users at its peak. The shift in consumer preferences from prerecorded CDs to music downloads was established, and although many websites have come and gone that facilitate music downloads, Apple was the pioneer in legal music downloads.

After his ouster from Apple in 1985, Steve Jobs bought the computer division of LucasFilms in 1986, renamed it Pixar, and was extremely successful in the animated movie business with hits like *Toy Story*. After returning to Apple as CEO in 1997 and sensing the slow growth in the Macintosh computer business, Jobs tried to reposition the iMac as the center of a consumer's digital lifestyle, with digital video editing software like iMovie for editing home movies. Excited by the Napster phenomenon in the late 1990s, he had an idea for a legitimate digital music business model that resulted in the launch of iPod and iTunes for Mac users in 2001.

The iPod wasn't the first digital music player. Offerings from Creative based in Singapore and iRiver based in South Korea existed before iPod (and in fact, Creative and Apple had competing patents to some of the underlying technology in their players); however, users of these other digital music players depended mainly on illegal music downloads.

The iPod captured the hearts and minds of consumers with its sleek, cool design and storage capacity. Apple designed it from scratch with a scrolling touch wheel and proprietary software. The first version of iTunes was used mainly to upload music from CDs to the Mac, which were then downloaded from the Mac into the iPod. Apple worked closely with suppliers on the components for the iPod, but the iTunes software was developed completely in-house.

In April 2003 Apple launched the iTunes Music Store, which had an initial catalog of 200,000 titles, each downloadable for an affordable 99 cents. The launch of iTunes capped a two-year effort during which Jobs had to personally convince the music labels to let him sell their music online, something they feared because of the threat of piracy. Jobs had to agree to some constraints to mollify the music labels: Music downloaded from iTunes was in a proprietary format, and it could be played only on iPods, not on any other digital music players; so iPod users could download music only from iTunes and not from any other website. Thus, although iPod/iTunes was a closed system, the legitimate business model hit the sweet spot of the market and catapulted Apple to a dominant market share of 76% of digital music players and 88% of all legal music downloads in 2006. Apple took a whopping 22% of all music sold in the U.S. by 2006.

Paradoxically enough, the music industry itself was the reason that Apple gained such a powerful presence. The only way Jobs was able to convince the major music labels to offer their

(continued)

music through iTunes was Apple's assurance that because its digital rights management system (FairPlay) meant that Apple downloads could play only on Apple products, it wouldn't affect a large percentage of the music market. The lack of interoperability combined with the iPod's dominance allowed Apple to essentially become a de facto monopoly. Indeed, Apple has come under scrutiny in the European Union for possible antitrust concerns. Yet, despite the popularity of Apple's iTunes, as late as 2006, research firm BigChampagne estimated that 90% of music downloads were still from underground file-trading programs such as eDonkey and LimeWire.

With the demonstrated success of the iTunes Music store, many competitors jumped into the fray with their own legal music download services. Roxio bought the rights to the Napster name and launched its legal Napster online music store. Real Networks launched its Rhapsody service. Yahoo! and Wal-Mart also launched their own online music stores in addition to Musicmatch and MusicNow. While iTunes downloads were owned by the consumer for the fixed price of 99 cents each, most of the other services rented music for a monthly subscription fee; the music expired if the fees were not paid. For example, Yahoo! Music Unlimited users paid $6.99 per month or $60 per year for unlimited rentals that could be downloaded to 10 portable players, which worked with the Microsoft digital music format. In this sense it was more open than iTunes. In 2004 Microsoft launched Windows Media Player and its MSN Music Service, which gave users access to online music downloads from Napster, Musicmatch, MusicNow, and Wal-Mart. These downloads would be compatible with more than 70 portable music players. In late 2006, Microsoft launched its Zune music player. These different music download services come with various antipiracy restrictions that prevent a song file from playing on some devices or from being shared more than a certain number of times.

The competing file formats for digital music (affecting both the music files themselves and the players) as well as the different business models means that the players and services that supply the music are incompatible with one another. Some refer to Apple's closed system as a "walled garden," a hallmark of Apple's product strategy since the Macintosh launch in 1984. By contrast Microsoft's system provides a software developer's kit to encourage developers to develop more applications for its products—although its Zune players are incompatible with others on the market (such as the Sansa player).

This chapter focuses on product and technology management considerations in the high-tech company. Based on guidance from a technology map, which provides a blueprint or road map for the company's various products and technology development projects (both incremental as well as breakthrough), the chapter delves into product development and management issues in a high-tech environment.

Effective technology development requires ongoing monitoring of technology trends to guide the development of a high-tech company's product road map. One major technology trend over the past 50 years has been the "digital revolution"—the transition from analog to digital technology. Analog devices use as input continuously varying real-world signals (such as pressure, sound, or heat) and convert them into electrical signals like current and voltage. During this conversion process, there is some signal loss or distortion. Digital devices use as input electrical signals coded as zeros and ones, manipulate them, and transmit output as electrical signals, quickly and efficiently. Therefore, digital systems (based mainly on digital devices) have greater speed and accuracy than analog systems. Currently, many countries are undergoing a government-mandated transition from analog to digital television broadcasts. For example, in the United States, all full-power television broadcasts will be exclusively digital from February 17, 2009. In Europe different countries are implementing the switchover in different years. However, the European Union is recommending to member countries that they complete this conversion by 2012 at the latest. In industries as diverse as automobiles, computers, music, movies, television, and telecommunications, the move toward digital technology has produced better quality for consumers and tremendous business opportunities and challenges for technology companies. Despite this well-pronounced trend, high-tech companies in many industries struggle with the migration. For example, Kodak waited so long in transitioning from analog to digital technology in the camera industry that it was faced with a painful restructuring of the organization.

Many companies are trying to understand, and profit from, the technology trend called **digital convergence**.[2] Digital convergence encompasses three phenomena: (1) previously stand-alone devices such as home appliances or office equipment are being connected by networks and software, significantly enhancing their individual functionalities; (2) previously stand-alone products are being converged onto the same platform, creating hybrid products such as camera phones, smart phones, multifunctional printers/copiers/fax machines, and the like; (3) companies are crossing traditional industry boundaries such as computers, telecommunications, consumer electronics, and entertainment to provide new sources of competition to existing players (Apple is a good example of this).

Digital convergence provides opportunities for innovation that offer consumers more value for less cost. Because consumer lifestyles are merging across work, play, and travel, and they need to increase their productivity through ubiquitous automation, digital convergence products and services fit well with current consumer lifestyles. Two other underlying reasons for the emergence of digital convergence include the emergence of global standards for digitization and communication, and the availability of cheaper, more powerful hardware, software and technology infrastructure.

Already many global companies are pursuing digital convergence opportunities, including Sony, Samsung, and LG electronics in Asia; Nokia in Europe; and Apple and Microsoft in the United States, to name a few. Global competition is putting pressure on the business models of such companies to lower costs. Further, companies need to expand their capabilities beyond existing competencies to offer consumers all the functionality they need. These challenges of digital convergence can be addressed through outsourcing (lower costs) and alliances for new product development (broaden existing product lines to offer whole solutions for consumers), as discussed in Chapter 5.

One purpose of technology mapping is to help a company anticipate technology trends and be poised to move as these change and evolve. The company's technology road map also informs decisions about how to develop new technologies. Firms may choose to develop a new technology in-house, with their own resources and skills; alternatively, high-tech firms may acquire new technologies either through licensing agreements or by purchasing new technology start-ups that have developed such technologies. (Partnering to develop new technologies and products was addressed in Chapter 5.)

High-tech marketers must also consider how much to develop the technology prior to offering it for sale in the marketplace. For example, a firm may decide to market and sell basic know-how (say, through a technology transfer or licensing arrangement) or develop a whole product (see Chapter 7), marketing and selling a ready-to-use product and everything that goes with it. (For an example of one organization's marketing decision to license its technology, see the accompanying Technology Solutions for Global Problems.) The decision about how close to final form the new product should be is not a clear-cut decision but one that warrants consideration.

Other product development/management issues in a high-tech environment that will be discussed in this chapter include the following:

- How should a firm conceptualize its product architecture in terms of its use of modularity, platforms, and derivatives?
- When or how should a firm halt investments and development in a new product whose success looks questionable?
- What is the role of the product manager in a high-tech company?
- What are the unique issues involved in the development of technology-related services?
- How should intellectual property rights be controlled?

By way of a brief review, other concepts related to product and technology management were introduced earlier. For example, technology life cycles (see Chapter 1) depict the relationship between

TECHNOLOGY SOLUTIONS TO GLOBAL PROBLEMS
The World's First Solar-Powered Hearing Aid Battery Charger
By Jamie Hoffman

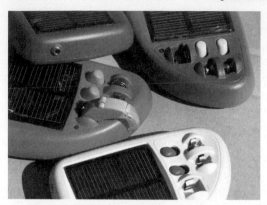

Photo reprinted with permission of Godisa Technologies Trust, Gabarone, Botswana.

One-tenth of the world's population suffers from a disabling hearing impairment; of the 600 million people experiencing hearing loss, 80% live in developing countries. Those affected by hearing loss routinely experience consequences beyond auditory impairment, involving social interactions, education, and employment. In developing nations, access to affordable hearing aids was impossible due to the expensive battery. To address this issue, Godisa Technologies Trust, a nonprofit enterprise spearheaded in Botswana, developed and distributes SolarAid, the world's first and only solar-powered hearing aid battery charger.

Although available in more than 30 countries, Godisa has opted to partner with organizations that can develop similar products for the hearing-impaired community. A complete SolarAid includes a behind-the-ear hearing aid, solar-powered charger, and four rechargeable batteries for approximately US$150 (standard hearing aids cost US$500). One battery can be reused 400 times, takes two to six hours to recharge, and once recharged, lasts from two to seven days, depending on hearing aid power demand. The charger accommodates most #13 and #675 rechargeable batteries. More than 7,000 units are presently in use.

Godisa envisions a world where hearing impairment is not a barrier and currently offers the product to developing and developed nations alike (including Canada, the United States, and the United Kingdom).

Sources: "SolarAid," Design for the Other 90%, online at http://other90.cooperhewitt.org/Design/solaraid; "About Godisa," Godisa Technologies Trust, online at www.godisa.org/about.html; Wightman, David, "Deaf Technicians Share SolarAid Success," 4HearingLoss News and Reviews, January 22, 2006, online at www.4hearingloss.com/archives/2006/01/deaf_technician.html; and personal correspondence with Modesta Nyirenda Zabula, Godisa/SolarAid, July 23, 2008.

investments in improving the underlying technology of a product and its performance–price ratio. This relationship is often depicted as an S-shaped curve: Initial investments in a new technology may show modest improvements in the product's performance–price ratio, but after some threshold, additional investments show a drastic improvement in the performance–price ratio, which then levels off at some point. Importantly, technology life cycles help managers to understand that when a new technology enters the market, the new and the former technologies may compete with each other for a period of time, until the new technology eventually supersedes the former. Technology life cycles allow a product manager in a high-tech environment to anticipate when new technologies may supersede existing technologies and highlight the need to always be on the cutting edge. Product managers must be willing to cannibalize their current-generation technologies so that new industry players do not disrupt them.

Other technology management concepts introduced earlier include the adoption and diffusion of innovations (Chapter 7), the timing of market entry (addressed in the section titled "Strategy Types" in Chapter 2), the use of teams in the new product development process (Chapter 4), and the role of partners in new product alliances (Chapter 5).

TECHNOLOGY MAPPING

An important tool used to monitor technology trends and systematically manage a company's resources is a **technology map**. A technology map defines the stream of new products, including both break-through and incremental products, that the company is committed to developing over some future time period. Companies that are most successful in defining next-generation products use the map to force decisions about new projects amid the technological and market uncertainty found in high-tech markets.[3] The use of a technology map can promote cohesion and commitment to new product development plans and can be used to clarify possible sources of confusion, to allocate resources, and to make trade-offs among various projects. Importantly, technology maps are not cast in stone but are updated and revisited regularly. Rather than being a tool that inhibits innovation and creates blinders for a firm, a technology map should serve as a flexible blueprint for the future, updated and revised regularly.

For example, the rivalry in the market for semiconductor chips between AMD and Intel has caused some changes in Intel's technology road map. In 2003 AMD beat Intel in coming out first with a 64-bit microprocessor, the Opteron. In 2005 AMD started stealing market share from Intel in the server market by emphasizing the energy efficiency of Opteron microprocessors. In fact, from 2003 to 2006, AMD made Intel scramble and change many of its product plans. In September 2006 Paul Ottelini, Intel CEO, shared his technology map with hardware and software developers at the Intel developer forum in San Francisco to stem the market share it had lost to AMD. Intel would launch a powerful, energy-efficient four-core microprocessor in November 2006 for the enthusiast or gamer segments, ahead of its 2007 schedule.[4]

Capon and Glazer offer the following steps in developing and managing technology resources[5] (also see Figure 8.1):

1. *Technology Identification.* Technology identification requires taking an inventory of the firm's know-how to find those ideas having the most value. Technology know-how can be found in products, processes, and management practices.

Recall from Chapter 1 that *product technology* is embodied in the goods and services that a company sells in the marketplace. Although most firms can easily identify the technology that forms

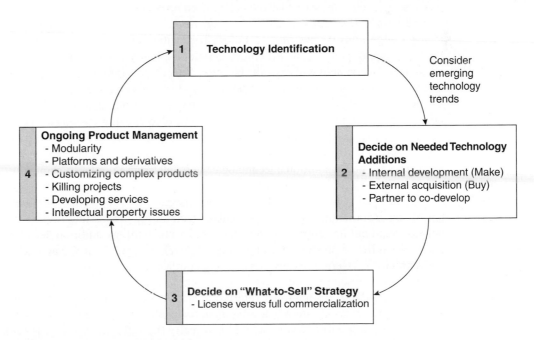

FIGURE 8.1 Technology Map

the basis of their products, identifying process technology that might have value outside the firm itself can be difficult. Recall that *process technologies* are used to facilitate the routines and procedures in the company; they are developed in the course of running the company. For example, process technologies might include manufacturing technologies the company has innovated, marketing innovations (say, a unique online customer interface for e-commerce or automated customer services), or lab equipment or scientific processes used in the R&D process itself that facilitate the discovery of underlying scientific properties (such as lab equipment to sequence genes). Cisco's engineers and network architects have developed their own internal tools to help them design, configure, optimize, and compare alternative network infrastructures for their clients.[6] For example, to determine the number of routers and switches to recommend to customers, Cisco runs sophisticated simulations. This process technology, developed in the course of managing their own business, has been repackaged as a customer tool, which customers have adopted to manage their own internal networks. In this case, the process technology has also become a product technology for sale in the market.

Management practices, embodied in organizational and managerial know-how and routines, are another source of technology that must be inventoried in the first step of the technology mapping process. As best-practices routines become more common—including superior skills in total quality management (TQM), e-commerce, customer service, and the like—such organizational and managerial know-how has become a revenue-generating asset. For example, although Amazon is still considered an e-tailer of many different categories of consumer products, about 27% of its annual revenue comes from B-to-B services such as website management, fulfillment services for companies like Target and Borders, Web hosting, and providing Web services for smaller sellers.[7]

At this first step of the technology mapping process, the firm is most concerned with its current technologies. Given the discussion on core competencies, core rigidities, and the innovator's dilemma earlier in this book, most companies need to exploit current core competencies as well as explore new competencies to leverage future business opportunities. One study found that exploiting current competencies is positively related to incremental innovation, while exploring new competencies is positively related to radical innovation.[8] Therefore, as companies "map" their technologies, they must be sensitive to the impact of current competencies and the development of new competencies, in order to avoid potential competency traps.

2. *Make Decisions about Technology Additions.* The technology identification conducted in step 1 highlights the existing technology base within the company. At this next step of the technology mapping process, the firm must consider explicitly what emerging technology trends (such as digital convergence) will affect its business. This consideration should explicitly evaluate disruptive technologies outside the current portfolio so the firm can escape the tyranny of the served market. Recall that these emerging technologies may come from areas outside the industry's existing boundaries—for example, new technologies that solve existing customers' needs in radically new ways—and so a company must scan the environment very broadly in order to avoid being myopic. The point of this step is to look for areas of weakness in the firm's technology portfolio, identifying areas where it needs additional technologies in its platform in order to compete effectively as technologies and markets evolve over time.

When the firm, through the combined insights from its technology identification process and its environmental scanning, recognizes technology arenas that require new skills or products to round out its offering, it must then make a technology addition decision. A company can add these skills or products through internal development (the R&D and new product development process), external acquisition (buying another firm that has the requisite technology or licensing the technology from another firm), or partnering. The decision about how to add needed skills and products is sometimes referred to as a "make-versus-buy" decision. "Make" refers to the decision to rely on internal development to develop new products and technologies, whereas "buy" refers to the decision to acquire externally the rights to a new technology developed by another firm.

The key issue in the decision about how to add technology is *development risk*. A firm would appropriately pursue *internal development* (i.e., make the product) if:

1. The R&D area is close to current corporate skills.
2. The firm wishes to keep its technological thrust confidential.
3. The firm's culture fosters the belief that the only good technology is developed internally.

The third point, the belief that the only good technology is developed internally, is sometimes referred to as the *NIH syndrome* ("not invented here"), as described in Chapter 6. Often, a company's technical personnel believe they can always "build a better mousetrap" (develop a more sophisticated technology) than people at other companies. However, this erroneous—even arrogant—belief can harm a company. Breakthrough technologies often result from knowledge spillovers and synergies through partnering and acquisitions; the NIH syndrome precludes these spillovers and synergies, and can create its own form of myopia.

Issues pertinent to internal product/technology development were discussed earlier in Chapters 2, 3, and 4—for example, the ideas of R&D–marketing interaction, product champions, and so forth. In some cases, internal R&D may be cheaper than external acquisition. Intel is building key portions of transistors in its chips from hafnium instead of silicon dioxide, an industry staple since the 1960s, in order to boost speed and reduce power consumption.[9]

On the other hand, *external acquisition* (i.e., buy the product) makes sense if:

1. Someone else has already developed the technology, and acquisition can save the firm time and effort.
2. The firm does not have all the necessary skills to develop the desired technology.
3. The firm wants to let others take big risks before participating in development of the new technology.
4. The firm needs to keep up with a competitor whose new technology is potentially threatening.
5. The firm wants to obtain technology for products that can use present brand names, distribution channels, and so forth (for example, Roxio's purchase of Napster in the opening vignette of this chapter).

Many companies acquire another company to gain access to its technology and people. There are pros and cons of such an acquisition. On the cons, a large financial acquisition may distract the management from their innovation goals, or the top talent of the target company may leave. These were risks faced in early 2008 by Microsoft's proposed acquisition of Yahoo!. Recent research shows that acquisition synergies can be achieved when the acquiring company has internal knowledge in particular technology areas that help it to better identify and assimilate knowledge from an acquired firm. The acquiring firm's internal knowledge should not be too similar or too different from the acquired firm's in order to boost innovation outcomes.[10] In another study, Sorescu, Chandy, and Prabhu show that firms with higher product capital (product development assets that the firm has built up over time) (1) select better acquisition targets with more innovation potential, (2) deploy the innovation potential of targets more extensively, and (3) create higher long-term shareholder value.[11]

As a middle ground between internal development and external acquisition, a firm may choose to form alliances with either competing firms or firms that provide complementary pieces to the product solution for new product development, as discussed in Chapter 5.

3. *Make Decisions about Commercializing, Licensing, and So Forth.* After acquiring or developing the desired technological know-how, the firm faces the commercialization decision. Here, *marketing risk* is the critical issue. The firm must decide exactly how far in the development process to proceed before marketing and selling the product. For example, as described by the accompanying Technology Expert's View from the Trenches, some companies decide to focus their efforts on developing underlying components of a product that is sold to other manufacturers, while other companies decide to focus on a turnkey, end-to-end solution. Because this decision is so critical to high-tech companies' success, this decision is explored in detail in the next section, "The 'What-to-Sell' Decision."

TECHNOLOGY EXPERT'S VIEW FROM THE TRENCHES

Overcoming the Challenges in Commercializing Disruptive Technology
BY BILL MCGLYNN
CEO, Memjet, Boise, Idaho

Memjet Technology is a new solution to printing color documents that will likely revolutionize the printing market. Memjet is a two-generation step beyond existing printing methodologies and is capable of printing 60 color pages per minute in a printer costing $300; ink supplies will cost the consumer 20% of existing prices. Just as cell phones, computer monitors, laptop displays, and photography have migrated to color, Memjet holds the promise of finally converting everyday printing to color as well. Memjet uses highly integrated silicon combined with micro-electro-mechanical (the "MEM" in Memjet) devices to produce a 20 mm × 0.72 mm chip with 6,400 nozzles. By aggregating these chips end to end, it's possible to build a very fast printer that squirts tiny ink drops producing high-resolution text and images that dry instantly. This means all printing in the future can be full, uncompromised color, including applications such as cash register receipts, shipping labels, baggage tags, and so forth.

There are five key challenges in bringing a new disruptive technology like Memjet to market:

1. *Committed Investors.* Technology disruptions come from invention, not from iterations of existing products. Therefore the normal ROI expectations on a rational product time line have to be suspended. Disruptions carry with them unpredictable obstacles. An obstacle we encountered was having to invent a new silicon etching process that eliminated the use of contaminating surface coatings, while maintaining an industry standard process to preserve low cost. A challenge like that can take months or years to resolve—time that neither corporate engineering timetables nor venture capital ROI expectations could withstand. Finding investors who understand the unpredictable nature of innovation and who also know that added pressure doesn't necessarily deliver better results can be the most difficult task in bringing disruptive technology to market. In our case, we have been fortunate to build relationships with two business leaders who have personally taken on innovation challenges and hence understand the challenges and the time required. The market opportunity also must be sufficiently large to make this investment worth the long-term investment commitment. In our case, revolutionizing a $100 billion-plus industry provides the incentive.

2. *The Entry Point.* A company must choose whether it is going to vertically integrate and brand the technology with its own brand, or sell the technology to brand partners. Branding your own technology sounds really exciting, until you consider how much it costs to establish a new brand. Some set the number at $1 billion for a new consumer electronics brand. If you don't have investors willing to invest that kind of money, you may want to consider either licensing your technology, or partially developing the final product by designing and building components. Licensing is the easiest path, but yields the lowest return. Manufacturing components is more difficult with a longer time to market, but yields a higher return. We chose to build components that brands could buy to build Memjet printers. In our case, this component approach has another advantage. It opens the market so that virtually anyone can enter and build a printer product. As a result, the business model becomes as disruptive as the technology, creating new opportunities for incumbents and new entrants alike.

3. *The Target Partner.* A company must attract the right partners or customers, ones that are capable and have the brand strength to be able to bring the technology to market with credibility. If it's truly disruptive, as in our case, just getting the partner to believe that the technology is *real* can be daunting. The incumbent partner with experienced engineering teams may, ironically, be the toughest to convince. Incumbents have been struggling in the trenches trying to evolve their existing technology, possibly for decades. When you show up with a technology that is two generations beyond their designs, there is, to say the least, some disbelief. These customers can use up tremendous amounts of your resources just trying to comprehend "how" you did it, rather than how they can "use" it. Your best choice could be a market-oriented partner, with resources to be a first mover. They aren't as interested in "how" you did it, as they are in "when" it will be available. Memjet sought out companies with strong consumer and business brands that didn't have strong printer businesses. They understood the market, and were not hesitant to adopt new technology to win market share. Much has been written about how market incumbents lose their position in markets where they have majority shares. Part of the reason this happens is this failure to quickly adopt new disruptive innovation that they didn't develop.

4. The Target Market. Once you have a brand partner, you need to help them choose the right end user target. Large corporations often look like the most lucrative customer, but that isn't always true. When it comes to information technology (IT) disruptions, many large corporations like to sit back and let the new technology be proven elsewhere before adopting. These companies also like standardizing on equipment to minimize the number of models and technologies to support. As a result, new technology usually gets its start elsewhere. Small business is a good place to target new technologies. They have a driving need to lower cost while increasing productivity and they often use new technology to help them do it. They don't have IT departments that set standards for products; nor do they have bureaucratic procurement procedures. They are frantically competing with larger competitors and are motivated to get an advantage. In Memjet's case, we have found that small businesses are the perfect end users of our printing solutions. Memjet is cheaper to acquire and much less expensive to operate, while printing at speeds that are 10 to 20 times faster than other printing products in the market, making a small business more productive and competitive.

5. The Right Product. Choose the right kind of product to launch your technology. Take into account the "immaturity" of your technology and manufacturing process, along with initial low production yields (as low as 50% in silicon processes), low margins, and a potentially slow manufacturing ramp. In our case, we chose an office printer for small business sold through office product channels as our initial product—a challenging but doable goal for us. As our manufacturing processes mature, production yields and quality—as well as costs—will improve, allowing us to enter higher volume markets with bigger brands, at lower costs and a more dependable supply chain. Starting with a very high volume application and segment of the market (in the Memjet case, it could be a corporate workgroup printer), could have some very difficult challenges, including lack of supply due to constraints in the manufacturing processes.

Memjet printers will be available in the market in mid-2009 and be offered through several brands simultaneously. We have made a conscious choice not to have our own branded Memjet printers, but to be the quiet component supplier to our brand partners. This will allow us to continue to focus on new technology and offer component enhancements to our partners that successively open opportunities in segments of the printing market for many years to come.

4. *Ongoing Management.* Finally, the firm needs to actively manage its technology asset base, including the issues of modularity, the development of product derivatives, whether to use product platforms in the strategy, when to "kill" new product development projects, how to augment product revenues with a services revenue stream, intellectual property management, and so forth. These issues are also discussed later in this chapter.

THE "WHAT-TO-SELL" DECISION

In a high-tech firm, technology itself either *is* the product (e.g., in a firm that licenses proprietary technology) or *gives rise to* the product (e.g., in a firm that chooses to commercialize products based on a new technology).[12] Firms that innovate technological know-how face a unique decision: Should they sell the knowledge itself or possibly license it? Should firms fully commercialize the idea—marketing, distributing, and selling a full solution including service and support? Or, given that final products can be "decomposed" into subsystems and components, should firms manufacture and sell some subsystem or component on an original equipment manufacturer (OEM) basis? Essentially, the decision about what to sell boils down to the basic issue of how to transform know-how into revenues.

Possible Options

As discussed next, firms can choose to sell (1) know-how only, (2) "proof-of-concept," (3) commercial-ready components to OEMs, (4) ready-to-use final products or systems, or (5) service bureaus that supply complete, end-to-end (integrated) solutions for the customers' needs.

The what-to-sell decision is based on an underlying dimension: the required expenditures by customers to derive the intended benefits, above and beyond their acquisition costs of the focal purchase.[13] For example, to derive the intended benefit of the product, funds must be spent for the core product, complementary items, services, and training, all of which can be necessary components. Customers who buy at the know-how end of the continuum must expend greater resources to realize the benefits of using the product, whereas customers who buy at the other end of the continuum (purchasing the complete product and all ancillary support services) can fully realize the benefits of the product with their purchase price.

- *Sell or license know-how only.*   The sale of know-how requires the greatest additional expenditures of funds by the customer after the transaction to realize the intended benefit. For example, chemical firms may sell (or license) the rights to a specific molecule to downstream producers.
- *Sell "proof-of-concept."*   The sale may include a prototype or pilot plant to establish that the know-how can indeed be made to work. Selling at this point on the continuum decreases the technological uncertainty for the buyer.
- *Sell commercial-grade components to OEMs.*   Firms may manufacture and sell components that are ready to use in another firm's manufactured product. As the Technology Expert's View from the Trenches illustrated, Memjet decided to sell components for printer engines non-exclusively to OEMs because that offered greater returns than licensing and leveraged the OEMs' marketing efforts in the end user market.
- *Sell final products or systems with all essential components, ready for use "out-of-the-box," to customers.*   Had Memjet chosen to sell at this point on the what-to-sell continuum, it would have sold a fully functional printer with its own brand name. Many technology companies mistakenly believe that this point is their best what-to-sell option. However, it can be complicated to work with end users. While Hewlett-Packard and Apple do a good job of offering an end product out of the box, other companies stumble. For example, flat panel TV displays often don't come with all the necessary cables for hooking up other components.
- *Sell a complete, end-to-end solution.*   The "whole product" solution delivers the intended benefits directly to customers with no need for them to incur additional expenditures on complementary items. Had Memjet operated at this level of the continuum, it might have established a type of printing service, where customers could send their print jobs electronically to use Memjet's technologies. As the chapter's opening vignette illustrates, Apple prefers to sell complete solutions to end users with its "walled garden" approach to product development, while Microsoft prefers to license out most of its software to be incorporated into end user products like PCs and cell phones.

What Decision Makes Sense?

The factors that affect a firm's decision about where along the continuum to generate revenues are shown in Figure 8.2 and discussed next.

In general, firms should lean toward the *selling know-how* end of the continuum when:[14]

- The technology does not fit with the firm's corporate mission.
- The firm has insufficient financial resources to exploit the technology.

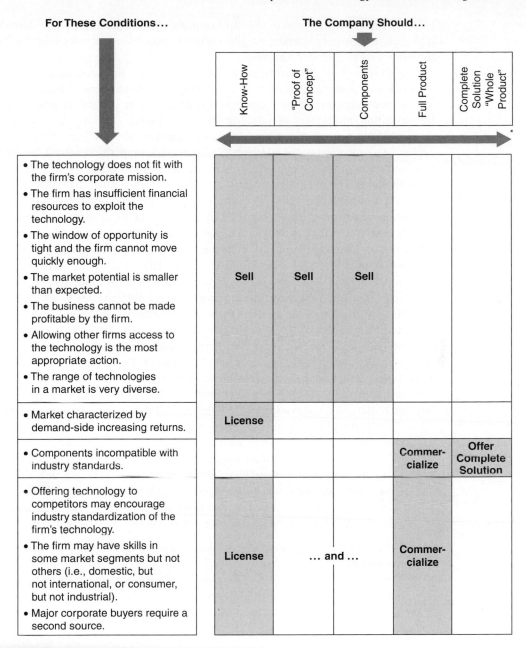

FIGURE 8.2 What to Sell

- The window of opportunity is tight and the firm cannot move quickly enough.
- The market potential for the end product is smaller than expected.
- The business cannot be made profitable by the firm.
- Allowing other firms access to the technology leverages their strength to develop the end user market, exploit network externalities, and discourage competing technologies.
- The range of technologies in a market is very diverse.

In this last case, it is difficult for firms to keep up with all the relevant technologies across all the components or subsystems of a final product. As a result, firms may find it desirable to compete closer to the know-how end of the continuum rather than the end product level.[15] This means that high-tech companies experience a pull toward licensing or "componentization" and promote their technology as an industry standard.

Allowing other firms access to technology (e.g., via licensing agreements) makes sense when the market is characterized by network externalities. Recall from Chapter 1 that many high-tech markets are characterized by a situation in which the more customers who adopt and use a particular innovation, the greater its value for all users. Making a technology available via licensing can discourage other companies from producing competing technologies. Although this protects the licensor's revenue stream, the licensees face increased competition among themselves and this tends to drive commoditization and price cuts in the end user market.

Firms tend to *sell more toward the end-product level* when they offer components that are incompatible with general industry standards. In general, customers want to be able to mix and match compatible components. Poor compatibility across the range of relevant technologies raises the customer's costs of putting together an acceptable system. In such a situation, selling know-how or individual components is difficult, and selling new, improved components is even more difficult. In the format wars over video discs between Blu-ray and HD-DVD, rivals Sony and Toshiba had to make both the players and disks for end users because the new disks could not play on existing DVD players. Making the end product with proprietary components ensures better optimization and control over the performance of the product, for which customers might be willing to pay a premium—indeed, this is a hallmark of Apple's product strategy. However, the lack of interoperability (also known as "plug-and-play" compatibility) for customers that would come from using more open platforms can potentially constrain the customer adoption process. Customers can also be frustrated by a sense that they are "held hostage" by proprietary standards. Moreover, they may find work-arounds and devise ways to make ostensibly proprietary technology compatible with other products in the market. Because of the risks, companies must be very careful if they choose to go it alone in this fashion.

Clearly, the preceding discussion highlights the strategic significance of selling more toward the component or the end product ends of the continuum, with major implications for branding strategies, the development of industry standards, as well as indirect network externalities and the availability of complementary products. Many would believe that Apple's walled garden approach invites more risks because it means that Apple not only has to develop all the necessary components of the solution, but it also stymies the development of both direct and indirect network effects.

As a result of the risk, some firms may *both commercialize and license technology* when:

- Offering technology to competitors may encourage industry standardization of the firm's technology.
- The firm may have skills in some market segments but not others (e.g., domestic but not international, or consumer but not industrial).
- Major corporate buyers require a second source.[16]

Doing both maximizes revenue streams while diversifying risks between business and consumer markets. Indeed, empirical evidence shows that firms that are good at product development processes (have low development times, can handle product changes, and have a higher percentage of revenue from new products) leverage these process to increase the vertical scope of their supply chain (make more of their inputs) as well as increased horizontal scope of products (offer a variety of different products).[17]

Rather than selling or licensing know-how, most firms have historically tended to sell close to the final user level. For example, royalty revenues (from licensing rights to know-how) have historically *not* been a major source of income compared to aggregate product sales revenue. Indeed, in many industries, royalty revenues are less than even the R&D expenditure itself.[18] But more and more high-tech firms are realizing revenues at multiple points along the what-to-sell continuum—say, by commercializing its technology in some markets and selling or licensing it in others.[19] One company that has transformed itself to utilize the whole range of these options is

IBM. IBM collaborated with Toshiba in the design of the cell microprocessor chip for Sony's PS/3 and has a five-year contract to manufacture and supply these to Sony; it makes final products like Netfinity server systems for enterprises; it offers global consulting services; and even operates data centers on a turnkey basis for companies.

One of the reasons high-tech firms are moving to sell more than just the products or components themselves is that, in a high-tech environment, sustainable long-term corporate growth depends on the continual development and leverage of a firm's technology. To maximize the rate of return on technology investment, a firm must plan for full market exploitation of all its technologies. These technologies may, but need not necessarily, be incorporated into that company's own products and services. Thus, a company's marketing strategy may (and probably should) provide for the sale of technologies for a lump sum or a royalty.[20]

For example, Qualcomm receives 58% of its revenue for chips that it designs and has manufactured by third parties, and 32% of its revenue as royalty payments from licensing agreements with 130 wireless equipment manufacturers.[21] As another example of a firm realizing revenues from different points on the continuum, Canon simultaneously sells printer subsystems on an OEM basis to both Hewlett-Packard and Apple, and ready-to-use laser printers to end customers. In a different vein, Toshiba historically licensed to other companies its underlying technologies for memory chips and disk drives. After new CEO Atsutoshi Nishida took the helm in 2005, he reversed course because the earlier strategy had not been profitable enough for Toshiba. Now it is making more final products on its own, including flash memory chips and gadgets like cell phones and music players powered by its proprietary fuel cells. It has also becoming more aggressive in filing lawsuits for violation of its patents.[22]

A final issue related to this what-to-sell decision pertains to *international markets*. In many cases, technology transfers to other countries may take place on the basis of standard turnkey deals,[23] in which a company transfers rights (e.g., via licensing) to use its technology to another company. Decisions to transfer technology to potentially low-cost producers must account for the effect on the company's own manufacturing plans. Many developing countries want to add value to their natural resources by buying sophisticated process technologies. Brazil, for example, wants to sell steel, not iron ore, and may be able to export steel relatively cheaply because it possesses key raw materials. But a non-Brazilian company licensing steel manufacturing technology to a Brazilian company has to consider the long-term effects of such a decision.

Although many of the preceding examples come from the computing industry, the pharmaceutical and biotech industries also rely heavily on the what-to-sell continuum to leverage their R&D investments.

Technology Transfer Considerations[24]

Technology can be transferred from small companies or inventors to larger companies with the resources and expertise in the commercialization process. One specific example of such a **technology transfer** is the transfer of know-how developed with federal funds (say, at research universities and government labs) to the private sector. Technology transfer agreements specify profit sharing from sales of the products or services with the inventor and the inventor's institution.

In the United States, a key stimulus for technology transfer from public research institutions and universities to the private sector, with the purpose of commercializing promising technologies, was the Bayh-Dole Act of 1980; it allows the transfer of technologies developed with federal funds (taxpayer dollars) to nonprofit institutions and gives small businesses the rights to inventions made with federal funds. Technology transfer has emerged as a specialized field, which requires legal, scientific, and business/marketing know-how. Its practice is proving to be one of the driving forces in economic development in the United States and other parts of the world.

A survey sponsored by the Association of University Technology Managers (AUTM) found that U.S. universities received about $45 billion in research and development funds from the federal government in 2006. U.S. universities filed 15,908 patent applications and received 3,255 patents in the same year. As a result of technology licensing by universities, 553 new start-up companies and 697 new products were launched in the United States in 2006.[25]

One of the major hurdles in the technology transfer process is establishing a realistic and accurate value on the technology to be transferred. Many methods have been developed for this process, and they are referenced in publications available from the AUTM. Inventors often overvalue their innovations and undervalue the investment risk the purchaser of the invention takes on. Most inventors, for example, do not appreciate that more than $200 million and seven to ten years may be spent in developing a drug or a vaccine with no assurance that it will get marketing approval or that it will be a commercial success if approved. Participation in the AUTM organization provides a network of colleagues that may be consulted about their opinions on contentious terms in an agreement under negotiation.

A second key issue in the technology transfer process is the protection of intellectual property rights, both for the inventor and the licensing/purchasing company. A company is unlikely to make the enormous investments required to develop a product for market unless it has some period of exclusivity during which it can recoup its investment and earn a fair return. A **patent** is a form of protection that gives the owner(s) the right to exclude others from making, using, offering for sale, or selling the product or process described in the patent for a specific time period. Not patenting an invention greatly reduces the incentive for a company to invest in its development, and thus (despite critics' concerns about the negative effects of patents) may delay or prevent application of the technology for public use.

Regardless of where on the continuum a firm chooses to sell, the development of high-technology products and innovations necessitates enormous R&D investments. To better leverage these investments, technology companies need to carefully design their product architecture, dealing with the issues of product modularity, platform products, and derivatives. This takes us to the fourth step in the technology development process: ongoing technology/product management, and the related issues of managing platforms and derivatives.

PRODUCT ARCHITECTURE: MODULARITY, PLATFORMS, AND DERIVATIVES

Modularity

Modularity is an approach to design that refers to building a complex product from smaller subsystems that are created independently yet function together as a whole.[26] Modularity in design requires information to be partitioned into visible and hidden components. The visible information consists of design rules on how the subsystems should work together. The hidden information is about how to design each subsystem independently while following the visible design rules on working together with other subsystems. Different companies can take responsibility for each of the modules with the assurance that their collective efforts will create value for customers. Companies in diverse industries such as automobiles, computers, and software use modularity because it provides many benefits. Each supplier company is able to focus on a module and make it better, thus accelerating the rate of innovation in the industry. The company enforcing the visible design rules (the architect company) gets the best subsystems due to supplier competition. And customers are able to mix and match modules to suit their specific needs. However, modularity requires the architect company to have deep product knowledge on every subsystem so that the design rules can be specified in advance.

Modularity reduces the uncertainty in product design and makes things more predictable for a technology company. Accordingly, research shows that modularity results in product standardization, lower barriers to entry for competitors, and more incremental rather than breakthrough innovations.[27] For a significant breakthrough in product performance, individual components and subsystems have to be highly interdependent or integrated. But an integrated product strategy carries more design risk because a change in one component affects all other components and the overall product. It seems prudent to have an integrated product strategy in the early stage of the life cycle when performance is important to customers. Later, a modular strategy may be used when customer needs shift to convenience, customization, flexibility, and price.

The software industry offers a good example of modularity in the product development process.[28] The subsystems for most software programs include the file access, editing, graphics formatting, and printing subsystems, all with internal subsystem interfaces and a graphical user interface. In the software world, the interfaces are particularly important, and their design and evolution can lead to long-lived systems and market domination. For example, Microsoft effectively guides the innovation of thousands of independent software companies by having developed and promoted as a standard the interface mechanisms that allow different programs to communicate with one another in a distributed computing environment. The resulting compatibility, or *interoperability,* for customers reduces learning time from package to package and allows sharing of data. Moreover, by establishing the de facto standard for an industry, Microsoft enables other companies to build modules that operate on Windows. The development of these third-party products reinforces the standard, yet Microsoft need not share the costs of development or marketing. These independent third-party developers, in addition to being software producers, become advocates for Microsoft Windows.

Google's Android software for mobile phones has modules including an operating system (based on Linux), a user interface, and Web-browsing software. This is being offered free to handset manufacturers who are part of Google's Open Handset Alliance. The first Android-based Google G1 phone, manufactured by HTC Taiwan and marketed by T-mobile, was launched in the United States for $179 on October 22, 2008.[29]

Platforms and Derivatives

A **product platform** is a common architecture based on a single design and underlying technology. New product platforms have enhanced performance benefits and involve significant investments compared to existing platforms, hence they are also called "next-generation" products. A product platform can be shared by a set of **derivative** products—spin-offs from the common underlying technology—whose features meet specialized needs of customers.[30] Derivative products include different models, brands, or versions of the platform product intended to fill performance gaps between the platform products. For example, each generation of Intel's microprocessor chips, such as the Intel Itanium 2 or the Intel Pentium D, is a platform product sharing the same underlying technology. After introducing the first version of a new chip, Intel introduces slower and faster versions of the chip within months, sometimes concurrently. Each company needs a technology map or product strategy encompassing its various platforms and derivatives.

WHY USE A PLATFORM AND DERIVATIVE STRATEGY?   At least two underlying reasons exist for using a platform and derivative strategy in high-tech markets. First, recall one of the common characteristics underlying high-tech markets—unit-one costs. High-tech marketers typically face a situation in which the cost of producing the first unit is very high relative to the costs of reproduction.[31] For example, the cost of burning an additional high-definition DVD is trivial compared to the cost of hiring specialists to produce the content in the first place. This underlying feature of technology markets makes a product platform strategy very attractive. If the incremental costs of developing derivative products are relatively small, compared to the platform product, then proliferating versions of a common design to reach various segments adds incremental revenue.

As shown in Figure 8.3, a second reason for using a platform and derivative strategy is that, when a firm introduces a breakthrough product, it will inevitably create "gaps" in the marketplace.[32] These gaps, or holes, exist in the customer's migration path from the old technology to the new technology. Importantly, a firm should not overlook these gaps, which would, in essence, allow

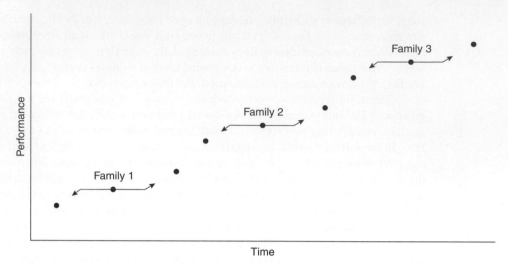

FIGURE 8.3 Product Platforms and Derivatives

competitors to come in with gap-filling strategies and possibly even dislodge a firm from the very market it creates.

A firm must understand who is buying current products and why they are buying, in order to make informed judgments about gaps in the market that a new product platform will create. Then, a firm must fill the holes with derivative products, which might include adding new features to the former model or scaling down versions of new products. The strategy addresses the needs of future customers while providing a migration path for current customers from the older to the newer product.[33]

Intel is a master at filling in the holes it creates by introducing new-platform products.[34] As shown in Figure 8.4, Intel has introduced platforms and derivatives to "gap-fill" both the high and

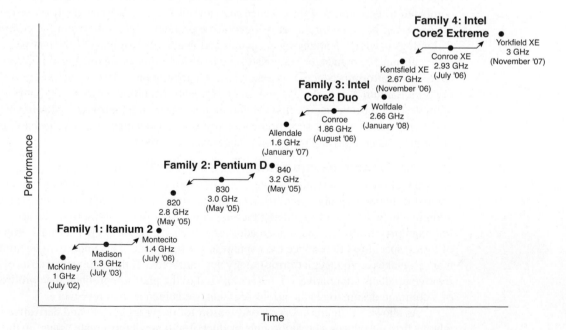

FIGURE 8.4 Product Platforms and Derivatives, in the Case of Intel

Source: www.wikipedia.org and Intel website.

low end of its products. For example, after launching the Core 2 Duo platform with the Conroe version in August 2006, Intel launched the slower Allendale version in January 2007 and the faster Wolfdale version in January 2008.

A company's product strategy encompassing platforms and derivatives could be either integrated or modular. If modular, the company can invite other players in the industry to develop derivatives. The modular approach may require the specification of interoperability design standards, which as noted previously are vital to the success of new technologies (Chapter 1).[35]

An interesting application of modularity, platforms, and derivatives to mobile handsets is offered by an Israeli company, modu© (no mistake here, this is how they spell it). Its platform is based on a single mobile phone module, 1.5 ounce, 2.8 inches long, with the guts of a phone—antenna, cellular radio, contact list, text messaging capabilities, 1 gigabyte of flash memory, and a battery. The derivatives are the modu jackets into which this module fits, and they can be customized to change the looks and functions of the device. The modu jackets are phone enclosures; by changing modu jackets, users can have a daytime phone with a full QWERTY keyboard, a more fashionable phone for the evening, and a still different appearance for niche applications like mobile TV. Service providers could offer a wide variety of modu jackets, increasing both frequency of customer visits to the store and revenue as well. Since the module would be certified to work with a carrier's network, new modu jackets could be brought to market quickly. Other consumer electronics manufacturers could get involved in bringing out modu jackets to serve a variety of needs.[36]

MAKING DECISIONS ABOUT PLATFORMS AND DERIVATIVES   How does a firm know what an appropriate common platform will be, so it can be "versioned" into multiple derivatives inexpensively and lucratively?[37] Rather than designing it to maximize its appeal to a specific segment, the platform should be designed for the high end of the user market and should incorporate as many of the desired features as needed for this segment.[38] Although the high end is not likely to be representative of the market as a whole, the large fixed development costs are more likely to be recovered

Photos reprinted with permission of modu Limited, Israel.

from developing a design with the attributes desired by the highest willingness-to-pay segment. Then, subsequent versions can be sold at much lower prices with only modest incremental cost. It is the *subtraction* of features that is a lower incremental cost activity, rather than the addition of new, higher-end features.

This recommendation is consistent with both the notion of crossing the chasm and the lead user process. In crossing the chasm, products developed for innovators and enthusiasts will typically incorporate more technologically advanced features than those versions desired by conservatives. Similarly, because lead users tend to be more sophisticated users, developing products with input from lead users may result in more technologically advanced products. Regardless, before delving into the development of these subsystems, designers must first engage in a careful study of the users' requirements and incorporate these discoveries into engineering initiatives.[39] Conjoint analysis can be a useful tool in assessing desired features in the platform (Chapter 6).[40]

The determination of how much better each new version should be, the time intervals between versions, and the positioning of versions relative to each other are complex issues that must be considered. Moreover, how to help OEMs and end users manage their migration choices (as covered in Chapter 7) must be part of these decisions.

Customizing Complex Products

In B-to-B relationships, customizing a vendor's offerings is of great value to the customer as it meets the latter's needs better; it is also of great value to the vendor because it earns customer loyalty and profits. Vendor firms vary widely with regard to the amount of customization they undertake. Some vendors, as in the PC industry, offer a menu of options from which a customer may self-select, but there is no proactive design of a customized solution by the vendor. On the other hand, installation of high-end enterprise software may require detailed customized implementation for each separate customer. A key issue is how much customization the vendor should undertake, relative to the customer's own customization efforts.

In deciding how much to get involved in customizing complex products for their customers, B-to-B vendors should consider the underlying technology and knowledge considerations—specifically modularity, technological unpredictability, and customer's knowledge.[41] Vendors can do a better job of customizing complex products when they and their customers have better product knowledge. Yet, in high-tech markets, technological unpredictability can make it hard for customers to be knowledgeable about the technology. Moreover, high-tech companies may find it difficult to have a customer orientation, learning and diffusing information about customers throughout their organizations. A recent study of 304 purchasing contracts from four industry sectors (industrial machinery, electronic equipment, transportation equipment, and instruments) found that in markets characterized by technological unpredictability, the vendor assumes control over customization because customers don't want to invest in knowledge that will soon become obsolete. On the other hand, vendors assume less control over customization when (1) industry products are more modular, because standard interfaces among components make it easier for the customer to undertake customization, and (2) when customers have a higher degree of product knowledge, because then they have the capabilities to undertake customization themselves.[42]

Further, the same study investigated how the vendor learns from customers and disseminates that knowledge about past customer engagements throughout its organization.[43] When vendors have structures and processes to learn from customers and disseminate that knowledge throughout the company, vendors can undertake more customization, even if product modularity is low and there are proprietary interfaces among components. Moreover, when vendors have the ability to learn and disseminate customer knowledge throughout their company, they will assume more customization even when buyers have more product knowledge. So, in order to reap the benefits of customization (improved customer loyalty and profits), vendors must do a good job of diffusing customer knowledge throughout their companies.

A CAUTIONARY NOTE ON ISSUES RELATED TO "KILLING" NEW PRODUCT DEVELOPMENT

The decision about when to pull the plug on a new product that is not doing well in the market, or on a high-tech development project that is not meeting its milestones, is extremely difficult. Managers often remain committed to a losing course of action in the context of new product introductions; such a scenario is often referred to as "good money chasing after bad." Decision makers have strong biases that affect "stopping" decisions. Product champions and technology enthusiasts are perennial optimists about the future viability of their pet projects. Furthermore, because they have a personal stake in these projects, they tend to persevere (sometimes irrationally) at all costs. This escalation of commitment is a major problem. Why does this happen?

Managers who tend to believe they can control uncertainties in their favor continue to escalate commitment to such projects. To justify their decisions, managers try to make data fit their desired expectations about the new product. So, if the decision maker is highly positive about the new product, then he or she will attempt to confirm this hypothesis by seeking out supporting data. For example, managers are much more likely to recall information that is consistent with prior (positive) beliefs, they tend to interpret neutral information as positive, and they will even ignore or distort negative information to support their desire to see their new pet project succeed.[44] Indeed, they may even interpret negative information as positive! Research shows that distorting negative information as positive occurs more than twice as often as distorting positive information as negative.

These biases are one reason that improving the information affecting a decision to withdraw a new product does not result in an actual withdrawal. Indeed, asking managers to set and commit to a stopping rule simply is ineffective. A change in direction will not be made unless managers are aware there is a problem in the first place. Problem recognition requires that managers attend to possible negative feedback. However, in light of distortion bias noted previously, such recognition can be extremely difficult.[45] Moreover, a "quit" decision requires that managers carefully reexamine the previously chosen strategy, both clarifying the magnitude of the problem and possibly redefining it. This is equally difficult because of conflicting information and differences of opinion. Some stakeholders have a vested interest in maintaining the status quo, whereas others exert pressure to change direction.

In light of these complications, managers should attempt to do the following:

1. Managers should search for an alternative course of action, attempting to obtain independent evidence of problems and identifying new courses of action. Creativity is vital in identifying a wide range of options, and the firm should foster a culture that encourages open questioning.[46] The enthusiasm of advocates may be contagious and could lead to collective blind faith, what social psychologists call "groupthink." Therefore, it may be a good idea, from the outset, to have naysayers or skeptics on the review committee. Just as a new product development team needs a product champion, the team also needs an "exit champion," someone who can pull the plug based on hard objective evidence.[47]

2. Managers should prepare key stakeholders for an impending change and attempt to manage impressions. Active attempts to remove the project from the core of the firm will help in this step.[48]

3. Biyalagorsky, Boulding, and Staelin suggest that managers should:[49]
   - Change the organization structure so that a new manager with no prior beliefs about the project makes the continue/stop decision.
   - Use stop rules based on objective data.
   - Put in place policies and procedures to check the biases that affect decisions.

A summary of the possible biases that can lead to escalation of commitment and some prescriptive advice on how to deal with these is presented in Table 8.1.

TABLE 8.1 Possible Biases Leading to Escalation of Commitment

Bias	Manifestation	Advice
Confirmation	Managers seek out information that supports their argument, discount information that does not.	Replace incumbent with new manager; hold managers accountable.
Sunk-cost fallacy	Justify project continuation because there has already been significant investment.	Set up contingencies to pull the plug when budgets are exhausted; use zero-based budgeting.
Anchoring and adjustment	Managers show tendency to adjust estimates insufficiently from an initial value.	Use independent evaluators.

Based on "When to exit a failing venture," *The McKinsey Quarterly,* August 2007.

THE ROLE OF PRODUCT MANAGEMENT IN THE HIGH-TECHNOLOGY COMPANY

Product managers face complicated terrain in high-tech companies. Often, they are squeezed between product development on the one side, and sales on the other. The scope of their responsibilities is often unclear, and their interactions with personnel from other departments can be fraught with complications: differing perspectives, conflicting objectives, and so forth. Although they may be involved in product development activities (providing market research and feedback to developers about desired feature sets, target markets, price targets, competitive intelligence, etc.), product managers typically take over from the product development team during the launch window. They coordinate with marketing communications personnel to manage trade shows, determine messaging for public relations, develop brochures and needed collateral, interface with field sales personnel and reseller organizations, organize technical support and customer service; the responsibilities are certainly daunting. Particularly in small organizations where there may be no other marketing personnel to do all these tasks, product managers may take it on themselves—or coordinate with external service providers to do so. The views of a Sun Microsystems product manager regarding successful product launches are captured in the accompanying Technology Expert's View from the Trenches.

DEVELOPING SERVICES AS PART OF THE HIGH-TECHNOLOGY PRODUCT STRATEGY

The process of developing services presents different challenges from developing physical products. Many firms involved in developing, manufacturing, and distributing high-tech products—like IBM, HP, and Agilent—have turned to services as a way to augment their revenue streams. Other companies exist solely to provide high-tech services; these would include many of the outsource providers of information technology and business services, both in the United States (e.g., Accenture, IBM, and HP) and in emerging markets (e.g., Infosys, TCS, Wipro). Indeed, the range of services related to high-tech products is wide. For example, the Bureau of Labor Statistics (BLS), in its review of employment opportunities in R&D-intensive industries, identifies a category called "R&D-intensive services." This category includes, for example, computer and data processing, engineering and architectural, and research and testing services. Recall from Chapter 1 that the BLS definition of *R&D-intensive* is "industries with 50% higher R&D employment than the average proportion for all industries surveyed." In other words, these service categories have a higher degree of technical workers employed than other service industries.

TECHNOLOGY EXPERT'S VIEW FROM THE TRENCHES

Managing a Successful Product Launch: Ensuring Effective Teamwork among Product Management, Sales, and Marketing

By Tami Syverson
Business Development Manager, Sun Microsystems, Hillsboro, Oregon

For most of my career at Sun Microsystems, I have held various product management, and marketing roles. Each of these assignments has yielded many challenges that technology companies face on an ongoing basis. For example, the product launch is a very challenging time for any technology company. After several years of product development, when a product is ready for customer availability, the steps required to bring the product to market need to be quickly executed.

There are two key success factors that are vital to a successful high-tech product launch: (1) managing an effective cross-functional launch team composed of sales, marketing, customer service personnel, and training; and (2) ensuring field readiness.

Step 1: Assembling and Managing the Cross-Functional Launch Team

Effective product managers assemble the cross-functional team and ensure that all the critical players are on board early. This not only includes coordinating the interface between sales and marketing communications ("marcomm"), but also ensuring that training, service and delivery personnel, and other key players who will be part of the launch activities are involved. One obstacle to an effective launch occurs when the launch team is not in place early. When key members are not engaged, this can significantly delay the product launch.

Sun's matrix organizational structure is composed of representation from the business unit of the new product (typically where product management and marketing are) and sales. Sales reports to one VP while product management and marcomm will report to another VP. The two VPs *must* work collaboratively for this dual-reporting structure to effectively function.

Once these key functions have representation on the launch team and, most importantly, *are engaged early* (i.e., before critical decisions are being made), the odds of a successful product launch are greatly enhanced. In many cases, service and support can be overlooked and often aren't thought of as critical to the launch team. However, because their role often comes so much later (i.e., after the product is ready) and because the focus pre-launch is sometimes still on the product's functionality and features, ensuring that key people are actually trained on the product is an afterthought.

Other key success factors for the cross-functional team are:

- Open communication: Regular weekly meetings are vital to ensure all players know their role and are aware of key activities.
- Use of a "dashboard" to track key events and accountability: Successful product managers effectively lead by holding people accountable for action items and time lines. The dashboard is an effective way to communicate with executive management. For example, color-coding shows where the launch was off schedule (red) and where activities were performed on schedule (green).
- Ensure that product supply is available throughout the key launch phases.

Step 2: Ensuring Field Readiness

Essential to an effective launch is ensuring that the field and distribution partners are ready to sell and deliver the new product. Product managers coordinate with marketing to ensure that the sales team is ready, getting the *right materials* to the *right audience*. In addition, effective product management requires working with sales to coordinate beta tests, which often serve as the source of customer references and testimonials.

The audience is composed of both internal and external personnel. Internal personnel include individuals from sales, service and support, and training. External stakeholders include partners (such

(continued)

as resellers, independent software vendors and other complementors, and service/delivery organizations), the press, analysts, and customers. Often times a funnel-type approach to communication is utilized, in which the launch team decides on the new product launch message internally (see preparation of materials, below), and then that information is handed to the internal audience, that is responsible for communicating with the external audiences.

With respect to the internal audience, salespeople must be informed early about when the new product is coming out and how it is positioned against other products. This communication is vital, particularly if replacement (and the associated obsolescence of their legacy products) is an issue for their customers; salespeople will need sufficient lead time to help their customers manage the transition and must be able to do so in a way that maintains their customers' trust and loyalty. Some companies believe that successful product launches are shrouded in secrecy—if the salespeople are informed too early, they will tell their customers, and the launch itself is undermined. Yet, my experience is that the majority of salespeople are professionals; they must know in advance about the product transition in order to effectively manage their pipeline. Trust is key to ensuring an effective communication strategy.

In addition, the internal service and support team must have sufficient documentation and training so that the help desk is ready. The launch time line must include sufficient time to get training materials together. People need to be trained; they need to know how to implement the product in the context of the customer's business and existing systems, and how to provide break/fix/warranty service. There are occasions where such documentation was given to internal and external audiences only one day before the launch of the product. Certainly, this is not ideal. Unless a very clear plan is in place for these key launch activities, the product manager will be pressed for time and they will not be completed for a successful launch.

Of course, top management must be fully informed, and provide support and involvement as needed.

With respect to the external audience, salespeople typically manage communication with the customers. In addition to communicating with customers about new products, communication with other key partners is also essential. VARs (Value-added Resellers), service delivery partners, ISVs (independent software vendors) and other parties who need to make modifications to their sales strategies and products need lead time to make these needed changes. For example, when a new line of servers are developed, we must ensure that the different operating systems will work on the new hardware. Similarly, new operating systems require compatible application software.

With respect to materials, marketing is typically responsible for generating necessary collateral materials, including needed sales training materials. By definition, neither product management nor marketing is as close to the customer as the sales team. Therefore, sales must be involved in reviewing collateral and training materials.

All too often, salespeople don't use what marketing creates, resulting in a significant waste of company resources. The most frequent criticisms of marketing materials and programs are that they are not relevant or appropriate, that the messaging is "off," and that they contain "too much marketing fluff." The only way to avoid this is for product managers to ensure that both marketing and sales are at the table in preparing materials.

Product managers must also ensure that products are available to give to customers prelaunch. Prioritizing which customers get trial products can be a contentious issue. A variety of metrics can be used to make this prioritization (e.g., customer revenue, growth potential, visibility). Handling these programs and the needed communication can be a job in itself!

With respect to customer references and testimonials, sales is responsible for the relationship with customers; they work with customers to serve as beta sites. The launch time line must allow roughly 90 days once the product is field ready to get it into the customer's hands, get feedback, and fine-tune it. From this effort, customer testimonials and case studies are built to market the product.

Of course, many other issues affect successful product management. For example, managing the actual logistics of the launch—staging events, managing the press and analysts, visiting customers. Moreover, managing decisions regarding end-of-life of products and end-of-service-life (timing of availability of obsolete products, managing the release of information about these decisions), and communicating with sales and marketing about these decisions, can be as important as the launch itself. Product management plays a role in each of these arenas.

Note: The information presented here represents the opinions of the author and not those of Sun Microsystems Inc.

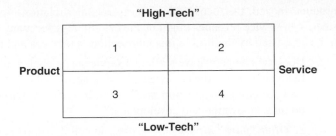

FIGURE 8.5 Intersection of Technology and Service

In conjunction with the increasing importance of services in the economy, a new paradigm in the marketing field, "the services-dominant logic of marketing," also emphasizes the role of services in companies' strategies.[50] Rather than a goods-dominant view focusing on tangible (physical) products, best-practices marketing now emphasizes that customer purchases are motivated less by ownership of a "thing" and more by the provision of value. In other words, rather than being an end in itself, the good or product is the vehicle that creates value. Consistent with the philosophy of competitive advantage (based on resources), and with a market-driven, customer-centric view of the firm, the service-centered dominant logic moves beyond viewing services as mere revenue-augmentation opportunities or pure service industries. Rather, the integration of goods and services emphasizes the importance of intangible resources (such as know-how), relationships with customers, and an extended network of partners. In this sense, marketing becomes a process of interacting with customers and partners, who become coproducers of value.

Figure 8.5 shows the range of ways that technology and services potentially interact, as follows:

1. *Augment high-tech product sales with services revenue.* In cell 1 of the figure, companies that sell a tangible high-tech product such as hardware or consumer electronics can augment their product sales with services revenue in several ways. First, a company might offer consulting services in conjunction with its product sales. This is one of the key reasons that high-tech companies have purchased (or attempted to purchase) some of the major consulting firms. For example, in 2008 HP purchased EDS in order to beef up the technology consulting services it can offer. Of course, like the issues the accounting industry faced in the early 2000s, it is imperative that the consulting arm of a company be viewed as credible and not merely a quasi sales force attempting to find ways to sell more of the company's own products.

Second, the company might offer training, repair, or maintenance contracts to supplement its revenues from product sales. The increasing popularity of a variety of industry trade groups attests to the viability of this strategy. For example, founded in 1989, the Service and Support Professionals Association is the industry trade group for technology service and support professionals; its membership includes nearly 200 member companies such as Microsoft, Dell, Hewlett-Packard, Oracle, Cisco, and IBM, as well as scores of innovative small and mid-sized companies. The goal of technology companies in cell 1 is to engage their development, service, and support organizations—the teams involved in delivering elements of the customer experience that lower the total cost of ownership, accelerate the business benefit of the product, and build valuable new product features that were not part of the original product purchase. A similar organization emphasizing research on services is the Center for Services Leadership at the W. P. Carey School of Business, Arizona State University. Members include a range of companies, both high-tech and otherwise, who share a commitment to customer-focused services and service excellence.

Consulting services and training, repair, or maintenance contracts to supplement product sales revenues can offer a company an ongoing revenue stream after the customer has purchased the product. Indeed, the focus on long-term customer relationships suggests that augmenting product sales with service revenues might be the best strategy for many high-tech companies. Critical issues to be addressed by the firm in this intersection of services and technology are:

- Does the company have sufficient service personnel, and are they appropriately trained, to effectively manage the customer's service needs and questions?
- Can the company develop proficiency in services, without losing its core competencies in product innovation and development?

2. Offer "pure" high-tech services. In cell 2, companies that offer high-tech services might include contract research firms, consulting companies that advise customers on implementation of technology solutions, and service providers that offer outsourced information technology or business process outsourcing services to corporate customers (as in on-demand computing). Again, industry trade associations such as the Technology Professional Services Association (TPSA) exist to offer excellence in this business domain.

Some of these companies offering high-tech services must rely on high-tech products to deliver their services. An example would be a firm's use of sophisticated imaging equipment to provide oil exploration services for its clients. More commonly, the issue faced by these high-tech services companies is how to leverage the customized developments made on behalf of one client to other revenue opportunities. Indeed, creating a platform for standardization of services is key to such leverage—but is also, to some extent, an oxymoron: How can a firm "standardize" customized services for resale to other customers? Yet, in order to manage rapid growth and to capitalize on immediate market opportunities, this is the challenge many high-tech services companies face.

Other issues that must be addressed are:

- Can the technical service personnel communicate in user-friendly ways with customers?
- Are the underlying technologies used to deliver the service reliable in terms of uptime? (Consider that Web-hosting services for e-tailers must ensure 24/7 access to shoppers.)
- Does the company invest continuously in upgrades to maintain cutting-edge technologies?
- Does the company invest adequately in training service personnel to be at the leading edge of practice?

3. Use technology to improve services in traditional products companies. In cell 3, for companies that offer more traditional products, issues related to the intersection of services and technology include the adoption of technologies to improve customer service or make the supply chain more efficient. For example, the use of retail self-scanning devices by consumers to check out their own goods and pay by credit card and the use of the Internet to check order and delivery status help improve customer service and reduce costs. As another example in cell 3, some companies that offer traditional products must rely on technology solutions within their companies in order to stay abreast of cutting-edge business practice. For example, Nike spent $500 million on a management system from SAP to automate its supply chain.

To be successful, technology providers that offer technology solutions to these traditional companies must develop an intimate understanding of these customers' industries. Whether the technology provider is selling an enterprise resource planning solution (such as Oracle or SAP), a sales force automation system (such as Salesforce.com), an automated customer service platform (such as RightNow Technologies), or a technology platform to harness the power of customer input (such as Grupthink.com), these technology companies will succeed only if they can communicate in the language of their customers' business.

4. Use technology to improve traditional services industries. Finally, in cell 4, companies that offer "low-tech" services also must use technology as part of their service-delivery strategy. Many traditional service industries—not R&D-intensive services—are affected by new technological innovations. The movement to self-service technologies (SSTs) has allowed many industries to automate previously labor-intensive operations.[51] The use of automated teller machines (ATMs) and online banking are examples in the banking industry. Another example is the adoption of DNA testing to apprehend and convict criminals, an improvement over traditional gumshoe detective work à la Sherlock Holmes.

Both cells 3 and 4 offer the potential for introducing **self-service technologies (SSTs)**, either to consumers or to employees. A critical issue for companies is how to get consumers or employees to adopt them. Recent research has shown that, in addition to innovation characteristics and adopter characteristics (discussed in Chapter 7), *adopter readiness* variables (such as their motivation and ability) play an important role in customer's willingness to try out these new technologies.[52] Managers of these self-service technologies can stimulate users' ability through training and "handholding" new customers, and they can stimulate users' motivation by emphasizing the benefits such as time or cost savings.

Unique Characteristics of Services: Implications for High-Tech Marketing

Regardless of the specific intersection of services and technology—by their very nature, services have some critical characteristics that make marketing them particularly challenging. Two of these are intangibility and inseparability.

INTANGIBILITY   First, there are some **intangible services**, meaning that they can be neither touched nor examined prior to the customers' purchase decision and consumption. Moreover, the customer of a service does not take home a tangible good. In other words, the intangibility exacerbates the anxieties a customer may experience in a high-tech purchase decision. The marketers' role here is to reduce the perceived risk of the high-tech offering through product demonstrations, free trials, money-back guarantees, extended warranty, training, or technical support.

INSEPARABILITY   Second, the production of a service cannot be separated from the consumption of a service. For example, when a high-tech company is providing training for a customer on its new technology implementation, the customer is simultaneously receiving the service experience. In this sense, the **inseparability of services** means that the company cannot produce/inventory the service in advance of customer demand. Thus there may be peaks and valleys in demand. One way to handle this is to configure service operations to be flexible or scalable. Additional service providers can be hired during peak periods and let go during lean periods. For example, online retailers may add additional customer service representatives to meet the increased workload at call centers during the holiday shopping season. Another way to manage the perishable nature of services is to price such that customers have an incentive to utilize services during low-demand periods. This can be found, for example, among cell phone service providers who provide a fixed number of weekday calling minutes at a monthly subscription fee and unlimited night and weekend minutes.

Inseparability also means that the quality of each service experience might vary considerably, from occasion to occasion and from service employee to service employee. For example, the SAP personnel assigned to manage Nike's supply chain management application might vary in expertise and customer communication skills from those assigned to manage Intel's SAP application of supply chain management software. This potential for inconsistency in the service-delivery experience means that companies that have service as part of their high-tech marketing strategy absolutely *must* train their customer-facing personnel so that there is a level of consistency and reliability, regardless of which employee team is assigned to which customer.

INTELLECTUAL PROPERTY CONSIDERATIONS

High-tech firms face an environment characterized by frequent innovation, high mortality rates, a high priority on R&D, and stiff competition in a race to the marketplace. Concerns about intellectual property and protection of trade secrets are important in many industries, but they are paramount in high-tech industries, in which the basis for competitive advantage is most likely to be superior technological know-how.

First, because know-how often forms the basis of a high-tech company's competitive advantage, protecting it is vital. Such protection can be especially important for companies such as pharmaceutical and biotechnology companies, whose R&D cycle is not only extraordinarily lengthy but also very costly. For example, in order to bring a blockbuster drug based on a new technology to market, a firm must spend $250–350 million.[53] As knowledge and intellectual property become more important than physical capital, businesses feel compelled to protect that intellectual capital by taking extraordinary steps. Indeed, the high-tech industry is littered with the sagas of many companies who have sued or been sued for infringement of intellectual property rights. In 2007, Vonage, the Voice-over-Internet Protocol (VoIP) company, lost patent infringement trials and had to pay AT&T $39 million, Sprint Nextel $80 million, and Verizon $66 million in licensing fees and damages for violating their patents covering Internet phone calls.[54]

Second, high-tech companies engage in strategic alliances with firms that may be potential competitors. In such an environment, the management of sensitive information is particularly critical.[55] One manager at a high-tech firm summarized the issue:

> Product life cycles are very short and development times are very long. That sets up a situation where having knowledge is extremely valuable. If one of our competitors were to acquire information, it would give them an edge. They could respond with their offerings in a stronger fashion than they might have.[56]

Protection of information can be particularly crucial when strategic alliances fail.[57]

Third, the globalization of trade requires some shared understanding of intellectual property rights. Outsourcing trends can exacerbate intellectual property considerations; high-tech companies must share their underlying trade secrets with outsourcing partners in order for them to do their work. Yet, depending on where those partners are located, legal standards and cultural values regarding intellectual property rights can vary.

Fourth, employees are highly mobile in today's information age. Almost anything a person creates, develops, or builds while on the employer's payroll can be considered the employer's property. Unless a person can prove the idea (or, say, the customer list) was developed separately and not as part of the company's work product, it belongs to the company. Because it can include important information on customers, even an employee's Rolodex® may belong as much to the employer as to the person.[58] Yet, when employees leave the company, their knowledge leaves with them; indeed, the *doctrine of inevitable disclosure* recognizes that people who possess sensitive competitive information may, in the course of doing a new job, use information from their former employers. Because this potentially makes the former company vulnerable, courts have been willing to ask a person to sit out of the industry for a time until the information he or she possesses is no longer as sensitive. Moreover, companies can protect against departing employees taking trade information with them by having employees return all company documents and disks before leaving, and by having the employee confirm the obligations of confidentiality during an exit interview.

Thefts of intellectual property are especially common in today's high-tech industries, in which the most prized assets can be stored on a disk or shared over the Internet. The Software and Information Industry Association reports that the software industry loses about $11–12 billion worldwide annually due to software piracy, with about half of that coming from Asia, about $3 billion from western Europe, and $2 billion in the United States.[59]

Information presents something of a double-edged sword to high-technology firms. On the one hand, as noted in previous chapters, high-tech companies want to be skilled in gathering and utilizing information—which includes information about their competitors—to gain advantage. On the other hand, each company wants its information to remain proprietary and its firm's boundaries to be impenetrable to the competitive intelligence-gathering efforts of its competitors. So, how is a company to manage this situation, which requires that it be simultaneously open and restrictive in information sharing? Knowledge of the various strategies to protect intellectual property is vital.

This section of the chapter focuses on issues in intellectual property rights. It begins with an overview of the types of intellectual property protection. It continues with a discussion of the reason for intellectual property rights in the first place. At its core, protection of intellectual property, by granting ownership to an inventor, is intended to give people an incentive to innovate. Yet, such protection must be balanced against society's right to benefit from the innovation. The tension between these two objectives is particularly acute in the areas of business method patents, patents for naturally occurring phenomena (such as gene sequences and "discoveries" in nature), and protection versus sharing of digital media such as music, films, and books (digital rights management). This section of the chapter also addresses how a high-tech company can effectively manage its intellectual property as a strategic asset—addressing issues such as what inventions to patent, in what countries to pursue patent protection, and when and how to form licensing agreements.

Types of Intellectual Property Protection[60]

Intellectual property refers to original works that are essentially creations of the mind. In the business arena, intellectual property may include not only creations from R&D but also general business information that a firm has developed in the course of running its business and that it needs to maintain as proprietary in order to remain competitive. *Intellectual property rights* refer broadly to a collection of rights for many types of information, including inventions, designs, and so forth. Table 8.2 reviews the options available to companies in protecting their intellectual property: patents, copyrights, trademarks, and trade secrets. For example, patentable subject matter includes any "new

---

TABLE 8.2 Types of Protection for Intellectual Property

Patent

Confers owner(s) the *right to exclude* others from making, using, offering for sale, or selling the invention claimed in the patent, or from importing the invention into the country granting patent protection, for a specific time period, usually 20 years from the date of filing an application for patent.

An invention must meet three requirements in order to obtain a patent:

- *Utility:* The invention must be useful; it must function as intended and provide some benefit to society.
- *Novelty:* The invention must be a new idea, one not previously identified.
- *Nonobviousness:* The invention must not be "suggested," either implicitly or explicitly, in prior literature on the subject matter. To assess the nonobviousness requirement, the teachings in multiple literature references may be combined; if the teachings, when collectively considered, show or suggest the invention, then the invention is not patentable.

The patent procurement process is procedurally complex, and Appendix 8.A provides supporting details.

Copyright

Copyrights protect the *form or manner in which the idea is expressed* (not the idea itself) and grant the owner exclusive rights to reproduce and distribute the copyrighted work as follows:

- For individual authors, the term for copyrighted works is the life of the author plus 70 years.
- For works created by employees for their employers—that is, corporate authorship—the copyright term lasts for 95 years from the date of publication or 120 years from the date of creation of the work, whichever occurs first.

Copyrights are used not only for artistic creations (music, literature), but also for products such as software that are mass-marketed (which makes the protection of the information difficult). For these types of products, although the underlying concept or generic capability gives rise to its expression in the product, only the tangible representation of the idea is subject to copyright.

Copyright infringement occurs most typically when someone other than the copyright owner engages in unauthorized reproduction or distribution of the copyrighted work.

(continued)

TABLE 8.2 (Continued)

Trademark

Trademarks protect words, names, symbols, or devices used to identify a specific manufacturer's goods and to distinguish them from others. Registration of a trademark protects a company against unscrupulous competitors who attempt to confuse, deceive, or mislead customers about the true identity of the producer of the goods.

Ownership of a trademark arises automatically under common law when a mark is used in trade. Registering a trademark with the U.S. Patent and Trademark Office (USPTO) can supplement the rights available under common law—for example, a right to assistance from the U.S. Customs Service in preventing importation of infringing goods, and in lawsuits. Even if a trademark has been registered, another company can use the mark on a different, unrelated product, or use a similar mark on similar product as long as consumers are not confused about the true origin of who is producing the product.

Trade Secret[*]

A trade secret is any concrete information that:

- Is useful in the company's business (i.e., provides an economic advantage or has commercial value)
- Is generally unknown (secret)
- Is not easily ascertainable by proper means
- Provides an advantage over competitors who do not know or use that information

Trade secrets are broadly defined and mean all forms and types of financial, business, scientific, technical, economic, or engineering information including patterns, plans, compilations, program devices, formulas, designs, prototypes, methods, techniques, procedures, programs, or codes; whether tangible or intangible, and whether stored physically, electronically, photographically, or in writing.

The requirements for trade secret protection are:

- The company must have a precise description of its trade secrets in order for the courts to recognize them as valid.
- The information must have been developed at some expense by the owner.
- The company must maintain a rigorous program to protect the information that forms the basis of its success, including the use of confidentiality agreements and other evidence that signals the company's efforts to protect its trade secrets.

---

[*] In the United States, trade secret law evolved as common law in each of the 50 states and was standardized in 1979 by the Uniform Trade Secrets Act. This law establishes rules for fair competition among businesses with respect to proprietary information.

and useful process, machine, article of manufacture, composition of matter, or any new and useful improvement thereof."[61] Appendix 8.A provides detail about the patenting process.

Because of the expense and public nature of patents, they are not universally desirable as a form of intellectual property protection. Table 8.3 examines the pros and cons of patents relative to trade secret protection, and recommends when one is more useful than the other.

Rationale for Protection of Intellectual Property

Intellectual property law seeks to balance the right of the inventors to benefit from their work with the right of the general public to benefit from the usefulness of a new discovery. By granting the right to exclude others from making or selling the same product, intellectual property rights generally, and patents in particular, foster creativity by providing an incentive to inventors to undertake the risks of innovation.

Because patents are public information (made public 18 months after filing an application, even if not granted), proponents argue that patents aid the "public good." When invention data are made public—a key requirement for conferring a patent is that sufficient information be provided so

TABLE 8.3 Patents versus Trade Secrets	

Patents

- Patent application information is made public 18 months after filing date.
- Patent's lifetime is 20 years from the date of filing the application.
- Certain types of business information do not qualify for patent protection (i.e., because they are not within the definition of patentable subject matter, are not novel, or are obvious).
- The patent owner must enforce the patent— that is, keep watch that competitors and others are not "infringing" the patent—an expensive undertaking.
- Grant of a patent confers a presumption, but not a guarantee, of validity.

When is patent protection preferred over trade secrets?

- When the product will have a long market lifetime, such as a drug or pharmaceutical composition
- When the product can be reverse engineered, thus permitting a competitor to fabricate or manufacture an identical product
- When it makes sense as a matter of corporate policy (i.e., as an indicator of financial viability, to enhance intangible corporate assets, or for the professional growth of employees by being named on a patent)
- When the patent covers a product or method that, if copied by a competitor, is detectable (i.e., a violation of the patent claims is detectable)

Trade Secrets

- A trade secret can be protected only as long as the company successfully prevents the secret from becoming widely known.
- A person or company who independently discovers or develops information identical to another's trade secret (as long as it wasn't through improper means) has no duty to another company holding the trade secret.

When are trade secrets preferred over patents?

- When the secret is not eligible for patent protection—for example, a customer list or a way of doing business (unless the latter is eligible for a business method patent)
- When the product life cycle is short (e.g., a computer chip with a life of one to two years)
- When the patent would be hard to enforce or would offer only narrow protection
- When the trade secret is not detectable in the product (i.e., a secret component, ingredient, or process of making cannot be discerned via reverse engineering of the end product, the notable example being Coca-Cola)

that others will know exactly how to "perform" the invention—others can build on that public knowledge.

Indeed, some companies find that the public nature of patents is a reason *not* to apply for one. Because patent applications are published 18 months from their filing date (prior to grant of the patent), competitors gain a window of insight into other companies' technology development efforts. One study found that 60% of patented innovations were "invented around" by other firms within four years.[62] Competitors may be able to design around the claimed invention by modifying a minor aspect of the invention and thus avoid infringing the patent.

Critics argue that by granting a 20-year period in which the patent owner has exclusive rights, it actually stifles competition and harms the public good. In May 2007, the U.S. Supreme Court ruled that the U.S. Court of Appeals for the Federal Circuit, the court with jurisdiction for patent cases, had been too generous toward patent holders, allowing them to claim a patent monopoly for mere incremental advances. The court promulgated a higher standard for granting patents, particularly in applying the "nonobvious" criterion.[63] Moreover, high-tech companies complain about **patent trolls,** organizations that license or acquire patents without ever commercializing products based on their discoveries. Patent trolls hope to make money by filing a lawsuit against a wealthy company whose products may be infringing their patents. For example, in March 2006 the patent troll NTP Inc. received $612 million from BlackBerry-maker Research in Motion (RIM) to settle such a patent infringement lawsuit.[64]

Three areas where intellectual property rights have been particularly contentious concern patents for business methods, patents for naturally occurring phenomena such as genes, and restrictions on sharing of digital media content.

BUSINESS METHOD PATENTS   Until the late 1990s, business information and other business know-how were not patentable; **business methods patents** were prohibited. However, on July 23, 1998, the U.S. Court of Appeals for the Federal Circuit held that a business method that uses a mathematical formula can be patented, as long as it meets the three traditional criteria for legal protection (useful, novel, and nonobvious) and yields a "useful, tangible and concrete result."[65] In particular, the Boston-based Signature Financial Group was granted a patent for a unique data-processing software program that cut down on the cost of crunching numbers.[66] In October 2008, the U.S. Court of Appeals for the Federal Circuit announced that for a business method to be patentable, the method must do more than produce a "useful, tangible and concrete result."[67] The court ruled that for a business method to be patentable it must be "tied to a particular machine" or must "physically transform an article into something different. This ruling effectively narrowed the types of business methods that qualify for patent protection.

The initial decision to patent business methods reflected an important shift in legal thinking and was quite controversial. Critics contend that those few who hold a patent to a business method will slow the spread of valuable commercial innovations, to the detriment of society. Although processes are patentable, a process of purely mental steps is not patentable, and any business method patent claims must include more than mental steps, but involve a machine or apparatus or transform an article (e.g., data) into a different state or thing. It is the fine line between a process of concrete steps and a process of mental steps that has allowed such business methods to be patented.

With the change in legal precedent that allowed business methods to be patentable, the U.S. Patent and Trademark Office was flooded with applications for business method patents on e-commerce methods with the hope that such patents could lock in competitive advantage for a few firms. During this rush, many e-commerce patents were granted, including that of CyberGold for a system of giving rewards to people who click on an advertising message. Another was Priceline's patent for being the first to innovate a buyer-driven e-commerce system. It was granted a broad patent on a method of auctioning goods and services on the Internet.[68] Amazon.com was granted a patent for its one-click shopping cart, and Barnes & Noble was excluded from using a one-click shopping cart on its website.

Critics say that because these methods are so common, they should not be eligible for patents. Patented processes are required to be novel, and many believe that the Internet patents are not novel. These critics argue that merely transferring a marketing technique to the Web does not necessarily constitute novelty.[69] Patents are also supposed to be nonobvious, but with so many people using similar strategies on the Internet, these critics also say they cannot be considered nonobvious. While patents are designed to give an incentive to innovate (by having a protected period to recoup R&D investments), Internet ventures often do not have the same start-up costs as, say, a new drug or a manufacturing process.

Moreover, although patents have historically been granted to protect the common good by providing an incentive to innovate, some believe that patents granted for broadly defined Internet business models do not protect the common good. Rather, they create protected profits for the few, to the detriment of many.[70] By inhibiting the diffusion of these useful innovations, patents in this area have potentially created inefficiencies in an area that had so much potential for economic efficiencies. The biggest issue in deciding whether to patent business methods on the Internet may be the financial value of such protection. With technology changing so quickly, some patents may be obsolete by the time they are awarded (usually two to three years after application).

The World Trade Organization's Agreement on Trade-Related Aspects of Intellectual Property Rights (TRIPS) does not address business method patents. Granting of business method patents in the European Union is controversial. Even in the United States, people like Amazon CEO Jeff Bezos have called for reform of the current law pertaining to business method patents.[71]

NATURALLY OCCURRING PHENOMENA   Patents have been granted for "naturally occurring" technological breakthroughs such as genes, prime numbers, and even lab animals (a Harvard mouse). In addition, some companies patent "new" therapeutic applications or forms of delivery of traditional medicinal plants or compounds. For example, Bristol-Myers Squibb patented its cancer treatment drug Taxol, which is derived from the Pacific yew tree.

Critics argue that such patents should not be granted in the first place. Using the analogy of Microsoft filing a patent on a series of 1's and 0's, the question is: How novel and nonobvious can these phenomena really be? However, although the patent process allows the grant of a patent for the unique application of these discoveries, it does raise profound ethical and social questions. For example, will patents on these phenomena stymie possible knowledge spillovers that could build on these developments in profound, but unknown, ways? Although patents are publicly available documents, will advances that could improve consumer welfare be restricted? For example, Monsanto, which patents genetically modified foods, faces an antitrust lawsuit from five U.S. farmers and one French farmer accusing the company of conspiring to control markets for corn and soybean seeds.

Proponents of these patents warn that if patent holders can't fully control their property, it might chill innovation. Moreover, they argue that more open information about intellectual property will actually enable inventors to experiment and innovate even better inventions. This issue is likely to heat up in the coming years.

DIGITAL RIGHTS MANAGEMENT (DRM)   Another area where friction between the rights of the owner and the "public good" is especially acute is in the area of digital information/media (music, films, books, etc.). Copyright laws grant copyright holders (often, these are the entertainment companies rather than the artists themselves) a monopoly over the right to reproduce and distribute their works during the term of the copyright. However, technology developments make it increasingly easy for consumers to access digital content (music, books, movies, TV shows) from a variety of devices—creating challenges for "content owners" to retain control over their copyrighted works.

This tension is seen not only in the ways in which entertainment companies try to restrict consumer access to digital media (through antipiracy restrictions and **digital rights management (DRM)**, but also in the very public acrimony between the technology and entertainment industries. Entertainment companies criticize technology companies for dragging their feet in establishing technical standards to prevent digital piracy. In a very strong lobbying effort by content providers (the media), lawmakers initially proposed that computer and consumer electronics manufacturers embed copyright protection technology in all of their products.

Take the case of the music industry. In 2006, the number of CDs sold worldwide fell 10%—the largest one-year drop ever, steeper than any of the post-Napster years from 2001 to 2004.[72] In 2007, sales of music on CD and other physical formats fell by another whopping 19% in the U.S. (more than anyone expected); 6% in Britain; 9% in Japan, France, and Spain; 12% in Italy; 14% in Australia; and 21% in Canada (sales were flat in Germany).[73] In the meantime, paid digital downloads grew rapidly, soaring by 40% in 2007. Tracks downloaded from the Internet generated revenues of $2.9 billion for the music companies, compared with $2.1 billion the year before.[74] However, legal downloads did not begin to make up for the loss of revenue from CDs. More worryingly for the industry, the growth of digital downloads appears to be slowing. Music companies blame piracy, and are taking great pains to make it tough for consumers to engage in illegal file sharing. Indeed, a mother of two who makes $36,000 a year was fined $222,000 in damages for having made copyrighted songs available on a file-sharing network.[75]

Table 8.4 provides an overview of the major events in the digital rights management controversy. At its heart, the questions are: How should the needs of the various stakeholders be addressed?

TABLE 8.4 History of U.S. Copyright Infringement Laws

Year	Event
1984	In the Betamax case (*Sony Corp. of America v. Universal City Studios*), Sony was *not* held to be liable for facilitating copyright infringement, even though it knew that some consumers would use its VCRs illegally to make copies of copyrighted materials. The critical issue was that because the VCR's primary uses were "non-infringing," the courts should not stifle potentially beneficial technologies before their usefulness might be fully understood.
1992	The Audio Home Recording Act granted consumers the right to unlimited private use of legally purchased music, videos, books, and other media content.
1998	The Digital Millennium Copyright Act made it a crime to circumvent copy protection. However, service providers have protection from liablity if their users post unauthorized material provided they have a mechanism to remove it immediately when a complaint is made.
1999	The Recording Industry Association of America (RIAA) successfully sued Napster for contributing to copyright infringement. After a lengthy trial and appeal process, Napster was shut down in July 2001. The courts held that the 1984 Sony ruling didn't apply to Napster. The VCR was a product that, once sold, Sony had nothing to do with any longer; as a result, Sony could not track what its users did. However, Napster had an ongoing relationship with its users and could track what users were trading.
2001	After Napster, companies like Morpheus, Grokster, Kazaa, Gnutella, LimeWire, and BearShare developed second generation peer-to-peer file-sharing software that didn't need a central server. They were providers of software tools that enabled users to search one another's PC hard drives and copy digital files, with no ability to know what people searched for and downloaded. Despite being sued in October 2001, Morpheus and Grokster were protected from liability by the 1984 Sony ruling.
2003	Unable to prosecute software developers, between September and December 2003 RIAA sued nearly 400 individuals for copyright violations. Initial evidence suggested that the lawsuits had the desired effect. Prior to the lawsuit, 29% of Internet users downloaded songs to their computers; in the months following RIAA's legal strategy, 14% of Internet users reported downloading songs to their computers.
2005	Sony introduces new DRM technology on its CDs, which installed DRM software directly onto a user's computer without notifying them and which created a security vulnerability. Sony was forced to recall millions of CDs and settled class action lawsuits with cash or album downloads free of DRM.
2006	Citing online theft of copyrighted works on campus computer networks as an enormous problem for the music community, the Recording Industry Association of America (RIAA) sent letters to 700 colleges across the country stating that students who ignore warnings and continue to engage in illegal downloading of music will be sued.
2007	The RIAA continues its efforts to identify and punish college students who download movies and music illegally, with penalties as high as $150,000 per work infringed. Students have 20 days to settle with the RIAA after being served, typically paying a $3,000 settlement.
	Radiohead pioneers a new business model, releasing its new album, *In Rainbows,* with a "pick-your-own-price" download model.
	More and more online stores selling music free of DRM.

Sources: "Digital Music: High Volume," *The Economist,* January 30, 2008, online at www.economist.com/displaystory. cfm?story_id=10598460; "From Major to Minor," *The Economist,* January 10, 2008, online at www.economist.com/ business/displaystory.cfm?story_id=10498664; "The Slow Death of Digital Rights," *The Economist,* October 13, 2007, pp.75; Byrne, David, "The Fall and Rise of Music," *Wired,* January 2008, pp. 124–129; and Mnookin, Seth, "The Angry Mogul," *Wired,* December 2007, pp. 202–213.

What are the rights and responsibilities of customers, artists, technology companies, and the media? Best-practices high-tech marketing argues for novel approaches to business models that are responsive to customer needs and technology trends.

Some believe that the entertainment industry's attempts to restrict information sharing, rather than to develop new business models, exhibit a desire to control intellectual property in ways that are incompatible with an information-based economy.[76] It is instructive to note that studios, fearful of new technology, initially fought the advent of the VCR, nearly asphyxiating the home video market that today provides up to 50% of their revenues from sales of videotapes and DVDs. Some believe the moral to the story is that, if the entertainment industry were to embrace the new file-sharing technology, it could end up making more money than ever before.[77] In other words, maybe it's time to accept technology and deal with the digital age.

Rather than searching for ways to defend their current business models, entertainment companies must reframe the issue as how to maximize revenues in the digital age and still protect the value of their intellectual property. As many say, "Content yearns to be free." Paradoxically, one way to maximize revenue may be to abandon efforts to protect digital rights at all costs. Rather, entertainment companies must begin to think nontraditionally about how to distribute content in order to increase their options for creating value. For example, instead of simply selling a book, CD, or movie via the Web, entertainment companies could focus on selling a variety of complementary products and services, such as merchandise or tickets to special events with artists and authors. With such a strategy, music companies could deliver many forms of additional services that their customers would be willing to pay for—and have a direct connection to the consumer.[78] Universal Studios has taken a step in this direction by allowing some songs to be DRM-free, through Amazon, Best Buy, Wal-Mart, and other online retailers; the files can be duplicated at will, but they do contain a watermark that allows Universal to see if they end up on peer-to-peer networks. Universal has also pioneered a Total Music subscription whereby, for a monthly fee attached to a cell phone, cable, or Internet bill, consumers have unlimited access to music from a particular label, as well as new recordings a week before their general release.[79] Universal's hope is to break Apple's presence in the industry and allow other music players and services a viable option. But, this service will require some form of DRM that doesn't allow what customers want: music to be played across any platform at any time. Until the music industry gets this, they will encourage customers to turn to illegal uses of technology to do it. Consistent with the notion of creative destruction, established incumbents at the top of the industry likely have to reinvent themselves—with some degree of pain—or face being "Amazoned."[80]

Missing from the intense rhetoric is the voice of the consumer. Is it possible to reframe this debate with a win-win-win lens for technology companies, entertainment companies, and consumers? Although strong protections for intellectual property are essential for promoting continued innovativeness, such protections also shouldn't stifle competition and access. A Consumer Technology Bill of Rights has been developed by DigitalConsumer.org, a consumer advocacy group formed by Silicon Valley businesspeople who oppose the erosion of consumer rights and of technological innovation.[81] The bill of rights proposes that once consumers have legally purchased digital content, they have the right to "time-shift" (record for later playback), "space-shift" (copy to blank CDs or portable players), make backup copies, use content on any platform (PC, MP3 player, etc.), and translate content into different formats. These ideas are also supported by legal scholars, such as Lawrence Lessig (see http://CreativeCommons.org).[82]

Managing Intellectual Property[83]

In today's successful high-tech companies, the management of intellectual property has become a core competence. Because intellectual assets rather than physical assets are the principal source of competitive advantage, unlocking the hidden power of these assets is often a key to success. One study reported that 67% of U.S. companies failed to exploit technology assets, and these companies let more than 35% of their patented technologies go to waste because they had no immediate use in products. Yet, active management of intellectual property assets is vital because:

- Patents can be tapped as a revenue source (e.g., via licensing).
- Costs can be reduced by cutting maintenance fees on unneeded patents (that could be donated to universities or nonprofits for a tax write-off).
- Patents can be repackaged to attract new capital and communicate an asset picture in a more attractive way to investors.

In addition to helping companies in the market by protecting their core technologies and business methods, patents can help a firm manage its product line. The potential strength of patents can help a firm establish R&D priorities. For example, Hitachi tries to develop only those products for which patents can help it establish market dominance. Similarly, in the biotech arena, Genetics Institutes says the strength of the potential patent position is a leading factor in deciding which research to pursue. Moreover, a patent strategy can help companies respond to shifts in the marketplace in an effective manner, by acquiring or partnering with firms that own patent rights to important developments.

Each of these issues points to the reality that intellectual property rights must be considered strategically in the product management process and not relegated solely to the realm of corporate and patent attorneys.

For any given company, its employee-scientists find numerous potentially patentable inventions each year. Because of the expense involved in procuring patent protection, careful thought must be given to discern which of those inventions best fit with the business strategy. Knowing with certainty what will be the best fit, however, is a bit like predicting the future. It is entirely possible that a seemingly unimportant invention becomes a cornerstone product for a company.

Further complicating the decision is the fact that in today's highly competitive, fast-paced world, filing for patent protection before a competitor is crucial. This means filing an application as soon as possible—and thus the luxury of doing a few more experiments to resolve one or two questions is often not possible. Decisions on whether to file for patent protection are often made prior to a full understanding of the commercial viability of the invention.

The core patent position of the company can effectively exclude other companies from producing copycat products.

Summary

This chapter has covered a wide range of topics that must be understood in order to effectively develop and manage high-tech products and services. Organized around the framework of steps in the technology/product management process, the chapter has addressed ways to manage a firm's products to maximize success. The first step is to take an inventory of a firm's know-how (product, process, management) and figure out how to use it to create and deliver value to customers. If there are gaps in technology identified in step 1, the next step is to figure out how best to add these technologies, whether by making them in-house, buying them on the market, or partnering to co-develop these. The next step is to decide on the firm's revenue-generating product offerings, whether simply to license its technology or go the whole way into commercializing end user products. The final step is the ongoing management of the new product development process, including decisions on the product architecture (modularity, platforms, and derivatives), the criteria for discontinuing projects, a services strategy to complement product offerings, and management of intellectual property issues.

Discussion Questions

Chapter Concepts

1. What is digital convergence? What are the opportunities and threats it poses for companies?
2. What is a technology map? How is it beneficial to a high-tech company?
3. What four steps does managing technology resources involve? Name and explain each step.
4. What are the possible sources of technology that a company should inventory in the technology identification step?

5. What is the "make-versus-buy" decision? What considerations influence the decision?
7. What is the NIH syndrome? How does it relate to myopia?
8. Describe the what-to-sell continuum. What is the underlying dimension of the what-to-sell continuum?
9. What factors affect a firm's decision about what to sell? What options are available to a firm?

10. How does the issue of international markets relate to the what-to-sell decision?
11. What is technology transfer? What are its benefits and barriers?
12. What is modularity? What are the pros and cons of following a modular approach to product design?
13. Is the mobile handset industry going to evolve along a path similar to the PC industry? Which companies are likely to make money in the future?
14. What is a product platform? What are the advantages to developers? To users?
15. When does a platform and derivatives strategy make sense? What product should be the company's first version under this strategy?
16. How does customization add value to a complex product? When is it beneficial to the vendor to customize?
17. What issues does a high-tech firm face in the decision to stop or kill a particular project? In light of these issues, how should a company manage the decision to end a project?
18. What is the range in which services and technology interact? Describe each dimension and the four resulting cells depicted in Figure 8.5.

19. What are the characteristics of services? How do these characteristics make the marketing of services difficult?
20. What factors affect the critical importance of intellectual property considerations in high-tech markets?
21. What are the three criteria for a patentable innovation? What are the steps in the patenting process?
22. What is a trade secret? What are the requirements for trade secret protection?
23. What are the pros and cons of using patents versus trade secrets to protect intellectual property?
24. What is the controversy over business method patents?
25. What is the controversy over patenting "naturally occurring phenomena" such as genes or traditional plant compounds?
26. What is the controversy over digital rights management? Be sure to address the issue from the perspectives of all affected stakeholders: the artists, the studios, the retailers, and consumers. What does best-practices high-tech marketing say is the best solution?
27. What are your thoughts on the use of patents in highly innovative markets? Do they encourage or stifle innovation?

Discussion Questions from the Appendix

28. Overview the process used to apply for a patent. What happens when a patent examiner's report is unfavorable?
29. What is the difference between patentability and freedom to operate?
30. What are the two types of patent applications in the United States?

31. How does a company file for international patent protection?
32. What are the typical fees associated with filing for patent protection?

Application Questions

1. Answer the following three questions based on the opening vignette in this chapter.
   a. What key factors led to Apple's success with iPod?
   b. From a technology management perspective, when does it makes sense for a company to build a complete, end-to-end solution on its own? When is it better to focus on various elements/components of the solution and let partners and complementors build on those components?
   c. Which of the two business models in digital music downloads—"walled garden" versus "open plain"—do you think will be successful in the long run? Why?

2. Answer the next two questions based on the Technology Expert's View from the Trenches feature, "Overcoming the Challenges in Commercializing Disruptive Technology."
   a. Which customer companies do you think Memjet should license its technology to for making printers? Why?
   b. Which end user market should Memjet's customers target these printers to and why?

3. Answer the next two questions based on the Technology Expert's View from the Trenches feature, "Managing a Successful Product Launch: Ensuring Effective Teamwork among Product Management, Sales, and Marketing."

   a. Would you like to apply for the position of product manager in a high-tech company like Sun? Why or why not?

   b. Tie Tami Syverson's essential success factors to concepts covered previously in this book.

4. Select and research a technology company of your choice. Assess this company's product strategy in terms of its technology map, including platforms and derivatives. Do you see any opportunities here for new product development?

5. Take the organization where you work for or study. Which cell of Figure 8.5 would you place it in with regard to the intersection of technology and service? What are the roles of technology and service in the product offering? Is technology being used for external or internal purposes? What are the issues the organization faces for increasing supply chain efficiency or enhancing customer value?

Glossary

business method patent   A patent granted for a way of doing business (versus a specific tangible invention), such as Amazon's one-click shopping cart; very controversial and being reexamined.

derivative   A spin-off product from a common underlying technology platform that includes either fewer or additional features to appeal to different market segments.

digital convergence   The technology trend of converting previously analog products (both consumer durables as well as media and entertainment content) into digital products.

digital rights management (DRM)   Attempts by companies to restrict the unauthorized sharing of copyrighted material such as music, movies, books, and TV shows.

inseparability of services   The fact that production of a service (e.g., the service provider trains a customer on its technology) cannot be separated from consumption of that service (e.g., the customer simultaneously experiences the service)—meaning that the company cannot produce/inventory the service in advance of customer demand.

intangible services   Services that can be neither touched nor examined prior to purchase and consumption.

intellectual property   Original works that are essentially creations of the mind—including inventions, designs, and so forth.

modularity   An approach to design that refers to building a complex product from smaller subsystems that are created independently yet function together as a whole; based on interoperability (plug-and-play compatibility).

patent   A form of protection for intellectual property; confers to the owner(s) the right to exclude others from making, using, offering for sale, or selling the product or process described in the patent for a specific time period, usually 20 years from the date of filing an application for patent with the U.S. Patent and Trademark Office (USPTO).

patent trolls   Organizations that buy and hold patents without ever commercializing products based on their discoveries, hoping to make money by filing lawsuits against wealthy companies whose products may be infringing their patents.

product platform   A common architecture based on a single design and underlying technology from which a stream of derivative products can be efficiently developed and produced.

self-service technologies (SSTs)   Technologies that facilitate customers' ability to perform service functions previously provided by a company—for example, checking delivery status on a website (versus calling a company to inquire) or checking out groceries with an automated scanner (versus relying on a human checker).

technology map   A blueprint that defines a company's stream of new products, both breakthroughs and derivatives, that it is committed to developing over some future time period.

technology transfer   The transfer of technological know-how/ideas from a small company or inventor to a larger company that has resources and expertise in the commercialization process; tech transfer agreements specify profit sharing among the company that commercializes the invention, the inventor, and the inventor's institution.

Notes

1. "Who's the Winner in the Tug-of-War between 'Walled Garden' and 'Open Plain' Strategies?" Knowledge@Wharton, September 5, 2007, online at http://knowledge.wharton.upenn.edu/article.cfm?articleid=1804; Mnookin, Seth, "The Angry Mogul," *Wired,* December 2007, pp. 202–213; Tam, Pui-Wang, Bruce Orwall, and Anna Wilde Matthews, "With Apple Stalling, Steve Jobs Looks to Digital Entertainment," *Wall Street Journal,* April 25, 2003, p. 1; and Delaney, Kevin, "Yahoo's Big Play in Music," *Wall Street Journal,* May 11, 2005, p. D5.

2. Yoffie, David, *Competing in the Age of Digital Convergence* (Boston: Harvard Business School Press, 1997); and Sengupta, Sanjit, Jakki Mohr, and Stanley Slater, "Strategic Opportunities at the Intersection of Globalization, Technology, and Lifestyles," in *Handbook of Business Strategy,* ed. P. Coate (Bradford, UK: Emerald Group Publishing, 2006), pp. 43–50.

3. Tabrizi, Behnam, and Rick Walleigh, "Defining Next-Generation Products: An Inside Look," *Harvard Business Review* 75 (November–December 1997), pp. 116–124.

4. Abate, Tom, "Intel Says New Chips Will Arrive This Year," *San Francisco Chronicle,* September 27, 2006, online at www.sfgate.com/cgi-bin/article.cgi?f=/c/a/2006/09/27/BUGGLLDA6J1.DTL&hw=Intel+technology+map&sn=002&sc=517.

5. Capon, Noel, and Rashi Glazer, "Marketing and Technology: A Strategic Coalignment," *Journal of Marketing* 51 (July 1987), pp. 1–14.

6. Schrage, Michael, "My Customer, My Co-Innovator," *Strategy+Business,* August 31, 2006, online at www.strategy-business.com/press/enewsarticle/enews083106.

7. Mangalindan, Mylene, "Who's Selling What on Amazon," *Wall Street Journal,* April 28, 2005, p. D1.

8. Atuahene-Gima, Kwaku, "Resolving the Capability–Rigidity Paradox in New Product Innovation," *Journal of Marketing* 69 (October 2005), pp. 61–83.

9. Clark, Don, "Intel Shifts from Silicon to Lift Chip Performance," *Wall Street Journal,* November 12, 2007, p. B7.

10. Prabhu, Jaideep C., Rajesh K. Chandy, and Mark E. Ellis, "The Impact of Acquisitions on Innovation: Poison Pill, Placebo, or Tonic?" *Journal of Marketing Research* 69 (January 2005), pp. 114–130.

11. Sorescu, Alina B., Rajesh K. Chandy, and Jaideep C. Prabhu, "Why Some Acquisitions Do Better Than Others: Product Capital as a Driver of Long-Term Stock Returns," *Journal of Marketing Research* 44 (February 2007), pp. 57–72.

12. Capon and Glazer, "Marketing and Technology."

13. John, George, Allen Weiss, and Shantanu Dutta, "Marketing in Technology Intensive Markets: Towards a Conceptual Framework," *Journal of Marketing* 63 (Special Issue 1999), pp. 78–91.

14. Capon and Glazer, "Marketing and Technology."

15. John, Weiss, and Dutta, "Marketing in Technology Intensive Markets."

16. Capon and Glazer, "Marketing and Technology."

17. Wernerfelt, Birger, "Product Development Resources and the Scope of the Firm," *Journal of Marketing* 69 (April 2005), pp. 15–23.

18. Thurow, Lester, "Needed: A New System of Intellectual Property Rights," *Harvard Business Review* 75 (September–October 1997), pp. 95–103.

19. Capon and Glazer, "Marketing and Technology."

20. Ford, David, and Chris Ryan, "Taking Technology to Market," *Harvard Business Review* 59 (March–April 1981), pp. 117–126.

21. Bigelow, Bruce V. "Patent Payoff," *The San Diego Union-Tribune,* May 14, 2006, www.signonsandiego.com/articlelink/fallbrook2/fallbrook2.html.

22. Hall, Kenji, and Peter Burrows, "Why Toshiba Is Clamming Up," *BusinessWeek,* December 19, 2005, online at www.businessweek.com/magazine/content/05_51/b3964044.htm.

23. Ford and Ryan, "Taking Technology to Market."

24. Our thanks to Jon A. (Tony) Rudbach, director of technology transfer, University of Montana, Missoula, Montana, for his assistance in the first edition with the material in this section. For additional information, see Steele, Thomas, W. Lee Schwendig, and George Johnson, "The Technology Innovation Act of 1980, Ancillary Legislation, Public Policy, and Marketing: The Interfaces," *Journal of Public Policy and Marketing* 9 (1990), pp. 167–182.

25. AUTM U.S. Licensing Activity Survey, FY 2006, Association of University Technology Managers, 2007, available at www.autm.net/surveys/dsp.Detail.cfm?pid=215.

26. Baldwin, Carliss Y., and Kim B. Clark, "Managing in an Age of Modularity," *Harvard Business Review* 75 (September–October 1997), pp. 84–93.

27. Fleming, Lee, and Olav Sorensen, "The Dangers of Modularity," *Harvard Business Review* 79 (September 2001), pp. 20–21.

28. Meyer, Marc, and Robert Seliger, "Product Platforms in Software Development," *Sloan Management Review* 40 (Fall 1998), pp. 61–75.

29. Vascellaro, Jessica E., and Amol Sharma, "Google's Android Has Phone Debut via T-Mobile, " *Wall Street Journal,* September 24, 2008, p. B3.

30. Tabrizi and Walleigh, "Defining Next-Generation Products."

31. John, Weiss, and Dutta, "Marketing in Technology Intensive Markets."

32. Tabrizi and Walleigh, "Defining Next-Generation Products."

33. Meyer and Seliger, "Product Platforms in Software Development."

34. Tabrizi and Walleigh, "Defining Next-Generation Products."

35. Ford and Ryan, "Taking Technology to Market."

36. Kim, Ryan, "Modular Unit Could Revolutionize Cell Phones," *San Francisco Chronicle,* February 7, 2008, online at www.sfgate.com/cgi-bin/article.cgi?f=/c/a/2008/02/07/BUNRUTB3L.DTL&hw=modu&sn=001&sc=1000; and personal correspondence with Matt Stewart (San Francisco) and Liton Ali (London) of Solid Ground, modu's marketing communications and branding agency, July 15, 2008.

37. Shapiro, Carl, and Hal Varian, "Versioning: The Smart Way to Sell Information," *Harvard Business Review* 76 (November–December 1998), pp. 106–114.

38. John, Weiss, and Dutta, "Marketing in Technology Intensive Markets."

39. Meyer and Seliger, "Product Platforms in Software Development."

40. Moore, William L., Jordan J. Louviere, and Rohit Verma, "Using Conjoint Analysis to Help Design Product Platforms," *Journal of Product Innovation Management* 16 (January 1999), pp. 27–39.

41. Ghosh, Mrinal, Shantanu Dutta, and Stefan Stremersch, "Customizing Complex Products: When Should the Vendor Take Control?" *Journal of Marketing Research* 43 (November 2006), pp. 664–679.

42. Ibid.

43. Ibid.

44. Boulding, William, Ruskin Morgan, and Richard Staelin, "Pulling the Plug to Stop the New Product Drain," *Journal of Marketing Research* 34 (February 1997), pp. 164–176.

45. Keil, Mark, and Ramiro Montealegre, "Cutting Your Losses: Extricating Your Organization When a Big Project Goes Awry," *Sloan Management Review* 42 (Spring 2000), pp. 55–68.

46. Ibid.

47. Rover, Isabelle, "Why Bad Projects Are So Hard to Kill," *Harvard Business Review* 81 (February 2003), pp. 48–56.

48. Keil and Montealegre, "Cutting Your Losses."

49. Biyalagorsky, Eyal, William Boulding, and Richard Staelin, "Stuck in the Past: Why Managers Persist with New Product Failures," *Journal of Marketing* 70 (April 2006), pp. 108–121.

50. Vargo, Stephen, and Robert Lusch, "Evolving to a New Dominant Logic for Marketing," *Journal of Marketing* 68 (January 2004), pp. 1–17.

51. See, for example, Meuter, Matthew L., Amy L. Ostrom, Robert I. Roundtree, and Mary Jo Bitner, "Self-Service Technologies: Understanding Customer Satisfaction with Technology-Based Service Encounters," *Journal of Marketing* 64 (July 2000), p. 50; and Bitner, Mary Jo, Amy L. Ostrom, and Matthew L. Meuter, "Implementing Successful Self-Service Technologies," *Academy of Management Executive* 16 (November 2002), pp. 96–109.

52. Meuter, Matthew L., Mary Jo Bitner, Amy L. Ostrom, and Stephen W. Brown, "Choosing among Alternative Service Delivery Modes: An Investigation of Customer Trial of Self-Service Technologies," *Journal of Marketing* 69 (April 2005), pp. 61–83.

53. Van Arnum, Patricia, "Drug Makers Look to New Strategies in Portfolio Management," *Chemical Market Reporter* 254 (November 1998), pp. 14–15.

54. Kardos, Donna, "Vonage, Nortel Settle Patents Dispute," *Wall Street Journal,* January 2, 2008, p. B3.

55. Mohr, Jakki, "The Management and Control of Information in High-Technology Firms," *Journal of High-Technology Management Research* 7 (Fall 1996), pp. 245–268.

56. Ibid.

57. MacDonald, Elizabeth, and Joann Lublin, "In the Debris of a Failed Merger: Trade Secrets," *Wall Street Journal,* March 10, 1998, p. B1.

58. The information in this section is drawn from Lenzner, Robert, and Carrie Shook, "Whose Rolodex Is It, Anyway?" *Forbes,* February 23, 1998, pp. 100–104.

59. "What Is Software Piracy: The Piracy Problem," Software and Information Industry Association, 2008, online at www.siia.net/piracy/whatis.asp.

60. Our sincere thanks to the efforts and insights of Judy Mohr, J.D., Ph.D., King & Spalding LLP, Redwood Shores, California, for her technical advice on the intellectual property sections and Appendix 8.A in this chapter.

61. Prior to June 8, 1995 (the effective date of the GATT-TRIPS legislation), the term of a U.S. patent was 17 years from the date the patent issued. The U.S. participation in the Uruguay Round Agreements included an agreement on Trade-Related Aspects of Intellectual Property (TRIPS) that harmonized the U.S. patent term with the rest of the world, by changing the term from 17 years from date of issuance to 20 years from the filing date. As a result, the present patent term for patent applications filed before June 8, 1995, is the greater of (1) 17 years from the date of issuance or (2) 20 years measured from the filing date of the earliest referenced application. For applications filed on or after June 8, 1995, the patent term is 20 years measured from the earliest claimed application filing date. *See also* 35 U.S.C. § 101.

62. Mansfield, E., M. Schwartz, and S. Wagner, "Imitation Costs and Patents: An Empirical Study," *Economic Journal* 91 (December 1981), pp. 907–918.

63. Bravin, Jess, "Patent Holders' Power Is Curtailed," *Wall Street Journal,* May 1, 2007, pp. A3.

64. Ibid.

65. *State Street Bank & Trust Co. v. Signature Financial Group, Inc.,* 149 F.3d 1368, (CAFC1998).

66. Updike, Edith, "What's Next—A Patent for the 401(k)?" *BusinessWeek,* October 26, 1998, pp. 104–106.

67. In re: Bernard L. Bilski and Rand A. Warsaw, case number 2007-1130, in the U.S. Court of Appeals for the Federal Circuit, October 30, 2008.

68. Ibid.

69. France, Mike, "A Net Monopoly No Longer?" *Business Week,* September 27, 1999, p. 47.

70. Gurley, J. William, "The Trouble with Internet Patents," *Fortune,* July 19, 1999, pp. 118–119.

71. Fisher, William, and Geri Zollinger, "Business Methods Patent Online," The Berkman Center for Internet and Society at Harvard Law School, June 22, 2001, online at http://cyber.law.harvard.edu/ilaw/BMP.

72. Mnookin, Seth, "The Angry Mogul," *Wired,* December 2007, pp. 202–213.

73. "Digital Music: High Volume," *The Economist,* January 30, 2008, online at www.economist.com/displaystory.cfm?story_id=10598460.

74. "From Major to Minor," *The Economist,* January 10, 2008, online at www.economist.com/business/displaystory.cfm?story_id=10498664.

75. "The Slow Death of Digital Rights," *The Economist,* October 13, 2007, Volume 385 Issue 8550, p. 93.

76. Byrne, David, "The Fall and Rise of Music," *Wired,* January 2008, pp. 124–129.

77. Ibid.

78. Mnookin, "The Angry Mogul."

79. Ibid.

80. Heft, Miguel, "Reports of the Death of the Dot Com Have Been Greatly Exaggerated," *Industry Standard,* November 6, 2000, online at http://findarticles.com/p/articles/mi_m0HWW/is_45_ 3/ai_66673076/pg_2.

81. Mossberg, Walter, "Consumers Must Protect Their Freedom to Use Digital Entertainment," *Wall Street Journal,* March 14, 2002, p. B1.

82. Black, Jane, "Lawrence Lessig: The 'Dinosaurs' Are Taking Over," *Business Week,* May 13, 2002, online at http://www.businessweek.com/magazine/content/02_19/b3782610.htm.

83. Rivette, Kevin, and David Kline, "Discovering New Value in Intellectual Property," *Harvard Business Review* 78 (January–February 2000), pp. 54–66.

APPENDIX 8.A

Details on the Patenting Process

By JUDY MOHR, J.D., PH.D.
King & Spalding LLP, Redwood Shores, California

This appendix outlines the basic steps in filing and obtaining a patent in the United States and, briefly, internationally. Most inventors use a patent agent or attorney to prepare and file the patent application and to act as their representative before the U.S. Patent and Trademark Office (USPTO) during the examination process.

A common first step in the process is to conduct a preliminary assessment of patentability prior to preparing or filing an application. The patentability assessment is performed by either the inventor or a patent attorney on behalf of the inventor in order to determine—to the extent possible—that the idea to be described and claimed in the patent application is *novel* and *nonobvious*. The novelty requirement is evaluated based on a search of "prior art," or the published literature— for example, a published patent application, patent, journal article, conference publication, or meeting abstract. A variety of databases, including patent databases* and literature databases relevant to the field of the invention,† are carefully scrutinized for documents that describe all or parts of the idea. If even a single published writing is found that describes the invention, either explicitly or inherently, or if it can be shown that the invention was known or used by others in the patenting country, the invention is not novel. If the idea is not identically described in any documents, a patent application is then prepared.

---

* Several patent databases are available on the Internet, such as the patent database at the USPTO website (www.uspto.gov/patft/index.html), the Google patent database (www.google.com/patents?hl=en) and an international patent database maintained by the European Patent Office (http://ep.espacenet.com).

† Literature databases available on the Internet include, for example, PubMed (www.ncbi.nlm.nih.gov/pubmed), the Google Scholar database (http://scholar.google.com/schhp?hl=en&tab=ws), and Scirus (www.scirus.com).

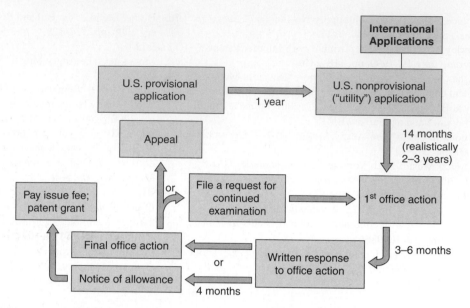

FIGURE 8A.1 U.S. Patent Application Life Cycle

In addition, prior to preparing or filing an application, the inventors should not have disclosed the idea to anyone; if the inventor has previously disclosed the idea, either orally or in writing (say, by giving a talk at a conference or publishing an abstract or a paper on the research before the patent application is filed) without a confidentiality agreement in place, then it might not be considered novel, and a patent may not be granted.

One important point is that grant of a patent does not automatically confer the right to "practice" that invention. A separate issue is "freedom to operate" and this issue must be researched as well. For example, suppose Company A has a patent that sets forth a claim to the chemotherapeutic drug paclitaxel and derivatives of paclitaxel for treatment of cancer. Company B has found that paclitaxel, when derivatized in a particular way, is significantly more effective in fighting colon cancer. So, Company B applies for and obtains a patent that claims this new derivatized paclitaxel for treating colon cancer. In this case, the patent issued to Company B falls within the broad claim to paclitaxel obtained by Company A, and hence, if Company B practices the invention, they would infringe Company A's patent. In this case, Company B can practice its invention only if a license to Company A's patent is obtained. Likewise, Company A is not free to practice the invention claimed in the patent to Company B. Although Company A has a broad claim to paclitaxel and its derivatives, Company A is excluded from using the paclitaxel derivative of Company B for treatment of colon cancer. Company A would need a license from Company B to practice the invention claimed in Company B's patent.

Types of Patent Applications

In the United States, a common strategy for obtaining patent protection on an idea is to initially file a provisional patent application and then a nonprovisional, or "utility," patent application.

With reference to Figure 8A.1, a *provisional application* is filed with the USPTO with a filing fee of $220[‡] and has a lifetime of only one year. Filing a provisional application secures a "filing

---

[‡] The dollar amounts indicated reflect the fees that took effect September 30, 2007. Fees are increased periodically; a current fee schedule can be found at the USPTO website (www.uspto.gov/main/howtofees.htm).

date," also referred to as a "priority date." During the one-year life, the application remains in the USPTO unexamined by a patent examiner. One year after filing, to avoid abandonment, the application must be refiled as a utility application (discussed in the next paragraph) for examination. A benefit of the provisional application is that the inventor has that one-year period to do further research on the invention prior to incurring the expenses involved in examination by the patent examiner, or to obtain investor monies to develop the idea.

Alternatively, if an inventor does not need time for further research or to obtain monies for development, or is anxious to obtain a patent, the application can be filed with the USPTO as a *nonprovisional or utility application,* along with the filing fee of $1,030. (Note that certain fees charged by the USPTO are reduced by 50% for so-called "small entities"—any person, nonprofit organization, or small business that has not assigned, granted, conveyed, or licensed, and is not under an obligation to do so, any rights in the invention to an entity having more than 500 persons.)

The patent application includes a description of the invention and any drawings necessary to fully and clearly convey the idea, and concludes with one or more claims. The claims are of critical importance, as they define the scope of legal protection being sought by the inventor. In drafting claims to cover an invention, a patent agent or attorney evaluates the invention and compares it to what is already publicly known to determine the correct legal language to cover the invention without capturing what is already publicly known.

After filing the nonprovisional application with the USPTO, it is assigned to an examiner, who will render an initial report regarding patentability, referred to in the United States as an "office action" (see Figure 8A.1). If that initial decision is unfavorable, the applicant can amend the claims and/or submit remarks in writing to the examiner addressing the unfavorable report. The examiner will then issue a final decision as to patentability of the claims. If the decision is favorable, the examiner issues a notice of allowance, and upon payment of an issue fee, a patent is granted (see Figure 8A.2).

If the examiner's final decision is unfavorable, the examiner issues a final office action rejecting the claims, and the patent application is denied. The applicant may appeal the examiner's decision to a three-person panel at the PTO, the Board of Patent Appeals and Interferences. Alternatively, the applicant can file a request for continued examination (for an $810 fee) with or without amended claims, to continue to try and convince the examiner of the patentability of the claims.

Representative costs for filing are shown in Table 8A.1 (page 312) on patenting fees.

International Patent Protection

The most common route used by companies today for obtaining international patent protection is to file an international application under the Patent Cooperation Treaty (PCT). By filing a PCT application, the applicant preserves the right to obtain patent protection in any country that is a member of the PCT. The 138 member countries of the PCT as of January 1, 2008, are shown in Table 8A.2, along with the two-letter code used by the PCT to refer to each country.

Under the PCT, an applicant can file an international application within one year of the filing date of a national application for patent (e.g., a U.S. provisional or utility application) and claim priority back to the filing date of the national application. The PCT enables an applicant to file one application in his or her home language and have that application considered for patent in any or all of the member states of the PCT. At the time of filing the international PCT application, the applicant has automatic and all-inclusive designation of all states that are bound by the treaty, thus preserving the right to seek patent protection in any or all of the member states.

Like U.S. applications, an application filed under the PCT is published 18 months from the priority date. For example, suppose an application, either a provisional application or a utility application, is filed in the United States, establishing a priority date. Within one year from the filing

The Commissioner of Patents and Trademarks

Has received an application for a patent for a new and useful invention. The title and description of the invention are enclosed. The requirements of law have been complied with, and it has been determined that a patent on the invention shall be granted under the law.

Therefore, this

United States Patent

Grants to the person(s) having title to this patent the right to exclude others from making, using, offering for sale, or selling the invention throughout the United States of America or importing the invention into the United States of America for the term set forth below, subject to the payment of maintenance fees as provided by law.

If this application was filed prior to June 8, 1995, the term of this patent is the longer of seventeen years from the date of grant of this patent or twenty years from the earliest effective U.S. filing date of the application, subject to any statutory extension.

If this application was filed on or after June 8, 1995, the term of this patent is twenty years from the U.S. filing date, subject to any statutory extension. If the application contains a specific reference to an earlier filed application or applications under 35 U.S.C. 120, 121 or 365(c), the term of the patent is twenty years from the date on which the earliest application was filed, subject to any statutory extension.

Bruce Lehman

Commissioner of Patents and Trademarks

Arnita Manley

Attest

FIGURE 8A.2a   Sample of Actual Patent

United States Patent [19]

Mitchell-Olds et al.

[11]   Patent Number: **5,770,789**

[45]   Date of Patent: **Jun. 23, 1998**

[54]   **HERITABLE REDUCTION IN INSECT FEEDING ON BRASSICACEAE PLANTS**

[75]   Inventors: **Storrs Thomas Mitchell-Olds; David Henry Siemens**, both of Missoula, Mont.

[73]   Assignee: **University of Montana**, Missoula, Mont.

[21]   Appl. No.: **496,016**

[22]   Filed: **Jun. 28, 1995**

[51]   **Int. Cl.^{6}** **A01H 5/00**; A01H 1/04; A01H 1/06; C12N 15/01

[52]   **U.S. Cl.** **800/200**; 800/230; 800/DIG. 15; 800/DIG. 17; 435/6; 435/172.1; 47/58; 47/DIG. 1

[58]   **Field of Search** .... 435/172.1, 6; 47/58, 47/DIG. 1; 800/200, 230, DIG. 17, DIG. 15

[56]   **References Cited**

PUBLICATIONS

Berenbaum et al., *Plant Resistance to Herbivores and Pathogens Ecology, Evolution, and Genetics*, Chapter 4:69–87 (1987). Univ. Chicago Press, Chicago.
James et al., *Physiologia Plantarium*, 82:163–170, (1991).
Koritsas et al., *Ann. Appl. Biol.*, 118:209–221, (1991).
Lister et al., *Plant J.*, 4:745–750, (1993) No. 4.
Magrath et al., *Heredity*, 72:290–299, (1994) No. 3 Mar.
Magrath et al., *Plant Breeding*, 111:55–72, (1993).
Richard Mithen, *Euphytica*, 63:71–83, (1992).
Mithen et al., *Plant Breeding*, 108:60–68, (1992) No. 1
Parkin et al., *Heredity*, 72:594–598, (1994).
Thangstad et al., *Plant Molecular Biol.*, 23:511–524, (1993).
Giamoustaris et al., *Ann. appl. Biol.* 126:347–363, 1995.
F.S. Chew, "Searching for Defensive Chemistry in the Cruciferae. or, do Glucosinolates Always Control with Their Potential Herbivores and Symbionts? No!," *Chemical Mediation and Coevolution*, Ed., Kevin C. Spencer, Academic Press, Inc., Ch. 4, pp. 81–112 (1988).

Lenman et al., "Differential Expression of Myrosinase Gene Families," *Plant Physiol.*, 103:703–7111 (1993).
Xue et al., "The glucosinolate–degrading enzyme myrosinase in Brassicacee is encoded by a gene family," *Plant Molecular Biology*, 18: 387–398 (1992).
Höglund et al., "Distribution of Myrosinase in Rapeseed Tissues," *Plant Physiol.*, 95: 213–221 (1991).
Ibrahim et al., "Engineering Altered Glucosinolate Biosynthesis by Two Alternative Strategies," *Genetic Enginering of Plant Secondary Metabolism*, Ed., Ellis et al., Plenum Press, New York, pp. 125–152 (1994).
Hicks, "Mustard Oil Glucosides: Feeding Stimulants for Adult Cabbage Flea Bettles, *Phyllotreta cruciferae* (Coleoptera: Chrysomelidae)," *Ann. Ent. Soc. Am*, 67: 261–264 (1974).
Reed et al. 1989 Entomol. exp. appl. 53(3):277–286.
Bartlet et al. 1994. Entomol. exp. appl. 73(1):77–83.
Butts et al. 1990. J. Econ. Entomol. 83(6):2258–2262.
Bodnaryk et al. 1990. J. Chem. Ecol. 16(9):2735–46.
Haughn et al. 1991. Plant Physiol. 97(1):217–226.
Mithen et al. 1987. Phytochemistry 26(7): 1969–1973.
Jarvis et al. 1994. Plant Mol. Biol. 24(4):685–687.

Primary Examiner—David T. Fox
Attorney, Agent, or Firm—Fish & Richardson, P.C.

[57]   **ABSTRACT**

A method for producing plants of the Brassicaceae family that have reduced feeding by cruciferous insects is disclosed. The method comprises selecting for the heritable trait of altered total non-seed glucosinolate levels or for the heritable trait of increased myrosinase activity. Selection may be performed on Brassicaceae cultivars, mutagenized populations or wild populations, including the species *Brassica napus*, *B. campestris* and *Arabidopsis thaliana*. Plants having such altered levels show reduced feeding by cruciferous insects, including flea beetle, diamond back moth and cabbage butterfly. Plants selected for altered levels of both glucosinolates and myrosinase also show reduced feeding by cruciferous insects.

25 Claims, 5 Drawing Sheets

FIGURE 8A.2b   Continued

date of that U.S. application, a PCT application must be filed in order to preserve the right to claim back to priority date. The PCT application is then published 18 months from the priority date. Because patent applications are published prior to examination, a company must be mindful that even if granting of a patent on the application is denied, others still have access to the information in the application.

Shortly after filing the PCT application, an PCT examiner in Europe, Korea, or in the United States will review the application, conduct a search of the prior art and issue an international search report that lists documents the examiner believes to be the closest prior art. The examiner will also prepare a preliminary, nonbinding written opinion on whether the claimed invention appears to meet the requirements for patentability in light of the closest prior art. Copies of the search report, the cited art, and the written opinion on patentability are sent to the applicant. The applicant can respond in writing to the examiner's opinion and amend the claims, if desired.

The fees for filing a PCT application depend on the number of pages of the application and the country selected to conduct the prior art search and initial assessment of patentability. A typical

TABLE 8A.1 Patenting Fees

U.S. Patent Costs		Foreign Patent Costs[1] in Selected Countries	
Searching/patentability	$3,000–$5,000	PCT filing fee:[1,2]	$2,000–3,750
Application preparation	$8,000–$18,000	Costs[1,3] associated with entry of the PCT application into selected national countries	
		Australia	$3,000
Filing fees	$1,030	Canada	$2,500
Examination/prosecution		China	$4,500
Attorney's time (per response)	$3,000–$6,000	Europe	$6,000
PTO fees for "late" responses	$120–$2,230	Japan	$7,000
Issue fee	$1,740	Costs[1,4] associated with examination of an application before selected national patent offices:	
Typical Total Range	$15,000–$30,000	Australia	$2,000–3,000
Maintenance fees		Canada	$3,000–5,000
3½ years	$930	China	$4,000–6,000
7½ years	$2,360	Europe	$3,000–6,000
11½ years:	$3,910	Japan	$4,000–8,000
		Costs[1,5] associated with validation and grant of a European Patent in five selected European countries:	
		Germany	$2,500
		France	$2,000
		Italy	$5,000
		Spain	$5,000
		United Kingdom	$1,500

[1]Costs are estimated using Global IP Estimator® from Global IP Net (www.globalip.com) and are based on an application with 30 pages of text, 2 pages of claims with 20 claims, and 3 pages of drawings. The costs are based on minimum fee schedules supplied by foreign associates and are for processing of straightforward applications. All costs are subject to currency fluctuations.
[2]PCT: Patent Cooperation Treaty
[3]National filing fee, U.S. attorney time, foreign attorney time, translation costs for China and Japan
[4]U.S. attorney time, foreign attorney time, translation of written responses filed with Chinese Patent Office and Japanese Patent Office
[5]Granting fees and translation into national language

range for filing fees for a PCT application having 25 to 35 pages is about $2,000 to $3,750 (see Table 8A.1). About 18 months after filing the PCT application, the applicant must decide whether to seek patent protection in any of the countries that are members of the PCT. That is, the PCT application itself does not mature into a patent, but must be filed in any country in which patent protection is desired. The search report and written opinion provided by the PCT examiner provide the applicant an evaluation of patentability before incurring the major costs involved in filing the application in individual member countries to obtain national patent protection. Specifically, the applica-

TABLE 8A.2 PCT Contracting States, January 2008

Albania – AL	Ghana – GH	Nicaragua – NI
Algeria – DZ	Greece – GR	Niger – NE
Angola – AO	Grenada – GD	Nigeria – NG
Antigua and Barbuda – AG	Guatemala – GT	Norway – NO
Armenia – AM	Guinea – GN	
Australia – AU	Guinea–Bissau – GW	Oman – OM
Austria – AT		Papua New Guinea – PG
Azerbaijan – AZ	Honduras – HN	Philippines – PH
	Hungary – HU	Poland – PL
Bahrain – BH		Portugal –PT
Barbados – BB	Iceland – IS	
Belarus – BY	India – IN	Romania – RO
Belgium – BE	Indonesia – ID	Russian Federation – RU
Belize – BZ	Ireland – IE	
Benin – BJ	Israel – IL	St. Kitts and Nevis – KN
Bosnia and Herzegovina – BA	Italy – IT	St. Lucia – LC
Botswana – BW		St. Vincent & Grenadines – VC
Brazil – BR	Japan – JP	San Marino – SM
Bulgaria – BG		Senegal – SN
Burkina Faso – BF	Kazakstan – KZ	Serbia –RS
	Kenya – KE	Seychelles – SC
Cameroon – CM		Sierra Leone – SL
Canada – CA	North Korea – KP	Singapore – SG
Central African Republic – CF	South Korea – KR	Slovakia – SK
Chad – TD		Slovenia – SI
China – CN	Kyrgyzstan – KG	South Africa – ZA
Colombia – CO	Lao People's Dem. Rep. – LA	Spain – ES
Comoros – KM	Latvia – LV	Sri Lanka – LK
Congo – CG	Lesotho – LS	Sudan – SD
Costa Rica – CR	Liberia – LR	Swaziland – SZ
Côte d'Ivoire – CI	Liechtenstein – LI	Sweden – SE
Croatia – HR	Lithuania – LT	Switzerland – CH
Cuba – CU	Luxembourg – LU	Syrian Arab Republic – SY
Cyprus – CY	Libyan Arab Jamahiriya – LY	
Czech Republic – CZ		Tajikistan – TJ
	Macedonia – MK	Tanzania – TZ
Denmark – DK	Madagascar – MG	Togo – TG
Dominica – DM	Malawi – MW	Trinidad and Tobago – TT
Dominican Republic – DO	Malaysia – ML	Tunisia – TN
	Mali – ML	Turkey – TR
Ecuador – EC	Malta – MT	Turkmenistan – TM
El Salvador – SV	Mauritania – MR	
Egypt – EG	Mexico – MX	Uganda – UG
Equatorial Guinea – GQ	Moldova – MD	Ukraine – UA
Estonia – EE	Monaco – MC	United Arab Emirates – AE
Finland – FI	Mongolia – MN	United Kingdom – GB
France – FR	Montenegro – ME	United States of America – US
	Morocco – MA	Uzbekistan – UZ
Gabon – GA	Mozambique – MZ	
Gambia – GM		Vietnam – VN
Georgia – GE	Namibia – NA	
Germany – DE	Netherlands – NL	Zambia – ZM
	New Zealand – NZ	Zimbabwe – ZW

tion is filed at the national patent office of the countries where patent protection is desired, for review and examination by an examiner in each national country. To facilitate this, a translation of the application into the home language of each country is required, along with the national filing fee for each country.

Costs to prepare multiple translations and to pay for the national filing fees, as well as the attorney fees for handling the application before each national patent office, quickly escalate to a range of $5,000 to $20,000 per country. The examiner in each country will consider the preliminary examination performed by the PCT examiner; in some countries the PCT examiner's recommendation is merely rubber-stamped, whereas in others the national patent examiner's review yields a different outcome.

International patent applications can also be filed directly in the patent office of those countries where patent protection is desired. This is the only way to obtain patent protection in countries that are not members of the PCT, such as Taiwan or Argentina.

Considerations in filing for international patents include the differing standards of enforceability in each country and the expensive translation costs. For example, it may be difficult to enforce patent rights in China or elsewhere. Translation costs, particularly for Japan, South Korea, and so forth, account for the rapid escalation in international patent costs for a U.S. company. In contrast, Canada and Australia are relatively inexpensive because no translation is needed.

Online Resources for Intellectual Property

IP Worldwide, the Magazine of Law and Policy for High Tech (www.ipmag.com)

PIPERS Virtual Intellectual Property Library (www.piperpat.com/VirtualIPLibrary/tabid/53/Default.aspx)

U.S. Patent and Trademark Office (www.uspto.gov)

Patent Office Sites around the World (www.pcug.org.au/~arhen)

Free copies of U.S. patents (www.pat2pdf.org)

Europe's Network of Patent Databases (www.espacenet.com)

World Intellectual Property Organization (www.wipo.int)

The following documents can be found free at the Organisation for Economic Cooperation and Development website (http://www1.oecd.org/):

Fiscal Measures to Promote R&D and Innovation

Foreign Access to Technology Programmes

Government Venture Capital for Technology-based Firms

Industry Productivity: International Comparison and Measurement Issues

The Knowledge-based Economy

National Innovation Systems

Patents and Innovation in the International Context

Policy Evaluation in Innovation and Technology

Regulatory Reform and Innovation

Technology Diffusion Policies and Programmes

Technology and Environment: Towards Policy Integration

Technology Foresight and Sustainable Development

Technology Incubators: Nurturing Small Firms

Venture Capital and Innovation

CHAPTER

9

Distribution Channels and Supply Chain Management in High-Tech Markets

---

Channel Decisions for Cisco Systems[1]

In 2000, Cisco Systems was the world's most valuable company, with a market capitalization in excess of $500 billion, sales of over $18 billion, and an elaborate structure of 6,000 channel members worldwide. Although it had originally started as a direct sales company, its explosive growth through the 1990s (due in part to the rise of the Internet) required the addition of channel partners to serve the large increase in customers across a wide variety of market segments. Indeed, with the rapid rise in the Internet and associated technologies, and caught up in its own "tornado," Cisco's channel strategy was one of "box pushing": Offering its products for sale through the largest number of channels allowed the company to grow very quickly.

Cisco's Historical Distribution Channel Strategy

Its various routes to market (distribution channels) can be seen in Figure 9.1 with the following breakdown:

- Cisco direct sales (through its own sales force to business customers): 10% of total sales
- Information technology (IT) consultants and "systems houses" (such as IBM and Accenture) who provided full-service consulting for IT needs and bundle a total solution for their business customers: 25–30% of total sales
- Value-added resellers (VARs) who added unique value in regional and niche markets, serving not only large enterprise customer accounts, but also small and medium enterprise (SME) accounts: 30–35% of total sales
- Distributors (such as Tech Data) who handled stocking, logistics support, and order fulfillment for smaller VARs: approximately 10% of total sales

Despite its "box pushing" approach, an important part of Cisco's strategy was collaborative selling between its 10,000-strong worldwide sales and field marketing organization and its VARs to generate demand, design solutions, and fulfill orders. Indeed, roughly 22% of Cisco's revenue was generated by channels prospecting for customers on their own, while the remaining 78% was generated from leads prospected by Cisco itself and then placed with resellers and other channel members.

In the late 1990s, Cisco categorized VARs according to both the volume of business they did with Cisco Systems and the level of technical support/certification of the VARs' employees. Each category received differential discounts from Cisco. Of the 3,000 VARs in the United States, approximately 2,200 were in one of three preferred categories, with about 5% in the Gold, 5% in the Silver, and 90% in the Premier categories.

(continued)

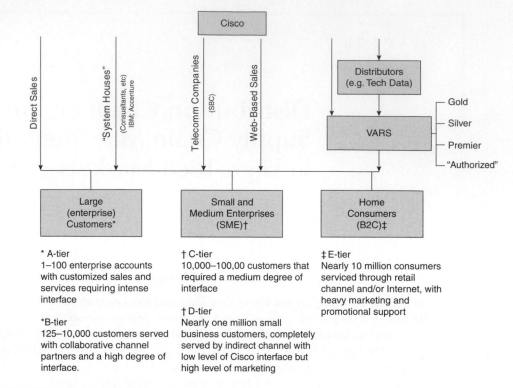

FIGURE 9.1 Cisco's Distribution Channels

However, with the technology slowdown in early 2001, Cisco found that dividing its diminishing total U.S. sales volume among 3,000 VARs caused extreme levels of *intrabrand competition* in its channels; this was competition among Cisco's own channel members selling Cisco products (in contrast to *interbrand competition,* which occurs between Cisco and competitive networking products). In order to retain the loyalty and commitment of VARs to its product line, it was faced with reducing the total number of VARs: "With brisk demand for routers and switches gone, Cisco VARs were increasingly at loggerheads with one another and with Cisco."[2]

The channel conflict was due not only to overcrowding in the shrinking market, but also to at least three other factors:

- Telecommunications providers who previously had been Cisco's end users began reselling Cisco products as part of their own services to their own customers. Because these telecommunications companies owned the wires that connected their customers' wide-area computer networks, getting into network solutions was a natural way for them to grow their own revenues and offer more services to their customers. Cisco could realize a large volume of sales through this channel, but it ended up commoditizing their products. For example, one telecommunication provider offered the Cisco products for free as an incentive to close the deal. Telecommunications companies could use this strategy because, rather than being a focal part of the telecommunication provider's solution to customers, the Cisco component was typically less than 5% of the total package being sold. VARs complained bitterly of these new competitors' "dumping" techniques (i.e., selling products below costs in order to maintain their customer base).

- In 2001, Cisco pioneered a direct online sales channel that bypassed or *disintermediated* the VARs and allowed Cisco to connect directly to the end customer. The goal was to sell preconfigured low-end equipment ("network in a box") that required no advanced networking skills to install to smaller end customers. Although Cisco maintained that this channel would "merely funnel away low-volume, low-dollar-amount transactions on a limited number of low-margin products so that resellers could focus on obtaining more lucrative business,"[3] the fact that Cisco's new wireless networking line, an area of high growth, was also available on this website seemed to undermine Cisco's pledge to resellers. Moreover, as one reseller said, "We have customers who would be buying these products, and our design services and implementation as well. This online site pegs the price for the market [limiting our pricing flexibility]. And, you can bet that big customers will be clicking there too."[4]

Cisco competitors moved rapidly to exploit the unhappiness of Cisco's VARs, offering them generous deals to carry their products instead. Clearly, Cisco faced a challenging channel situation.

2001 Channel Redesign

In a radical approach to channel management, Cisco completely changed the criteria for VARs to participate at the Gold, Silver, or Premier levels; it eliminated the sales volume requirement, and instead implemented a point system in which VARs were categorized based on the value they added to the Cisco solution. Value-added points could be earned by:

- *Technology specialization:* Selling products and building expertise in Cisco's new product initiatives
- *Engineering expertise (certification of employees):* A previous criterion that was retained, but made more stringent
- *Customer satisfaction levels:* Utilizing surveys conducted by an independent research provider to award points for exceeding numerical targets for customer satisfaction, including pre-sales and post-sales support, responsiveness, communications, systems engineer availability and skill levels, and the ability to diagnose problems

This recategorization now allowed small but highly skilled VARs to compete with larger ones based on value added, since the discount they received from Cisco was exactly the same. Moreover, the inclusion of customer satisfaction allowed small regional VARs to show the value they provided to Cisco end users.

Cisco continued with its collaborative approach between its own sales force and the VARs, including the resellers earlier in the sales cycle in pursuing Cisco-generated leads. In addition, Cisco provided sophisticated advice to its VARs about how to enhance their own profitability, offering consulting about, for example, how to craft sophisticated service offerings.

On the basis of this channel redesign, the number of VARs fell from 6,000 worldwide to 3,000, with about 1,200 dropping out in the United States alone. The overall effect was to decrease competition among resellers. Cisco now had 1,000 partners going after $9 billion in U.S. business, instead of 2,200. Now, Gold, Silver, and Premier partners constituted 15%, 10%, and 75%, respectively, of the total number of VARs, and—importantly—Cisco represented up to 70% of a specific VAR's overall revenues.

Many VARs were pleased with the new strategy, saying that the field was now competing on the basis of value and solutions, rather than price discounts.

Channel Challenges Based on New Technologies and Products

Cisco's product strategy focused on identifying high-growth markets with a forecast of $1 billion per year in sales. For example, early in 2003, Cisco identified the home consumer networking market, with a growth rate of nearly 40%. Although this market was only $800 million in 2003, it was projected to be a $4 billion-a-year business in 2008. So, Cisco purchased Linksys, a

(continued)

company that made home networking equipment, and added a B-to-C (home consumer) channel serviced by traditional consumer electronics retailers such as Best Buy.

Another new market was VoIP, Voice-over-Internet Protocol, which allowed people to make phone calls over the Internet using their broadband lines. This technology, in effect, made land-lines obsolete, significantly cutting the cost of telephony for large-enterprise customers and home consumers as well. A benefit of the VoIP product market was that it also generated demand for Cisco's core products, creating a virtuous cycle/growth engine for Cisco. These products were targeted at the $750 billion telecommunications and equipment market, expected to grow from $3.5 billion in 2003 to $10.5 billion by 2008. The Cisco products allowed a single network infrastructure to transmit voice, data, and video communications over one network rather than three. Internet-based phone systems are cheaper than traditional phone switches for customers; however, Cisco's traditional VARs were "data VARs" that focused on the networks for data; "voice VARs" were different businesses altogether.

Cisco also faced new competitors in these new markets, including Nortel, Siemens, and Avaya (a Lucent spinoff), that had established relationships with voice VARs. Channel margins for voice products were in the 20%-plus range, compared to the data market where intense channel competition resulted in margins of only 12–20%. Further complicating the channel strategy was the fact that voice VARs were less interested in the new VoIP technology because—despite the compelling benefits to their end customers—it cannibalized their existing product line of PBX switches that earned a higher margin for them. So, Cisco had to decide whether to try to work with new voice VARs (and the resulting complications), or to expect its existing data VARs to pick up new product/customer knowledge to sell the new VoIP equipment.

Ultimately, despite the new tensions with its existing data VARs, Cisco chose to go with voice VARs in selected markets. In other areas, Cisco's data VARs who were either qualified or willing to invest in learning were also allowed to carry the new VoIP products. To harmonize the channel margins, VARs were awarded points and aggressive discounts *after* their transactions with customers were complete and value/satisfaction assessed.

D istribution channel productivity is a critical success factor for business. Indeed, the importance of distribution channel strategy development and effective management of channels is growing, as many businesses move to a more complex channel design with multiple direct and indirect traditional channels as well as expansion of electronic (Web-based) channels. Clearly, channel strategies are complicated. Managing them well requires effective use of best-practices decision frameworks and concepts, the focus of this chapter. Consider the following:

- More and more companies are offering "multi-channel" options to their customers, in which a customer can buy a product directly from the company's own website, indirectly from either a brick-and-mortar retailer (e.g., Best Buy or Circuit City) or an e-tailer (e.g., Vann's), or from some other distribution channel (e.g., warehouse clubs such as Costco). For example, consumers can purchase a TiVo digital video recorder (DVR) directly from TiVo.com, or from their cable company.
- In 2007 Dell decided that its direct-only sales model (through the Dell catalog or Internet sales at Dell.com) should be augmented with an indirect distribution channel. The indirect channel partner it added? Mass merchant Wal-Mart!
- At the extreme, with increased penetration of broadband Internet access, some industries are completely eliminating physical (brick-and-mortar) distribution channels and delivering digital products (say, music or movies) to customers over the Internet (e.g., by downloading a newly released movie rather than releasing it in movie theaters).
- The major issue affecting business in many developing countries is the lack of distribution infrastructure. For example, because electric power is not widely available in many parts of

the developed world, the XO laptop from the One Laptop Per Child (OLPC) project has a hand crank to recharge the battery for short periods of time.

- Technologies such as supply chain management software programs, electronic marketplaces including virtual private networks and reverse auctions, and RFID chips (radio frequency identification; see the opening vignette from Chapter 7 for an overview) continue to streamline distribution channels and supply chains.

Distribution channel activities include traditional *logistics and physical distribution functions* (e.g., inventory, transportation, order processing, warehousing, and materials handling decisions) as well as the *activities used to structure and manage distribution channel relationships* (e.g., the selection and management of distribution channel structure and intermediaries). The goal of channel management is to coordinate the various logistics and distribution processes to provide value to the end customer effectively and efficiently. Typically, optimal distribution decisions involve trade-offs among inventory carrying costs, transportation costs, and risk management (say, demand and supply uncertainty). In addition, distribution channels are used to establish brand identity and preference in the marketplace. In consumer electronics, for example, even when shoppers come into the store knowing what product they want to buy, many of them have not decided which brand to buy. Indeed, nearly 60% of final brand purchase decisions are made inside the store.[5] Hence, channel members are crucial to the marketer's success.

However, distribution channels can be inefficient because suppliers and manufacturers, as well as manufacturers and distributors, often work at odds with each other; they may have conflicting goals and objectives and often don't think in terms of joint problem solving. For example, depending on the particular type of intermediary, many channel members are more interested in creating a unique identity and position for their own store in a local market, while manufacturers generally have less interest in the particular store where a customer buys its goods, so long as the customer chooses the manufacturer's brand over competing brands.

Moreover, because of the high value of many technologically sophisticated products, and because of the rapid pace of market evolution, high-tech marketers experience serious incentives to minimize the number of products held in inventory in the channel. Effective distribution channels allow a firm to identify redundancies and inefficiencies in the system, to develop relationships and alliances with key players, and to achieve both cost advantages and customer satisfaction.

This chapter focuses on managing the complexities in high-tech distribution channels. The chapter first presents a simple outline of the issues involved in channel strategy, including channel structure, channel management, and channel performance. This section addresses, among other things, complexities in direct channels, including direct sales, Web-based sales, and company-owned retail outlets, as well as the evolution of channel structure over time. In addition, it overviews coordination mechanisms and legal issues such as tying and exclusive dealing, as well as the ways managers evaluate channel performance.

The chapter then continues with a focus on **multi-channel marketing**—marketing that includes a mix of direct and indirect distribution channels as well as Web and offline (brick-and-mortar) channels. A key issue in multi-channel marketing is minimizing the conflict between the various channels. To create harmonized, or integrated, multi-channel systems, a contingency framework for hybrid channel design and management is presented.

The third section of the chapter explores emerging trends and concerns in high-tech distribution, including the "long tail" (in which the combination of broadband Internet penetration and the digitization of information goods has resulted in the ability to distribute profitably a wider variety of "less popular" content such as books, movies, etc.); gray markets (unauthorized distribution), piracy, and restricted exports (say, for technology with potential military applications); and the unique distribution needs in base-of-pyramid markets.

The final section of the chapter broadens the focus to the management of the incoming components used in the manufacturing process itself. Whereas distribution channel management is concerned about delivery of and support for manufactured goods *after* they leave the producer's hands, supply chain management is concerned about matching the inflow of supplies and other materials used at every stage of the manufacturing process to the actual demand exhibited by customers in the marketplace. Hence, the final part of the chapter examines best-practices supply chain management that high-tech companies use to mitigate both demand and supply uncertainties. This section of the chapter also overviews technologies used to maximize supply chain efficiency, such as supply chain management software, e-procurement, reverse auctions, and RFID chips, as well as the trend toward ecologically friendly supply chains.

ISSUES IN DISTRIBUTION CHANNEL DESIGN AND MANAGEMENT

Table 9.1 provides a brief outline of the issues in distribution channel strategy. In addition to providing a framework for evaluating distribution strategies, the focus is on issues that are particularly pertinent for high-tech distribution channels. Additional coverage of general issues in distribution channel management can be found in more traditional sources.[6]

TABLE 9.1 Outline of Issues in Distribution Channel Strategy

I. *Channel Structure:* The set of companies involved in the flow of product from producer to end user
  A. Direct channels
    1. Company sales force
    2. Company website
    3. Company-owned retail outlets
  B. Indirect channels
    1. Types of intermediaries
    2. Number of intermediaries (coverage; penetration)
  C. Hybrid channels, dual channels, or concurrent channels
    1. Multi-channel marketing

II. *Channel Management:* The ongoing control and leadership of channel members to manage conflict, and to encourage cooperation and coordination
  A. Governance mechanisms
  B. Legal issues
  C. Managing gray markets (unauthorized distribution)
  D. Managing conflict in multi-channel marketing

III. *Channel Performance:* The monitoring and assessment of performance of the overall channel as well as individual channel members; metrics include:*
  * Reseller's contribution to supplier profits
  * Reseller's contribution to supplier sales
  * Reseller's contribution to growth
  * Reseller's competence
  * Reseller's compliance
  * Reseller's adaptability
  * Reseller's loyalty
  * Customer satisfaction with reseller

---

*Kumar, Nirmalya, Louis Stern, and Ravi Achrol, "Assessing Reseller Performance from the Perspective of the Supplier," *Journal of Marketing Research* 29 (May 1992), pp. 238–253.

Channel Structure

Channel structure refers to the number of levels and companies involved in the flow of product from producer to end user. A *direct* channel structure is one in which a manufacturer sells directly to the customer, say with its own sales force, through company-owned stores, or via the Web. Because they are owned, managed, and controlled by the manufacturer, direct channels are sometimes referred to as "vertically integrated." An *indirect* channel is one in which a manufacturer uses intermediaries to market, sell, and deliver products to customers. Because they are composed of independent companies, indirect channels are sometimes referred to as "market-based" channels. In the mid-range between vertically integrated or market-based channels are contractually administered channels, such as franchises and cooperatives.

Some people argue that by using a direct channel, the manufacturer can eliminate the intermediary and lower the price of the product. However, a careful examination of the logic of distribution channels shows the fallacy of this argument. Although intermediaries can be eliminated, the functions they perform—providing assortments for customers in the amount and variety they desire, providing service and other facilitating functions, communicating with end users, and so forth—cannot. If a firm were to use a direct channel, either the manufacturer or the customer must assume responsibility for those functions. And, in either case, although the price of the product may be lower to customers, the costs for one or the other channel members have increased because of the additional functions performed.

In addition, **contact efficiencies** arise from using intermediaries. In a market composed of, say, 10 vendors and 10 customers, for each customer to evaluate each vendor's offering would require 10×10, or 100 interactions. However, with the addition of a distributor, each of the 10 vendors and each of the 10 customers can efficiently deal through just one intermediary, cutting the number of market transactions from 10×10 to $10 + 10$, or merely 20. This notion of contact efficiency is a key reason that even on the Internet, where transactions are presumably "costless," *cybermediaries* (online intermediaries) have sprung up. For example, even online, customers don't necessarily want to check 10 different websites to compare products and features; it is much easier to have that information aggregated by one site.

Moreover, distribution and logistics require a rather unique set of core competencies that differ significantly from manufacturing companies' competencies in product development and marketing. And, companies who perform distribution functions can gain scale economies that are not available to direct channels.

Direct and indirect channels are not mutually exclusive; a firm may use some combination of them to get its products to customers. Using a combination of direct and indirect channels goes by several different terms including *hybrid channel, dual channel, concurrent channel*, and *multi-channel marketing*. Figure 9.2 shows a multi-channel system that might be used for high-tech products.

Research on channels serving business customers has found concurrent channels are more prevalent when:[7]

- Market size and growth are strong.
- The offering is perceived as less standardized.
- Customers don't form buying groups to increase their bargaining power.
- Customers' needs and buying behavior are stable across purchasing occasions. (If customers' needs varied across purchase occasions, they would invite more bid proposals from the two competing channels, which would result in channel conflict and possible cannibalization.)

Intrabrand competition, or different channels competing to sell a company's products to the same target markets, is more acute when a firm deploys a direct (company-owned) channel in addition to an indirect channel. Indirect channel members feel the company will favor its own employees, while company employees feel they have to perform better than the indirect channel to justify the company's investment in fixed overhead cost. Thus, the concurrent use of a direct and indirect channel makes sense only under certain conditions. Because of their prevalence and more complicated management, we later devote a full section of this chapter to multi-channel marketing.

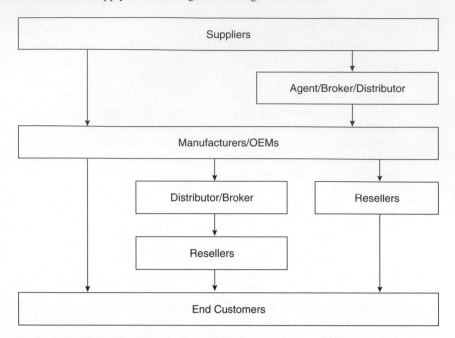

FIGURE 9.2 Multi-Channel System

In designing the channel structure, high-tech marketers have to assess what determines customers' selection of different channels—their channel preferences—and the likely outcomes of customers' choices. For example, channel attributes that affect customers' channel selection include, among other things, ease of use, price, search effort, service information, quality, aesthetic appeal, convenience, assortment, and enjoyment.[8]

DIRECT CHANNELS High-tech firms face specific issues with respect to three types of direct channels: (1) direct sales and the interface between sales and marketing, (2) sales over a company website, and (3) sales through company-owned retail outlets. As always, the focus here is selective; additional resources are recommended for interested readers in each section.

Direct Sales. Although using a sales force as a mode of distribution can be expensive, companies gain control over the selling process as well as control over customer relationship management. Many companies use a sales force to reach key customer/market segments such as enterprise accounts or key industry verticals. In addition, even when companies use an indirect channel, they frequently rely on direct sales (personal selling) as the major mode of interacting with distributors and other channel members.

Sales force management is a highly specialized field in its own right. In addition to navigating the complications between a direct sales force and indirect channels, effective sales force management must also address the interface between the Sales Department and the Marketing Department. Because these two groups are typically separate functions in an organization, they often work toward separate goals and experience conflict. However, when sales and marketing departments work well together, companies experience improved performance including shorter sales cycles, lower sales costs, and improved customer relationships.[9] For example, at IBM, sales and marketing used to work independently of one another: Marketing's new product announcements often came when the sales department was not prepared to capitalize on them; marketers did not link advertising expenditures to actual sales, so the sales department couldn't see the value of marketing efforts; sales people focused on fulfilling product demand, not creating it. To overcome some of these problems, IBM created a new function called Channel Enablement, whose purpose was to integrate the sales and marketing activities. Box 9.1 provides a

BOX 9.1

Sales–Marketing Interactions

"When sales are disappointing, Marketing blames the sales force for its poor execution of a brilliant plan. But, Sales claims that Marketing sets prices too high and uses too much of the budget, which would be better spent hiring more salespeople or paying sales reps higher commissions."[1]

"Sales people believe that marketers are out of touch with what's really going on with customers. Marketing believes that Sales is too focused on short-term results."[2]

Types of Sales–Marketing Relationships

As shown in Table 9.2, the relationship between Sales and Marketing takes different forms in different companies at different life-cycle stages. Many small companies may not have a formal marketing function; if small businesses have a marketing person, often that person's position is to support the sales staff (e.g., develop collateral materials, qualify leads). As companies grow and become successful, they typically realize the need for competencies in segmentation, targeting, and positioning; people are hired to perform these tasks. At this point Marketing can compete with Sales for funding, and as Marketing takes on more tasks, disagreements arise over who does what. Moreover, Marketing begins to work with other departments, such as strategic planning, product development, and manufacturing. The company may begin to think about a branding strategy (beyond selling products). People start to say that the organization should be transformed into a marketing-led company (not the same as *market oriented*, mind you).

Table 9.2 shows conditions under which the various types of Sales–Marketing relationships can work well, and conditions under which they should change.

Causes of the Disconnect

Why do Marketing and Sales experience conflict? The friction derives from two sources: economic issues and cultural differences.

Economic friction arises from competition over resources and how Marketing spends its share of the pie. For example, Marketing wants the sales force to sell on value, while salespeople may favor lower prices. While Marketing establishes suggested retail prices, Sales often negotiates trade discounts and promotions. The VP of Sales often goes directly to a COO or CFO for financial negotiations, leaving Marketing out. Moreover, salespeople often say that new products lack the features, style, or quality their customers want. Finally, because the sales force's productivity is both more tangible (easier to measure) and more immediate (short-term) than Marketing's activities, CEOs are sometimes more biased toward Sales. When compensation systems between Sales and Marketing are not aligned (say, salespeople are evaluated on quotas/sales volume, while marketing personnel are evaluated on profitability), conflict is also exacerbated.

Cultural differences in the "thought worlds" between Sales and Marketing also cause friction. For example, salespeople are often more strongly oriented toward customers and short-term results, while marketers may be more strongly oriented toward products and long-term results. In addition, Sales and Marketing often have different competencies, with salespeople often more knowledgeable about markets (both customers and competitors) and marketing personnel often more knowledgeable about products and internal processes to support those products.

Interestingly, although the differences in both the orientations and competencies have a negative effect on the quality of the cross-functional collaboration between Sales and Marketing, the differences in orientation have a positive impact on the market performance of the firm. Therefore, the solution is not to eliminate their differences, but rather to balance their roles in order to improve the outcomes.

More specifically, differences between Sales and Marketing in both the customer (versus product) orientation and time frame (short-term versus long-term) are positively related to market performance. This positive effect likely arises from the fact that each side tempers the orientation of the other, bringing some

---

[1] Kotler, Philip, Neil Rackham, and Suj Krishnaswamy, "Ending the War between Sales and Marketing," *Harvard Business Review* 84 (July–August 2006), pp. 68–78, quote p. 68.

[2] Ibid., p. 68.

(*continued*)

balance. In addition, reasoned arguments between the two (say, when marketing argues that short-term price cuts may have a negative long-term effect) result in more profitable business decisions. Hence, when one side champions a product orientation, while the other side champions a customer orientation, improved market performance results—despite the lower quality of team integration.

However, differences in the competency sets between Sales and Marketing have a negative impact both on the quality of the interaction between the two groups as well as on market performance. When the two groups have different skills, it is more difficult to share and assimilate the knowledge and perspective of the other, posing an "interpretive barrier that precludes the exchange, understanding, and synthesis of ideas, and ultimately, optimally decisions."[3]

Facilitating Shared Competencies to Enhance Integration between Sales and Marketing

Some of the ways in which a company can facilitate more integrated Sales and Marketing functioning are:

- Encourage communication between the two groups. More communication is not necessarily better. Rather, communication should be relatively formal, meetings should have an agenda, and a systematic process for sharing information.
- Joint assignments in which Marketing and Sales are given opportunities to work together can help them become more familiar with each other's work. For example, product managers could be sent on sales calls; salespeople can be involved in reviewing ads and campaigns, as well as developing marketing plans.
- Co-location in which Sales and Marketing groups reside in physical proximity to each other can facilitate interactions.
- Common metrics for evaluation should be used wherever possible, including compensation systems.
- Encourage shared databases for CRM and monitoring the results of the marketing and sales cycles.
- Develop a systematic way for the sales force to share its knowledge of customers and the market, and vice versa.
- In terms of organizational structure, it can be helpful to have both functions report to a C-level executive.

Finally, as described in Chapter 4, companies can use the following set of six items (assessed on a 7-point scale from "Strongly Agree" to "Strongly Disagree") to assess the degree to which the two groups function collaboratively:

In our company, Marketing and Sales:

- Have frictionless collaboration
- Act in concert
- Coordinate their market-related activities
- Have few problems in cooperation
- Achieve common goals
- Trust each other

Sources: Kotler, Philip, Neil Rackham, and Suj Krishnaswamy, "Ending the War between Sales and Marketing," *Harvard Business Review* 84 (July–August 2006), pp. 68–78; and Homburg, Christian, and Ove Jensen, "The Thought Worlds of Marketing and Sales: Which Differences Make a Difference?" *Journal of Marketing* 71 (July 2007), pp. 124–142.

[3] Homburg and Jensen, "The Thought Worlds of Marketing and Sales: Which Differences Make a Difference?", quote p. 134.

brief overview of some ways that companies have facilitated effective cross-functional collaboration between sales and marketing, and Table 9.2 charts four types of sales–marketing relationships. In addition, Chapter 11 provides a very brief overview of the role of personal selling in a company's advertising and promotion strategy.

Sales over the Company Website. Many companies offer their products for sale through their own websites. When a company-direct website is used, in addition to traditional offline channels, the strategy is frequently referred to as a **bricks-and-clicks** model of

TABLE 9.2 Four Types of Relationships between Sales and Marketing			
Type	**Characteristics**	**Useful When**	**Needs to Change When**
"Undefined" (Independent)	Lack of knowledge about what the other is doing—until conflict arises	Company is small Informal relationships are productive Marketing's role is to support sales	Conflict occurs regularly Duplication of effort or tasks falling through cracks
"Defined" (Clear Roles)	Clear processes, boundaries, and responsibilities; rules for contentious areas, such as "how is a lead defined?" Some collaboration on events such as trade shows	Products are simple Traditional roles work in the market	The industry changes (e.g., products are commoditized or customers need customization); technology changes are accelerating; Sales may need new skills
"Aligned" Cooperate	Joint planning; salespeople understand marketing terminology (i.e., value proposition, brand image); marketers confer with salespeople on important accounts	Sales cycle is short Sales process is straightforward Marketing and sales report separately	A common process and shared responsibility can generate more revenue
"Integrated" Full Collaboration	Shared structures, systems, and metrics/rewards; marketers assist with key account management		

Source: Kotler, Philip, Neil Rackham, and Suj Krishnaswamy, "Ending the War between Sales and Marketing," *Harvard Business Review* 84 (July–August 2006), pp. 68–78.

distribution. Factors that need to be considered before moving to a bricks-and-clicks model include:[10]

- Does the company have the competencies and capabilities to innovate and operate an industrial-strength e-commerce website that provides an excellent user experience and makes money on a full accounting basis?
- Does the company currently sell products through a catalog?
- Are the company's products simple in nature (e.g., no configuration required, not integrated with products from other manufacturers)?
- Is the sales process clear-cut and non-consultative?
- Are the products easy to install and maintain?
- Does the company have an existing infrastructure to support direct sales (e.g., order fulfillment, returns, customer service)?
- Do customers usually know what they want when they are ready to buy, or do they need information about competitors' products and product benefits from a third party?
- Is the company willing to promote the website enough to attract sufficient prospects?

If a company can answer "yes" to all of these questions, then a Web-direct sales model can make sense. If, however, the answer to any of the questions is "no," then a Web-direct sales model may be problematic.

In addition to considering the factors above, companies must also consider the backlash from their existing channels (both direct as well as indirect). Many companies experience resistance from their other channels when they add a Web-based sales channel. The resistance arises from cannibalization, or displacement, of sales from other channels. Indeed, this conflict, when coupled with the cost overhead associated with fulfilling Web-based orders (including the costs of maintaining a strong e-commerce platform and associated technology costs) as well as the competencies needed to execute this model well, means that **disintermediation**, or the elimination/bypassing of intermediaries due to direct Web sales, has not taken off on a widespread scale. Although going direct on the Web may work for some companies (particularly those that can gain from online transaction efficiencies and/or increased market reach), the inherent complexities can mean that a middle-ground approach may be preferred. For example, companies may have a direct marketing channel with direct mail and telemarketing, an Internet-direct channel, and a sales force for business accounts, as well as indirect channels with retailers, wholesalers, and mass merchandisers and traditional agents. Conflict management in multi-channel marketing is discussed in a dedicated section later in this chapter.

Company-Owned Retail Outlets. One type of direct channel that high-tech companies have utilized with mixed success is the company-owned retail store. IBM opened its first product center for the personal computer in New York City in April 1982. The number of IBM stores had grown to 84 by 1986 when they were sold off to NYNEX due to poor performance.[11] During the personal computer industry boom, Gateway expanded aggressively into company-owned stores, but it had to shutter all 188 of them in 2004.[12] In July 1991, Sony launched its Gallery store in Chicago, the first in the United States to showcase the company's entire consumer electronics product line to the public. In 1996 Sony updated its retail strategy with the launch of the Sony Style showrooms in the United States, which had expanded to 57 by June 2008.[13]

Notwithstanding these mixed success stories, during the past few years many more high-tech consumer electronics companies have jumped on this bandwagon. Apple opened its first company-owned store in McLean, Virginia, in 2001. By June 2008, due to a phenomenal track record of success, Apple operated 215 retail stores in six countries including Australia, the United States, the UK, Japan, Canada, and Italy,[14] with these stores contributing nearly 20%—and growing—of Apple's revenue.[15] Many companies have tried to imitate Apple's retail strategy. Palm has opened its own gadget shops in airports and shopping malls. In 2006 Dell opened two retail stores, in Dallas and New York City, to let consumers gain hands-on experience with Dell computers, notebooks, widescreen TVs, and other products. Dell also has about 160 product kiosks at shopping malls all over the United States.[16] Nokia now has flagship stores in Chicago, New York, London, Helsinki, Moscow, and Shanghai, with a goal of reaching 18 worldwide.[17]

Why have some companies like IBM and Gateway failed and others like Apple succeeded with this direct retail channel strategy? One possible explanation is offered by channel evolution theory (discussed shortly). In the early days of the PC industry, before the chasm was crossed, early adopters may not have been comfortable going to a single-brand retailer like IBM or Gateway because they needed to explore more alternatives. Later, as the market matured and more mainstream consumers came into the market, they had enough prior information from other sources to feel confident going to a single-brand store.

Another possible explanation is that Apple had created so much buzz with its iPods since 2001, and their stores were so attractive, that mainstream consumers were drawn in to see product demonstrations and had a wonderful experience inside the store. In addition, the company apparently got into the retail business because it was not happy with the service and support being provided by its indirect channels.[18] The retail strategy was well suited to Apple's transformation from a computer company to a consumer electronics company. Direct retail channels provide high-tech companies with full control over the execution of their marketing strategy and provide a performance benchmark for indirect channels to aspire to. As noted previously, a problem affecting many high-tech companies is that the sales and marketing functions are not well aligned in their systems, processes, metrics, and reward systems, leading to sub-optimal performance.[19] A direct retail channel enables companies like Apple to fully integrate retail sales into their marketing strategy for superior results. However, as with any multi-channel marketing efforts, such company-owned direct channels can cause conflict with intermediaries in indirect channels.

INDIRECT CHANNELS Recall that indirect channels rely on one or more intermediaries to assist the manufacturer with a variety of channel functions, including sales and marketing, customer service and support, installation, training, and maintenance, and warehousing. For example, in 2007 computer maker Acer was the fourth-ranked PC company globally, behind Hewlett-Packard, Dell, and Lenovo. *BusinessWeek* attributes its growth to a combination of its low-price computers and "unconventional distribution." It sells only through an indirect distribution model and, despite its low prices, offers retailers a higher margin than rivals.[20]

Two of the key issues when using indirect channels are the type of intermediary to use and the number of intermediaries to use.

1. *Types of intermediaries:* Indirect channels can rely on a variety of different types of intermediaries. *Distributors* usually buy directly from the manufacturer and sell to other intermediaries such as resellers or retailers who, in turn, sell to end users. For example, Tech Data Corporation based in Clearwater, Florida, distributes a variety of information technology products and services to customers ranging from small retailers to large corporations. *Resellers,* who typically operate locally, have a closer relationship with end users by providing products and services matched to their needs. Many resellers of high-tech products are referred to as **value-added resellers (VARs)**, or *value-added dealers (VADs).* These are resellers who purchase products from one or several high-tech companies, add value through their own expertise, and usually market the bundled solutions to particular vertical (i.e., industry-specific) markets. For example, Meridian IT Solutions, based in Schaumberg, Illinois, is a VAR for Cisco Systems and other manufacturers. **Systems integrators** are specialized resellers who typically manage very large or complex projects involving hardware, software, or services from different vendors and take responsibility for customized implementation for specific customers. For example, Science Applications International Corporation (SAIC), a *Fortune* 500 company based in San Diego, provides IT systems integration services to many government customers. While VARs and systems integrators effectively serve the needs of business customers, high-tech marketers serve end-consumer markets through a number of indirect channels including offline and online retailers, catalogs, and kiosks.

2. *Number of intermediaries to use:* When using an indirect channel, a company must decide how many intermediaries to use within each region or territory, trading off the *degree of coverage* and the *degree of intrabrand competition.* As the opening Cisco vignette indicated, companies typically want as much market coverage as possible, and as a result they may sell through as many intermediaries as they can. Although the company might gain additional market coverage, it comes at a cost. When a firm has many dealers in any area, each dealer competes with others in its territory in selling the company's products—known as **intrabrand competition**. Competition among different brands or products in the marketplace, referred to as **interbrand competition**, is healthy. However, intrabrand competition can cause conflict and price erosion

When selling the same brand (intrabrand competition), resellers often rely on price competition. Not only can this be damaging to the manufacturer's reputation and perceived quality in the market, but the dealers themselves often end up making a lower margin on those manufacturer's sales. As a result, they find it difficult to support the level of service and training that the high-tech product often requires. So, too much coverage can actually lead to problems over the long term, in which the product is neither supported nor valued by channel intermediaries and end users in the way a firm might desire. Hence, in deciding on the degree of coverage in a market, a firm must maintain a balance between too limited coverage and too much coverage (which invites intrabrand competition). Vertical or territorial restrictions (discussed subsequently), in which a select distributor is granted exclusive rights to a particular territory, can be used to inhibit intrabrand competition and promote harmony among channel members.

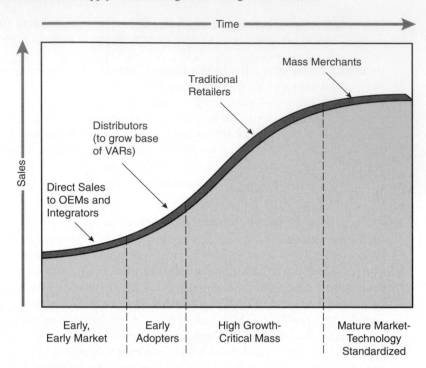

FIGURE 9.3 Evolution of High-Tech Channels

EVOLUTION IN CHANNEL STRUCTURE OVER THE TECHNOLOGY LIFE CYCLE   The type of channel a firm uses typically changes over the course of a technology's life cycle,[21] shown in Figure 9.3. When new technologies first appear in the market, sales strategies are naturally focused on OEMs, independent software vendors, and integrators as the products struggle to gain support and a presence in the industry. Once the technology gains a toehold in the market and the target moves to the early adopters, technically astute VARs join integrators as a key sales channel. VARs and early adopters determine whether a new technology establishes a presence in the market. And to leverage full efficiencies of the indirect channel, vendors work with distributors to expand their reach and grow their base of VARs.

As the technology approaches critical mass and enters a high-growth phase, a fairly traditional dealer channel is necessary. National distributors, value-added dealers, and traditional retailers (such as computer superstores) each serve to add coverage in a growing market. At this point in the technology life cycle, the earlier channel members (such as the VARs) may transition to newer technologies and opportunities, even while maintaining their presence in the current technology as a convenience to their customer base.

Once technology reaches maturity as a standardized technology, the distribution cycle shifts again and mass market channels become increasingly important. As the technology reaches maturity, increased use of mass merchants, consumer electronics, and office products stores may be seen.

Of course, there are exceptions to the model depicted here. Some products' first success might be with a retail channel. For example, if the product has an application in a home environment, then vendors must sell where this customer buys, which is typically at a retail store or via the Internet. Or, if the market is composed of business customers with an installed base of existing technology, customers face headaches in migrating to a new technology and so may be more conservative and require more personal selling.

Geoffrey Moore believes that a direct sales channel is the most effective one to create demand for a new product and to cross the chasm.[22] But he argues that volume and predictability of revenues determine whether a direct-sales model is even viable. To support a single consultative salesperson requires a revenue stream of anywhere from $500,000 to several million dollars, depending on the amount of pre- and post-sales support needed. For a quota of $1.2 million per year, the salesperson must close $100,000 per month. If the sales cycle is six to nine months, and if the close rate is one of every two opportunities, then 12 to 18 prospects of $100,000 each must be in the pipeline at all times, or some smaller number of significantly larger deals.[23]

A retail channel model may also be successful for a mainstream market, but it is not ideally suited for crossing the chasm. A retail channel does not create demand (rather, it is suited for situations in which customers are looking for a channel to fulfill demand), and it does not help develop the whole product.[24]

Channel Management

Channel management generally deals with the ongoing control and leadership of channel members to manage conflict and to encourage cooperation and coordination. Because multi-channel marketing is increasingly the norm for most high-tech companies, the next section has extensive coverage on how to manage these hybrid channels to minimize conflict and maximize performance. In this section, we address governance structures, or the coordination mechanisms used in channel management, as well as two legal issues that have recently received increased attention in high-tech distribution channels: tying and exclusive dealing. Other channel management topics, such as gray markets and export restrictions, are covered later in the chapter.

GOVERNANCE MECHANISMS   Manufacturers must use *coordination mechanisms,* or governance structures that specify the terms, conditions, systems, and processes to manage, guide, and monitor ongoing interactions between marketers and channel intermediaries.[25] Recall from Chapter 5 that the two approaches to coordinating relationships are authoritative (unilateral) controls and bilateral controls. **Authoritative (unilateral) controls** reside in one firm's ability to develop rules, give instructions, and, in effect, impose decisions on the other. **Bilateral controls** originate in the activities, interests, or joint input of *both* firms. Also known as relational norms, the spirit of bilateral controls is reflected in a commitment to flexibility and adaptation under market uncertainty, mutual sharing of benefits and burdens, and collaborative communication.[26]

Coordination mechanisms are usually put in place at the beginning of the relationship. As the relationship unfolds, marketers use influence strategies to obtain channel member compliance. *Coercive influence* is based on promises and threats, whereas *noncoercive influence* is based on information and persuasive arguments. Companies must use caution in using coercive influence attempts, as research has found that coercive influence is effective at obtaining channel member compliance only when the latter's dependence on the marketer is high. The most effective noncoercive influence strategy to obtain channel member compliance is a rational argument, which includes three elements: make a claim, provide evidence, and then exhort the channel member to act.[27]

The fact is, distribution channel members (whether direct or indirect) are economically motivated. Therefore, the margin structure is key to aligning the channel member's goals with the manufacturer's. In addition, the relative power balance between the channel members—which in some cases is more strongly oriented toward large, powerful intermediaries—also has a large effect on how channel members behave.

LEGAL ISSUES   **Tying** occurs when a manufacturer makes the sale of a product in high demand conditional on the purchase of a second product. Among other things, the U.S. Justice Department and the European Union accused Microsoft of tying its operating system to its Internet browser in 1997. Tying may also come in the form of bundled rebates that give buyers discounts for hot products if they also purchase the company's other products.[28] Because the Nintendo Wii has had insufficient supply to meet demand, Nintendo has had to allocate units to the different members of its retail channels. One rationale for its allocation has been whether or not the channel member has sold a sufficient number of the higher-margin Wii games. As a result, mass merchants like Wal-Mart have told customers that they must buy at least two Wii games in order to obtain the right to purchase a Wii game console, the classic definition of tying.

Exclusive dealing arrangements restrict a dealer to carrying only one brand of a product. Exclusive dealing can also involve licenses of technology and availability of components/supplies in the supply chain. For example, Toshiba exclusively sells large-capacity 1.8-inch hard drives to Apple Computers.[29] Ostensibly, such arrangements are put into place to ensure adequate service, but antitrust implications arise if large companies use their dominance to restrict customers' access to competitors' products. In 2000, Amazon and Toys "R" Us signed a 10-year distribution agreement in which Amazon granted Toys "R" Us exclusivity to be the sole provider of toys and baby products on Amazon's website. Over the years, other merchants (e.g., Target) began to sell toys on Amazon's website. In 2004 Toy "R" Us filed a lawsuit against Amazon for breach of contract of the exclusive agreement. In 2006 a judge sided with Toys "R" Us and permitted the company to exit the agreement without penalty.[30]

Channel Performance

Effective channel strategy includes assessing the performance of the various channels as well as individual channel members. As shown previously in Table 9.1, both quantitative and qualitative performance indicators are pertinent.[31] *Quantitative* indicators include sales volume moving through a particular intermediary or the manufacturer's market share in the intermediary's relevant territory. *Qualitative* indicators include the dealer's satisfaction with and commitment to the manufacturer and the dealer's willing coordination of activities with the manufacturer's national programs.

For example, the channel performance of a dual distribution system (e.g., a channel that includes a mix of company-owned and franchised retail outlets) shows a complicated pattern of interactions between the firm's distribution strategy and its intangible value (i.e., value beyond physical assets, not captured in a firm's accounting data alone, but captured in forward-looking stock market metrics, such as Tobin's q).[32] In some cases, dual distribution interacts with other firm strategies, such as advertising and financial liquidity, in its impact on intangible value, but in other cases, either company-owned (vertically integrated) or indirect (market-based) channels perform better. Given that no one-size-fits-all distribution strategy works, a contingency theory of distribution channel design and management is warranted—as discussed in the next section.

MANAGING HYBRID CHANNELS: EFFECTIVE MULTI-CHANNEL MARKETING

As with most marketing issues, companies must keep the customer in mind when designing channel strategy. What benefits can various channels deliver to customers? How can multi-channel marketing be used to enhance customer value? Consumers generally prefer to interact with multiple channels for the same purchase or on different purchase occasions.[33] For example, some consumers like the convenience of ordering and paying on the Web, but they want to avoid shipping and handling charges. When a deadline looms, it may be faster to order online and pick up offline rather than wait for home delivery. Consumers may also want to physically examine the goods before purchase or use the offline channel to exchange or return merchandise they are not satisfied with. Evidence indicates that multi-channel customers buy more and are more profitable than single-channel customers.[34] For example, one study found that shoppers who use a combination of channels—say, the Internet, catalogs, and physical stores—spend an average of 30% more per year in stores than single-channel shoppers.[35] The Direct Marketing Association refers to these shoppers as MVPs: Most Valuable Purchasers. Therefore, it is imperative that high-tech marketers offer multiple channels in large geographic markets and have tight coordination across these channels to provide a seamless customer experience.

Yet, managing multiple channels is complex. Indirect channels are subject to less management authority than are direct channels, posing a control issue. Harmonizing objectives and driving results across the different channels are challenges. Further, as in the opening Cisco vignette, channel conflict arises from multiple sources, including too many resellers in a particular geographic territory, conflict between a manufacturer's direct channels and its indirect channels, and gray markets (in which the products destined for one market are diverted and end up in another market, discussed in a subsequent section of this chapter).

Using multi-channel marketing might invite conflict, but is consistent with best-practices distribution strategies. In addition, using multiple distribution channels is consistent with the notion of creative destruction, in which a company is willing to cannibalize existing business models to take advantage of new technologies and business models that customers desire. Companies that make decisions based on keeping the peace with their existing distribution channel likely will not succeed in the new world. As the experts say: Cannibalize before there is nothing of value left to cannibalize.

The objectives of a hybrid channel are to (1) increase market coverage while (2) maintaining cost efficiency and (3) minimizing conflict.[36] Increased coverage and lower costs can create a competitive advantage for firms that understand how to implement and manage a hybrid channel effectively. The two-step strategy to achieving these objectives is described next.

Step 1: Gather Market Data

Effective management of a hybrid channel requires solid data. Analytical rigor is important. The assessment of market opportunity, coverage models, channel-specific cost–benefit analysis, and other data are needed to determine channel resource allocation. But, more importantly, logic and quantification are needed to communicate to the organization the justification for the distribution strategy. The numbers and facts behind the logic make channel changes compelling and understandable.

One of the most critical pieces of necessary data is the assessment of the magnitude of conflict. This includes not only an assessment of the market segments that multiple channels will be simultaneously pursuing, but also an assessment of the degree of cannibalization (sales that used to flow through one channel that will now be displaced to another channel). With no revenue in conflict, channel members may become complacent and not pursue their tasks effectively enough. On the other hand, with too much revenue in conflict, dysfunctional effects distract channel members from performing their jobs with clarity and enthusiasm. One guideline is that having more than roughly 10% to 30% of a firm's revenue in conflict between multiple channels provides a dysfunctional amount of conflict and can result in angry feedback from both customers and marketing personnel.[37] Without clear evidence that incremental revenues will be generated, the financial costs as well as morale costs of adding new channels may be unwarranted.

Step 2: Work toward Harmonization Following Contingency Theory

From a general perspective, concurrent channels can be harmonized by differentiating the offerings carried by each channel; by having a system in place that clarifies which channel has ownership of each lead, inquiry, or order; or by offering compensation to each channel for its role in a sale that went to the other channel.[38]

A useful model to design hybrid channels follows the familiar contingency model, depicted in Figure 9.4. Contingency theory implies that no one channel can be used for optimal performance; rather, the type of channel used must be matched to particular "contingent" factors in order to optimize outcomes. In this case, the contingent factors are:

- The tasks, or functions, that the specific channel will perform
- The target, or the market segment the channel will focus on

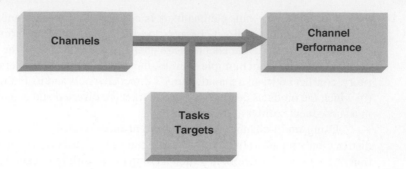

FIGURE 9.4 Contingency Approach to Developing Hybrid Channels

The key steps in effectively implementing this contingency theory to manage hybrid channels follow.[39]

a. ***Identify customer target segments.*** For channel purposes, useful segmentation variables are the size of customer, geographic region, products purchased, or buying behavior/needs. For example, high-tech companies frequently use a direct sales model for their largest customers, for those with complex buying needs, or in key geographic regions (either larger metropolitan areas or certain international markets). In addition, high-tech companies frequently bifurcate their product line so that certain products are available through certain channels, as a way to minimize channel conflict. The channel strategy must be integrated with the total company strategy of what products go to what markets. Even the best distribution strategy will fail without well-founded product and target decisions.

b. ***Delineate the tasks or functions that must be performed in selling to those segments.*** As noted previously, channel members must perform a variety of functions. One useful way to apply the contingency theory for channels is to delineate the tasks in terms of the selling cycle (or sales process): lead generation, sales prospect qualification, presales activities, closing-the-sale activities, post-sales service and support, and ongoing account management. Hybrid channel strategy should include not only the "front end" of sales and sales support, but also the "back end" of order processing, inventory fulfillment, and customer care. One cannot make channel decisions in isolation of these back-office functions.

c. ***Allocate the best (i.e., most efficient and effective) channels to those tasks.*** In multi-channel marketing, companies have a range of options to perform channel functions and to reach market segments. Effective multi-channel marketing recognizes that channels must be harmonized to best meet customer needs. In addition, channels should be combined to optimize costs and coverage relative to the tasks they are performing for various customer segments. For example, direct sales, company-owned retail stores, independent retailers, distributors, dealers/VARs, the Internet, and so on, can be used in some combination.

With respect to a Web-based sales channel, a company might decide to use its website only to disseminate product information and to generate leads; potential buyers may then be directed to existing dealers. Alternatively, to differentiate it from other channels, the company may use its website for sales only for certain products or for certain order sizes. In addition, to the extent that prices can be harmonized across channels (say, the company's pricing on its website is consistent with its suggested retail selling prices), cross-channel conflict can be minimized.

The idea behind the channel contingency theory is to establish clear boundaries for who owns which customer.[40] The hybrid channel strategy delineates customers on the basis of size, order size, decision process, industry, geography, or product line (low end, mid-range, or high end). Using such heuristics to identify which channels can pursue which customers will keep the level of conflict in a manageable, functional range. For example, a major manufacturer of computer-aided manufacturing systems (CAD/CAM) sells its offerings in the United States and Europe through a direct sales force; in Japan, it uses an exclusive distributor. Because the channels and customers are physically separated, little conflict occurs (except in global accounts). Similarly, Xerox used product boundaries when it entered the personal copier market. It sold mid-range and high-end machines through a combination of direct sales and dealer distribution and low-end machines exclusively through retail channels (e.g., electronics and appliance stores, mass merchants).

Additional Considerations: Channel Relationship Quality, CRM, Compensation, and Communication

Effective multi-channel marketing also must consider:

- *The tenor of the relationship:* Business relationships can be characterized on a continuum as either more relational (based on trust, commitment, and a shared sense of collaboration) or more adversarial. Companies that have a history of working collaboratively with channel partners have more flexibility in adding new channels, because their partners have the expectation that the increased market opportunities will be equitably allocated.
- *CRM practices:* The company's multi-channel strategy must be supported with a customer relationship management (CRM) system that allows customers to be effectively tracked. Multi-channel marketing requires a fully integrated experience supporting the different channels, and although traditional information technology has been able to provide this, the evolution of the technology is to *service-oriented architecture* (SOA). SOA spending for multi-channel retailing will reach \$9.5 billion by 2012, according to market analyst Datamonitor.[41]
- *Compensation and communication:* One of the key ways to get buy-in to changes in channel strategy is consistent reinforcement of the strategy delivered by sales leadership and a compensation structure supportive of that strategy. Even for indirect channels, company salespeople establish relationships with channel intermediaries. If the salespeople do not buy into the changes, they may sabotage or undermine the changes. Existing channels must be compensated for the potential loss of income that flows to the new channel, or the new channel may be sabotaged. Communication with all channel members (new and old) about the logic behind the strategy can go a long way toward keeping the focus on ultimate objectives.

The accompanying Technology Expert's View from the Trenches explains Cisco's approach to multi-channel marketing.

TECHNOLOGY EXPERT'S VIEW FROM THE TRENCHES

Bucking the Conventional Wisdom on Channel Programs

SURINDER BRAR

Worldwide Channel Strategy & Programs
Cisco Systems Inc., San Jose, California

Cisco's 2001 redesign of its channel program from volume-based compensation to value-based compensation is still unique in the technology industry and was very timely: Over the next few years the dot-com bubble burst, but the smaller partners in the Cisco channel were not driven out of business by larger partners because they competed on a level discount basis.

In addition to introducing the value-based model in 2001, Cisco introduced the requirement that, because all channel partners are a direct extension of the Cisco brand, all channel members be measured for customer satisfaction through an independent third-party survey. Partners are required to send a survey to their Cisco end-customers and meet a minimum target on a scale of 1 to 5. This program has been extremely successful; the score through channel partners has increased from 4.06 to 4.59 over the past eight years.

(continued)

By 2006 Cisco had about almost 2,800 Gold, Silver, and Premier Certified partners worldwide, with each level representing an increased capability around Cisco technologies. But the market had evolved and Cisco needed to drive additional channel capability in two dimensions: (1) integrated technology partners who could integrate multiple Cisco advanced technologies (like Unified Communications, Security and Wireless LAN) onto the network, and (2) deeply specialized partners who could provide a level of depth in Unified Communications and Security that was far deeper than the integrated partner's capability to support complex deployments in these technologies.

Cisco also needed to drive new advanced technologies through channel partners and create a new set of partners that could focus on covering the small and medium business (SMB) customers. Thus, in 2006 Cisco enhanced its channel program to address these gaps by:

1. Repositioning the Gold, Silver, and Premier tiers as "integrated partners" by changing their requirements so that these partners now had capability across multiple Cisco advanced technologies
2. Creating a new brand called Select, which was positioned below Premier and required these partners to take two days of training through an SMB specialization that focused on small and medium business customers
3. Formalizing Cisco's technology training framework into specialization tiers called Express, Advanced, and Master. This also allowed the company to create a new Master brand for Unified Communications and Security to provide recognition for the most deeply specialized partners

Cisco allowed partners two years to transition to the enhanced program and this was extremely well received by partners. By June 2008, Cisco increased the channel community to 10,000 certified partners including 7,000 SMB-focused partners. In addition to this, Cisco has another 36,000 registered partners who have typically not invested in Cisco capabilities, but have access to Cisco products (through distributors) that are easy to sell and deploy.

In Cisco's Channel Program, there is no incentive for volume at all, avoiding destructive price competition among VARs. Rather, incentives are designed to encourage value-added activities by partners. The company identified four key behaviors that it wanted to reward partners for: hunting for new business, upgrading the installed base, selling Cisco technology wrapped in business solutions, and building practices in key technologies like Unified Communications and Security. These incentives have also been very well received by the channel partners and the company now provides almost $1 billion through these programs to reward partner behavior in these areas.

Cisco focuses on two primary metrics to measure channel success: partner profitability and customer satisfaction. The former is important because it is key to channel loyalty and allows Cisco to ensure that its channel partners make sufficient margin on their Cisco business. And customer satisfaction is critical to ensure that the partners are successfully representing the company's brand.

Through its channel program, Cisco has both expanded channel capacity and increased the value-add that partners provide to end customers over the past eight years. During this period, the Cisco Channel Program has also been recognized as the best program in the industry by VARBusiness ARC and by the CRN Champions awards every year. The company today does over 80% of its total business through the channel, or over $30 billion in revenue per annum.

Cisco's experience with its channel program bucks the conventional wisdom that a B-to-B technology company needs to depend more on direct rather than indirect channels, or that an indirect channel needs to be managed on a volume-based reward structure instead of a value-based reward structure.

EMERGING CONSIDERATIONS IN DISTRIBUTION CHANNELS

Distribution for "Digital" Goods: The Long Tail

The deployment of broadband technology coupled with the digitization of information has profound implications for distribution in many industries, both high-tech and traditional. For example, the distribution of software is moving to an online business model, as is distribution of other digitized products, such as information (books, news, research), entertainment (music, TV shows, movies), and other goods that can be transformed electronically (advertising, banking, travel, financial planning, stock trading). Across this broad swath of industries, the Web enables economies of scope that cannot be duplicated offline, with many implications that go well beyond distribution to include pricing as well. We explore one specific implication here: the notion of the "long tail."

In the offline world composed of physical inventory, companies tend to follow the 80/20 rule to promote efficiency: Stock the 20% of the most popular items that will bring in 80% of the revenue. However, with information goods distributed electronically over the Web, the 80% of slow-moving items—what is called the **long tail** of a frequency distribution of consumer preferences,[42] or the large number of items for which only a few consumers express a preference—can also be sold profitably (in addition to the 20% of the most popular items).

Because the Web provides tremendous geographical reach of consumers with very little inventory and fulfillment cost, even individual, slow-moving, low-demand items—when aggregated—bring in sufficient revenue that exceeds their inventory and fulfillment costs, thus creating a viable business model. For example, in the movie rental business, Blockbuster's traditional retail business model was constrained by the number of movies it could physically stock on DVD in each of its stores. Moreover, it had to stock a large number of the newest hit movie releases—which raised its costs—to satisfy its customers in each store. In contrast, the Netflix DVD-by-mail business model allows Netflix to distribute more independent movies, in the aggregate, at lower cost and higher margin. No wonder, then, that Blockbuster's traditional business model has faltered and it has belatedly embraced the Netflix business model with its Total Access service.[43]

There is an interesting controversy between the blockbuster strategy of focusing on hit products versus the long tail strategy of focusing on niche products. Chris Anderson, the original proponent of the long tail theory, predicted that not only will the number of products in the "long tail" exceed the number of products in the "head," but also the total sales from the "long tail" products would exceed the total sales from the "head" products.[44] Anderson's theory is based on the argument that offering products that better meet the needs of niche customers will shift the demand curve in their favor, benefits which continue to be debated.[45] A recent study examined data from the online music store Rhapsody and Quickflix, an Australian DVD-by-mail service.[46] The analysis suggested that companies should not abandon the blockbuster strategy even in the case of digital information products and services and ought to proceed cautiously in pursuit of the long tail.

Understanding Gray Markets

In managing and controlling distribution channels, manufacturers want to ensure that the channels they use reach the appropriate customer segments with the combination of products and services each segment desires. In some cases, however, some channel members, rather than selling and marketing the firm's products to legitimate customers, may sell to unauthorized distributors or markets. For example, if a manufacturer offers large volume discounts, a distributor may feel compelled to buy a large quantity to take advantage of the discount; however, rather than inventorying those goods to be sold at a later date to legitimate customers and resellers, the distributors may "divert" them to unauthorized distributors, exporters, or other markets. Alternatively, if a large price differential exists between export markets (say, due to tariffs or other conditions), channel intermediaries in one market might try to take advantage of the price differential and sell to unauthorized distributors in those markets.

Known as the **gray market**, such unauthorized distribution refers to the selling of goods at discounted prices through resellers not sanctioned by the producer[47] and results in a lack of control over the product, distribution, and services. It is considered a legal violation for firms to sell trademarked goods without the manufacturer's approval. In the case of international markets, the appropriate customs service can levy cease and desist orders or prevent the import of offending goods.[48] A study published by KPMG (a tax and accounting consultancy) and the Anti-Gray Market Alliance (AGMA, a trade association made up of major technology firms) found that roughly $40 billion of computer-related products move through a gray market every year, resulting in nearly $5 billion of lost profits for OEMs.[49]

A case in point is the Apple iPhone. Apple reported that sales of iPhones worldwide for 2007 were about 3.7 million units. During 2007, iPhones were available through authorized channels only in the United States, France, and Germany. After accounting for France and Germany and units that may have been unsold in distributors' inventory, the U.S. market accounted for sales of about 3 million units. iPhones were available in the United States through Apple direct and through AT&T. However, AT&T, the exclusive provider of iPhone service in the U.S. market, reported service activations for only 2 million units. Industry analysts were confounded by this unaccounted inventory of 1 million units.[50] It turns out that people were buying iPhones in the U.S. market, having them unlocked with easily available hacker software from the Web, and taking them to the overseas gray market, where they were sold in countries like China, where Apple had no authorized service provider. When unlocked, the iPhone could work with any local service provider.[51]

Although some may believe that the gray market offers a way to move product through another channel, it can cause serious problems. Even when only a small amount of total sales is diverted, legitimate channel members and resellers may become confused and angry over the unfair advantage the unauthorized distributors gain in buying and selling at a lower price point. Both the legitimate outlets and the firm's own sales force may lose business to the gray marketers. Gray marketers typically don't provide the sales and service that authorized distributors do. In turn, legitimate channel members become less motivated to support the firm's products. In some cases, the company's products may wind up competing across different channels, lowering the price point.

Ultimately, the negative backlash can end up hurting the firm's brand and position in the marketplace. A recent study from Taiwan finds that authorized and gray goods have a significantly different impact on brand equity for high involvement products such as consumer electronics. Consumers evaluate authorized goods as having higher perceived quality than gray goods. Therefore, manufacturers and authorized dealers should try to customize offerings to local needs in different countries to minimize the impact of competition from gray goods.[52]

CAUSES   In order to address the issue of gray markets, a firm must have a solid understanding of its root causes. As shown in Table 9.3, the most commonly cited cause of gray markets is a firm's pricing policies.[53] Manufacturers tend to structure discount schedules in favor of large orders, which causes distributors and other customers to buy more than they can sell or use (known as *forward buying*) and then to resell the rest to unauthorized resellers. This problem can be exacerbated if distributors and customers must commit to purchases far in advance, with penalty clauses for canceling orders. Although manufacturers may have strong production reasons for such conditions, it contributes to the gray market problem.

TABLE 9.3 Causes and Solutions for Gray Markets

Causes	Solutions
Volume discount price policies	Eliminate sales to the source of the gray market
Differentials in exchange rates	Eliminate the arbitrage problem: one-price policy
Different resellers' cost structures	Increase market penetration
Highly selective distribution	Gather information on gray market problem
Producers performing many marketing functions	Institute consistent performance measures internally
Inconsistent internal policies	

A second cause is the price arbitrage opportunity that arises from differentials in exchange rates in international markets. Termed *parallel importing,*[54] this type of gray market activity occurs when goods intended for one country are diverted into an unauthorized distribution network, which then imports the goods into another country. Globalization has increased this type of gray market opportunity.

A third cause arises from the cost differences between different resellers' cost structures.[55] For example, full-service resellers have a higher cost structure than do discounters. Full-service resellers tend to provide functions including advertising, product demonstrations, post-sales service support, and so forth. Gray marketers can free-ride on these functions. Resellers may also use some products as *loss leaders,* selling products at or below cost in order to attract traffic. Loss-leader tactics are common among gray marketers of popular brand-name products. Full-service resellers look to brand-name products to perform a very different role in their businesses; the product is expected to generate sufficient profit to cover the costs associated with providing full service.

Fourth, gray markets may be fostered when suppliers practice highly selective distribution. Territorial restrictions, exclusive dealing, and so forth may lessen intrabrand competition, but can draw unauthorized dealers into the market if demand is strong.[56] This was the case with the Apple iPhone.

Fifth, gray markets can also develop when producers assume a range of marketing functions that might otherwise be provided by full-service dealers. When such services are available from the manufacturer, the buyer's risk in buying from minimal-service dealers is reduced. Moreover, heavy advertising by the producers may build total demand and brand image and decrease buyers' dependency on resellers' reputations.[57]

Sixth, inconsistent and incompatible policies regarding a manufacturer's own departments can contribute to the gray market problem.[58] Plant managers may view gray markets positively when they contribute to the ability to operate plants at full capacity. Sales personnel may overlook gray market activity if the volume contributes to the quota in their territory. Such problems make it difficult to establish a clear solution to the gray market problem.

SOLUTIONS   In light of the causes, the solutions are many and varied. A basic question for firms dealing with the gray market problem is whether they should resort to legal enforcement and punitive action against offenders. For example, a firm may use serial numbers on products to track the source of units sold to gray marketers. Then, the firm can cut off the offender and is legally justified in doing so. Although the firm may lose some sales in the short term, it may sell more units through its remaining authorized distributors and mitigate price erosion. Such a move sends a strong signal of commitment to the authorized distribution channel. However, this solution can be costly in terms of the time and administrative burden of identifying the offending dealers. Moreover, cutting off a distribution network may be satisfying to the authorized dealers but may actually cut off a market where the firm may have some competitive advantage.[59] Finally, empirical evidence shows that enforcement severity (or raising the magnitude of the punitive response to gray marketers through fines, litigation, and termination), by itself, does not result in decreased gray market activity.[60] Rather, enforcement severity must be coupled with a high likelihood of detection ability as well as speedy retaliation. However, these actions require significant investments in monitoring and are affordable only for larger firms.

A second solution is to eliminate the source of the arbitrage and offer a one-price policy with no quantity discounts.[61] This strategy, although useful in eliminating one of the root causes of the problem, forecloses valid price differentiation opportunities among different types of customers who have different transaction costs and receive different benefits from the product. It doesn't reward the larger, full-service dealers in the network, who may have other options available to them.

A third solution is to move toward increased penetration in the market,[62] balancing the potential of attracting unauthorized distributors with restrictive distribution against the increased intrabrand competition from too-intensive distribution. Some companies opt for a differentiated product line, in which only some versions of their products are available through certain channels.

To minimize the gray market problem for the iPhone, Apple in 2008 signed up authorized service providers to distribute the next generation iPhone 3G in 73 countries.[63] Surprisingly, China wasn't on this list as of November 2008. For the U.S. market, Apple made one big change: iPhone 3G would be available for purchase only with service activation, to minimize loss of service revenue from device sales on the gray market.

Above all, it is important to have information on the extent to which gray markets exist in the distribution system, coordinated pricing brackets, and consistent performance measures.

Black Markets, Piracy, and Restricted Exports

Counterfeit, high-quality knockoffs pose yet another problem that is endemic to high-tech industries. Given the unit-one cost structure, in which the cost of producing the very first unit is high (due to R&D investments) relative to the reproduction costs of subsequent units, pirated copies of software and related items can be made relatively easily and inexpensively. A 2006 study revealed that 35% of the software installed on PCs worldwide was obtained illegally, amounting to losses of about $40 billion due to software piracy. Progress was reported in curbing software piracy in China and Russia, where the piracy rate over three years dropped by 10% and 7%, respectively.[64]

In addition to being aware of potential counterfeit problems, firms selling strategic high-tech products, such as satellites, must also be aware of **export restrictions**. For example, U.S. chip makers must submit applications to sell microprocessors to certain countries, such as the former Soviet Union states or China. Sales of "dual use" products—nonmilitary items with military applications—are restricted to some countries. Even when restricted items are sold legally, problems can arise. For example, in March 2007, ITT, the U.S. military's largest supplier of night-vision goggles, pleaded guilty to two violations of U.S. export law for selling night-vision technology to a company in China for more than two decades. The company pleaded guilty to exporting without a license and making false statements in connection with transferring technology to the Chinese company through a contractor in Singapore. It agreed to forfeit $100 million and was barred from shipping devices to "specific parties" for at least a year. The company attributed the violations to the actions of a few rogue employees.[65]

The purpose of rules against technology sales is to protect U.S. security interests abroad. But the policy is fraught with problems. Some industry experts say that it is impossible to prevent the goods sold to friendly countries from ending up in restricted-access countries. Intel spokesperson Bill Calder says, "We ship chips to thousands of distributors all over the world, who aren't prevented from selling to these countries. There's a disconnect there."[66] Moreover, some believe that the controls actually undermine the U.S. position as a technology leader. For example, the U.S. share of the international satellite market dropped from 73% to 53% during the 1999–2000 time frame; and French, Canadian, and German firms picked up the satellite contracts that the U.S. firms couldn't fill due to export restrictions.[67] Others believe that, rather than restricting access to such products, it may actually serve U.S. strategic interests to have countries such as China use U.S. technology—the United States will understand the technology being used. Moreover, strict restrictions could drive these countries to other suppliers, which might put even more information in the hands of other countries.

Manufacturers of high-tech goods must be aware of issues surrounding export controls in order to ensure that their distribution channels operate legally. And manufacturers must take proactive steps to protect their goods against counterfeiting.

Unique Distribution Requirements for Developing (BOP) Markets

As global technology companies address the needs of base-of-the-pyramid (BOP) markets, they must recognize that the distribution channel infrastructures they take for granted in developed markets may not be available in emerging markets. Some of the challenges they face are:

- Large segments of the population are poor and illiterate, so product design has to take into account affordability and rely more on experiential learning rather than documentation in manuals.

- Large segments of the population live in rural areas where transportation access is difficult and advertising media are not well developed, so physical distribution and product promotion are constrained.
- Electric power is not widely available, so innovative ways to recharge batteries for devices or alternative energy sources must be thought of.
- Independent businesses to perform distribution channel functions are not often available, so companies have to choose between investing in direct channels or motivating and training channel members from scratch.

Constraints and trade-offs are so unique that products usually have to be designed from scratch rather than adapting products that were successful in developed markets. Designers and marketers have to spend time observing and learning about these consumers and markets through systematic research efforts; the payoff can come from the sheer volume and growth in these markets.

Technology companies need innovative business models to serve BOP markets. For example, in Bangladesh, Grameen Telecom, a wholly owned subsidiary of Grameen Bank, founded by Nobel laureate Muhammad Yunus, offers cell phone service to 40,000 mainly rural consumers. The pilot project involved providing cell phones to about 950 village women via microcredit. These women each acted as the public call center in the village and resold cell phone service to consumers by the minute for a modest profit of $50 to $100 per month.[68] The program was on track to expand to 40,000 village women by 2004. Other companies such as Selco (see case at the end of the book) and Envirofit are designing creative distribution models for their products (solar lighting and two-stroke motor engine retro-fit kits, respectively) in BOP markets. This is such a new area that very little research and few frameworks exist on best-practices distribution for developing countries—although social entrepreneurs, governmental agencies, non-governmental organizations, and corporate philanthropy efforts likely all will play a role as these markets continue to be an important focus for business models of the future. The accompanying Technology Solutions for Global Problems offers one example, and Chapter 13 further explores this topic.

SUPPLY CHAIN MANAGEMENT

The final section of the chapter broadens the focus beyond just the distribution of the product from the manufacturer to downstream customers, to include the logistical management of all of the incoming components and pieces used in the manufacturing process of a particular product, also known as **supply chain management**. Whereas distribution channels trace the flow of product after it leaves the manufacturer, through the various intermediaries and institutions that add value along the way to the end user, supply chain management is concerned about matching the inflow of supplies and other materials at every stage of the manufacturing process to the actual demand exhibited by customers in the marketplace. The goal of the supply chain manager is to partner effectively to design an entire supply chain from raw material to end customer that meets all the goals of service levels, availability, flexibility, and cost. Ultimately, intelligent supply chain design requires accurate forecasts, flexibility, a focus on the customer, and effective collaboration across both intra- and interorganizational boundaries.[69]

High-tech product life cycles are getting shorter and shorter, which means the production cycles ramp steeper and steeper, both during new product introduction and at the end of the product's life. Also, the days of steady, reliable forecasts from large OEM customers are gone. Demand is constantly changing, causing tremendous pressure on materials supply chains to support the customer's needs without taking unwarranted risk or incurring unnecessary liability. Moreover, when the company outsources the production/manufacturing of the high-tech product, supply chain management is even more challenging, and requires cross-functional integration across procurement, planning, operations, outsourcing, and marketing.

TECHNOLOGY SOLUTIONS FOR GLOBAL PROBLEMS
Bike Technology Solutions for "the Other 90%"
By Jamie Hoffman

Photos reprinted with permission of Ed Lucero, International Bicycle Development, Santa Fe, New Mexico.

East African residents have long relied on bicycle-taxis, called boda-bodas, for personal and commercial transportation in the same way that developed countries utilize trucks and school buses. In Uganda alone, more than 200,000 men are employed as boda-boda drivers. Unfortunately, the majority of the bikes are ancient roadsters, engineered in the 1930s, which have serious maintenance, safety, and capacity issues resulting in accidents and chronic health problems for the drivers. Due to the roadsters' dangerous qualities, locals call them "Black Mambas" (after the very deadly snakes found in Africa).

The U.S. nonprofit organization Worldbike takes an open-source and partnering approach to solving these problems, innovating modifications to bike frames to allow greater stability, carrying capacity, and safety. It offers blueprints, instructions, and support for all its designs on its website (www.worldbike.org). In addition, it works with partners in local communities for manufacturing and distribution, creating income-generating opportunities for the world's poor. In May 2007, Worldbike was selected by the Smithsonian's Cooper-Hewitt National Design Museum to have two load-carrying bicycles—the Big Boda and the Worldbike—in their "Design for the Other 90%" exhibit.

For example, the Big Boda (see photo above) is a simple add-on bike frame that extends the bike's wheelbase, allowing two adults, three children, or bulky cargo to ride with greater stability. To achieve this, several alterations are necessary: Long rear and side steel rods are added at the back of the frame, the back wheel is pushed back, and the wheel base is extended. The Big Boda sold for around $110 in 2006; although 40% higher than the price of a typical bike, the increased capacity allowed the bike to pay for itself in only 3 to 4 months.

Newer developments by Worldbike include a new process of modifying a mountain bike (see photo at left), rather than adding on to the 1930s-era Black Mamba. This utility bicycle modification is less expensive and faster/easier to build since the extension is welded directly to the existing mountain bike with fewer materials and less labor. Mountain bikes have many benefits compared to Black Mambas, such as safe brakes, more gears, and a stronger frame.

Sources: Freedman, Paul, "Final Report: Big Boda Load-Carrying Bicycle Trial Market in Kisumu, Kenya," Worldbike.org, 2006, online at www.worldbike.org/files/Final-report.pdf; "Big Boda Load-Carrying Bicycle," Worldbike.org, 2007, online at www.worldbike.org/technologies/big-boda-load-carrying-bicycle; and personal correspondence/interviews with Kristna Evans, executive director, Worldbike, July 25, 2008, and Ed Lucero, International Bicycle Development, July 23, 2008.

Given the characteristics of high-tech products (high per-unit costs, high obsolescence, and competitive volatility), firms are concerned about efficient supply chain management, and frequently, standard practices are ill-equipped to deal with the risks and uncertainties in these markets.[70] For example, problems in its supply chain resulted in numerous delays of the Boeing Dreamliner jet (see case at the end of this book). Also, Nintendo's low availability of its Wii game console at launch resulted in huge opportunity costs from lost revenue. Samsung follows the "sashimi" theory of inventory and pricing: Supply should be maximized at launch because that is the time that a company can charge the highest prices for electronics components; later, price declines as inventory tends to become obsolete.[71]

For example, a company may set a goal of ensuring that no one component will result in production delays; other goals may be flexibility and cost. A goal of high availability can be achieved by building a very large amount of product, far above what a company expects to sell. However, in doing so, the likelihood is high that a company will have to scrap a large amount of inventory once the product reaches end of life. This is neither efficient nor cost-effective.

Effective supply chain management uses a demand-backward approach, starting with an accurate forecast of customer demand for individual products and probable lead times. Then, companies can set finished goods inventory levels and channel locations to best support the highest demand items with the most inventory closest to the customer. For the remaining items, companies may choose lower inventory levels and support product requests with timely factory builds. After looking at customer demand, the company will address component inventories both in the factory and in the pipeline. By considering cost, lead time, number of suppliers, and overall industry conditions, the firm can determine the inventory level for each component, where in the pipeline that inventory should be held, and when it should move from one stage to the next.

In supply chain networks, the vertical relationship between any two firms in the network is affected by the other relationships that the firms have. Indeed, firms in supply chains are highly dependent on each other for dealing with uncertainty at the end-consumer level. Consider a supplier like Intel providing microprocessors to HP, whose laptops and notebooks are sold through retailer Best Buy to end-consumers. HP's ability to deal with end-consumer uncertainty depends, in large part, on its relationship with Intel. If consumer preferences shift quickly in the direction of more powerful notebooks with longer lasting batteries, HP will be able to respond to these needs only if Intel accommodates HP's requests for more powerful processors that consume less power. A manufacturer's flexibility in accommodating requests of the downstream customers is closely related to the relationship with its upstream suppliers. In particular, manufacturers exhibit more flexibility in responding to downstream customer needs when: (1) the manufacturer invests in prior supplier qualification and (2) the supplier and manufacturer both make symmetric idiosyncratic investments in their relationship.[72]

Traditional supply chain management texts address topics such as warehousing, logistics, and transportation;[73] the focus here is on how high-tech companies can effectively manage their supply chains to deal with the high level of uncertainty in their markets, as well as on the technologies used to enable effective supply chain management. Consistent with the theme in this book, the final topic in this section addresses how companies are making their supply chains more environmentally friendly.

Matching Supply Chain Strategies to Uncertainty

Supply chains must be designed to deal with two kinds of uncertainty surrounding the product—demand uncertainty and supply uncertainty. *Demand uncertainty* is related to the difficulty of predicting end-consumer demand for the product. Supply chains have to deal with the challenge of the *bullwhip effect*:[74] Even though the end-consumer demand may be quite stable, market signals get distorted up the supply chain so that demand fluctuations at the manufacturer or supplier stage are typically larger than end-consumer demand fluctuations. Demand uncertainty varies from low for functional products familiar to end consumers to high for innovative products perceived as risky by end consumers. The way to minimize the stresses of demand uncertainty on the supply chain is for supply chain members to share information about market demand. A variety of information technology solutions and supply chain management software are available to help reduce demand uncertainty.

Supply uncertainty is the unpredictability associated with obtaining the right quality and quantity of raw materials, components, infrastructure, supplies, and services to offer a finished product to the market. The variation in supply uncertainty can vary from low for a stable supply process with mature technology, to high for an evolving supply process with changing technology and unknown supplier base. Supply uncertainty may be reduced by early design collaboration and joint product development with suppliers, and later on by sharing information about product transitions and end-of-cycle issues. In addition, manufacturers may participate in online marketplaces (also known as *electronic hubs* or *exchanges*) for synchronized planning with their suppliers.

By aligning supply chain strategies with the underlying dimensions of demand and supply uncertainty, contingency theory provides a useful model for supply chain planning. Depending on the nature of the demand and supply uncertainties, four different types of supply chains are possible:[75]

Efficient supply chains. When demand uncertainty is low (for familiar, functional products) and supply uncertainty is also low (for stable supply processes with mature technology), supply chains are geared toward economies of scale, reduction of non-value-added activities, and optimization of forecasting and inventory by sharing accurate information speedily with suppliers. Supply chains of Wal-mart and Costco operate largely in this manner.

Risk-hedging supply chains. When demand uncertainty is low and supply uncertainty is high (an evolving supply process with changing technology and suppliers), supply chains are geared toward pooling and sharing of resources to avoid disruptions in the chain. Cultivating second sources, maintaining extra inventory, and having manufacturing facilities in alternate locations are all intended to hedge risks of disruption. To the extent that firms can share these resources with others in the industry, they can be efficiently utilized as well. Military supply chains operate largely in this manner.

Responsive supply chains. When demand uncertainty is high (for innovative products perceived as risky by end users) and supply uncertainty is low, supply chains must be flexible in meeting the changing needs of customers; they rely on accurate order information and mass customization to be truly effective. Dell is a master practitioner of the responsive supply chain strategy.

Agile supply chains. When both demand and supply uncertainty are high, supply chains should combine the characteristics of the risk hedging and responsive supply chain strategies. Working with alternate suppliers on different technologies, sharing these resources with others in the industry, and striving to be flexible in mass-customizing products for customer needs characterize agile supply chains. Amazon's ability to work with a number of distributors or wholesalers for inventory and its utilization of a number of different shippers to get customized assortments in an order shipped to consumers at the time they want and at the price they are willing to pay exemplifies an agile supply chain.

Supply Chain Management Technologies

The Internet and other technologies have greatly improved supply chain management, enabling several kinds of efficiencies to be realized.[76] Information sharing between suppliers and customers helps in better matching supply and demand conditions. The Internet enables suppliers to increase their reach and find new customers at lower cost. Dynamic pricing, as in auctions, helps suppliers clear unsold inventory and helps customers procure goods and services at lower cost. In addition, technologies such as collaborative software and RFID chips that can track and harmonize supply chain management activities across supply chain members can provide further efficiencies.

ONLINE PLATFORMS   During the dot-com boom, *electronic hubs* or *exchanges* proliferated. These online marketplaces, whether vertical (industry specific) or horizontal (across industry boundaries, such as those used for advertising space or maintenance, repair, and operating supplies) brought business buyers and sellers together and provided Web services ranging from e-commerce (sales of products) to document sharing, bill presentment, order tracking, contract negotiations, and credit services. Some of them were independent public exchanges, open for participation to any company that wanted to join and was willing to pay a membership fee. Others were sponsored by leading players in an industry. After the dot-com bubble burst, the number of electronic marketplaces dwindled due to their inability to attract a critical mass of users. Many closed and others consolidated. For example, Covisint, a vertical exchange for the auto industry, sold its exchange assets to Freemarkets, which was acquired by Ariba. Virtual private networks (VPNs), or private exchanges, are open only by invitation to trusted partners. For example, a VPN allows channel members to check inventory levels across all geographic markets, and to place orders in a secure online environment.

Because public, open exchanges easily allow for price-based competition (using reverse auctions and dynamic pricing, for example), they are well suited for short-term transactions of commodity products between businesses. However, such tools can seriously undermine the tenor of more collaborative, trusting relationships where parties may have worked together over a longer time horizon to jointly develop innovations or open new markets.[77] Conversely, because private exchanges allow trusted business partners to share sensitive information and open new lines of business (e.g., by reaching new customer segments), they are more suited to collaborative relationships; in more collaborative relationships (unlike short-term, transaction-oriented situations), firms can be more confident that their information and knowledge will not be unfairly used by their business partners.[78]

The latest incarnation of these Web-based supply chain practices is called **e-procurement**, which is a large Web-based marketplace encompassing many vertical industries. Ariba now offers software, services, and expertise to help companies analyze their procurement spending. It can also connect companies with its supplier network of more than 140,000 global suppliers of a wide range of commodities to realize savings.[79] E-procurement enables Web-based procurement services such as sourcing or finding suppliers, tendering or requesting information on prices and specifications from suppliers, placing purchase orders and receiving goods, reverse auctions (see below), and transfer of payments. Being Web-based, e-procurement solutions are also Software as a Service (SaaS), which means rather than residing on the customer's own computers, the software can be accessed over the Internet through a Web browser on an as-needed basis.

REVERSE AUCTIONS   A **reverse auction** is a method of e-procurement where a company puts its purchasing requirements for suppliers to see on a website, and then asks them to competitively bid against each other for the company's business over a predefined time period. The notion of "reverse" captures two facts: (1) the seller bids for the right to supply the buyer's purchasing needs, and (2) the lowest price bid wins. Although reverse auctions can save on procurement costs for buyers, they are not very attractive for suppliers who have to engage in destructive price competition. Reverse auctions can be complicated; they vary according to factors such as the number of suppliers invited to bid, the size of the purchase contract, the number of lots (similar products grouped for common production and delivery), price visibility (the degree to which bidders can view competitive bids), how the winning bid is determined (sometimes based only on price, but in other cases the buyer may consider other factors such as quality or supplier reliability), and the magnitude of price savings the auction generates.[80]

Research by Sandy Jap on the impact of reverse auctions on suppliers shows mixed effects. On the one hand, as the number of bidders increases, suppliers become more suspicious of a buyer's motives, suspecting the buyer of providing false information, making hollow promises, and essentially being a poor partner.[81] On the other hand, as the size of the purchase contract increases, suppliers' overall satisfaction with and commitment to the relationship increases. By having about six to seven lots per auction (neither too few nor too many), suppliers express higher continuity expectations for the relationship. Partial price visibility (providing only rankings of competitors instead of exact price levels) has a better impact on supplier satisfaction than full price visibility. The higher the price drops during the auction, the greater the negative impact on the relationship. Thus, if buyers want to effectively leverage reverse auctions, they must carefully balance the factors that drive cost efficiencies against the factors that may negatively impact supplier relationships. In general, buyers should strive for auctions with large contract values, partial price visibility, and auction-determined award rules to safeguard their relationships. This provides the financial motivation with some level of pricing confidentiality without sacrificing bidding aggressiveness. Moreover, to avoid creating distrust in the relationships, buyers should have neither too few nor too many suppliers or lots in the auction.

SUPPLY CHAIN MANAGEMENT (SCM) SOFTWARE While supply chain management (SCM) software has some of the functionality of e-procurement discussed in the previous section, it is different in two ways: (1) It goes beyond procurement and includes manufacturing and distribution planning systems, forecasting, inventory management and warehouse management modules, transportation management systems, and international trade compliance and documentation; and (2) rather than being on a Web platform controlled by a third party, the software is typically installed on computers located within the company's facilities and those of its supply chain partners. For both of these reasons, installing SCM software is more expensive than Web-based e-procurement. SAP and Oracle are the big suppliers of SCM software and it is often offered as part of more comprehensive enterprise resource management/planning (ERM/ERP) packages. Other players include i2 Technologies, JDA Software, Infor Global Solutions, and IBM, among others. "Best of breed" software vendors (those who specialize in supply chain management applications, such as LeanLogistics and RouteView Technologies, for example) also are strong players in SCM software. In addition, companies are migrating to service-oriented architecture (SOA) platforms, jumping to an entirely new technology stack. Collaboration is the name of the game in spending on supply chain technologies, with an emphasis on sales and operations planning (S&OP) as well as vendor-managed inventory (VMI).[82]

"Supply chain management software revenues were $1.6 billion in 2005, and are expected to reach $2.5 billion worldwide by 2010, according to market analyst DataMonitor. The growth will be driven by small-to-medium-size manufacturers as well as larger manufacturers looking to replace legacy systems or start new implementation."[83] Advanced technologies, such as RFID, which rely on a supply chain management software suite, will also drive adoption as manufacturers seek to improve their overall business performance. For high-tech industries, these investments are designed to increase revenue growth as much as to reduce costs.[84]

RFID The opening vignette of Chapter 7 discussed the slow adoption of RFID technology by companies for supply chain management. With leadership from large retailers like Wal-Mart, Target, and Best Buy, technology companies and university researchers all over the world are working to improve the technology and lower the cost. Currently RFID readers can read the cheaper, passive RFID tags within a 30-foot radius. A Los Angeles start-up, Mojix, is working on technology that could potentially increase the coverage area of RFID tags and readers by a factor of 100, at 25% of the cost of current technology.[85] Such a breakthrough would accelerate the adoption and diffusion of RFID technology and realize huge supply chain efficiencies for the big retailers.

Outsourcing

High-tech companies have outsourced production to contract manufacturers in low-cost countries for many years. Now, other functions are also being outsourced. U.S. companies are relocating a greater number and range of functions to foreign countries, including call centers, stock analysis, accounting, tax return and insurance claims processing—and even software programming and other tech jobs. Clearly, this trend is fueled by the need to lower costs in a competitive, slow global economy. But there are downsides as well in terms of losing control over the supply chain and having to invest more in training and monitoring of outsourced contractors. A new trend is the move by technology companies to do *offshoring,* wherein they locate their own facilities and employees in low-cost countries rather than utilize contractors. For example, out of a global workforce of around 127,000, IBM employs 53,000 people in India.[86] An expanded discussion of outsourcing is in Chapter 5.

The "Greening" of the Supply Chain

Think about a customer's choice for an environmentally friendly consumer electronics gadget. It's all very well to select a product that was made with biodegradable plastics and environmentally friendly (nontoxic) manufacturing methods. But, if the product was air-freighted and cannot be recycled, then what's the point? Best-practices supply chain management is increasingly concerned about the environmental impact of supply chain activities. From sourcing, transportation, and production to end-of-life disposal and recycling considerations, companies can find ways to cut costs and gain competitive advantage through environmentally responsible, or "green," supply chain management practices. Companies' concerns are driven not only by the costs of doing business, but also by customers' increased desire to purchase from ecologically sensitive companies and by governmental regulations. For example, as much as 75% of a company's carbon footprint comes from transportation and logistics alone.[87] Moreover, the import requirements of the European Union countries follow ISO 14000 standards, and are much tougher with respect to environmental impacts than the U.S. market.[88] Companies such as Semitool, which have developed environmentally friendly manufacturing processes and products that meet ISO 14000 standards, have gained a first-mover advantage over other companies that have been slower to respond to these needs.

Companies are responding to this trend by having supply chains that deliver products and services with a minimal impact on the environment, and environmental organizations are holding them accountable. Since 2006, Greenpeace has been ranking global technology companies on how eco-friendly they are. In 2007 out of 14 companies ranked, the top five companies were Lenovo, Nokia, Sony Ericsson, Dell, and Samsung, while the bottom five companies were Toshiba, Sony, LG Electronics, Panasonic, and Apple.[89] The rankings were based on factors such as the use of toxic chemicals in the production process and how the companies manage recycling their goods when customers have stopped using them.

The scope of green supply chain management includes:[90]

- Green product design, or designing products with environmental impact in mind
- Green purchasing and materials sourcing
- Green manufacturing processes, including waste management, energy utilization, and minimal use of toxins in production
- Green delivery of the product, including transportation and minimal waste in packaging
- Recycling of e-waste (both in production and at end-of-life of product use) and adoption of reverse logistics

As companies continue their quest for cost efficiencies, environmentally friendly practices, and competitive advantage through differentiated strategies, the greening of the supply chain is a "natural" choice.

Summary

Managers must design and manage channel systems and supply chains strategically to achieve competitive advantage. This chapter focused on issues in designing and managing high-tech distribution channels and supply chains. A framework for developing channel strategies, including channel structure, management, and performance, helps managers think through the myriad issues that must be addressed. For example, relying on direct channels—such as a direct sales force, a company-owned website, or company-owned retail outlets—requires significant investments. Indirect channels can include a host of types of intermediaries, and the coverage, or number of intermediaries used, affects the level of intrabrand competition. Channels need to be adapted over the product life cycle to reach different segments of the population. Managing these channels requires careful attention to their governance or coordination mechanisms, as well as complex legal issues such as tying and exclusive dealing. Assessment of channel performance includes evaluating not only the various channels a company uses, but also the individual channel members.

In addition, multi-channel marketing (including bricks-and-clicks channels) was addressed. Concurrent channels must be integrated, or harmonized, so that customers can order products online and use physical stores for return centers, for example. While multiple channels increase market coverage for companies and provide convenience for customers, they need to be actively managed to minimize the conflict between alternate channels. Companies need to think through the pros and cons of adding new channels, particularly Web-based channels, because of the likely conflict with and cannibalization of its existing channels.

Other complexities in high-tech distribution channels include the new business possibilities that are cre-ated through electronic distribution of information and the "long tail," as well as the unique distribution consid-erations for base-of-the-pyramid markets and the con-comitant need to innovate new channels in these emerging markets. Gray markets, piracy, and export re-strictions are thorny problems that most successful global technology companies have addressed.

The need for supply chain strategies to be aligned with underlying demand and supply uncertainty is ad-dressed with a contingency framework providing useful managerial guidelines. Finally some recent trends in supply chain management were discussed, such as e-procurement, reverse auctions, outsourcing, RFID technology, and eco-friendly supply chains.

Discussion Questions

Chapter Concepts

1. What do distribution channel activities and strategies encompass? What is the goal of channel strategy?
2. Why are distribution channels hard to coordinate?
3. What is channel structure? Describe and give an ex-ample of each of the following structures:
   a. Direct channel structures
   b. Indirect channel structures
   c. Hybrid (dual, concurrent, or multi-channel) chan-nel structures
4. Are prices of goods in direct channels generally more or less expensive than in indirect channels? What are the sources of efficiencies that indirect distribution channels can provide?
5. Do the efficiencies available on the Internet mitigate the notion of contact efficiencies? Explain and/or give an example.
6. Is intrabrand competition exacerbated or lessened with multi-channel marketing? Why?
7. What affects customers' selection of which distribu-tion channel to use?
8. What are some of the reasons a company uses a direct sales channel?
9. Why do companies experience conflict between their Sales and their Marketing Departments? What are the benefits a company can experience when Sales and Marketing work well together?
10. From Box 9.1 and Table 9.2:
    a. What are the four types of Sales–Marketing rela-tionships? Under what conditions is each type of relationship viable, and under what conditions should each be "upgraded"?
    b. What are the two major sources of the friction be-tween Sales and Marketing? Explain each with sufficient detail to capture the complexity.
    c. What are the differential effects of Sales' and Marketing's different orientations and compe-tencies on the quality of the collaboration and on the firm's market performance? What are the implications?
    d. What are seven ways that companies can work to-ward more collaborative Sales and Marketing rela-tionships?
11. What is a bricks-and-clicks distribution model? What factors should a company consider before adding a Web-based sales channel?
12. What is disintermediation? What are the factors that have limited or facilitated the extent of disintermedia-tion?
13. What are the factors that appear to support a com-pany's successful use of company-owned retail stores? Based on these factors, what companies do you predict will open company-owned retail stores in the next 3 to 5 years?
14. What are the two key issues a company must address in using indirect channels?
15. What are the various types of intermediaries com-monly used in high-tech channels?
16. What is the nature of the trade-off between the *degree of coverage* and the *degree of intrabrand competition?*
17. What is the typical evolution pattern for channels in high-tech markets?

18. What is channel management?
19. What are the primary coordination mechanisms and influence strategies used to guide channel member interactions?
20. What is tying? What are its legal implications?
21. What is an exclusive dealing arrangement? What are its legal implications?
22. How can channel performance be measured?
23. What are the opportunities and challenges for technology companies in using multi-channel marketing?
24. What are the objectives of a hybrid (multi-channel) channel system? What are the steps in designing harmonized hybrid channels?
25. What is the contingency theory of hybrid channel design?
    a. What are the common segmentation variables for designing hybrid channels?
    b. What are the channel functions that the channel design should consider?
    c. How does a company "map" the various targets and tasks to optimize multi-channel design?
26. What other issues must be considered to effectively execute multi-channel marketing?
27. What is the long tail? What are its implications for profitable business models? What is its relationship to distribution?
28. What is the gray market? Why is it a problem? What are its causes and solutions?
29. Why are high-tech markets particularly susceptible to counterfeits (the black market) and piracy?
30. What is the purpose of export restrictions? Are these restrictions justifiable for U.S. companies?
31. What are the unique channel needs in base-of-the-pyramid markets? What are the implications?
32. What is supply chain management? Why is supply chain management complicated in high-tech markets?
33. What is demand uncertainty? How is it related to the bullwhip effect? What is supply uncertainty?
34. How does contingency theory help a company mitigate demand and supply uncertainties? What are the four types of supply chains that arise from this contingency model?
35. What types of online platforms have facilitated supply chain efficiencies?
36. What is the difference between e-procurement and SCM software? When should these different tools be used by what kinds of companies?
37. What are reverse auctions? What are their pros and cons? How should they be designed?
38. What are the various types of supply chain management software?
39. How are RFID chips used to streamline supply chains?
40. How has outsourcing affected supply chain management?
41. What are the various components involved in "green" supply chain management or eco-friendly supply chains?

Application Questions

Based on the opening vignette and the Cisco Technology Expert's View from the Trenches, answer the following questions:

1. What were the drawbacks of Cisco's distribution channel strategy prior to 2001?
2. What factors led to Cisco's channel redesign in 2001? How did these address the earlier drawbacks?
3. What factors led to Cisco's decision to redesign the VAR program yet again in 2006?
4. What were the new features of the 2006 VAR redesign? Do you think these were appropriate? Why or why not?
5. Conventional wisdom says that B-to-B technology companies should have more of a direct rather than indirect channel. Yet Cisco's experience goes against this. How can you explain Cisco's success in the light of this conventional wisdom?
6. Find a current example of a gray market. Offer your ideas for the causes and solutions.
7. Find an example of multi-channel marketing. Offer your insights about the degree of conflict, how the targets and tasks might be allocated, and whether the company is managing the channel well.

Glossary

agile supply chains Supply chains that combine the characteristics of risk hedging and responsive supply chain strategies; they encourage working with alternate suppliers on different technologies and flexibility in mass-customizing products for customer needs.

authoritative (unilateral) controls In coordinating channel relationships, the controls that reside in one channel member's ability to develop rules, give instructions, and in effect, impose decisions on the other.

bilateral controls In coordinating channel relationships, the controls that originate in the activities, interests, or joint input of *both* channel members and include mutual sharing of benefits and burdens, information sharing, and so forth.

bricks-and-clicks A distribution model that offers the firm's products through both retail stores ("bricks") and an online channel ("clicks").

contact efficiencies The decrease in the number of transactions (contacts) that must occur between vendors and customers when an intermediary is added.

disintermediation The elimination or bypassing of existing intermediaries that occurs when a firm adds an online distribution channel.

efficient supply chains Supply chains that are geared toward economies of scale, reduction of non-value-added activities, and optimization of forecasting and inventory by sharing accurate information speedily with suppliers.

e-procurement A Web-based marketplace offering supply chain services such as sourcing or finding suppliers, tendering or requesting information on prices and specifications from suppliers, placing purchase orders and receiving goods, holding reverse auctions, and transferring payments.

exclusive dealing arrangements Arrangements that restrict a dealer to carrying only one brand of a product, or that involve licenses of technology and availability of components/supplies in the supply chain.

export restrictions Government regulations prohibiting companies from selling certain products to other specified countries.

gray market Unauthorized distribution of goods at discounted prices through resellers not franchised or sanctioned by the producer.

interbrand competition Competition among different brands or makes in the marketplace.

intrabrand competition Competition among dealers who are selling the same manufacturer's brand.

long tail Term for the 80% of slow-moving items in a firm's inventory that usually bring in only 20% of total revenue; refers to the "long tail" at the end of a frequency distribution for consumer preferences.

multi-channel marketing Marketing that uses a combination of direct and indirect distribution channels; also referred to as *hybrid channels, dual channels,* or *concurrent channels.*

responsive supply chains Supply chains that are flexible in meeting the changing needs of customers.

reverse auction A method of e-procurement where a company puts its purchasing requirements for suppliers to see on a website and then asks them to competitively bid (via lower prices) against each other for the company's business over a predefined time period.

risk-hedging supply chains Supply chains that are geared toward pooling and sharing of resources to avoid disruptions in the chain.

supply chain management The logistical management of all of the incoming components and pieces used in the manufacturing process of a particular product.

systems integrator A type of dealer who manages very large or complex computer projects, typically creating customized solutions for customers by bundling and reselling different brands of equipment.

tying A practice whereby a manufacturer makes the sale of a product in high demand conditional on the purchase of a second product.

value-added resellers (VARs) Smaller channel intermediaries who purchase products from one or several high-tech companies and add value through their own expertise, to meet their target markets' needs.

Notes

1. Rangan, V. Kasturi, "Cisco Systems: Managing the Go-to-Market Evolution," Harvard Business School, 2006, Case #9-505-006, pp. 1–20.
2. Ibid, p. 8.
3. Ibid, p. 10.
4. Ibid.
5. Hutchinson, Brian, "In-Store Marketing in the Consumer Electronics Industry," Brandchannel.com, December 2005, online at www.brandchannel.com/papers_review.asp?sp_id=1224.
6. Coughlan, Anne, Erin Anderson, Louis Stern, and Adel El-Ansary, *Marketing Channels,* 7th ed. (Upper Saddle River, NJ: Prentice Hall International Series in Marketing, 2006); Rangan, V. Kasturi, and Mari Bell, *Transforming Your Go-to-Market Strategy: The Three Disciplines of Channel Management* (Boston: Harvard Business School Press,

2006); and Pelton, Lou, David Strutton, and James Lumpkin, *Marketing Channels,* 2nd ed. (New York: McGraw Hill, 2001).

7. Vinhas, Alberto Sa, and Erin Anderson, "How Potential Conflict Drives Channel Structure: Concurrent (Direct and Indirect) Channels," *Journal of Marketing Research* 42 (November 2005), pp. 507–515.

8. Neslin, Scott A., Dhruv Grewal, Robert Leghorn, Vekatesh Shankar, Marije L. Teerling, Jacquelyn S. Thomas, and Peter C. Verhoef, "Challenges and Opportunities in Multichannel Management," *Journal of Service Research* 9 (November 2006), pp. 95–112.

9. Kotler, Philip, Neil Rackham, and Suj Krishnaswamy, "Ending the War between Sales and Marketing," *Harvard Business Review* 84 (July–August 2006), pp. 68–78.

10. Kirsner, Scott, "Channel Concord: The Web Isn't Just for Alienating Partners Anymore," *CIO Web Business,* November 1, 1998, pp. 32–34.

11. Yang, Frank, "IBM's Big Retail Sale," *Datamation,* July 15, 1986, pp. 46–49.

12. Wingfield, Nick, "How Apple's Store Strategy Beat the Odds," *Wall Street Journal,* May 17, 2006, online at http://online.wsj.com/article/SB114783026556754978-search.html?KEYWORDS=apple+store+strategy&COLLECTION=wsjie/6month.

13. "Sony Style Retail Stores," 2008, online at www.sonystyle.com/webapp/wcs/stores/servlet/SYStoreLocatorMainView?storeId=10151&catalogId=10551&langId=-2&SR=nav;service_support:electronics:sonystyle_store_locations:ss.

14. "Apple's First Retail Store in Australia Opens in Sydney on Thursday, June 19," press release, June 18, 2008, online at www.apple.com/pr/library/2008/06/18retail.html?sr=hotnews.

15. Hafner, Katie, "Inside Apple Stores, a Certain Aura Enchants the Faithful," *New York Times,* December 27, 2007, online at www.nytimes.com/2007/12/27/business/27apple.html.

16. Thurrott, Paul, "Dell Copying Apple's Retail Strategy," *Connected Home Tech Blog,* May 24, 2006, available at www.connectedhomemag.com/Blog/Articles/Index.cfm?ArticleID=50396.

17. Chao, Loretta, "Nokia Takes 'Wow' Factor to Shanghai," *Wall Street Journal,* October 29, 2007, online at http://online.wsj.com/article/SB119340270057372852.html.

18. Wingfield, "How Apple's Store Strategy Beat the Odds."

19. Kotler, Rackham, and Krishnaswamy, "Ending the War between Sales and Marketing."

20. Einhorn, Bruce, "A Racer Called Acer," *BusinessWeek,* January 18, 2007, online at www.businessweek.com/globalbiz/content/jan2007/gb20070118_038922.htm?chan=search.

21. "How Technology Sells," 1997, Dataquest, Gartner Group, and CMP Channel Group (CMP Publications, Jericho, NY).

22. Moore, Geoffrey, *Crossing the Chasm: Marketing and Selling Technology Products to Mainstream Customers,* rev. ed. (New York: HarperCollins, 2002), Chapter 7.

23. Ibid., p. 173.

24. Ibid.

25. Mohr, Jakki, Christine Page, and Greg Gundlach, "The Governance of Inter-Organizational Exchange Relationships: Review and State-of-the-Art Assessment," working paper, University of Montana, Missoula, Montana, 1999.

26. Sahadev, Sunil, and S. Jayachandran, "Managing the Distribution Channels for High-Technology Products: A Behavioral Approach," *European Journal of Marketing* 38 (2004), pp. 121–149.

27. Payan, Janice M., and Richard G. McFarland, "Decomposing Influence Strategies: Argument Structure and Dependence as Determinants of the Effectiveness of Influence Strategies in Gaining Channel Member Compliance," *Journal of Marketing* 69 (July 2005), pp. 66–79.

28. France, Mike, "Are Corporate Predators on the Loose?" *Business Week,* February 23, 1998, pp. 124–126; and Stremersch, Stefan, and Gerrard J. Tellis, "Strategic Bundling of Products and Prices: A New Synthesis for Marketing," *Journal of Marketing* 66 (January 2002), pp. 55–72.

29. Sherman, Erik, "Inside the Apple iPod Design Triumph," Electronics DesignChain (2002), www.designchain.com/coverstory.asp?issue-summer02.

30. Mangalindan, Mylene, "Court Rules against Amazon in Toys Dispute," *Wall Street Journal,* March 3, 2006, online at http://online.wsj.com/article/SB114130712302487542-search.html?KEYWORDS=amazon&COLLECTION=wsjie/6month.

31. Kumar, Nirmalya, Louis Stern, and Ravi Achrol, "Assessing Reseller Performance from the Perspective of the Supplier," *Journal of Marketing Research* 29 (May 1992), pp. 238–253.

32. Srinivasan, Raji, "Dual Distribution and Intangible Firm Value: Franchising in Restaurant Chains," *Journal of Marketing* 70 (July 2006), pp. 120–135.

33. Weinberg, Bruce D., Salvatore Parise, and Patricia J. Guinan, "Multichannel Marketing: Mindset and Program Development," *Business Horizons* 50 (September 2007), pp. 385–394.

34. Venkatesan, Rajkumar, V. Kumar, and Nalini Ravishanker, "Multichannel Shopping: Causes and Consequences," *Journal of Marketing* 71 (April 2007), pp. 114–132.

35. Chandler, Michele, "Companies Find Multichannel Marketing Key," Knight Ridder Newspapers, January 1, 2006.

36. Moriarty, Rowland, and Ursula Moran, "Managing Hybrid Marketing Systems," *Harvard Business Review* 68 (November–December 1990), pp. 146–155.

37. Ibid.

38. Vinhas and Anderson, "How Potential Conflict Drives Channel Structure."

39. Moriarty, Rowland, and Ursula Moran, "Managing Hybrid Marketing Systems," *Harvard Business Review* 68 (November–December 1990), pp. 146–155.

40. Ibid.

41. Kenworthy, James, "Rise in Multi-channel Retailing Drives Growth of SOA Deployment in Retail Sector," *Database and Network Journal* 37 (June 2007), pp. 10–11.

42. Anderson, Chris, "The Long Tail," *Wired,* October 2004, online at www.wired.com/wired/archive/12.10/tail.html; Anderson, Chris, *The Long Tail, Revised and Updated Edition: Why the Future of Business Is Selling Less of More* (New York: Hyperion, 2008); and Anderson, Chris, "Does the Long Tail Work for Mobile Music?" Wired Blog Network, October 17, 2008, online at http://longtail .typepad.com/the_long_tail.

43. "Blockbuster to Close 282 Stores This Year," Associated Press, June 28, 2007, online at www.msnbc.msn.com/id/ 19489372.

44. Anderson, "The Long Tail."

45. Elberse, Anita, "Should You Invest in the Long Tail?" *Harvard Business Review* 86 (July–August 2008), online at http://harvardbusinessonline.hbsp.harvard.edu/hbsp/ hbr/articles/article.jsp?ml_action=get-article&articleID =R0807H&ml_ page=1&ml_subscriber=true.

46. Elberse, "Should You Invest in the Long Tail?"

47. Duhan, Dale, and Mary Jane Sheffet, "Gray Markets and the Legal Status of Parallel Importation," *Journal of Marketing* 52 (July 1988), pp. 75–83; and Corey, E. Raymond, Frank V. Cespedes, and V. Kasturi Rangan, "The Gray Market Dilemma," in *Going to Market: Distribution Systems for Industrial Products* (Boston: Harvard Business School Press, 1989), pp. 169–186.

48. Myers, Matthew, and David Griffith, "Strategies for Combating Gray Market Activity," *Business Horizons,* November–December 1999, pp. 2–8.

49. KPMG and the Anti Gray Market Alliance, "The Grey Market," Spring 2003, white paper, available at www .agmaglobal.org/press_events/whitepapers.shtml.

50. Kharif, Olga, and Peter Burrows, "On the Trail of the Missing iPhones," *BusinessWeek,* January 31, 2008, online at www.businessweek.com/magazine/content/08 _06/b4070025757036.htm?chan=search.

51. Burrows, Peter, "Inside the iPhone Gray Market," *BusinessWeek,* February 12, 2008, online at www .businessweek.com/print/technology/content/feb2008/ tc20080211_152894.htm.

52. Chen, Hsiu-Li, "Gray Marketing and Its Impact on Brand Equity," *Journal of Product and Brand Management* 16 (2007), pp. 247–256.

53. Corey, Cespedes, and Rangan, "The Gray Market Dilemma."

54. Duhan and Sheffet, "Gray Markets and the Legal Status of Parallel Importation."

55. Corey, Cespedes, and Rangan, "The Gray Market Dilemma."

56. Ibid.

57. Ibid.

58. Ibid.

59. Ibid.

60. Antia, Kersi D., Mark E. Bergen, Shantanu Dutta, and Robert J. Fisher, "How Does Enforcement Deter Gray Market Incidence?" *Journal of Marketing* 70 (January 2006), pp. 92–106.

61. Corey, Cespedes, and Rangan, "The Gray Market Dilemma."

62. Ibid.

63. "iPhone 3G: Coming to Countries Everywhere," Apple website, 2008, online at www.apple.com/iphone/countries.

64. Fourth Annual BSA and IDC Global Software Piracy Study, 2006, online at http://w3.bsa.org/globalstudy.

65. Epstein, Keith, "ITT to Plead Guilty in Export Case," *Business Week,* March 27, 2007, online at www .businessweek.com/investor/content/mar2007/pi20070327 _964132.htm?chan=search.

66. Cohen, Adam, "When Companies Leak," *Time,* June 7, 1999, p. 44.

67. Gay, Lance, "U.S. Satellite Controls Have Backfired," *Missoulian,* June 1, 2000, p. A7.

68. Richardson, Don, Ricardo Ramirez, and Moinul Huq, "Grameen Telecom's Village Phone Programme: A Multi-Media Case Study," TeleCommons Development Group, Asia Branch Poverty Reduction Project, Canadian International Development Agency, March 2000, online at www .telecommons.com/villagephone/index.html.

69. Slone, Reuben, John Mentzer, and J. Paul Dittmann, "Are You the Weakest Link in Your Company's Supply Chain?" *Harvard Business Review* 85 (September 2007), pp. 116–127.

70. Spearman, Mark, "Realities of Risk," *Industrial Engineer,* February 2007, pp. 36–41.

71. Quelch, John, and Anna Harrington, "Samsung Electronics Company: Global Marketing Operations," Harvard Business School Publishing, 2005, Case #9-504-051.

72. Wathne, Kenneth H., and Jan B. Heide, "Relationship Governance in a Supply Chain Network," *Journal of Marketing* 68 (January 2004), pp. 73–89.

73. Lambert, Douglas, *Supply Chain Management: Processes, Partnerships, Performance* 3rd ed. (Sarasota, Florida: Supply Chain Management Institute, 2008).

74. Lee, H. L., V. Padmanabhan, and S. Whang, "The Bullwhip Effect in Supply Chains," *Sloan Management Review* 38 (Spring 1997), pp. 93–102.

75. Lee, Hau L. (2002), "Aligning Supply Chain Strategies with Product Uncertainties," *California Management Review,* 44 (Spring 2002), pp. 105–119.

76. Jap, Sandy, and Jakki J. Mohr, "Leveraging Internet Technologies in B2B Relationships," *California Management Review* 44 (Summer 2002), pp. 24–38.

77. Jap, Sandy, "An Exploratory Study of the Introduction of Online Reverse Auctions," *Journal of Marketing* 67 (July 2003), pp. 96–107.

78. Jap and Mohr, "Leveraging Internet Technologies in B2B Relationships."

79. Supplier List, Ariba website, 2008, online at www.ariba .com/network/supplierlist.cfm.

80. Jap, Sandy, "Online Reverse Auctions: Issues, Themes, and Prospects for the Future," *Journal of the Academy of Marketing Science* 30 (September 2002), pp. 506–525.

81. Jap, Sandy, "The Impact of Online Reverse Auction Design on Buyer–Seller Relationships," *Journal of Marketing* 71 (January 2007), pp. 146–159.

82. Fontanella, John, and Eric Klein, "Supply Chain Technology Spending Outlook," *Supply Chain Management Review,* April 2008, pp. 14–20.

83. "Looking to Supply Chain Management: Manufacturers Up Their Software Investment," *Industrial Engineer,* October 2006, p. 13.

84. Fontanella, John, and Eric Klein, "Supply Chain Technology Spending Outlook," *Supply Chain Management Review,* April 2008, pp. 14–20.

85. Clark, Don, "Enhanced Tracking Technology May Propel Adoption of RFID," *Wall Street Journal,* April 14, 2008, p. B4.

86. Lohr, Steve, "At IBM, a Smarter Way to Outsource," *New York Times,* July 5, 2007, online at www.nytimes.com/2007/07/05/business/05outsource.html?fta=y.

87. Regan, Dominic, "Forge a Green Supply Chain to Drive Growth," *Computer Weekly,* February 26, 2008, pp. 20–23.

88. Koh, S. C. Lenny, Frank Birkin, Linda Lewis, and Adrian Cashman, "Current Issues in Sustainable Production, Eco-supply Chains, and Eco-logistics for Sustainable Development," *International Journal of Global Environmental Issues* 7 (2007), pp. 88–101.

89. "Lenovo Tops Eco-Friendly Ranking," BBC News, April 4, 2007, online at http://news.bbc.co.uk/2/hi/technology/6525307.stm.

90. Srivastava, Samir, "Green Supply-Chain Management: A State-of-the-Art Literature Review," *International Journal of Management Reviews* 9 (March 2007), pp. 53–80; and Mongelluzzo, Bill, "Green for 'Green,'" *Journal of Commerce,* June 16, 2008, pp. 46–47.

10 Pricing Considerations in High-Tech Markets

The iPhone Pricing Dilemma

On January 9, 2007, at the MacWorld conference in San Francisco, Steve Job pre-announced that in June Apple would launch the iPhone—a product that would be a combination cell phone, iPod music player, and Internet appliance. It would be available in two versions: a 4 GB version and an 8 GB version, retailing for $499 and $599 respectively. It would be available only in Apple's on-line store, in Apple-owned retail stores, and (through an exclusive partnership) in AT&T (formerly Cingular Wireless) stores. At that event Jobs also announced that the next-generation Macintosh operating system, Leopard, would be released in June as well—and he announced a name change for the company, from Apple Computer Inc. to Apple Inc.

The iPhone product demonstration that Jobs delivered wowed the MacWorld audience. The iPhone had a sleek black design, was only a half-inch thick, and had a 3.5-inch touch screen. By flicking a finger across the screen, users could scroll through music in their iTunes playlists. Photos could be zoomed in for closer viewing by touching thumb and forefinger at a spot and then separating them while being in contact with the screen. Photos or album covers could be viewed in portrait or landscape mode by turning the iPhone 90 degrees. Visual voice mail allowed users to touch any listed message and listen to it directly without sequentially listening to messages recorded earlier. The iPhone was WiFi-enabled for high-speed Internet access in the presence of a hotspot. The interface was easily navigable. The battery life had a telephone talk time up to 8 hours, audio playback time up to 24 hours, and Internet use time up to 6 hours.

Steve Jobs announced that Apple's objective for the iPhone was to get a 1% global market share of the 1-billion-unit cell phone handset market, or to sell 10 million iPhone units by the end of 2008. Apple's stock price went up 8% on the day of the pre-announcement.

The pre-announced iPhone had some deficiencies. Cell phone service through AT&T's EDGE network (2.5G) was rather slow compared to 3G network service available from Verizon and Sprint. The iPhone would not work with any carrier's service except AT&T, which required a two-year contract from iPhone users. The iPhone had no physical keyboard, only a virtual keyboard, which required users to tap characters one by one on the touch screen. The iPhone's e-mail program would not interface with secure, corporate e-mail servers from BlackBerry, Microsoft, and Motorola. Soon after the pre-announcement, Cisco sued Apple for copyright infringement, having owned the brand name iPhone since 2000. It took both companies about six weeks of negotiation to reach a settlement whereby Cisco permitted Apple to use the iPhone brand name in return for ensuring interoperability of the iPhone with Cisco's products and services.

The January pre-announcement was out of character for Apple, which heretofore did not make product announcements until they were ready to be shipped. It set in motion a frenzy of media hype. For the next six months, newspapers, magazines, radio and TV stations, and blogs were abuzz. Some analysts lauded Apple's move to protect its iPod dominance, increasingly under attack from smartphones (Internet-enabled cell phones) with integrated music players offered by Nokia, Motorola, Samsung, Sony Ericsson, and LG. Within days of Apple's pre-announcement, LG announced that its LG Prada handset would be available in Europe in February, where it would cost Euro 600 ($775), and in Asia in March. It would have a touch screen, music player, and Internet capability..

Other analysts were of the opinion that the combination of iPhone features at pre-announced price points would not offer value to mainstream U.S. consumers. In February, a survey done by Changwave Research of Rockville, Maryland, showed that one-third of people not interested in the iPhone cited high prices as a reason. Analysts from iSupply analyzed the expected components of the iPhone and estimated the cost of making the $499 version as $245.93 and the cost of making the $599 version as $280.83. This showed that Apple had a lot of flexibility on iPhone pricing.

Ryan Kim wrote in the Tech Chronicles blog of the *San Francisco Chronicle* that price would be a stumbling block for the iPhone. The pre-announced iPhone prices made them comparable to a Sony PlayStation 3, a powerful machine with the new Blu-ray optical drive. Most smartphone users had a reference price around $250 for a music-enabled smartphone.

In April 2007, Apple announced it was pushing back the release of the new Macintosh operating system, Leopard, so that more product development resources could be allocated to the iPhone for its timely release in June. By this time, AT&T had received inquiries from about one million people who asked to be alerted when the iPhone went up for sale.

In early June, Apple announced the iPhone launch date as June 29 and began a TV advertising campaign with commercials featuring product demonstrations of the cool new product.

In a product review just prior to launch, the *Wall Street Journal* said:

> Expectations for the iPhone have been so high that it can't possibly meet them all. It isn't for the average person who just wants a cheap, small phone for calling and texting. But, despite its network limitations, the iPhone is a whole new experience and a pleasure to use.

People started camping out at Apple retail stores several days before June 29 to be sure they got the product on the first day. Apple reported 270,000 iPhones were sold during the first 30 hours of the first weekend, while AT&T reported that 146,000 iPhone buyers signed up for its service during that period. The Apple online store gave wait times of 2 to 4 weeks for those placing an order. Apple said it expected iPhone sales to cross 1 million units by the end of September.

In late August, Sony announced a challenge to Apple, a new Walkman player and an online store that would allow Walkman users to download music and video content onto their players. Nokia also announced an Internet download service for music and games, under the brand name Ovi, which would provide content for its multimedia handsets and terminals.

On September 5, 2007, at an invitation-only event for journalists and analysts at the Moscone Center in San Francisco, Steve Jobs announced that the $499 iPhone was being discontinued. Further, the $599 iPhone would be reduced to $399 so that Apple could make a big push with shoppers during the holiday season. This surprised many analysts because iPhone seemed to be meeting sales targets and satisfying customers even at the higher price during the first two months. Further, the price cut irritated some loyal iPhone buyers who had bought within the first two months. In response, an open letter to iPhone customers was posted on Apple's website, in which Steve Jobs apologized for the price cut. Any customers who bought an iPhone within 14 days of the September 5 announcement were offered a $200 cash refund. Other iPhone customers were offered a $100 credit toward purchase of future Apple products and services.

On September 10, Apple reported that it had sold 1 million iPhones in 74 days—3 weeks ahead of schedule by its own estimates.[1]

THE HIGH-TECH PRICING ENVIRONMENT

What forces impinge on high-tech pricing decisions? As shown in Figure 10.1, the forces are varied and strong. Many high-tech firms might find it desirable to price at a high level, in order to recoup investments in R&D and to signal high product quality. However, many factors conspire to push prices down.

High-tech firms face an environment characterized by ever-shortening product life cycles, with the inevitable rapid pace of change and potential obsolescence of products. Moore's Law[2] operates unforgivingly: Every 18 months or so, improvements in technology double product performance at no increase in price. Stated a different way, every 18 months or so, improvements in technology cut price in half for the same level of performance. So introductions of product versions with better price–performance ratios are a given, creating downward pressure on prices.

Moreover, as identified in Chapter 1, network externalities and unit-one costs affect pricing in high-tech markets. Recall that network externalities exist when the value of the product increases as more users adopt it; some examples include the telephone, portals on the Internet, and social networking sites. Unit-one costs refer to the situation in which the cost of producing the first unit is very high relative to the costs of reproduction for subsequent units. For example, the costs of distributing software either over the Internet or on CDs are trivial compared to the costs of hiring programmers and specialists to develop, test, and debug the software program itself. Both of these factors create pressure to acquire a critical mass of users through lower price structures.[3]

Furthermore, customer perceptions of the costs versus benefits of the new technology affect pricing strategy. Customer anxiety may cause delays in adoption. For example, as firms introduce one new and improved version after another, consumers may postpone purchases in the hope (fear?) that prices eventually will come down and performance will improve substantially.[4] For example, prices for solar energy have fallen about 90% since the mid-1980s—and fell 40% annually during the past five years. In addition to cost efficiencies arising from economies of scale in production, advances in technology will lead solar electricity prices to be competitive with conventional electricity by 2014.[5] In such a situation, marketers may need to lower the prices of newer technologies aggressively to reduce possible switching costs, to offer special deals for upgrades, to generate manufacturing economies, or to entice customers switching from a competing application.[6]

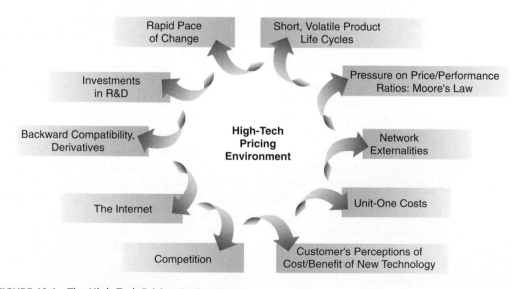

FIGURE 10.1 The High-Tech Pricing Environment

Competitive volatility means that new entrants in a market may enter with disruptive innovations and disruptive business models. For example, using Voice-over-Internet Protocol (VoIP) technology, Skype offers free long-distance phone calling anywhere in the world to anyone who has downloaded the free software from its website and has access to a broadband Internet connection. Established telecommunications companies are hard pressed to compete against a "free" service such as this, with their heavy investment in telecommunications infrastructure.

Moreover, the Internet allows both individual consumers and organizational customers to compare prices and negotiate for lower prices to a much greater degree than in the past. The Internet creates **cost transparency**, which allows buyers to more easily find information about manufacturers' costs and prices, providing them more leverage in making product choices. For example, the Internet lowers prices of new cars for consumers because it reduces the search cost of information.[7] Consumers can find information on dealer invoice prices on the Internet or they can get referrals from online buying services to affiliated lower priced dealers. The Internet, thus, makes a buyer's search more efficient. Reverse auctions (see Chapter 9), in which suppliers make lower and lower bids in order to "earn" the right to sell a manufacturer supplies for its business, allow customers to identify suppliers' *price floors,* or the lowest price at which they are willing to sell a product or service. Moreover, due to the transparency of online pricing information, the Internet makes it more difficult for a firm to engage in different pricing strategies in different markets—something that was commonly done in international markets in the past. And the availability of low-priced or free offers on the Internet makes customers more sensitive to prices.

Issues of backward compatibility (with older versions of the product), support for existing products, changing industry standards, pricing for product derivatives, and so forth, all must be considered in pricing strategy.

Pricing even in conventional marketing contexts is a very complex decision; this overview of the high-tech pricing environment shows that it is doubly complex for high-technology products and services. As in prior chapters, this chapter does not go over pricing basics (e.g., calculating payback periods, return on investment, break-even points, leasing, competitive bidding, price elasticity, penetration versus skimming strategies); interested readers should consult other resources to learn the basics. Rather, this chapter addresses how to make pricing decisions that incorporate and address many of the complications in the high-technology context, mentioned previously. The chapter begins by examining the three major factors that all marketers must systematically consider when setting prices: costs, competition, and customers.

THE 3 C'S OF PRICING

The 3 C's of pricing—costs, competition, and customers—are analogous to a three-legged stool, shown in Figure 10.2. Stools with only two legs are unbalanced and likely to topple over. Similarly, setting price on the basis of only one or two of the 3 C's results in an unstable situation. Solid pricing strategy must systematically consider all three factors.

Competition

Costs   Customers

FIGURE 10.2   The 3 C's of Pricing

Costs

Costs that a company incurs to produce its goods provide a *floor,* generally below which marketers ought not to price. Companies that position on a low-price basis should not do so unless (1) they have a strong, *inimitable* cost advantage in the industry, and (2) that inimitable cost advantage is *sustainable* (in other words, it is unlikely to disappear with future generations of technology). For example, a cost advantage based on economies of scale (large volume) on an existing technology may not translate to a cost advantage when a new generation of technology comes down the pike. Sony's PlayStation 2 was selling at $250 when the PlayStation 3 (PS/3) was introduced in November 2006—at a price of $500 for a 20-gigabyte console. Sony's justification was that the PS/3 had more expensive technology including powerful graphics processors and the new high-definition Blu-ray DVD drive. Even at this price, experts speculated that Sony would lose money on each unit. Indeed, Sony needed sales of 20 million units before per-unit costs would decline enough for the product to make a positive margin.[8]

In addition to break-even pricing and payback pricing,[9] an important cost-based pricing concept for high-tech companies is based on a company's **experience curve**. As shown in Figure 10.3, an experience curve captures the per-unit cost declines in production every time the company's accumulated manufacturing volume doubles. These cost savings arise from the combined effects of learning, volume, and specialization: Employees become more efficient, purchasing costs fall, production lines run more smoothly. The figure shows that learning effects are stronger earlier on and flatten as volume increases. Plotting the unit costs and cumulative volume on a logarithmic scale results in a linear experience curve. So, for example, using a logarithmic scale to quantify the experience curve effect, an 85% experience curve means that every time accumulated volume doubles, costs decline by a fixed and known amount (15%).

One possible pricing strategy arising from experience curve effects is for a firm to willingly take a per-unit loss early in the production process, pricing aggressively either to attract new customers (grow the installed base, generate network externalities if appropriate, and/or establish an industry standard through a first-mover advantage) or to deter competitors. A second pricing possibility arising from experience curve effects is to lower the price of the product as experience curve effects kick in; this modify-over-time strategy is consistent with price-based market segmentation where later adopters are more price sensitive than initial adopters. Toyota appears to have priced its hybrid-engine Prius using a combination of these factors. Despite the initial cost outlay in development and manufacturing, Toyota set a sales price in which it subsidized the cost of purchase.

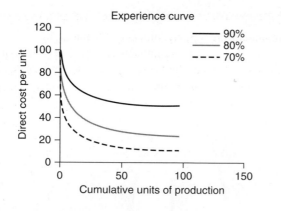
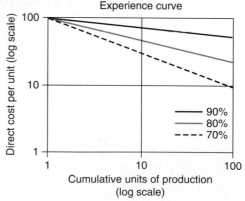

FIGURE 10.3 Experience Curve

Every time accumulated volume doubles, costs decline by a fixed (known) amount; that amount depends on each individual company's production processes, scale economies, and learning.

Source: http://en.wikipedia.org/wiki/Image:Experience_curve.gif#file

To reach profitability on this new technology, Toyota set an aggressive sales target of 1 million hybrid units by 2010—a goal it actually reached in 2007! While the other car makers will struggle to achieve the scale economies Toyota has attained, Toyota is further hoping to leverage the experience curve by cutting the costs of hybrid engines in half.[10]

Despite its appeal, a firm that bases its pricing strategy primarily on its own cost structure (i.e., on a markup, cost-plus, or target-return basis) often fails to recognize the impact that market factors have on profitability. Overlooking the impact of market factors, such as competition and demand, on pricing and profitability can be a fatal mistake in high-tech markets.

Competition

Competition provides a benchmark against which to evaluate prices. A firm might let competitors set prices and then establish its price below, equal to, or above those of competitors, depending on its desired positioning in the market. While Dell tries to position itself as the low price leader, Apple tries to differentiate itself with innovative products and premium pricing.

As stated earlier in this book, in the high-tech arena, a firm that introduces a radical innovation to the marketplace often (wrongly) believes that, because its innovation is so new, there is "no" competition. However, this belief is not necessarily the case from the customer's perspective. A customer can always choose not to adopt new technology, instead solving problems based on former solutions (which provide the competitive benchmark for radical innovations). Indeed, one executive from Motorola said, "Our biggest competitor isn't IBM or Sony. It's the way in which people currently do things."[11]

In addition, some high-tech companies myopically define competition as anything that uses a similar underlying technology to solve the customer's needs. Viewing competition a bit more broadly, in terms of alternative ways of satisfying the same customer need, helps a firm to understand competitive threats and opportunities. For example, as the price of oil goes up, it makes competing fuel technologies (hydrogen fuel cells) that previously seemed pricey more attractive. Economists refer to this as the **cross-price elasticity of demand**, or the percentage change in one product's sales due to a percentage change in a price of another product. Products are considered competitive substitutes when sales of one product decline (increase) as the other product's price drops (increases).

Conventional wisdom is that prices decline as more competitors enter the market. While this is true in many industries, the reverse may occur in the special case of information-based products (e.g., consultants' reports or analysts' forecasts). When information products are of unknown quality, products from different sellers are seen as complements: Using multiple products from multiple providers increases the consumers' confidence in the information. In such a situation, a larger number of sellers can result in increased rather than decreased prices.[12] This explains why product-market forecasts from companies like Forrester, IDC, and others can cost several thousand dollars each—even though they differ widely in their estimates of market size.

Customers

Customers' perceptions of value provide a *ceiling* above which marketers should not price. Simply, customers balance the benefits of a purchase against its costs. High-tech marketers often find it difficult to understand fully the customer's perceptions of benefits and costs. The innovating firm may find the new technology so compelling, so sophisticated, or so "innovative" that it assumes the benefits are obvious to users. Similarly, the innovating firm may not fully appreciate the customer's perceptions of costs, which may include switching costs, learning effort, and so forth.

Product benefits as perceived by customers might include the following:[13]

- *Functional benefits:* The utilitarian aspects that might be attractive to engineers or technology enthusiasts

- *Operational benefits:* The product's reliability and durability, and the product's ability to increase efficiency for the customer
- *Financial benefits:* Credit terms, leasing options—and for business customers, whether the product helps boost revenues
- *Personal benefits:* The psychosocial satisfaction from being an early adopter, purchasing a well-known brand to avoid risk, and being professionally rewarded for making good organizational buying decisions

The costs a customer perceives are similarly diverse and might include these:[14]

- *Monetary costs:* The price paid, as well as transportation, installation, and operating costs
- *Nonmonetary costs:* The risk of product failure, risk of obsolescence, obsoleting of a prior piece of equipment or related products, switching costs, learning costs—and for business-to-business goods, factory downtime for repair and maintenance of machinery

Reference price is a standard against which the purchase price of a product is judged by a customer. It is based on a customer's prior experience and current purchase environment. However, for durable technology products where the interpurchase cycles are long, information on past purchases may not be a salient benchmark. Instead, a reference price based on current competitors' prices is more relevant.[15] Further, reference prices of technology durables are also affected by customers' perceptions of the cost of critical components, "guesstimates" of the expected size of the installed base, availability of complementary products, and anticipated macroeconomic conditions and price trends.

The **total cost of ownership** (or *life cycle costing*) is one way to look at customers' costs; it reflects the total amount a customer expends in order to own and use a product or service. Total cost of ownership includes the initial price paid (including financing fees), as well as delivery or installation costs, maintenance and repair costs, power costs to run the equipment, supplies, and other operating costs over the life of the product. Using the total cost of ownership in pricing strategy can help a firm position its products relative to those of competitors. Showing that a product's total cost of ownership is lower than a competitor's can be a compelling benefit to a customer—despite an initially higher outlay for the product or an investment in new/updated additional equipment. Similarly, showing a customer that buying the cheapest product available might result in the most expensive product to own can also provide a compelling purchase rationale. Because the total cost of ownership concept is so important to a company's value proposition, two examples follow.

1. *Energy-efficient products:* One of the biggest difficulties people have in buying energy-saving technologies is that they may cost more than traditional technologies. For example, the replacement price for an electric water heater might range from $300 to $400, while purchasing solar technology to heat the water might cost $3,000. Similarly, one can replace a lightbulb for 39 cents for an incandescent bulb or $6 for a compact fluorescent bulb. To calculate the total cost of ownership of these energy-saving devices, one must adjust the purchase price for the annual energy savings over the estimated lifetime of the product. Financially savvy customers will further adjust this cost of ownership by a discount factor, to recognize that a dollar (or euro, or similar monetary equivalent) today is more valuable than one in the future. In calculating the total cost of ownership, one may find that the energy costs alone used to operate a product may exceed the product's initial purchase price. Take a new refrigerator. Most refrigerators last between 15 and 20 years. A new energy-efficient model will cost $850; a conventional model will cost $750. If the energy-efficient model will save $25 a year in energy costs (estimates are that most energy-efficient appliances will save more than this), then over 15 years the energy savings will be $375—more than three times the incremental cost of the energy-efficient product.[16] Similarly, that incandescent bulb will run for 1,000 hours while the CFL bulb will run for 10,000 hours. When coupled with its energy efficiency (a CFL bulb uses a fourth of the electricity of an incandescent bulb), the longer-lived yet more expensive bulb actually saves the user money in the long run.[17] Hence, for many environmentally friendly and/or energy-efficient products, a total cost of ownership approach is a sound strategy.

2. *Software as a service:* Many business innovations can also benefit from a total-cost-of-ownership strategy. For example, take the new delivery method for software based on "software as a service," or SaaS in which customers access the software program over the Internet via an online delivery model rather than using a client-server model in which the software is sold as a packaged application and hosted at the customer's site; SaaS is also called *cloud computing,* whereby a customer signs up to use the software that is "hosted" by the software vendor on its servers. The pricing strategy for SaaS is typically a monthly subscription fee rather than a one-time ownership fee. Consultants at McKinsey compared the total cost of ownership of CRM software for a customer who purchased and owned the software (installed for 200 users on the company's premises) versus a customer who accessed the software in a subscription, SaaS model. Components of total cost included procurement costs, implementation and deployment, training, customization and business process changes, data center facilities, unused licenses and unscheduled downtime. The study showed a 30% lower total cost of ownership for CRM software when delivered under the SaaS model compared with a conventional sales/licensing model. According to McKinsey & Company, SaaS has a lower total cost of ownership because:[18]

- It does not require customer infrastructure and application testing.
- It requires only limited customization.
- It lowers training requirements through simpler user interfaces and self-training/self-service capabilities.
- It does not require ongoing business process change management.
- It reduces unused licenses by 20%.
- It provides 99.9% server availability/reliability (vs. 99% for in-house ownership).

Consistent with the innovator's dilemma, many established industry players initially dismissed the idea of software as a service. Siebel allowed Salesforce.com to steal a lead in the online delivery of CRM applications. Even business process outsourcers may be faced by disruption from the SaaS companies.

Similar pricing arguments based on a total-cost-of-ownership strategy can be found for open source software, which sometimes requires large up-front customer investments in new hardware but actually results in lower total operating costs. For example, Morgan Stanley's Institutional Securities Division decided to replace 4,000 high-powered servers running traditional software with cheaper machines running Linux (a free, open source operating system); the projected five-year savings from the switch: $100 million![19] This value proposition is part of the reason that the market share of Linux (in the market of server operating systems) has grown from 0% in 2000, to 13.7% in 2003, to 33% in 2007.

Consolidating the 3 C's into a Successful Strategy

In summary, solid consideration of costs, competitors, and customers is vital to establish a successful pricing strategy. Focusing on costs alone can be myopic and can cause problems. Similarly, focusing on competition can be hard in high-tech markets, when the competition for a breakthrough innovation might be the customer's current behavior pattern. Both of the drawbacks in focusing solely on costs or competition point to the value in taking a customer perspective in pricing. Taking a customer perspective in pricing helps the marketer to realize that the firm's costs to manufacture a product and its investments in R&D are relatively unimportant to the customer's perceived value. Moreover, the customer tends not to care about the firm's costs so much as his or her own costs in buying and using the product. For example, the LG Internet Refrigerator was launched in the U.S. with a whopping retail price of $8,000 in 2002. Later, in 2005, LG reduced the price of the full-featured product to $3,999 with the idea that consumers would perceive the combined value of roughly $2,000 for the refrigerator and $2,000 for its computer/TV aspect (appearing as an LCD display on the door of the refrigerator). LG soon learned that rather than using all the functions of the computer, consumers were accessing only a few Internet functions such as weather, photos,

TECHNOLOGY SOLUTIONS FOR GLOBAL PROBLEMS
The New Public Call Office

Telecommunications service providers are bringing cellular phone service to isolated low-income communities in Asia and Africa. In Algeria, Orascom Telecom Holding SAE has a pilot project to bring cell phone service to remote villages that border the desert and include watering holes and bathing tents for nomadic travelers. The rugged wireless phones seem old-fashioned but are designed to withstand extreme temperatures. Orascom looks for an entrepreneur to buy the phone in each local area, who then rents it out to neighbors for a usage fee based on the number of minutes used. The 10 phones used in the pilot are being used for about 40 minutes a day, a level that is profitable for the entrepreneur. The reaction of local communities to this new service has been phenomenal. Now Orascom plans to roll out 500 phones, which cost the operator $570 each, in similar areas all over Algeria.

Source: Cassell, Bryan-Low, "New Frontiers for Cellphone Service," *Wall Street Journal,* February 13, 2007, pp. B1.

recipes, calendar, date, and time. So LG offered these dedicated functions on the LCD screen for $3,499 in 2006. LG also eliminated the TV LCD screen capability and offered an LCD-free version of the product for $2,499.

Because of the importance of a customer orientation in pricing, and because of the benefits to high-tech marketing of being customer focused, this leg of the stool deserves additional consideration. But first, the accompanying Technology Solutions to Global Problems offers another example of a creative customer-based pricing strategy.

CUSTOMER-ORIENTED PRICING

Steps in Customer-Oriented Pricing[20]

In order to price products based on the value that customers perceive, marketers can use the steps shown in Table 10.1.

1. *A firm must understand exactly how the customer will use its products.* Customer-oriented pricing requires that a marketer completely understand how customers apply and use the products they buy from the firm. Each end-use of a product may have a different cost–benefit analysis. For example, a customer who purchases a Quicken tax program to do tax preparation and consulting as a small business would place a different value on the product from a person who purchases the same program to do his or her personal taxes. Because of the varying ways in which customers use products, marketers may need to segment on an end-use (usage occasion) basis.

In the business market, such end-use segments are referred to as *vertical markets.* The idea is to examine how different customer segments in different industries use a product and to price accordingly. Because of the varying requirements in their end-to-end (or total) solution, customers in different vertical markets evaluate costs and benefits of a specific product in terms of a complete

TABLE 10.1 Steps in Customer-Oriented Pricing

1. Understand exactly how the customer will use a firm's products.
2. Focus on the benefits customers receive from using the products.
3. Calculate all relevant customer costs, and understand how a customer trades off costs versus benefits in the purchase decision.

usage system, and not just in terms of an isolated part of that system. For example, if a small business decides to use a Web-based solution for its business processes (e.g., customer relationship management, supply chain management, customer service and billing), it must also have an Internet service provider, a Web-hosting service, and technical support (either in-house or outsourced). Evaluating the costs/benefits of, say, the Web-hosting service, really cannot be considered in isolation of the total value to be gained from the Web-based business process. (Recall the discussion of the whole product from Chapter 7.) And, obviously, for critical applications, customers will perceive greater value. Therefore, a company like IBM could charge higher fees for its Linux services for corporate customers, than, say, public university customers.

2. *A firm must focus on the benefits customers receive from using its products.* The various types of benefits a customer can obtain were previously discussed and include functional, operational, financial, and personal benefits. In analyzing benefits, firms must not fall into the trap of focusing on product features at the expense of benefits. A familiar example is that the person who buys a quarter-inch drill bit does not want the drill bit itself, but wants the *capability* to drill quarter-inch holes. Customers buy benefits, not features. High-tech firms often mistakenly stress the cool technical wizardry of their inventions and are hard pressed to identify the real benefits customers receive. Additionally, the benefits that the technical/development personnel find compelling are often confusing or not clearly important to the customers. Focusing on customer needs is a good way to overcome this problem.

For example, in marketing computers, advertisements frequently discuss terms such as *megahertz, megabytes, pixel resolution,* and so forth. Although customers might know that greater numbers on each of these categories are presumably better, they might not know what the "improved performance" really delivers. Speaking in terms of processing speed (less wait time for functions to be performed), greater storage capacity (for the ever-increasing size of software programs), and greater clarity of the screen can help customers translate technical jargon into understandable benefits.

Another example of understanding customer benefits is exhibited in the refinement of the pricing strategy of Parker Hannifin, a diversified manufacturer of motion and control technologies for a variety of commercial, mobile, industrial, and aerospace markets. After Donald Washkewicz became CEO in 2001, he questioned the flat markup of 35% that the company added to the cost of each product to arrive at a list price. After bringing in consultants to do detailed pricing studies, he introduced revised prices that took into account what a customer was willing to pay instead of just what it cost to make each product. This new approach boosted the company's return on invested capital from 7% in 2002 to 21% in 2006.[21]

3. *A firm must calculate customer costs*—including information search costs, product purchase, and other relevant costs (discussed previously) such as transportation, installation, maintenance, training, and nonmonetary costs—*and must understand how a customer trades off costs versus benefits in the purchase decision.*

For example, in considering the purchase of a high-definition TV, U.S.-based customers must now factor in the February 18, 2009, deadline for converting all television signals to digital. This means that the roughly 19 million households that do not have the capability to receive digital broadcast signals will either need to upgrade to an HDTV, purchase a digital-ready device such as a DVD player that contains a digital tuner, or purchase a digital-to-analog converter box—or their analog televisions will effectively go dark. Given that their existing TV sets will become obsolete with the new broadcasting standards, the "cost" of missing their favorite programming may be a compelling incentive to upgrade to a new digital TV. But, customers will still trade off the benefits against the purchase-price outlay, and installation and setup costs including expensive cables. Many experts have argued that, because consumers who already subscribe to cable or satellite services will not have to make any changes, the cable and satellite companies should step in and offer a cohesive message about high-definition programming and installation, and offer an easy installation package at the time of HDTV purchase. Moreover, because of worries about negative consumer backlash, the U.S. government offered each household the opportunity (through March 31, 2009) to apply for up to two $40 coupons to defray the cost of converter boxes, which sell for about $50 to $70 apiece.

An example of step 3 in the business-to-business arena is seen in hospitals' adoption of RFID (radio-frequency identification) chip technology.[22] A common problem in many hospitals is finding the necessary equipment to use for a particular procedure. For example, many medical devices are on rolling carts that can be wheeled from room to room. One hospital found that nurses were stashing away patient-controlled anesthesia pumps. It wasn't that the hospital lacked sufficient equipment; rather, the problem was finding the equipment at the right time: Nurses felt that by hiding the pumps, they would be able to locate one when they needed it. So, the hospital's emergency department put RFID tags on equipment and used an asset-tracking application to tell hospital personnel where to find intravenous pumps, ventilators, and other devices. The $50 RFID tags contained a battery and a transmitter, using the hospital's high-speed wireless network to broadcast the location of the equipment. The results showed that health care workers spent an average of 20 fewer minutes per day looking for equipment. The results also showed a reduction in equipment losses ($600,000 annually), some of which was attributable to theft, but more often to poor handling (for example, when a $1,000 portable monitor is rolled up in a sheet and thrown down a laundry chute). When such errors happened, the RFID tag signal could be read, and within five minutes, a clerk could find and retrieve the item. Another benefit was that equipment that was not being stashed for a particular worker's station would get maximum use and didn't need to be overstocked. The hospital estimated savings of $1–2 million by not buying extra devices—the kind of investment that pays for itself within a year.

Implications of Customer-Oriented Pricing

The implications of these steps in customer-oriented pricing assist marketers in the following ways. First, this analysis helps marketers to realize that *pricing considerations should not be made after a product is developed and ready for commercialization, but early in the design process.* Treating price as a design variable helps the firm to understand the relevant cost/benefit trade-offs involved for the customer.[23] Recall that conjoint analysis (Chapter 6) is a useful tool in this regard. Smart firms take a customer-oriented perspective on pricing early in the design process, and then develop the product around the relevant price point. For example, Hewlett-Packard, in its initial foray into the digital photography market in the mid-1990s, had research showing that a $1,000 price point was the maximum a consumer would be willing to pay for a scanner and printer for digital photography needs. (Yes, by today's comparison standards, this $1,000 seems exorbitantly high, but this was the going rate for a digital photography package in the mid-1990s! As noted earlier, Moore's Law operates relentlessly in most high-tech markets.) As a result, HP worked its price analysis backward from the customer value point, through the retail channel, subtracting out the margin that retailers would take, ending with a **target cost** figure that HP had to meet in product design and manufacturing. It then did the sourcing and manufacturing around this target cost. Similarly, in the HDTV example, working more diligently with other industry players early in the design process on the whole product—which would include not only programming considerations, but also installation issues, cable/satellite access, and any needed set-top boxes—might help to tip the balance more positively toward higher benefits versus costs.

Second, this analysis shows that *different customers in different segments will value the same product differently.* Prices must account for both the perceived value of the product to customers and the cost to serve a particular customer account. Understanding that different customers value the product differently, and that different customers require distinct levels of service, means that the profitability of different customer accounts can vary widely—and differentially affect the profitability to the firm. Customer-oriented pricing requires that companies manage their customers based on profits, not just sales.[24] High-tech firms must be attuned to the costs of serving customers and filling orders, which can vary significantly by customer, depending on the sales support, design or applications engineering, and systems integration required. Costs to serve customers can include presales costs (e.g., sales calls, applications engineering), production costs, distribution costs, and post-sales service costs. Unfortunately, the price paid by a particular customer often does not correlate with the costs to serve that customer.

With the adoption of activity-based costing practices[25] and CRM software from companies like SAP, Oracle, or Salesforce.com, businesses can now track profitability at the level of each individual customer. This analysis can provide more useful pricing insights than segment-level profitability analysis. For example, a study of customers in four industries (a U.S. high tech corporate services provider, a U.S. mail-order company, a French retail food business, and a German direct brokerage house) found that loyal customers can cost more to serve and pay lower prices than newer customers.[26] If loyal customers turn out to be unprofitable, prices may need to be revised upward. This implication of customer-oriented pricing (focusing not just on sales, but on profits) is consistent with and reinforces the customer relationship management strategies identified in Chapter 5. As noted in that chapter, a key implication for pricing is: *Firms should track the profitability of different customer accounts.*

In analyzing the profitability of customer accounts, companies may actually decide *not* to serve some customers[27]—unless there are mitigating reasons for doing so (e.g., the lifetime value of a particular account is likely to be positive, or ancillary products and services might be sold at a profitable level). Recall how Tech Data from Chapter 5 uses activity-based costing to calculate customer profitability. Unprofitable customers are identified and encouraged to order more efficiently or take their business elsewhere. Using customer-based pricing, Tech Data reduced expenses to 3.5% of sales, compared to the industry average of 5%.[28]

Similarly, using vast data warehouses of customer information and sophisticated hardware and software, FedEx divided its customers into the "good," the "bad," and the "ugly" with respect to profitability.[29] The good (profitable) customers are carefully monitored and receive regular customer service follow-up to prevent defections. The bad customers who spend much but are expensive to serve may be charged higher prices. And the ugly customers, who don't spend much and are expensive to serve, are not targeted with marketing communication at all.

Customers are attuned to issues of fairness in pricing. Some pricing practices—for example, rising gasoline prices, prohibitive prescription drug prices for life-threatening illnesses, and charging higher prices to existing customers than new customers—may result in perceptions of unfairness.[30] When consumer groups voice their protests publicly, companies face a potential public relations crisis. To handle such situations, the company should provide as much information as possible on the reasons for the higher prices. One study found that consumers are more willing to accept price differentials based on cost arguments. They need high levels of interaction and customer service to neutralize their negative emotions. Marketers can charge different prices to different customers and avoid perceptions of unfairness if they can show that customization and differentiation do result in better experiences for consumers.

The accompanying Technology Expert's View from the Trenches gives a customer-oriented view on pricing software as a service.

PRICING OF AFTER-SALES SERVICE

Many manufacturers of durable high-tech products earn significant revenue from after-sales service. As discussed in Chapter 8, services have the potential to provide higher margins and competitive differentiation to sellers. Pricing of services poses a unique challenge because the benefits are often intangible to customers, and companies lack data on unit production costs. As a result, many companies default to pricing service contracts by intuition. Some use uniform pricing based on a fixed percentage of the sales price of the equipment. This technique is too simplistic, because service costs can vary by accessibility of the customer, age of equipment, usage, and operating conditions.

TECHNOLOGY EXPERT'S VIEW FROM THE TRENCHES

Pricing Based on Customer Value: Insights from the CRM Software Industry

BY JASON MITTELSTAEDT

Vice President of Marketing, RightNow Technologies, Bozeman, Montana

Software pricing has evolved as the industry has evolved, migrating from simple heuristics based on, say, the number of computers or users, to a model based on understanding customer value.

RightNow Technologies provides customer relationship management (CRM) software to organizations that want to improve the service they provide to customers, while simultaneously reducing associated costs. Our global client base* spans a wide range of industries, including consumer electronics, higher education, media and entertainment, public sector, and retail. RightNow's software is delivered via a "software as a service" (on-demand) model, where the software itself resides on company servers, and clients access it over the Internet. Our software allows clients to automate many customer-related processes, which in turn, enables them to better serve their customers. For example, many RightNow clients use our software as the customer service agent's desktop in their contact center and to provide self-service options to customers through their websites. Almost immediately after implementing RightNow's software, organizations typically see dramatic reductions in the volume of incoming calls and e-mails. In addition, customer service employees have more time to provide personalized service and address complex customer issues.

Just 10 years ago, pricing for e-commerce software—the software used to transact business on a website—was typically based on the number of central processing units (CPUs) used by a customer's organization; it was not linked in any way to the tangible value it provided for those customers. Today, e-commerce software tends to be priced as a percentage of the incremental revenue driven by the solution. For example, if a company generates an incremental $100M in revenue per year by using the software, the software provider could collect a percentage of that amount, say 5% or $5 million.

Similarly, a decade ago, software pricing for CRM software was based on a customer's revenue or number of end users, neither of which had any relation to the value garnered from use of the software. CRM software pricing is now based on explicit usage during a specified time period—for example, the number of call center agents using the software or the number of end consumer interactions within a 12-month period.

Matching a company's price to the actual customer value its products deliver can be complicated, but it is the optimal pricing strategy. I offer the following recommendations about how to manage this pricing challenge, based on my own experiences:

1. *Embrace change.* In high-tech industries, pricing models change rapidly—as the technology matures, as more competitive offerings are available, and as customers more clearly understand the value added. Although changes to pricing models can fundamentally change your business model, resisting the change may result in outdating your solutions, your business model, and ultimately your competitiveness in the market. In my industry, pricing has changed from licensing fees paid up front by a customer for lifetime ownership, to a rental or subscription model where customers pay for a particular time period of use. One benefit of this "software as a service" model is that customers no longer have to pay huge start-up fees to begin using the software. For software companies, it enables a much more stable and predictable revenue stream while also holding the vendor accountable for delivering ongoing value to the customer.

2. *Align pricing as closely as possible to the measureable value it delivers.* If a CRM system enables a customer to complete 2 million successful interactions with their end-customers during

---

* We often refer to RightNow's customers as "clients," as this helps to distinguish between our customers and our customers' customers.

a month, the customer will be able to strategically and financially justify the value of the purchase of that software. An important corollary of this insight is that sales people won't be put in the awkward role of trying to justify a pricing model that is not clearly tied to the benefits the customer receives.

3. *Structure the contract to allow for easy usage level adjustments.* If a customer's usage needs expand, you should be able to easily add incremental usage at a predetermined pricing level without having to go through an additional sales cycle. This flexible pricing structure lowers your cost of sales to existing customers and accelerates the speed and ease of service that you can provide your client. Conversely, if your customer is not using the full amount of technology they purchased, you should quickly adjust their pricing accordingly. Maintaining a satisfied customer at a lower usage and pricing level is better than losing the customer due to overselling.

4. *Allow your clients to try your product risk free.* Technology skepticism is at an all-time high, particularly with software, thanks to decades of unfulfilled promises. Offering a risk-free trial, limited to a predetermined quantity and time frame, allows the client to experience the business benefit in a real-world environment. Actual experience with the product can help showcase its value, helping to remove any skepticism—as well as increase the likelihood of selling broader usage and minimizing discount levels.

5. *Make sure your products deliver tangible value.* This should go without saying, but if your company's products don't deliver quantifiable value, the pricing strategy is pointless. The root of most pricing challenges is that the product doesn't deliver tangible value to the customer. Open, honest interactions with customers can help highlight what customers are looking for in a solution and how to ensure that value is delivered. The open, honest dialogue can then be used to ensure a pricing strategy that matches the value the customer is receiving.

We at RightNow are committed to providing our clients with products that quickly deliver real value, in terms of improved customer service, satisfaction, and loyalty, as well as substantial cost savings. Many of our clients begin with a small project and rapidly realize significant benefits. They then use the savings from the initial project to expand their use of RightNow. We're proud of the fact that our clients are consistently recognized for the outstanding service they provide to customers and win awards for greatest returns on their investment in RightNow—a sure sign that we are accomplishing our mission.

At the other extreme, some companies have a bewildering array of special contract terms negotiated with each customer. These may be costly to negotiate and can be perceived as unfair to customers. Technology companies, thus, can end up losing money on services.

A better approach, and one that is consistent with the steps in customer-oriented pricing, is to price services based on careful segmentation of customer requirements. Customer needs for service usually include one or more of the following: technical support, training, maintenance, response times, parts coverage, after-hours availability, and add-on services. McKinsey & Company has found that most firms' service customers can be segmented into three categories:[31]

- "Basic needs customers" who want a standard level of service with basic inspections and periodic maintenance
- "Risk avoiders" who want to avoid big bills but don't care as much about response times
- "Hand-holders" who need high levels of service, often with quick and reliable response times, and are willing to pay for the privilege

The three types of service pricing approaches, which will vary for the three categories of customers, are (1) fixed price contract, (2) time and materials, and (3) full coverage. On the basis of this segmentation, it makes sense to offer the basic-needs customers a fixed-price, well-defined, limited service contract, while the hand-holders should be happy to invest in a full-coverage contract. The risk avoiders' needs may be met best with a combination of fixed price plus a time and materials add-on option. This type of service pricing strategy, based on customer needs and provider costs, has a better chance of profitability than either the more simplistic or complicated strategies.

THE TECHNOLOGY PARADOX

One of the most significant factors high-tech marketers face is the rapid pace of price declines. The inexorable march of technological improvements (via Moore's Law) and competition continues to force down prices in products ranging from semiconductor chips to finished personal computers to bandwidth and storage. This situation requires huge gains in volume if a firm is to maintain sales revenues, let alone profitability. Falling prices can help a firm or an industry sell more units—some believe that the demand for digital resources is almost infinitely elastic[32]—and increasing volumes can allow for more price cuts. But the cycle is spinning ever faster, and companies have to scramble to keep up.

During 2006 despite falling prices of mobile handsets, Nokia was able to retain its position as global market leader by focusing on rising demand in emerging markets like China and India. In the fourth quarter of 2006, average selling price of Nokia handsets declined from Euro 99 (US$136) to Euro 89 (US$123). However, sales rose 13% and profit rose 19% in the quarter amid tough conditions.[33]

Known as the **technology paradox**,[34] businesses can prosper at the same time that their prices are falling—if they understand how to thrive in such an environment. At a minimum, the situation requires exponential growth in the marketplace, such that volume grows faster than prices decline. However, at its extreme, technology is virtually free and companies cannot count on volume to provide profits when they are literally giving the products away. The question of how to be competitive when technology is free requires a whole new paradigm for profitability.[35]

In such an environment, there is no single set of rules, as value can be found in several solutions.[36] For example, some companies will thrive by charging a premium for their products (e.g., Intel and Microsoft). Others can make money by selling products like commodities (e.g., disk drives). But, in the middle, companies must be inventive with their pricing strategies, as the solutions in Table 10.2 suggest.

Solutions to the Technology (Pricing) Paradox

Obviously, one solution to the technology paradox is that high-tech companies must know how to *keep costs falling faster than prices*. Companies must redefine value in an economy driven by unit-one costs.

TABLE 10.2 Solutions to the Technology Paradox

1. Squeeze out cost inefficiencies.
2. Avoid commodity markets.
   • Maintain a steady stream of innovation.
   • Differentiate offerings.
3. Have agility and speed in getting products to market.
4. Find new revenue streams.
   • Find new uses for existing products.
   • Offer whole product (end-to-end solution).
   • Offer product bundles.
   • Find new, less price-sensitive, segments.
   • Offer product derivatives under a price lining strategy.
5. Develop long-term relationships with customers.
   • Focus on revenue from complementary products and services (such as captive product pricing or advertising revenues).
6. Use smart (dynamic) pricing.

Second, *technology companies must make every effort to avoid getting stuck making commodity goods.* Maintaining a steady stream of innovations is highly desirable. Commodity markets compel companies to follow supply/demand dynamics, and pricing power dissolves altogether. When products become near-commodities, firms must focus on giving customers something that provides value above and beyond the competition's offerings. Such value might be found in *differentiation* through, say, customization, 24-hour tech support or maintenance agreements, or a strong brand name (Chapter 12). **Mass customization,** or serving mass markets with products that are tailored to individual customers, can be a compelling source of competitive advantage and provides knowledge of individual customer tastes and preferences. This is what Dell offers with its build-to-order business model.

Third, *firms must have agility and speed.* If a firm can't get to market on time, it might have missed its chance for profitability, because the price point will have moved. Relatedly, engineers must focus less on the best possible solution and more on the *best solution possible* in the fastest time frame. In a market where prices decline rapidly, efficient design and systems are probably less important than getting the product to market quickly. As noted in Chapter 6, Guy Kawasaki refers to this as Rule #2 for revolutionaries: "Don't worry, be crappy."[37] His inflammatory rhetoric means that it is sometimes acceptable to strive not for perfection, but for the minimum level of market acceptability with the first generation of a radical new product. This is also consistent with the notion of expeditionary marketing from Chapter 3.

Fourth, *companies can strive to find additional revenue streams.* There are many alternatives to consider here. One source could come from *new uses* for existing products. NTT DoCoMo in Japan encouraged users to send text messages to compensate for declining revenues from voice calls. Additional revenue could also be generated by providing more of the *whole product* that the customer needs. For example, Nokia has unveiled a new, multimedia digital content product strategy that will let its handset users play music, take photos, and watch video in addition to making calls or sending text messages.[38] Additionally, a **product bundle** could be offered where revenues and margins are higher than offering stand-alone products. Former telephone companies are currently dueling with former cable TV companies to offer consumers telephone, Internet, and video entertainment services (triple-play) at home. *New segments,* especially less price-sensitive ones, could be targeted. Google offers its search engine free to consumers but charges companies when they license the software or servers for their web sites. **Price lining,** or *versioning,* follows the practice of offering derivative products and services at various price points to meet different customers' needs. One extreme form of this is to provide varying levels of services, with the lowest level being a "free" version. For example, for many digital products, the ratio of free to paid subscriptions follows the 1% rule: 1% of users support all the rest; 99% get basic free version.[39]

Fifth, rather than being found in selling hardware or software, a real source of value is found in *developing long-term relationships with customers.* When the cost of manufacturing one more unit is negligible (unit-one costs), the goal of the firm changes from making a high margin on each product sold to building relationships with customers. This calls for a high level of responsiveness to demands for service. Many companies are recognizing this as they move away from focusing on sales of hardware or software to providing ongoing services that are a sustainable source of revenue—and competitive advantage. For example, IBM has steadily moved away from being a provider of computer equipment and software to a provider of information-technology–related services.

One business model that relies on establishing long-term relationships is based on the existence of an installed base of customers. An installed customer base often may have loyalty to a product and/or may experience switching costs. As a result, these existing customers are likely to purchase upgrades of a product line or migrate to a new technology platform within the same company. For this reason, some companies are willing to incur higher customer acquisition costs at the outset and plan on a longer-term revenue stream from the installed base over multiple generations of a product. Another business model built on an installed customer base is known in traditional marketing as **captive product pricing.** Also called *razor-and-blade pricing* (pioneered by Gillette), this model appeals to customers who view the product acquisition cost as salient. Thus, companies may price the initial base product (the razor) at a fairly low price (and make correspondingly lower margins), but then earn a longer-term revenue stream on the complementary products (the blades) over the customer's lifetime usage of the initial product. For example, Hewlett-Packard inkjet printers are fairly inexpensive (in the sense that HP takes only a modest markup on them), but the company makes higher margins on the consumable ink cartridges. This model is also useful in markets driven by indirect network externalities (where the value of the base product is derived from complementary products).

An additional strategy based on a customer acquisition model can be found in the advertising revenue model followed by many websites. Rather than focusing on the revenue generated by the customer for the use of the product, the company focuses on amassing a large number of customers; the goals are not only to generate network externalities for the site itself (also an entry barrier to future competitors), but also to acquire a sufficiently large customer base ("eyeballs") to sell to advertisers. Getting big fast, gathering enough consumer eyeballs, and acquiring knowledge about those consumers' shopping habits are keys to success. MySpace became the premier social networking site for teens and young adults by providing a user-friendly platform to post customized Web pages containing photographs, blogs, and music. While registration is free for users, MySpace has an advertising-based business model offering advertisers a chance to connect with a large, desirable demographic.

Sixth, **smart pricing** (or *dynamic pricing*) uses data on customer shopping habits to adjust prices in real-time on the Internet. This allows a company to identify customer preferences and gauge how sensitive certain customers might be to price differentials.[40] Web pricing systems (based on sophisticated software and data analysis) go hand in hand with data mining and one-to-one marketing techniques, which allow marketers to target individual customers geared to their profitability and volume.

A book by Chris Anderson, editor of *Wired* magazine, provides a provocative perspective on the technology paradox. Reprised in the March 2008 of *Wired,*[41] Anderson's 2009 book, *Free: The Past and Future of a Radical Price*, describes two key forces at work: the impact of Moore's Law on the technologies used to serve up bandwidth (as well as computing storage and processing), and the digitization of goods. Although companies have long subsidized "free" products with sales of other, complementary products, increasing digitization and Moore's Law are opening up radical new business models based on the notion of "free." He develops a taxonomy of six different business models based on these forces, including the price versioning, advertising, and captive product pricing models discussed previously.

The final section of this chapter explores other pricing issues germane to high-tech markets.

ADDITIONAL PRICING CONSIDERATIONS

The role of pricing in any market is to transfer rights of the product to the buyer, in exchange for some form of payment. High-tech products are valuable because of the know-how embedded in them. Recall from Chapter 8 that revenues can be generated by selling the know-how in multiple forms: Firms can sell the know-how itself; they can sell components to OEMs; they can sell complete systems in a ready-to-use form; or they can operate a service bureau, providing complete, hassle-free solutions to customers (as IBM does with its e-business and Web-hosting solutions).

The embedded nature of know-how makes it difficult to price high-tech products at different levels on the what-to-sell continuum. A firm can price for a complete transfer of rights, whereby the buyer fully owns the product and related know-how and operates without restrictions, or it can use highly restrictive licensing arrangements that specify volume, timing, and purpose of usage.[42] The issue for a firm is how to maximize profits by choosing the "right" amount of property rights to transfer. This section examines three options: (1) outright sale versus licensing agreements, (2) licensing restrictions for a single use versus multiple uses, and (3) price promotions.

Outright Sale of Know-How versus Licensing Agreements[43]

An outright sale of know-how assumes that the net present value of the technology over the relevant time horizon can be estimated. However, it is hard to assess the value of the technology at the time of transfer, so outright sales of know-how can be difficult to consummate. On the other hand, short-term licenses require an estimation of value over specific fields of use, which can be more readily estimated. When compared to an outright transfer of rights—for which buyers will presumably pay more—short-term licenses may reduce the revenue stream. However, because of the difficulty in valuating know-how, firms might be more willing to use short-term licenses. Rather than undervaluing the know-how, they yield a guaranteed stream of revenue for a specified time and use.

Usage Restrictions

Usage restrictions may include individual versus multiple users (site licenses), one-time use versus multiple uses over a specific time period, and per-use pricing versus subscription pricing.

When licensing technology products, especially software, to enterprise (corporate) customers, a major pricing issue is whether a company should continue its licensing policy for individual users or offer a site license for multiple users. In general, a discounted site license for multiple users provides more value to enterprise customers than the individual license policy.

For many technology products, especially software, individuals are given a license to use the product. Sometimes the license is given free for one-time use or for a specified time period. When a fee is charged for a license, additional conditions may apply, including restrictions on the transferability of the license, time period of use, number of users permitted, or number of physical products on which the software may be used. At the Apple iTunes Music Store, music downloads at 99 cents per title permit users to export, burn, and copy music for personal noncommercial use only, and an audio playlist may be burned no more than seven times.

An additional consideration is whether customers should pay per use or a subscription fee.[44] *Subscription plans* charge one fee, regardless of usage, for a time period—usually monthly or annually. If the technology is new and unfamiliar to consumers, *pay-per-use* (also called *micropayment*) encourages trial at lower risk. However, as consumers become familiar with the technology, beyond a certain usage volume the subscription plan will offer more value to consumers. In 2007, AT&T's data messaging service in the U.S. offered pay-per-use at 20 cents per text message or 50 cents per multimedia message for light users, and monthly subscription plans starting at $4.99 for text-heavy users, or $9.99 for multimedia-heavy users.

The economics of direct billing for micropayments is challenging because the fixed costs of billing infrastructure may not be recovered from the low margins on the micro-transactions. NTT DoCoMo did the billing on behalf of third-party providers of iMode services but charged them 9% of revenue for the service. Credit card payments are the standard way for most service providers to accept micropayments. But credit card processing fees do eat into margins. And many consumers are not comfortable using credit cards. One alternate way to handle micropayments is through eBay's PayPal service, though it has yet to gain sufficient traction outside the United States. A South Korean start-up, Danal, has come up with an innovative micropayment solution for consumers who don't want to use credit cards for online purchases. While making an online purchase, a consumer can send an SMS (short message service) text message on a mobile handset to the wireless carrier requesting approval of the purchase. The carrier responds by SMS, providing an approval code which the consumer enters into the checkout form to complete the transaction. The purchase is then billed to the consumer by the carrier on his or her monthly bill.[45] More innovations in the area of online micropayments are likely to be forthcoming.

In setting subscription prices, one has to consider the impact on consumption. Evidence suggests that a large up-front fee such as an annual subscription paid in advance may have a negative impact on consumption and result in lower usage.[46] Therefore, monthly subscription fees, rather than annual fees, may better encourage technology utilization.

Industry experts predict that the future of the software industry rests on moving away from the traditional licensed model to subscription pricing. This change is driven by the new model of software delivery, "software as a service," described earlier. The monthly subscription fees give the customer access to regular upgrades, and (many believe) lower switching costs in the case of dissatisfaction.[47]

Price Promotions

Price promotions or temporary discounts are sometimes offered to induce trial or to overcome consumer resistance to adoption of breakthrough products or services. Anderson and Simester's study of consumer behavior for a catalog retailer of books and software between 1999 and 2001[48] found that deep discounts were successful in attracting new customers who eventually bought more over time. Current customers also bought more initially when deep discounts were offered. However, over time, current customers bought fewer and less expensive items, probably because they had already stocked up on what they needed or were predisposed to value to begin with. Sellers need to distinguish between prospective and existing customers and consider the long-term impact of their promotions, especially since these could have a negative impact on brand equity.

Summary

This chapter has addressed salient issues in pricing in high-tech environments. After examining the factors that create a complex pricing environment, the chapter presented an overview of the 3 C's of pricing (cost, competition, customers), a framework for the issues that must be simultaneously considered to set prices. Because of the vital importance of a customer orientation in the high-tech arena, the chapter delved more deeply into customer-oriented pricing, including analyzing the profitability of individual customers and revising pricing as a result of this analysis. High-tech marketers need to pay special attention to the pricing of after-sales service.

One of the most significant factors that high-tech marketers face is the unavoidable decline in prices over time; therefore, special attention was given to strategies that generate revenue in light of price declines. Known as the *technology paradox,* businesses who use these strategies can be profitable despite falling prices.

Finally, special pricing considerations—such as sale of know-how versus licensing, usage restrictions (e.g., pay-per-use versus subscription pricing), and price promotions—were briefly addressed. For the many reasons outlined, pricing decisions are difficult; despite this difficulty, marketers must systematically address the issues presented here in order to minimize the odds of making a mistake. Importantly, success is difficult to guarantee.

Discussion Questions

Chapter Concepts

1. What are some of the factors that complicate high-tech pricing decisions?
2. What are the 3 C's of pricing strategy? Describe the importance of each.
3. What is the experience curve? How does it affect pricing strategies?
4. What is cross-price elasticity of demand? How does it help a company understand competition?
5. What are the relative costs and benefits of the purchase of a high-tech product from the customer's perspective?

6. What is reference price? How does it affect pricing in high-tech markets?
7. What is the total cost of ownership? What is its pricing implication?
8. What is customer-oriented pricing? What are the steps in customer-oriented pricing? What are the implications of understanding this approach to pricing?
9. Why is it important for high-tech firms to identify the benefits a customer receives from product features? What type of benefits can the customer receive?

10. What are the various options for a company in pricing after-sales service?

11. Name and describe the three categories of service customers.

12. What is the technology paradox in pricing? What six strategies can a firm use to stay profitable, despite the downward pressures on price, or even free products?

13. What is captive product pricing? Give an example.

14. What is smart pricing?

15. When should a firm use the following pricing strategies?
- Outright sale of know-how versus licensing
- Licenses versus subscription plans versus pay per use
- Price promotions

Application Questions

1. Based on the opening vignette of this chapter, answer the following questions:

a. Do you think Apple set the price appropriately for the iPhone at launch in June 2007? Why or why not?

b. Was the $200 price cut for iPhone on September 5 a good move by Apple? Why or why not?

c. How do you think Steve Jobs handled the response to loyal iPhone customers' anger? How (if at all) would you do this any differently?

2. Based on the chapter's Technology Expert's View from the Trenches, answer the following questions:

a. Identify the shifts in pricing trends for e-commerce software. Give an example of a company other than RightNow that illustrates each trend.

b. Explain the idea of pricing based on customer value. Give an example of a company other than RightNow that practices this kind of pricing along with how it sets prices based on customer value.

c. One of the recommendations of the Technology Expert is that software marketers should allow their clients to try their product risk free. Pick three software products that you are familiar with. Use the Web to research these products, and summarize how and to what extent they implement this policy.

Glossary

captive product pricing A strategy of either giving away or charging a low price for a base or foundation product that requires use of a corollary revenue-producing product; also called *razor-and-blade pricing*.

cost transparency The ability of buyers to use the Internet to obtain information about manufacturers' costs and prices, providing them greater leverage in purchase decisions.

cross-price elasticity of demand The percentage change in one product's sales due to a percentage change in a price of another product.

experience curve A graph that captures the per-unit cost declines in production every time the company's accumulated manufacturing volume doubles.

mass customization Use of technology to serve mass markets with products that are tailored to individual customers.

price lining The strategy of offering different versions of a base product or service at various price points to meet different customers' needs; also called *versioning*.

product bundle Grouping of products to be sold together as a package, rather than selling each one as a stand-alone product.

reference price A standard against which a customer judges the purchase price of a product; an implicit "reasonable" amount based on a customer's prior experience and current purchase environment.

smart pricing The use of sophisticated software that allows analysis of data on customer shopping habits to adjust prices in real-time; also called *dynamic pricing*.

target cost The amount a company is willing to spend to produce an item; determined by working backward from the suggested list (market) price and deducting the company's desired profit margins and any channel margins.

technology paradox The notion that high-tech businesses can thrive at the very moment their prices are falling the fastest—if they use the right strategies.

total cost of ownership The total amount of money a customer expends in order to own and use a product or service, including the initial price paid, delivery or instal- lation costs, maintenance costs, supplies, and other oper- ating costs over the life of the product; also called *life cycle costing.*

Notes

1. Wingfield, Nick, and Li Yuan, "Apple's iPhone: Is It Worth It?" *Wall Street Journal,* January 10, 2007, p. D1; Boslet, Mark, "Apple iPhone Faces Price Hurdle," *Wall Street Journal,* February 21, 2007, p. B11; Kim, Ryan, "Apple Considering a Rebate for the iPhone?" Tech Chronicles, April 17, 2007, online at www.sfgate.com/cgi-bin/ blogs/sfgate/detail?blogid=19&entry_id=15505; Mossberg, Walter S., and Katherine Boehret, "Testing Out the iPhone," *Wall Street Journal,* June 27, 2007, p. D1; and Kim, Ryan, "Apple CEO Jobs offers early iPhone buyers a $100 credit," *San Francisco Chronicle,* September 7, 2007, http://www.sfgate.com/cgi-bin/article.cgi?f=/c/a/2007/09/ 07/BUNBS0HJ0.DTL&hw=apple+iphone& sn=001&sc= 1000.

2. This phrase was coined by Gordon Moore, cofounder of Intel.

3. Smith, Michael F., Indrajit Sinha, Richard Lancioni, and Howard Forman, "Role of Market Turbulence in Shaping Pricing Strategy," *Industrial Marketing Management* 28 (November 1999), pp. 637–649.

4. Dhebar, Anirudh, "Speeding High-Tech Producer, Meet the Balking Consumer," *Sloan Management Review* 37 (Winter 1996), pp. 37–49.

5. Davidson, Paul, "Forecast for Solar Power: Sunny," *USA Today,* August 26, 2007, online at www.usatoday.com/ money/industries/energy/environment/2007-08-26-solar_ N.htm.

6. Smith, Sinha, Lancioni, and Forman, "Role of Market Tur- bulence in Shaping Pricing Strategy."

7. Zettelmeyer, Florian, Fiona Scott Morton, and Jorge Silva- Risso, "How the Internet Lowers Prices: Evidence from Matched Survey and Automobile Transaction Data," *Journal of Marketing Research* 43 (May 2006), pp. 168–181.

8. Chadwick, Alex, "Sony's PlayStation Challenge," National Public Radio, May 9, 2006, online at www.npr .org/templates/story/story.php?storyId=5393492.

9. Monroe, Kent B., *Pricing: Making Profitable Decisions,* 3rd ed. (New York: McGraw-Hill, 2003).

10. "Battery Assault: How Toyota Has Seized the Initiative," *The Economist,* September 24, 2005, p. 79.

11. Martin, Justin, "Ignore Your Customer," *Fortune,* May 1, 1995, p. 122.

12. Christen, Markus, and Miklos Sarvary, "Competitive Pric- ing of Information: A Longitudinal Experiment," *Journal of Marketing Research* 44 (February 2007), pp. 42–56.

13. Shapiro, Benson, and Barbara Jackson, "Industrial Pricing to Meet Customer Needs," *Harvard Business Review* 56 (November–December 1978), pp. 119–127.

14. Ibid.

15. Mazumdar, Tridib, S. P. Raj, and Indrajit Sinha, "Refer- ence Price Research: Review and Propositions," *Journal of Marketing* 69 (October 2005), pp. 84–102.

16. Sheinkopf, Ken, "Energy Efficient Appliances Save in the Long Run," *Orlando Sentinel,* February 5, 2004, online at www.accessmylibrary.com/coms2/summary_0286-6131692 _ITM.

17. Moore, Michael, "Incandescent Light Bulbs Fading out of Sight," *Missoulian in Business Monthly,* June 2007, p. 10.

18. Dubey, Abhijit, and Dilip Wagle, "Delivering Software as a Service," *McKinsey Quarterly,* June 2007, online at www.mckinsey.de/downloads/publikation/mck_on_bt/2007/ mobt_12_Delivering_Software_as_a_Service.pdf.

19. Greene, Jay, "The Linux Uprising," *BusinessWeek,* March 3, 2003, pp. 78–86.

20. Shapiro and Jackson, "Industrial Pricing to Meet Cus- tomer Needs."

21. Aeppel, Timothy, "Seeking Perfect Prices, CEO Tears Up the Rules," *Wall Street Journal,* March 27, 2007, p. 1.

22. King, Rachael, "Radio Shipment-Tracking: A Revolution Delayed," *BusinessWeek,* October 9, 2006, online at www .businessweek.com/technology/content/oct2006/tc20061009 _438708.htm.

23. Shapiro and Jackson, "Industrial Pricing to Meet Cus- tomer Needs."

24. Shapiro, Benson, V. Rangan, R. Moriarty, and Elliot Ross, "Manage Customers for Profits (Not Just Sales)," *Harvard Business Review* 65 (September–October 1987), pp. 101–108; and Myer, Randy, "Suppliers—Manage Your Customers," *Harvard Business Review* 67 (November– December 1989), pp. 160–168.

25. Cooper, Robin, and Robert S. Kaplan, "Profit Priorities from Activity Based Costing," *Harvard Business Review* 69 (May–June 1991), pp. 130–136.

26. Reinartz, Werner, and V. Kumar, "The Mismanagement of Customer Loyalty," *Harvard Business Review* 80 (July 2002), pp. 86–94.

27. Bishop, Susan, "The Strategic Power of Saying No," *Harvard Business Review* 77 (November–December 1999), pp. 50–61.

28. Cruz, Mike, "Tech Data Adds Pricing Tiers," *Computer Reseller News,* May 4, 2001, online at www.crn.com/ it-channel/18835636.

29. Judge, Paul C., "Do You Know Who Your Most Profitable Customers Are?" *BusinessWeek,* September 14, 1998, on- line at www.businessweek.com/1998/37/b3595144.htm.

30. Xia, Lan, Kent B. Monroe, and Jennifer L. Cox, "The Price Is Unfair! A Conceptual Framework of Price Fairness Perceptions," *Journal of Marketing* 68 (October 2004), pp. 1–15.

31. Bundschuh, Russell G., and Theodore M. Devzane, "How to Make After-Sales Services Pay Off," *McKinsey Quarterly,* November 2003, online at www.mckinseyquarterly.com/How_to_make_after-sales_services_pay_off_1343_abstract.

32. Gross, Neil, and Peter Coy with Otis Port, "The Technology Paradox," *BusinessWeek,* March 6, 1995, pp. 76–84.

33. Thomas, Daniel, and Cassell Bryan-Low, "Nokia Profit Rises 19% Even as Prices of Cellphone Handsets Decline," *Wall Street Journal,* January 26, 2007, online at http://online.wsj.com/article/SB116851095357673752-search.html?KEYWORDS=nokia+profit+rises&COLLECTION=wsjie/6month.

34. Gross and Coy, "The Technology Paradox."

35. Anderson, Chris, "Free! Why $0.00 Is the Future of Business," *Wired,* February 25, 2008, online at www.wired.com/techbiz/it/magazine/16-03/ff_free.

36. Gross and Coy, "The Technology Paradox."

37. Kawasaki, Guy, and Michele Moreno, *Rules for Revolutionaries* (New York: Harper Business, 1999).

38. Bryan-Low, Cassell, "Music to Nokia's Ears," *Wall Street Journal*, September 28, 2006, online at http://online.wsj.com/article/SB115940457667876166-search.html?KEYWORDS=nokia&COLLECTION=wsjie/6month.

39. Anderson, "Free! Why $0.00 Is the Future of Business."

40. Keenan, Faith, "The Price Is Really Right," *BusinessWeek,* March 31, 2003, pp. 62–67.

41. Anderson, "Free! Why $0.00 Is the Future of Business."

42. John, George, Allen Weiss, and Shantanu Dutta, "Marketing in Technology Intensive Markets: Towards a Conceptual Framework," *Journal of Marketing* 63 (Special Issue 1999), pp. 78–91.

43. Ibid.

44. John, Weiss, and Dutta, "Marketing in Technology Intensive Markets."

45. Buckman, Rebecca, "Just Charge It—To Your Cellphone," *Wall Street Journal*, June 21, 2007, online at http://online.wsj.com/article/SB118236880977842343-search.html?KEYWORDS=mobile+payment&COLLECTION=wsjie/6month.

46. Gourville, John, and Dilip Soman, "Pricing and the Psychology of Consumption," *Harvard Business Review* 80 (September 2002), pp. 90–96.

47. Dubey and Wagle, "Delivering Software as a Service."

48. Anderson, Eric T., and Duncan I. Simester, "Long-Run Effects of Promotion Depth on New versus Established Customers: Three Field Studies," *Marketing Science* 23 (Winter 2004), pp. 4–20.

11

Marketing Communication Tools for High-Tech Markets

People-Ready Software[1]

In March 2006, Microsoft launched an integrated marketing communications campaign, "People-Ready Software," to be launched later in the year. Promoting Windows Vista and Office 2007 to business customers, the campaign theme emphasized that this collection of software helps companies reduce costs, increase worker productivity, and accelerate innovation by enabling collaboration, searching company databases, conducting business intelligence, and managing customer relationships. The campaign also intended to take market share away from IBM by suggesting that as a service company, IBM's expensive consultants do projects for clients and then depart. The claim was that Microsoft's software is so familiar to users that consultants are not required. Microsoft needed to generate excitement among business users for these new products because there hadn't been a major release of Windows since Windows 2000, nor of Office since Office 2003. Some of the new products' capabilities include phoning in for e-mail and having messages read out through voice recognition and translation technology, creating virtual workspaces for syncing documents and calendars for team members, and analyzing worldwide sales patterns from an Excel spreadsheet.

The campaign began with eight-page advertisements in the *Wall Street Journal,* the *New York Times,* and other newspapers. There was also a major launch conference in New York City, where Bill Gates and Steve Ballmer appeared onstage alongside several computer manufacturers like Dell, HP, Sony, and Toshiba, as well as semiconductor manufacturers like Intel and AMD, who were all featured as Microsoft partners committed to Vista. Chief Information Officers (CIOs) from major corporations were invited to this event to see product demonstrations. In addition to newspapers, ads appeared on TV, in technology magazines, and on websites. A specific tag line for Vista in these ads was "The Wow Starts Now." It was estimated that Microsoft would spend $500 million over the coming year on this campaign, created by the ad agency McCann Erickson in San Francisco, a unit of Interpublic Group of Companies.

Because business customers have long decision cycles to upgrade software throughout their organization, the Microsoft "People-Ready Software" campaign preceded the actual launch of the products by about nine months.

A solid advertising and promotion mix is as important in high-tech markets as in traditional markets. This importance is backed by research that found higher expenditures on both advertising and R&D lower a firm's systematic risk, thereby increasing stock returns across a variety of publicly listed U.S. companies.[2] Some of the key marketing communications tools include traditional advertising (in both mass media as well as trade journals), trade shows, sales promotions (contests, incentives), public relations (event sponsorships, etc.), publicity (articles in the news media), direct marketing (mail, telemarketing), and personal selling. In addition to these traditional media, marketers are increasingly augmenting their advertising and promotion strategies with **new media**—a variety of Internet-based marketing tools including paid (sponsored) search advertising, online display ads, Web 2.0 techniques such as blogging and social networking, marketing in virtual worlds such as Second Life, and mobile marketing (marketing to cell phones via text messaging services), to name a few. Moreover, most (if not all) companies today have a website that is a key element in their marketing communications efforts.

This chapter discusses how high-tech marketers can use these marketing communications tools in a synergistic fashion. The chapter begins with the "advertising and promotion pyramid," a useful device to coordinate and leverage the various marketing communications tools. It continues with a focus on new media and how high-tech marketers are using new media to complement their traditional advertising and promotion strategies. Because a company's website often is one of the foundational elements of its marketing communications efforts, this chapter also presents an overview of website design and management. Appendix 11.A at the end of the chapter provides details on monitoring website traffic, an increasingly important topic given that so many companies' marketing efforts are focused on driving traffic to their websites.

The discussion of high-tech marketing communications continues in the next chapter as well, covering topics such as strategic brand management, ingredient branding strategies, and new product pre-announcements.

ADVERTISING AND PROMOTION MIX: INTEGRATED MARKETING COMMUNICATIONS

The advertising and promotion (A&P) pyramid,[3] shown in Figure 11.1, positions advertising and promotion tools based on two dimensions:

- The degree of coverage, or reach, of the target audience
- Cost efficiency

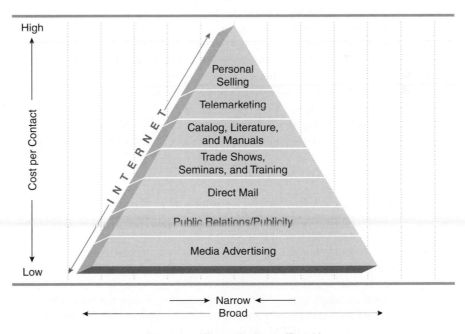

FIGURE 11.1 The Advertising and Promotion Pyramid

Source: Adapted from Ames, B. Charles, and James D. Hlavacek, *Managerial Market for Industrial Firms* (New York: Random House, 1984), p. 253.

One useful way to assess cost efficiency is based on **cost per thousand (CPM)**, where M is the Roman numeral for 1,000. CPM is computed as follows:

$$CPM = \frac{\text{\$ Cost of the advertising and promotion tool (say, an ad in a particular trade journal)}}{\text{Number of people the tool reaches}} * 1,000$$

CPM allows advertisers to compare the relative cost efficiency of various media tools. By "amortizing" the total cost of an advertisement in a specific media outlet by the number of viewers the specific advertisement will reach, CPM allows an "apples-to-apples" comparison of disparate media across media categories. Also, as discussed later in this chapter, CPM is one of the major pricing models for online advertising.

At the base of the A&P pyramid are tools that have wider coverage of the target audience and lower cost on a per-contact basis. Tools at the top of the pyramid generally have narrower coverage and higher cost on a per-contact basis. While the pyramid implies a trade-off between coverage and cost per contact using traditional communication tools, the Internet as a new communication tool is scalable, extremely cost efficient and can be used at all levels in the pyramid.

The pyramid can be used as a coordinating device for advertising and promotion. A firm should not use the tools at the highest level of the pyramid in isolation of the tools at lower levels. The role of the tools at the lower levels of the pyramid is to create product and brand awareness, and to "warm up" prospects prior to using the more expensive, narrower tools.

For example, a new product may be announced with free product reviews in trade journals or blogs. Leads from the announcement (that come in via either a website or a toll-free number) can be sent a direct-mail piece or an e-mail follow-up, consisting of, say, a brochure or other collateral material. As prospects continue to express their interest, they can be contacted more personally via telephone, further e-mail, or a salesperson's visit. Importantly, the use of higher-level tools in the pyramid is leveraged on the efficiency and effectiveness of the lower-level tools and is geared to the prospect's continuing interest. Using all the promotional tools together in an orchestrated campaign to deliver a consistent message to the target audience constitutes what is known as **integrated marketing communications**.

Matching the advertising and promotion tools to the appropriate task is based on the strengths and weaknesses of each tool, relative to the objectives each is to perform. The tool used must be matched (based on effectiveness and efficiency considerations) to the task at hand. By necessity, this discussion of pertinent issues for each of the elements in the advertising and promotion mix is brief. Because A&P requires a high degree of specific knowledge, we encourage high-tech marketers to avail themselves of other resources that cover the more advanced nuances of A&P tools and issues.[4]

Media Advertising

Advertising is a paid, nonpersonal form of communication using either mass-media channels such as print, radio, television, billboards, and so forth, or more targeted trade journals. At the base of the pyramid, high-tech companies use media advertising in both mainstream media and trade journals. Trade journals might include *vertical* (industry-specific) publications as well as *horizontal* publications (often specific to a particular job title/function) that cut across industry boundaries. Critical issues in selecting the specific media are (1) the degree to which the audience of each particular media vehicle overlaps with the firm's target market, where the goal is to minimize *wasted coverage* (people who are exposed to the ad in the media vehicle but are not in the company's target market) (2) the cost efficiency (CPM) of the vehicle, (3) the fit of the editorial content of the media vehicle with the brand message, and (4) the size and frequency of the ads that will be run.

How can high-tech firms identify appropriate media outlets? An advertising agency is a good start. For example, Godfrey Q and Partners in San Francisco is an example of an advertising agency that focuses exclusively on high-tech clients.[5] If a firm does not utilize the services of an ad agency, the Standard Rate and Data Service (SRDS), a part of the Nielsen Company, is a useful resource to search by industry, audience, and so forth for appropriate media vehicles. Moreover, CMP Media, a media services company, has events, websites, and trade publications serving a wide spectrum of builders, sellers, and users of technology worldwide. Finally, media kits, available either online or from the advertising sales representative of any media outlet, provide useful information about not only the specific media vehicle but also its audience, ad rates, editorial calendar with upcoming special issues, other publications in the industry, and pertinent industry statistics.

Identifying appropriate media options is only half of the equation in using media advertising. A firm also must decide on an effective message that breaks through the competitive clutter of the many ads vying for the viewer's attention, while simultaneously reinforcing a brand message in a quick and easy manner. Striking a balance between gaining attention and reinforcing a brand message can be tough. Some firms err on the "gaining attention" side, using cute techniques (e.g., humor, sex, babies) to gain attention in a way that actually has little to do with the brand message, and so may even distract the viewer from the brand message. Other firms err on the side of providing so much technical detail—in jargon, which is, at best, uninteresting to the audience, and, at worst, unintelligible—that attention is quickly lost.[6] The best ads will both break through the clutter and use an attention-getting device that quickly and effectively delivers the key benefit without being lost in details—details that would be better provided in the next contact with a customer. For example, Apple's "Mac Guy versus PC Guy" humorous TV ads have been effective in creating a hipper, cooler image for Macs over Windows PCs. In fall 2008, Microsoft began an image campaign as well with its "I'm a PC" TV ads featuring people from different walks of life.

Public Relations/Publicity

Public relations (PR) includes the activities a firm undertakes to develop goodwill with its customers, the community in which it does business, stockholders, and other key stakeholders (e.g., government regulators, industry trade associations). These activities might include sponsoring events, such as sporting events or charitable causes, cause-related marketing (aligning with a nonprofit organization), corporate advertising regarding the firm's position on critical issues or its philosophy of doing business, or other outreach activities (e.g., speeches by company executives, tours). *Publicity* refers to any coverage the company receives in the news media (print or broadcast) regarding its products or activities. Firms can attempt to gain favorable news coverage by holding press conferences, sending press releases, or staging events. Harnessing the power of bloggers can also generate publicity, and the use of these "new media" tools are covered later in this chapter. The accompanying Technology Solutions for Global Problems is an example of corporate social responsibility in the pharmaceutical sector.

Companies can use the Web to distribute their press releases to a number of media outlets. PRNewswire is a membership-based organization that allows companies (or their PR agencies) to look up the editorial calendars of many media outlets and then upload press releases to the PRNewswire website (www.prnewswire.com). Journalists are allowed to register free on this website and get immediate access to all uploaded press releases. PRNewswire keeps track of journalist views and which outlets carry each story and reports these back to the client.

High-tech firms should not ignore the value that comes from maintaining a positive public image—and maintaining good relations with the media is key in nurturing a positive image. When bugs are discovered in new high-tech products after launch, the media determine whether to make a big issue of it or to relegate it to a few paragraphs on an inside page of a newspaper or magazine. In 2006 when Sony's lithium-ion batteries overheated and caught fire, affecting 8 million laptop users, the news grabbed headlines all over the world. But news of Apple CEO Steve Jobs's alleged involvement in backdating stock options in 2007 attracted hardly the same media attention.

High-tech companies should think strategically about their PR efforts. For example, some industry experts believe that Google was a bit late to the PR game with respect to its efforts in

TECHNOLOGY SOLUTIONS FOR GLOBAL PROBLEMS
Mass Drug Administration and Rapid Testing to Eliminate Lymphatic Filariasis in Tropical Regions
By Jamie Hoffman

In 2000, one billion inhabitants of the world's tropical regions risked acquiring lymphatic filariasis (LF), commonly referred to as *elephantiasis*. Transmitted by infected mosquitoes, LF afflicted 120 million people that year, one-third of which exhibited serious disfigurements and suffered from incapacitation. LF causes grotesque swelling of the legs, arms, head, genitals, breasts, and vulva; those infected also suffer from intense social stigmas that hinder employment and marital opportunities.

In 1997 the World Health Assembly resolved to eliminate LF by 2020; since 2000, the World Health Organization (WHO) has partnered with pharmaceutical companies GlaxoSmithKline (GSK) and Merck, various private and governmental agencies, and the academic sector to do just that. After six years of mass drug administration (MDA) in 44 of 83 endemic countries, WHO lowered the at-risk population to 1.3 million people by annually co-administering single doses of two drugs: albendazole and diethylcarbamazine (DEC). GSK has pledged to donate 5 billion doses of albendazole for 20 years. Merck has donated Mectizan.

To stop the spread of LF transmission, WHO first identified those regions in which LF is considered endemic (10–15% of men and 10% of women affected.) Utilizing a new technology, the whole blood immunochromatographic (ICT) filariasis test, to rapidly detect *Wuchereria bancrofti* antigens, infected individuals were quickly identified by a simple finger-prick blood sample that doesn't require a laboratory facility. In communities deemed to be endemic, the entire at-risk population is treated with both DEC and albendazole over a span of 4 to 6 years.

The results have been astounding. After 2 to 6 rounds of MDA of the two drugs, most areas were able to decrease LF to below 1% of the population. The countries unable to implement MDA faced lack of funds or lack of drugs.

Sources: World Health Organization, "Weekly Epidemiological Record," October 19, 2007, online at www.who.int/wer/2007/wer8242.pdf; and World Health Organization, "Lymphatic Filariasis," Fact Sheet, September 2000, online at www.who.int/mediacentre/factsheets/fs102/en.

Washington, D.C., particularly over the net neutrality controversy.[7] **Net neutrality** (which Google supports) is the idea that all websites, content, and platforms should be treated equally for their broadband needs, in contrast to a "tiers of service" perspective advocated by some of the telecom companies, which states that some Internet companies could or should pay more to ensure high-speed transmission of their data. The analogy would be that some companies would pay a "toll" to access a high-speed traffic lane while those that don't would be relegated to the slow lane. Net neutrality advocates argue that broadband carriers should not be allowed to use their market power to discriminate between large and small Internet-based companies, which could potentially stifle new start-ups and innovations. (Net neutrality opponents argue that companies such as YouTube that send extremely large amounts of data over the Internet ought to pay proportionately for the requisite broadband infrastructure that is a key foundation for their operations.) As Google's business has expanded into so many new domains that intrude upon traditional industries' businesses (e.g., its move into user-generated content with YouTube places it in the sights of huge entertainment companies and studios) and as its traditional search-based advertising business amasses user data with its attendant privacy concerns, having a lobbying presence in Washington is critical. "Opening an office in Washington, D.C., really shows they are becoming a grown-up company," says Fred von Lohmann, senior intellectual-property attorney at the Electronic Frontier Foundation. "The Internet companies are belatedly realizing that they can't ignore Washington."[8]

Direct Mail

Because direct mail can generally be targeted more precisely than advertising or public relations, it is placed higher in the pyramid than those two tools. Lists can be obtained from a number of list brokers, including Dun and Bradstreet or the mailing lists of many trade publications, to name just two sources. Additional information on possible lists can also be found in the Standard Rate and Data Service. Costs are usually determined on a per-name basis, with more targeted lists that are frequently updated costing more. Firms must decide how many pieces to mail, as a function of the potential size of the target market, as well as the frequency of mailings. One mailing is typically insufficient to achieve results. The online equivalent of direct mail is an online newsletter or targeted e-mail communications. Again, lists can be purchased for such purposes.

Trade Shows, Seminars, and Training

Trade shows, seminars, and training reach yet a narrower group of customers at a proportionately higher cost than tools lower in the pyramid. However, research has documented that exposure to a product at a trade show accelerates the progression of business buyers through the selling cycle after the trade show.[9] The international Consumer Electronics Show (CES) held in Las Vegas every January is the industry launchpad for new products and innovative technologies, as well as a key environment to compare, contrast, and test-drive these products. CEBIT is an international trade show for office and information technology held every spring in Hanover, Germany.

Trade shows can be quite expensive for exhibitors, including expenses for exhibitor fees, design and setup of a booth, personnel to staff the show (with the attendant travel, meals, and entertainment costs), and so forth. Trade show costs vary by show, industry, and location. For CEBIT 2008, a special package to first-time exhibitors was priced at Euro 5,000 (about US$7,300) for a 15-square-meter (161 square feet) row stand. At CES, the minimum price of a 200-square-foot booth is over $6,000.

Especially in large shows such as CES or CEBIT, firms must trade off the large exposure potential from exhibiting at the show against the investment needed to break through the competitive clutter. Large players tend to dominate the space, and unless attendees have a compelling reason to visit a particular booth, a firm may not realize a return on its costs of attending the show. At a minimum, a company must run an accompanying advertising and promotion campaign just to generate traffic at its booth. Additionally, some attention-getting device at the show itself can be helpful, to the extent it ties in with the product's message. Follow-up on the leads generated and a post-show evaluation are also critical. Despite the costs and risks involved, trade shows are effective in the marketing of technology products and services because face-to-face interactions help to reduce technology, market, and competitor uncertainty.

Online trade shows and *webinars* (Web-based seminars that allow users to participate from their own locations) are also a way to leverage this level of the A&P pyramid.

Catalogs, Literature, and Manuals

Companies must have brochures or other collateral material of some sort to provide customers with additional information. The material must build on earlier communications with customers and showcase the key benefits of the products in terms customers can understand. Technology companies sometimes err in calling their collateral materials "data sheets." Although supporting technical detail is appropriate in these later follow-ups with customers, such details must not come at the expense of clear communication of the product's benefits and a clear, compelling value proposition. Issues such as relative advantage (costs versus benefits), compatibility/interoperability, scalability, service, and warranties should be addressed. Especially for new technologies that may cost more than existing solutions, communicating benefits in terms of total cost of ownership can be effective.

Often, tech companies prepare a *white paper,* which is a technical paper detailing the technology they have developed and the benefits it provides to customers, to disseminate to prospective customers. A white paper gives potential customers more information and provides credibility to the company.

Telemarketing

Telemarketing can be done on an outbound or inbound basis. *Outbound* calls are made by a company's personnel to existing customer accounts or on a cold-call basis. *Inbound* calls are made by customers or prospects calling into the company's call center, typically via a toll-free number. Having the opportunity to maintain customer relationships with a person dedicated to particular accounts can be an efficient use of resources. Telemarketers can help provide support to the field salespeople, answering questions, maintaining contact with accounts, and staying abreast of account changes in between visits from field sales personnel. In October 2003, the U.S. Congress passed legislation creating a national "Do Not Call" list for those who don't want to be bothered by telemarketers' calls. By 2007 about 145 million phone numbers had been registered on this list.[10] This has been effective in reducing the number of unwanted calls to consumers. It has also resulted in the reduced effectiveness of a communication channel for marketers to make cold calls.

Personal Selling

At the highest level in the pyramid is the use of personal selling, where each salesperson can generally reach just one customer at a time. One implication derived from the advertising and promotion pyramid is that small, resource-constrained high-tech companies generally should *not* use their company founders and salespeople to call on prospects *unless the prospects have first been contacted using less expensive, broader-reach tools.* Although small firms may say that they lack the resources to fund PR or a direct-mail campaign, the real issue is whether they can afford to use their existing resources in an inefficient manner. By leveraging the tools at lower levels in the pyramid, firms ensure that the value of their high-expense tools is maximized.

Do not misunderstand this message. It is *not* meant to imply that personal selling and wooing customer accounts with top executives is unimportant—or that having top managers and technical personnel involved in, say, a customer visit program as part of a company's market-oriented culture is not valuable. For example, EMC Corporation, maker of computer storage devices, heard from its customers that sending its top sales executive to its customers' headquarters was a key to winning business away from competitors.[11] Similarly, Semitool, maker of high-end semiconductor manufacturing and test equipment, relies extensively on personal selling to its large customers (the semiconductor manufacturers)—but it also relies extensively on trade shows and on very detailed brochures to communicate with prospects first. The lesson to be learned is that the value of such resource-intensive tools is maximized by ensuring that customers are appropriately primed to receive the company's message.

INTERNET ADVERTISING AND PROMOTION

High-tech and more traditional companies alike are increasing the percentage of their advertising and promotion expenditures spent on *new media,* a term that refers to technology-enabled advertising and promotion tools, typically via the Internet. Display (banner-type) ads, branded sponsorships, classifieds, search ads, affiliates, Web 2.0, viral marketing, widgets, and mobile advertising are all cost-efficient, Internet-based tools that can complement many traditional advertising and promotion tools.

Indeed, more and more companies are embracing online advertising.[12] The Internet's share of global advertising spending was 5.8% 2006 with a total value of $18.7 million; it included:

- Paid search ads, $7.8 million
- Display ads, $7.0 million

- Classified ads, $3.3 million
- Other forms of online ads, $0.6 million

The Internet's share of global advertising was projected to increase to 7% in 2007, 8% in 2008, and 8.7% in 2009.[13] A survey of 410 marketing executives from around the world found that the most used online advertising techniques, in decreasing order, were e-mail ads, display ads, search ads, branded sponsorship, referrals from other web sites, video ads and podcasts.[14] On the other hand, the same executives said that the most efficient online ads in reaching company goals (in decreasing order of efficiency) were search ads, e-mail ads, branded sponsorship, referrals from other websites, video ads, podcasts and display ads.[15] There is thus a surprising asymmetry in the case of display ads. They are viewed as inefficient and yet executives seem to be using them a great deal. A possible explanation is that display ads are used more for long-term brand building rather than generating immediate sales.

Regardless of the specific advertising tools used, best-practices online marketing follows the philosophy of **permission-based marketing**[16]—asking customers or prospects to *opt in* to receive marketing messages based on their interests in a particular topic. Permission-based marketing is in contrast to nearly all conventional advertising and promotion, which might be characterized as "interruption marketing." Not only does most traditional advertising "interrupt" consumers from the tasks they are engaged in (say, TV viewing); if a consumer doesn't want to receive a company's ads, he or she must "opt out" by telling the company to remove his or her name from its mailing list.

Furthermore, permission-based marketing takes a very direct approach in marketing to customers, rather than the more veiled persuasion inherent in most traditional marketing attempts. The Internet is a natural environment for permission marketing because of e-mail communication and its growing ability to close sales and, in the case of purchasing informational or digital products, to fulfill sales. In a comparative study of the ROI of e-mail marketing versus direct response TV and direct mail conducted by the Direct Marketing Association, e-mail marketing came out on top.[17] Among e-mail marketing campaigns, those targeted to existing customers through in-house databases did dramatically better than those intended to prospect for new customers.

Once a company has received the customer's permission to receive e-mail information and product offers, a business should follow *relationship marketing* strategies (see Chapter 5). Rather than pouring marketing dollars into prospecting for new customers, a business should try to up-sell and cross-sell to existing customers. Customers who trust the vendor to give them reliable products and services at a fair price are receptive to such marketing. In relationship marketing, the true value of any customer is based on the lifetime value of his or her future purchases across all the product lines, brands, and services offered by a firm.

This section discusses the different types of Internet ads, their relative effectiveness, and cost implications. As with traditional media, companies must delineate the role of each of these tools in the overall communications mix so as to create an effective integrated marketing communications campaign with a consistent message. For example, research indicates that e-commerce enabled websites can enhance the effectiveness of salespeople by relieving them of mundane tasks so they can focus on high-value opportunities. However, such websites also increase job insecurity of salespeople. Thus, managers in multi-channel environments have to provide better training to salespeople to take advantage of these new technologies.[18]

Display Ads

Display ads or **banner ads** include graphics as well as text, are placed adjacent to editorial content on a Web page and, when clicked, take the user to the advertiser's website. They may be static or involve rich media such as animation, audio, and video. Click-through rates on display ads have been decreasing over time. However, display ads, like traditional advertising, are better at long-term brand building rather than generating short-term clicks or sales. Research has shown that display ads increase advertisement awareness, brand awareness, and purchase intent. Evidence suggests that exposure to display ads increases the purchase probabilities of existing customers. Display ad campaigns for current customers, therefore, should be designed to deliver a consistent, creative message across highly trafficked Web pages across many websites.[19] They act as a reminder to current customers to buy again.

Further, the effectiveness of display ads increases with appropriate behavioral and demographic targeting. **Behavioral targeting** refers to a technique that allows "supersmart, supertargeted display ads" to be exposed to an individual based on the person's past online behavior (via cookies, log files, and other online data); ads placed using behavioral targeting not only do a better job of getting a Web surfer's attention, but also can be tracked with "laserlike precision."[20] For example, in 2007 Yahoo! had about 131 million monthly unique visitors to its sites. By dropping cookies onto every Web browser that looks up one of its sites, Yahoo! analyzes this information and combines it with data about what people are doing on its search engine. Its sophisticated model can then be used to predict consumer behavior. In one campaign, Yahoo! found that visitors who saw the ad for one specific brokerage were 160% more likely to search in that category ("online brokerages") over the next three weeks. Most importantly, the visitors who previously saw the ad overwhelmingly clicked on a display for this brokerage when it appeared in Yahoo!'s paid search results. The benefit is a user profile that goes well beyond a particular search episode (what search string, for example), and integrates the data with a host of other surfer behaviors. For example, a person's cookie profile may show that he spent time at Yahoo! Auto evaluating cars on fuel efficiency, and then clicked over to Yahoo!'s Green Center to read about alternative fuels, and then looked at cars on eBay (a Yahoo! partner).[21] Yahoo!'s behavioral targeting program can predict with 75% certainty which of the 300,000 monthly visitors to Yahoo! Auto will actually purchase a car within the next three months. And, the next time this person visits Yahoo! Sports, he will see an ad for hybrid cars. Indeed, based on this analysis, Yahoo! is finding that ads on sites that seemingly have nothing to do with them (where the content seems irrelevant) can perform very well, because they are based on an elaborate analysis of a user's complete Internet behavior (and not merely a group of search terms).[22] Similarly, on a social networking site such as Facebook, a display ad featuring tickets for the New England Patriots American football team would have maximum impact if it were served to users whose Facebook profile shows they are interested in football and attend Harvard University.[23] Yahoo! also developed new SmartAds technology that lets advertisers mix different backgrounds, images, and text in real time to be able to target different users based on interest and location.

The importance of the online display advertising market can be seen in the flurry of acquisitions made in late 2007 and early 2008 by companies such as Google and Microsoft. In late 2007, Google made a bid for DoubleClick, an online display ad platform that operates at the intersection between advertisers and publishers (websites where the ads appear), valued at $3.1 billion. Yahoo! made a bid for ad-serving platform called Right Media. Microsoft made several acquisitions, including an offer of $6 billion to acquire aQuantive, which owns another dominant display ad-serving platform named Atlas; a 1.6% equity stake in Facebook to better target display ads to Facebook users; and in early 2008 an aggressive bid to buy Yahoo! outright. All of this activity in display ad space indicates it will be a hot growth area as more companies reallocate budgets from offline to online display advertising.

Branded sponsorship is undertaken by an advertiser seeking to associate its brand with a popular Web page using a variety of large-format, rich-media display ads. This is the online equivalent of a branded offline location such as Qualcomm Stadium in San Diego.

Classified advertising is text-based advertising, typically in a particular local market area, similar to its offline counterpart found in print newspapers. Probably the biggest, free online classified website is Craigslist.org. In 2007, Google launched a similar service called Base, which continued in beta test form in 2008.

Search Ads

In contrast to display ads (which are served up by a third-party service such as DoubleClick), search ads are served up based on the keywords a user enters into a search engine. A search engine like

Google or Yahoo! sends out automated software "spiders" or "crawlers" that index the content of billions of pages on the Web and include the pages as listings in their database. After users enter keywords for the information they are seeking, these keywords are matched against the content of the pages in the database and a listing of the pages is produced in order of decreasing relevance.

Any website that wants to receive a high ranking from search engines must design its pages around the keywords that are most relevant to the content of its pages. This practice of designing Web pages so they will rank high in relevance when users search with certain keywords on a search engine is called **search engine optimization**.[24] Natural search rankings ("organic search") that result from search engine optimization are generally unpaid, although new websites may get indexed faster by the payment of inclusion fees to select search engines. Box 11.1 provides an overview of considerations in search engine optimization.

In 1998 Overture (now part of Yahoo!) introduced the concept of *paid search*. Overture invited advertisers to bid for certain search keywords in real time on its home page. For example, IBM could buy the word *computer* on Yahoo! for a defined time period. Every time someone conducted a Yahoo! search using that term, an IBM Web page would come up as the first sponsored link on the page. The advertiser would be charged the bid price only if a user actually clicked on the link. So paid search is also called *pay per click*. Paid search includes boxed text ads appearing horizontally on the top of the user screen identified as "sponsored links" or "advertisements," as well as the links appearing vertically along the right side of the screen in decreasing order of their bids. Paid search ads are being embellished with additional features such as "click to call," which enables users to click an icon and speak to a company representative right away; such ads cost more than conventional pay-per-click ads. Over time, the algorithms used by search engines have become more sophisticated so that the rank of a search ad is determined by historical click-through rate and number of external links pointing to the page in addition to relevance and bid price per click. Google's paid search ad service is called AdWords.

Contextual ads are a form of paid search on various websites rather than the search engine's home page. Google, with its AdSense service, acts like an ad agency on behalf of smaller advertisers. Websites that register with AdSense allow Google's search technology to monitor the content of their sites. If Google finds editorial content on a registered website that is of interest to one of its advertisers, it will serve up an ad on that website to match the content.[25] Each time a visitor clicks on these ads, the advertiser pays Google, which splits the revenue with the website that hosts the ad.

Pricing Models for Online Advertising

Payment for online advertising follows either a CPM or cost-per-click pricing model. CPM (cost per thousand, see page 376) means that the advertiser pays the stated CPM amount (which varies from site to site, based on the type of viewers the website attracts) every 1,000 times the ad is served up. *Cost per click* refers to the ad rate charged only if the surfer clicks on an ad (as in paid search). This type of payment structure requires measuring how many surfers click on the ad. Whether an advertiser is willing to pay on a CPM or cost-per-click basis likely depends on its online advertising goals. If the advertising is designed to support broad-based awareness and brand familiarity, a CPM pricing model might make sense, as for display ads. On the other hand, if the goals of the campaign are to develop a database of possible customers and move toward a direct sale, then paying on a cost-per-click basis might make more sense, as in paid search.

As with any advertising techniques, online advertising also has problems and limitations. Furthermore, as with any new technology, online advertising faces a number of barriers to adoption including insufficient metrics to measure impact, insufficient internal capabilities to fully utilize them, and resistance from management. The barrier of insufficient metrics may seem counterintuitive because most people believe that online advertising is more measurable than traditional advertising such as TV, print, or radio. For example, monitoring the action people take online after being served up an ad can be fairly easy to assess through click-stream analysis. However, even search

BOX 11.1

Optimizing a Website for Search Engine Rankings

Businesses must know how to effectively design their websites so search engine rankings will give them maximum visibility. However, because search engine algorithms change regularly and because search engine companies form and dissolve partnerships frequently, search engine optimization is a moving target. Although each search engine has a unique algorithm for ranking websites, in general, two major factors affect how sites rank:

1. *Keyword relevance: Relevance* captures how often a particular keyword appears in the Web page, as well as the placement of those keywords (e.g., in major headings). Keywords should be chosen to replicate what a user would enter into a search engine to find that particular site—including any possible misspellings. Resources to help identify relevant keywords and search terms include WordTracker, a subscription service that provides related keywords and search counts for the major search engines; some search engines offer a "Search Term Suggestion Tool" that shows related keywords and the number of actual user searches for each term.

Then, the company's Web pages should be designed with those keywords in mind by using "tags" in the HTML, as listed below. Because search engines use these tags to sort and rank Web pages, the HTML code should be optimized for the keywords the company has selected in the tags.

- The *title tag* is the line of text that appears at the top of the Web browser when the window is open. Be sure the title includes relevant keywords.
- *Meta tags* keywords tags and the page description tag—provide keywords and a description of the page contents. The search engine results typically provide the description just below the displayed link; the description should not only include keywords, but should also offer an explanation of what the website contains.
- *Link tags* are URL addresses to other websites. Many search engines follow the links to other pages and find out how many times the keyword comes up on the corresponding Web pages. By following the hyperlinks to other pages, search engines can ensure that the website's content is devoted to the specific keyword that is entered into the search engine.

2. *Link popularity:* When website A creates a link to website B, search engines define that link to be a vote for site B. Beyond the sheer number of links to a website, search engines place greater importance on the quality of a link. If site A has high link popularity, the link from site A to site B will create additional link popularity for site B. Also, if site A is relevant to the search term, it is determined to be a more valuable link.

One way to create link reciprocity is to visit a major search engine and perform a search on the target keywords. The resulting websites that are highly ranked by the search engines are all prime candidates for quality links. With a list of prime link candidates, the next step is to visit each site individually and request a link. The most successful approach for soliciting links from these sites is by requesting a "reciprocal link." Each site will create a link to the other, a win-win solution that improves the visibility for both websites.

There are many other creative ways to create quality links with other websites. A company should ensure that links are included with its content or services shared with other sites on the Web. For instance, if a company writes press releases or articles for syndication, including a link back to the company's website can provide quality link opportunities. Similarly, affiliate programs have become very popular on commercial websites. These programs provide a network of affiliate websites that present a prime opportunity to build links.

Simply designing a website and optimizing it for search engine positioning does not ensure that it will be picked up by the search engines as well. Use of a resource such as SearchEngineWatch can help to stay abreast of this evolving field.

Source: Ferrini, Anthony, "Search Engine Placement," in *Marketing of High-Technology Products and Innovations,* 2nd ed., J. Mohr, S. Sengupta, and S. Slater (Upper Saddle River, NJ: Prentice Hall, 2005), pp. 350–361.

ads, the most efficient form of online advertising, have measurement problems. Advertisers usually pay on a cost-per-click basis, although clicks aren't necessarily easily converted into purchases. Moreover, the problem of *click fraud* poses another concern for online advertisers. Click fraud occurs when a company maliciously clicks on a competitor's sponsored ad (say, on a search engine page) to exhaust the competitor's budget. Or, click fraud occurs when a website owner clicks on a contextual ad on one of its own pages repeatedly so that it earns a portion of the revenue the advertiser pays the search engine. At the extreme, people are paid to engage in click fraud by unscrupulous companies in distant countries. Search engines try to monitor for click fraud and provide credit to advertisers when this is detected. Further, search engines are beginning to offer search ads to advertisers on a "pay-per-action" basis, where action could be defined by the advertiser as a verifiable purchase or a registration. However, cost-per-action search ads cost more than pay-per-click ads.

Web 2.0 Technologies

Web 2.0 refers to a collection of Internet-based software tools that emphasize users generating and viewing content in a proactive social setting—hence these tools are also called *social media*. In contrast, first-generation World Wide Web technologies (Web 1.0) allowed websites to generate their own content. Users could surf to these sites and gain useful information or engage in e-commerce, but the technology did not allow the level of interactivity and user input that the Web 2.0 technologies do. Based on increased connectivity and communications among users, these second-generation Web-based tools include, for example:

- *Social networking* sites (such as MySpace, Facebook, or LinkedIn) and websites whose members invite contacts and friends from their own personal networks to join the site. New members repeat the process, growing the total number of members and links in the network. Sites then offer features such as automatic address book updates, viewable profiles, the ability to form new links through "introduction services," and other forms of online social connections. MySpace, for example, builds on independent music and party scenes, and Facebook was originally designed to mirror a college community (though now it is open to anyone; in fact the fastest growing segment of members are people aged 26–44). The newest social networks on the Internet are becoming more focused on niches such as travel, art, tennis, football (soccer), golf, cars, dog owners, and even cosmetic surgery. Other social networking sites focus on local communities, sharing local business and entertainment reviews, news, event calendars and happenings. Social networks can also be organized around business connections, as in the case of LinkedIn.
- *Blogs* (short for "Web log") are online journals on a variety of topics from technology to politics; typically updated daily, blogs often reflect the personality of the author. Some of the popular technology blogs are Engadget (www.engadget.com) and Slashdot (http://slashdot.org); other useful tech blogs are Tech Crunch (www.techcrunch.com), focusing on technology start-ups particularly in the Web 2.0 sector, and AlwaysOn (http://alwayson.goingon.com).
- *Podcasting* is the audio analog of a blog; not only can people can record their own thoughts and distribute them to the others through Web-based subscription services such as iTunes, but any Web publisher can make an audio recording and share its content via podcasting, which can be very useful for busy executives who want to learn about new topics.
- *Video sites* (e.g., YouTube; BrightCove) enable users to upload and share their own videos with others while sites like Flickr (http://flickr.com) allow the sharing of photographs; another tagging or labeling website is Delicious (http://del.icio.us).
- Second Life (www.secondlife.com) and other *virtual worlds* are virtual economies in which companies can buy real estate, build their own virtual facilities, and have the opportunity to expose their brands to the avatars of potential consumers (see also Zwinky.com and Webkinz.com as other examples).

Some other Web 2.0 technologies are shown in Table 11.1.

TABLE 11.1 Some Web 2.0 Technologies

AJAX (Asynchronous JavaScript and XML): A programming technique for websites whose data are regularly refreshed by the user. It allows the website to exchange small amounts of data with the server behind the scenes (rather than reloading the entire Web page each time the user requests an update), resulting in enhanced interactivity, speed, functionality, and usability.

RSS (Real Simple Syndication): This technology allows people to sign up for news articles, blog posts, or audio interviews/podcasts from their favorite websites to be sent directly to their computers—essentially, the syndication of Web content. A website that wants to allow other sites to publish some of its content creates an RSS document and registers the document with an RSS publisher. A user that can read RSS-distributed content can then read content from a different site. Syndicated content can include data such as news feeds, events listings, news stories, headlines, project updates, excerpts from discussion forums, or even corporate information.

Twitter: A Web service that allows users to send "updates" ("tweets") about what they are doing at a particular moment in time via text messages (SMS), instant messaging, or e-mail to the Twitter website; also called *micro-blogging* because of the short nature of the frequently updated posts, these updates can also be displayed on the user's profile page and can be delivered instantly to other users who have signed up to receive the updates. Twitter look-alikes include country-specific services (e.g., Germany's Frazr) or sites that combine micro-blogging with other functions such as file sharing (e.g., Pownce).

Wiki: A collaborative website composed of the collective work of many authors. A wiki allows anyone using a browser interface to edit, delete, or modify content that has been placed on the website, including the work of previous authors. In contrast, a blog, typically authored by an individual, does not allow visitors to change the original posted material, only add comments to the original content. The term *wiki* refers to the web-hosted software used to create the site. *Wiki* means "quick" in Hawaiian.

Source: Ferrini, Anthony, and Jakki Mohr, "Uses, Limitations, and Trends in Web Analytics," in *Handbook of Web Log Analysis*, eds. B. J. Jansen, A. Spink, and I. Taksa (Hershey, PA: IGI, 2008), pp. 124–142.

All Web 2.0 websites have the potential to host display or search ads provided they are successful in attracting large audiences. For example, Google has now begun to serve ads in the bottom fifth of the YouTube video screen. Other videos and podcasts may contain embedded advertising before or after the content that is of interest to the user. Ad spending on social networking sites was projected to reach $1.38 billion or 5.0% of total online advertising spending in 2008.[26]

Web 2.0 technologies can be extremely cost efficient. It costs nothing for a small business to post a 10-minute video about its products and services on YouTube. The lack of a big budget for traditional or online advertising can be compensated by creativity and a personal touch. An analysis of the most viewed marketing videos on YouTube indicates the following key success factors: (1) be funny, (2) tap into current events, (3) find a partner (for their creativity and production capability), (4) provide useful information for users, and (5) get users involved in creating or spreading the message. BlendTec, a maker of high-end food blenders, uses comedy to deliver its product's value proposition. In a series of videos labeled "Will It Blend?," BlendTec CEO Tom Dickson uses the blender to grind up a number of unusual objects including credit cards, razor blades, and marbles. In one year since the series launched in 2006, the videos collectively received 60 million views, brought the company name recognition, and retail sales shot up 500%.[27]

Web 3.0 Technologies

In addition to effectively leveraging Web 2.0 technologies in their A&P mix, marketers must also monitor the Web 3.0 technologies coming down the pike. **Web 3.0** technologies, still only loosely defined, generally refer to emerging technologies such as the Semantic Web, based on the use of meta-tags and other identifiers coded into the data itself, which will enable computers to perform searches and other activities independently of human intervention. Other visionaries suggest that increases in Internet connection speeds, modular Web applications, or advances in computer graphics will play a key role in the evolution of the World Wide Web.

At the Seoul Digital Forum in May 2007, Eric Schmidt, CEO of Google, speculated that

[Web 3.0 is] a different way of building applications. . . . My prediction would be that Web 3.0 will ultimately be seen as applications which are pieced together. There are a number of characteristics: the applications are relatively small, the data is in the cloud, the applications can run on any device, PC or mobile phone, the applications are very fast and they're very customizable. Furthermore, the applications are distributed virally: literally by social networks, by e-mail. You won't go to the store and purchase them. . . . That's a very different application model than we've ever seen in computing.[28]

Web 3.0, then, is likely to include some combination of the Semantic Web, and an increasingly mobile Web that is driven by increased bandwidth and speed to handle the increasingly dense applications of streaming media. How marketers will leverage this in their future marketing activities remains nebulous at this time.

Viral Marketing

Viral marketing refers to making marketing offers so compelling that people voluntarily pass them around to their friends. It takes advantage of the power of contacts and shared interests to stimulate word of mouth using technology such as e-mail, mobile messaging, and Web 2.0 tools. Viral marketing is sometimes thought to be a subset of what marketers refer to as **buzz marketing**,[29] or word-of-mouth communications to generate interest and excitement in a company's product. This technique is often used in a subtle (even covert) fashion by marketers seeking out trendsetters in particular areas, who are then compensated in some fashion to "talk up" the product to friends. Buzz strategies can be so effective that they result in a critical mass of product adoption, reached at "the tipping point"[30] after which the product becomes a market success. Nissan launched a massive viral marketing campaign leading to the announcement of its new GT-R sports coupe at the Tokyo Motor Show in October 2007. It posted sneak previews of the car masked in a black rubber cover on blogs and YouTube. It struck product placement deals with Sony's videogame unit and Electronic Arts, to feature the product in downloadable games. Initially downloadable versions of Sony's Gran Turismo game allowed users to drive a masked version of the GT-R, while downloadable versions of Electronic Arts' Need for Speed game let users drive a prototype version of the car. Coinciding with the Tokyo Motor Show, users could download new versions of both games featuring the actual GT-R. Nissan estimated the cost of this viral campaign was about half what it would cost to launch a mass-media campaign using traditional media.[31]

Widgets or **gadgets** are Web-based software applications provided free to Web users by companies in return for exposing them to their online ad. These downloadable, interactive icons allow users to perform a task from their desktop without opening a Web page. Typical widgets include applications like news feeds, stock tickers, clocks, and calculators. Such freebies provide more of an incentive for users to pay attention to the ad than a standard display ad. Widgets, such as chat boxes and music players, may be uploaded by users to place on their own MySpace pages or other websites, giving the ad message a viral element that encourages sharing. For example, the San Francisco–based interactive agency AKQA created a weather widget to promote Microsoft's Flight Simulator X that allows users to fly virtually to any airport to find out the weather report there, giving users a feel for the Xbox videogame. The widget was downloaded 150,000 times in two months with users spending an average of 23 minutes checking out the simulator.[32]

Affiliates are websites that drive traffic and sales to another website by carrying their display ad or link. A website compensates an affiliate by sharing revenue from each referral. Amazon runs a popular affiliate program in which any website can post ads for books and other items that Amazon sells; through a unique tracking code, the owner of the affiliate site will then receive a portion of the revenue Amazon makes through any sales this affiliate referred back to Amazon. Another benefit of an affiliate program is that getting other websites to link to the company's own website on a reciprocal basis is one step to a higher rank in natural (organic) search listings.

Mobile Advertising

Mobile advertising, also known as *mobile marketing* or *M-commerce,* uses the mobile phone platform as an advertising tool. In the United States, nearly 70% of mobile phone owners use text messaging; among these users, 44% send text messages daily. Consumer participation in mobile marketing efforts jumped from only 8% in 2005 to 29% in 2006.[33] Worldwide spending on mobile advertising was about $2 billion in 2006, and it is forecasted to grow to about $15 billion in 2011.[34]

Mobile marketing has exploded earlier and faster in areas of the world outside the United States, where the cell phone culture is much more prevalent—partially because other countries such as Finland and South Korea had the necessary infrastructure in place long before the United States did and because people in those countries had to pay much more for landline phones than Americans, giving them added incentive to use mobile technologies.[35]

The most prevalent form of mobile advertising involves sending messages, usually SMS or text messages, directly to an individual's mobile phone. Customers must sign up and provide their mobile phone numbers to receive mobile messages from marketers. If customers also provide demographic and preference information at the sign-up stage, the messages can be customized and better targeted. Like their Web-based search applications, companies such as Google, Yahoo, and MSN are making their search applications available on mobile devices. In turn, this development provides additional opportunities for advertisers to serve paid search ads on mobile screens on a pay-per-click basis. NeoMedia's Qode technology is designed to let mobile users click an image of an icon in an ad (similar to a barcode on consumer products) that will take them directly to that company's website on their mobile devices. This new medium lets advertisers target their messages to individuals, especially younger consumers, who carry mobile devices with them most of the time. By requesting message recipients to participate in some kind of activity on their mobile devices, mobile advertising can be interactive. In addition to being able to easily measure the response rate to mobile marketing campaigns, the effectiveness of traditional media advertising such as TV or print or billboards can also be tested. For example, by having a call to action to customers involving their mobile devices, such as calling a number or sending a text message to participate in a vote or contest, marketers can test the effectiveness of their traditional campaigns.

Marketers are also sending coupons to consumers on their mobile phones. A company called Cellfire based in San Jose, California, has business agreements with 200 marketers including entertainment and food companies such as Virgin Megastores, Hollywood Video, Domino's Pizza, and Quizno's to send their coupons to the mobile phones of about 1 million consumers. These companies pay Cellfire a fixed up-front fee for this service plus an additional incentive per coupon redeemed by consumers. Cellfire gets consumers to sign up for this free service on its website or by sending a text message, offering them coupons and discounts worth over $100. To redeem the coupon, a consumer shows her mobile phone screen with the coupon displayed to a cashier at the checkout. Virgin Megastores has tested Cellfire-distributed coupons on DVDs at seven of its stores in New York and California and found redemption rates of about 15%, higher than the 5% redemption rate they get on paper coupons. To allay privacy concerns, Cellfire says that it does not share individual consumers' phone numbers with other organizations except with prior consent.

Mobile advertising opportunities will expand even more when embedded chips using Global Positioning System (GPS) technology make possible the customization of **location-based marketing**. For example, while walking past a Best Buy store, a customer could get an SMS text message via cell phone to come in for a special offer on a newly released computer game or electronic device. Advertisers need to strictly follow the principle of opt-in permission marketing so that this does not go the way of spam in e-mail marketing or invite regulation as in the Do Not Call list for telemarketing.

In many areas of the world, mobile phones are used to facilitate product purchases for customer convenience. For example, people can use their phones to make purchases from vending machines, for movies, and of other convenience items. Although the technologies involved vary, essentially mobile users can sign up for a subscription service or enter a text number that adds the price of the good they've purchased to their cell phone bill, much like a credit card purchase.

Many experts predict that mobile marketing is the technology wave of future advertising,[36] referring to it as the "third screen" (with television having been the first screen, and the computer monitor the second screen). Marketing in the future will leverage advertising on all three screens synergistically.[37] The up-and-coming developments in the mobile arena explain why companies such as Apple with its iPhone, and Google with its Android platform, are paying so much attention to a field previously dominated by relatively staid telecommunications giants. The 2.5 billion mobile phones around the world reach a much bigger audience than the billion or so personal computers.[38] Moreover, the number of mobile phones is growing much faster than the number of computers, particularly in developing countries. And, unlike their televisions and computers, most people carry their phones with them everywhere. However, the industry does face barriers—including privacy concerns; a lack of agreement for revenue sharing among advertisers, network operators, and other intermediaries; as well as a lack of a common format for mobile platforms.[39]

Marketing in Virtual Reality Environments

Many companies are exploring the possibilities of marketing in virtual worlds such as Second Life. For example, IBM has a "pavilion" in Second Life and holds real meetings with employees worldwide in this online world. Dell, Toyota, Sun Microsystems, and Sony are also experimenting with virtual marketing as well.[40] Companies can use it either for internal purposes (meetings) or for connecting with customers, suppliers, and business partners. Yet, the effectiveness of this online tool is hotly debated. On one side are those who believe that despite the relative uncertainty, connecting with customers via their online avatars is an opportunity that should not be passed up.[41] Others believe that companies are simply jumping on a fad with no clear direction or objectives. But, the truth is somewhere in the middle. To use Second Life or any other virtual world effectively, marketers should:[42]

- *Engage:* Cisco has an in-world campus with meeting rooms and areas where user groups get together; it says the real value of Second Life is the opportunity for spontaneous customer interaction. With 650 employee avatars registered in Second Life and six locations ("islands" or "sims"), the company sponsors events about once a week.
- *Add value to the online community:* Rather than following the mind-set of "build it and they will come," success in Second Life requires some sort of value to the people who are online. Pontiac has an online pavilion called Motorati, which is not used for marketing messages, but rather as place for users to interact, including an area called AskPatty.com for female car enthusiasts.
- *Be smart about policing:* Despite the negative press about Second Life being "a wasteland of freaks and perverts" and "griefers" (Second Life jargon for vandals who wreak havoc in the virtual world), companies can take a leadership position in their areas by setting a civilized tone and keeping their distance from areas where inappropriate behavior is rampant. Griefers can be ejected when their malicious intent is obvious.

Although predicting where marketing in this area will head is difficult, half a million people do use Second Life regularly. Although it is not a mass-market tool, those who participate are a niche market—including software developers and IT managers. So, companies should continue to keep an eye on this arena.[43] Importantly, there are many other types of virtual worlds that reach a variety of different demographics. For example, PenguinClub and Webkinz reach the "tween" market (children aged 8-12 who are older than "kids" but not yet teenagers). This is an area of online marketing that will continue to see experimentation as companies learn how to capitalize on it effectively.

Dealing with Disruption

New media technologies are significantly disrupting traditional media.[44] TV commercials face disruption not only from online and mobile advertising but from digital video recording (DVR) technology as well. For example, DVR technology, as offered by TiVo for example, lets viewers fast-forward through commercials. Traditional media are taking proactive steps to figure out how to survive and thrive in this brave new world, and are responding to these threats in several ways.

First, TV programs are being modified and made available on the Web and mobile platforms in shorter form, and these provide new advertising opportunities. Content in traditional print media is being embellished with additional audio and video content on the Web, which also provides opportunities to embed advertising. Popular TV shows are engaging viewers more by letting them use the Web and mobile devices to participate in games and contests of various kinds, providing more value from the traditional medium while letting people use the newer digital media.

To counter the threat of video content available from sites like YouTube, the traditional TV companies are taking legal action against YouTube over copyright infringement as well as signing deals with alternate video websites like Joost and Hulu. A new start-up from the originators of Kazaa and Skype, Joost has access to legally licensed, long-form original content from TV companies, is free to consumers, and is supported by advertising. It is positioned as next-generation TV on the Web.

Because DVR technology enables TV viewers to fast-forward through commercials, TV studios and advertisers are cooperating to embed brand and product plugs into the script of TV shows. Product placement fees can compensate for lost advertising revenue to some extent. Some advertisers are having TV companies create whole shows around their products, called "branded entertainment." TV companies are also buying viewer data from DVR companies like TiVo to get a better understanding of who is watching their show and who is skipping through which commercials to enable better targeting. TiVo offers new technology called *interactive tags,* in which consumers see a tag with the sponsor company's name while they are fast-forwarding through commercials and they can click on this tag for more information if they like.[45] Although viewed by fewer people than traditional ads, these ads are more effective overall.

While traditional media are taking laudable steps toward offering more value to advertisers and consumers, the most effective advertising campaigns in the future must exploit the synergy between traditional and new digital media in the true spirit of integrated marketing communications. The accompanying Technology Expert's View from the Trenches explains some of the new challenges and opportunities for high-tech marketers.

THE WEBSITE'S PART IN ADVERTISING AND PROMOTION STRATEGY

As noted in the Technology Expert's View from the Trenches, a company's website must be integrated with its advertising and promotion (A&P) strategy. Indeed, a company's Web presence is as powerful an embodiment of the company as is its products or its retail presence. The website must be more than "brochure-ware" or "gratuitous digitization"; in other words, a company must do more than simply place its existing information online. Website visitors must have a meaningful online experience that helps reinforce the company's value proposition and brand meaning. Because of the importance of a company's website as part of its overall A&P strategy, issues in creating an effective website are explored next. Designing an attractive, professional, useful website is only the first step in leveraging a company's Web presence. It must also build site traffic and evaluate the effectiveness of its investment in the website itself.

TECHNOLOGY EXPERT'S VIEW FROM THE TRENCHES

Moving beyond a Website: Building a Powerful Web Presence
By Mario Schulzke
Director of New Media, Respond2 Communications, Inc., Portland, Oregon

Consider the following statistics:

- 41% of consumers research products online before purchasing
- 80% of all Web traffic begins at search engines
- 66% of all consumers regularly use TV and the Internet simultaneously

Companies that pay attention to such facts reap the greatest rewards from their online strategies, doing two things extraordinarily well. First, they integrate their online and offline marketing efforts in a synergistic fashion. Second, they enlist the energy and voices of a variety of constituencies (via Web 2.0 technologies) to effectively engage their audiences. We use a three-step process to optimize our client's Web presence, ensuring that their expenditures in offline media yield maximum effectiveness:

Analyze → Optimize → "Particize"

Analyze

The first step in building a powerful Web presence is to analyze a company's current Web presence. Indeed, before making any changes to the Web strategy or spending a single dollar on focus groups, companies should use a few of the following techniques:

- Set up automatic (RSS) feed searches for the company, products, and competitors' names on sites such as Google News and Feedster.
- Stay up to date on what customers are saying about products and services on popular review sites such as Amazon.
- Analyze Web traffic and figure out what keywords people are using to find the website, products, or services.
- Read relevant posts or comments on blogs.
- Monitor popular forums relevant to the category.
- Assess any videos relevant to the company's products or categories on YouTube or video sites.
- Assess any pictures of the company's products on Flickr or similar sites.

This first step provides useful insights about the company's online presence (e.g., how easy it is to find; whether the information is positive, negative, or neutral; unintended links to the company or its products) and monitors existing conversations about the company and its products.

Optimize

After analyzing its existing Web presence, the company must then optimize its existing site for search engine positioning, with the goal of making the site easy to locate and navigate. Building a search-engine-friendly site usually translates seamlessly into a user-friendly site; the search engine's primary goal is to provide Internet users with reputable, relevant, and user-friendly search results.

Before launching a search engine optimization campaign, companies need to conduct a keyword analysis to determine what words people search for and how many other websites are optimizing for those keywords. The goal of the keyword research phase is to choose the best 20 to 50 primary keywords based on two factors: First, select keywords that will yield the highest number of potential searches (Google, Yahoo!, and other sites like WordTracker provide tools to research this); and second, balance those popular keywords against the number of other websites that are also optimizing for those same keywords. The point here is that a certain keyword that might yield

(*continued*)

only 1,000 searches per month, but has only a handful of sites optimizing for it, might be a better keyword selection than one that yields 5,000 searches but has 100 sites also optimizing for it. The balance requires both the quantitative assessment as well as judgment and experience.

Three ways to optimize a website for search engines are to (1) work on content (using target keywords), (2) establish links (both within the website and with incoming links from other reputable and relevant websites), and (3) use navigation techniques that can be read and followed by search engine spiders (for example, some Flash navigation techniques still cannot be read up by search engine spiders).

Other optimization tactics include:

- Develop a blog to communicate with prospective customers, current customers, and all stakeholders in an authentic fashion. Because search engines look for keyword content, the text-heavy nature of blogs can also be a great asset in a search engine optimization campaign.
- Optimize press releases with relevant keywords, then publish them through newswires.
- Create a PDF guide (for example, a white paper or other document that provides useful information to potential website visitors) and use it as an incentive to get people to sign up for an e-mail newsletter. Creating these guides will also help with search engine optimization because other websites may link back to them if they are truly useful, and search engines will also pick up html content based on keywords.
- Last (but not least), set up an e-mail newsletter for regular communication with interested parties.

Properly executing the optimization phase results in a website that derives a majority of its traffic from organic search engines (in contrast to traffic from paid search) and converts that traffic at a higher rate into prospective and new customers. While optimization is a necessary step in building a powerful Web presence, it should also be considered a necessity in any Web development project.

"Particize"

Once a company analyzes its Web presence (to identify and understand ongoing conversations) and optimizes its existing website for search engines, it's time to *participate and publicize*—or as I like to call it, *particize*. Here are some tactics that we frequently employ to help our clients particize:

- Encourage and reward current customers who contribute to online conversations and reviews.
- Establish a board of customers (heavy users, fans, bloggers, partners and journalists) to engage in product development and marketing efforts.
- Plan and implement an online PR program.
- Answer questions relevant to the business on Yahoo! Answers.
- Film short instructional videos about products and upload them on YouTube.
- Upload high-quality product images on Flickr.
- Set up and manage discussion groups about topics relevant to the company's business, category, and products (e.g., Apple starting a Yahoo! discussion group about iPhones).
- Implement a blogger relations program:
  - Point bloggers to content that would interest their readers.
  - Leave comments on other blogs and point back to resourceful content.
  - Communicate!

The purpose of this third step is to help companies participate in ongoing conversations and help publicize their contributions to the Web community. Of course, many more opportunities to build a Web presence are specific to an industry, products, or services; companies will uncover these opportunities in the analysis phase of this process.

Outcomes

Using this process has enabled our clients not only to build their online presence, but also to increase the effectiveness of their traditional media investments in print media and direct response TV. For example, the hair restoration industry leader Bosley used a combined strategy of running a consumer-focused blog called BattleAgainstBald in combination with a YouTube channel and other social media postings to give prospective customers a firsthand perspective on the results of undergoing a hair restoration procedure. BattleAgainstBald has thousands of monthly readers, ranks highly on all relevant search engine keywords, was featured in PR pieces in the *Boston Globe* and on CNBC's "The Big Idea," is cited on Wikipedia as an information resource for baldness treatments, and has significantly increased its effectiveness across all advertising channels.

Similarly, Vonage created a viral publicity site called FreeToCompete.com to raise awareness of the anticompetitive behavior of a large telecommunications provider and to communicate Vonage's desire to compete in the marketplace. The site was created instead of a traditional PR campaign, as most of the target audience included influencers and first adopters who used the Web as their primary source of information. FreeToCompete.com had content relevant to the issue, which was also explained through an entertaining flash animation, and gave visitors the opportunity to forward the site to their friends, send a petition to their state representative, or e-mail the VP of Public Affairs of the large telecommunications provider. Upon launch, the site was leaked to business and technology bloggers, who wrote about the site and issue and whose audiences engaged in avid discussions about the topic. Based on this blog buzz, traditional media publishers such as *Business Week,* the *Wall Street Journal, Wired,* and the *New York Times* published articles about FreeToCompete and its issue. More than 17,000 petitions were signed and the large telecommunication provider's VP of Public Affairs received 11,405 e-mails—more than he ever had before. Thanks to the campaign's success, FreeToCompete.com was actually purchased by a major telecommunications lobbying association and is now a leading industry voice for competition within the telecommunications industry.

Building a meaningful Web presence is a strategic priority for every marketer, offline or online. Much of the interest created by television, print, outdoor advertising, or PR will result in interested consumers doing Web searches and participating in online conversations, and companies need to capture that interest—because if they don't, their competition will.

Website Design

Effective websites are designed with clear objectives of what the company hopes to accomplish with its Web presence. For example, different types of websites have different objectives. For some websites, attracting a large number of visitors—and getting them to return and spend time at the site—is the name of the game. For others, attracting users who take a particular action at the site is the goal—say, asking users to join a particular cause, to play an online game, or to make a purchase. Other companies might want to leverage their websites to gain cost efficiencies in performing tasks that would otherwise be done offline, such as allowing customers to get customer service inquiries answered or to find answers to frequently asked questions. Thinking about objectives prior to design is paramount.

In addition, effective websites are designed with the user's experience in mind. Conceptualizing, implementing, and successfully managing any website requires an intimate understanding of customer behavior, and how the Web will deliver compelling value to those customers. As shown in Figure 11.2, a well-designed online customer interface consists of context, content, community, connection, customization, communication, and commerce,[46] each of which are described here.

Context refers to a website's design and layout, sometimes termed "look and feel." Some sites are more aesthetic, intended to create a mood or image with lots of graphics, sound, or video. Others are functional, largely text-based, like Google. The context of a website must be designed with an eye to the target customers and the value proposition.

Content includes all the digital material on a website including information, products, and services. Content pertains to what is presented, whereas context is about how it is presented. It is imperative that a website include the relevant content a user will need, in a design and layout that is user-friendly and easily navigable.

Community refers to the ability of users on a website to interact with each other, build bonds, and share common interests or values. The community itself is made up of the people who participate

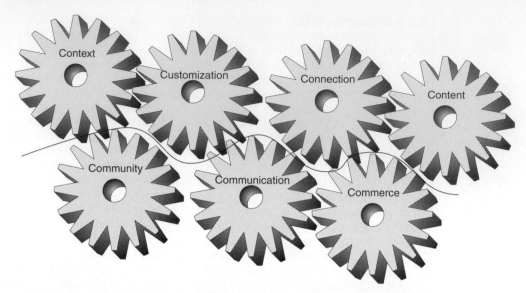

*All elements have reinforcement (consistency) and fit (meet needs of the target market)

FIGURE 11.2   Elements of Effective Website Design

in the activities of the site on a regular basis and are willing to share information with those who share their interests. Community is a useful tool for a website to build relationships with and among its users: eBay has a vibrant community of sellers who help each other with tips and information and provide useful feedback on company policies; Amazon has an active community of book, music, and other product reviewers.

Customization is the ability of a website to present flexible information tailored to different users or to be modified by users to suit their unique needs. For example, MyYahoo! allows users to choose what types of information they would like to receive when they log into this portal. Customization is an important feature that delivers personalized value to individual visitors. The ability to customize depends largely on being able to build a database of useful information on customers' tastes and preferences, demographics, and actual purchases.

Communication refers to the means by which the website permits interaction between itself and its users—via live chat, e-mail, or telephone. This communication could be one-way (website to the user) or two-way (website to the user, and the user back to the website); studies show that two-way communication contributes to a more satisfying user experience.

Connection refers to how well the website is linked with other useful websites. If a user cannot find what is needed on a website, it adds value to the user experience to be referred to a site where the needs can be met. The desire to build connection explains why Amazon has integrated a Google search box into its website. Building connections also presents opportunities to earn revenue from referrals to affiliated websites.

Commerce refers to a website's capability to execute commercial transactions between buyers and sellers. It requires features such as a shopping basket, invoicing, secure credit card processing, and order acknowledgment.

Effective websites exhibit both reinforcement and fit.[47] *Reinforcement* refers to the degree of consistency between the seven characteristics described above. If a website wants to build a vibrant community but does not provide easy two-way communication tools, it would be low on reinforcement. *Fit* refers to the match between the website design and the needs of the target market. For example, Yahoo!'s site has an attractive look and feel, as well as useful content to draw users and keep them; therefore, it is high on fit.

Building Site Traffic

Once the website is designed, it must be promoted to attract visitors. The interface between website design and search engine optimization was previously discussed. In addition, all of the various tools discussed previously in this chapter can be used to build site traffic.

Many companies follow a "get big fast" strategy for their websites (i.e., grow the number of visitors as quickly as possible), focusing on customer acquisition as the key to success. However, such a strategy makes sense only under certain conditions. Three conditions support a "get big fast" strategy:[48]

- *Network effects:* Recall that network effects exist when adoption by a customer increases the value of that product for other customers already using it. For example, because most shoppers do not communicate with each other, network effects are minimal for e-tailers. Although chats and customer recommendations do benefit from network externalities, these features are only of secondary importance for most e-tail sites. Portals, on the other hand, as well as market aggregators (such as eBay), benefit from network effects and so can justify a "get big fast" approach. Websites might also follow a "get big fast" approach when driven by indirect network externalities, in which the number of customers provides an attractive opportunity for advertisers.
- *Scale economies:* When a company can spread its fixed costs over a growing revenue base, it makes sense to get big fast. Some Internet companies have only fixed costs (i.e., portals), so operating margins can increase dramatically with growth. However, online retailers have a large share of variable costs (i.e., cost of goods sold) and are less likely to benefit from scale economies.
- *Customer retention rates:* The real payoff from "get big fast" customer acquisition strategies is found in customer lock-in. Customer retention rates tend to be high in the presence of network effects, solid competitive position, and high customer switching costs. For e-tailers, these factors can be highly variable. Despite e-tailers' assertions that they can build loyalty based on customer service or knowledge of customer preferences, retention is highly variable by segment. As a result, excessive expenditures on customer acquisition to grow site traffic may not be justifiable for some companies.

Evaluating Website Effectiveness

Effective website management requires a way to track not only the traffic (number of visitors), but also what those visitors are doing at the particular site. Importantly, effective website management requires a way to map the behavior of the visitors against the particular objectives and purpose of the site.

The indicators used to measure website performance, known generally as *Web metrics,* are available from the server (the computer) on which the website is hosted, or "served up," on the Internet. In particular, the server records data for every time a browser hits a particular Web page, and includes information for every action a visitor at that particular website takes; these data, known as *log files,* include a host of information, as shown in Table 11.2. They can show how a website visitor navigates through the various pages of a given website, or the *click trail,* also known as *clickstream data. Log file analysis,* also known as *Web log analytics* or more simply **Web analytics,** is the study of the log files from a particular website to assess the performance of the site. Google Analytics can also be used to examine traffic at a particular website. By embedding a specific html code into each page of the website (through directions given by Google Analytics), the company can then access the free Google Analytics program to see the data about its website traffic.

Web analytics can be used in a wide variety of ways. For example, marketers can assess the effectiveness of search engine listings by analyzing which search engines visitors came from and which *search queries* (keywords typed into the search engine) they used. Often, the best keywords to use (both for search engine positioning and paid search) are not always obvious. A popcorn chain

TABLE 11.2 Types of Data Used in Website Analysis

Hit: A count of Website traffic for each element of a Web page that is downloaded to a viewer's Web browser (such as Internet Explorer, Firefox, or Safari); hits do *not* correspond in any direct fashion to the number of pages viewed or number of visitors to a site. For example, if a viewer downloads a Web page with three graphics, the Web log file will show four hits: one for the Web page and one for each of the three graphics.

Unique visitors: A count of Website traffic for the actual number of viewers to the website that came from a unique IP address (see *IP address* below).

New/return visitors: A ratio comparing the number of first-time visitors to the site to the number of returning visitors.

Page views: A count of Website traffic based on the number of times a specified Web page has been viewed; shows exactly what content people are (or are not) viewing at a site. Every time a visitor hits the page-refresh button, another page view is logged.

Page views per visitor: A ratio of the number of page views divided by the number of visitors; indicates how many pages viewers generally look at each time they visit a specific website.

IP address: A numeric identifier for a computer. (The format of an IP address is a 32-bit numeric address written as four numbers separated by periods; each number can be anything from 0 to 255. For example, 1.160.10.240 could be an IP address.) The IP address can be used to determine a viewer's country of origin; it also can be used to determine the particular computer network a website's visitors are coming from.

Visitor location: The geographic location of the visitor.

Visitor language: The language setting on the visitor's computer.

Referring pages/sites (URLs): Indication of how visitors get to a website (i.e., whether they type the URL, or Web address, directly into a Web browser or whether they click through from a link at another site).

Keywords: The keywords (search string) that the visitor used to reach the website, if the referring URL is a search engine.

Browser type: The type of browser software a visitor is using (Firefox, Safari, Internet Explorer, etc.).

Operating system version: The specific operating system on the visitor's computer.

Screen resolution: The display setting for the visitor's computer monitor.

Java or Flash enabled: Whether or not the visitor's computer allows Java (a programming language for applications on the Web) and/or Flash (a software tool that allows Web pages to be displayed with animation, or motion).

Connection speed: The speed at which visitors are accessing the website, whether a slower dial-up connection, high-speed broadband (DSL or cable modem), or T1.

Errors: The number of errors recorded by the server, such as a "404-file not found" error; can be used to identify broken links and other problems at the website.

Technical information regarding server load and any unusual activity: Assesses load on the site (to determine whether the bandwidth can accommodate the expected traffic) and monitors abnormal activity such as repetitive login attempts, unusually large numbers of requests from a single IP address, and so forth.

Visit duration: Measure of the average time spent on the site (i.e., the length of time the visitor stays on the site before leaving). Sites that retain visitors longer are referred to as "sticky" sites.

Visitor paths (navigation): The way visitors navigate the website—by specific pages, most common entry pages (the first page accessed) and exit points (the page from which a visitor exits the site), and so forth. For example, if a large number of visitors leave the site after looking at a particular page, the analyst might infer that they either found the information they needed, or alternatively, there might be a problem with that page (e.g., is it the page where shipping and handling fees are posted, which might be large enough to turn visitors away?).

Page elements: The elements (images, text, etc.) of the Web page that visitors are utilizing.

Bounce rate: The percentage of visitors who leave the site after the first page; calculated by the number of visitors who visit only a single page divided by the number of total visits. The bounce rate is sometimes used as another indicator of "stickiness."

in New Jersey had been using keywords like "gourmet popcorn" and "popcorn tins." But, when it started using Web analytics, the company learned that more people were searching by "chocolate popcorn" and "caramel popcorn," so it boosted the use of those phrases, both in the site content as well as in its marketing efforts.[49] Moreover, it found that most visitors were typing "kettle corn" as two words rather than the single word that the site was using, so it added a two-word version in its strategies as well.

Metrics to assess marketing effectiveness can include:

- *Cost-per-click:* The total online expenditure divided by the number of click-throughs to the site
- *Conversion rate:* The percentage of the total number of visitors who make a purchase, sign up for a service, or complete another specific action
- *Return on marketing investment:* A measure of the benefits of a marketing expenditure; calculated as the total revenue generated from the advertising expense divided by the advertising expense
- *Bounce rate:* The number of users that visit only a single page divided by the total number of visits; this ratio is one indicator of the "stickiness" of a Web page

Trends in Web analytics allow marketers to assess the link between online advertising and sales, analyzing the entire marketing process from a user clicking an advertisement through to the actual sale of a product or service. This information helps to identify not merely which online advertising is driving traffic (number of clicks) to the website and which search terms lead visitors to the site, but which advertising is most effective in actually generating sales (conversion rates) and profitability. This integration of the Web log files with other measures of advertising effectiveness, such as that provided by Google Analytics or Salesforce.com, is critical to guiding further advertising spending. Appendix 11.A at the end of this chapter provides more information on evaluating website traffic.

Summary

This chapter addressed a range of tools used in high-tech companies' advertising and promotion (A&P) strategies. Beginning with the idea of the advertising and promotion pyramid, this chapter initially addressed how a firm can leverage a variety of traditional A&P tools in a cost-efficient manner. Next the chapter introduced the new media A&P tools including display advertising, search engine optimization, paid search advertising, and the opportunities provided by Web 2.0 technologies such as social networking, tagging, blogs, podcasting, user-generated video, virtual worlds, widgets, and mobile advertising. New media are potentially disruptive technologies for traditional media, so the chapter also discussed the implications of this trend for advertising and promotion. Finally, the chapter discussed the website's role in advertising and promotion and ways to design an effective Web site. The chapter closed with a discussion of how Web analytics can be used to track and measure website effectiveness.

Advertising and promotion issues are particularly salient in the high-tech marketing environment. What can work particularly well is an integrated marketing campaign that makes use of a mix of offline and online advertising and promotional tools in a coordinated fashion to maximize impact.

Discussion Questions

Chapter Concepts

1. What is the logic behind using the advertising and promotion pyramid as a coordinating device?
2. What is integrated marketing communication? How does a company achieve it? What are some of the tools used to implement it?
3. What are some of the critical issues concerning each of the tools in the advertising and promotion mix?
4. Explain permission-based marketing. What type of strategies should a company pursue once it has received a customer's permission? Why?

5. Find examples of the various types of online ads. Examine their strengths and weaknesses.

6. What is behavioral targeting? Why is it effective?

7. What is search engine optimization? How is it achieved?

8. What are the two pricing models used for online advertising? Which makes sense under which conditions?

9. Describe some of the problems and limitations of online advertising.

10. What is Web 2.0? What are some technologies available to Web 2.0? How does Web 2.0 affect online advertising?

11. What is mobile advertising? How does location-based marketing fit into mobile advertising?

12. How are traditional media dealing with the disruption caused by new media?

13. What must a company consider in order to create an effective website? Why is design alone not effective in marketing the company?

14. List and describe the elements of a well-designed website user interface.

15. How does a company build site traffic? Under what conditions does it make sense for a website to follow a "get big fast" strategy?

16. What tools are available to analyze website traffic? Why are Web analytics important?

Application Questions

1. Based on your reading of the opening vignette in this chapter, answer the following questions:
   a. What were the objectives of Microsoft's advertising campaign?
   b. Given the types of media and the messages that were used in this integrated marketing campaign, do you think Microsoft will be able to achieve its objectives?

2. Pick your favorite technology company. Go to its website and find the section where its press releases are published (usually in the investor or media or news sections of the site). These are examples of what is submitted to publications and media outlets for the firm's PR efforts. They usually portray the company and its products in a positive light, since the company or its representatives prepare them. Now surf the Web or use other library research tools to find an example of negative publicity for the same company. Who wrote this and where was it published?

What do these contrasting communications tell you about PR/publicity?

3. Which specific Web 2.0 sites do you use (social networking, blogging, video, etc.)? Are there any attempts by technology companies to post commercial messages on these websites? Research one such attempt by a technology company and justify why it is targeting that specific site.

4. Have you participated in helping spread a commercial message as part of a viral marketing campaign? What was the company, product, or service? What was your incentive in participating in this campaign and who (number of people) did you pass on the message to?

5. Some examples of location-based mobile advertising have been discussed in this chapter. What specific companies, products, or services do you think you could give permission to, to send you these type of messages?

Glossary

advertising   A paid, nonpersonal form of communication using mass-media channels such as print, radio, television, and billboards, as well as targeted trade journals.

affiliates   Independent, third-party websites that drive traffic and sales to another website by displaying its ad or link; the affiliate earns a commission on any sales derived from traffic that the affiliate sends to the host site.

banner ads   Internet ads that include graphics and text, placed next to editorial content on a Web page, which, when clicked, take the user to the advertiser's website; also called *display ads*.

behavioral targeting   A technique that allows a company to place "supertargeted" online ads that are matched to a particular Web surfer's cookie profile, log file data, and other online behavior.

branded sponsorship   An advertiser's sponsorship of a popular Web page, achieved by placing a variety of its ads prominently on the page, thus associating its name with that page.

buzz marketing   Harnessing word-of-mouth communications to generate interest and excitement in a company's product.

cost per thousand (CPM) The relative cost of reaching 1,000 people through an advertising medium; a good tool for comparing the cost efficiency of different media. Computed by dividing the total cost of the ad or promotion by the total number of people reached, and then multiplying that number by 1,000.

display ads *See* banner ads.

gadget A Web-based software application provided free to Web users by companies in return for exposing them to their online ad; also called a *widget*.

integrated marketing communications Marketing that combines all the promotional tools—including various forms of advertising, public relations, events, and the Internet—in a planned, coordinated campaign to deliver a clear and consistent message to a target audience.

location-based marketing The ability to send content-specific ads to a person's mobile phone, based on the person's location, using GPS technology.

mobile advertising Also known as *mobile marketing* or *M-commerce;* uses the mobile phone to deliver advertising (coupons and SMS text messages).

net neutrality The notion that providers of Internet content and platforms should be treated equally in terms of their broadband needs—in contrast to a "tiers of service" perspective, advocated by some telecom companies, that some Internet companies could or should pay more to ensure high-speed transmission of their data.

new media Technology-enabled advertising and promotion tools—specifically, Internet-based options such as sponsored search advertising, online display ads, ads on blogging and social networking sites, marketing in virtual worlds, and marketing to cell phones via text messages.

permission-based marketing An approach to communicating with customers or prospects that asks them to *opt in* to receive marketing messages based on their interests in a particular topic, as opposed to more traditional "interruption marketing."

public relations (PR) The activities a firm undertakes to develop goodwill with its customers, the community in which it does business, its stockholders, and other key stakeholders.

search engine optimization (SEO) The practice of designing Web pages around the keywords that are most relevant to the content of a company's website, so the pages will rank high when users do a Web search; link reciprocity is also part of SEO.

viral marketing Use of the Internet to spread word-of-mouth influence among members of a target audience; a subset of *buzz marketing.*

Web 2.0 A collection of Internet tools based on communications and connectivity among users in which the users themselves generate, view, and share content in a proactive social setting; includes technologies such as blogging, podcasts, video sites, and social networking.

Web 3.0 The emerging technologies such as the Semantic Web, which will enable computers to perform searches and other activities independently of human intervention, as well as new technologies to enable mobile Web access at greater speeds/bandwidth.

Web analytics The study of the log files (data collected by a server to measure every action a visitor takes at a website) to assess the performance of the site.

widget *See* gadget.

Notes

1. Lohr, Steve, "Microsoft Reveals Plan to Take Business from IBM," *New York Times,* March 17, 2006, online at www.nytimes.com/2006/03/17/technology/17soft.html.

2. McAlister, Leigh, Raji Srinivasan, and Min Chung Kim, "Advertising, Research and Development, and Systematic Risk of the Firm," *Journal of Marketing* 71 (January 2007), pp. 35–48.

3. Ames, B. Charles, and James D. Hlavacek, *Managerial Market for Industrial Firms* (New York: Random House, 1984).

4. O'Guinn, Thomas, Chris T. Allen, and Richard J. Semenik, *Advertising and Integrated Brand Promotion,* 4th ed. (Mason, OH: Thomson/South-Western, 2006).

5. Raine, George, "New Ad Agency to Focus on Tech," *San Francisco Chronicle,* April 24, 2003, online at www.sfgate.com/cgi-bin/article.cgi?f=/c/a/2003/04/24/BU67295.DTL&hw=New+Ad+Agency+to+Focus+on+Tech&sn=001&sc=1000.

6. Bellizzi, Joe, and Jakki Mohr, "Technical versus Nontechnical Wording of Industrial Print Advertising," in *AMA Educators' Conference Proceedings,* eds. R. Belk et al. (Chicago: American Marketing Association, 1984), pp. 171–175.

7. Helm, Bert, "Google Goes inside the Beltway," *BusinessWeek,* October 10, 2005, online at www.businessweek.com/technology/content/oct2005/tc20051010_0156_tc024.htm.

8. Ibid.

9. Smith, Timothy M., Srinath Gopalakrishna, and Paul M. Smith, "The Complementary Effect of Trade Shows on

Personal Selling," *International Journal of Research in Marketing* 21 (March 2004), pp. 61–76.

10. Levitz, Jennifer, and Kelly Greene, "Marketers Use Trickery to Evade No-Call Lists," *Wall Street Journal,* October 26, 2007, p. 1.

11. Auerbach, Jon, "Cutting-Edge EMC Sells the Old-Fashioned Way: Hard," *Wall Street Journal,* December 19, 1996, p. B4.

12. Delaney, Kevin J., "Once-Wary Industry Giants Embrace Internet Advertising," *Wall Street Journal,* April 17, 2006, p. A1.

13. Marketing News, *Global Media Statistics in Marketing Fact Book,* July 15, 2007, pp. 28–29.

14. Bughin, Jacques, Christoph Erbenich, and Amy Shenkan, "How Companies Are Marketing Online: A McKinsey Global Survey," *McKinsey Quarterly,* September 2007, online at www.mckinseyquarterly.com/How_companies _are_marketing_online_A_McKinsey_Global_Survey _2048

15. Ibid.

16. Godin, Seth, *Permission Marketing: Turning Strangers into Friends, and Friends into Customers* (New York: Simon & Schuster, 1999).

17. Parker, Pamela, "House Lists Generate Best E-Mail ROI," EarthWeb News, October 15, 2003, online at http://news .earthweb.com/IAR/article.php/3092211.

18. Johnson, Devon S., and Sundar Bharadwaj, "Digitization of Selling Activity and Sales Force Performance: An Empirical Investigation," *Journal of the Academy of Marketing Science* 33 (December 2005), pp. 3–18.

19. Manchanda, Puneet, Jean-Pierre Dube, Kim Yong Goh, and Pradeep K. Chintagunta, "The Effect of Banner Advertising on Internet Purchasing," *Journal of Marketing Research* 53 (February 2007), pp. 98–108.

20. Sloan, Paul, "The Quest for the Perfect Online Ad," *Business 2.0,* March 2007, p. 88

21. Ibid.

22. Ferrini, Anthony, and Jakki Mohr, "Uses, Limitations, and Trends in Web Analytics," in *Handbook of Web Log Analysis,* eds. B. J. Jansen, A. Spink, and I. Taksa (Hershey, PA: IGI, 2008), pp. 124–142.

23. Holahan, Catherine, and Robert D. Hof, "So Many Ads, So Few Clicks," *BusinessWeek,* November 12, 2007, online at www.businessweek.com/magazine/content/07_46/ b4058053.htm.

24. Mangalindan, Mylene, "Playing the Search-Engine Game," Special Report: E-Commerce, *Wall Street Journal,* June 16, 2003, p. R1.

25. Borzo, Jeanette, "On Point: New Services Promise to Deliver Ads to Web Sites That Are a Lot More Relevant— and a Lot More Lucrative," Special Report: E-Commerce, *Wall Street Journal,* October 20, 2003, p. R4.

26. Sullivan, Elisabeth, "Be Sociable: 2008 Is the Year That Social Marketing Will Reach Critical Mass," *Marketing News,* January 15, 2008, pp. 13–16.

27. Flandez Raymond, "Lights! Camera! Sales! How to Use Video to Expand Your Business in a YouTube World," Journal Report on Small Business, *Wall Street Journal,* November 26, 2007, p. R1.

28. Schmidt Eric, speech in Seoul Digital Forum (May 2007), www.youtube.com/watch?v=T0QJmmdw3b0.

29. Rosen, Emanuel, *The Anatomy of Buzz: How to Create Word-of-Mouth Marketing* (New York: Currency, 2000); Khermouch, Gerry, "Buzz Marketing," *Business Week,* July 30, 2001, pp. 50–56; and Dye, Renée, "The Buzz on Buzz," *Harvard Business Review* 78 (November–December 2000), pp. 139–146.

30. Gladwell, Malcolm, *The Tipping Point: How Little Things Can Make a Big Difference* (Boston: Little, Brown, 2000).

31. Chozick, Amy, "You've Played the Videogame, Now Buy the Car," *Wall Street Journal,* October 24, 2007, p. B1.

32. Steel, Emily, "Web-Page Clocks and Other 'Widgets' Anchor New Internet Strategy," *Wall Street Journal,* November 21, 2006, online at http://online.wsj.com/article/ SB116406464019128830-search.html?KEYWORDS= web-page+clocks+widgets&COLLECTION=wsjie/ 6month.

33. Enright, Allison, "Third Screen Tests: Mobile Growth Offers Targeted Connection Opportunities," *Marketing News,* March 15, 2007, pp. 17–18.

34. Steel, "Web-Page Clocks and Other 'Widgets' Anchor New Internet Strategy."

35. Christensen, Tyler, "Text-Message Marketing Makes Debut," *Missoulian,* February 6, 2007, pp. A1, A5.

36. Sharma, Chetan, Joe Herzog, and Victor Melfi, *Mobile Advertising: Supercharge Your Brand in the Exploding Wireless Market* (New York: Wiley, 2008).

37. Enright, "Third Screen Tests."

38. "Mobile Advertising: The Next Big Thing," *The Economist,* October 6, 2007, pp. 73–74.

39. Ibid.

40. Enright, "Third Screen Tests."

41. Rose, Frank, "How Madison Avenue Is Wasting Millions on a Deserted Second Life," *Wired,* July 24, 2007, online at www.wired.com/techbiz/media/magazine/15-08/ff_ sheep?currentPage=all.

42. Wagner, Mitch, "Five Rules for Bringing Your Real-Life Business into Second Life," *Information Week,* August 21, 2007, online at www.informationweek.com/story/ showArticle.jhtml?articleID=201500141.

43. Hemp, Paul, "Avatar-Based Marketing," *Harvard Business Review* 84 (June 2006), pp. 48–57.

44. McBride, Sarah, "TV + The Web = ?" *Wall Street Journal,* May 15. 2006, p. R1.

45. Kang, Stephanie, "NBC to Use TiVo's Viewership Data," *Wall Street Journal,* November 27, 2007, p. B8.

46. This section is based on the 7 C's framework in Rayport, Jeffrey F., and Bernard J. Jaworski, *Introduction to e-Commerce,* 2nd ed. (New York: McGraw-Hill/Irwin, 2004).

47. Ibid.

48. Eisenmann, Thomas, "Online Portals," Note, Product # 9-801-305, Harvard Business School Publishing, December 15, 2000; and Eisenmann, Thomas, "Online Retailers," Note, Product # 801306, Harvard Business School Publishing, December 11, 2000, both available for purchase at www.hbsp.com/hbsp/index.jsp;jsessionid=HYMWNI0SUQX AIAKRGWCB5VQBKE0YOISW?_requestid=266387.

49. Spors, Kelly, "Small Business: For the Web, Choose Your Words Carefully," *Wall Street Journal.com,* May 13, 2007, p. A1.

APPENDIX 11.A

Web Analytics: Monitoring the Traffic at a Website

Web analytics refers to the assessment of the performance of a given website. Web analytics uses data that are recorded by the computer server (the computer on which a website is hosted, or "served up" on the Internet) every time a browser hits a particular Web page. Known as *log files,* these data include information for every action a visitor at that particular website takes, showing how the visitor navigates through the various pages of a given website, or the *click trail,* also known as *clickstream data.* Clickstream data can show which goods a customer looked at on an e-commerce site, whether the customer purchased those goods, which goods a customer looked at but did not purchase, which ads generated many click-throughs but few purchases, and so forth. Software called *log analysis software* (such as that available from WebTrends, Web Side Story, Urchin Web Analytics, or Google Analytics) pulls data from the server log files and presents the information in a variety of useful templates.

Uses of Web Analytics

Because the details in log files give clues as to which website features are successful and which are not, they assist website designers in the process of continuous improvement by adding new features, improving upon current features, or deleting unused features. Then, by monitoring the Web logs for user reaction (increased or decreased usage of the site's features), and making adjustments based on those reactions, the website designer can improve the overall experience for site visitors on a continuous basis.

Web logs can also be used to track the amount of activity from offline advertising, such as magazine and other print ads, by utilizing a unique URL in each offline ad that is run. Unlike online advertising, in which the referring website can be tracked via log file analysis, offline advertising requires a way to track whether or not the ad generated a response in the viewer. One way to do this is to use the ad to drive traffic to a particular website. So, many advertisers place a unique URL in each offline ad that they run; each unique URL directs viewers who saw the ad to a different Web address from the website's regular URL. Web marketers can create a unique URL by buying a completely new domain name (Web address) or by using a subdomain, such as *subdomain.domain.com,* or by creating unique pages within the current site structure, such as *www.domain.com/unique.* Any visitor traffic that enters the website via the unique URL is assumed to have been driven there by the offline ad—the only way a visitor could have discovered the specific URL. So, by tracking the number of visitors to each unique URL, the advertiser can evaluate the effectiveness of different offline ads.

Limitations of Web Analytics

Despite the wealth of useful information available in log files, the data also suffer from limitations. The limitations of Web log files generally arise because certain types of visitor data are not logged, such as information about the person visiting the site rather than just the computer visiting the site, and some of the data that are logged may be incomplete, such as visit duration or unique visitors. For example, log files cannot track visitor activity from cached pages because the Web server never acknowledges the request. As a result, conclusions based on these data may lead to unsound business decisions.

Page tagging is a technique that allows marketers to record visitor information in a database, identifying the most valuable customers (typically defined as those who account for a significant volume of purchases or Web-based activities). Because of its increased flexibility (compared to traditional Web analytics based on server log files), most of the innovation in Web analytics is coming from page tagging. This method easily adapts to the rapidly changing Web environment and allows new ways to capture, manipulate, and display visitor information, as discussed next.

New Techniques in Web Analytics: Site Overlays and Geo-Mapping

Site overlays are a visual representation of the click activity on a specific page of the website. The complete Web page is displayed as seen by the user in a browser, with the addition of the percentages of click activity for each link on the Web page. This overlay feature is a useful addition to previous Web analytics software. Rather than reviewing a numerical Web log report for the most popular links and paths through a site, the site overlay provides a detailed visual representation of each individual Web page, with all click activity represented. One benefit of a site overlay is that it provides an easy way to quickly identify what features visitors are clicking. Moreover, it gives a more complete picture of the activity on a specific Web page, as compared to traditional Web analytics, which is usually limited to a simple list of the most popular click paths. Web developers and marketers can utilize a site overlay to analyze a specific Web page, and even each individual link within a Web page.

Geo-mapping relies on new mapping technologies being made available by services like Google Earth and Microsoft Virtual Earth to display Web analytics with a geographic perspective. In the past, most Web analytics reports provided a list of visitor countries (and number of visitors from each country) with little additional detail. Improvements in Web analytics and mapping software provide more detail on visitor locations. In addition to providing country of origin, geo-mapping provides detail on visitors' specific cities of origin, and creates a visual representation of all the visitors on a world map. This technique can be useful for tracking the penetration of a website in a particular geographic region or for tracking the effects of marketing activities in a specific city.

Web 2.0 Considerations

Web 2.0 technologies pose new complications for Web analytics. First, some Web 2.0 technologies make it difficult to count website traffic. If a person wants to determine how many readers are reading a certain blog, it becomes complicated when the blog is shared, say, via an RSS feed. In addition to monitoring traffic at the blog itself, the person would have to measure how many people access the blog via the RSS feed. The page views of the blog that occur in GoogleReader, Bloglines, LiveJournal, or any place the blog is syndicated are nearly impossible to track and count.

Second, new technologies such as AJAX and widgets make it difficult to count site traffic. AJAX technology that allows a page to update itself without reloading creates a problem for counting "page views." When visitor hits a page using AJAX, only the first page view is recorded, no matter how long that person stays and interacts with the page. As a result, more emphasis is being placed on newer metrics such as visit duration and user interaction. In addition, although AJAX does provide some capability for tracking refreshed page views through a tagging and "call back" to the server, most experts today find AJAX problematic for the mainstream, commercial analytics software that most companies use.

More important than the problems in counting site traffic per se are the metrics themselves. In the Web 2.0 environment, traditional metrics used to evaluate website performance are called into question. Prior to Web 2.0, most visitor activity could be tied to simple page views. However, some argue that, at the extreme, "page views are obsolete" and that "there will come a time when no one who wants to be taken seriously will talk about their Web traffic in terms of 'page views' any more than one would brag about their 'hits' today."* In many cases, the sheer number of visitors to a particular site matters less than how engaged the visitors are: "Most bloggers would rather be read by a handful of key influencers who provide thoughtful commentary rather than by legions of regular Joes."[†] Or, the bloggers are interested in the thoughtfulness of a handful of responses to their blogs rather than merely the number who read the blog. As one person stated on blog:

> I would much rather have 100 focused people reading my site than 100,000 people mindlessly wandering through. With a strong, well-defined niche, I can advertise to it, pull advice and knowledge from it, and learn a lot. This might be [only] a handful of page views. The analogy would be an airline company that brags about how many millions of people it is moving every day. If the quality of the interaction is low and people don't have a reason to come back, bragging about some number you counted up doesn't capture the reality of the situation.[‡]

Therefore, Web 2.0 presents a challenge for measuring Web activity because much of the key user activity is more complicated than simply viewing a page. Because user activity on Web 2.0 sites can involve watching a video, listening to a podcast, subscribing to RSS feeds, or creating rather than just viewing content, new metrics must be considered. For example, Web analytics of rich-media content might include, say, metrics such as the number of times a video has been played, the average duration of viewing, and completion rates. Or, in an interactive user environment, the quality of the user base may be more important than the quantity per se. Quality might be captured by visitors who stimulate word-of-mouth, for example.

Unfortunately, the dominant Web analytics companies provide little functionality to track these more nuanced issues posed by Web 2.0 technologies. However, new companies are springing up to address these issues. While there really isn't a comprehensive application to track all of the various Web 2.0 content, an assortment of new companies can provide information on the effectiveness of Web 2.0 sites. For example, TubeMogul provides information on various video websites. The service will even aggregate the video comments and ratings from the various sites. Viewership is plotted over time, which allows users to monitor spikes and trends. FeedBurner (purchased in June 2007 by Google) can provide insight on the popularity of various blogs and analysis of RSS feeds and podcasts as well. This service allows users to determine the number of subscribers, where subscribers are coming from, what they like best, and what they are downloading. In much the same way as TubeMogul tracks video, FeedBurner overcomes the analytics challenge presented by blogs and other types of feeds by offering a solution to track content that is no longer contained in a single website, but rather is distributed to other sites and feed readers across the Web.

As these two examples show, new companies are springing up to handle measurement and monitoring of new websites based on Web 2.0 technologies. Although complications still exist, the evolving nature of the Internet implies that Web analytics will continue to evolve as well, providing better tools to manage such complications.

Source: Ferrini, Anthony, and Jakki Mohr, "Uses, Limitations, and Trends in Web Analytics," in *Handbook of Web Log Analysis,* eds. B. J. Jansen, A. Spink, and I. Taksa (Hershey, PA: IGI, 2008), pp. 124–142.

*Zawodny, Jeremy, "Hit Counter 2.0, or 'Web 2.0 Metrics,'" posted October 9, 2006, at http://jeremy.zawodny.com/blog/archives/007665.html.

†Ibid.
‡Ibid.

Strategic Considerations in Marketing Communications

Samsung's Global Branding Strategy[1]

For almost three decades Samsung Electronics was a low-priced manufacturer and supplier of dynamic random access memory (DRAM) chips to OEM companies like Dell and Nokia. In 1997 Samsung was in deep trouble following the Asian financial crisis, with a huge debt and a brand associated with cheap, me-too TVs and microwave ovens. A new CEO, Yun Jong Yong, started out by cutting 24,000 jobs and selling $2 billion in noncore businesses. He sensed the opportunity in electronics that the technology shift from analog to digital would create and decreed that Samsung would sell only high-end products. Yun hired new blood from outside the company. Eric Kim, who moved to the United States from Korea at age 12 and had worked in many U.S. tech companies, was hired in 1999 as executive vice president of global marketing, based in Seoul. Together, Yun and Kim developed and executed the global branding strategy.

For a start, Kim severed the unwieldy relationships Samsung had with 55 advertising agencies and consolidated these into one, awarding Madison Avenue's (the famed New York City street of high-powered ad agencies) Foote, Cone, and Belding a $400 million contract to build a global brand to rival Sony. The global ad campaign featured ethereal models equipped with Samsung gadgets. They advertised heavily in the 2002 Winter Olympics in Salt Lake City and sponsored the World Cup Soccer tournament in Seoul in 2002 and are now a regular sponsor of the Olympic Games. An electronic billboard in New York City's Times Square prominently featured Samsung advertising.

The tagline "Samsung DigitAll" is becoming quite familiar to U.S. consumers. The DigitAll campaign was the first time Samsung linked its global advertising to a Hollywood film, *The Matrix*. Samsung believed that the breakthrough visual effects and prestige of the film resonated with its core audience of high-tech enthusiasts: stylish trendsetters, aged 17 to 39, who wanted to own the latest gadgets. The tools Samsung used in this promotional campaign included the development of a limited-edition Matrix-themed wireless telephone for in-film use by the lead characters. Samsung received good exposure for this product in video releases of *The Matrix Reloaded* and also the *Enter the Matrix* videogame. Samsung developed ads featuring the film's stars with 10 Samsung electronics products. Participating global regions were encouraged to develop relevant campaigns in their territories, utilizing advertising media, an online microsite for the campaign, point-of-purchase displays, banners, wireless marketing tie-ins, public relations and buzz marketing, and guerrilla marketing. Samsung had corporate screenings of the film's release, where it held additional events such as on-site displays and product sampling. From this campaign alone—on which Samsung spent $100 million on global media and promotions—Samsung generated more than 1 billion advertising impressions, and the website traffic on Samsung.com increased 65%.

In addition, Kim created an office of Global Marketing Operations, improved coordination with the business units in the different geographies, and reallocated marketing resources to support premium products such as cell phones, flat panel displays, and consumer electronics. With these brand initiatives, Samsung's rank as a global brand went up from 34 in 2002, to 25 in 2003, and 21 in 2004. Meanwhile, rival Sony's global brand rank, which was 21 in 2002, was dropping. By 2005 Samsung, at global rank 20, had overtaken Sony, at rank 28—a remarkable achievement.

In September 2004, Eric Kim left Samsung to take a position at Intel. In 2005 Samsung agreed to pay a $300 million fine to the U.S. government for settling a price-fixing conspiracy charge with its DRAM chip competitors. Affected customers included Dell, HP, Apple, and others. Having focused historically on premium mobile phones, Samsung missed an early opportunity to offer low-end handsets in India and China, which Nokia and Motorola capitalized on during 2005 and 2006. In 2007, the South Korean government launched an investigation into allegations that Samsung bribed public officials to overlook some of its management practices. All of these factors have had an impact on Samsung's global brand ranking. After maintaining its global rank at 20 in 2006, Samsung's global brand ranking dropped to 21 in 2007, while Sony has picked up after 2005 and increased its rank to 26 in 2006 and 25 in 2007.

This chapter continues the focus on high-tech company's advertising and promotion strategies. While Chapter 11 covered specific advertising and promotion tools in the high-tech company's marketing communications arsenal, this chapter deals with two overarching strategic issues that affect a company's use of advertising and promotion tools: strategic brand management and new product pre-announcements.

Fear, uncertainty, and doubt often plague the customer's buying decision for high-tech products. In such situations, customers rely on heuristics to help them make safer, easier decisions—and a solid brand name is one such heuristic. Many high-tech companies spend a significant amount of money on marketing to develop brand equity. Yet as the opening vignette illustrates, global branding needs a sustained commitment and constant vigilance on several business fronts.

The second strategic issue covered in this chapter is when and how a high-tech company pre-announces new products that are not yet available in the market. The timing of new product announcements can be vitally important in the high-tech arena. Pre-announcements help customers know what new products are coming down the pike and can delay them from buying a competitor's product in anticipation of another one coming in the near future. However, they often cue competitors to a firm's specific strategies, and in that sense present something of a double-edged sword. The pros and cons of pre-announcing new products must be considered carefully, and the second half of this chapter provides insight into the trade-offs.

BRANDING IN HIGH-TECH MARKETS

The importance of building strong brands in high-tech markets, accomplished by efforts to establish and nurture a corporate identity, can be seen in many examples. Microsoft, Intel, Hewlett-Packard, Apple, Cisco, IBM, Dell, and Yahoo! are just a few of the names that come to mind. These companies use TV, print, radio, billboards, and the Web to reach a large audience to define their brands. For 90 years the Matsushita Electric Industrial Company bore the name of its founder and became a global consumer electronics company, manufacturing and marketing products including camcorders, TV sets, microwave ovens, and DVD players. In January 2008, the company announced that effective October 1 it would take the name of its best-selling brand, to become Panasonic Corporation, because of the greater global brand equity in the latter name.[2]

Every year, *Business Week* magazine reports the rankings of the world's top 100 brands, as determined by the consulting firm, Interbrand. The Interbrand method for valuing brands examines brands through the lens of financial strength, importance in driving consumer selection, and the likelihood of ongoing branded revenue. In 2008, the top 10 global brands, in order of decreasing brand value, were Coca-Cola, IBM, Microsoft, GE, Nokia, Toyota, Intel, McDonald's, Disney, and Google.[3] It is interesting to note that at least six of the top 10 global brands could be classified as high-tech brands.

Despite these high-profile examples, Kevin Keller, one of the premier experts on strategic brand management, states that many high-tech companies—often managed by "technologists"—struggle with branding.[4] They often lack any kind of brand strategy and sometimes see branding as simply naming their products. Yet, as shown in earlier chapters, financial success in high-tech markets is not driven by product innovation alone or by the latest "gee whiz" product specifications and features. Rather, marketing skills are fundamental to the success of high-tech products, and one of these critical marketing skills is mastering the intricacies of strategic brand management.

This section first defines what the rather nebulous term *brand* means, and what is meant by a *branding strategy*. This is followed by a discussion of why strategic brand management is important, as well as specific strategies used for creating a strong brand. Because so many high-tech products are used as components, or ingredients, embedded within a finished product, one specific branding strategy, *ingredient branding,* is discussed in some detail.

What Is a Brand?[5]

At its simplest level, a **brand** is a name, term, sign, symbol, design, or a combination of them, used to identify and differentiate the goods and services of one seller from another.[6] Branding as a means to distinguish the goods of one producer from another has existed for centuries; indeed, the Old Norse word *brandr,* which means "to burn," was the means by which livestock owners marked their animals to identify them. When used in this manner, a brand allows a company legal protection for its products, to protect against counterfeiting and other methods that unscrupulous competitors might use to ride unfairly on its reputation. Further, a brand can be protected through registered trademarks, and these intangible assets provide additional value to the company that owns them. From an accounting perspective, when a firm is acquired the difference between the "book value" of its tangible assets and the price paid represents goodwill, or the value attributed to the brand and related intangibles.

Yet, when marketing managers use the word *brand,* they generally are referring to much more than the simple term or name used to denote a company's products. When marketers talk about "a brand," they typically mean that the specific brand has gained a certain amount of awareness and prominence in a market, and the company's reputation usually is well known in terms of the value proposition it offers to customers. Examining the value of a brand in this manner is directly related to a customer-oriented perspective on branding.

From a customer perspective, a brand is a pledge or a promise that a product will perform in certain ways and provide customers with consistent performance across all *touch points*—or "moments of truth"—with the company (distribution channels, customer service, pricing, warranties, etc.). In this sense, the brand represents an emotional connection with the company and its products. Moreover, the brand represents a bond or a pact whose special meaning frequently transcends the functional or utilitarian benefits of the product. Strong brands serve as symbols of customers' self-image, allowing those who use a brand to communicate to others—or even to themselves—the type of person they envision themselves to be.

Many different factors can affect a company's brand image in the market, some of which are under their control and others which are not. **Branding strategies** refer to the specific activities that a firm undertakes in its strategic brand management program, with the objective of creating awareness and prominence in its market. At a basic level, branding strategies might include the choice of *brand elements* (e.g., logos, symbols, slogans), the *usage experience* (including marketing activities and supporting programs that create strong, favorable, and unique associations for the brand), and *associations with other entities* (e.g., country of origin, distribution channels, co-branding, sponsorship of events, celebrity endorsements) to help build the brand. In addition, as discussed subsequently, branding strategies must also include a range of other issues, such as a company's innovation road map, its internal communication with employees, and even its investments in customer service and support.

Aaker and Jacobson's study of high-tech firms in the computer industry identified drivers of changes in brand attitude, including:

- *Changes in a company's innovation road map:*   Introductions of great new blockbuster products enhance brand attitudes, while a failure to innovate erodes them.
- *Changes in top management:*   IBM's induction of Lou Gerstner in 1993 or the reinvolvement of Steve Jobs with Apple in 1997 were associated with improving brand attitude. Indeed, many of the world's top technology companies have highly visible CEOs; in these cases, the CEO's identity and persona are inextricably woven into the fabric of the brand.[7]
- *Competitor actions:*   Hard-hitting, direct-comparison advertising can have a negative impact on brand attitude.
- *Legal actions:*   Actions like the Justice Department case against Microsoft, which gathered momentum in the fourth quarter of 1997, can have a negative effect on brand attitude.

These drivers of brand attitude highlight the breadth of the factors a company must consider in its strategic brand management.

When a firm's branding strategies are successful, they create **brand equity**. Keller defines brand equity "as a tool to interpret the potential effects of various brand strategies."[8] Branding endows products and services with specific meaning, and therefore results in differential customer response to marketing activities. Three key drivers of brand equity are customer brand awareness, customer brand attitudes, and customer perceptions of brand ethics. Importantly, brand equity yields key benefits for a company. For example, the higher price premiums and increased levels of loyalty to brands generate incremental cash flows.

Some high-tech companies may be a bit uneasy discussing their innovations in the language of branding. To be sure, talking about psychosocial benefits and symbolic value may seem rather far-fetched for some high-tech products or breakthrough scientific inventions, such as a pacemaker for a heart patient or the use of DNA testing in law enforcement. However, as the next section demonstrates, the benefits of a branding strategy are universally generalizable regardless of a company's industry or customer base. As a matter of fact, Kevin Keller and other experts note (see below) that branding strategies may be more important for high-tech companies than for other types of companies.

The Benefits and Risks of Branding Strategies

As shown in Table 12.1, strategic brand management confers a variety of benefits, to both the companies that undertake branding strategies and the customers to whom they sell.

COMPANY BENEFITS   Strong brands possess important advantages in a competitive marketplace. Well known brands generally are priced at a premium, resulting in higher margins to the companies that sell them. Strong brands are used as a badge or emblem that bestows credibility; they signal the quality level that customers can expect, and the brand name usually connotes particular attributes or associations. Brands can also attract attention in a new market, be it a new country, a new category, or a new industry. Hence, brands can reduce risks companies face in introducing new products, because customers trust that the new introduction will be consistent with the company's efforts in the past; in that sense, customers may be less vigilant about examining the specifics of the new offering.

Probably the single largest benefit for a company's investment in branding is the positive return in financial performance and stock market valuation. Strong brands create value for their shareholders by yielding higher returns than the overall market.[9]

In addition, brand equity results in many ancillary benefits (beyond the differential customer response to marketing activities), such as attracting higher quality employees, eliciting stronger

TABLE 12.1 Advantages and Disadvantages of Strong Brands

	Advantages	Disadvantages
For Companies	• Command premium prices • Earn higher margins • Bestow credibility • Send a signal of product quality • Facilitate entry into new markets • Reduce risk with new initiatives • Can have positive impact on financial performance and market valuation • Can have strong impact on consumer behavior, especially for breakthrough innovations • *Ancillary benefits:* Attract high-quality employees; elicit support from partners; gain licensing opportunities	• Can be expensive • Can trigger backlash if company actions are inconsistent with brand message • Require a long-term commitment
For Customers	• Provide a decision-making heuristic; simplify complex decisions • Provide a signal of safety, or a pledge of quality	• Can become angry if company doesn't live up to promises

support from channel and supply chain partners, and gaining additional licensing opportunities; each of these ancillary benefits represent "options" that might be exercised in the future, based on a company's branding efforts in the current marketing environment.

CUSTOMER BENEFITS   From a customer's perspective, strong brands serve as a heuristic, providing a simple, safe shortcut in decision making. Strong brands stand out as a beacon to the harried customer, a safe haven from the daily cacophony of technologies, new products, and media clutter around them,[10] serving as a pledge of quality and a promise that the company will back up its product. A sizable segment of customers tend to migrate to the familiar in a cluttered and confusing world.

Some believe that in high-tech markets where products change rapidly, a strong brand name is even more important than in the consumer packaged goods industry[11]—an industry that "wrote the book" on developing strong brands. Indeed, "many high-tech companies have learned the hard way the importance of branding their products and not relying on product specifications alone to drive their sales."[12] High-tech companies need well-designed and well-funded marketing programs to create brand awareness and a strong brand image. Corporate images associated with innovation and trustworthiness have a greater positive impact on consumers' product evaluations when the perceived risk of the purchase is high (rather than low); this implies that branding is more important for high-tech products and innovations—and even more so for breakthrough innovations compared to incremental innovations.[13]

Even high-tech services companies, such as those offering IT consulting and IT outsourcing (e.g., KPMG, Accenture, Wipro, TCS), must consider the customer benefits of strategic brand management. High-tech service providers face the double-whammy of "high-tech" (unfamiliar, risky, fear, uncertainty, and doubt) as well as *intangibility:*

> One of the challenges in marketing services [any service, not only high-tech services] is that they are less tangible than products and more likely to vary in quality, depending upon the particular person or people providing them. For that reason, branding can be

particularly important to service firms as a way to address intangibility and variability problems. Brand symbols may also be especially important because they help to make the abstract nature of services more concrete. A brand can help to identify and provide meaning to the different services provided by a firm. . . . Branding a service can also be an effective way to signal to customers that the firm has designed a particular service offering that is deserving of its name. Branding has clearly become a competitive weapon for services.[14]

Therefore, high-tech services providers stand to benefit disproportionately from undertaking a strategic branding campaign.

Despite these benefits, branding campaigns carry a certain amount of risk as well. Certainly, branding strategies can be expensive—although we note later that this is not always the case. Moreover, by creating an implicit promise or pledge of quality to customers, companies also must ensure that their brand promises are kept. When problems arise that are inconsistent with the brand promise, brand equity can be eroded and the investment a company has made in creating a strong brand is potentially wasted. For example, BP has worked for the last few years on creating a brand image of caring for the environment. Yet, leaks in pipelines and oil spills have undermined its brand-building efforts.

Similarly, from a customer's perspective, a brand represents that customer's trust in the company. Trust can be misplaced, and a company may not live up to its promises. For example, Dell had a brand promise to deliver the best customer service in the industry. Yet, so many customers were finding the Dell customer support organization unwieldy and ineffective that there was a backlash of negative online publicity and blogging—embodied in the phrase "Dell hell," the moniker coined by blogger Jeff Jarvis at BuzzMachine.com. Although brands are supposed to be a pledge of quality and commitment, companies can encounter difficulties in fulfilling their promises, meaning that consumers must remain vigilant, even when relying on brands as a symbol of quality.

Branding strategies are best viewed as long-term investments. Even successful brands lose their luster with consumers after a while, necessitating changes in brand strategies. For example, when Eric Kim joined Intel after leaving Samsung in 2005, Intel had had a 14-year successful run of its "Intel Inside" and "Pentium" brands, which gave it the fifth rank among the top global brands. However, consumer research revealed that "Intel Inside" was redundant because consumers knew that microprocessors go inside PCs. Further, the Pentium brand was too closely associated with PCs and Intel wanted to diversify its product portfolio to include chips for consumer electronics and mobile devices.[15] So it modified its logo and tagline, and it also launched some new brands. The "Intel Inside" brand was replaced with "Intel" with a circular swirl around it followed by the words "Leap Ahead." A new low-powered microprocessor was launched under the brand "Core" and it came in two versions, "Core Solo" and "Core Duo." Intel launched another brand, "Viiv" (rhymes with "five"), a family of chips and software for bringing Internet content to other devices like TV sets. Unfortunately Intel's 2005 branding efforts did not immediately bear fruit. In the 2008 *Business Week* survey, Intel's ranking among global brands had declined to seventh.[16]

Building brands in the high-tech arena, with its rapid change and ambiguity, requires careful consideration of the strategies and tools discussed next.

Developing a Strong Brand

So, how does a firm develop a strong brand? Strategies for branding in the high-tech environment are shown in Table 12.2.

CREATE A STEADY STREAM OF INNOVATIONS WITH A STRONG VALUE PROPOSITION Major new product introductions (dramatic and visible blockbuster innovations) with strong advertising support, like the Apple iPhone, create positive brand attitudes; conversely, failure to innovate erodes brand attitude.[17] For example, Motorola's failure to follow its Razr handset with other hit products

TABLE 12.2 Strategies for Branding in the High-Tech Environment
Create a steady stream of innovations with strong value proposition.
Emphasize advertising (rather than price promotions) to create awareness and brand image.
Effectively harness Web 2.0 technologies and new media.
"Influence the influencers" and stimulate word of mouth.
Think strategically about corporate social responsibility efforts.
Brand the company, platform, or idea.
Rely on symbols or imagery to create brand personality.
Engage in internal branding to enlist the support of all company personnel.
Effectively manage all points of contact ("touch points") with customers.
Work with partners (co-branding and ingredient branding).

has led to an erosion in brand attitude. Customers expect marketers of strong brands to supply a steady stream of innovations in exchange for their loyalty. And it is important that strong brands deliver the value they promise. The price–performance ratio must not be perceived as inequitable for the exchange, whether the customer is an OEM, an enterprise (business) user, or a consumer.

EMPHASIZE ADVERTISING TO CREATE AWARENESS AND A BRAND IMAGE   Advertising with a strong brand message focused on brand positioning and value (versus price deals and promotions) is a vital ingredient in the branding mix. Traditional sales promotions, with their focus on price discounting, tend to erode brand equity. Chapter 11 delved extensively into the wide range of advertising tools that can be used. The point here is that great brands have high awareness levels in their markets, and an astute advertising campaign is a critical ingredient in accomplishing that. Great brands also require a savvy positioning strategy, and the advertising message must create the right brand associations to create the brand image.

For example, Cisco Systems became a Silicon Valley giant by making and selling network equipment (routers, hubs, and switches) to IT officers in large enterprises that wanted to make better use of the Internet. The technology bust of 2001 adversely affected Cisco's performance due to its overdependence on this market segment. In 2003, Cisco acquired Linksys to get into home networks, giving consumers the ability to share broadband Internet connections, printers, and digital files over a wired or wireless local area network. In 2006 Cisco acquired Scientific-Atlanta for $6.9 billion to get into the TV set-top business. Cisco wanted to enter consumers' living rooms with home networking equipment, wirelessly networked DVD players, and video-on-demand.[18] Reaching out to this wide range of segments from large enterprises to home consumers with a unifying, consistent message was a challenge for Cisco's marketing and branding strategy. Cisco appointed a new marketing chief, Susan Bostrom, an eight-year Cisco veteran, to craft and deliver this message. The company adopted a new logo (vertical bars of varying lengths that together look like a bridge) and started a new TV ad campaign with the unifying message, "Welcome to the Human Network." The new message was more consumer-friendly than the earlier campaign messages, "This is the Power of the Network. Now" (in 2003) and "Are You Ready" (in 2000). The new branding campaign included product placement in movies and TV shows, and billboard ads at sporting events. The brand was also discussed on blogs in social networking sites. Keeping a brand relevant to its target markets and consistent with its products and services is important for a fast-paced high-tech firm and advertising plays a big part in this.

EFFECTIVELY HARNESS WEB 2.0 TECHNOLOGIES AND NEW MEDIA   As discussed in Chapter 11, the percentage of companies' advertising budgets spent on traditional media is declining as they experiment with new media and online advertising techniques. At a minimum, an effective branding strategy needs coordinated online and offline campaigns. For example, Nike has a partnership with Apple whereby a sensor embedded under the insole of a Nike shoe can send runners' data to an iPod. The online community for this new product (http://nikeplus.nike.com/nikeplus) allows runners to send data to their friends and challenge them to virtual contests. The payoff to Nike is that 40% of the users in this community have become converts to Nike shoes.[19]

One potential concern in today's new media environment is that companies have less control over their brand messages compared to traditional advertising and public relations. Consumers and users participate actively in these online environments, so most high-tech companies need to think about how to influence the brand messages on these sites. One important decision is whether a company should host its own blog for consumers or participate in publicly available blogs. A company-owned blog gives a marketer a little more influence over conversations but requires additional resources, and skeptical users may hesitate to participate. In either case, the guiding principles for marketers to successfully leverage new media for branding purposes are that they must (1) be honest in their communications, (2) demonstrate positive action in responding to users' complaints, and (3) provide real value.[20]

"INFLUENCE THE INFLUENCERS" AND STIMULATE WORD OF MOUTH[21]   Because brands in the high-tech arena are built up not over decades but over months, brand strategies designed to quickly stimulate word of mouth, and public relations and publicity, are crucial. Many high-tech companies rely on an "influence the influencers" program to generate publicity and favorable word-of-mouth endorsements among technology experts, enthusiasts, and opinion leaders. This is also a form of buzz marketing where a group of enthusiasts who are eager to talk about the product with their peers generate "buzz," or favorable word of mouth. With its viral nature, the Internet can further accelerate word-of-mouth influence through viral marketing (discussed in Chapter 11). Public relations personnel must work hard to identify and court the experts that influence the masses. For example, some companies give products away for free or at substantially reduced prices to these technology experts, enthusiasts, and opinion leaders, particularly journalists and bloggers who specialize in reviewing technology products. Their reviews have a big impact on new product sales. As a result, these experts can become advocates for some products, with corresponding credibility greater than advertising. Third-party endorsements from top companies, leading industry or consumer magazines, or industry experts may help to achieve the necessary perceptions of product quality.[22] To gain such endorsements requires demonstrable differences in product performance, suggesting the importance of innovative product development over time.

THINK STRATEGICALLY ABOUT CORPORATE SOCIAL RESPONSIBILITY   Best-practices business strategy in the 21st century explicitly acknowledges the responsibility that companies have to society, engaging in business strategies that benefit society as well as the firm by balancing economic profit with what is known as "social profit." As discussed at length in Chapter 13, companies can use a variety of strategies to "give back" to society and to demonstrate their commitment to corporate social responsibility (CSR). The role of strategic brand management is to consider proactively how to leverage investments in CSR for brand building efforts. CSR should be used strategically to help create positive brand associations in key target markets and provide an edge to the company in differentiating itself from competitors. It should work to enhance customer loyalty through a point of differentiation that builds on product features, yet is distinct from them.

Toyota's chairman, Eiji Toyoda, a member of the founding family, had an unusual obsession with saving energy. Under his leadership, in 1993 a team of 100 engineers was assembled to develop a car that could provide a 50% fuel efficiency improvement over current ratings and somehow be "the car of the 21st century." What was unusual about this was that oil was around $15 a barrel then, and people were buying big trucks and SUVs. Four years later, after a $1 billion investment, the hybrid Prius was launched in Japan and in 2000 in the United States. Analysts predict that all cars in the near future will use some form of hybrid gasoline-electric technology due to the spectacular success of the Prius.[23] The Toyota Prius example illustrates that CSR can result in big corporate payoffs over the long term.

TECHNOLOGY SOLUTIONS FOR GLOBAL PROBLEMS
Fighting Poverty with Agricultural Technologies
By Jamie Hoffman

Reprinted with permission of International Development Enterprises–India, New Delhi, India.

Subsistence farming affords the survival of many of the world's poorest people, often on less than one-tenth of an acre of marginal land. Regions where subsistence agriculture thrives are also plagued by variable and seasonal rainfall, restricting farmers' abilities to provide a consistent food supply for their families throughout the year. Over the last 25 years, International Development Enterprises' (IDE) has improved the lives of 2 million families through a 2-pronged approach: (1) developing and marketing affordable agricultural technologies and (2) connecting farmers to the world markets. One of IDE's most successful products has been the bamboo treadle pump.

IDE's bamboo treadle pump provides farmers of marginal land access to a controlled supply of irrigation water through an affordable, easy-to-use, and durable product. Available in Nepal, Bangladesh, and India, the product costs approximately US$9.50, and consists of two parts: the pump and the operating device. The pump lifts groundwater from shallow depths or open water using two independent metal cylinders, both connected to a suction pipe at the base and a spout at the top. The operating device involves two bamboo foot pedals attached to a bamboo superstructure with hand-holds. A user of the bamboo treadle pump merely shifts body weight from one pedal to the next to discharge 0.8–1.5 liters of water per second. As each pedal is compressed, water is drawn into the attached cylinder, where non-return valves on the lower plate protect the inflow and non-return valves on the upper plate prevent the water from being forced down the suction pipe. As the plunger moves upward, the water within the cylinder is lifted and released through the spout while the next volume of water is simultaneously drawn in.

The bamboo treadle pump creates a range of benefits for producers and users alike. The pump provides farmers with an opportunity to grow crops year-round and double their annual income. The product's simplicity allows children, elders, and women to take a larger role in farming activities. The pump establishes development within the local economy by creating a market for rural mechanics with average metalworking skills and limited access to materials. Demand for local distributors, dealers, installers, and maintenance is subsequently created.

Sources: Zenia, Tata, "Fighting Poverty with Profits," *eJournal USA,* August 2005, online at http://usinfo.state.gov/journals/ites/0805/ijee/tata.htm; and "Bamboo Treadle Pump," Food and Agricultural Organization of the United Nations, Technology for Agriculture, 2006, online at www.fao.org/sd/Teca/search/tech_dett_en.asp?lang=en&tech_id=1200.

BRAND THE COMPANY, PLATFORM, OR IDEA[24]   Given the rapid obsolescence of high-tech products, branding new products with new names is not only prohibitively expensive, but it is also difficult for customers to develop product or brand loyalty. Therefore, high-tech firms should consider what is known as the *brand hierarchy,* which includes the company/corporate name (say, Microsoft), the particular technology platform (say, Windows), and then the individual products within that platform (say, Windows Vista).

At the corporate level, high-tech companies benefit from branding the company itself.[25] A strong corporate brand is vital in the technology industry to provide stability and help establish a presence on Wall Street and in other global financial markets. Moreover, when enterprise customers purchase high-tech products, they are committing to a long-term relationship. This is another reason high-tech companies should establish a strong corporate brand that will endure over time.

The rapid pace of innovation in high-tech markets puts a premium on creating a corporate or family brand with strong credibility associations. Because of the often complex nature of high-tech products and the continual introduction of new products or modifications of existing products, consumer perceptions of the expertise and trustworthiness of the firm are particularly important. In a high-tech setting, trustworthiness implies longevity and "staying power." Yet unfortunately, many technology companies are so enamored with their new product innovations that they brand the product name to the detriment of a corporate brand identity.

High-tech companies may rely on *family names* or *umbrella brands* taken from either the company or the underlying technology platform. With this strategy, names for new products are given modifiers from existing products—Windows Vista, for example, or Microsoft Office 2007. Alternatively, the company might choose to *brand the idea* behind the product, as in "Powered by Google" or Apple's "Think Different" campaign. Branding the company, platform, or idea is more durable and cost-efficient than branding individual products.

In fact, research shows that multiproduct firms benefit disproportionately from strong customer loyalty that arises from the company's distinct market position based on its constellation of products. This customer loyalty drives adoption of the company's products, even when a single product offered by a competitor better matches a customer's stated preferences.[26] This suggests that companies should use distinct brands for different product platforms.

Despite the benefits of a corporate or umbrella branding strategy, because of the rapid introduction of new products, tech companies exhibit a tendency to "overbrand" their offerings—meaning that they brand individual products. This is a mistake because it taxes consumers' memory and causes confusion. When a company applies distinct brand names to too many products in rapid succession, the brand portfolio becomes cluttered and consumers may lose perspective on the brand hierarchy.[27] Apple used to overbrand its laptops (e.g., Apple Macintosh iBook, Apple Macintosh PowerBook 17) but now has simplified branding to MacBook and MacBook Pro for its two lines of laptops.

Relatedly, high-tech marketers should use branding selectively to differentiate major new product introductions from minor extensions. For example, after launching several extensions of its music player—including iPod mini, iPod Shuffle, and video iPod—when it wanted to launch a new category of hybrid phone and music player, Apple chose the brand name iPhone to differentiate these two product lines.

RELY ON SYMBOLS   Associations related to brand personality or other imagery may help establish a brand identity, especially in near-parity products.[28] For example, the open source Linux operating system has a cute Penguin mascot that conveys comfort and contentment. Napster's use of a silent, inscrutable cat—part of its attempt to build a hip, rebellious, edgy image—was successful, and its new owner (Roxio) stated its desire to continue using that icon given its high awareness level and positive associations with the Napster brand.

UTILIZE EFFECTIVE INTERNAL BRANDING   A company must ensure that all its personnel are informed about its brand message and campaigns prior to their public release. For example, when Intel modified its logo and brand strategy to the "Leap Ahead" in 2006, it accompanied its external, public efforts with a 26-page color brochure that was distributed internally to all Intel personnel. Intel realized that the company's public image is a function of its internal decisions and behaviors. Hence, for a major change in branding to be successful, all company personnel had to first be informed about the change, the reason for the change, and their roles in successfully executing the new strategy. This leads directly to the next important element of a branding strategy: managing points of contact.

MANAGE ALL POINTS OF CONTACT   Ultimately, a brand is a promise of quality to a customer. So, customer service is as important in building a strong brand as any advertising campaign or new product innovation initiative. If customers do not receive high-quality service across all their touch points with the company, it negatively affects the brand.

This point cannot be overemphasized. Far too many high-tech companies think they are ready for a brand-building campaign. They invest significant sums of money in creating a logo, generating a tagline, creating expensive ads, and buying media placement. Yet, what if customers cannot find the product on the shelf when they go to a retail store? Or, what if the retail salesperson cannot adequately explain the new product's features? What if a customer's first Google search results are a negative blogger's rants about the company? What if the customer's attempts to reach customer service via an online inquiry go unanswered? Or, what if a phone call to the 1-800 number results in a mechanized routing system that creates frustration and more headaches than the original problem? The point is that a company should not undertake a branding campaign until its internal house is in order; in other words, it has thought proactively about all the touch points that prospective, new, and existing customers will have with the company and has ensured that each and every interaction will support the underlying brand promise.

WORK WITH PARTNERS High-tech branding strategies must consider their impact on not only customers and the financial community, but also alliance partners. The effectiveness of a branding strategy is only as effective as its partners' commitment. For example, when a company outsources its customer service, it must ask whether its partner has the same commitment to customer excellence as it expects. Or, when a company relies on a channel partner, it must be certain that that channel member will support the company's brand promise. Certainly, working with partners means relying on another company to deliver and execute effectively. Yet, as discussed in Chapter 5, working with partners means giving up some control and depending on the capability of others. When a company has invested in brand building, partners need to be informed and committed as well.

A special type of branding strategy that relies on partners is known as **co-branding**, in which two companies jointly brand their product offering. For example, when Yahoo! co-brands certain areas of its website with its content partners' names and logos, it is using a co-branding strategy. Co-branding is based on the idea of synergy: The value of two companies' brands, when used together, is stronger than that of one brand alone. Intel has relied on a variation of this strategy extensively, based on the idea of creating a brand identity for a component used in the customer's end product, or *ingredient branding,*[29] discussed subsequently.

Additional guidelines and insights for high-tech companies to improve their brand strategies include the following.[30]

1. *Have a brand strategy that provides a roadmap for the future.* Technology companies too often rely on the false assumption that the best product based on the best technology will sell itself. As the market failure of the Sony Betamax illustrates, the company with the best technology does not always win.
2. *Know that brands are owned by customers, not engineers.* In high-tech firms, CEOs in many cases work their way up the ladder through the engineering divisions. Although engineers have an intimate knowledge of products and technology, they may lack the big-picture brand view. Compounding this problem that high-tech CEOs may lack branding capabilities is the fact that technology companies typically spend less on consumer research compared with other types of companies. As a result of these factors, tech companies often do not invest in building strong brands.
3. *The rapidly changing environment demands that companies stay in tune with their internal and external environments.* The rapid pace of innovation in the technology sector dictates that marketers closely observe the market conditions in which their brands do business. Trends in brand strategy and marketing change almost as rapidly as the technology.
4. *Invest the time to understand the technology and value proposition and don't be afraid to ask questions.* Technology marketers must ask questions in order to educate themselves and build credibility with the company's engineering group and with customers. To build trust among engineers and customers, marketers must strive to learn as much as they can about the technology.

Ingredient Branding

An **ingredient branding** strategy pulls demand from end users through the distribution channel back to the original equipment manufacturers (OEMs), who feel pressure to use the branded ingredient in the goods they make. Business-to-business marketers will recognize this strategy as one designed to stimulate **derived demand**, whereby demand for the component, or ingredient, at the upstream level of the supply chain is derived from the end customers' demand (at the downstream level of the supply chain) for the products in which the components are used:

Supplier of ingredient or component →	OEM manufacturer →	Retail →	Customers
Intel	Hewlett-Packard	CompUSA	You

With an ingredient brand strategy, typically the ingredient supplier provides cooperative (co-op) advertising dollars to OEMs, who feature the ingredient in their own ads. Using this campaign, Intel's awareness level increased from 22% in 1992 to 80% in 1994.[31] Intel continued this approach for its Centrino product in 2003 and the Viiv product in 2005. To use the Viiv logo on packaging and advertisements, PC makers had to use the new Intel Core Duo chip along with some other communication chips and software in a bundle. In return, the PC maker received marketing subsidies from Intel. Digital content providers certified to work with the multimedia Viiv product could also qualify for these subsidies.[32]

Co-op advertising dollars are typically provided as a percentage of revenue for product purchased. For example, at one time Hewlett-Packard's program set aside 3% of a dealer's/retailer's purchases of HP products into a co-op advertising account. Then, when the dealer ran an ad that also featured an HP product, the dealer would submit a copy of the ad to HP, along with the invoice from the media in which the ad was run, and HP would then cut the dealer a check from his or her co-op account.

Ingredient branding seems to make the most sense when the supplier's "ingredient" is integrally related to the performance capabilities of the end-product in which it embedded. For example, Intel's chip plays a key role in the performance capabilities of computers.

ADVANTAGES AND DISADVANTAGES OF INGREDIENT BRANDING As shown in Table 12.3, ingredient branding has advantages and disadvantages for both the ingredient supplier as well as the OEM customer. On the upside, the brand awareness created by the strategy creates a competitive advantage in the marketplace for the ingredient supplier. It establishes brand preference among end users, allowing the firm to fend off growing competition and stake out its own turf or identity. However, developing a strong brand name for an ingredient and providing the necessary co-op advertising dollars to make it work can be very costly. Moreover, if one of the OEMs that is participating in the branding program experiences product/performance problems, a *halo effect* occurs in which the supplier's reputation is tarnished also. (Note that this halo effect can work both ways: If the supplier's

---

TABLE 12.3 Pros and Cons of Ingredient Branding

	Pros	Cons
Supplier	Creates competitive advantage	Costly
		Possible risk if OEM has product problem
		Conflict with large OEMs
Large OEM		Erodes ability to differentiate
		Risky if supplier's product has performance problems
		If it doesn't co-brand, consumers might question product quality
		Worry about supplier forward integrating
Small OEM	Lends credibility to its product	Risky if supplier's product has performance problems
	Gets advertising support	

ingredient has performance problems, the OEM's image will suffer.) And, as discussed subsequently, this strategy can create conflict with a supplier's large OEM customers.

From a small OEM's perspective, using this type of a co-branding strategy can lend credibility to its brand, making it more competitive with stronger brands in the market. Smaller companies also receive a more significant benefit (compared to larger OEMs) from the use of co-op advertising funds that share advertising expenses with the ingredient supplier.

However, for large OEMs with top-end brands in the market, say IBM, an ingredient branding strategy can erode their ability to differentiate their products. Customers in the marketplace often assume that industry leaders have something special in their products, but, with an ingredient branding strategy, customers begin to realize that the industry leader's product has the same ingredients as lesser-known brands. This effect can cause conflict between the ingredient supplier and key OEMs in the marketplace, making the strategic relationship difficult to manage.

An ingredient branding strategy can also cause problems when the ingredient supplier and the OEM have different goals in the marketplace.[33] For example, the ingredient supplier, say Intel, would like end users to demand the latest chips—those that sell at the highest margin. To drive this engine, ingredient suppliers typically invest huge expenditures in R&D. And the best scenario from the ingredient supplier's perspective is to have a large number of OEMs all competing on the basis of price. But the OEM may have a very different strategy. An OEM may be less interested in working with a supplier that has developed a strong brand for its ingredient—and that charges a premium price for it. Large OEMs may actively try to cultivate ties with other suppliers, to avoid being so dependent on the branded-ingredient supplier only. Dell for many years was an exclusive OEM for Intel's microprocessors. But since 2006 it has started buying microprocessors from AMD as well.

An OEM that sells products without the ingredient's logo (e.g., "Intel Inside") might arouse suspicion among consumers in the marketplace. Last, if OEMs fear that the ingredient supplier will attempt to enter their marketplace in the future, they won't want to participate in the co-branding strategy, which will only serve to develop a strong brand for a company that may become a future competitor. Reconciling these tensions may be next to impossible, but at least being aware of them helps firms to make more educated choices.

Branding for Small Business

Small businesses often face resource constraints in building a brand. A low-cost way of entering any market is with a no-brand strategy: The small business can find one or more large business customers to resell the product under their own brands. This is a good way for the small business to get started. Samsung started out like this in the semiconductor business (see the chapter's opening vignette). If the small business wants to build and own direct relationships with end users, however, then branding is critically important. Creativity in marketing programs has to compensate for a low budget.[34] A small business needs to focus on one or two brands with one or two key associations. Creative messages can attract attention and get people to try the product, but product quality and the customer experience are critical in building brand preference or loyalty over time.

Small businesses may also piggyback on the brand names of larger companies by leveraging business relationships. Small suppliers to larger companies could seek testimonials from customers and publicize them on their websites or in press releases. In e-business, smaller sellers get more credibility by having their websites hosted by larger e-tailers like Amazon, eBay, or Yahoo!.

Resources permitting, brand names can also be acquired or licensed. Lenovo was a relatively unknown company when it acquired IBM's PC business in 2005 for $1.25 billion. Having obtained the exclusive rights to the IBM ThinkPad brand name until 2010, Lenovo used this brand on its products to become the third largest manufacturer of PCs worldwide, after HP and Dell. In November 2007, ahead of schedule and with a newfound confidence, Lenovo announced it was discontinuing the use of the IBM corporate brand and renaming its products to Lenovo ThinkPad.[35]

To learn how a small technology start-up went about introducing a new brand, see the accompanying Technology Expert's View from the Trenches.

TECHNOLOGY EXPERT'S VIEW FROM THE TRENCHES

Ingredient Branding: Foveon X3*

BY ERIC ZARAKOV
Former Vice President of Marketing, Foveon Inc., Santa Clara, California

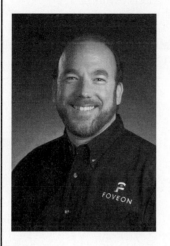

Between 1999 and 2002, the scientists and engineers at Foveon made a breakthrough in the fundamental design of an *image sensor*. Image sensors are the semiconductor chips that detect light (and replace film) inside a digital camera. The new sensor had multiple benefits over the incumbent image sensor technology, including significantly sharper images, elimination of color artifacts, and lower cost through a more efficient use of silicon. A cornerstone of the company's marketing strategy was to establish an ingredient brand for its image sensors in order to develop a premium for its products and to distinguish its technology above incumbent products. The goal was to establish a relationship between the Foveon ingredient brand and the assurance of the highest level of image quality regardless of the host brand of the digital camera.

Foveon launched its new image sensor in February 2002 and received worldwide visibility. The press coverage extended to almost all of the available photography trade media, key business press, and mainstream consumer press. Overall, the introduction resulted in more than 100 feature stories about Foveon and its X3 technology. The coverage included the front page of the *New York Times* business section, front page of the *Wall Street Journal* business section, *Time* magazine, *Newsweek* magazine, 15 seconds of prime time television coverage on *ABC World News Tonight* with Peter Jennings, and exposure to 150 million people through a radio interview with the BBC. The media attention brought enormous visibility to Foveon. Through this news-based media introduction, Foveon created interest in its technology from potential customers (digital camera makers) and sparked the foundation for developing end user pull that would reach through the digital camera makers to generate demand for new Foveon products.

As I look back to that time, I believe the success of Foveon's launch can be attributed to a number of factors. First, the timing of the story aligned with the explosive growth of digital photography. A second contributing factor was the visibility of Dr. Carver Mead, the company's founder. Dr. Mead is a world-famous physicist who co-developed the very first methods for designing computer chips using Very Large Scale Integration (VLSI). He is well known and followed by key industry analysts due to his association with many breakthrough technologies. Lastly, the visibility was high due to the obvious breakthrough nature of the technology and products in a domain that many people relate to: photography. While all of the above were significant contributions, we would not have achieved this visibility if we had not planned our strategy and prepared carefully for the product launch.

Planning the Launch Strategy

As a foundation for the launch, the new technology needed a name, a logo, and simple eye-catching visuals that we could use to develop a brand identity. After considering many alternatives, our final decision was to use "X3." The "X3" name had multiple benefits. It was two syllables, which made it short and memorable. It reinforced the fact that our technology has three layers compared to the competing image sensor technology that has just one layer. The X3 name also worked easily as a suffix, which would minimize the potential identity tension between X3 as an ingredient brand and the product host brand with which we would co-exist. For example, if Nikon developed a digital camera with our technology, a potential name such as "Nikon D100 X3" sounded good without imposing on the well-established Nikon brand.

(continued)

Having selected a name, we needed to develop a logo. The current design was selected because of its simplicity, scalable output size, color match to our corporate identity, and its ability to work as a logo on the front of a camera. The final X3 logo included our corporate name, Foveon, to bridge the familiarity that Foveon had achieved through previous technology announcements. The logo "X3" was designed, however, to eventually work without the corporate name. Color versions of the logo and images can be found at www.foveon.com.

In addition, we developed a single dominant visual that was easy to understand, would catch attention, and would communicate the essence of the technology. This single illustration was used consistently throughout our product launch and is still used in virtually all of our current marketing communications.

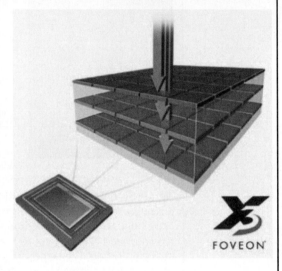

The third element of the visual portfolio was a 15-second video that animated the basic concept of X3. The animation shows light separating into color layers. A Foveon X3 logo watermark was positioned in the bottom right corner of the animation to support the brand association. This effort paid off dramatically when we received a phone call from *ABC World News Tonight* saying they wanted to run a news report on the Foveon X3 sensor. We offered the animation, and they ran it as a backdrop, with the visible X3 brand, while Peter Jennings gave a 15-second overview. The animation was also used on local TV stations throughout the country. The ABC news story alone gave us exposure to more than 10 million prime time viewers. The video cost us less than $5,000.

While we had no certainty that the product and technology would receive the visibility that it did, we had made sure we were prepared. Based on this experience, the following themes emerged from our successful launch of the brand campaign:

1. ***Get crystal clear on product positioning.*** Without this clarity, the foundation of the marketing communications plan will not hold together and the messaging has the potential to fragment. Positioning drove strategy, marketing communication asset development, and press communications.
2. ***Spend the time to work out the details.*** We spent countless hours on details of the illustrations, wording, and messages. We knew it was worth the time to get it right. The foundation communication elements would ultimately been seen by millions of people, and once it was launched there was no going back.
3. ***Work small.*** Ultimately, all of the key marketing communications work was done with a very small internal team of two to three people. Large agencies cost a lot of money, need education on the company's products (at the company's expense), and take time to get up to speed. Make sure your team has balanced points of view so that you do not steer yourself down a blind alley of enthusiasm. Seek out specialized consultants if needed. This approach requires more hands-on management—-but you get what you need.
4. ***The press is your friend.*** Make their job as easy as possible. They constantly get pitched by companies to cover self-serving stories. Use your specific elements to make your story easy to understand and have a catch. Try to anticipate what might be needed—-visuals, sound bites, testimonials. These will be golden assets when you need them, so have them ready.
5. ***Create mystique.*** Most people like to be on the inside of a secret. Creating mystique gets people intrigued and creates an atmosphere of excitement and inside knowledge.

6. ***Create critical mass in the press.*** I never to this day turn down an interview, no matter how small. You never know where it might lead. Once a story starts to spread, the press start coming to you and you need to keep it alive to build momentum. Each exposure generates more inquiries.

7. ***Keep a secret.*** Of all the product launches in which I have participated, the best results were achieved when *no* news leaked out in advance. News that leaks out through dribbles has no impact when the formal announcement is made.

Author's note: While I headed and directed the launch of the X3 technology, I worked side-by-side with a coworker, Brian Behl, who was a key contributor to our brand identity. Brian's relentless perfectionism, keen eye for design, and multifaceted skills significantly influenced the elegant look and feel of Foveon's branding and communications.

NEW PRODUCT PRE-ANNOUNCEMENTS

Many high-tech products and innovations are announced before they are ready for the market. More formally referred to as **pre-announcements**[36]—or formal, deliberate communications before a firm actually undertakes a particular marketing action (say, shipping a new product)—they are a form of signaling that conveys information to competitors, customers, shareholders, employees, channel members, firms that make complementary products, industry experts, and observers of the firm's future intentions. Because of their versatility, pre-announcements are a very appealing tool for strategic marketing communications.

High-tech firms routinely pre-announce new products. In a break from its prior strategy of not using new product pre-announcements, on January 9, 2007, Apple pre-announced the iPhone at MacWorld in San Francisco, saying it would be available in June. For the next five months, the amount of worldwide publicity that the Apple iPhone received was so great that it caused some operational difficulties for Apple. Apple had to postpone the launch of its new Leopard operating system (also pre-announced for June) to October and reassign software engineers from that project to the iPhone. It barely kept its word, releasing iPhone on June 29, the last weekday of the month. But the enormous buzz generated by the pre-announcement resulted in consumers lining up several days earlier outside Apple and AT&T stores so they could be among the first to buy the product.

Many companies pre-announce products and then encounter difficulties in delivering the product on the promised date. Microsoft began work on Windows Vista in May 2001 (known at the time by its code name "Longhorn")—which was a full five months prior to the release of Windows XP, the next-generation Microsoft operating system at that time—and originally expected to ship it sometime late in 2003 as a minor step between Windows XP and a product temporarily code named "Blackcomb." After many changes, development delays (caused in part by feature creep), and extensive beta testing (including its official naming in July 2005), the Vista feature set was frozen in February 2006. While Microsoft had originally hoped to have the consumer versions of the operating system available worldwide in time for Christmas 2006, it was announced in March 2006 that the release date would be pushed back to January 2007, to give the company and its partners time to prepare device drivers. Despite speculation of further delays, the product was released in November 2006, completing Microsoft's lengthiest operating system development project.[37]

Similarly, in the highly competitive world of computer gaming, Sony pre-announced plans for its PlayStation 3 just prior to the major Electronic Entertainment Expo (E3) industry trade show in May 2005. Its release window was set for spring 2006. Yet, delays in the development of the Blu-ray drive and other concerns led the company to push the release out to fall 2006. In September 2006, Sony experienced further delays. The product finally hit the North American and Japanese markets in early December 2006, but wasn't available in Europe until early 2007. Indeed, Sony's stock price declined 7.3% over six months. As one industry commentator stated, "It's like starting a race and Microsoft and Nintendo have already completed one lap and Sony hasn't even started."[38]

At the extreme, if products are announced early, and are slow to materialize—or in some cases, do not ever appear on the market, they are called **vaporware**, software or hardware that is announced by a developer prior to its release, but which then fails to emerge in the market, either with or without a protracted development cycle. The term implies unwarranted optimism, or sometimes even deception—that is, the announcer *knows* that product development is in too early a stage to support responsible statements about its completion date, feature set, or even feasibility.[39] Vaporware undermines the credibility of the high-tech firm and could even earn negative publicity. As the Acme Vaporware website says: "Yesterday's technology tomorrow."[40] Every year in December, *Wired* magazine has a feature article identifying the top 10 vaporware products of that year. In 2007, this list included the videogame *Spore* from Electronic Arts and the Tesla Roadster electric car.[41]

Thus pre-announcements can have both advantages and disadvantages. If (and when) a company decides to pre-announce new products, the potential advantages must be carefully weighed against the disadvantages.

Advantages and Objectives of Pre-Announcements

Firms choose to pre-announce new products for many reasons. In order to maximize their value, firms must have a clear intent for pre-announcements. As shown in Table 12.4, by pre-announcing, firms can potentially reap a *pioneering advantage,* creating entry barriers to later entrants. By announcing products before they are fully available, a firm can preempt competitive behaviors. Research shows that new product pre-announcements create market anticipation or an industry-wide buzz, which makes the firm allocate more resources to new product development resulting in greater new product success.[42]

This pioneering advantage is related to another driver of a firm's propensity to pre-announce: its *pursuit of a high-profile leadership position* within its industry.[43] Indeed, this may be one reason why Apple pre-announced the new iPhone with such fanfare.

Pre-announcements can also *stimulate demand.* By helping develop word of mouth and opinion leader support, pre-announcements can accelerate the adoption and diffusion of innovation when the product does hit the market. In addition to building interest for the product among channel members and customers, another factor related to consumer behavior is to *encourage customers to delay purchasing* until the announcing firm's new product is available. This latter reason is primarily used for big-ticket items that are purchased rather infrequently.

Businesspeople say that announcing products before they are ready for market is a "valuable tradition" in the software industry, divulging future product plans to customers. Such pre-announcements can be beneficial when they *help customers to plan for their future needs,* or when they *allow customers to have input* in order to develop a more useful product. Users often need to know a software

---

TABLE 12.4 Pros and Cons of Pre-Announcements

Pros	Cons
Pioneering advantage: Preempt competitors, pursue a leadership position	Cue competitors
Stimulate demand	Product delays damage reputation or jeopardize firm survival
Encourage customers to delay purchase of competitor's products	Cannibalize current products
Help customers plan	Confuse customers
Gain customer feedback	Create internal conflict
Stimulate development of complementary products	Generate antitrust concerns
Gain access to distribution	
Increase shareholder value	

maker's plans early, because the software is so critical for their own businesses. So, alpha and beta versions (prototypes used to test and refine a program) may be sent to key customers months before the program is made commercially available. This gives customers, channel members, and OEMs time to prepare their operations for the new product and gives the manufacturer valuable market feedback. Indeed, companies are more likely to use pre-announcements for new products that entail switching costs for customers to give them enough time to plan for the switch.[44]

Pre-announcements are also used to *stimulate the development and marketing of complementary products.* In markets characterized by indirect network externalities, where the value customers receive from a specific product is dependent on the availability of complementary products, companies must do some sort of pre-announcing in order to inform developers of the complementary products. For example, Microsoft pre-announced Longhorn, the internal code name for what later became Vista, several years before the 2006 launch in order to give application software providers ample time to have new products ready for the new operating system. In addition, when Sony pre-announced the PlayStation 3 in the spring of 2005, it also announced the major game developers who were onboard to deliver their next-generation games for the PS3. By including these major game developers in its pre-announcement publicity, Sony sent a signal to the market stimulating indirect network externalities; the pre-announcement gave ample time to other game developers to have their products ready by the time the PS3 hit the market. The inclusion of game developers in the pre-announcement publicity also reassured end users that games would be available.

Pre-announcements can also help *gain access to distribution.* If a well-known company pre-announces its intent to enter a new market, this could help generate inquiries from prospective channel partners.

Recent research has linked the use of pre-announcements to *increased value to shareholders.* High-technology hardware and software companies that use pre-announcements increase shareholder value about 13% after a year compared to those that do not.[45]

Disadvantages of Pre-Announcements

The benefits of a pre-announcement strategy must be balanced against the costs of pre-announcing. Pre-announcements can *cue competitors* to what is coming down the pike, allowing them the opportunity to react to the new product.[46] For example, within a week of Apple's January 2007 pre-announcement of the iPhone's June 2007 release in the United States, featuring a touch screen, LG pre-announced that it was launching the LG Prada phone with a touch screen in Europe. Indeed (and despite the desire to use pre-announcements to reap a pioneering advantage), the risk of cuing competitors is one reason why a firm's propensity to be a pioneer is *negatively* related to a firm's use of pre-announcements.[47] Concerns about competitive retaliation and the risk of delays combine in such a way that firms seeking a pioneering advantage may actually avoid pre-announcements. In fact, one study found that companies are more likely to use pre-announcements when they face more attractive competitive environments (in other words, when the competitive environment is not too intense).[48]

Because the development process of high-tech products is so complex, *delays* are sometimes inevitable—which can damage the firm's reputation. One study of delays in launching products after pre-announcements in the computer hardware, software, and telecommunications industries found that delays are due to (1) pre-announcements being a hasty reaction to competitors' actions, (2) the fear of cannibalization of existing products, and (3) the innovativeness or complexity of the product.[49] On the other hand, delays are *less likely* to occur when (1) a partner firm has high power that it can use to dictate deadlines (e.g., Apple's suppliers are less likely to delay delivery and risk Apple's wrath), (2) the firm has effective interfunctional coordination, and (3) top management emphasizes meeting deadlines.

Pre-announcements can result in *cannibalization* of the firm's current product line, caused when customers delay purchases of current products in anticipation of the new ones. Some analysts expressed concern that Apple's pre-announcement of iPhone in January 2007 would cannibalize sales of its blockbuster iPod line, but this did not come to pass as the two were well differentiated and positioned for different consumer segments. Other disadvantages include the risk that pre-announcements might also *confuse customers* who try to buy the product, thinking that it is already available. Pre-announcements might also *cause internal conflict* between departments. For example, engineering might want secrecy, but financial officers may want to send signals to the market early.

Finally, *antitrust concerns* can lead a firm to avoid pre-announcing strategies, particularly for dominant firms. Vaporware can have detrimental effects in the marketplace when a firm has no intention of following through on its announcements, which are used merely as a marketplace tactic to harm competitors. In an early investigation of Microsoft, the U.S. Justice Department decried the frequent use of vaporware when businesspeople know it is deceitful.[50] When the pre-announcement is for the sole purpose of causing consumers not to purchase a competitor's product, then the predatory intent is inferred to be anticompetitive and subject to regulation under antitrust laws. Some believe that dominant players in a market—such as Microsoft—must be held to a higher legal standard because of the likely harm of their pre-announcements to the healthy competitiveness of a market.[51]

Tactical Considerations in the Pre-Announcement Decision

In making decisions about new-product announcements, firms must consider several tactical factors, shown in Figure 12.1. These include the timing of the announcement, the nature and amount of information, the communication vehicles used, and the target audience(s).[52]

TIMING   The timing of pre-announcements must consider many factors, including the advantages and disadvantages of pre-announcing relative to the innovativeness and complexity of the new product, the nature of customer switching costs, the length of the buying process, and the timing of final determination of the product's attributes.

Earlier pre-announcements (further away from the actual product launch) are particularly useful when complements to the new product are necessary to its success, for highly novel or complex products that will engender buyer uncertainty, for products that have a long buying process, or for those in which buyer switching costs are high. *Later pre-announcements* (closer to the actual product launch) make more sense when the firm needs to keep information about the new product from potential competitors, when product features are frozen late in the process, and when the firm seeks to minimize risks of cannibalization. Regardless, pre-announcements must be *timed to coincide with the purchase cycle of customers*. For example, if customers take approximately six months to decide on a new purchase, a pre-announcement with a six-month lead time would be acceptable.[53]

INFORMATION   Firms must consider how much and what kinds of information to include in the pre-announcement. Some new product pre-announcements contain information on the attributes of the new product, how the product works, and how it compares to existing products in the market, whereas others contain very limited information. Information about pricing and delivery date may also be important. Communication vehicles might include trade shows, advertisements, press releases, or press conferences. Target audiences for the information might include customers,

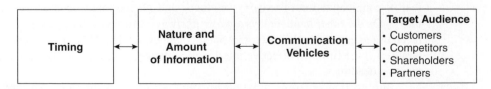

FIGURE 12.1   Tactical Considerations in Pre-Announcement Decisions

competitors, shareholders, or partners. Sorescu, Shankar, and Kushwaha found that pre-announcements containing specific information do create shareholder value in the short term.[54]

OTHER CONSIDERATIONS Pre-announcements are useful for firms with low market dominance, those that face lower cannibalization risks, and smaller firms that face fewer antitrust concerns.[55] When the firm making the pre-announcement is more reliable (in keeping its promises), pre-announcements have a greater impact on short-term and long-term shareholder value.[56]

Summary

This chapter covered the topics of branding and new product pre-announcements as part of the strategic marketing communications efforts of a business.

It defined what a brand is and the benefits of branding to a company and its customers. It discussed specific initiatives for developing a strong brand, including the customer's usage experience, brand elements (logos, slogans), and associations with other entities. Two specific branding programs popular with technology companies are co-branding and ingredient branding, which is a special case of co-branding. The advantages and disadvantages of ingredient branding were discussed, and an

implementation of this by a small company, Foveon, is presented as the Technology Expert's View from the Trenches.

New product pre-announcements are used to deliver early signals to a market and the objectives, risks, and tactical considerations associated with these were discussed.

High-tech companies must ensure that their investments in marketing communications tools are part of a planned strategy. Branding campaigns and pre-announcing practices have strategic implications that a firm must consider wisely.

Discussion Questions

Chapter Concepts

1. Explain the different meanings of the term *brand*. Why is a customer perspective on branding so important?
2. What are the advantages of a strong brand to firms? To customers? What are the risks?
3. What are the arguments for and evidence of the idea that strategic brand management is more important for high-tech companies than for other types of companies?
4. What is brand equity? When does a company create brand equity?
5. How does a firm develop a strong brand? Name and describe the strategies for branding.
6. What are some disadvantages of using Web 2.0 technologies in branding? How can Web 2.0 be used effectively in a branding strategy?
7. Explain best-practices branding for a brand hierarchy: company, technology, product.
8. Explain the role of corporate social responsibility in branding strategies.
9. Why is it important to manage all points of contact, including internal points of contact?
10. What is the logic behind ingredient advertising (also known as *co-branding* and *derived demand advertising*)?

11. Under what conditions might ingredient advertising be most useful or most likely?
12. What are cooperative (co-op) advertising dollars? Explain their function in ingredient branding.
13. What are the advantages and disadvantages of ingredient advertising? Be sure to consider the viewpoints of both the supplier and the OEM.
14. What methods are available to small businesses to build brands, which often face resource constraints?
15. What are pre-announcements? What are the pros and cons?
16. What is vaporware? What are the ethics involved? (Be sure to consider the perspectives of the various constituencies, including customers, competitors, and the firm.)
17. What factors affect the timing (early or late) of pre-announcements?
18. When is pre-announcing most likely to be effective (from the company's perspective)? When is it most likely to cause problems?

Application Questions

1. In a small group, brainstorm all of the associations you have with the brand name Samsung. Organize these into positive and negative associations.
2. After this exercise reread the opening vignette of this chapter. Are the associations summarized from your group above consistent with the brand messages that Samsung is trying to convey as discussed in the vignette? What does this tell you about the effectiveness of Samsung's global branding campaign?
3. Why did Samsung consolidate its advertising agencies from 55 to 1 worldwide?
4. Evaluate all the contributions that Eric Kim made to Samsung.

5. Based on your reading of the Technology Expert's View from the Trenches feature, answer the following questions:
   a. The Foveon X3 received a lot of free publicity during its launch. What were the reasons it was able to get this amount of free publicity?
   b. Discuss the pros and cons of the ingredient brand name X3.
   c. Eric Zarakov offers seven practical rules for successful launch of a new brand. If you had to prioritize them from most to least important what would the order be? Why?

Glossary

brand   A name, term, sign, symbol, design, or a combination of them, used to identify and differentiate the goods and services of one seller from another.

brand equity   The intangible value of a brand, earned by the success of a firm's branding strategies; results in positive consumer response to the brand.

branding strategies   The specific activities a company undertakes in order to create awareness and prominence of its brand in its market.

co-branding   A branding strategy in which two companies jointly brand their product offering; based on the idea that the value of the two companies' brands when used together is greater than that of either brand alone.

derived demand   Demand for a component, or ingredient, in a product (e.g., computer chips) that is derived from the end customers' demand for the products in which the components are used (e.g., computers).

ingredient branding   The strategy of creating a brand identity for a component or ingredient used in the customer's end product; usually stimulates derived demand.

pre-announcement   A formal, deliberate communication before a firm actually undertakes a particular marketing action (such as shipping a new product), to convey information to competitors, customers, shareholders, and industry experts.

vaporware   Software or hardware that is announced prior to its release, but then never materializes.

Notes

1. Based on Keller, Kevin Lane, *Strategic Brand Management: Building, Measuring, and Managing Brand Equity*, 3rd ed. (Upper Saddle River, NJ: Prentice Hall, 2008); Solomon, Jay, "Samsung Vies for Starring Role in Upscale Electronics Market," *Wall Street Journal*, June 14, 2003, p. 1; Edwards, Cliff, Moon Ilwahn, and Pete Engardio, "The Samsung Way," *Business Week,* June 16, 2003, pp. 56–64; Quelch, John, and Anna Harrington, "Samsung Electronics Company: Global Marketing Operations," Harvard Business School Publishing, 2005, Case 9-504-051; "The Top 100 Brands," *Business Week,* August 7, 2006, pp. 60–66; "The Top 100 Brands," *Business Week,* August 1, 2005, pp. 90–94; and Ramstad, Evan, "Samsung Suffers as Gadget Boom Ends," *Wall Street Journal,* October 11, 2007, p. A7.

2. Maxwell, Kenneth, "Matsushita to Become Panasonic in Branding Push," *Wall Street Journal,* January 11, 2008, p. B4.

3. "The Top 100 Brands," *Business Week,* September 29, 2008, Special Report, pp. 59–60.

4. Keller, *Strategic Brand Management.*

5. This and subsequent sections draw heavily from Keller, *Strategic Brand Management.*

6. American Marketing Association definition, online at www .marketingpower.com/_layouts/Dictionary.aspx?dLetter=B.

7. Tickle, Patrick, Kevin Lane Keller, and Keith Richey, "Branding in High-Technology Markets," *Market Leader* 22 (Autumn 2003), pp. 21–26.

8. Keller, *Strategic Brand Management,* p. 37.

9. Madden, Thomas J., Frank Fehle, and Susan Fournier, "Brands Matter: An Empirical Demonstration of the Creation of Shareholder Value through Branding," *Journal of the Academy of Marketing Science* 34 (March 2006), pp. 224–235.
10. Morris, Betsy, "The Brand's the Thing," *Fortune,* March 4, 1996, pp. 73–86.
11. Ibid.
12. Keller, *Strategic Brand Management,* p. 649.
13. Gürhan-Canli, Zeynep, and Rajeev Batra, "When Corporate Image Affects Product Evaluations: The Moderating Role of Perceived Risk," *Journal of Marketing Research* 41 (May 2004), pp. 197–205.
14. Keller, *Strategic Brand Management,* pp. 15–17.
15. Clark, Don, "Intel to Overhaul Marketing in Bid to Go Beyond PCs," *Wall Street Journal,* December 30, 2005, p. A3.
16. "The Top 100 Brands," *BusinessWeek.*
17. Aaker, David A., and Robert Jacobson, "The Value Relevance of Brand Attitude in High-Technology Markets," *Journal of Marketing Research* 38 (November 2001), pp. 485–493.
18. White, Bobbie, "Expanding into Consumer Electronics, Cisco Aims to Jazz Up Its Stodgy Image," *Wall Street Journal,* September 6, 2006, online at http://online.wsj.com/article/SB115751189903254778-search.html?KEYWORDS=cisco+consumer&COLLECTION=wsjie/6month; Borden, Jeff, "Cisco Humanizes Technology and Connects the World," *Marketing News*, September 1, 2008, pp. 14–18.
19. Atal, Maha, and Conrad Wilson, "Feeling Trashed on the Web?" *BusinessWeek,* August 30, 2007, online at www.businessweek.com/innovate/content/aug2007/id20070830589214.htm?chan=innovation_special+report+-+inside+innovation_inside+innovation.
20. Atal and Wilson, "Feeling Trashed on the Web?"
21. Winkler, Agnieszka, "The Six Myths of Branding," *Brandweek,* September 20, 1999, p. 28.
22. Keller, *Strategic Brand Management.*
23. Collier, Robert, "Behind Toyota's Hybrid Revolution," *San Francisco Chronicle,* April 24, 2006, online at www.sfgate.com/cgi-bin/article.cgi?f=/c/a/2006/04/24/MNG3JIE6DK1.DTL&hw=toyota&sn=001&sc=1000.
24. Winkler, "The Six Myths of Branding."
25. Tickle, Keller, and Richey, "Branding in High-Technology Markets."
26. Anand, Bharat N., and Ron Shachar, "Brands as Beacons: A New Source of Loyalty to Multi-Product Firms," *Journal of Marketing Research* 41 (May 2004), pp. 135–150.
27. Tickle, Keller, and Richey, "Branding in High-Technology Markets."
28. Keller, *Strategic Brand Management.*
29. Arnott, Nancy, "Inside Intel's Marketing Coup," *Sales and Marketing Management,* February 1994, pp. 78–81.
30. Tickle, Keller, and Richey, "Branding in High-Technology Markets."
31. Morris, "The Brand's the Thing."
32. Clark, Don, "Intel Signs Partners for 'Viiv' Brand," *Wall Street Journal,* November 30, 2005, online at http://online.wsj.com/article/SB113331518180209754-search.html?KEYWORDS=viiv&COLLECTION=wsjie/6month.
33. Kirkpatrick, David, "Why Compaq Is Mad at Intel," *Fortune,* October 31, 1994, pp. 171–176.
34. Keller, *Strategic Brand Management.*
35. Spencer, Jack, "Lenovo to Drop the IBM Logo; Profit Surge Foretells Growth," *Wall Street Journal,* November 2, 2007, p. B5.
36. Except as noted, this section is drawn from Eliashberg, J., and T. Robertson, "New Product Preannouncing Behavior: A Market Signaling Study," *Journal of Marketing Research* 25 (August 1988), pp. 282–292; Lilly, Bryan, and Rockney Walters, "Toward a Model of New Product Preannouncement Timing," *Journal of Product Innovation Management* 14 (January 1997), pp. 4–20; and Calantone, Roger, and Kim Schatzel, "Strategic Foretelling: Communication-Based Antecedents of a Firm's Propensity to Preannounce," *Journal of Marketing* 64 (January 2000), pp. 17–30.
37. "Windows Vista," Wikipedia, online at http://en.wikipedia.org/wiki/Windows_Vista.
38. "PlayStation 3 Delay Raises Fears for Sony," *International Hearld Tribune,* September 12, 2006, online at www.iht.com/articles/2006/09/07/business/sony.php.
39. "Vaporware," Wikipedia, online at http://en.wikipedia.org/wiki/Vaporware.
40. Acme Vaporware, online at www.acmevaporware.com.
41. Calore, Michael, "Vaporware 2007: Long Live the King," *Wired,* December 20, 2007, online at www.wired.com/gadgets/gadgetreviews/multimedia/2007/12/YE_Vaporware?slide=1&slideView=9.
42. Schatzel, Kim, and Roger Calantone, "Creating Market Anticipation: An Exploratory Examination of the Effect of Preannouncement Behavior on a New Product's Launch," *Journal of the Academy of Marketing Science* 34 (Summer 2006), pp. 357–366.
43. Calantone and Schatzel, "Strategic Foretelling."
44. Eliashberg and Robertson, "New Product Preannouncing Behavior."
45. Sorescu, Alina, Venkatesh Shankar, and Tarun Kushwaha, "New Product Preannouncements and Shareholder Value: Don't Make Promises You Can't Keep," *Journal of Marketing Research* 44 (August 2007), pp. 468–489.
46. Additional detail about incumbent competitors' reactions to new product pre-announcements can be found in Robertson, Thomas, Jehoshua Eliashberg, and Talio Rymon, "New Product Announcement Signals and Incumbent Reactions," *Journal of Marketing* 59 (July 1995), pp. 1–15.
47. Calantone and Schatzel, "Strategic Foretelling."
48. Eliashberg and Robertson, "New Product Preannouncing Behavior."
49. Wu, Yuhong, Sridhar Balasubramanian, and Vijay Mahajan, "When Is a Preannounced New Product Likely to Be Delayed?" *Journal of Marketing* 68 (April 2004), pp. 101–113.

50. Yoder, Stephen Kreider, "Computer Makers Defend 'Vaporware,'" *Wall Street Journal,* February 16, 1995, p. B1.

51. Ibid.

52. Lilly and Walters, "Toward a Model of New Product Preannouncement Timing"; and Calantone and Schatzel, "Strategic Foretelling."

53. Eliashberg and Robertson, "New Product Preannouncing Behavior."

54. Sorescu, Shankar, and Kushwaha, "New Product Preannouncements and Shareholder Value."

55. Eliashberg and Robertson, "New Product Preannouncing Behavior."

56. Sorescu, Shankar, and Kushwaha, "New Product Preannouncements and Shareholder Value."

13

Strategic Considerations for the Triple Bottom Line in High-Tech Companies

---

Fighting a Global Pandemic: The Lilly MDR-TB Partnership

Tuberculosis (TB), is a deadly and highly infectious disease of the lungs, second only to HIV/AIDS (human immunodeficiency virus/acquired immunodeficiency syndrome) as a leading killer of adults among infectious diseases.[1] A chronic, lingering, wasting disease that can be transmitted with tiny droplets from a cough or a sneeze, TB infects 2 billion people worldwide—nearly a third of the global population. Of these, 1 in 10 will develop active TB. TB kills 2 million people every year—5,000 per day—mostly adults between 15 and 54 years of age.[2] Given increased air travel, migration, and urban crowding, control efforts are difficult. Indeed, once TB enters a country, it is difficult to get rid of, says Dr. Pieter Van Maaren, a World Health Organization regional adviser for TB.[3]

Moreover, TB is the leading cause of death among people who are HIV-positive.[4] In fact, people with both HIV and TB are much more likely to develop active TB. In Africa, HIV is the single most important factor contributing to the increased incidence of TB since 1990.

If TB is detected early and fully treated, people become noninfectious and eventually cured. However, of the 9 million new TB cases each year, 400,000 are *multidrug-resistant TB (MDR-TB)*, meaning that the disease does not respond to standard TB drugs. The reasons for MDR-TB are interrupted, inconsistent, or incomplete treatment of standard TB drugs; patients do not take all their medicines regularly for the required period because they start to feel better, because doctors and health workers prescribe the wrong treatment regimes, or because the drug supply is unreliable—problems that are most prevalent in the developing world, where trained health care staff and drug supplies are limited.

Patients with MDR-TB are 4 times more likely to die. The average MDR-TB patient infects up to 20 other people in his or her lifetime. Rates of MDR-TB are particularly high in some countries (China, India, Russia, and South Africa). Given its combination of drug resistance and highly infectious nature, MDR-TB threatens TB control efforts worldwide.

The treatment for MDR-TB is long and complex, requiring extensive chemotherapy (up to 2 years) with second-line anti-TB drugs, which are more costly than first-line drugs. Indeed, the drugs to treat MDR-TB are about 100 times more expensive than those for normal TB. These second-line drugs produce adverse drug reactions that are more severe, though manageable. These problems often result in poor patient compliance, leading to the development of further drug resistance, or *extensively drug-resistant TB (XDR-TB)*. This particularly deadly new strain of TB killed 52 of 53 patients within 25 days in South Africa in 2006.[5]

These developments pose a serious threat to TB control globally, and the World Health Organization is working to address this threat through its Stop TB strategy launched in 2006, and the

(continued)

Global Plan to Stop TB. A key part of the strategy is a "directly observed treatment" program (DOT) where patients *must* take their drugs under supervision. In addition, the strategy emphasizes training of health care workers, appropriate use of diagnostic tests, and the availability of a reliable supply of affordable second-line drugs for MDR-TB. The WHO Green Light Committee approves programs for second-line anti-TB drugs at reduced prices.

The Lilly MDR-TB Partnership

In October 1997, two young physicians at Harvard, Dr. Paul Farmer and Dr. Jim Yong Kim, desperately needed batches of two drugs developed nearly 50 years earlier by Eli Lilly and Company—capreomycin and cycloserine—to fight a TB outbreak in Peru. (Tracy Kidder's 2004 book *Mountains beyond Mountains* documents Paul Farmer's pioneering efforts in battling the TB pandemic.) Their mentor, Howard Hiatt, the former dean of the Harvard School of Public Health, called a friend, Dr. Gail Cassell, Lilly's vice president for infectious disease research and clinical development. Dr. Cassell not only ensured that the two doctors received free drugs for their program, but also invited them to meet Lilly's CEO, Sidney Taurel.

So began the program to fight the rapidly growing threat of multidrug-resistant TB—a pioneering multipronged, philanthropic, global initiative that Lilly formalized in 2003. The Lilly MDR-TB Partnership is an alliance of 14 public and private organizations including businesses, humanitarian organizations, academic institutions, and professional health care associations. This partnership was funded with $70 million dollars from Eli Lilly and includes the diagnosis, treatment, and monitoring of patients; training of doctors and nurses; and community support, patient advocacy, and antistigma efforts.[6] That effort was expanded in 2007 with the commitment of an additional $50 million for training health care workers in the developing world,[7] and again with another $15 million for additional research[8]—a total of $135 million. This partnership has been commended by Nobel Peace Laureate Archbishop Desmond Tutu, as well as former U.S. Ambassador to the United Nations Richard Holbrooke.

One of Lilly's main goals is to increase the supply of quality, affordable, second-line drugs to fight MDR-TB (cycloserine and capreomycin) for WHO. These are older-generation TB drugs that Lilly had planned to stop manufacturing. To achieve this goal, not only does Lilly heavily discount its MDR-TB drugs for WHO, but it has also engaged in a technology transfer agreement to share its knowledge with four pharmaceutical companies in the four countries hardest hit by MDR-TB to increase the availability of low-priced local drug supplies: Aspen Pharmacare in South Africa, Hisun Pharmaceutical in China, Shasun Chemicals and Drugs in India, and SIA International in Russia. Not only did Lilly turn over its formulation for the drugs, but it also spent $37 million to help these companies buy the necessary equipment for their factories, trained their workers, and then taught health care workers how to administer treatment. Lilly also donated $33 million in drugs.[9]

Certainly, such programs face their challenges, and Lilly's has not been immune from that. The four partners needed significant assistance with quality control and safety, which Lilly also provided. These technology transfer arrangements are facilitated by the Chao Center for Industrial Pharmacy at Purdue University in Indiana, which provides the necessary manufacturing and quality assurance expertise in addition to providing contract manufacturing of ceromycin.

Another key partner in the fight against MDR-TB is the Harvard Medical School and its association with Partners in Health (PIH). With Lilly's support, PIH has established a Center of Excellence in Tomsk, Russia, to train doctors in treating MDR-TB. Lilly also has collaborated with the International Council of Nurses (ICN) to develop international TB and MDR-TB training guidelines and programs that are fine-tuned for nurses, an overlooked constituency in many countries. Lilly has also provided support to the International Federation of Red Cross and Red Crescent Societies (IFRC) to develop MDR-TB awareness and outreach programs, and has worked with the World Medical Association (WMA) to craft an MDR-TB Internet training course for private physicians. Lilly is also working with the International Hospital Federation

(IHF) to control the spread of TB in hospitals. Lilly collaborates with the World Economic Forum (WEF) to engage corporations in the prevention and treatment of TB in the workplace; for example, the WEF recently spearheaded the Business Alliance to Stop TB in India.[10]

While Eli Lilly's efforts to fight MDR-TB are notable on their own merits, they stand out because pharmaceutical companies frequently have been criticized over the price and availability of cutting-edge drugs. The only other pharmaceutical gift similar to Lilly's is Merck & Company's fight against river blindness (see Appendix 13.A at the end of this chapter). Merck has donated medicine worth about $2.7 billion to that program, started in the late 1980s. Lilly is using a similar approach for its diabetes treatment programs.

Ongoing innovations to diagnose and treat MDR-TB are also necessary. Toward this end, a rapid saliva test with results in two days' time can be used to determine whether a patient with TB can be treated with first-line antibiotics. Previous tests required the saliva to be incubated for up to 60 days in order to determine whether or not the TB strain was MDR.[11] In addition, the National Institute of Allergy and Infectious Diseases (NIAID), part of the National Institutes of Health, has a biomedical research program to combat "the ominous and accelerating rate" of drug-resistant TB.[12]

Other Eli Lilly Corporate Citizenship Initiatives

Lilly follows a triple-bottom line approach to its corporate strategy, realizing that in addition to economic (financial) performance, social and environmental performance matter as well. Its approach to global health challenges is shaped by its desire to bring breakthrough medicines to patients and to improve access to health care. Lilly believes that corporate citizenship includes:[13]

- Researching, developing, and marketing its products in an ethical and transparent manner
- Manufacturing its products to protect patients, the environment, and the health and safety of its workers and the communities in which its facilities are located
- Using resources and influence to strengthen the local and global communities in which it operates
- Looking beyond its traditional operations to understand and address global health needs
- Building engaged philanthropy through gifts of cash, products, and in-kind donations to charitable causes
- Facilitating employee volunteerism
- Creating a diverse workforce, including gender and ethnic diversity
- Incorporating "green chemistry" principles to minimize health, safety, and environmental impacts from the use of hazardous material purchases
- Investing in more efficient energy usage through 16 energy conservation projects at the nine sites that account for 84% of Lilly's global energy use
- Investing in employee health and safety, including a colon-screening program and an ergonomic workplace assessment that have cut and reduced employee injuries
- Protecting the privacy of employees' genetic information to ensure it cannot be used to make employment and benefits eligibility decisions
- Developing an anticounterfeiting strategy to secure the integrity of its supply chain
- Partnering with governments to strengthen and enforce piracy laws

Lilly was the first pharmaceutical company to publicly disclose online the results of all its clinical trials, as well as all its grants to U.S. nongovernmental organizations, research institutions, individual researchers, and others. It has worked with the U.S. Food and Drug Administration to improve direct-to-consumer advertising.

Consistent with a triple-bottom line philosophy, Lilly prepares its Corporate Citizenship Report following the Global Reporting Initiative (GRI) Guidelines (www.globalreporting.org/ReportingFramework/G3Guidelines/).

Marketing in all companies must be attuned to emerging trends that affect business strategy and profitability. Indeed, astute business planning proactively scans the external environment for important trends that will pose future opportunities and/or threats. Trends are most often found in six key domains in both domestic and international markets:

1. *Economic:* Changes in the macroeconomic environment of a market or region, typically reflected in inflation rates, unemployment rates, savings/debt rates, exchange rates, and other factors affected by federal/government economic policy decisions.
2. *Competitive:* Changes in competition at both the selective demand level (brand competition; e.g., between the Apple iPhone and the Samsung Instinct) and the primary demand level (product form competition; e.g., between PDAs and cell phones).
3. *Regulatory/political:* Changes in the laws and regulations that affect companies, society, markets, and regions.
4. *Sociocultural/sociodemographic:* Changes in widely held values and beliefs by members of society and demographic trends, in terms of the composition of various markets/regions.
5. *Technological:* Changes in technology that potentially obsolete existing ways of doing business or present opportunities for new ways of doing business. Note that new technologies can also be a competitive trend (product form competition).
6. *Natural resources:* Changes in the availability of natural resources that affect a company's business strategy; often seen in changes in supply/pricing of the raw materials used as supply inputs, such as iron, copper, steel, and petroleum. In addition, concerns about global climate change also affect companies' strategies.

For example, with respect to this last domain, 60% of 2,192 executives from around the world regard global climate change as a significant issue affecting their business strategy.[14] Table 13.1 provides an overview of other predominant trends affecting high-technology businesses in 2008.

The focus of this chapter is the constellation of several of the most profound trends affecting business today: society's expectations for business to be both socially and environmentally responsible, coupled with pressing global problems (such as global population growth and global warming, to name just two) that are potentially catastrophic. As a result, companies today are using expanded criteria to develop their business strategies and to measure their success—economic, environmental, and social criteria (or, alternatively, "profits, planet, and people"[15]). Often referred to as the **triple bottom line**,[16] companies today must concern themselves not only with their economic profits, but also with social and environmental profits. This includes corporate social responsibility initiatives and corporate philanthropy, as well as efforts toward environmental sustainability.

The notion that business has an obligation to society is not new. Indeed, some of the earliest writing in this area can be traced back more than half a century, to management guru Peter Drucker:[17]

> Economic purpose does not mean that the corporation should be free from social obligations. On the contrary, it should be so organized as to fulfill, automatically, its social obligations in the very act of seeking its own self-interest. An individual society based on the corporation can function only if the corporation contributes to social stability and to the achievement of social aims independent of the goodwill or the social consciousness of individual corporate managements.[18]

> Economic change can be made into the most powerful engine for human betterment and social justice.[19]

Reasons for today's increasing prevalence of a triple bottom line approach to business are multidimensional, but a key driver is responding to changes in society's values and beliefs. For example, a growing number of students—41% in 2008, up from 37% in 2007—are looking to spend their money on socially responsible brands.[20] A 2007 survey conducted by Edelman, a global public relations firm, found that "85% of consumers around the world are willing to change the brands

TABLE 13.1 Predominant Trends Affecting High-Technology Businesses, 2008

- **Economic**

  Increasing inflation rates in the United States, due primarily to rising energy, fuel, and food costs

  Lowering value of the U.S. dollar relative to foreign currencies

- **Competitive**

  Digitization; convergence across industries

  Increased global competition with disruptive business models

- **Regulatory/political**

  Laws regarding recycling, pollution, and carbon emissions, including:

  - U.S. CAFE (corporate average fuel efficiency) targets that automotive manufacturers' fleets must meet.

  - Regulations in the state of California requiring that from 2008 to 2016, greenhouse gas emissions from new cars be reduced by 30%; California passed legislation to reduce total emissions to 1990 levels by 2020.

  - Twenty states require utilities to obtain a percentage of the power they sell from renewable sources, and 208 U.S. cities have adopted programs to reduce emissions.

- **Sociocultural/sociodemographic**

  Changes in the composition of society, such as the aging (or "graying") of the population (Italy and Japan are the "oldest" countries in the world by average age and composition)

  Developments at the base of the pyramid regarding factors such as distribution of wealth and access to education

  Changes in people's expectations about the role of business in contributing to societal welfare

  People's expectations that companies should act in an environmentally responsible fashion

- **Technological**

  Digitization, as it continues to affect many industries and their business models

  New (alternative) energy technologies, which are creating opportunities and threats

- **Natural resources**

  Continued rising costs for fossil fuels (driven primarily by supply/demand imbalances)

  Depletion of natural resources

---

they buy, or their consumption habits, to make tomorrow's world a better place. Over half (55%) would help a brand 'promote' a product if there was a good cause behind it."[21] In addition, companies can find recruiting advantages in being socially responsible, and employee morale and retention is higher in companies that augment their profit-driven agenda with a social mission.

These trends have been gathering steam for about 15 years, but the current constellation of global factors gives them added urgency, visibility, and priority for businesses. Indeed, regardless of one's views on the issues, the culmination of these trends has been an expanded role of business in addressing social and environmental problems. Eighty-four percent of 2,687 executives from around the world agree that "making broader contributions to the public good should accompany generating high returns to investors; only 16% believe that high returns to investors should be a company's sole focus."[22]

A triple bottom line approach is increasingly important for all businesses, but high-technology businesses in particular have taken a leadership role in this area. Not only is access to technology a key driver for economic prosperity, but technological innovations are also a key part of the solution for many problems facing society today. Moreover, because of the need to innovate new technologies to solve pressing global problems, high-tech companies have the opportunity to more directly link their business strategies to the triple bottom line. In turn, this may provide them greater opportunity for differentiation from socially responsible actions than companies in more traditional environments.[23] For example, three companies that have been cited extensively for their leadership in the social and environmental arenas are:

- *Salesforce.com:* Marc Benioff, the founder of Salesforce.com, integrated philanthropy into the company's corporate culture from its inception following a 1-1-1 model (equity, profits, time): 1% of the company's shares are put into a foundation for social causes, 1% of the company's profits go to community needs, and 1% of employee working hours are devoted to community service. For example, the two-day new-employee orientation includes a half day devoted to volunteerism to inculcate this value as a priority in new employees, to continue this philosophy as part of the company's DNA. Benioff documented this approach to business in his book, *Compassionate Capitalism: How Corporations Can Make Doing Good an Integral Part of Doing Well.*[24]

- *Intel:* Intel, named one of the "100 Best Corporate Citizens" by *Business Ethics* magazine and one of the "World's Most Socially Responsible Companies" by *Global Finance,* has a focus on environmental stewardship, safety, diversity, community building, and education. Indeed, its commitment to education (in particular, strengthening math, computer science, and engineering programs) is its top priority. As Chairman Craig Barrett says, "We don't do golf tournaments, we don't sponsor football games. We think that doing the right thing is helping to prepare young people—wherever they are—to become contributing members of society."[25] Its philosophy of corporate social responsibility pervades the internal culture, and Intel has a consistent ranking on *Fortune*'s list of 100 Best Companies to Work For. Finally, Intel has also been a leader in environmental responsibility. As early as 2004, it had a goal to reduce emissions of perfluorocarbons (PFC), chemicals used in chip making that are potent greenhouse gases with toxic by-products. They continue to innovate in this area, as highlighted later in the chapter.[26]

- *General Electric:* In 2003, GE used the Greenhouse Gas Protocol to construct an emission inventory, it quantified its risks of regulatory restrictions based on its emissions, and it launched the ambitious "ecomagination" initiative to develop clean technologies for the transportation, energy, water, and consumer products sectors. Its goals were to increase its investments in clean technologies to $1.5 billion by 2010, and to generate $20 billion in revenue from products that offer customers measurable environmental performance advantages. Its R&D program has paid off, with revenues of $10.1 billion in 2005. GE's chairman and CEO Jeff Immelt says, "Our customers have made it clear that providing solutions to environmental challenges like climate change is essential to society's well-being and a clear growth opportunity for GE. Companies with the technology and vision to provide products and services that address climate and other pressing issues will enjoy a competitive advantage."[27] Moreover, GE Healthcare has worked to create modifications of its medical devices for use in developing areas of the world, like rural areas in India. Engineers have adapted components and functioning so that the devices are portable, durable, and low-cost.[28] In other words, GE will do not just well, but better, by doing good.[29]

Many other high-tech companies—too numerous to single out individually—ought to be mentioned here as leaders of similar initiatives. And these are only the large, multinational corporations. Many emerging technology businesses have built their entire business models around serving the needs of impoverished consumers as well as larger societal needs. These social and environmental entrepreneurs are also addressed later in this chapter.

In this spirit, this chapter addresses how high-tech companies are adopting triple bottom line strategies in their businesses, including corporate social responsibility and environmental initiatives. As shown in Figure 13.1, these include, for example, business strategies designed to serve customers in base-of-the-pyramid markets as well as strategies to ensure wide access of technology to all segments of society. As with the prior 12 chapters in this book, we highlight best practices and pitfalls that high-tech companies encounter in implementing triple bottom line strategies.

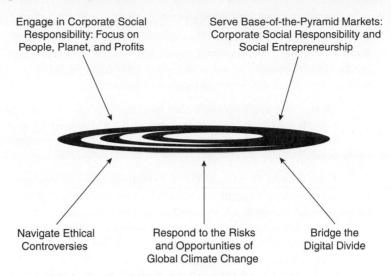

FIGURE 13.1 Triple Bottom Line Considerations in High-Tech Companies

CORPORATE SOCIAL RESPONSIBILITY

Corporate social responsibility (CSR), also called *corporate responsibility,* or *corporate citizenship,* refers to companies' explicit consideration of the impact of their activities on all stakeholders affected by their business decisions. Consistent with the concept of the triple bottom line, corporate social responsibility broadens a company's focus beyond shareholders to include anyone who is influenced or affected—either directly or indirectly—by the actions of the firm. These stakeholders include customers, suppliers, employees, shareholders, communities—indeed, society as a whole. CSR efforts focus on social problems (e.g., health care and education), as well as environmental issues (e.g., concerns about the use of toxic chemicals in production and the carbon footprint of a supply chain).

The United Nation's Global Compact, the world's largest global corporate citizenship initiative, focused on building legitimacy for the role of business in social issues, consists of 10 universally accepted principles in the areas of human rights, labor, the environment, and anticorruption. For example, with respect to the environment, businesses should undertake initiatives to promote greater environmental responsibility and encourage the development and diffusion of environmentally friendly technologies.[30] The UN Global Compact provides a useful framework for businesses that are committed to aligning their operations and strategies with a CSR philosophy. Cisco and Toshiba are two high-tech companies that have signed on to these principles. The UN's goals are for participating companies to engage in partnership projects and policy dialogues, catalyzing actions in support of broader UN goals, such as the Millennium Development Goals (MDGs).

This section on Corporate Social Responsibility addresses why CSR continues to be somewhat controversial. It also covers the various domains of and models for CSR, as well as how companies monitor the results from their CSR initiatives. The section concludes with tips for best-practices CSR for high-tech companies.

Debates over and Criticisms of CSR

Despite its increasing prevalence, the practice of CSR is subject to much debate and criticism. Two key issues include:

- Should firms engage in CSR?
- Is it truly CSR if companies benefit? Does it matter? If so, why?

SHOULD FIRMS ENGAGE IN CSR? This debate is often couched in terms of the role of business in society. On the one hand are the critics of CSR who argue that the primary role of business is to serve a company's stockholders; any distraction from the making of profits is inconsistent with the business's obligations. These critics often invoke the Nobel-winning economist Milton Friedman, who spoke against any social obligation of business:

> There is one and only one social responsibility of business—to use its resources and engage in activities designed to increase its profits so long as it stays within the rules of the game, which is to say, engages in open and free competition without deception or fraud.[31]

According to Friedman, the only responsibility of business is to maximize wealth for the firm's stockholders. Typically espoused by business conservatives, the logic is that business's concern for social causes leads to spending and expenses that do not contribute to profitability. These critics argue that when companies donate company funds to charities, for example, they are giving away shareholders' money. If the funds are not going directly into the business, the shareholders should decide how to spend them.

Critics also like to point out that many CSR initiatives are nothing more than superficial window-dressing, designed to burnish a company's reputation without doing any real good for society. Finally, critics also say that CSR is merely an attempt to preempt the role of governments as a watchdog over powerful multinational corporations. If companies can say they are doing things that benefit society and protect the environment, then maybe they can stave off costly regulations that would actually require more stringent behaviors.

On the other side of the argument are people who support the role of business in serving social needs. They point out that companies benefit in multiple ways by operating with a perspective broader and longer than their own immediate, short-term profits. In other words, business economics and social responsibility are not mutually exclusive—but rather, businesses that are socially responsible are actually more successful. The logic is that socially responsible companies will reduce their risks, energize their employees, and build a stronger emotional connection with their customers and investors by taking on social causes and issues that matter. Social responsibility advocates strongly argue that, if reputation and brand matter, companies with a mission that goes beyond making money will do better; the alternative—a focus on short-term profits that come from polluting the environment (or avoiding investments in sustainability) or mistreating workers—can be profitable in the near term but may prove costly over the long term.

Moreover, business and society need each other if both are to thrive. Healthy societies expand demand for business as more human needs are met and aspirations grow. Finally, companies that ignore their social and environmental roles run the risk of protests and boycotts.[32] For these reasons, many believe that it is directly in a business's interests to attack social problems.

Of course, one way to address this debate is to assess the impact of CSR on company performance. If companies that adopt a proactive CSR stance exhibit better performance than those that do not, the point is potentially moot. We turn to this topic later in this chapter.

IS IT TRULY CSR IF COMPANIES BENEFIT? Some argue that when businesses address social causes that dovetail with their business strategy—say, when GE modifies an expensive medical device to be affordable for rural doctors in India—it is not corporate social responsibility, it is simply good business: "If the company must insert itself into certain social causes due to business opportunities and threats, so be it, but to claim it is somehow more than a legitimate business expense is dishonest."[33] In other words, "strategic CSR is hardly distinguishable from good business."[34] These critics state that, rather than cloaking their motives under the sheath of CSR, businesses should just call them the strategic business decisions that they are.

At the extreme, these critics ask whether companies' CSR initiatives are merely a blissful façade in a pathological search for profits.[35] For example, when a company "greens" its business and lowers its energy and operational costs in doing so, can this be considered CSR? What about companies like GE or Toyota that offer green products for sale? What about companies that profit from serving impoverished customers in base-of-the-pyramid markets? Similarly, companies may engage in seemingly hypocritical CSR activities. Is BP acting in a socially responsible manner by striving to address global climate change through the development of alternative energy sources, or is it being hypocritical because it continues to produce fossil fuels?

Other criticisms of CSR also exist, including unintended results from CSR, the potential to use corporate resources inefficiently, the potential vulnerability of social causes that become dependent on the vagaries of corporate donations, and the potential for arbitrary or capricious selection of which social causes to support[36]—criticisms which really can be levied against any source of support, corporate or otherwise, for social and non-profit causes.

Despite these criticisms, this chapter takes the position that a triple bottom line philosophy is consistent with best-practices business strategy. As stated by Rochelle Lazarus, the CEO of Ogilvy and Mather Worldwide, one of the world's largest advertising agencies (with respect to corporate philanthropy in particular, but her comment applies equally well to corporate social responsibility generally):

> The debate over whether corporate philanthropy is right for the bottom line is over. The market has spoken and said it cares whether companies are good citizens; these issues do in fact matter to society. Engaging your company in corporate philanthropy is now part of being a good CEO.[37]

Moreover, to some extent, the debate over "Is it CSR if a company benefits?" is spurious. One prominent author, John Elkington, uses a metaphor drawn from Polish poet Stanislaw Lec, who asked, "Is it progress if a cannibal uses a fork?" Elkington applied the same question to 21st century capitalism in his book, *Cannibals with Forks*, pondering whether holding corporations accountable to a triple bottom line of economic prosperity, environmental quality, and social justice constitutes progress. He argues that there are seven dimensions of—or revolutions leading to—a sustainable future, and that it most assuredly is progress when companies work toward sustainable development.[38]

Identifying win-win solutions does not make the causes or initiatives a company pursues any less noble; to accuse business of being disingenuous when its socially responsible actions also benefit the business itself overlooks the fact that if a business benefits, the action is more likely to last, which is good all the way around. Companies are frequently "damned if they do, and damned if they don't" with respect to their CSR initiatives. So, in the spirit of proactive business, we explore what best-practices CSR strategies consist of, which in today's society means being responsive to social and environmental trends: Get with it or get out of the way—and potentially be left behind. As GE's Vice Chairman David Calhoun said in the "early days" of environmental thinking—2005—with respect to that specific domain of CSR:

> Environmental responsibility is something that sensible customers will eventually want. Therefore, let's stop putting our heads in the sand, dodging environmental interests, and go from defense to offense.[39]

Therefore, the CSR debate might be more appropriately reframed in terms of how companies select the CSR initiatives they want to support/pursue. How do they measure the impact of their initiatives? How can organizations prosper from addressing social and environmental challenges? The topics that follow address these questions.

Desired Outcomes from CSR

Businesses seek desired outcomes from CSR in two broad areas: business goals and social/environmental goals.

Business goals include the following (in order of frequency of citation by business executives):[40]

- Enhance corporate reputation or brand
- Build employee/leadership capabilities
- Improve employee recruitment and retention
- Differentiate from competitors
- Manage current or future risk
- Build knowledge about potential new markets
- Inform areas of innovation
- Meet industry norms

For example, whether through the increased sale of premium-priced products that are environmentally friendly, or through new markets—say, in developing countries—companies can drive revenue growth with their CSR initiatives.[41] As discussed subsequently, companies that pursue base-of-the-pyramid markets with their CSR initiatives can also stimulate economic development in those markets. For example, assistance that allows people to enter the labor force can raise living standards and allow people to earn income for consumer spending. (Of course, CSR initiatives are criticized for this seemingly crass market goal, as acknowledged earlier.)

Similarly, companies can use knowledge gained from CSR projects to build new competencies while simultaneously improving social or environmental conditions. Social responsibility can support a company's pursuit of breakthrough technologies,[42] which are needed to address social and environmental challenges.

Social goals generally include the desire to improve quality of life and to build capacity in disadvantaged communities.[43] Capacity-building initiatives might include, for example, development of infrastructure, training of local workers to take ownership of initiatives, and investments that are self-sustaining after the company's CSR initiative comes to a close. Companies can address causes that benefit the community in which they are located, disadvantaged groups in society, or other causes that a company or its managers believe in, such as the arts. In addition, given expanding societal expectations, companies also address more expansive goals, such as mitigating the negative effects arising from offshoring, obesity, debt, environmental degradation, governance problems in resource-rich/income-poor nations, and so forth.

CSR Domains: People and Planet

Following the triple bottom line approach, CSR initiatives can be broadly categorized into two domains: "people" and "planet" (recall that profit was the third domain).

PEOPLE "People," broadly construed, can include the employees of the company, its suppliers, its customers, people in the local communities in which it operates, or more generally, society as a whole. To avoid being viewed as hypocritical (were it to treat people outside the company better than those inside the company), before a company turns its CSR focus outside of its own boundaries, it must practice fair, ethical treatment of its own employees, including fairness in hiring and personnel practices. A triple bottom line business does not knowingly use child labor, pays fair salaries to its workers, maintains a safe work environment and tolerable working hours, and does not exploit a community or its labor force. It may also take steps to improve the quality of life for employees and their families.

In addition, companies adopting a CSR philosophy engage in responsible supplier relations, and they ensure that their suppliers follow fair and ethical treatment in their own businesses. Moreover, the "upstreaming" of a portion of profit from the marketing of finished goods back to the original producer of raw materials—a farmer in fair trade agricultural practice—is not an unusual feature. Because a triple bottom line philosophy is inconsistent with exploiting or endangering any stakeholder group,[44] companies should ensure (as discussed in the "Planet" section that follows) that they do not degrade the physical environment in the community in which they operate.

Language surrounding corporate social responsibility has grown more expansive as the trend has gained increased prominence, and as companies have developed different versions and flavors of how they leverage it in their business models. So, in addition to the ethical and responsible treatment of the business stakeholders noted above, a company may address larger social issues, such as Eli Lilly's TB initiative. Indeed, CSR initiatives that are the most visible address, in some large-scale way, large societal problems like hunger, health care, poverty, or education. Respect for indigenous groups, fair competition (antibribery/anticorruption), and accountability (transparency in performance reporting) also fall under the rubric of CSR.

One specific type of CSR is **corporate philanthropy** (or *benevolence*), in which a company donates resources in some form (cash, in-kind donations of products or services, employee time/volunteerism, or other company resources such as the use of a corporate jet to transport sick children to hospitals and summer camps) to support social causes (health, education, poverty, hunger, recycling) or nonprofit organizations. One early form of this is *cause-related marketing,* in which a company donates a portion of its sales during a specified time period to a social cause. Cause-related marketing campaigns were often viewed as promotional tools to generate customer interest and loyalty by linking a company's sales to causes of interest to its customers.

When companies engage in corporate philanthropy specifically, the following issues are addressed (in order of frequency of companies addressing):[45]

- Education (75%)
- Community (58%)
- Economic development (52%)
- Environment (52%)
- Civic/public affairs (51%)
- Heath, social services (48%)
- Culture/arts (47%)

Because committed and engaged employees are a key driver of business prosperity, they are the most influential of all stakeholder groups on the selection of causes to which their companies give. Local communities are second, consumers and shareholders are tied at third, and personal interests of the CEO and/or board members is fifth.[46] As one example of how a company's engaged citizenship philosophy creates loyalty among employees, Anne Mulcahy of Xerox stayed with the company, in part, because of its culture of citizenship that included fair treatment of people, customers, suppliers, and communities. Similarly, Medtronics (featured in the opening vignette in Chapter 2) sponsors annual parties at which employees meet patients whose lives have been improved by their products; this can be a defining moment in their career, when they come face-to-face with a patient whose story deeply touches them.[47] A 2006 survey of 2,100 MBA students by Net Impact revealed that 59% of respondents planned to seek socially responsibly work upon graduation and 79% said at some point in their careers.

Notably, companies cite neither "alignment with business needs" nor "ability to leverage companies' existing capabilities" as the basis for selecting the causes they donate to. This lack of tie to business issues can be both beneficial and detrimental. Only 54% of respondents who rated their philanthropy programs as less effective align their programs with trends relevant to their business, while 71% who rated their programs as very or extremely effective do align philanthropy with business trends and needs.[48]

PLANET   This domain of the triple bottom line refers to engaging in environmentally responsible, sustainable business practices. The definition of **sustainability**, first developed in 1987 by the UN's Brundtland Commission (led by the former Norwegian Prime Minister Gro Harlem Brundtland), refers to business practices that meet the needs of the current generation without compromising the ability of future generations to meet their needs. Thus, to engage in sustainable business practices, a company must take into account the rights of future generations to raw

TECHNOLOGY SOLUTIONS FOR GLOBAL PROBLEMS
The Solar Revolution: Nanosolar Is the Answer!
By Jamie Hoffman

Photo reprinted with permission of Nanosolar, San Jose, California.

For decades, the viability of solar power as an alternative energy source remained elusive due to its long and costly manufacturing process, high shelf price, and low return on investment. Thanks to technological advances made in the nanotechnology sector, solar power is about to become much more attractive, economical, and attainable for the average energy consumer. *Nanotechnology* refers to engineering on a microscopic scale, between 1 and 100 nanometers. (A nanometer is equal to one billionth of a meter.) The recent surge in nanotechnology is attributed to the fact that, thanks to new powerful microscopes, scientists can actually see the particles they are manipulating, down to the atomic level.

By coupling nanotechnology with solar power, Nanosolar is on the cusp of revolutionizing the way the world acquires electric power. Located in Palo Alto, California, Nanosolar used nanotechnology to create Power Sheet™—a cheaper, lighter, and more flexible, albeit less efficient, alternative to silicon-based wafer solar cells. Nanosolar's genius is in its unique use of Copper Indium Gallium diSelenide (CIGS)—high-performing, durable solar cells made from flexible plastic semiconductors. By mixing CIGS and a nanoparticle ink, Nanosolar has been able to lock in the exact distribution ratios necessary to make CIGS work, while also allowing the mixture to be "printed" onto sheets of aluminum foil. Power Sheet consists of five layers: a clear zinc oxide semiconductor atop a semiconductor that doesn't absorb light, CIGS, an electrode, and aluminum foil. Using continual manufacturing processes similar to printing presses, Power Sheet can be produced quickly and efficiently without the costly vacuum distribution techniques required by other competitors. Nanosolar claims that Power Sheet will cost US$1 per watt of capacity, a third of the current solar market cost, and comparable to the cost of coal.

Nanosolar considers Power Sheet to be the third wave of solar technology; the first and second waves were silicon wafer cells (photovoltaics) and CIGS thin-film on glass. Power Sheet thin-film on foil allows for cost savings not only by using less material than silicon wafer cells, but also by requiring cheaper, continuous processing that is 100 times faster than traditional thin-film on glass. Thin-film on aluminum foil, as compared to thin-film on glass, utilizes at least 50% more of the materials, provides a high panel current, cell matching, and robust process yield. Where thin-film on glass took 1.7 years to achieve energy payback, thin-film on aluminum foil achieves energy payback in less than a month.

After building two massive manufacturing centers in 2007—one in San Jose, California, and the other in Germany—Nanosolar has hit the ground running and its first solar panels rolled off the line in December 2007. If Nanosolar delivers power as cheaply as promised, it may usher in an era where power generation is removed from the hands of big-industry juggernauts and put back in the hands of the people.

Sources: Nanosolar Inc., online at www.nanosolar.com; Markoff, John, "Start-Up Sells Solar Panels at Lower-Than-Usual Cost," *New York Times,* December 18, 2007, online at www.nytimes.com/2007/12/18/technology/18solar.html?_r=2&adxnnl=1&oref=slogin&ref=business&adxnnlx=1208644801-5JyWOiTbWpsc5EkeTOjNMg; Firmage Jr., Ed, "Revolutionary Solar Technology Is Set to Transform Energy Generation," *Salt Lake Tribune,* April 3, 2008, online at www.nanosolar.com/cache/slt.htm; and LaMonica, Martin, "Nanosolar 'Prints' First Flexible Solar Cells," CNET News, December 18, 2007, online at www.news.com/greentech/8301-11128_3-9835241-54.html?tag=nefd.top.

materials and vital ecosystems. Following a triple bottom line philosophy, a company avoids ecologically destructive practices, such as depletion of natural resources, and it should at the least do no harm and curtail environmental impact. In the best case, it should strive to benefit the natural environment as much as possible.[49] Moreover, it should not produce harmful or destructive products such as weapons, toxic chemicals, or batteries containing dangerous heavy metals, for example. For an example of how one company is revolutionizing the way the world acquires electric power, see this chapter's Technology Solutions for Global Problems.

Hence, companies can reduce their ecological footprints, carefully manage consumption of energy and nonrenewable resources, reduce manufacturing waste, render waste less toxic before disposing of it in a safe and legal manner, and plan for end-of-life disposal of products. Cradle-to-grave—or indeed, *cradle-to-cradle*[50]—thinking is uppermost in the thoughts of triple bottom line businesses; this means that they conduct a life cycle assessment of products to determine their true environmental cost, from the growth and harvesting of raw materials to manufacture, distribution, and eventual disposal by the end user. Proactive consideration of how to recycle or dispose of a product and its components is a key aspect of cradle-to-grave, or cradle-to-cradle, thinking. In the triple bottom line approach, an enterprise that produces and markets a product that will create a waste problem should not be given a free ride by society. It would be more equitable for the business that manufactures and sells such a product to bear part of the cost of its ultimate disposal—rather than residents near the disposal site, or governments, as is usually the case.

Arguments that it costs more to be environmentally sound are often specious when the course of the business is analyzed over a period of time. Indeed, environmentally sustainable practices can be more profitable than environmentally damaging practices.[51] Moreover, as the Sierra Club's David Brower has stated, "A company can't do business on a dead planet." Because of the timeliness and importance of this topic, companies' responses to the risks and opportunities of global climate change will be explored later in this chapter.

Importantly, although corporate social responsibility makes good business sense, it is more than just "good business." Companies that engage in standard good business practices would, for example, treat employees, suppliers, and the community well; run an operationally efficient business; and produce products that are well positioned to capitalize on key trends. Essentially, these are companies that "do the right things" by engaging in standard good business practices, but they don't take on larger social challenges. As companies become more engaged in social causes, they move beyond serving existing customers and actively seek out ways in which they can do good for others. These companies are more proactive about "doing good" and initiating programs that move beyond mere standard business processes.

Models of and Approaches to CSR

Different companies have different motivations for CSR and take different approaches to fulfilling their social responsibility. Some companies' motives may be economic in nature, relating to the financial goals of the business; in other cases, the goals may be primarily noneconomic, with financial goals of only secondary concern. Depending on the degree to which CSR is embedded in a firm's day-to-day culture, activities, and processes, and which type of goals a company seeks, three distinct profiles of CSR practices are possible,[52] as shown in Figure 13.2:

1. *Business case model:* This approach to CSR is driven primarily by the ability of CSR initiatives to create positive business results. Because serving shareholders is paramount, a strong tie to economic outcomes drives CSR investments. Companies following this approach to CSR may also be responding to other external drivers, such as the threat of regulation, activists targeting the business, or other negative events. In such businesses, CSR is viewed as relevant only when it translates unambiguously to competitive advantage. As a result, companies following this model of CSR generally look for the nexus of business opportunity and social responsibility. The

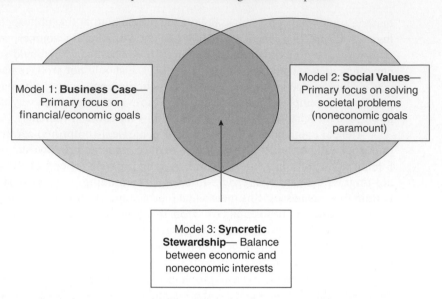

FIGURE 13.2   Models of and Approaches to Corporate Social Responsibility

idea is to align CSR with the company's capabilities/mission, and CSR becomes an extension of its business strategy in terms of "fit" with target market's preferences. Importantly, CSR activities should not result in a higher cost of doing business, which would raise the price of the company's products.

In these companies, justifications for CSR expenditures are often grounded in statistics about the size of the CSR-sensitive market. This approach to CSR should allow a company to leverage its core competencies in some social domain that will grow its market beyond a narrowly focused, socially conscious niche to a broader, more mainstream consumer market. In order for this market growth to be a high-probability event, CSR activities must clearly influence consumers' decisions—and metrics for CSR must be available. Metrics are useful both so consumers are not confused and so the company can gauge CSR's impact on the financial bottom line.

Because of its desire to achieve competitive advantage through its CSR initiatives, companies following a business case model do not want their competitors to follow their actions in this regard; such companies want to maintain CSR as a unique point of differentiation.

Employees of companies with this approach to CSR may view the firm's commitment as less than genuine; because it is not central to the organizational culture or its processes, CSR is also not a source of retention or motivation for employees. Moreover, employees aren't necessarily engaged in volunteerism as part of the company's CSR efforts. More commonly, the company's approach to employees' volunteerism is "hands-off": What employees do personally is fine, and their own acts of charity are not the company's responsibility.

Large public companies often follow this approach to CSR.

> It comes down to the business case. . . . We live in the short run . . . that is the reality. We must meet investor expectations. . . . There has to be a business case. . . . Within the world we live in, you've got to make money. As a public company, there is not an appetite for the benefits of CSR without the business case.[53]

2. *Social values model:* This approach to CSR defines a particular social issue as "the force" that drives—even consumes—the organization. CSR is the organization's lifeblood and purpose, and it is integrated into the organizational fiber in every way; visible symbols of the cause can be found everywhere in the company. As a result of the values-led approach, employees who believe in the issue are drawn to such companies. Their convictions, values, energies, and innovativeness about the social issue lead CSR initiatives in these companies to be approached with evangelical zeal—solving the problem is an imperative. In these companies, noneconomic criteria are dominant, and economic/external market factors are of secondary importance. The philosophy is: The more we prosper, the more society prospers.

Companies following this CSR model welcome competitive firms to adopt their models as well. Consistent with their evangelical approach, they are willing to share their business models, use an open source approach to their innovations, and "spread the word." Moreover, they abhor firms that use superficial support of a cause to position themselves as socially responsible.

Certainly, this CSR approach is more likely to be found in smaller, private companies. Known as **social entrepreneurship**, these small businesses are founded by people who recognize a social problem and use entrepreneurial principles to create and manage their organization to foment social change (often following a for-profit business model, but not always—see the topic titled "Social Entrepreneurship" later in this chapter). However, large companies may also follow this model. For example, using a microcredit approach to financing, the Grameen Bank founded by Muhammad Yunus in Bangladesh has revolutionized the socioeconomic status of rural women entrepreneurs.

3. *"Syncretic stewardship" model:* This third approach to CSR, with the rather unwieldy moniker—*syncretize* means "to unite and harmonize"—combines, integrates, negotiates, and balances the competing claims of varied stakeholders. Such companies recognize their broad obligations to serve multiple stakeholders, but they also recognize that doing so is a juggling act. Companies that follow this approach find their business decisions rife with paradoxes, contradictions, and interdependencies; they tend to be large, publicly held firms with extensive international operations.

These companies embrace economic objectives, but view the market as more nuanced and complex than firms following the business case model. These companies also use noneconomic criteria that are integrated with economic goals, in a triple bottom line approach. Here, as with the social values approach, CSR represents a management philosophy that guides its overarching approach to business. CSR is fully integrated into these companies' missions, values, and decision making. Moreover, the CSR efforts enhance the company's culture to create a cooperative, congenial atmosphere; employees are motivated by knowing they make a difference. Often, employee retention and satisfaction is a key reason for CSR in these companies, and the causes selected are often based on employees' interests. Company hiring is based on the belief that individuals who are motivated by a company's CSR efforts are more likely to exhibit other valued traits such as empathy and integrity.

Finally, as with model 2 companies, model 3 companies encourage competitors to practice CSR and legitimate stakeholder engagement.

These three models implicitly assume a company does engage in some form of CSR. Certainly, other business models are possible—say, a single-minded focus on profits with no CSR, or engaging in CSR in a hypocritical fashion. However, these models are inconsistent with best-practices business strategy in the 21st century, for the reasons noted previously.

Measuring the Outcomes and Effectiveness of CSR Initiatives

At a very simple level, companies can evaluate the business impact of triple bottom line strategies by gauging their impact on either revenues (finding new business opportunities, new markets, new customers) or costs (say, decreasing fixed costs or variable costs, depending on the specific initiative). For example, GE's "green is green," or "ecomagination," initiative had goals to double its revenue from 17 clean-technology businesses—ranging from renewable energy (wind turbines) and hydrogen fuel cells, to water filtration and purification systems, to cleaner aircraft and locomotive engines—as well as to cut its own energy-related costs through astute energy management. It made a solid business case that it grew revenues and cut costs with these measures, contributing to enhanced profitability. Hence, at a simple level, evaluating revenue and cost impacts can help a company gauge the profitability of its CSR decisions.

However, triple bottom line performance measurement goes well beyond this simple approach. At least two questions arise with respect to the measurement issues:

1. How does a company evaluate its performance for the triple bottom line?
2. How does a company's social performance relate to its financial performance?

EVALUATING TRIPLE BOTTOM LINE PERFORMANCE With respect to the first question, quantifying the triple bottom line is relatively new, problematic, and often subjective. Although many companies publish an annual report on their corporate social performance, challenges arise. Unlike reporting of financial results, in which generally accepted accounting principles (GAAP) guide the reporting process and auditors verify the results, only recently have reporting standards been developed for a company's noneconomic (social and environmental) performance. In addition, these standards have been developed by many different organizations, and each uses a different measurement approach; therefore, depending on the approach used, the results can vary widely.

For example, AccountAbility and CSR Network ranked the 100 largest global corporations by the quality of their commitment to social and environmental goals; they used 100 metrics grouped into four categories:

- *Strategy:* The extent to which a company includes social and environmental goals in its business decisions
- *Governance:* The processes used to hold executives and board members accountable for nonfinancial goals
- *Stakeholder involvement:* How well a company responds to employees, local communities, and activist groups
- *Follow-through:* As evaluated by metrics such as the output of carbon per dollar of revenue and public controversies a company was involved in

The 2007 rankings published in *Fortune* magazine[54] showed that British Petroleum was ranked #1, with a score of 75.2 (out of a possible 100). Other high-tech companies in the top 50 included:

Vodaphone (#5, with a score of 66.3)	Deutsche Telekom (#33)
GE (#13, with a score of 59.1)	Sony (#34)
Telefonica (#23)	Toyota (#38)
HP (#24)	Siemens (#41)
Matsushita (#29)	Fiat (#45)
Toshiba (#30)	France Telecom (#46)

As a sector, computers and electronics scored only 42 out of 100 points, the fifth industry behind energy, refining, automotive, and industrial. The pharmaceutical industry scored only 17 points, despite the high-profile efforts of some companies to widen access of medications to poor people.

On the other hand, Innovest, an international research and advisory firm, prepared a list of the world's 100 most sustainable corporations based on 120 different factors such as energy use, health and safety records, litigation, employee practices, and dealing with supplier problems.[55] Companies published on its Global 100 list in *Business Week,* including Nokia and Ericsson for example, did not appear on the 2007 AccountAbility ratings.

Ratings are also conducted for companies' environmental responsibility. Climate Counts, an environmental group that ranks major corporations on their tracking, reporting, and reductions of greenhouse gases, released a scorecard in 2007 where Apple came in dead last, with a score of 2 out of 100.[56] Despite the wild success of the iPhone that year, poor marks damaged Apple's reputation as being cutting-edge and cool. Apple also faced negative effects from Greenpeace's "Green my Apple" campaign, which lambasted the company for its "iWaste." CEO Steve Jobs posted a letter pledging the company would become "a greener Apple."

Further compounding the reporting problem is that companies like to portray their CSR efforts in the most favorable light possible, at the extreme engaging in mischaracterizations of their efforts, such as **greenwashing**, in which a company inaccurately describes its products or business practices as beneficial to the environment, when in fact they are not. Greenwashing will be discussed more fully later in the chapter.

Another complicating factor is that many organizations are strong in some areas of CSR yet weak in others.[57] As noted above, BP was rated by AccountAbility as #1 in 2007. Yet, does BP exhibit CSR by striving to address global climate change through the development of alternative energy sources, or is it irresponsible for continuing to produce fossil fuels? Was it only a marketing ploy to change its name from BP (British Petroleum) to bp (beyond petroleum)? Similarly, does HP's community development initiatives in developing countries make it socially responsible, or does its abandonment of its long policy of job security for employees mean that it is not?

Despite these measurement complications, companies must evaluate their social and environmental performance. Rather than waiting for an industry watchdog group to single it out as being a poor performer, a company should proactively monitor its own performance, using one of the generally accepted frameworks listed below. Moreover, as cited in Chapter 2's section on performance measurement, a company's efforts to measure performance are actually associated with enhanced performance;[58] the area of the triple bottom line also shows this effect. Knowing which assessment criteria are used to evaluate a firm's CSR performance actually helps it score higher.[59]

In this regard, **social accounting** is used to document and communicate a company's impact on social and environmental concerns.[60] A number of reporting guidelines or standards have been developed to serve as frameworks for social accounting, auditing, and reporting, including:[61]

- AccountAbility's AA1000 standard (www.accountabilityrating.com)
- Verite's Monitoring Guidelines to assess sweatshop labor and labor abuses (www.vcritc.org)
- Social Accountability International's SA8000 standard for human rights (www.sa-intl.org/index.cfm?&stopRedirect=1)
- The FTSE4Good Index, an evaluation of CSR performance of companies (www.ftse.com/Indices/FTSE4Good_Index_Series)
- GlobeScan's Corporate Social Responsibility Monitor (www.globescan.com/csrr_overview.htm).

The idea is to develop guidelines to enable corporations and nongovernmental organizations (NGOs) alike to comparably report on the social impact of a business.

Generally, sustainability reporting metrics are better quantified and standardized for environmental issues than for social ones. A number of respected reporting institutes and registries exist including:

- The Global Reporting Initiative's Sustainability Reporting Guidelines (www.globalreporting.org)
- CERES (www.ceres.org)
- Institute 4 Sustainability (www.4sustainability.org/international/index.htm)
- Green Globe Certification/Standard (www.greenglobeint.com)
- The ISO 14000 environmental management standard (www.iso.org)

For example, CERES ranks 100 global corporations on their strategies for curbing greenhouse gases. There are standards that vary by industry as well, such as those that monitor green mutual funds, clean energy, green construction, and green building (the LEEDs Standard).

Even the measurement of financial performance following a triple bottom line approach requires special attention. Rather than being limited to the internal profit made by a company or organization, within a sustainability framework, "profit" includes the lasting economic impact the organization has on its broader economic environment. Therefore, despite its convenience and prevalence, merely using traditional corporate accounting *plus* an assessment of a firm's social and environmental impact does not fully capture the notion of a triple bottom line.[62]

THE RELATIONSHIP BETWEEN A FIRM'S CORPORATE SOCIAL PERFORMANCE AND ITS FINANCIAL PERFORMANCE   With respect to the second measurement challenge, many companies want to know how their investments in corporate social responsibility translate into financial performance. Does CSR detract from financial performance through unnecessary expenditures that not only raise the firm's costs, but also detract from its focus on its for-profit business? Or, does it better enable the firm to attract resources, including quality employees and customer loyalty, as well as create new opportunities that provide additional leverage in its business?

Before addressing this question, it is important to note that strong adherents to a triple bottom line approach argue that the question itself is spurious: The essential point of CSR is that interests of equity holders may need to be set aside in favor of interests of the firm's other stakeholders. Firms engage in activities that benefit employees, suppliers, customers, and society at large, even if those activities reduce the present value of the cash flows generated by the firm.[63] Moreover, equity holders of these firms often have interests besides simply maximizing their financial wealth. Hence, they argue that addressing whether or not firms should engage in CSR by looking at the effect on the "bottom line" is inconsistent with a strong triple bottom line philosophy.

At first blush, a company's scores on social performance generally do not correlate well with financial performance. *Business Week*'s Innovest Ranking of the "Global 100" cites numerous examples, including:

- Companies that received relatively high scores for corporate social responsibility (including BP and Sony) but whose financial (stock market) values did not perform well
- Companies that received lower scores for corporate social performance (including Apple and Nintendo) but whose stock market prices did perform well[64]

Says *Business Week:* "Results [of studies that look for] direct relationships between a company's social and environmental practices and its financial performance are mixed."[65] *Fortune*'s Account-Ability ranking also noted the lack of a link between corporate social performance and financial performance—with a few notable exceptions, such as GE.[66]

Several explanations exist for the lack of a direct relationship between corporate social performance and financial performance. First and foremost, the studies lump together all types of companies, without differentiating between companies that neither link their CSR efforts directly to their business strategies nor execute their CSR programs as well as others. Moreover, there is most certainly a time lag between when a company implements a social or environmental initiative and when it actually translates into business performance. Sometimes, this lag may be years, as many environmental initiatives in particular will play out only after industry and market trends have become well established, and the more nimble, proactive companies are separated from those who pursued more short-term endeavors. Finally, as noted in the various models of CSR approaches, some companies simply don't have a "business case" approach to their CSR initiatives.

Academic studies show somewhat similar results. A compilation of early findings (pre-2003) show a positive association, and very little evidence of a negative association, between a company's social performance and its financial performance.[67] More recent studies show differential effects of corporate social performance on financial performance, depending on either the degree of social responsibility or the degree of innovativeness of the company. A 2006 study of the performance of a panel of 61 socially responsible investing funds found that the strongest financial returns accrue to low and high levels of social responsibility, with significantly lower returns to moderate levels of social responsibility.[68] A second study (2008) also shows that "corporate social performance can be achieved without negative effects on financial performance." More importantly, because corporate social performance can be used to differentiate a company from its competitors, the 2008 study shows that the link between corporate social performance and financial performance is actually stronger for:

- Less innovative companies (as measured by a three-year average of R&D spending)
- Commodity industries where companies' products are less differentiated (say, through less emphasis on advertising and brand positioning)[69]

Rather than inferring that technologically innovative companies will not benefit from CSR, the authors caution: Companies can creatively combine corporate social performance and innovation, particularly because many innovative firms' CSR initiatives involve new innovations. Moreover, because many CSR initiatives allow firms to gain new insights about new markets, improved learning capabilities can translate into financial performance in the future.[70]

A final consideration is how customers respond to a company's CSR initiatives. Many studies show a positive effect on consumers' attitudes toward a firm and its brands, as well as higher intentions to purchase and actual purchase behavior.[71] What are some of the reasons for this effect? First, consumers who believe that there is a strong fit between the company and the cause are more likely to believe that the company is proactive rather than defensive. Moreover, consumers who have a personal connection to the CSR initiative are quite likely to have a favorable attitude toward the company. Consumers are more likely to identify with companies whose values, as represented in its CSR initiatives, are consistent with their own values. Finally, a company's CSR initiative can enhance its customers' sense of well-being. However, before consumers' attitudes toward a company or a brand can be influenced, the consumer must be aware of and knowledgeable about CSR activities.[72]

Although the triple bottom line has clearly caught the attention of many companies, their shareholders, customers, employees, and other stakeholders, the challenge—as with all strategies and initiatives—is to develop a measurement system that enables the company to assess its progress and to take corrective action as necessary. In practice, there is no commonly accepted "right way" to identify, measure, and report on the nonfinancial inputs or outcomes in the triple bottom line.

However, just as the company's marketing strategy should guide the selection of metrics for its marketing dashboard (see Chapter 2), its triple bottom line priorities must guide the development of appropriate metrics. Without an appropriate measurement, control, and adjustment system, triple bottom line may become another management fad. Notwithstanding the current problems of measuring CSR discussed above, companies, nonprofit organizations, and academic scholars are all working to improve metrics in this area, which should bear fruit in the near future.

Best-Practices CSR for High-Tech Companies

Clearly, the field of CSR has grown increasingly complex and nuanced, and so a simple recipe for success is simply not feasible. Despite "consumers' growing expectations of companies [that] make corporate philanthropy more important than ever," companies' philanthropy programs aren't fully meeting their social goals or stakeholders' expectations for them.[73] In this spirit, this section offers some tips and best practices that can be useful, knowing that a one-size-fits-all approach is not possible (see Figure 13.3).

LINK CSR TO BUSINESS STRATEGY Some 90% of executives seek business benefits from their CSR programs in addition to social goals.[74] Therefore, for firms that adopt a business case approach to CSR, the first priority must be to make CSR "strategic" in the sense that it is linked clearly to the business. Companies who report that their CSR programs are very or extremely effective at meeting both social goals and stakeholder expectations select social and political trends that are relevant to their business, and are influenced by both community and business needs.[75] For example, a

FIGURE 13.3 Best-Practices CSR Considerations

telecommunications company's investments in broadband access to rural communities make sense, because it is part of the company's raison d'etre. However, based on this factor, such a company's sponsorship of planting trees in these communities may make less sense, because this activity is unrelated to its mission.

Michael Porter and Mark Kramer[76] suggest that companies' CSR investments should occur in the following four domains in order to strengthen competitive position through pursuit of social causes:

1. *Supply/input conditions:* Investments in social causes that either directly or indirectly develop a company's human resources, capital resources, physical infrastructure, natural resources, scientific or technological infrastructure, and so forth, will deliver both economic and social benefits. For example,[77] to cultivate needed talent, Microsoft pursued an initiative to bolster information technology education infrastructure, including $47 million to help the American Association of Community Colleges and $15 million for the African American collegiate community. Similarly, GlaxoSmithKline needs to attract top scientists, and it finds that its initiatives to improve access of poor people to medications and vaccines draw such talent.

2. *Demand/customer conditions:* CSR investments that can develop local markets, improve the capabilities of local customers, provide insights into needs of emerging customers, or develop product standards will be in a firm's strategic interests. Investments in base-of-the-pyramid markets, discussed in the next main section, are illustrative of this approach. Philips tailors its offerings to the local infrastructure and economy of developing countries, providing traveling medical vans to bring health care to isolated villages in rural India via telemedicine, low-cost water purification systems, and smokeless wood-burning stoves to reduce death due to exposure to cooking smoke.[78]

    Also, CSR investments that attract the business of consumers who want a socially and environmentally responsible lifestyle, such as the segment referred to as "Lifestyles of Health and Sustainability" (www.LOHAS.com), would fall under this domain.

    Finally, companies can select causes based on their target markets' interests. For example, if a telecommunications company's primary target market, new users of telecommunications services, was women-owned small businesses, then sponsorship of a breast cancer awareness campaign or support of educational programs for young girls in math and sciences might be a good fit. In this case, attempting to connect to the target market and to build competitive advantage via causes in which the target is interested may take the company into socially responsible activities somewhat removed from its primary business.

BOX 13.1

Cisco's Network Academy

Cisco's Networking Academy provides an example of the powerful links that exist between a company's philanthropic strategy, the four domains that can be affected by or targeted with such investments, and the resulting economic and social benefits. Cisco (maker of networking equipment and routers used to connect computers to the Internet) found that many of its customers faced a chronic shortage of qualified network administrators. Although Cisco was already engaged in a type of cause-related marketing (including donations of networking equipment to a high school near its headquarters), the company decided to formalize not only the donation of equipment, but also a program to train teachers (and students) at the schools on how to build, design, and maintain the networks. This program grew into a Web-based distance-learning curriculum. At the suggestion of the U.S. Department of Education, the company began to target schools in "empowerment zones," areas designated by the federal government as the most economically challenged communities in the country. The program was expanded to include developing countries as well. Cisco has added a worldwide database of employment opportunities for academy graduates. Other companies have joined the effort, donating Internet access and other needed computer hardware and software. And, rather than reinventing their own training infrastructure, other companies such as Sun Microsystems and Adobe Systems have expanded the academy's curriculum by sponsoring courses in other areas. As of December 2002, only five years after it began, the academy was operating 9,900 programs in all 50 states and 147 countries—and it continues to grow rapidly. Cisco's $150 million investment has brought technology careers, and technology itself, to people in some of the most economically depressed areas of the world. More than 115,000 students have graduated from the two-year program, and half of the 263,000 students were outside the United States. Cisco has found a pool of talented employees, improved the sophistication of new customers, attracted international recognition, generated pride and enthusiasm among its employees and partners, and is known as a leader in corporate philanthropy.

Source: Porter, Michael, and Mark Kramer, "The Competitive Advantage of Corporate Philanthropy," *Harvard Business Review* 80 (December 2002), pp. 5–16.

3. *Competitive context:* Any investments a firm makes to facilitate policies to reduce corruption, encourage fair competition, protect intellectual property, and in general support an attractive business environment can be beneficial to both society and the individual business. For example, the opening vignette mentioned Eli Lilly's efforts to address counterfeiting and piracy in the pharmaceutical industry.

4. *Supporting infrastructure:* Investments that bolster supporting industries (e.g., services, suppliers) can foster the development of vibrant industry clusters, which can become an engine for economic development, in line with the spirit that "a rising tide floats all boats."

Box 13.1 provides an illustration of these four domains for Cisco.

MEASURE THE IMPACT OF CSR   Impact should be assessed based on the measurement guidelines outlined earlier; it should include assessing impact on both the stakeholders addressed by the CSR initiative as well as on a broad set of company indicators (not just revenue/cost/profit, but also employee morale, innovativeness, etc.). Indeed, because of the measurement difficulties, some caution companies to be wary of philanthropic spending where value is difficult to assess.[79] However, these areas could also offer the most opportunity for long-term impact and innovativeness.

BALANCE COMPETING STAKEHOLDER INTERESTS   Companies following a syncretic stewardship model know that they serve multiple stakeholders, but are faced with the need to balance the stakeholders' competing interests. Indeed, some believe that existing stakeholder models are inadequate because they do not offer insights on how to do so.[80] One approach is to treat such competing interests as an ethical controversy, and to use the framework provided later in this chapter to evaluate carefully all the stakeholders affected by the decision. Another is to appoint devil's advocates to argue the viewpoints of specific stakeholders.

COLLABORATE WITH RELEVANT PARTNERS   Companies who report that their corporate philanthropy programs are very or extremely effective at meeting social goals and stakeholder expectations should collaborate with other companies to achieve the goals they have set for their CSR programs.[81]

BEWARE OF UNINTENDED REPUTATION EFFECTS (HYPOCRISY)   Socially responsible behavior is pursued in part, to enhance a company's reputation. However, the single-minded pursuit of goodwill and enhanced reputation can degrade moral capital—particularly if a company's CSR activity is viewed as insincere. Similarly, overpromoting CSR can create cynicism.[82] Hence, reputation-related CSR must be tempered with two caveats.

First, reputation-related CSR goals can lead a company's CSR initiatives into areas unrelated to its business. For example, a telecommunications company might donate money to a local food bank in some of its primary communities. In this case, the cause is not directly tied to its mission, nor specifically targeted to its primary customers. Companies that exhibit social responsibility in areas more closely related to their businesses, and that are genuine in their efforts, are less likely to be accused of hypocrisy. (A possible benefit of such CSR, even without a direct link to the business, is the opportunity to create goodwill. Further, to the extent the CSR initiative is marketed or televised, or if the company plans on advertising its affiliation to the cause, then the company may also gain some positive publicity.)

Second, companies that focus primarily on reputation benefits run great risk if they use socially responsible behaviors in one area to compensate for socially irresponsible behaviors in others. When corporate social responsibility initiatives are mere window dressing, a public relations ploy designed to "divert attention from corporate rapacity and corruption,"[83] CSR will backfire—possibly not in the short term, but certainly in the long term.

EMBED CSR IN SENIOR LEADERSHIP AND IN ORGANIZATIONAL REWARDS[84]   For any organizational initiative to have staying power, senior executives must be vested with authority for and believe in the initiative. Moreover, unless compensation systems are in some way tied to CSR, the initiatives will languish.

LINK CSR TO INNOVATION   Through their CSR initiatives, companies are developing innovative business models that bring needed products and services to base-of-the-pyramid markets, and they are developing new technologies to solve entrenched problems. CSR initiatives are thus a source of market learning and breakthrough innovations.

SERVING BASE-OF-THE-PYRAMID MARKETS: CORPORATE SOCIAL RESPONSIBILITY AND SOCIAL ENTREPRENEURSHIP

As shown in Figure 13.4, of a total global population of roughly 6 billion people, the world economic pyramid varies widely in terms of income distribution. At the top of the pyramid is the small percentage of the world's population that lives in relative affluence. Found primarily in mature/developed markets, these people enjoy the highest standard of living and per capita income. At the base of the pyramid (BOP) are the roughly 4 billion people who live at a bare-bones, subsistence level. Indeed, 3 billion of them live on less than $3 a day, and another 1 billion live in extreme poverty, on only $1 a day. BOP markets tend to be concentrated in India and China, as well as certain other Asian countries, and Africa as well. Roughly two-thirds of these people are concentrated in urban areas, and one-third in rural areas.[85]

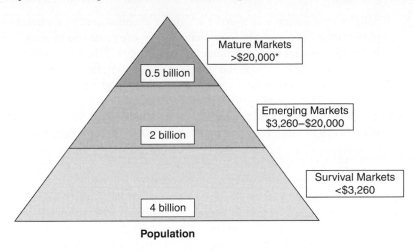

Population

FIGURE 13.4 The World Economic Pyramid *Source:* World Resource Institute
*Individual annual income, in 2005 $US, purchasing power parity

Poverty and its related ailments are not just a phenomenon of the developing world. Even in developed countries, marginalized groups experience extreme poverty and the deprivations that go with it. For example, according to the U.S. Census Bureau, California has more than 5 million people living below the poverty line, some in Third World situations without clean water, lights, sanitation, or jobs. Los Angeles County alone has more than 1 million people living below the poverty line. Dire situations are also found on most of the Indian reservations in the United States.

BOP problems can be addressed by taking a public assistance/aid approach, or a market-based approach, arrayed on the following continuum:

1. Government/ NGO assistance   2. Nonprofit organizations   3. Corporate donations/ philanthropy*   4. "Enlightened CSR"/ social entrepreneurship

*Often distributed through partnerships with government agencies, nongovernmental organizations (NGOs), or nonprofit organizations

The label on the fourth category in the continuum above is not meant to imply that the other three approaches are not also "enlightened"; rather, the term is meant to differentiate the idea of mere "giving" from capacity building and empowerment. Market-based approaches to poverty reduction differ from traditional approaches in that, rather than thinking in terms of aid, charity, and public assistance, the focus is on using business models and tools.[86] The logic is multidimensional and rests on the foundation that aid programs to date have not demonstrated they can effectively solve large, seemingly intractable social problems. As importantly, public assistance may not be sustainable over the long term. Only sustainable solutions can scale to meet the needs of 4 billion people.

> Ironically, where government programs have been ineffective or faltered, corporations have increasingly stepped up to the plate to tackle thorny global challenges, ranging from climate change to poverty. Notable among these recent corporate initiatives has been the quest to reach the base of the pyramid—the more than four billion people globally with per capita incomes below $1,500 (purchasing power parity).[87]

In a survey conducted by GlobeScan's CSR Monitor asking people "the best way for a company to make a positive contribution to society," philanthropy was the last (of three) choices. First and foremost, people want companies to act ethically (e.g., develop safer products), and second, people want companies to have a broader commitment to corporate citizenship that goes beyond philanthropy programs to those that take a more proactive approach, including direct involvement in tackling social and environmental challenges.[88] In other words, investments in infrastructure, develop of human capital and other self-sustaining investments are key.

Furthermore, rather than assuming that people in poverty are victims, unable to help themselves, market-based approaches believe these people are a key part of solving seemingly intractable problems. As individuals, the people living in these situations bring excellent insights into what needs to be different and what will work; in that sense, BOP markets are an ideal setting for co-creation (discussed previously in Chapter 6).

Enlightened BOP strategies are less about serving an existing market, and more a developmental activity that requires new and creative approaches to address poverty. Social entrepreneurship in particular can be "a curative potion for the poverty malaise," and requires that the process start with respect for people. Solutions that better meet the needs of impoverished people should increase their opportunities to earn income and empower their entry into the formal economy.[89] In turn, this will unleash economic mobility that will transform the world economic pyramid to a diamond, with a larger, more prosperous middle class.

A final argument used to support the business case for serving people in the base of the pyramid is found in the sheer size of the market. In Asia and the Middle East there are 2.86 billion people with a total of $3.47 trillion; in Eastern Europe, $458 billion; in Latin America, $509 billion; and in Africa, $429 billion. This is a total of $5 trillion![90] Even a 1% penetration of 4 billion people results in a market of 40 million customers! If per capita income can be raised even marginally—say, if 1 billion people make $1 more per day—that represents a $365 billion market.[91] Hence, the phrase "the fortune at the bottom of the pyramid"[92] captures this notion: "[Although] the motivation for exploring base-of-the-pyramid markets may be to bring needed goods and services to poor people, in the case of more commercially-minded, it is to make money in unlikely circumstances."[93]

However, these markets are either unserved, or underserved,[94] and the fact remains that serving people in these markets is fraught with challenges. Distribution systems are few and fragmented. If a business wanted to reach even five customers per day, it would need 550,000 stores or salespeople to reach 1 billion people.[95] Moreover, the nature of the informal economy means there is little financing for purchases over even a few cents. However, creative solutions are being found every day, whether it is microcredit or treadle pumps, franchises, co-ops, or "Tupperware" models in which home-based selling occurs.

This section presents the latest thinking about BOP interventions, which are at some level market driven, based on a co-production/co-creation model where the people whom the initiatives are meant to serve are empowered to take ownership of the solutions developed. As mentioned previously, the opportunities for high-tech companies in particular arise because technological innovations are necessary to solve many problems in BOP markets.

Domains for Intervention

The problems faced by people living in poverty, when viewed through the lens of entrepreneurship and capacity-building corporate social responsibility, provide the underlying basis for imagining the solutions. Two approaches to frame these problems are presented here.

First, as shown in Figure 13.5, the problems and ailments facing society can be characterized on at least two dimensions:[96]

- Whether the problem exists in top-of-the-pyramid markets or in base-of-the-pyramid markets
- Whether the problem is precipitated by a crisis (e.g., hurricane or tsunami), or whether the problem is chronic and ongoing (e.g., poverty, illiteracy, malnutrition)

In times of crisis (cells 1 and 2 of the figure), any form of assistance—whether government aid, corporate giving, or volunteerism—is required for both BOP and TOP markets. For chronic problems (cells 3 and 4), best practices would be to engage in capacity building rather than solely donations.

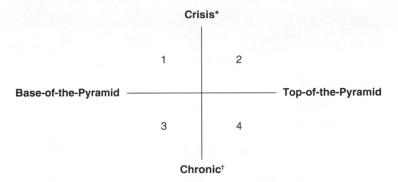

FIGURE 13.5 Domains for Intervention   *Source:* Developed by Paul Hudnut (Co-Director of Colorado State University's Global Innovation Center for Energy, Health & Environment and a founder and director of Envirofit International; see his blog at BOPreneurs.com). This framework highlights the spheres for charity, social entrepreneurship, and for-profit business ventures for environmental and public health concerns.

*Tsunami, earthquake, hurricanes, etc.

†Poverty and its associated ailments, including hunger, chronic illness, illiteracy, etc.

Second, Table 13.2 provides another way to dimensionalize the chronic problems facing mainly BOP markets. These 10 "great divides" require a radical rethinking of typical business-case arguments for venturing into new, untested markets. At the extreme, it requires "unreasonable" thinking: "Unreasonable entrepreneurs find opportunity in the chasms that separate the fortunate from the less fortunate."[97] It also requires raising expectations for the poorest of the poor.

TABLE 13.2 Markets of the Future: BOP Issues as Business Opportunities

1. **Demographic divide:** Business opportunities based on solving problems caused by demographic trends, including population explosion, the aging of the population, and differential problems related to urban versus rural poverty

2. **Financial divide:** Business opportunities based on minimizing the inequities in wealth distribution—a source of economic injustice that can sow the seeds of rebellion and revolution

3. **Nutritional divide:** Business opportunities based on mitigating famine, hunger, and poor nutrition; may include agricultural innovations

4. **Natural resource divide:** Business opportunities based on minimizing shortages of things that are essential to modern living, such as energy, clean air, and water

5. **Environmental divide:** Business opportunities based on creating more environmentally sustainable practices in all countries; may include addressing the disparity in environmental degradation between resource-rich (developing) and resource-poor (developed) nations, as well as issues related to sanitation, waste, and toxins

6. **Health divide:** Business opportunities based on providing access to health care and medicines, as well as eliminating the uneven distribution of illness including HIV/AIDS, malaria, and diarrhea; for example, in 2006, 11 million children under 5 years of age died from preventable causes, 5 million in their first month of life

7. **Gender divide:** Business opportunities based on creating equality for women

8. **Educational divide:** Business opportunities based on creating learning opportunities for all

9. **Digital divide:** Business opportunities based on creating access to information communications and technology

10. **Security divide:** Business opportunities based on bridging the financial divide, because poverty is a threat to world peace

---

Source: Reprinted by permission of Harvard Business School Press from Elkington, John, and Pamela Hartigan, *Power of Unreasonable People: How Social Entrepreneurs Create Markets That Change the World* (Boston: Harvard Business Press, 2008). Copyright © 2008 by the Harvard Business School Publishing Corporation. All rights reserved.

TABLE 13.3 Contrasting Perspectives on Base-of-the-Pyramid Strategies: BoP 1.0 versus BoP 2.0

	BoP 1.0 "Selling to the Poor"	BoP 2.0 "Business Co-Venturing"
View of People	As potential consumers	As business partners
Model of Engagement	Deep listening	Deep dialogue
Business Solution	Reduce price points	Expand imagination
	Redesign packaging	Build shared commitment
	Extend distribution	Marry capabilities
Nature of Relationships	Arm's length, mediated by NGOs	Direct, personal relationships facilitated by NGOs

Source: Used with permission of Erik Simanis and Stuart Hart (with Justin DeKoszmovszky, Patrick Donohue, Duncan Duke, Gordon Enk, Michael Gordon, and Tatiana Thieme), "The Base of the Pyramid Protocol: Towards Next Generation BOP Strategy," 2nd edition (Ithaca, NY: Center for Sustainable Global Enterprise, Johnson School of Management, Cornell University, 2008).

Business Models and Approaches to Solving BOP Problems: Enlightened CSR and Social Entrepreneurship

Entrepreneurial solutions to the world's greatest challenges are being developed in major global businesses pursuing enlightened CSR and in small-scale social enterprises. Although the major characteristics of these newer approaches are presented below, they focus on capacity-building models with the goal of creating lasting solutions with the requisite infrastructure and training to be self-sustaining over time.

"ENLIGHTENED" CSR   The Base-of-the-Pyramid Protocol Initiative was launched in 2003 by the Center for Sustainable Enterprise at Cornell University's Johnson School of Management to develop and refine a corporate innovation process geared for the unique challenges of sustainably serving BOP markets. It differentiates between BOP 1.0 (first-generation approaches to BOP markets) and BOP 2.0 (second-generation approaches).[98] As shown in Table 13.3, companies' initial efforts to pursue base-of-the-pyramid markets ("BoP 1.0" in the table) included efforts to repackage existing products at lower price points to tap into new markets. Or, they were based on a philanthropic model with an idea toward developing markets for the longer term. However, often these projects did not achieve corporate targets and were discontinued. For example, HP's e-inclusion strategy (c. 2000–2005) was canceled when "corporate convulsions" got in the way—business prospects didn't materialize fast enough to satisfy a new CEO and, ultimately, Wall Street.[99]

> First-generation BOP corporate strategies have failed to hit the mark. . . . BOP 1.0 strategies have implicitly imposed a narrow, consumption-based understanding of local needs and aspirations. . . . [They] represent nothing more than veiled attempts to sell to the poor, as though simply turning the poor into consumers will address the fundamental problems of poverty and sustainable development.[100]

Further compounding these problems is that when companies simply adapt environmentally unsustainable products to sell in the BOP "mass market," this path will lead to environmental oblivion: If 6.5 billion people consume at levels of today's typical American, we would need 3 to 4 planet Earths to supply the raw materials, absorb the waste, and stabilize the climate![101]

To remedy these deficiencies, second-generation corporate BOP strategies take a different approach. As shown in Figure 13.6, these "enlightened" strategies are based on an embedded process of co-invention and business co-creation that bring business into close, personal partnerships with BOP communities. By marrying companies' and communities' resources, capabilities, and energies, BOP 2.0 strategies create enduring community value, establish a foundation for long-term corporate

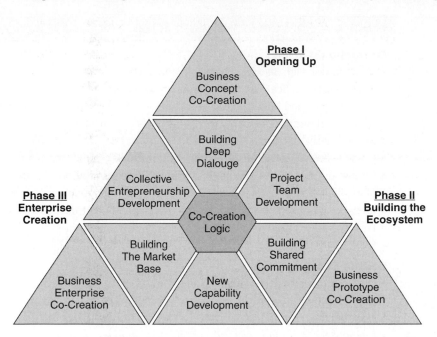

FIGURE 13.6 Co-Creation Logic for BOP 2.0 Solutions
Source: Used with the permission of Simanis, Erik, and Stuart Hart, "The Base of the Pyramid Protocol: Toward Next Generation BoP Strategy," 2nd ed., (Ithaca, NY: Center for Sustainable Global Enterprise, Johnson School of Management, Cornell University, 2008), p. 1, online at www.johnson.cornell.edu/sge/docs/BoP_Protocol_2nd_ed.pdf.

growth and innovation, and require a new strategic process and corporate capability.[102] Table 13.4 itemizes the operating guidelines and code of conduct for the BOP Protocol Initiative.

The SC Johnson Company and the Solae Company (a subsidiary of DuPont) have been pioneers in developing and utilizing this co-creation model. For example, SC Johnson partnered with youth groups in a Nairobi slum to create a community-based waste management and cleaning company. They provide home cleaning, insect treatment, and waste disposal services for residents of the slum.[103] Similarly, after extensive on-site interactions, the Solae Company co-developed business concepts in a Hyderabad, India, slum and in a rural village three hours outside of Hyderabad to address malnutrition and food security through the sale of fresh, prepared foods using locally sourced ingredients. An initial business prototype was put into action at the end of 2007, with two full-time Solae employees based in each site working closely with 20-plus women team members from each community to evolve and expand the business prototypes. The thinking is that only locally based initiatives can be truly culturally appropriate and embedded in the local community—both necessary ingredients for success.

SOCIAL ENTREPRENEURSHIP   A second approach to developing solutions to BOP markets does not rely on large multinational corporations, but rather is based on the power of small business to create innovative solutions. Recall from earlier in the chapter that *social entrepreneurship* refers to the process of applying the principles of business and entrepreneurship to address social problems; *environmental entrepreneurs* apply the principles of business and entrepreneurship to address environmental problems. Because so many social entrepreneurship ventures are focused on base-of-the-pyramid markets, one creative blog refers to them as "BOPreneurs"—entrepreneurs focused on improving the lives of the world's poorest people.[104]

Social entrepreneurs can be called crazy, even unreasonable,[105] people who, paradoxically, seek profit in unprofitable pursuits. Social and environmental entrepreneurs attack intractable problems, taking huge risks and forcing people to look beyond the edge of what seems possible. In seeking to

TABLE 13.4 BOP Protocol Business Principles

Operating Guidelines

- Suspend disbelief. Be willing to admit ignorance.
- Put the last first. Seek out the voices seldom heard.
- Show respect and humility. All parties have something important to contribute.
- Accept and respect divergent views. There is no one best way.
- Recognize the positive. People who live on $1 per day must be doing something right.
- Co-develop solutions. Creating a new business takes mutual learning by all partners.
- Create mutual value. All parties must benefit in terms important to them.
- Start small. Begin with small pilot tests and scale out in modular fashion.
- Be patient. It takes time to grow the ecosystem and win trust before the business takes off.
- Embrace ambiguity. The greatest opportunities often arise from unplanned events and circumstances.

Code of Conduct

- Design businesses that increase earning power, remove constraints, and build potential in the BOP.
- Ensure that wealth generated is shared equitably with the local community.
- Use only the most appropriate and sustainable technologies.
- Promote the development of the affected communities as broadly as possible in ways defined by the local people themselves.
- Track the "triple bottom line" impacts associated with the entire BoP business system.
- Monitor and address any unintended negative impacts associated with the business model.
- Share best practices with local partners to the extent possible.
- Report transparently and involve key stakeholders in an ongoing dialogue.
- Commit to increase community value regardless of the business outcome.

Source: Used with permission of Erik Simanis and Stuart Hart (with Justin DeKoszmovszky, Patrick Donohue, Duncan Duke, Gordon Enk, Michael Gordon, and Tatiana Thieme), "The Base of the Pyramid Protocol: Towards Next Generation BOP Strategy," 2nd edition (Ithaca, NY: Center for Sustainable Global Enterprise, Johnson School of Management, Cornell University, 2008).

transform systems whose dysfunctions create or aggravate major socioeconomic, environmental, or political problems, they end up disrupting established industries while creating new paths for the future. To do so, they delight in coming up with new products and services, or new approaches to delivering products and services to previously undiscovered markets. Motivated less by "doing the deal" and more by "achieving the ideal," they are insanely ambitious, propelled by emotion, and can communicate a clear vision of how the future can be different: "[They] can create a powerful alternative to the orthodoxy of capitalism: a social-consciousness-driven private sector, created by social entrepreneurs."[106] As one social entrepreneur, David Green, founder of Aurolab, stated:

> I could apply my talents to making lots of money, but where would I be at the end of my lifetime? I would much rather be remembered for having made a significant contribution to improving the world into which I came than for having made millions.[107]

Table 13.5 describes some of the other defining characteristics of successful social entrepreneurs.

Models of social entrepreneurship vary. One of the earliest organizations in this space was Ashoka, whose name is derived from a Sanskrit word meaning "the active overcoming of sorrows." This organization, founded by Bill Drayton in 1980 in Washington, D.C., continues to be a pioneer in social entrepreneurship. With a network of 200 staff members that supports 2,000 social entrepreneurs by integrating them into a network of other entrepreneurs, corporations, and business projects,

TABLE 13.5 Noteworthy Qualities of Social Entrepreneurs

1. Try to shrug off constraints of ideology or conventional wisdom
2. Identify solutions to social problems that combine innovation, resourcefulness, and opportunity
3. Primary focus is on value creation, and therefore, are willing to share their innovations and insights for others to replicate
4. Jump in before they are fully "resourced"
5. Have unwavering belief in everyone's innate capacity, regardless of education, to contribute meaningfully to economic and social development
6. Exhibit dogged determination that pushes them to take risks that others wouldn't dare
7. Balance their passion for change with a zeal to measure/monitor their impact
8. Display a healthy impatience for bureaucracy

Source: Reprinted by permission of Harvard Business School Press from Elkington, John, and Pamela Hartigan, *Power of Unreasonable People: How Social Entrepreneurs Create Markets That Change the World* (Boston: Harvard Business Press, 2008), p. 5. Copyright © 2008 by the Harvard Business School Publishing Corporation. All rights reserved.

they collaborate, share know-how, and partner. Their model is to improve economic and environmental conditions for the impoverished by creating a multiplier effect, leveraging a global movement of change-makers (www.changemakers.net) via a website that allows social organizations around the world to connect and share innovations. The organization follows an open source model that allows the best social solutions to surface and be adopted and modified by others.

Another early player in this area was Dr. Muhammad Yunus, who pioneered the microcredit business model in Bangladesh in the 1970s, and empowered rural women to be entrepreneurs by extending small loans to help them get set up. Another organization, Kiva.org, is a person-to-person website that empowers individuals to make microloans to entrepreneurs in developing countries.

Despite their roots, social entrepreneurs do not necessarily lean toward nonprofit models for their ventures. Indeed, government and philanthropic funding sources on which nonprofits rely create a dependency which, in turn, can create uncertainty. Hence, although social entrepreneurs might structure their organizations as nonprofit ventures, they might also structure them as for-profit businesses, or hybrid "blended value" operations.[108]

Leveraged nonprofit ventures are sometimes the only option for tackling many kinds of societal woes ("market failures"). Operating where "the air is too thin for business to even think of venturing," they are modern day alchemists. With minimal financing, they leverage the power of communities to transform a grim daily existence; they deliver public goods (safe drinking water, education, health, housing) to the most economically vulnerable who neither have access to nor can afford the services needed. These nonprofit ventures require multiple external partners to provide financial, political, and "in kind" (with donations) support. In addition to funding, they must leverage resources in ways that effectively tackle the nature and scale of the problems, especially when a crisis has precipitated a situation that other emergency responses have not addressed. Ideally, the goal of such ventures would be to enable the direct beneficiaries of the aid to assume ownership of the initiative, to enhance long-term sustainability.

One well-known venture structured under a nonprofit business model is Nicholas Negroponte's One Laptop Per Child organization (www.laptop.org), which offers a low-cost (desired target of $100) laptop computer for schoolchildren through the governments in developing countries. The laptop has many innovative features including a network to enable collaboration between classmates, a hand crank to charge the battery for short periods of time, and a dust-free, waterproof casing. Another nonprofit social entrepreneurship venture is International Development Enterprises (IDE), an organization that develops and/or markets products (such as the treadle pumps described in Chapter 12's Technology Solutions for Global Problems) to help people lift themselves out of poverty. This organization's mission is to develop low-cost tools to help subsistence farmers become small-scale commercial producers. IDE products are designed to pay for themselves in the first year through the farmer's increased productivity. IDE also nurtures private-sector supply chains, which are essential for economic development. For example, the organization has a network designed to offer training for well drillers to install the pumps and to educate farmers interested in irrigating to produce higher-value crops.[109]

Hybrid nonprofit ventures are an imaginative blending of nonprofit and for-profit strategies that create social or environmental value. As with leveraged nonprofit ventures, they deliver goods or services to disadvantaged populations that have been excluded or underserved by mainstream companies, and they also develop a plan to ensure that the targeted population can access the product or service. Given their hybrid nature, these enterprises recover a portion of their costs through the sale of goods and services, and in the process often identify new markets. However, because these sales cannot support the organization, the entrepreneur must also be an astute fund-raiser. Some of the investors may push the hybrid nonprofit venture to evolve to a for-profit business.

One creative hybrid nonprofit venture is Envirofit, featured in Box 13.2. Aravind Eye Care is another hybrid nonprofit, featured in Chapter 4's Technology Solutions to Global Problems.[110]

Because they operate in emerging markets, leveraged nonprofit and hybrid nonprofit initiatives may provide early indicators of where business could head in the future. The potential for services, products, and technologies created at this level to leapfrog back into the developed world markets can present both opportunity and threat to established businesses.

Social business ventures, the third category of social entrepreneurial ventures, are set up as for-profit businesses from the outset—although they tend to think about what to do with any profits very differently from mainstream businesses. They are typically set up with a mission to drive social and/or environmental change. Rather than maximizing financial returns for shareholders, the goal is to benefit financially low-income groups and to reinvest profits to grow the venture, enabling it to reach more people; however, managing this balance can create internal tension in the organization. Backing comes from investors interested in coupling economic and social returns. Because of its business model, these businesses can scale (grow) more readily because they can more easily take on debt and equity. Evidence indicates that environmental entrepreneurs tend to pursue for-profit ventures.

Experts continue to debate whether problems such as poverty can be best alleviated by market-based mechanisms or by governmental aid.[111] For example, with respect to technology access (or the digital divide, discussed at length in the next main section), One Laptop Per Child is based on a partnership between a nonprofit organization and governmental agencies, avoiding traditional marketing and distribution and instead selling to governments, which in turn allocate the laptops to schools using existing textbook distribution channels. However, governmental demand has not been forthcoming, and the initiative has struggled. Intel's Classmate PC project, on the other hand, has taken a market-based approach. While new models and approaches are still being evaluated, the governmental aid/nonprofit model can be the only choice, say, if pricing makes needed products unaffordable. On the other hand, creative market-based approaches have certainly filled a void. As detailed in the next section on the digital divide, "No one thinks that mobile phones would have spread so fast if the task had been left to nonprofit providers."[112] Before turning to that topic, however, two final topics with respect to BOP markets are addressed: ongoing challenges and keys to success, and criticisms of BOP strategies.

Ongoing Challenges and Keys to Success for BOP Strategies

Successful approaches to BOP markets often defy conventional wisdom. For example, rather than being fearful of new technology, base-of-the-pyramid consumers can be eager adopters of advanced technologies. They can also be brand conscious and sophisticated about value. However, illiteracy makes many standard approaches to products and marketing problematic. Many developing countries struggle with corruption challenges and antiquated governmental regulations. This section addresses

BOX 13.2

Social Entrepreneurship in Action: *The Case of Envirofit*

Envirofit provides retrofit kits for the dirty (pollution-generating), carbureted two-stroke engines which power more than 100 million "two-wheelers" (motorcycles, scooters) and "three-wheelers" (tricycle taxis, tuk tuks) across much of Southeast Asia. Each of these engines produces the pollution output of 50 modern cars, contributing the pollution equivalent of more than 5 billion cars! For example, in Manila (Philippines), 40% of the air pollution is attributed to these forms of transportation, which causes nearly 5,000 premature deaths per year in Manila alone (12% of all premature deaths). Taxi drivers are too sick to work roughly seven days each month, contributing to lost income. Schoolchildren are particularly vulnerable to, and disproportionately affected by, this pollution.

Envirofit started in 2003 as a student entrepreneurship project at Colorado State University, and became a nonprofit company later that year. The first prototype for the retrofit was built in 2004, and the Philippines was selected for field testing: It had more than 1.3 million registered two-stroke tricycle taxis, and more than 80% of the country's population depends on the tricycle transport sector in some form. In 2007, the company saw its first customers and hired its tenth employee.

Using a direct in-cylinder fuel injection retrofit kit, composed of a fuel injector, an air injector, a cylinder head, a fuel pump, and an oil pump, each retrofit reduces roughly one ton of CO_2 per year, based on an 89% reduction in hydrocarbons, a 76% reduction in carbon monoxide, a 50% reduction in oil consumption, and a 35% reduction in fuel consumption.

Value Proposition for the Taxi Owner

Driving a dirty two-stroke engine, a cab owner will earn roughly $9.25 in daily fares, of which fuel costs consume $4.25, leaving earnings of $5 a day, for an annual income of roughly $1,500. The purchase of an Envirofit kit costs roughly $400, which is financed over a 365-day time horizon. The cab owner still earns $9.25 per day, but his fuel costs are now only $2.55 (35% reduction in fuel consumption), his financing fees for the retrofit purchase are $1.15 a day for the first year, leaving an income of $5.55 a day for Year 1, and after the kit is paid for, a daily income of $6.70 a day. Hence, the retrofit not only results in cleaner air, but the taxi owner sees a 34% increase in income to $1,650–2,000 a year.

What do people at the base of the pyramid do with such a significant increase in income?

"After 6 months of using the Envirofit retrofit kit, my extra income helped me save for a matching house grant. I rebuilt my home and my neighbor's home, which provided housing for 6 families."　　　*—Rolando Santiago, taxi driver and early Envirofit customer*

"With his savings from using Envirofit, Vigan City taxi driver Conrado Valdez can now afford to send his children to school."　　　*—Philippines People's Journal, February 18, 2008*

Envirofit has won a slew of international awards: in 2005 for "Technology Benefiting Humanity," in 2006 as a "Top 10 Innovative Technology" by Stanford's Social Innovation Review, in 2007 for one of "20 Companies That Will Change the World" as well as the "World Clean Energy Award," and in 2008 as one of *Newsweek*'s "Top 10 Fixes for the Planet." It now is taking its successful model for the two-stroke retrofits and applying it to cook stoves.

Source: Paul Hudnut, "Technology and Marketing in the Base of the Pyramid," Presentation at the Academy of Marketing Science, Vancouver, BC, May 29, 2008; and Envirofit International, online at www.envirofit.org.

five key areas that BOP market initiatives must proactively consider. Overarching considerations are to think in terms of social transformation and capacity building, unconventional partnering, scaling issues (whether the business model solution will grow to accommodate large numbers of people), and new business models with a completely "reimagined" approach. The areas discussed here are not mutually exclusive, and can be used in creative combinations.

DISTRIBUTION   Poor roads, high transportation costs, information asymmetries between buyers and sellers, and insufficient infrastructure pose complications for distribution in BOP markets.[113] For example, as Eli Lilly found out, distributing TB drugs was problematic, and those problems have led to the development of MDR-TB.

Best practices in socially responsible distribution include localizing value creation, such as through franchising or agent strategies that build local ecosystems of vendors and other forms of creative collaboration. Investments in capacity building and training are also necessary.

PRICING   Pricing can also create challenges in BOP markets. Some say that the price of Western products must be slashed by at least 90% to be affordable for BOP markets. Stanford University has a program called "Entrepreneurial Design for Extreme Affordability" to address this challenge.[114] For example, the initial target price of the One Laptop Per Child initiative was $100, which led to using a "skinny" Linux operating system, and avoiding retail sales, marketing, and distribution that account for roughly half of the price of a typical laptop.[115] Other pricing strategies include using "surplus revenue" from wealthy consumers to subsidize access for poorer consumers, as in the Aravind Eye Care example in Chapter 4. The Envirofit model of tying daily payments to increased income for BOP consumers is another pricing option. Finally, community-based pricing and distribution models, in which, say, a village (rather than an individual consumer) is the focal unit, can make products affordable. Grameen Telecom has used this model for its community-based phones. Prepaid voice and text messages also make mobile services affordable.

CREATIVE OPEN SOURCE AND LICENSING STRATEGIES   Many social values-led companies and social entrepreneurs believe in the power of sharing their innovations. As a result, BOP strategies often rely on open source models and licensing strategies to reuse, adapt, modify, and tap into previously developed solutions. For example, Eli Lilly, in the opening vignette, used a licensing strategy to give low-price access to its drugs for the most in-need markets. GlaxoSmithKline has worked with local companies to produce generic versions of its patented AIDS medication, selling vaccines at volume discounts to developing countries, and dedicating researchers for malaria vaccines. Cambia is based on an open source approach to the life sciences. For example, PatentLens, a platform based on its Biological Open Source (BiOS) Initiative, offers new technology, royalty-free, for further research or to create new products to any party agreeing to freely share improvements with other license holders, even if it has been patented.[116]

RESPONDING TO FAILURE   Companies that operate in these markets can expect failure. As stated in Chapter 2, though, innovative companies accept such failure as part and parcel of success in high-tech markets. Indeed, they should fail early and fail fast.[117] When companies fail in interesting ways, and when they are willing to learn from their failures, the seeds of success are sown. In this sense, BOP markets create learning labs and experiments that can provide new insights and opportunities even for established markets. As always, a key ingredient is creativity. Moreover, mainstream businesses should learn from the successes and failures of social entrepreneurs.

TACKLE MEASUREMENT CHALLENGES   As with other CSR initiatives, BOP business strategies require measurement. However, the measurement challenges are even greater in these markets. One tool that can be used is the BOP impact assessment framework, a holistic guide for BOP ventures to assess and enhance their poverty alleviation impacts.[118]

Criticisms of BOP Strategies

As with many business models, base-of-the-pyramid business models are also criticized. One criticism is that the data indicating a "fortune" at the bottom of the pyramid have been misconstrued, and that in fact the market is actually very small.[119]

Others find the notion of profiting from the poor to be morally reprehensible. This criticism may be focused primarily on BOP 1.0 initiatives, premised on trying to sell products to poor people as a way to alleviate poverty. These critics argue that the only viable solution to alleviate poverty and its ailments is to find ways for the poor to become producers and earn their own income.[120] One possible solution is found in microcredit business models. Another is helping local artisans find global markets for their crafts.[121]

THE DIGITAL DIVIDE

The term **digital divide** refers to the disparity in technology access between technology haves and have-nots. Disparity in access is an important issue; as stated by the former UN Secretary-General Kofi Annan, "Today, being cut off from basic telecommunications [Internet] services is a hardship almost as acute as other deprivations, and may reduce the chances of remedying them."[122] Similar to issues related to corporate social responsibility and base-of-the-pyramid markets, "if the technology revolution leaves some behind, all will suffer."[123]

In 2006, statistics indicated 91 cell phones and 51 landline phones per 100 people in developed countries, compared to 32 cell phones and 14 landline phones per 100 people in developing countries. Internet penetration was 59% in developed countries compared to 10% in developing countries.[124]

Although the technology gap is a symptom of wider inequality, technology is a key enabler to close these other inequalities. Many believe that providing technology access for developing countries could stimulate a push to democracy and economic development. Based on the belief that technology provides access to more diffuse communications, information, and economic opportunity, the UN sponsored a conference in Tunis in 2005—World Summit on the Information Society—with a goal of improving access to information and communication technology (ICT) by 2015.[125]

The disparity in access to technology can manifest itself in myriad ways:

- Between different socioeconomic groups (e.g., affluent versus poor)
- Between different geographic areas (e.g., urban versus rural users, inner-city versus suburban users)
- Between different ethnic groups (e.g., between Caucasians and African Americans or Hispanics)
- Between different countries (e.g., developed versus developing versus least developed)

In the United States, the digital divide is manifested as differential *access* between different income and education levels, between different ethnic groups, and between rural and suburban/urban areas. Table 13.6 shows gaps in broadband adoption between affluent and less affluent people (76% for the wealthiest compared to 30% for the poorest); between college graduates and lower education levels (70% versus 21%); between rural and urban dwellers (52% versus 31%); and in ethnicity (48% adoption for Caucasians compared to 40% adoption for African Americans).[126] Some reasons cited for the disparities include lack of infrastructure (e.g., in rural areas) as well as low overall Internet *usage* in the low-adoption subgroups in the United States. In that sense, another digital divide issue is people's ability to utilize technology.

> Overall 73% of Caucasians use the Internet at least occasionally from any location, compared with 62% of African-Americans. The relatively lower incomes and relatively lower average levels of educational attainment for African-Americans contribute greatly to this gap in Internet usage, since individuals with low incomes and education levels (regardless of race) are generally much less likely to use the Internet.[127]

With only 47% of Americans having a broadband connection,[128] the U.S. has a bit of a digital divide of its own compared to other developed countries. Although the U.S. broadband adoption

TABLE 13.6 Disparities in Broadband Adoption in the U.S. Market

Percent with Broadband at Home	2005	2006	2007
All adult Americans	30%	42%	47%
Gender			
Male	31	45	50
Female	27	38	44
Age			
18–29	38	55	63
30–49	36	50	59
50–64	27	38	40
65+	8	13	15
Race/Ethnicity			
Caucasian	31	42	48
African American	14	31	40
Hispanic*	N/A	29%	N/A
Education			
Less than high school	10	17	21
High school grad	20	31	34
Some college	35	47	58
College +	47	62	70
Income			
Under $30k	15	21	30
$30k–50k	27	43	46
$50k–75k	35	48	58
Over $75k	57	68	76
Community Type			
Urban	31	44	52
Suburban	33	46	49
Rural	18	25	31

Source: Horrigan, John, and Andrew Smith, "Home Broadband Adoption 2007," Pew Internet & American Life Project, June 2007, online at www.pewinternet.org/pdfs/PIP_Broadband%202007.pdf.

*The report did not include the data for Hispanics, and instead reported data from a separate survey for 2006 only.

rate increased 5% in one year (2006-2007), it is still low—25th to be specific—compared to other countries: South Korea at 93%, the Netherlands at 74%, Canada at 65%, and Finland at 60%, to name just a few.[129] Moreover, in Japan, 90% of broadband subscriptions are through 3G networks.[130]

Solutions to Bridging the Digital Divide

Although technology companies have been accused of discriminating against those living in remote or rural areas because providing them service is not "economically viable," the causes and solutions to this dimension of the digital divide are varied. Even if technology is available, people worry about affordability and ability to use (education-related). The solutions to bridging the digital divide are many, and a variety of organizations are spearheading these initiatives, including Everyone

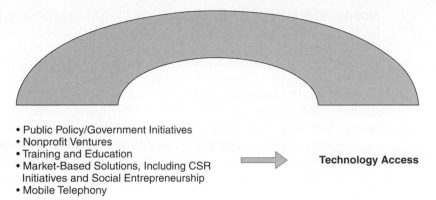

- Public Policy/Government Initiatives
- Nonprofit Ventures
- Training and Education
- Market-Based Solutions, Including CSR
  Initiatives and Social Entrepreneurship
- Mobile Telephony

Technology Access

FIGURE 13.7   Solutions to Bridging the Digital Divide

(www.everyone.org), Internet Evolution (www.internetevolution.com), Digital Divide Network (www.digitaldivide.net), InterConnection (www.interconnection.org), Digital Alliance Foundation's ICT Education for All (www.ictefa.org), Digital Opportunity (www.digitalopportunity.org), Close the Gap (www.close-the-gap.org), and Digital Dividend (www.digitaldividend.org). The solutions shown in Figure 13.7 and discussed here are not mutually exclusive, but can be used effectively in combinations.

PUBLIC POLICY   Government efforts to facilitate access to technology are one possible way to bridge the digital divide. For example, in the United States, ongoing debates occur about the role of the government in providing broadband Internet access. Similar to access to education, some believe that access to technology is an "essential facility," from which all should have the opportunity to benefit. If the economics of the situation do not support businesses developing digital access for, say, rural communities, the government has a responsibility to its citizens to develop a widely available network of computer technology. Therefore, as in the development of the interstate highways, phone lines, and electricity, the government may choose to play a similar role in the development of technology access and availability. Many believe that, because the long-term health and survival of rural America is, to a large extent, dependent on its ability to attract and retain economic development activity—which, in turn, requires access to current technologies—this type of government support is vital.

NONPROFIT VENTURES   As discussed previously in the section on social entrepreneurship, nonprofit ventures can be used to bridge the digital divide. For example, several of the organizations involved with digital divide initiatives are nonprofits.

TRAINING AND EDUCATION   Although providing access to technology for disadvantaged populations is an important part of bridging the digital divide, simply having access will not integrate technology into people's lives. An overlooked factor is the willingness and ability of these people to fully embrace and utilize the technology to their advantage.[131] Hence, training and education efforts on how to effectively utilize technology also must be initiated.

MARKET-BASED SOLUTIONS, INCLUDING CORPORATE SOCIAL RESPONSIBILITY INITIATIVES AND SOCIAL ENTREPRENEURSHIP   Digital technology can spread at an incredible rate among the poor when pursued as a for-profit activity. Many technology-based businesses offer affordable computing products and Internet access to disadvantaged populations. Intel's Classmate PC was designed to target schoolchildren in developing countries.

MOBILE TELEPHONY   While some digital divide initiatives focus on providing laptops (One Laptop Per Child and Intel's Classmate PC), others focus on mobile telephony. Given that 80% of the world has mobile coverage, and only 25% is accessing it,[132] companies are building phones at lower costs and creating innovative business models to get more phones in people's hands. Through its innovative product development, Nokia offers low-priced handsets in developing countries. It has 46% of the market share in Asia, 57% in India, and 66% in Africa.[133] It is highly efficient, and has operating profit margins of 25% on handset sales (fourth quarter 2007). Samsung and Sony Ericsson are working to challenge Nokia in these markets, but bigger threats may come from local start-ups offering ultra-cheap phones. Reliance Communications in India sells a model for $19, and China's ZTE offers a $30 handset, compared to Nokia's cheapest model at $32. Statistics indicate that most new subscribers will be on low incomes and spend less than US$5 per month on mobile communications.

Despite their limitations, mobile phones may be more suited to life in developing areas than computers, as reported on BBC News:

> The problem with PCs is they last for maybe one or two years and then start to break down; schools have no money to fix equipment. Moreover, computers still require some level of know-how such as typing or reading; mobile requires voice only. Mobile telephones are preferred for crossing the digital divide due to cost, convenience, and no need for literacy.[134]

Not only are they cheaper and easier to use than other forms of ICT, mobile phones consume less power, and they can be used by people without prior training in technology.

Moreover, while much of the developed world faces the legacy anchor of 100-year-old landline telecommunications networks to deliver broadband Internet, emerging markets are set to leapfrog such encumbrances using new mobile broadband-based technologies. Indeed, landline phones are now outnumbered 3 to 1 generally, and by 7 to 1 in least developed nations—and 9 to 1 in sub-Saharan Africa.

Data indicate the digital divide with respect to mobile telephony is shrinking.[135] Indeed, the figures for monthly mobile phone subscriber growth in many developing countries are astounding. Between 2000 and 2005, mobile subscribers in developing countries grew more than fivefold—to nearly 1.4 billion. Growth was fastest in sub-Saharan Africa. Nigeria's subscriber base grew from 370,000 to 16.8 million in just four years. Every month, China and India together add 13 million new subscribers, Pakistan adds more than 2 million, and Egypt more than 1 million. Mobile phones provide access to jobs, to medical care, to market prices, to family members living away from home and the money they can send, and increasingly to other financial services. The strong value proposition for low-income consumers has translated into financial success for creative mobile business models.[136]

Once penetration of mobile phones reaches 2%, the uptake of related mobile services rises more quickly in emerging markets than in the developed world. One reason for this rapid uptake is that because people in these areas don't have access to computers, mobile phones are their first and only gateway to the Internet. Moreover, the pioneering applications built on mobile platforms for these developing areas are improving the lives of millions.

One study estimates that the average developing nation experiences a 0.6%–1.2% rise in economic growth for every 10% increase in the number of mobile phone subscribers.[137] The total economic impact of all wireless activity on these developing countries is estimated to be up to 4 times the value to the wireless operators alone—with much of it coming from productivity gains due to the variety of ways in which these new wireless customers use their phones.[138]

How does mobile telephony raise incomes for the world's poorest? Take the case of fishermen in Kerala, India, who now use mobile phones to price their daily catch of sardines. Prior to the phones, on a good day too much supply would result in lower prices. If the fishermen chose

TABLE 13.7 Examples of Mobile Telephony in BOP Markets

- **Tradenet in West Africa:**

  Helps farmers find the most profitable markets for their agricultural products. Using simple SMS messages, the system provides farmers in their fields with information on supplies and pricing across the region.

- **Safaricom in Kenya (in conjunction with the GSM Association):**

  Has provided 100 physically disabled Kenyans with specially adapted wheelchairs that double as GSM-enabled payphones. Safaricom says, "This is not a charity: Selling phone calls this way is more cost-effective than a roadside kiosk."

- **The Gramijyoti Rural Broadband Project in India's Tamil Nadu region:**

  Uses mobile broadband applications for online access to government application forms, interactive educational classes, and telemedicine.

- **SMART and Globe Telecom in the Philippines:**

  Offer SMS services that allow Filipino workers abroad to deposit money with partnering banks and send it to SMART or Globe subscribers back home. The recipient picks up the money at a partner institution, and the fee (maximum of 1% commission per transaction) is less than traditional wire services.

- **Nokia's BridgeIT in the Philippines:**

  Provides more than 200 schools with Nokia set-top boxes with GSM mobile, a Sim card, and a guide to interactive math and science lessons. When teachers want to access lessons, they use their mobile phone to text a unique code and the interactive material is downloaded to the box; they then access that material in their lessons via a TV; parents began coming to the schools in the evening to do the lessons also. The boxes cost $250, the mobiles were under $30, and the Sim card is prepaid.

Source: Twist, Jo, "Pocket Answer to Digital Divide," BBC News, November 18, 2005.

to motor to a market farther down the coast, prices may or may not have been better, but they incurred additional fuel costs—and if the fishermen made the wrong choice, the catch would spoil. On average, 5%–8% of the total catch was wasted, even though there were willing buyers at markets in other places along the coast. Cell phones have allowed the fishermen to call around the region to find the best prices. Indeed, after the phones were introduced, the proportion of fishermen who sold their catch beyond their local market increased from 0% to 35% as soon as coverage became available. At that point, no fish were wasted, and prices converged to a single rate for sardines along the coast. Although the average price of sardines fell by 4%, there was an 8% rise in the average fisherman's profits—the mobile phone paid for itself within the first two months.[139] Table 13.7 provides other examples of the value proposition of mobile telephony in BOP markets.

Certainly, barriers to using mobile telephony as a solution to the digital divide must be addressed. These include the cost of ownership for operators, and the complicated issues of government subsidies, regulation, and taxation. For example, calling tariffs remain very high, and cheap calls are not consistently available. In 2004, there were 48 mobile subscribers in India for every 1,000 people, compared to 367 in Brazil. Yet Brazilians had to pay an average of US$18.90 per month while India subscribers paid just US$3.20. The fact is, services need to be available and it does require investment in infrastructure. Therefore, national strategies must be linked to other initiatives to build a "hybrid infrastructure."[140] In addition, the development of more affordable handsets, more efficient network equipment, new business models, and a more enlightened approach to the telecomm market are bringing down the barriers.

RESPONDING TO THE RISKS AND OPPORTUNITIES OF GLOBAL CLIMATE CHANGE

According to the 2007 Intergovernmental Panel on Climate Change—whose committee members shared the 2007 Nobel Peace Prize with Al Gore—greenhouse gas emissions are a key contributing factor to global warming.[141] Relatedly, environmental issues soared to the top of the agenda in executive suites around the world, with executives predicting in 2007 that "the environment will attract more public and political attention and affect shareholder value far more than any other societal issues."[142] A February 2008 McKinsey survey showed that 46% of 721 executives thought that "environmental issues, including climate change" will have the most impact on their companies over the next five years—nearly twice as many as for the next two most important issues (health care/employee benefits, cited by 27%, and privacy/data security, cited by 24%).[143] Finally, of the most important ways large companies harm the public good, the top one cited by 65% of executives was "polluting and damaging the environment" (followed by "putting profits ahead of people" cited by 40%, and "exerting improper influence on governments" cited by 30%).[144]

These surveys indicate that attention to the environmental impact of a firm's strategies is a business mandate. As Michael Porter and Forest Reinhardt stated in 2007:

> While individual managers can disagree about how immediate and significant the impact of climate change will be, companies need to take action now. Companies that persist in treating climate change solely as a corporate social responsibility issue, rather than a business problem, will risk the greatest consequences.[145]

Experts from the World Resources Institute also state:

> Even people skeptical of the dangers of global warming are recognizing that simply because so many others are concerned, the phenomenon has wide-ranging implications. Investors are discounting share prices of companies poorly positioned to compete in a warming world. Many businesses face higher raw material and energy costs as governments around the world enact policies placing a cost on emissions. Consumers take into account a company's environmental record when making purchase decisions. . . . According to GlobeScan, 76% of Americans believe global warming is a serious problem—and 50% believe it is very serious. (Although these numbers are high, all other countries surveyed except Kenya and South Africa reported even greater concern.) . . . Even in the United States, which has lagged the rest of the developed world in the regulation of greenhouse gas emissions, the debate is rapidly shifting from whether climate change legislation should be enacted to when and in what form.[146]

For example, big investors are demanding disclosure from companies in this area. The Carbon Disclosure Project, a coalition of institutional investors representing more than $31 trillion (yes, that's a "t") in assets, annually requests information from large multinational companies about their climate-risk positioning. In other words, crafting a strategy for environmental issues is a strategic imperative and not just a CSR initiative.

Companies who worry about the cost of a sustainable business program can usefully reorient their thinking in terms of the cost of not having one. More importantly, why run the risk of catastrophic changes when important steps can be taken now?

The strategic nature of environmental issues is reflected in a McKinsey study of 2,192 executives from around the world in December 2007, which found that 60% of global executives view climate change as an important consideration in their company's overall business strategy, including its impact on the company's reputation, new product development, purchasing, and supply chain management.[147] The varying responses by executives in different countries was striking, with the Asia-Pacific region, China, Europe, India, Latin America, and North America, in that order, viewing climate change as important. However, despite its acknowledged importance, climate change issues are not executed well. Only 36% of executives said that climate change is explicitly addressed in setting business strategy, and 55% said that their companies do only "somewhat well" or "not at all well" in taking climate change into strategic consideration.

The study also showed that high-tech companies specifically (compared to companies in other industries) are more likely to take climate change issues into consideration in response to concerns about company reputation and competitive pressures. Dell was an early mover in this regard. It was the first company with a computer recycling goal in 2004, and as part of its "Soul of Dell" initiative, instituted a culture of sustainability and responsibility to reduce environmental impacts through product design, manufacturing, operations, and end-of-life considerations (i.e., a goal of 50% increase in product recovery).[148]

High-tech companies also innovate new technologies in response to environmental considerations, and they "eat their own dog food"[149]—meaning that if they position their products on an environmentally friendly message, then their internal business operations better be environmentally friendly as well, often utilizing the very same technologies they sell.

Moreover, consumers may be particularly sensitive to the environmental friendliness of technology: "The very idea of 'green' technology is somewhat problematic. How can products full of electronics, wiring, and complex components be environmentally friendly?"[150] Information technology (IT) products are major energy drains for companies, just like building energy costs and transportation costs. For example, different computer chips are more or less energy intensive. In addition, the majority of energy use from computers and other consumer electronics comes from passive energy use from computers that remained plugged in when not in use (e.g., by people who leave their computers on at work to receive network upgrades at night). The Environmental Protection Agency estimated in 2004 that in U.S. companies alone, more than $1 billion a year is wasted on electricity for computer monitors that are turned on when they shouldn't be.[151] The energy used in tech-heavy offices "full of PCs and power-guzzling devices" (monitors, printers, copiers, faxes) is expected to grow at double the rate of energy consumption in offices in general. As a result, corporate buyers and customers want to know about the energy efficiency of the products they buy. Reducing energy-related operating expenses is the main reason for pursuing sustainable IT operations, above "doing the right thing for the environment."[152]

Because of these reasons, high-tech companies have been leading the way for at least 10 years in this arena.[153] The World Economic Forum in January 2008 in Davos, Switzerland, featured the founders and/or chairmen of Microsoft, Dell, Cisco, and Intel, pledging to coordinate their efforts regarding sustainable business practices. In addition, the Green Electronics Council introduced the Electronic Product Environmental Assessment Tool (EPEAT) to allow companies to gauge the sustainability of high-tech products.

The Kyoto Protocol

The Kyoto Protocol, an international treaty which took effect in 2005 and was signed by about 170 countries, places limits on greenhouse gas emissions (as of 2008 the United States had not signed the treaty). For the five-year compliance period 2008–2012, countries that emit less than their quota can sell credits to countries that exceed their quota. This has given rise to the concept of **cap and trade**, whereby companies have a cap, or limit, on the amount of greenhouse gases they are allowed to emit. If they exceed this cap, they can buy credits from companies that emit less. This gives rise to trading in carbon credits. Trading exchanges have developed, such as the European Union Emission Trading Scheme for greenhouse gases. The Kyoto Protocol also established the Clean Development Mechanism (CDM), which validates and measures projects to ensure they produce authentic benefits and are genuinely "additional" activities to limit emissions that otherwise would not have been undertaken. Over time, the cap on companies and countries may be lowered by negotiation and political agreement to reduce pollution. Box 13.3 provides more detail on these agreements.

BOX 13.3

Cap and Trade

Concerns about global warming are fostering creative, and controversial, strategies for reducing emissions of heat-trapping gases such as carbon dioxide (CO_2). One such strategy is a "cap-and-trade" system. Under a *cap-and-trade* system, the government sets a cap on emissions by a designated group of polluters (e.g., utilities, oil refineries, steel companies) *and* distributes carbon credits, or permits to pollute, to participants. Over time, the caps become stricter, allowing less and less pollution, until the ultimate reduction goal is met. If participants pollute above their cap, they must decide whether it is cheaper to buy credits or to invest in cutting emissions.

Since those permits give companies a right to pollute, they become an asset and have a financial value. Companies that emit less than their permit amount allows can then sell that additional right to emit to other companies. So, those companies that are able to reduce emissions at a low cost can sell their extra permits to companies facing high costs (which will generally prefer to buy permits rather than make costly reductions themselves). According to a World Bank report released in May of 2008, the global carbon market grew from $31 billion in 2006 to $64 billion in 2007. New Carbon Finance, a New York-based research firm, estimates that the U.S. carbon market could reach $1 trillion by 2020.

The argument in favor of cap-and-trade systems is that they will exert constant pressure on polluters to reduce emissions while allowing them flexibility in determining how to accomplish the reduction. This will encourage companies to meet (or exceed) their emission targets in the most innovative and cost-effective way possible. By promoting innovation, cap-and-trade systems could help slow the pace of global warming while stimulating the development of new technologies.

A counter-argument is that, if innovation does not reduce emissions, the cap will suppress energy production—and the economy requires energy to grow. When demand exceeds supply, prices will increase. The Congressional Budget Office estimates that a 15% cut in emissions would raise average annual household energy costs by roughly 3% of income for the bottom four-fifths of the population. In this case, a cap-and-trade system would represent an implicit tax on consumers. Furthermore, in February 2008, the Congressional Budget Office said, "A tax on emissions would be the most efficient incentive-based option for reducing emissions and could be relatively easy to implement."

Sources: "Carbon: A Hot Commodity?" *Canadian Business*, June 16, 2008, 81 (10), p. 7. Mathers, Jason and Michelle Manion (2005), "How it Works: Cap-and-Trade Systems," *Catalyst* 4 (Spring 2005), available at www.ucsusa.org/publications/catalyst/catalyst-past-issues.html. Samuelson, Robert J. "Let's Just Call It 'Cap and Tax,'" *Newsweek*, June 9, 2008, p. 39.

In the smaller voluntary market, individuals or companies can buy **carbon offsets** to mitigate their own greenhouse gas emissions from transportation, electricity use, and other sources. Carbon offsets are financial instruments; the entities selling them invest the money in carbon-reducing projects, such as alternative energy generation or tree planting initiatives. Offsets may be cheaper or more convenient alternatives to reducing one's own fossil-fuel consumption, but some deals do not deliver true reductions. When traced to their source, these dubious offsets often encourage climate protection that would have happened regardless of the buying and selling of paper certificates.[154] One danger of largely symbolic deals is that they may divert attention and resources from more expensive and effective measures. This is a legitimate concern, as the market for offsets in the United States alone is estimated to be $100 million.[155] Such concerns will continue to be raised and addressed as this field develops.

Best-Practices Environmental Strategy: A Four-Step Approach

As shown in Table 13.8, best-practices environmental strategies take a four-pronged approach and seek to find new sources of competitive advantage in effective risk management.[156]

1. *Quantify the company's carbon footprint.* Using available reporting standards (e.g., see the Greenhouse Gas Protocol developed with the World Business Council for Sustainable Development, www.ghgprotocol.org), prepare an inventory that provides a true account of the company's green-

TABLE 13.8 Four Steps to a Best-Practices Environmental Strategy

1. Quantify the company's carbon footprint.
2. Assess carbon-related risks and opportunities.
3. Adapt the company's business strategy.
4. Stay ahead of competitors.

Source: Lash, J., and F. Wellington, "Competitive Advantage on a Warming Planet," *Harvard Business Review* 85 (March 2007), pp. 20–29.

house gas emissions, including direct and indirect (e.g., firm's energy consumption, employee business travel) emissions. First and foremost, this assessment of the carbon footprint sends the signal that the company is serious about environmental responsibility; companies that don't report their carbon footprint are assumed to have problems. It also sets the stage for the second step (to assess risks and opportunities).

The importance of this first step cannot be overstated: McKinsey reports that 60% of executives whose companies consider managing environmental issues to be at least somewhat important report that their companies don't have defined corporate emissions targets for greenhouse gases—and 15% don't even know if their companies do or not. However, given that more than 80% of these executives expect some form of climate change regulation in the next five years, they will need a measurement program in place, and knowing a company's carbon footprint is the only place to start.

2. *Assess carbon-related risks and opportunities.* Carbon-related risks and opportunities can arise from regulatory trends (e.g., mandatory emissions-reduction legislation) or from competitive and technology trends (e.g., competitors developing climate-friendly products). In addition, consumer or shareholder backlash or lawsuits can cost the company in litigation or damaged reputation. Supply chain issues could include suppliers who pass on higher energy costs or damage from extreme weather. Risks may be found in availability problems for energy or water, the reliability of infrastructure or supply chain, or even the prevalence of infectious diseases.

With respect to regulation risks in particular, many U.S. states and cities are not waiting for the federal government to act. California has enacted state regulations requiring that by 2016, greenhouse gas emissions from new cars must be reduced by 30%; California also passed legislation to reduce total emissions to their 1990 levels by 2020. Twenty states require utilities to obtain a percentage of the power they sell from renewable sources, and 208 U.S. cities have adopted programs to reduce emissions. For most businesses, a comprehensive federal policy concerning climate change is preferable to a patchwork of state and local regulations. More than 40 *Fortune* 500 companies have announced that they favor mandatory federal regulation of greenhouse gases. They would prefer to know the rules soon rather than be surprised by sudden political urgency; regulations would help with long-term predictability.

In terms of opportunities, companies can find opportunities to enhance or extend positioning, and new sources of revenue, through new environmentally friendly products. For example, many companies featured throughout this book have found that "green" business and products are growth opportunities. The Toyota Prius, GE, BioPower Systems, and the Biomimicry Institute have all experienced revenue growth due to environmentally friendly products and approaches. Companies can also find opportunities by lowering their carbon footprint, cutting operational costs through energy-efficiency measures. Although these measures may require an up-front investment, they typically lower variable operating costs (e.g., energy and water utilization) in the long run.

3. *Adapt the company's business strategy.* The company should identify ways it can reduce energy consumption and carbon emissions, as well as develop new products or business models to seize new opportunities. The minimum adaptation to a business strategy is to seek operational effectiveness in terms of carbon emissions. Even before global warming became more broadly accepted, the cover article in *Business Week* (August 16, 2004) noted that "many companies have discovered, often to their surprise, that [environmentally friendly strategies] save money and spur development of innovative technologies."[157] For example, DuPont cut its greenhouse gas emissions by 65% between 1990 and 2004, and saved hundreds of millions of dollars in the process.

A firm that understand its emissions costs—say, through a simple ratio of profits to total emissions in the company's value chain—may find that just-in-time inventory management may not be optimal in a world where emissions costs rise. Moreover, e-commerce, with its proliferation of small shipments, may also face limits. Offshored manufacturing that drives up emissions by lengthening transportation hauls may be supplanted by "nearshoring."[158] These costs would be missed without an environmental strategy.

4. *Adapt to global climate change better and faster than competitors.* As Lash and Wellington (from the World Resources Institute) say, "Doing well by doing good isn't good enough."[159] Companies who find competitive advantage in dealing with global warming manage their risks better and find new business opportunities faster than competitors.

One company that was an early mover in adopting a proactive environmental philosophy was Intel, as highlighted in this chapter's Technology Expert's View from the Trenches.

TECHNOLOGY EXPERT'S VIEW FROM THE TRENCHES

Innovation and Environmental Leadership at Intel

J. McGregor Agan
Director of Marketing, Corporate Affairs Group
Intel Corporation, Hillsboro, Oregon
With contributions from John Harland, Intel Principal Engineer

Intel has a long history of environmental leadership in our products, operations, and advocacy around the world. We have a passion for innovation that we apply not only to developing new technologies, but also to managing our environmental impact. From responsible product design to corporate recycling programs, Intel strives to be a leader in environmental sustainability. Our culture of environmental excellence has been instilled since our founding 40 years ago.

Today it is not only important to "be" an environmentally responsible company, but to "be perceived" as one as well. This perception increasingly contributes to the value and trust attributed to our company by governments, investors, customers, consumers, and employees. Intel's operations and market presence span 70 countries served by more than 80,000 employees. Every day the environment is a consideration in the wide range of decisions we make. The implications of these decisions can affect our ability to operate, sell, and ultimately create value in the marketplace.

Building awareness of Intel corporate responsibility and establishing a credible relevance is something never to be taken lightly. Intel's perception as an environmentally responsible leader has in large part been built through a transparent approach to reporting over many years, direct community relationships, and observations communicated by others tracking environmental performance in the industry.

Increasingly our customers and purchasing consumers are requiring a known level of "greenness" in what they buy. In addition, there is a desire to understand the environmental impact related to the development, manufacturing, and distribution of products they consider. Responding to this, Intel has begun to call out the inherent environmental benefits in our products, such as energy efficiency and removal of lead and halogen materials in our products.

Manufacturing facilities in particular face obvious environmental challenges with regard to energy efficiency, water recycling, air quality, and materials recycling. Today, as materials science is pushed to its physical boundaries, the environmental challenges for fabricating cutting-edge materials grow more complicated. The key question for progress is: How does a manufacturer push those technological boundaries without causing environmental problems? Intel takes environmental protection seriously, and is committed to incorporating environmental performance goals into product design, manufacturing processes, and facilities use. The bottom line is that doing the right thing for the environment means doing the right thing for business. This is both a top-down and bottom-up ideal of environmentalism at Intel and, as the company has reaffirmed again and again, it is not a solely altruistic process. There are significant benefits to maintaining a strong, pro-environment policy.

A recent example of this policy in action was the decision to establish a major chip fabrication plant (Fab 68) in Dalian, China. Intel is committed to designing, building, and operating facilities to minimize the impact to the environment, no matter where we operate. In this case the design standards for Fab 68 often exceeded the local requirements in many areas such as water, energy, and chemical waste management. Intel is applying the same world-class design and construction standards in Fab 68 that we apply everywhere in the world.

We apply that discipline in every Intel location and we are proud to bring that culture to F68, the Dalian community, and China. A clear win-win is formed as we establish a strategic manufacturing location serving the growing Chinese consumer demand for PCs and Intel technology-based products. China wins though economic development and shares in the benefits of environmental practices that can influence and be shared with other companies and future development.

Every year, Intel spends enormous resources on R&D, and very large sums of capital to turn innovation into real-world components. For Intel the goal is to build environmental health and safety attributes into the design and development process. Delivering on this goal ensures alignment to supporting the "be" and "be perceived as" requirements of a global eco-responsible industry leader.

Natural Capitalism

Companies can find additional advantage in environmentally friendly business strategies through **natural capitalism**,[160] or the application of a proper valuing of natural resources so that business strategies accurately account for their impact on the environment. Natural capitalism demonstrates that business can be good for the environment, and its strategies include the following.[161]

- **Dramatically increase the productivity of natural resources.** By taking a "whole systems" approach to reducing wastefulness (rather than piecemeal), energy saving, productivity-enhancing improvements can be achieved at lower cost, say, by piggybacking them onto periodic renovations that all buildings need. For example, windows with special glass that lets in cool, glare-free daylight can be combined with more efficient lighting to reduce air-conditioning by 75%. In turn, saving on renovating an air-conditioning system can pay for the new windows. Dow Europe cut the use of its office paper by 30% in six weeks by discouraging unneeded information (by asking people to respond electronically if they actually read the document sent, and then trimming distribution lists). AT&T set default modes on office printers/copiers to double-sided, reducing paper costs by 15%.
- **Shift to biologically inspired production models.** "Closed-loop" manufacturing can create new products and process that can prevent waste. For example, when Motorola need to replace chlorofluorocarbons for cleaning printed circuit boards after soldering, rather than finding an alternative it redesigned the whole soldering process so they needed no cleaning at all. Xerox saved $700 million from remanufacturing and expects to save another $1 billion by remanufacturing its new reusable line of "green" photocopiers.
- **Engage in "solutions-based business" thinking.** Rather than building business models around selling products, companies that practice natural capitalism think in terms of solving customer problems. For example, rather than selling carpets as a product, Interface is an innovative company that provides floor-covering as a service, in its "Evergreen Lease." This minimizes carpeting material by 80%, with the company replacing only the worn parts of a carpet as needed. DuPont leases its dissolving services because the same dissolving solvent can be reused scores of times, whereas when individual customers bought the product, they might use it once and then throw away the remainder.
- **Reinvest in natural capital.** Companies actively think about ways to restore and expand ecosystems and biological resources. This can be tough to do, because many business practices reward companies for wasting natural resources: "Even though the road seems clear, the compass that companies use to direct their journey is broken."

Challenges for Environmentally Responsible Business Practices

Certainly, much will be learned in the coming years about best-practices environmental strategies. In the meantime, three challenges are highlighted here.

BEWARE OF UNINTENDED CONSEQUENCES[162] Often, a company's initial efforts to be environmentally friendly may have negative consequences. For example, the push to use biofuels has shown that the source of biomass matters. Corn as a source of biofuel is not only *not* environmentally friendly (the energy to produce it and its relative energy inefficiency means that it does not really cut emissions), but it has driven up the price of food. However, second-generation biofuels, say from cellulose or jatropha, are more environmentally friendly. Carbon trading also shows unintended consequences. For example, the EU's emissions-trading program could result in a financial windfall—to the tune of 34 billion euros—for Germany's coal-fired power generators.

BEWARE OF GREENWASHING A series of industry watchdog groups have stepped up to monitor the consistency of a company's marketing messages and its actual activities. For example, with respect to environmental claims, such groups include Greenwashing Index (www.greenwashingindex.com), Stop Greenwash (www.stopgreenwash.org), and Greenwashing.net (www.greenwashing.net). The term *greenwashing* (a combination of "whitewashing," meaning to gloss over or cover up, and "green," meaning environmentally friendly) was coined in 1986 to refer to the hotel industry's practice of placing placards in each room, promoting the reuse of towels and linens, ostensibly to lessen the use of water, washing, energy, and related environmental impacts. However, in most cases there was little or no effort toward actual waste recycling, and in fact the "green campaign" was designed primarily to increase profits.[163] As used today, *greenwashing,* or *green sheen,* refers to organizations' efforts to inaccurately describe their products or business practices as beneficial to the environment, when in fact they are not. Simply changing the name or label of a product—say, by putting an image of a forest on a bottle of harmful chemicals—may capitalize on the sociocultural trend of customers wanting to buy green products, but the company runs a serious risk in damaging its reputation.

In December 2007, the environmental marketing company TerraChoice found that 99% of 1,018 common consumer products randomly surveyed were guilty of greenwashing. The study identified six "greenwashing sins"[164] listed in Table 13.9. Companies must be careful about how they advertise and promote their greenness.[165] If a company spends significantly more money or time to advertise *being* green than it spends on environmentally sound practices, it is at risk. The most egregious greenwashers include the energy and automotive industries, with bottled water close behind. For example, one report stated that half of all hybrid vehicles on the market in 2008 were no more fuel-efficient than their nonhybrid versions, and "the other half are phony, or 'hollow hybrids,'" meaning they have neither hybrid technology nor fuel efficiency.[166] The consumer electronics and renewable energy industries fair better.[167]

FAIRNESS AND EQUITY FOR DEVELOPING NATIONS Environmental challenges also arise from the growth in China, India, and other developing nations and the accompanying increase in consumption of coal, gasoline, and other fossil fuels. In 2004, developing countries emitted just over a third of the world's greenhouse gases—less than one-fifth as much per person as industrialized nations. But, as their citizens buy more cars and consume more energy, by 2100 these countries will emit 2 to 3 times as much as the developed world.[168] These countries need incentives to invest in advanced technology, which will leapfrog standard energy practices.

A FRAMEWORK FOR NAVIGATING ETHICAL CONTROVERSIES

In addition to ethical dilemmas arising from corporate social responsibility initiatives, many ethical dilemmas also arise from technological breakthroughs themselves. Indeed, technological breakthroughs often outpace society's capacity to deal with the ethical dilemmas they pose.

TABLE 13.9 Six Greenwashing Sins

Sin	Frequency of Occurrence	Example
Sin of the Hidden Trade-Off	57% of all environmental claims	"Energy-efficient" electronics that contain hazardous materials
Sin of No Proof	26%	Shampoos claiming to be "certified organic," but with no verifiable certification
Sin of Vagueness	11%	Products claiming to be 100% natural when in fact they include many naturally occurring hazardous substances like arsenic and formaldehyde
Sin of Irrelevance	4%	Products claiming to be CFC-free, even though CFCs were banned 20 years ago
Sin of Fibbing	<1%	Products falsely claiming to be certified by an internationally recognized environmental standard like EcoLogo, Energy Star or Green Seal
Sin of Lesser of Two Evils	<1%	Organic cigarettes or "environmentally friendly" pesticides

Source: Paul Schaefer, "The Six Sins of Greenwashing—Misleading Claims Found in Many Products," Environmental News Network, December 3, 2007, online at www.enn.com/green_building/article/26388.

High-tech firms cannot simply ignore the controversy that inevitably surrounds many technological developments.

The unfortunate reality faced by companies wrestling with ethical dilemmas is that they often find themselves in a catch-22: They are "damned if they do and damned if they don't." Even if a company acts in a way that ultimately reflects some level of ethical responsibility, it can be criticized by those who disagree. This is the nature of business. At a minimum, a company should avoid things that would cause embarrassment, were they to appear on the front page of the news.

One framework for addressing ethical dilemmas is shown in Table 13.10. Appendix 13.A at the end of the chapter demonstrates the use of this framework in the context of the pharmaceutical industry, with the situation faced by Merck when it discovered a drug that would potentially cure "river blindness," a disease afflicting many young adults in developing countries. The systematic approach advocated by the framework considers all stakeholders involved and their relative stakes in the decisions, and uses a solid basis to prioritize their relative needs; this leads to a sense that the company's decision is not a hasty one, but rather explicitly considers all parties' perspectives.

The benefits of using a systematic framework to guide the resolution of ethical dilemmas are threefold. First, the framework makes explicit the various issues with which the company will wrestle, regardless of the resulting outcome. For example, in the Merck case, the framework makes explicit the need to manage possible shareholder concerns about a loss of profitability against the need

TABLE 13.10 Framework for Addressing Ethical Dilemmas

1. Identify all stakeholders who are affected by the decision.
2. For each stakeholder group, identify its needs and concerns, both if the decision *is* implemented and if the decision *is not* implemented.
3. Prioritize the stakeholder groups and perspectives.
4. Make and implement a decision.

to manage scientists' incentives for discovery. Second, the framework brings into stark relief the various perspectives of the affected stakeholders. By highlighting these perspectives and their respective "stakes," a firm gets a better sense of the magnitude of the controversy it will encounter, regardless of the decision it makes. Third, the framework leads to a heightened sense of commitment to the resulting outcome. In an ethical dilemma, a company is exposed to criticism, regardless of which side the resulting decision supports. By using the framework, company spokespeople can feel greater confidence in their decisions that come under attack. In communicating the issues to concerned third parties (e.g., the media or boards of directors), the thoroughness of the process used to make the decision can be reassuring.

Summary

This chapter addressed triple bottom line considerations for high-tech companies. Given trends in today's business environment, companies must adopt a triple bottom line approach or they will be out of step with societal needs and expectations.

Consistent with this view, the chapter presented a broadened perspective on corporate social responsibility (CSR). Although CSR is an increasingly prevalent part of a company's business strategy, it is not without its controversies. The chapter presented the arguments on both sides of the CSR debate, and concluded that best-practices businesses today adopt CSR as an overarching mandate. Companies adopting such a perspective pursue CSR initiatives that focus on people and societal needs/problems, as well as environmental needs/problems. Rather than a one-size-fits-all approach to CSR, different companies adopt different models. Some focus primarily on economic/financial outcomes, with a strong tie to business strategy, while others are more focused on noneconomic criteria and enhancing social welfare. The measurement of a company's corporate social and environmental performance is particularly challenging, and the methods used require further development.

The second major section in this chapter focused specifically on base-of-the-pyramid (BOP) markets. Second-generation CSR and social entrepreneurship BOP strategies bring a solution-based, capacity-building focus to solving problems caused by extreme poverty. Best-practices models in this area focus on empowerment of the impoverished people themselves, and a co-creation model of development. By understanding the key challenges and success factors, companies' efforts in BOP markets can be enhanced. Even when companies follow best-practices strategies, BOP strategies may still be criticized.

The third section of the chapter focused on the digital divide, or the disparity in access to technology between different groups of people in society. This problem manifests itself both in the U.S. as well as in developing countries. As with the other topics in this chapter, the solutions to bridging the digital divide are many and varied and require creative, market-based interventions. When implemented successfully, mobile telephony can be a key solution to empowering people in developing markets to raise their standard of living and gain access to education and health care, as well as other needed services.

Fourth, given the increasing prevalence of a triple bottom line approach to managing business, addressing environmental impacts is vital. Companies that manage and mitigate their exposure to climate change risks while seeking new opportunities for profit will generate a competitive advantage over rivals in a carbon-constrained future. High-tech companies have taken a leadership role in pursuing environmental sustainability. Companies that are serious about reducing their impact on the environment follow a four-step model presented in this section of the chapter. Cap-and-trade systems and carbon offsets provide other mechanisms to minimize greenhouse gas emissions.

Finally, technology products carry with them inherent ethical controversies. The chapter concluded with a framework to address such controversies; Appendix 13.A provides a thorough illustration of how to use the framework.

CONCLUDING REMARKS: REALIZING THE PROMISE OF TECHNOLOGY

Many of the vital technologies in the world today were literally inconceivable only a short time ago. Technological development requires creativity, resources, perseverance, and serendipity. Innovators must be tireless crusaders regarding their inventions, as the time lag between the invention and market acceptance may be significantly longer than expected. The promises that new technologies offer can be realized only if companies take careful steps in the commercialization and marketing of their ideas. Smart marketing based on systematic consideration of critical issues is necessary to allow innovations to benefit society in the ways they are intended. Without effective marketing of high-technology products and innovations, the benefits such innovations can yield will remain elusive.

Discussion Questions

Chapter Concepts

1. What are the six key domains that pose opportunities and threats for companies? How do the trends in these domains relate to companies' pursuit of the triple bottom line?
2. What is the triple bottom line?
3. What is corporate social responsibility (CSR)?
4. What is the United Nations Global Compact? Look up the 10 principles online and explain how a high-tech firm would use them in its business approach.
5. Why is CSR controversial? For the major issues, what are the two sides of the debate? Show a sophisticated understanding. State your own position.
6. What outcomes do businesses seek with their CSR initiatives?
7. What are the various domains included in the "people" aspect of CSR?
8. What is corporate philanthropy? How do companies select the causes they give to?
9. What are the various ways a company can exhibit social responsibility with respect to the planet?
10. What is sustainability?
11. What is meant by "cradle-to-grave" or "cradle-to-cradle" thinking?
12. Describe the three different approaches companies take to CSR. Be thorough.
13. Why is a simple "Profit = Revenue − Cost" model insufficient for evaluating a company's triple bottom line performance? How do triple bottom line companies view "profit" compared to companies that don't follow a triple bottom line philosophy?
14. What are the challenges in assessing triple bottom line performance (also known as *corporate social performance* and *environmental performance*)?
15. Why is it important that companies measure their social and environmental performance? How can they best do so?
16. What is the relationship between a company's social performance and its financial performance? Why would companies with a strong CSR philosophy find this question superfluous?
17. What explanations exist for the lack of a direct relationship between a company's social performance and its financial performance? Be sure to be thorough in your answer.
18. What is the impact of a company's CSR initiatives on consumer response?
19. Identify the various tips for best-practices CSR strategies. With respect to these best practices:
    a. What are the four domains that guide a company's CSR initiatives to its business strategy?
    b. How can companies balance competing stakeholder interests?
    c. How might unintended reputational effects occur? How should a company mitigate against this possibility?
20. Describe the tiers of the world economic pyramid. Are BOP markets found only in developing countries?
21. Why are high-tech companies particularly well suited to solving BOP problems?
22. What is the continuum of approaches used to address BOP problems? How do market-based approaches differ from traditional approaches? What is their logic?
23. Why are BOP markets hard to serve?
24. What are the two approaches presented that categorize BOP problems/opportunities?
25. What were some of the problems with BOP 1.0 strategies? How do BOP 1.0 strategies differ from BOP 2.0 strategies? What are the key elements of BOP 2.0 strategies?
26. What is social entrepreneurship? What are some of the characteristics of social entrepreneurs?
27. What are the three types of ventures that social entrepreneurs can form? What is the debate regarding nonprofit versus business ventures?
28. What are the criticisms of BOP strategies? Do you think they are valid? How can they be addressed?
29. What is the digital divide? Why is access to technology considered a problem?
30. How does the digital divide manifest itself in the U.S.? Globally?
31. What are the various solutions that can be used to bridge the digital divide?

32. Why is mobile telephony increasingly being viewed as a solution to the digital divide? How do mobile phones solve problems that create poverty? What are the barriers to mobile telephony as a solution to the digital divide?
33. Why are executives increasingly concerned about global climate change? Why are high-tech companies playing a leadership role in environmental strategies?
34. What is the Kyoto Protocol? How do cap-and-trade systems work? What is the Kyoto Protocol's Clean Development Mechanism (CDM)?
35. What are carbon offsets? What concerns do they raise?
36. What is the best-practices four-step approach for a company's environmental strategy?
37. Why would companies prefer regulation of greenhouse gas emissions to the alternative?
38. What are some other strategies that companies can use in their pursuit of "natural capitalism"?
39. What are three challenges that companies will face with their environmental strategies?
40. What is greenwashing? What are the six ways companies greenwash their claims?
41. What are the steps used to address ethical dilemmas? What are the benefits of this four-step framework?

Application Questions

1. Examine Eli Lilly's corporate citizenship initiatives.
   a. Which of its initiatives fall into which areas of the triple bottom line?
   b. How should the performance of Lilly's MDR-TB initiative be evaluated?
   c. Which of the characteristics of best-practices CSR does it meet?
   d. What suggestions do you have for Lilly with respect to its efforts to run a triple bottom line business?
2. Pick a high-tech company and assess how the six key domains in the external environment will affect it. Be thorough and specific in terms of those that pose opportunities and those that pose threats.
3. Evaluate one high-tech company with respect to its approach to the triple bottom line.
4. Select a high-tech company and offer three suggestions for corporate social responsibility initiatives based on linking CSR to business strategy.
5. Identify a social entrepreneurial business enterprise. Describe its mission and approach. Evaluate it on the keys to success for pursuing BOP markets.
6. Select a company that could benefit from an environmental strategy. Apply the four-step model to create a strategy for it.
7. Select an ethical dilemma faced by a high-tech company. Use the four-step framework presented to analyze the issue and make a recommendation.

Glossary

cap and trade   A method of addressing greenhouse gas emissions whereby companies have a cap, or limit, on the amount they are allowed to emit; if they exceed this cap, they can buy credits from companies that emit less.

carbon offsets   Financial instruments that individuals or companies can buy to compensate for their own greenhouse gas emissions from transportation, electricity use, and other sources; the entities selling the instruments invest the money in carbon-reducing projects, such as alternative energy generation or tree planting initiatives.

corporate philanthropy   Charitable donations by a company to solve societal problems and needs, such as education, health care, and hunger.

digital divide   The disparity in access to technology between the haves and have-nots in society.

greenwashing   False claims by a company that its products or business practices are environmentally friendly; also called *green sheen*.

natural capitalism   The application of a proper valuing of natural resources so that business strategies accurately account for their impact on the environment.

social accounting   The methods used to document and communicate a company's triple bottom line performance; often relies on intangible value and assets.

social entrepreneurship   The use of entrepreneurial principles to create and manage an organization in order to solve social problems; when used to solve environmental problems, it is called *environmental entrepreneurship*.

sustainability   The quality of business strategies that meet the needs of the current generation without compromising the ability of future generations to meet their needs by using renewable resources and accounting for environmental impacts.

triple bottom line   A business philosophy that measures corporate success using three criteria: economic ("profit"), societal ("people"), and environmental ("planet").

Notes

1. GMA Network, "World at Risk for Multi-Drug Resistant TB—WHO Official," GMANews.TV, July 21, 2008, online at www.gmanews.tv/print/108249.
2. Eli Lilly and Company, "The Lilly MDR-TB Partnership," 2008, online at www.lillymdr-tb.com.
3. Ibid.
4. World Health Organization, "Tuberculosis," Fact Sheet #104, March 2007, online at www.who.int/mediacentre/factsheets/fs104; and World Health Organization, "Drug- and Multidrug-Resistant Tuberculosis (MDR-TB)," 2008, online at www.who.int/tb/challenges/mdr.
5. Donnelly, John, "Doctors' Call Spurs Massive Pledge to Fight TB," *Boston Globe,* April 10, 2007, p. A1.
6. Eli Lilly and Company, "The Lilly MDR-TB Partnership."
7. Donnelly, "Doctors' Call Spurs Massive Pledge to Fight TB."
8. Business Roundtable, "'SEE'ing Change: 2008 Progress Report," online at http://seechange.businessroundtable.org/PDF/SEEing_Change_2008_Progress_Report.pdf.
9. Donnelly, "Doctors' Call Spurs Massive Pledge to Fight TB."
10. Lilly MDR-TB Partnership, Questions and Answers, http://www.lillymdr-tb.com/question_answers.html; and PRNewswire, "Lilly Partnership Helps with Multi-Drug Resistant Tuberculosis in South Africa," November 9, 2007, online at www.prnewswire.com/cgi-bin/stories.pl?ACCT=104&STORY=/www/story/11-09-2007/0004701760&EDATE=.
11. ABC News, "WHO Unveils Multidrug-Resistant TB Test," July 1, 2008, online at www.abc.net.au/news/stories/2008/07/01/2290543.htm.
12. National Institute of Allergy and Infectious Diseases/National Institute of Health, "NIAID Describes Research Priorities to Fight Drug-Resistant Tuberculosis," NIH News, April 22, 2008, online at www3.niaid.nih.gov/news/newsreleases/2008/tb_priorities.htm.
13. Lilly Corporate Citizenship Report 2006/2007, "Providing Value to Patients and Society," Executive Summary, available at http://www.lilly.com/pdf/citizenship_report_0607.pdf.
14. Enkvist, Per-Anders, and Helga Vanthournout, "How Companies Think about Climate Change: A McKinsey Global Survey," *McKinsey Quarterly*, February 2008, available at www.mckinseyquarterly.com/how_companies_think_about_climate_change_a_mckinsey_global_survey_2099.

15. "People, Planet, Profit" phrase was coined for Shell by SustainAbility (www.sustainability.com); see also Hart, Stuart, *Capitalism at the Crossroads: Aligning Business, Earth, and Humanity,* 2nd ed. (Upper Saddle River, NJ: Wharton School Publishing, 2007).
16. Elkington, John, "Towards the Sustainable Corporation: Win-Win-Win Business Strategies for Sustainable Development," *California Management Review* 36 (Winter 1994), pp. 90–100; Elkington, John, *Cannibals with Forks: The Triple Bottom Line of 21st Century Business* (Gabriola Island, BC, Canada: New Society Publishers, 1998); and Brown, D., J. Dillard, and R. S. Marshall, "Triple Bottom Line: A Business Metaphor for a Social Construct," Portland State University, School of Business Administration, March 3, 2006, online at www.sba.pdx.edu/faculty/darrellb/dbaccess/MIM/TBL.pdf.
17. See also Mohr, Jakki, and Shikhar Sarin, "Building on Drucker's Insights: Implications for Market Orientation and Innovation in High-Technology Marketing," *Journal of the Academy of Marketing Science* 37 (Winter 2009, forthcoming).
18. Drucker, Peter, *The Concept of the Corporation* (New York: John Day Company, 1946), p. 17.
19. Drucker, Peter, *The Practice of Management* (New York: Harper & Row, 1954), p. 4.
20. Bush, Michael, "Students Rank Social Responsibility," *Advertising Age,* August 4, 2008, available at http://adage.com/article?article_id=130037.
21. Brandsizzle.com, "Global Survey Says Consumers Will Pay More to Support Socially Responsible Causes," November 26, 2007, online at www.brandsizzle.com/blog/2007/11/global-survey-s.html.
22. Bonini, Sheila, Jiefh Greenye, and Lenny Mendonca, "Assessing the Impact of Societal Issues: A McKinsey Global Survey," *McKinsey Quarterly,* November 2007. online at www.mckinseyquarterly.com/assessing_the_impact_of_societal_issues_a_mckinsey_global_survey_2077.
23. Mohr and Sarin, "Building on Drucker's Insights."
24. Benioff, Marc, and Karen Southwick, *Compassionate Capitalism: How Corporations Can Make Doing Good an Integral Part of Doing Well* (Franklin Lakes, NJ: Career Press, 2004).
25. Benioff, Marc, *The Business of Changing the World* (New York: McGraw-Hill, 2007), pp. 151–152.
26. Carey, J., "Global Warming," *Business Week,* August 16, 2004, cover story, pp. 60–69.

27. Lash, J., and F. Wellington, "Competitive Advantage on a Warming Planet," *Harvard Business Review* 85 (March 2007), pp. 20–29, quote from p. 28; see also Elkington, John, and Pamela Hartigan, *Power of Unreasonable People: How Social Entrepreneurs Create Markets That Change the World* (Boston: Harvard Business Press, 2008).

28. McGregor, Jena, "GE: Reinventing Tech for the Emerging World," *Business Week,* April 17, 2008, p. 68.

29. Lash and Wellington, "Competitive Advantage on a Warming Planet."

30. "The Ten Principles," United Nations Global Compact, 2008, online at www.unglobalcompact.org/AbouttheGC/TheTENPrinciples/index.html.

31. Friedman, Milton, "The Social Responsibility of Business Is to Increase Its Profits," *New York Times Magazine,* September 13, 1970, pp. 122–126.

32. Heslin, P. A., and J. D. Ochoa, "Understanding and Developing Strategic Corporate Social Responsibility," *Organizational Dynamics* 37(April–June 2008), pp. 125–144; see also Benioff, Marc, *The Business of Changing the World* (New York: McGraw-Hill, 2007).

33. Sasse, Craig, and Ryan T. Trahan, "Rethinking the New Corporate Philanthropy," *Business Horizons* 50 (January–February 2007), pp. 29–38, quote on p. 34.

34. Ibid.

35. Ibid.

36. Ibid.

37. Benioff, *The Business of Changing the World*, p. xii.

38. Elkington, *Cannibals with Forks.*

39. "A Lean, Clean, Electric Machine: The Greening of General Electric," *The Economist,* December 8, 2005, pp. 77–79.

40. "The State of Corporate Philanthropy: A McKinsey Global Survey," *McKinsey Quarterly*, February 2008, available at www.mckinseyquarterly.com/the_state_of_corporate_philanthropy_a_mckinsey_global_survey_2106.

41. Heslin and Ochoa, "Understanding and Developing Strategic Corporate Social Responsibility."

42. Mohr and Sarin, "Building on Drucker's Insights"; London, Ted, and Stuart Hart, "Reinventing Strategies for Emerging Markets: Beyond the Transnational Model," *Journal of International Business Studies* 35 (September 2004), pp. 350–371; Hart, Stuart, and Clayton Christensen, "The Great Leap: Driving Innovation from the Base of the Pyramid," *Sloan Management Review* 44 (Fall 2002), pp. 51–57; and Kirkpatrick, David, "Two Ways to Help the Third World," *Fortune,* October 27, 2003, pp. 187–196.

43. Benioff, *The Business of Changing the World.*

44. "Triple bottom line," Wikipedia, 2008, online at http://en.wikipedia.org/wiki/Triple_bottom_line.

45. "The State of Corporate Philanthropy: A McKinsey Global Survey."

46. "The State of Corporate Philanthropy: A McKinsey Global Survey."

47. Heslin and Ochoa, "Understanding and Developing Strategic Corporate Social Responsibility."

48. "The State of Corporate Philanthropy: A McKinsey Global Survey."

49. "Triple bottom line," Wikipedia.

50. McDonough, William, and Michael Braungart, *Cradle to Cradle: Remaking the Way We Make Things* (New York: North Point Press, 2002).

51. Esty, Daniel, and Andrew Winston, *Green to Gold: How Smart Companies Use Environmental Strategy to Innovate, Create Value, and Build Competitive Advantage* (New Haven, CT: Yale University Press, 2006).

52. Berger, I. E., P. H. Cunningham, and M. E. Drumwright, "Mainstreaming Corporate Social Responsibility: Developing Markets for Virtue," *California Management Review* 49 (Summer 2007), pp. 132–157.

53. Ibid., p. 139.

54. Demos, Telis, "Accounting for Accountability," *Fortune,* November 12, 2007, pp. 66–69.

55. Engardio, Pete, "Beyond the Green Corporation," *Business Week,* January 29, 2007, pp. 50–64.

56. Esty, Daniel, "Transparency: What Stakeholders Demand," in Forethought Special Report: Climate Business/Business Climate, *Harvard Business Review* 85 (October 2007), pp. 5–7.

57. Heslin and Ochoa, "Understanding and Developing Strategic Corporate Social Responsibility."

58. O'Sullivan, Dan, and Andrew Abela, "Marketing Performance Measurement Ability and Firm Performance," *Journal of Marketing* 71 (April 2007), pp. 79–93.

59. Bansal, P., C. Maurer, and N. Slawinski, "Beyond Good Intentions: Strategies for Managing Your CSR Performance," *Ivey Business Journal,* January–February 2008, pp. 1–8.

60. "Corporate social responsibility," Wikipedia, online at www.wikipedia.org/wiki/Corporate_social_responsibility.

61. Ibid.

62. "Triple bottom line," Wikipedia.

63. Mackey, Alison, Tyson Mackey, and Jay Barney, "Corporate Social Responsibility and Firm Performance: Investor Preferences and Corporate Strategies," *Academy of Management Review* 32, (2007), pp. 817–835.

64. Engardio, "Beyond the Green Corporation."

65. See the full listing in "The Global 100," online at bwnt.businessweek.com/interactive_reports/g100.

66. Zadek, Simon, "Inconvenient but True: Good Isn't Always Profitable," *Fortune,* November 12, 2007, p. 70.

67. Margolis, J., and J. Walsh, "Misery Loves Company: Rethinking Social Initiatives by Business," *Administrative Science Quarterly* 48 (2003), pp. 268–305; and Orlitzky, M., F. Schmidt, and S. Rynes, "Corporate Social and

Financial Performance: A Meta-Analysis," *Organization Studies* 24 (May–June 2003), pp. 403–441.

68. Barnett, M. L., and R. M. Salomon, "Beyond Dichotomy: The Curvilinear Relationship between Social Responsibility and Financial Performance," *Strategic Management Journal* 27 (November 2006), pp. 1101–1122.

69. Hull, C. E., and S. Rothenberg, "Firm Performance: The Interactions of Corporate Social Performance with Innovation and Industry Differentiation," *Strategic Management Journal* 29 (July 2008), pp. 781–789; see also Waddock, S., and S. Graves, "The Corporate Social Performance–Financial Performance Link," *Strategic Management Journal* 18 (April 1997), pp. 303–319; and McWilliams, A., and D. Siegel, "Corporate Social Responsibility and Financial Performance: Correlation or Misspecification?" *Strategic Management Journal* 21 (May 2000), pp. 603–609.

70. Hull and Rothenberg, "Firm Performance."

71. Brown, Tom, and Peter Dacin, "The Company and the Product: Corporate Associations and Consumer Product Responses," *Journal of Marketing*, 61 (January 1997), pp. 68–84; and Sen, Sankar, and C. B. Bhattacharya, "Does Doing Good Always Lead to Doing Better? Consumer Reactions to Corporate Social Responsibility," *Journal of Marketing Research* 38 (May 2001), pp. 225–243.

72. For an excellent overview, see Bhattacharya, C. B., and S. Sen, "Doing Better at Doing Good: When, Why, and How Consumers Respond to Corporate Social Initiatives," *California Management Review* 47 (Fall 2004), pp. 9–24.

73. "The State of Corporate Philanthropy: A McKinsey Global Survey."

74. Ibid.

75. Ibid.

76. Porter, Michael, and Mark Kramer, "The Competitive Advantage of Corporate Philanthropy," *Harvard Business Review* 80 (December 2002), pp. 5–16.

77. Heslin and Ochoa, "Understanding and Developing Strategic Corporate Social Responsibility."

78. Ibid.

79. Sasse and Trahan, "Rethinking the New Corporate Philanthropy."

80. Berger, Cunningham, and Drumwright, "Mainstreaming Corporate Social Responsibility."

81. "The State of Corporate Philanthropy: A McKinsey Global Survey."

82. Sasse, and Trahan, "Rethinking the New Corporate Philanthropy."

83. Gunther, Marc, "Tree Huggers, Soy Lovers, and Profits," *Fortune,* June 23, 2003, pp. 98–102.

84. Heslin and Ochoa, "Understanding and Developing Strategic Corporate Social Responsibility."

85. NextBillion.net, Development through Enterprise, World Resources Institute, online at www.nextbillion.net; see also Hammond, Allen, et al., *The Next 4 Billion: Market Size and Business Strategy at the Base of the Pyramid* (Washington, DC: World Resources Institute, IFC, and the World Bank, 2007), available at www.wri.org/publication/the-next-4-billion.

86. Jeffrey Sachs, *The End of Poverty: Economic Possibilities for Our Time* (New York: Penguin, 2005).

87. Simanis, Erik, and Stuart Hart, "The Base of the Pyramid Protocol: Toward Next Generation BoP Strategy," 2nd ed., Cornell University, 2008, p. 1, online at www.johnson.cornell.edu/sge/docs/BoP_Protocol_2nd_ed.pdf.

88. GlobeScan Incorporated, "Corporate Social Responsibility Monitor," 2007, online at www.globescan.com/csrm_overview.htm.

89. Hammond et al., *The Next 4 Billion.*

90. Ibid.

91. Hudnut, Paul, "Technology and Marketing in the Base of the Pyramid," presentation at the Academy of Marketing Science, May 29, 2008, Vancouver, BC.

92. Prahalad, C. K., and Stuart Hart, "The Fortune at the Bottom of the Pyramid," *Strategy + Business* 26 (First Quarter 2002), pp. 54–67; Prahalad, C. K., *The Fortune at the Bottom of the Pyramid: Eradicating Poverty through Profits* (Upper Saddle River, NJ: Pearson Education/Wharton School Publishing, 2005); and Mahajan, Vijay, and Kamini Banga, *The 86 Percent Solution: How to Succeed in the Biggest Market Opportunity of the 21st Century* (Upper Saddle River, NJ: Pearson Education/Wharton School Publishing, 2005).

93. Elkington and Hartigan, *Power of Unreasonable People,* p. 42.

94. Hammond et al., *The Next 4 Billion.*

95. Hudnut, "Technology and Marketing in the Base of the Pyramid."

96. This framework was developed by Paul Hudnut, "Technology and Marketing in the Base of the Pyramid," who used it originally to highlight the spheres for charity, social entrepreneurship, and for-profit business ventures for environmental and public health concerns.

97. Elkington and Hartigan, *Power of Unreasonable People.*

98. Simanis and Hart, "The Base of the Pyramid Protocol."

99. Elkington and Hartigan, *Power of Unreasonable People,* p. 114.

100. Simanis and Hart, "The Base of the Pyramid Protocol," p. 2; see also Karnani, Aneel, "The Mirage of Marketing to the Bottom of the Pyramid: How the Private Sector Can Help Alleviate Poverty," *California Management Review* 49 (Summer 2007), pp. 90–110.

101. Simanis and Hart, "The Base of the Pyramid Protocol," p. 5.

102. Simanis, Erik, and Stuart Hart, "Beyond Selling to the Poor: Building Business Intimacy through Embedded Innovation," working paper, University of Michigan, 2008.

103. Gunther, Marc, "Chasing the 'Base of the Pyramid,'" *Fortune,* November 15, 2006, online at http://money.cnn.com/2006/11/14/magazines/fortune/guntherkenya.fortune/index.htm.

104. "What's a BOPreneur?" 2008, online at http://bopreneur.blogspot.com.

105. Elkington and Hartigan, *Power of Unreasonable People.*

106. Ibid., p. 17.

107. Ibid., p. 4.

108. This section on the three types of social entrepreneurship ventures is drawn from Elkington and Hartigan, *Power of Unreasonable People.*

109. Scanlon, Jessie, "Giving the Poor a Means to Work," *BusinessWeek,* February 22, 2008, online at www.businessweek.com/innovate/content/feb2008/id20080222_960476.htm?chan=innovation_innovation+%2B+design_service+innovation; see also Polak, Paul, *Out of Poverty: What Works When Traditional Approaches Fail* (San Francisco: Berrett-Koehler, 2008).

110. See also Elkington and Hartigan, *Power of Unreasonable People,* pp. 40–42.

111. "The Laptop Wars," *The Economist.Com,* January 8, 2008, http://www.economist.com/business/displaystory.cfm?story_id=10489616.

112. Ibid.

113. Vachani, Sushil, and N. Craig Smith, "Socially Responsible Distribution: Distribution Strategies for Reaching the Bottom of the Pyramid," *California Management Review* 50 (Winter 2008), pp. 52–84.

114. "Entrepreneurial Design for Extreme Affordability," Stanford University Institute of Design, 2008, online at http://extreme.stanford.edu/index.html.

115. "The Laptop Wars," *The Economist.Com.*

116. Elkington and Hartigan, *The Power of Unreasonable People,* pp. 152–153.

117. Ibid.

118. See bio sketch of Ted London, director of the Base of the Pyramid Initiative, William Davidson Institute at the University of Michigan, online at www.wdi.umich.edu/About/People/TedLondon.

119. Karnani, Aneel, "The Mirage of Marketing to the Bottom of the Pyramid: How the Private Sector Can Help Alleviate Poverty," *California Management Review* 49 (Summer 2007), pp. 90–110.

120. Karnani, "The Mirage of Marketing to the Bottom of the Pyramid"; see also Landrum, Nancy, "Advancing the 'Base of the Pyramid' Debate," *Strategic Management Review* 1 (2007), pp. 1–12, online at www.strategicmanagementreview.com/ojs/index.php/smr/article/viewFile/12/16.

121. Arnould, Eric, and Jakki Mohr, "Dynamic Transformation of an Indigenous Market Cluster," *Journal of the Academy of Marketing Science* 33 (Summer 2005), pp. 254–274.

122. Cited in Sallis, Edward, and Gary Jones, *Knowledge Management in Education*, London, UK: Routledge, 2002, p. 99.

123. Crockett, Roger, "High Tech's Next Big Market? Try the Inner City," *Business Week*, December 20, 1999, p. 48.

124. "World Telecommunications/ICT Development Report, 2006: Measuring ICT for Social and Economic Development," Geneva, Switzerland: International Telecommunications Union.

125. "The Digital Divide at a Glance," World Summit on the Information Society, 2005, online at www.itu.int/wsis/tunis/newsroom/stats.

126. Horrigan, John, and Andrew Smith, "Home Broadband Adoption 2007," Pew Internet & American Life Project, June 2007, online at www.pewinternet.org/pdfs/PIP_Broadband%202007.pdf.

127. Ibid., p. 6.

128. Ibid.

129. Gartner Research, "Dataquest Insight: Consumer Broadband, Global Penetration Rates and Growth Prospects," July 2008, data cited in "17 Countries to Surpass 60% Broadband Penetration into the Home by 2012," available at: www.wirelessdesignasia.com/article-9016-17countriestosurpass60broadbandpenetrationintothehomeby2012-Asia.html; see also "US Falls to 25th in Broadband Penetration," WebSiteOptimizaion.com., 2008, online at www.websiteoptimization.com/bw/0704.

130. Twist, Jo, "Pocket Answer to Digital Divide," BBC News, November 18, 2005.

131. Albert, Terri, and Charles L. Colby, "The Technology Readiness of Vulnerable or Impacted Groups and Public Policy Considerations: A Cross-Cultural Research Program," paper presented at the 2003 AMA Public Policy Conference, Washington, DC, May 2003; and Albert, Terri, and Jakki Mohr, "Technology Readiness of Vulnerable Consumers: Implications for the Digital Divide," working paper, University of Montana, Missoula, 2004.

132. Twist, "Pocket Answer to Digital Divide."

133. Ewing, Jack, "Mad Dash for the Low End," *BusinessWeek,* February 18, 2008, p. 30.

134. Twist, "Pocket Answer to Digital Divide."

135. "Calling an End to Poverty: Mobile Phones and Development," *The Economist,* July 9, 2005, pp. 51–52; "The Real Digital Divide," *The Economist,* March 12, 2005, p. 11; "Calling Across the Divide," *The Economist,* March 12, 2005, p. 74; and "Behind the Digital Divide," *The Economist,* March 12, 2005, pp. 22–25.

136. Hammond et al., *The Next 4 Billion,* p. 7.

137. Waverman, Leonard, and the London Business School and GSM Association, as cited in "Can Mobile Communications Close the Digital Divide?" Ericsson white paper, October 2007, online at www.ericsson.com/technology/whitepapers/can_mobile_communications_close_digitaldivide.pdf.

138. Enriquez, Luis, Stefan Schmitgen, and George Sun, "The True Value of Mobile Phones to Developing Markets," *McKinsey Quarterly,* February 2007, www .mckinseyquarterly.com/public_sector/the_true_value_of_ mobile_phones_to_developing_markets_1917.

139. Cited in "To Do with the Price of Fish," *The Economist,* May 10, 2007; the original research was conducted by Robert Jenson, "The Digital Provide: Information (Technology), Market Performance, and Welfare in the South Indian Fisheries Sector," *Quarterly Journal of Economics,* August 2007.

140. Shanmugavelan, Murali, "Technology Tuesday: Market Forces Alone Won't End Digital Divide," posted March 11, 2008, at EndPoverty blog, http://endpovertyblog.org/ node/1131.

141. Intergovernmental Panel on Climate Change, online at www.ipcc.ch.

142. Bonini, Greenye, and Mendonca, "Assessing the Impact of Societal Issues."

143. "The State of Corporate Philanthropy: A McKinsey Global Survey."

144. Bonini, Greenye, and Mendonca, "Assessing the Impact of Societal Issues."

145. Porter, Michael, and Forest Reinhardt, "A Strategic Approach to Climate," in Climate Business/Business Climate (Forethought Special Report), *Harvard Business Review* 85 (October 2007), pp. 1–17, quote on p. 1.

146. Lash and Wellington, "Competitive Advantage on a Warming Planet," p. 20, p. 26.

147. Enkvist and Vanthournout, "How Companies Think about Climate Change."

148. Benioff, *The Business of Changing the World.*

149. Rather than focusing on environmental strategies specifically, the phrase "eat your own dog food" is used more generally to convey the idea that if a company markets a high-tech solution, say, an enterprise resource planning software package, then it ought to use its own product in its business.

150. "'Greener Gadgets' Isn't an Oxymoron," interview with Allan Chochinov, *Business Week,* February 11, 2008, online at www.businessweek.com/innovate/content/feb2008/ id20080211_832566.htm.

151. Arnold, Chris, "Computer Energy Waste a Major Cause of Pollution," *NPR Morning Edition,* June 15, 2004, online at www.npr.org/templates/story/story.php?storyId=1960428.

152. Jana, Reena, "Green IT: Corporate Strategies," *BusinessWeek,* February 11, 2008, online at www.businessweek.com/ innovate/content/feb2008/id20080211_204672.htm.

153. Hart, Stuart, "Beyond Greening: Strategies for a Sustainable World," *Harvard Business Review* 75 (January 1997), pp. 66–76.

154. "Another Inconvenient Truth: Behind the Feel-Good Hype of Carbon Offsets, Some of the Deals Don't Deliver," *BusinessWeek,* March 26, 2007, online at www.businessweek .com/magazine/content/07_13/b4027057.htm.

155. Ibid.

156. Lash and Wellington, "Competitive Advantage on a Warming Planet"; see also Heslin and Ochoa, "Understanding and Developing Strategic Corporate Social Responsibility."

157. Carey, "Global Warming."

158. Porter and Reinhardt, "Grist: A Strategic Approach to Climate."

159. Lash and Wellington, "Competitive Advantage on a Warming Planet," p. 26.

160. Hawken, Paul, Amory Lovins, and Hunter Lovins, *Natural Capitalism: Creating the Next Industrial Revolution* (Boston, MA: Back Bay Books, 2000).

161. Lovins, A. B., L. H. Lovins, and P. Hawken, "A Road Map for Natural Capitalism," *Harvard Business Review* 77 (May–June 1999), pp. 145–158, 211.

162. Ellison, Jesse, "Save the Planet, Lose the Guilt," *Newsweek.Com,* June 28, 2008, www.newsweek.com/id/ 143701/page/1.

163. "Greenwash," Wikipedia, online at http://en.wikipedia .org/wiki/Greenwashing.

164. Schaefer, Paul, "The Six Sins of Greenwashing—Misleading Claims Found in Many Products," *Environmental News Network,* December 3, 2007, online at www.enn.com/ green_building/article/26388.

165. Ibid.; see also "Sustainability," Dictionary of Sustainable Management, online at www.sustainabilitydictionary.com.

166. Ellison, "Save the Planet, Lose the Guilt."

167. Ibid.

168. Carey, "Global Warming."

169. "Merck & Co., Inc.," in David Held, *Property, Profit, and Justice,* The Business Enterprise Trust, 1991.

170. Ibid.

171. "World Bank and Merck & Co., Inc., Announce US$50 Million Funding Initiative to Eliminate River Blindness in Africa," press release, online at http://web.worldbank.org/ WBSITE/EXTERNAL/COUNTRIES/AFRICAEXT/0,, contentMDK:21586793~menuPK:258658~pagePK:286 106~piPK:2865128~theSitePK:258644,00.html; see also the following Web pages: "Onchocerciasis (river blindness)," at www.who.int/blindness/causes/priority/en/ index3.html; "African Programme for Onchocerciasis Control (APOC)," at www.who.int/apoc/en; "Onchocerciasis Control in the WHO African Region," at www.afro .who.int/rc57/documents/AFR-RC57-5_Onchocerciasis_ control_in_the_WHO_Region_final.pdf; "The Merck MECTIZAN Donation Program," at www.merck.com/ about/cr/mectizan; and "The Carter Center River Blindness (Onchocerciasis) Program," at www.cartercenter.org/ health/river_blindness/index.html.

172. "River Blindness Parasite Becoming Resistant to Standard Treatment," *Science Daily,* September 5, 2007, online at www.sciencedaily.com/releases/2007/08/070829212815 .htm.

APPENDIX 13.A

Application of a Framework to Address Ethical Controversies: Merck, Ivermectin, and River Blindness[169]

In 1978, a disease called "river blindness" plagued at least 85 million people throughout Africa and parts of the Middle East and Latin America. The cause of the disease is a parasitic worm carried by a tiny black fly that lives and thrives along fast-moving rivers. When the flies bite people, the larva of the parasitic worm enters the human body, eventually growing to more than 2 feet in length at adulthood and causing grotesque but relatively innocuous nodules in the skin. The health problems caused by the worms begin when the adult worm reproduces, releasing millions of microscopic off-spring that swarm through the body tissue, causing terrible itching, so terrible that some victims have committed suicide. After several years, these microscopic larvae cause lesions and depigmentation of the skin. Eventually, they invade the eyes, often causing blindness.

Indeed, the disease was so prevalent in some areas that the children assumed blindness was simply part of growing up. The World Health Organization labeled river blindness as a public health and socioeconomic problem: In their attempts to avoid flies, people abandoned fertile ground near rivers, moving to poorer land with decreased food production.

In Merck's labs, a scientist stumbled upon the possibility that one of its drugs used to eliminate insect-borne parasites in cows—ivermectin—might actually have properties that would eliminate the parasite that caused river blindness (onchocerciasis is the disease's medical name). However, from Merck's perspective, this was not necessarily good news. The process to determine which discoveries the company should pursue was difficult. For every pharmaceutical compound that becomes a "product" candidate, thousands of others fall by the wayside. If Merck made the decision to invest in further research for this drug, including conducting field trials in remote areas of the world, it would be faced with the following business problems. First, the population that would benefit from this discovery was relatively small. Second, the population lacked the means to pay for the medication. Third, there was no infrastructure in place to deliver the medication and oversee the administration of the series of treatments that was required. Fourth, if a human derivative of ivermectin proved to have any adverse health effects when used on humans, these adverse human effects might taint its reputation as a veterinary drug. Fifth, there was concern that a human version of the drug distributed to the Third World might be diverted to the black market, undercutting sales of its veterinary product. Clearly, Merck faced an ethical dilemma in how to proceed with this scientist's discovery. How would the use of the four-step framework from Table 13.10 (p. 471) help resolve this dilemma?

Step 1: Identify All Stakeholders Who Are Affected by the Decision

In Merck's case, the stakeholders might include the following:

- Shareholders
- The public at large
- The Third World population affected
- The government
- Employees

Step 2: For Each Stakeholder Group, Identify Its Needs and Concerns

This step requires identifying the results both if the decision *is* implemented and if the decision *is not* implemented. What were each stakeholder group's needs and concerns if the decision to explore the drug further were to proceed, and if the decision to explore the drug further were curtailed?

From the *shareholders' perspective,* shareholders were concerned about the likely negative impact on profits if this discovery were pursued. Studies at the time showed that, on average, it took 10 years and $200 million to bring a new drug to market. So, pursuing this development would be costly, regardless of the outcome. And, even in the most positive scenario, the expense of development and drug trials, coupled with the lack of a target market's ability to pay for the drugs, indicated that the company might lose money in proceeding further with this discovery.

On the other hand, the company might face negative publicity if it withheld useful medications from people. Possible repercussions from such negative publicity might include competition gaining advantage.

From the *public at large's perspective,* if the drug were further pursued the company might be viewed favorably for its efforts and gain more loyal customers. If the drug were shelved, the public might question Merck's sincerity in developing drugs to address human suffering. In choosing not to develop drugs that have the potential to significantly enhance the quality of life for people, the resulting skepticism could cause customer relationship problems. Moreover, given the presence of knowledge spillovers in high-tech markets, where innovations and developments in one area have the potential to lead to new innovations in other areas, a legitimate consideration was what other discoveries might never be made if the drug were not pursued.

From the *perspective of people afflicted with the disease,* if the drug were developed, and if it proved to be effective in combating river blindness, the quality of life of this population would be greatly enhanced. This could lead to other improvements. For example, support currently provided by relief agencies or local governments to assist this population with ongoing sustenance and survival could be redirected to other needy populations. On the other hand, there was also the risk that if the drug were developed, some unknown side effects might occur (which hopefully would be uncovered during additional development and prior to administration of the drug to this population). The company would have to weigh (as with any drug) the benefits the drug could deliver relative to possible side effects.

If the drug were not developed, not only would this population continue with the grim situation, but the placement of profits over people could lead to a general backlash against corporations in developing countries, with, at a minimum, ensuing negative publicity or possibly even local protests with resulting violence.

From the *government's perspective,* if the drug were further pursued, the U.S. government might choose to expedite the lengthy approval process in order to hasten Merck's ability to relieve human pain and suffering. If the drug were not pursued, although the U.S. government typically does not legislate drug development, a sufficient outcry might cause intervention, requiring Merck to make its formularies for the bovine medicine available on a licensing basis to other competitors so they could pursue its development for river blindness. Issues related to infrastructure and drug administration ultimately also would be government concerns.

From the *employees' perspective,* if the drug were developed and created financial losses, Merck employees might experience possible belt tightening and layoffs. Still, the scientists and people involved would feel confident that their work had purpose and meaning. If the drug weren't developed, employees might not face negative revenue implications, although the decision not to pursue potentially lifesaving discoveries might be demoralizing.

Step 3: Prioritize the Stakeholder Groups and Perspectives

How did Merck prioritize the various stakeholders? It is vital that this step not be simply a debate over people versus profits. Unfortunately, this over-simplification happens frequently with ethical business dilemmas, and the ensuing "discussion" typically is based on whose view is "right" or which views come from more powerful people in the firm. However, the simplistic approach does not lead to insights about priorities, nor does it address the relationship between the dilemma and the company's mission and its long-term perspective. Rather, when this step to prioritize stakeholder groups is done well, it clarifies organizational values and the implicit assumptions that underlie decision making in a company.

In this case, Merck had, over the years, deliberately fashioned a corporate culture to nurture the most creative, fruitful research. Its scientists were among the best paid in the industry and were given great latitude to pursue intriguing leads. Moreover, they were inspired to view their work as a quest to alleviate human disease and suffering worldwide. Employees found inspiration in the words of George W. Merck, son of the company's founder and its former chairman, that formed the basis of Merck's overall corporate philosophy:

> We try never to forget that medicine is for the people. It is not for the profits. The profits follow, and if we have remembered that, they have never failed to appear. The better we have remembered it, the larger they have been.[170]

At this step, a company can also look for creative solutions that reframe the ethical debate, looking for win-win solutions that do not require pitting different stakeholder groups' needs against one another. It is vital that a company remain in touch with the underlying ethical dilemma, however, and not attempt to resolve the debate in a moral vacuum.

Step 4: Make and Implement a Decision

In wrestling with the dilemma, Merck explicitly recognized that its success in the pharmaceutical market was, in large part, due to the efforts of its scientists in making discoveries just like this one. If its scientists did not believe that their discoveries would be used to their fullest capabilities, then not only would they possibly become demoralized, but Merck might also have difficulties recruiting, attracting, and retaining the best scientists. On this basis, it decided to proceed with further study on the drug, which was eventually released for human purposes in 1987.

Some may say that Merck's decision, based on its desire to keep talented, driven scientists, was fundamentally based on profits and self-serving; therefore, Merck could be criticized for a lack of moral grounding. However, it is important to recognize that the most obvious decision to support profitability in this case would have been for Merck to forego the drug, rather than to proceed. Indeed, given the lack of a profitable market opportunity, a decision based solely on profits would have led to a very different outcome.

Epilogue: The Gift of Sight

After a seven-year clinical research program, Merck & Company, in collaboration with the World Health Organization (WHO), demonstrated that a single oral dose of Mectizan, taken once a year, could prevent blindness and alleviate skin disease. In 1987, Merck decided to donate Mectizan free of charge to all people affected by river blindness, for as long as necessary. Merck approached Dr. William Foege, then executive director of the Carter Center, for assistance with the global distribution of Mectizan. Together, they created the Mectizan Donation Program (MDP) and housed it at the Task Force for Child Survival and Development, an independent partner of the Carter Center. MDP acted as the liaison among Merck, nongovernmental development organizations (NGDOs), affected countries' ministries of health, and UN organizations such as WHO.

The statue of a boy leading a blind man (found at the headquarters of the WHO, the headquarters of Merck & Co, the World Bank, and the Carter Center) has become recognized as a symbol of the fight to eliminate river blindness. The statue commemorates the success of three programs led by WHO and the Pan-American Health Organization: the Onchocerciasis Control Programme in West Africa (OCP), operating in 11 countries; the African Programme for Onchocerciasis Control (APOC), covering 19 countries outside West Africa; and the Onchocerciasis Elimination Program for the Americas (OEPA), present in 6 countries. The disease was largely eliminated in 10 of the 100 West African countries that were targeted with a large-scale aerial spraying program, which has enabled resettlement and cultivation of more than 25 million hectares of land, boosted farm productivity and rural incomes, and prevented 600,000 people from going blind. As of December 2007, its 20th anniversary, the Mectizan Donation Program had donated more than 530 million doses of Mectizan valued at roughly US$2.7 billion. As the longest-running medicine donation commitment, the program reaches about 70 million people each year in 120,000 remote communities for treatment.[171]

Merck reaffirmed its commitment to this program on December 4, 2007, pledging $25 million toward the World Bank's $70 million needed to support it through 2015. This is important, as the disease shows signs of becoming drug resistant,[172] much like TB.

CASES

This section includes seven cases that span a wide range of technologies and business models, as well as business-to-business and business-to-consumer applications, as follows:

- **Skype:** A Web-based software company offering Voice-over-Internet Protocol (VoIP) services
- **TiVo:** Digital video recorder hardware combined with superior software functionality, advertising services, online and retail distribution
- **Xerox:** Information technology/hardware
- **ESRI:** A software provider for geographic information systems (GIS)
- **Boeing vs. Airbus:** Transportation/aeronautics
- **Goomzee Mobile Marketing:** A mobile marketing technology company offering text-message marketing services to real estate professionals
- **SELCO–India:** A solar home systems provider in India with a unique business/distribution model

These cases will allow students to apply their insights and knowledge of high-technology marketing based on material from the various chapters. The matrix in Table C.1 provides guidance on which cases might be best used with the various chapters' concepts. Of course, readers may also find other creative applications of the case material as well.

Other opportunities to apply chapter material to hands-on examples can be found in each chapter's opening vignette, in the Technology Expert's View from the Trenches features, and in the Technology Solutions for Global Problems features. For example, students could answer questions such as the following for these application materials in each chapter:

- How does the company in this box embody best-practices high-tech marketing?
- What insights can you offer for this company's strategies for future success?

Table C.2 provides a summary of each chapter's opening vignette, Technology Expert's View from the Trenches, and Technology Solutions for Global Problems features.

IS THERE MORE TO SKYPE THAN HYPE?

Skype was founded in 2003 by Niklas Zennström and Janus Friis (also the original founders of Kazaa, the music file-swapping service). Skype allows users to make telephone calls over the Internet to other Skype users free of charge and to landlines and cell phones for a fee. Skype uses a proprietary Voice-over-Internet Protocol (VoIP) that does not work with open source protocols. An objective of the proprietary strategy may have been to create user lock-in. Skype is available in 28 languages and is used in almost every country around the world. Skype generates revenue through its premium offerings such as making and receiving calls to and from landline and mobile phones, as well as voice mail and call forwarding.

In late 2005, eBay purchased Skype for $2.6 billion in an effort to help the company leverage its base of loyal users and extend its reach well beyond e-commerce, and to meld the Skype technology with its auction listings and PayPal payment service. Yet, on October 1, 2007, eBay announced that it had overpaid for Skype by nearly $1 billion and that the business had not achieved the targets for users, revenue, and profits set in the 2005 buyout agreement. On the same day, Michael van Swaaij, eBay's chief strategy officer, was appointed interim CEO of Skype following the departure of Niklas Zennström. While Zennström and other early Skype investors received a payout of about a half-billion dollars, that sum was substantially less than the $1.7 billion they would have received from eBay if Skype had met the targets set in the 2005 agreement.

TABLE C.1 Case Ties to Chapter Concepts

Case:	Skype	TiVo	Xerox	ESRI	Boeing vs. Airbus	Goomzee Mobile Marketing	SELCO–India
Chapter 1 Characteristics of High-Tech	X	X		X	X		
Network Externalities	X	X					
Disruptive Innovations	X			X			
Chapter 2 Strategy	X	X	X	X			X
Chapter 3 Culture of Innovativeness		X	X	X			
Chapter 4 Market Orientation/Teams/Cross-Functional Collaboration			X	X		X	
Chapter 5 Partnerships and Alliances/ CRM	X		X	X	X		
Chapter 6 Research, Forecasting			X				
Chapter 7 Customer Adoption Factors		X					X
Chasm	X	X		X		X	
Segmentation, Targeting, Positioning						X	
Chapter 8 Product/Tech Management	X	X		X	X		
Platforms/Derivatives					X		
Chapter 9 Distribution and Supply Chains		X	X		X		X
Chapter 10 Pricing	X	X		X		X	X
Chapter 11 Advertising and Promotion/New Media Website Considerations	X		X			X	X
Chapter 12 Strategic Branding/Pre-announcement			X			X	X
Chapter 13 CSR/Social Entrepreneurship/BOP Markets							X

TABLE C.2 Featured Companies

	Opening Vignette	Tech Expert	Technology Tidbits
Chapter 1: Intro to High-Tech	Cars of the Future	N/A	One Laptop Per Child
Chapter 2: Strategy	Medtronic	Auroras/IPTV	Dignity Toilets
Chapter 3: Culture	Google	ESRI	Star Sight
Chapter 4: MO, R&D–Marketing Interaction	Buckman Labs	Hewlett Packard, Xilinx	Aravind Eye Hospital
Chapter 5: Partnerships, CRM	Apple iPhone	REI	The Sling Shot
Chapter 6: Research	IDEO	Grupthink	bioWAVE
Chapter 7: Customer Behavior/Segmentation	RFID chips	Panasonic Mobile/Japan	Manila Water
Chapter 8: Product/Tech Management	Apple iPod	Memjet Sun MicroSystems	Godisa SolarAid
Chapter 9: Distribution/Supply Chain Mngt.	Cisco	Cisco	Big Boda WorldBike
Chapter 10: Pricing	Apple iPhone	RightNow Tech.	Orascom Telecom
Chapter 11: A&P	Microsoft	Respond2	Lymphatic Filariasis
Chapter 12: Branding	Samsung	Foveon	IDE/Bamboo Treadle
Chapter 13: CSR	Eli Lilly	Intel	Nanosolar

By early 2008, Skype's 276 million registered users around the world had talked with one another for more than 100 billion minutes using free Skype-to-Skype voice and video calls. But Skype had yet to figure out how to get more of its users (most of whom make computer-to-computer calls for free) to pay for premium services, and was generating revenues of only $100 million per quarter. On average Skype's users paid just 12 to 13 cents per month, compared to $28.38 a month for subscribers to Vonage's Internet-based phone service.

Contributing to Skype's poor performance, the market for making calls with a computer over a broadband Internet connection has gotten increasingly crowded since Skype blazed the trail. Free calling has become a standard feature of instant-messaging services from the likes of Yahoo! and Microsoft as well as an array of start-ups. In addition, Skype faced a setback in August 2007 when it experienced a systemwide failure after a routine Microsoft upgrade. Because of the distributed (peer-to-peer) nature of its network, the failure was difficult to troubleshoot.

What could the new management team do to turn the business, now profitable but generating less than $100 million in sales a quarter, into the moneymaking machine that eBay envisioned?

Three Alternatives

A New Business Model? Some analysts suggested that eBay needed to dramatically alter Skype's business model, perhaps focusing on advertising rather than user fees. One obstacle to that strategy is that Skype's seemingly rapid user growth—registered users nearly doubled between 2006 and 2007—may be deceptive. Stephan Beckert, research director at consultant TeleGeography, estimated that fewer than 20% of Skype's 220 million registered users (as of the second quarter of 2007) were actively using the site, compared to about 30% of Skype's 95 million registered users in early 2006. That would mean the active-user base grew only from 28.4 million to 44 million even as the total number of registered users more than doubled—casting doubt on Skype's ability to reach a large audience and generate substantial revenue from advertising.

Photo of Skype phone reprinted with
permission of Skype Limited, New York, NY.

A Mobile Model?   In October 2007, Skype announced that it would launch a customized cell phone developed jointly with 3 Mobile, a wireless carrier in Europe, Asia, and Australia. The Skype cell phone, developed with a software company named iSkoot, is equipped with multimedia capabilities and high-speed data for mobile Web browsing. But its most prominent feature is a big button right above the regular keypad that activates Skype's popular service for long distance and international calls. A press on that button triggers an iSkoot-developed application that brings up a list of the user's Skype "buddies" and regular phone contacts. A click on any entry in that list dials the call.

Calls on the Skype cell phone will cost the same as on a computer or Skype cordless phone: free when speaking to other Skype users, pennies per minute to dial regular phone numbers in most countries. People can also download Skype software into their cell phones. Called "Skype for Your Mobile," no special phone is needed.

A cheap international connection could prove to be a potent draw for wireless users. Currently, few mobile phone subscribers are willing to pay the rates charged by cellular companies for international calls. Skype is betting that easy mobile access to its service could spur more overseas call traffic, a revenue-producing business where growth has slumped sharply. And, because eBay owns the online payment service PayPal, success with the Skype phone could provide a springboard for using a cell phone or other handheld device to pay for items, as if it's a charge or debit card.

New Pricing?   A third alternative is for Skype to start charging for calls between Skype PC-to-PC users. In the five-plus years since its founding, Skype has established its brand and its service and has a loyal base of heavy users. The length of a user's "buddy list" may be a switching cost for such users. If they see real value in the service, however, they should be willing to pay a few pennies per minute at least for international calls. This could be a viable alternative to the advertising-supported business model. Especially for the newer mobile business, Skype should consider putting charging a fee for calls because the service provides users the additional benefit of ubiquity.

Challenges

The jury is still out on which mobile VoIP architecture will prevail. The purest form of mobile VoIP travels over IP links all the way to the handset. Usually this requires a 3G-speed cellular data service (and a cellular carrier that doesn't prohibit VoIP). It also requires software that works on the particular handset in question, which at present limits the number of handsets that can be used. Along with Skype, two other companies—Fring and Raketu—provide this functionality. An alternative is a handset that enables calls to be made through a WiFi hotspot. Truphone, a free piece of software that is downloaded to a mobile phone, is a high-profile example of this approach. Many argue that this approach does not qualify as true mobile VoIP, though. GrandCentral Mobile, by contrast, is mainly an alternate method of accessing the main GrandCentral service. That service provides the Web-based call management features and flexibility of VoIP, but all calls start and end on the public wire line or wireless networks. The mobile version merely lets users access key features through a handset, and it requires no client software.

Another major challenge for Skype and other VoIP companies is that the major wireless carriers such as Verizon, Cingular, Sprint, and T-Mobile have little incentive to allow the VoIP companies on their networks. In its PC-to-PC incarnation, Skype is a competitor, luring away landline customers. If Skype succeeds in the mobile phone arena as a major player, the company could also lure away customers from cellular operators, slashing the wireless carriers' revenue streams.

Skype's future as part of eBay will hinge largely on the success of these initiatives. In 2007 Scott Sleek, an industry analyst at Pike & Fischer, predicted: "This is really going to be a make-or-break year." What will it take for Skype's new management team to get the business back on track?

Discussion Questions

1. Describe the environment in which Skype competes based on the three characteristics of high-tech markets (Chapter 1).
2. Given the technological uncertainty present in the VoIP mobile market, would Skype be likely to emerge as the leader? Explain your reasoning.
3. Would VoIP be considered a disruptive technology by wireless carriers? Explain.
4. Collect current statistics regarding the number of VoIP users. At what stage of the adoption-and-diffusion process is this technology?
5. What would be an ideal beachhead for Skype to focus its efforts on?
6. What are the elements of a whole product solution for VoIP? What are the implications for Skype in terms of crossing the chasm?

7. What are the pros and cons of Skype's selection of a proprietary Voice-over-Internet Protocol?
8. What are the characteristics of a business that can successfully implement an advertising-based business model? Does Skype possess those characteristics?
9. a. Analyze the pros and cons of the three alternatives discussed in the case for Skype to generate revenue.
   b. What other strategies are available to Skype to generate revenue? Assess their pros and cons as well.
   c. Make a recommendation for a viable business model for Skype to succeed.
10. Is Skype likely to become the moneymaking machine that eBay envisioned? Why/Why not?

Sources: "3 Launches New Skype Mobile Phone," BBC News/Business, October 29, 2007, online at http://news.bbc.co.uk/2/low/business/7066271.stm; Browne, Marcus, "Australia to Wait until December for Skype 3 Mobile," CNET.com Australia, October 30, 2007, online at www.cnet.com.au/mobilephones/0,239025893,339283337,00.htm; Dannen, Chris, "Technology: The Skype Mobile Phone Will Blow Your Mind," December 18, 2007, online at http://blog.fastcompany.com/archives/2007/12/18/technology_the_skype_mobile_phone_will_blow_your_mind.html; Skype website, online at www.skype.com; Kharif, Olga, "Skype's 'Make-or-Break Year,'" Business Week, November 30, 2007, p. 25; Kharif, Olga, "eBay's Skype Bubble Bursts," Business Week, October 2, 2007, p. 1; Kharif, Olga, "Skype's Mobility Problem," Business Week, September 29, 2006, p. 12; Meyerson, Bruce, "Skype Takes Its Show on the Road," Business Week, October 29, 2007, p. 38; Meyerson, Bruce, "Skype Goes Mobile," Business Week, October 19, 2007, p. 27; and Poe, Robert, "Which Mobile VoIP to Choose?" VoIP News, May 15, 2007, online at www.voip-news.com/feature/which-mobile-voip-to-choose-051507.

THE FUTURE OF TIVO?

Personal video recorders (PVRs), also known as digital video recorders (DVRs), allow users to record and play back TV shows at their convenience. In addition, they allow users to easily bypass the advertising within such shows. Developments in the technology over time have increased the functionality, allowing users to create "Wish Lists" that enable them to search for and record by actor, director, keyword, or category; to create "Season Passes" that allow them to record their favorite shows each time the program airs regardless of the time, while automatically skipping repeats and without the use of tapes or timers; to record two shows at once; and to link with new technologies such as the Internet to record shows from Web browsers, download movies and songs to the TiVo box, and so forth. These devices even allow users to play MP3s and view digital photos on their TV screens.

TiVo Background

TiVo pioneered this product category in 1999. Indeed, the TiVo brand name in the United States has become synonymous with the category as a whole—and in fact, "TiVo" has become a verb: Television viewers today commonly ask their friends, "Are you going to TiVo that?"

In order to get these product benefits, the user must buy a piece of hardware that looks much like a DVD player. Prices of these "boxes" vary widely (including being bundled free with a cable or satellite broadcasting service) depending on the functionality and where the user buys it (discussed more fully below). The TiVo products available in 2007 included:

- The TiVo Series 2: Able to record two shows at once and record up to 80 hours; works with basic cable, digital cable, or satellite services; priced at $99.99 (with rebates available)
- The TiVo HD DVR: Able to record 20 hours in high-definition format or up to 180 hours in standard format; can control live HDTV and record two digital cable shows at once; priced at $299.99
- The TiVo Series 3 HD DVR: Able to record 32 hours in high-definition format or up to 300 hours in standard format; priced at $599.99 (again, with rebates available)

In addition, the user must subscribe to a service to access the programs/shows being broadcast. In addition to TiVo, common providers for DVR service include cable and satellite companies (such as Comcast, DirecTV, and Dish Network) whose customers pay a monthly fee to receive their programming (via cable or satellite) and then pay an incremental amount for DVR functionality (to record and play back programming at their convenience). Again, subscription prices vary widely.

The service plans available directly from TiVo in 2007 included:

1-year monthly plan	$12.95/month
1-year prepaid plan	$129.00 ($10.75/month)
2-year prepaid plan	$249.00 ($10.38/month)
3-year prepaid plan	$299.00 ($8.31/month)

Current Issues

Customers have experienced a fair amount of difficulty in using the products. TiVo's internal records showed that it took customers an average of six phone calls to Technical Support just to install the product correctly (hooking up the required cables between the box and the existing TV, programming it correctly, and so forth).

Margins on the devices are razor thin. To control its costs, TiVo uses the free Linux operating system. In addition, it has outsourced its manufacturing—and even efficient contract manufacturing firms have found it difficult to make their required margins. The hard disk alone accounts for 33% of the cost of manufacturing the device. When added to the required marketing (customer acquisition costs) and customer service expenses, selling stand-alone boxes has been a tough business model.

Cable and satellite operators frequently use DVR functionality (including giving the box away for free) as a bundled feature in their plans to differentiate their offerings from competitors. According to James McQuivey of Forrester Research, "TiVo has been fighting for years and has only sold about 1.7 million units of its boxes compared to [the] cable and satellite companies, who looked at TiVo and said, '1.7 million units? We can do that.'" Today, Comcast, DirecTV, Dish Network, Time-Warner Cable, and others, are offering DVR technology to their customers. These companies have indicated they consider DVR technology a competitive tool to help differentiate their services by offering their customers more programming features. Despite TiVo's superiority, only about 1.7 million units had been sold through 2007 compared to the 25 million units shipped by cable and satellite companies.

Table C.3 shows a comparison of the TiVo DVR product/service bundle compared to that available via cable or satellite providers.

In 2007, TiVo's *subscription acquisition cost*—the amount of money TiVo spends for each additional customer it adds, generally in the form of marketing and product development costs—was $294, while its annual revenue per subscriber was $107 and the cost to serve that subscriber was $25. That means it took 3 years and 7 months for TiVo to break even on a new subscriber. And, from 2004 to 2007, TiVo's *churn rate* (the percentage of service terminations over a given period) was also on the rise, making the need to gain new customers greater even while subscriber acquisition costs were rising and revenue per subscriber was falling. As a result, each new customer was less lucrative than the last, yet more expensive to get. However, in the first quarter of 2008 TiVo seemed to have turned a corner, driving its acquisition cost per subscriber to $116 by reducing marketing expenditures. This enabled TiVo to more than quadruple its net income compared to the first quarter of 2007.

The Stakeholders

In order to fully understand the business model for a company like TiVo, one must consider the many stakeholders involved in the value delivery process. Because the product includes the ability to skip all ads, advertisers initially were worried about the impact of TiVo on their ability to reach viewers through broadcast and cable programming. Therefore, TiVo offered a package to advertisers based on a proprietary technology. This advertising plan featured "Showcases," an interactive advertising platform allowing the delivery of special offers to TiVo viewers through a different format than embedded commercials in the TV programming. While 70% of television commercials are fast-forwarded in DVR households, the TiVo technology offers advertisers a way to engage TiVo viewers with their brands. After TiVo viewers have watched a recorded program, in addition to the options of deleting or archiving the program, they are invited to enter a showcase area that shows ads, program promotions, and long-form video. An additional benefit of these advertising showcases is the ability to measure the impact and results of the ads. The TiVo audience measurement service allows second-by-second analysis of viewing patterns within ads and shows. Moreover, viewers who watch TiVo-embedded ads can send their personal information to advertisers, requesting a brochure or other information. TiVo's fees for the services run a few hundred thousand dollars to place ads over a three- to six-month time period; TiVo's objective is to move to a cost-per-viewer fee (to reflect size of subscriber base).

Other affected stakeholders included:

- Broadcasters: They were worried about the impact of TiVo on their ad rates.

- Broadcasters and producers: They were concerned about the intellectual property rights of the shows, once they had been recorded for later playback.

- Category-killer retailers such as Best Buy, as well as smaller specialty electronics retailers: They frequently sold the stand-alone boxes; channel pricing and margin considerations needed to be factored into TiVo's business model.

TABLE C.3 Comparison of TiVo DVR Bundle to Cable and Satellite Providers			
	TiVo Series2 & Series3 boxes	**Leading cable service DVR***	**Satellite DVR****
Record from multiple sources:	Yes: Combine satellite and cable, depending on product	No: Digital cable only	No: Satellite only
Dual Tuner: Record 2 shows at once[1]	Yes	Yes	Yes
Easy Search: Find and record shows by title, actor, director, genre, or keyword	Yes	Titles only; browsing only	Title, subject, and actor only
Built-In Ethernet: Broadband-ready right out of the box to connect to your home network	Yes	No	No
Online Scheduling: Schedule recordings from the Internet	Yes	No	No
Movie and TV Downloads: Purchase or rent thousands of movies and television shows from Amazon Unbox and have them delivered directly to your television[2]	Yes	No	No
Home Movie Sharing: Edit, enhance, and send movies and photo slideshows from your One True Media account to any broadband connected TiVo box[3]	Yes	No	No
Online services: Yahoo! weather, traffic, and digital photos; Internet Radio from Live365, podcasts, and movie tickets from Fandango	Yes	Limited	Limited
Transfer Shows to Mobile Devices and DVD: Transfer shows to portable devices, laptops, or burn them to DVD[3, 4, 5]	Yes	No	Limited
Home Media Features: Your digital photos and digital music on your TV[3]	Yes	No	No
Transfer Shows between DVR Boxes: Record shows on one TV and watch them on another[3]	Yes	No	No

* Time Warner/Cox Communications Explorer® 8000™ DVR and Comcast DVR
** DISH Network 625 DVR
[1] On the TiVo® Series2™ DT DVR, users can record two basic cable channels, or one basic cable and one digital cable channel, at once.
[2] Requires broadband cable modem or DSL connection.
[3] Requires TiVo box to be connected to a home network wirelessly or via Ethernet.
[4] In order to burn TiVoToGo transfers to DVD, customer must purchase software from Roxio/Sonic Solutions.
Source: "Shop for TiVo DVRs, accessories & more," November 2007, online at https://www3.tivo.com/store/home.do#.

- Other DVR manufacturers: TiVo functionality can be found in other consumer electronics manufacturers' branded digital video recorders, including Sony, Pioneer, and Toshiba, to name a few. Based on an up-front license fee of several million dollars and annual "maintenance fees" at roughly 15–20% of the up-front fee, this revenue stream required TiVo engineers to work to make its technology compatible with these other companies' products.

- Third-party software vendors: TiVo also offered tools and resources for software developers to build new applications and services.

Revenues from advertisers and from other DVR manufacturers required TiVo to incur additional engineering expenses. TiVo was not originally built to be an extensible system that could easily be adapted to different hardware and software; it was not modular.

TiVo also offered a TiVo Affiliate Program, an online network of approved affiliate sites that have partnered with TiVo. TiVo provides official links and banners to these sites, which direct Web traffic back to the TiVo.com online store. Once the links have been placed live on the affiliate site, TiVo pays the affiliate site owner a commission for each TiVo box plus service plan purchased by a consumer coming from that affiliate's site. TiVo's hope was that e-tailers or websites that offered entertainment-oriented Web content might find it useful to recommend TiVo to their customers.

Meeting the Challenges

Further complicating TiVo's business strategy were emerging technology trends. Consumers increasingly sought alternative ways to access television content, whether over their mobile phones, through IPTV (TV delivered over the Internet to computers), or downloading movie content directly from the Internet.

Faced with these challenges, as TiVo looked to the future, its mission as a company—to make the TiVo DVR the focal point of the digital living room, a center for sharing and experiencing television, movies, video downloads, music, photos, and more—seemed increasingly challenging.

Discussion Questions

1. Draw a supply chain (or value net) that traces the various stakeholders involved in the TiVo value chain and their respective interactions. From this, what insights do you get about the relative value that each stakeholder adds in this process?
2. For TiVo, describe the six factors that affect customers' purchase decisions (see Chapter 7, Table 7.1). Which are most salient for TiVo adopters?
3. What are TiVo's core competencies? Is the company effectively leveraging these in its current strategy?
4. To what extent is the TiVo business model affected by network externalities? What are the implications for its business model?

5. Who are the key competitors facing TiVo? Given this competition, what are the pricing implications?
6. Does TiVo's pricing strategy make sense? Why or why not?
7. Do you believe that TiVo's business model is sustainable over the long term? If so, why? If not, why not? If not, what recommendations do you have for this company?
8. Why has TiVo apparently not been successful in "crossing the chasm"? What will it take for TiVo to penetrate the mass market?

Sources: Yoffie, David B., Pai-Ling Yin, and Barbara J. Mack, "Strategic Inflection: TiVo in 2005," Harvard Business School, Case #9-706-421; "TiVo DVRs," TiVo website, at www3.tivo.com/store/boxes.do; Reisinger, Don, "Does TiVo Have a Chance?" NewTeeVee.com, posted March 11, 2008, at http://newteevee.com/2008/03/11/does-tivo-have-a-chance; TiVo Inc., Form 10-K, Annual Report for fiscal year ended January 31, 2008, online at http://investor.tivo.com/secfiling.cfm?filingID=1193125-08-81777; and "TiVo Affiliate Program," 2008, online at https://www3.tivo.com/abouttivo/affiliateprogram/index.html.

CHARTING A NEW COURSE FOR XEROX: STRATEGIC MARKETING PLANNING

In December of 1998, the value of a share of Xerox stock reached $60.81. By December 31, 2000, that share was worth $4.63, a decline of 92%! With Chapter 11 bankruptcy looming after six consecutive years of losses, more than $17 billion of debt, and a debt-to-equity ratio of 9:3, the board of directors appointed Anne Mulcahy to the CEO position. Mulcahy was a 24-year Xerox veteran but had little hard-core executive experience. An English and journalism major, she had spent 16 years in sales and many of her remaining years as the head of human resources and the chief of staff for former CEO Paul Allaire—not exactly the kind of resume Wall Street could count on to stop Xerox's free fall. Xerox's stock dropped 15% on the day of the announcement. Having a revered brand and a well-known technology is no guarantee of survival in the brutal world of technology. Could Mulcahy guide Xerox to develop a strategy for success?

Xerox Background

Xerox was founded in 1906 as "The Haloid Company," which originally manufactured photographic paper and equipment. The company subsequently changed its name to "Haloid Xerox" in 1958. The company came to prominence in 1959 with the introduction of the first plain paper photocopier using the process of *xerography*. Xerox has long been recognized as a technology pioneer. Its Palo Alto Research Center (PARC) was at the leading edge of the development of laser printing (1971); object-oriented programming (1972); a personal computer, the Alto, that incorporated the first what-you-see-is-what-you-get (WYSIWYG) editor, a mouse, and a graphical user interface (GUI) (1973); a system that allows the assembly of an integrated office network in which users can electronically create, process, file, print, and distribute information (1980); and Colab, a meeting room that provides computational support for collaboration in face-to-face meetings (1987).

Today, Xerox Corporation engages in the development, manufacture, marketing, servicing, and financing of document equipment, software, solutions, and services worldwide. It serves commercial customers (primarily small to medium sized, but large customers as well), as well as government, education, and other public sector customers. Xerox markets its products through its direct sales force, as well as through a network of independent agents, dealers, value-added resellers, and systems integrators.

Mulcahy's First Steps

Faced with an immediate financial crisis, Mulcahy had no choice but to act boldly. Through a back-to-basics approach and a renewed focus on operational efficiency, Xerox cut its capital expenditures by 50%; reduced its sales, general, and administrative expenses by a third; and slashed its total debt in half. However, Mulcahy knew that companies can't shrink their way to greatness. After gaining some breathing room, her next challenge was to reignite Xerox's growth. Xerox's 2001 revenues were below those of 1998, a situation that worsened in 2002.

Return to Growth

Xerox's shrinking revenue base was primarily the result of losing market share to competitors. Not surprisingly, the years between 2000 and 2002 were fairly quiet in the marketing arena. The money just wasn't there to spend—but worse, there was no coherent message. After the turnaround took hold, Mulcahy was ready to spotlight the Xerox brand again. Xerox chose a subdued, customer-centric focus.

The base campaign, launched in 2003, was so successful that Xerox continues to add iterations. The core print ads feature a giant red Xerox logo and short case studies built around clients including Microsoft, Office Depot, and Enterprise car rentals. By early 2004, 60 of those print ads had been created and run in publications such as *Business Week, Forbes,* and the *Wall Street Journal.* In 2003, Xerox spent $53.5 million on advertising, according to TNS Media Intelligence/CMR, up significantly from the $27.3 million it spent in 2001.

In March 2007, Xerox announced an advertising campaign designed to bolster the fastest growing segment of its business: consulting services. The ad campaign featured print, broadcast, and online ads designed to position Xerox Global Services as the "premier consulting and outsourcing" partner for document management. For example, one print ad spotlighted a team of Xerox consultants under the banner "We Find Millions," and focused on document assessment services that yielded millions of dollars of cost savings for companies such as Owens Corning and InterContinental Hotels Group. Other print ads and television commercials revolved around the theme of affordable color for businesses small to large.

According to Michael C. MacDonald, Xerox's chief marketing officer, the ads are part of a broader push to build the value of the Xerox brand and better serve customers. "When it comes to the customer experience, we need to think big and small. That means paying equal attention to 'big brand' efforts like advertising and 'little brand' experiences that define the way customers interact with us every day," MacDonald said in remarks prepared for a keynote presentation at the 2007 Conference on Marketing in Las Vegas.

Of course, clever advertising will fail in the face of subpar products. To maintain an innovative product line, Xerox continued to spend about $1 billion annually on R&D. In May 2006 Xerox launched a line of digital color printers, presses, and copiers. Xerox's color line, already the most extensive in the industry, was expanded with five additional systems, including a new midlevel digital production color press and color printers, digital copiers, and multifunction products aimed at offices small to large. "The breadth of Xerox's digital color portfolio surrounds the competition, with options in every speed range, and offers superior value for customers who want to harness the proven results that color can bring to their operations," said Jim Firestone, president of Xerox North America, in announcing the new products. In 2007, some 44% of Xerox's total equipment sales came from color printing, up 12 points from 2005. The color focus was a big driver of revenue growth.

Xerox also began to involve customers in the early stages of product design. Xerox Chief Technology Officer Sophie V. Vandebroek called the process "customer-led innovation." The process played a key role in the design of the company's new dual-engine Nuvera 288 Digital Perfecting System. Brainstorming, or "dreaming with the customer," is critical, she says. The goal: "Involving experts who know the technology with customers who know the pain points." Promoted to the top technology spot in January 2006, Vandebroek expects scientists and engineers to meet face-to-face with some of the 1,500 to 2,000 customers who visit showrooms at the company's four global research facilities each year. Others work on-site for a week or two with a customer, observing how they behave with the product. A team of ethnographers is also charting customer behavior.

Also in 2006, Xerox detailed two major consulting services contracts: a $36 million seven-year deal with the University of Calgary, and a $17 million four-year contract with InterContinental Hotels Group. The signings included enterprise-wide document management services combined with advanced office and high-end production print technology to streamline the flow of information and reduce costs. Xerox also announced a number of software offerings and business development tools that help customers simplify work processes and boost performance.

"Xerox is intently focused on its customers, offering them the industry's broadest choice of hardware, software, and services so they can indeed work smarter," said Angele Boyd, group vice president of IDC, a global information technology research group. "As a result, customers in diverse environments—from the office to the print shop—can turn to Xerox to help them be more productive."

Another marketing priority for MacDonald in 2006 was to improve the global customer experience. "We are doing more one-to-one marketing with our largest global customers," he said. For example, in 2006 Xerox restructured its premier global account organization to put senior-level client managing directors on the company's largest accounts. In parallel with these efforts, Xerox leveraged a proprietary software tool called Sentinel to monitor and boost customer satisfaction. The tool automatically contacts customers and offers an easy way to provide immediate feedback and solve problems in real time. Use of Sentinel nearly doubled in two years, expanding to 14 countries worldwide and 124 major accounts.

These initiatives returned Xerox to solid financial footing. In 2006 Xerox had a profit margin of 7.75% and an ROE of 17.2%—and the value of a share of Xerox stock more than doubled since

Mulcahy took the reins. However, Xerox's revenues have remained flat and the company faces intense competition from Canon, Ricoh, Toshiba, and Konica Minolta.

Discussion Questions

1. What kind of strategic planning process (bottom-up or top-down) did Xerox follow for its turnaround? What inference can you make about the effectiveness of this approach?
2. While Xerox PARC made a number of breakthrough innovations during the 1970s and 1980s, it doesn't seem to have benefited competitively from these. What inferences can you make about the culture of innovation at Xerox in the 1970s and 1980s?

3. Identify and prioritize the key success factors for Xerox's turnaround. Link them clearly to high-tech marketing concepts.
4. What specific steps did Xerox take to restore its brand equity?
5. Does Xerox have a foundation for competitive advantage? Why or why not?

Sources: "Innovation Milestones," at www.parc.com/about/history; Vollmer, Lisa, "Mulcahy Took a No-Nonsense Approach to Turn Xerox Around," *Stanford Graduate School of Business News,* December 2004, online at www.gsb.stanford.edu/news/headlines/vftt_mulcahy.shtml; Snyder Bulik, Beth, "Rejuvenated Xerox Doubles Its Investment in the Brand," *Advertising Age,* May 3, 2004, p. 18; "Xerox Executes Growth Strategy with Five New Color Systems, $53 Million in Document Management Contracts," *Business Wire,* May 16, 2006, available at http://findarticles.com/p/articles/mi_m0EIN/is_2006_May_16/ai_n26862818; and Byrnes, Nanette, "Xerox' New Design Team: Customers," *BusinessWeek,* May 7, 2007, pp. 72–73.

ENVIRONMENTAL SYSTEMS RESEARCH INSTITUTE (ESRI)

By Bryant Ralston, ESRI's strategic account executive for Idaho and Montana.

ESRI, the Environmental Systems Research Institute, is a privately held company of 5,000 employees founded in 1969. The early mission of ESRI focused on organizing and analyzing geographic information. The firm carefully managed project work to ensure growth without the need for venture capital or going public. During the 1980s, ESRI devoted its resources to developing a core set of application tools that could be applied in a computer environment to create a geographic information system (GIS). In 1982, ESRI launched its first commercial GIS software called ArcInfo. Its 25-year flagship GIS software product, ArcInfo, is a full-featured professional GIS product that supports the full range of GIS data collection, import, compilation, editing, query and analysis, visualization, and analytical "geoprocessing." Over the years, ArcInfo has evolved to a modern, component-based graphical user interface (GUI) application. As the world's leading vendor of geographic information systems software in terms of revenue, market share, and innovation, ESRI has been called "the classic high-end GIS vendor."

Key factors contributing to ESRI's longevity have been its use of the classic smart business practices of listening to its customers, investing in new technology and aligning it to real customer needs, introducing sustaining innovations into its product architecture, working with an extensive network of channel partners to enhance the "whole product," and keeping an eye on its competitors. As the GIS market grew, so did ESRI's revenues and the company as a result.

Over its 35-year history, ESRI added software products based on its own innovation and research, technology trends, and user-driven needs. In the late 1990s, ESRI re-engineered its entire product line into a family of products, or a "platform," called ArcGIS. Consisting of four fundamental classes of GIS deployment options—desktops, servers, embedded solutions, and mobile applications—it is an example of how an integrated platform can command a very strong industry position.

GIS technology can be used for a variety of industry applications including resource management, environmental impact assessment, urban planning, cartography, criminology, marketing, and logistics. For example, GIS might allow emergency planners to easily calculate emergency response times in the event of a natural disaster; GIS might be used to find wetlands that need protection from pollution; or GIS can be used by a company to site a new business location to take advantage of a previously underserved market.

Industry History

The GIS systems of the past were primarily the domain of specialists, utilizing stand-alone workstations running monolithic specialized GIS software to manipulate volumes of diverse geographic information. In the early days of GIS, large government agencies often responsible for nationwide mapping projects implemented these large-scale and relatively expensive systems. At this time, two GIS vendors were clearly dominant: Huntsville, Alabama–based Intergraph Corporation, which had its roots in the defense and rocketry business, and Redlands, California–based ESRI, whose roots were in environmental consulting and land use projects.

In the 1980s, GIS vendors began to see demand coming from state and local governments (for land records and planning management) and from utilities (for automated mapping and facilities management, or AM/FM), as well as continued growth from the federal sector. Primarily, the original mainframe monolithic GIS applications—notably ArcInfo from ESRI and the Modular GIS Environment (MGE) from Intergraph—dominated the commercialization of GIS during this period; at that time, these applications were UNIX-based programs.

The early 1990s marked the rise of desktop GIS, which coincided with the "desktop wars" in which Microsoft eventually triumphed. For a time, both ESRI and Intergraph clung to their arcane and unwieldy software formats and high-end GIS products. However, as demand grew, it was clear that these new desktop GIS users were different from high-end users in that they lacked the tolerance for specialized, hard-to-use software; they demanded a simplified GIS system if they were going to harness the power of GIS. This eventually led to both ESRI and Intergraph launching their desktop GIS technologies. In addition, completely new desktop GIS industry entrants, like MapInfo and Strategic Mapping, capitalized on these new markets being created by desktop users. However, ESRI was able to adapt quickly and protect its market position by introducing its successful ArcView product that continues to dominate the desktop GIS market even today.

The mid- to late 1990s was characterized by the adoption of the Internet into the world of GIS. This included the launch of several new Internet-based GIS systems such as MapGuide from a small Canadian company called Argus (subsequently purchased by Autodesk), Arc-Internet Map Server by ESRI, and GeoMedia Web by Intergraph; it also included the rise of GIS data-sharing portals for searching, discovering, downloading, and consuming GIS data in a "clearinghouse" fashion.

Modern Industry Trends and the Emerging "GeoWeb"

In the early 21st century, new investments in Internet-based virtual globes by large mature IT vendors—notably Google and Microsoft—have challenged the traditional industry model of market segmentation. Unlike traditional GIS, these new virtual globes operate with a midrange personal computer plus a broadband Internet connection coupled with modern search engines; moreover, they do not require any existing GIS data to be stored on a user's own computer. Free versions are usually the norm with more advanced versions (particularly of Google Earth) available for a fee.

These offerings have proved to be extremely popular, experiencing hundreds of millions of downloads. They have attracted large, active online communities of users who post their own personalized "geo-tagged" content—including vacation and travel photos, restaurant and lodging reviews, and even the locations of past Bigfoot sightings—all linked to their appropriate location on the Earth's surface. This "architecture for participation" functionality is an example of how the Web 2.0 world recognizes and values user-contributed content (UCC) that promotes greatly expanded interaction on the Web. The use of these virtual globes in this participatory manner has recently been referred to as *neogeography*. Also referred to as *Volunteered Geographic Information (VGI),* much of the current user-contributed content on the Web represents casual and unverified observations or assertions about a place. Because of this, it is generally not as valuable a source of information as data collected by traditional "authoritative source" organizations utilizing traditional GIS solutions. However, these user-driven efforts serve as a model for the GIS community to publish and share more sophisticated and useful services in this dynamic, new online environment.

Another new development is the "GeoWeb," a "relatively new term that implies the merging of geographical (location-based) information with the abstract information that currently dominates the Internet." In essence, the GeoWeb is a modern, open, and interoperable Web environment for sharing and using a broad range of geographically related information, again using the underlying participatory Web 2.0 architecture. Expectations are that the future GeoWeb will be capable of supporting many interesting and diverse geographic information applications all over the globe that are driven by the convergence of Web 2.0, the explosive growth of consumer virtual globes fused with traditional GIS-created content, the recent growth in geo-referencing content (often from online "neogeographers"), and the ability to examine and share this content with easy-to-use search tools and "mashup" technology.

All these developments affected the traditional GIS community by raising GIS user expectations in terms of ease of use, performance, user interface design, data collection and integration, and application interoperability. In addition, virtual globes have redefined traditional GIS for two principal reasons:

1. Much of the new demand from consumers of geographic information comes from individuals who do not have any educational or conceptual background in GIS. Therefore, they do not necessarily have high-end analytical needs that traditional GIS solutions provide. These new users are satisfied with an intuitive, simple, fast, and "cool" interface to geospatial data. Previously, these types of users were served by Internet sites such as MapQuest or Yahoo! Maps, but virtual globes have enough "cool factor" currently to create many new users (and resulting hype) and create new demand for the most up-to-date geographic information available.

2. These virtual globes are coming from the hottest companies in the mainstream IT world today. The tremendous amount of geographic data (traditionally available only to stand-alone GIS systems and specialists) now available via a modern Web interface offers a new platform for GIS that is, essentially, free to end users. In addition, Microsoft and Google do not make the bulk of their revenue from mapping or GIS, but rather from selling advertising and software. For both companies, the traditional GIS market is not the target for their virtual globe offerings; rather, their interests lie in driving ad revenues, increasing competitive pressure on each other, enhancing their stock prices, and improving their positions in the marketplace.

ESRI Response

Early on, when virtual globes began to gain mainstream popularity, ESRI managers held conflicting views about the degree to which they posed a competitive threat. Many inside ESRI thought that "it's not *real* GIS so it's not our core business." Others in ESRI saw these as largely a competitive threat (they were from Google and Microsoft after all), while still others saw them as an opportunity to grow awareness of the value of geospatial data and analysis—which, it was argued, would help grow the market for traditional GIS. The company's value network, organizational structure, and business model tended to reinforce a rather rigid perspective. Because these new users of virtual globes were not necessarily existing ESRI customers, no part of the organization was ultimately responsible for bringing this new perspective on the value of geographic data and applications to ESRI's management.

These virtual globes were not positioned by Google and Microsoft as direct competitors of traditional GIS products; indeed, both companies (as well as traditional GIS industry vendors like ESRI) seemed to take care to emphasize and differentiate virtual globes in the context of traditional GIS market segmentation. Some industry participants also suggested that virtual globes were ultimately complementary to traditional GIS solutions. This "peaceful coexistence" perspective suggests that the mainstream exposure of Google Earth benefits traditional GIS vendors by helping to raise awareness of the value of GIS data and analysis—that visualizing and analyzing data in a geocentric way can be extremely compelling and benefit many types of sophisticated scientific analysis. Indeed, ESRI President Jack Dangermond has said that Google Earth was "the most fantastic thing I've ever seen" and that its popularity contributed to the fact that traditional GIS was "booming."

Another topic debated internally at ESRI was whether the new virtual globes offered a potential opportunity for ESRI to enhance and leverage its partnerships with Google and Microsoft. Presumably, both Google and Microsoft could benefit from activities that allowed their virtual globe users a path to "scale up" to more functionally rich and powerful, high-end GIS tools. ESRI had experience partnering with both firms in the past. In particular, it worked closely with Google (and other firms like IBM) for the successful implementation of the U.S. government's GeoSpatial One-Stop GIS portal project following 9/11. ESRI and Google teamed up in 2008 to provide attendees of the Where 2.0 Conference with a vision of how traditional GIS and Google Earth users can leverage one another's efforts. In regard to partnering with Microsoft, ESRI rebuilt its core GIS platform on the Microsoft Component Object Model (COM) software specification; ESRI also had close ties with Microsoft's SQL Server team in its Redmond, Washington, headquarters as they architected their "abstract data types" for SQL Server by testing the way GIS users utilize their new data types for storing and managing geographic data.

The degree to which virtual globes were expected to disrupt the traditional GIS world affected the way ESRI, as an industry incumbent, addressed the resulting threats and opportunities of such a potentially disruptive change.

In 2008, ESRI took a major step to integrate the emerging GeoWeb capabilities into its flagship product, ArcGIS Server 9.3. ArcGIS Server allows GIS professionals (where appropriate) to publish searchable "metadata" about their services, as well as maps, data, and related GIS services for others to view and use. *Metadata* are essentially data about data—information that contributes to the data's usefulness and allows the data to be found more easily via Internet-based searches. With these advancements in traditional GIS, the GeoWeb will now be able to go far beyond the simple visualization of unverified VGI by neogeographers; it will support integration of full GIS-based services representing "authoritative source" information created from users of traditional GIS. It will greatly enhance the geographic content available to the masses and GIS professionals alike. It will also provide the ability to access and manipulate layers of information and carry out spatial analysis supporting a host of Web applications. By engineering interoperability between GIS and many types of other Internet-enabled IT business systems, it will allow GIS professionals to distribute their own data and applications to a much wider audience. Both GIS professionals and casual Web users will be able to leverage one another. Web users will be able to access and use maps and GIS analysis applications authored by GIS professionals. GIS users will also benefit from greater access to geo-referenced data available from a variety of sources including casual observations and assertions. In addition, the availability of new and "authoritative" data sources will help ensure that the data requested are the most recently available and of the highest quality possible.

Discussion Questions

1. Of the three types of uncertainty that characterize high-tech markets, which type is ESRI experiencing most acutely? Elaborate and provide strategic implications for ESRI.

2. What are the characteristics of the newest competitors that have allowed them to gain traction in the GIS industry?

3. To what extent does ESRI face "disruption," in the classic (Chapter 1) sense of the word?

4. Should ESRI explicitly target the new GIS customer (mainstream, nonprofessional, low-end)? If so, what organizational changes might this require (Chapter 2, Chapter 3)?

5. How would you characterize GIS diffusion in terms of the categories of adopters? Has it crossed the chasm?

6. How can companies such as ESRI compete in this era of "free" (Chapter 10)?

7. What should the traditional GIS vendors such as ESRI do, in light of the new industry trends?

Sources: "Company History," www.esri.com/company/about/history.html; Longley, Paul, Michael F. Goodchild, David J. Maguire, and David W. Rhind, *Geographic Information Systems and Science* (New York: Wiley, 2005), p. 165; and Butler, Declan. 2006. "The Web-Wide World: Life happens in three dimensions, so why doesn't science? Declan Butler discovers that online tools, led by the Google Earth virtual globe, are changing the way we interact with spatial data." *Nature* magazine. Volume 439, February 16, 2006. Pp.776–78.

VISION OF THE FUTURE: AIRBUS 380 OR BOEING 787 DREAMLINER?

History of Boeing and Airbus

Boeing was incorporated in Seattle, Washington, by William E. Boeing in 1916 as "Pacific Aero Products Company" and renamed the "Boeing Airplane Company" in 1917. Boeing has consistently been at the forefront of commercial aircraft innovation, introducing the first truly modern airliner, the Boeing 247, in 1933; the world's first pressurized-cabin transport aircraft, the Stratoliner, in 1938; and the first commercial U.S. jet airliner, the 707, in 1958. The Boeing 747, or "Jumbo Jet," was launched in 1970 as the first wide-body commercial aircraft with a capacity of about 360 passengers. It was the world's largest-capacity passenger aircraft for more than 30 years and dominated the industry in sales and deliveries with more than 1,400 built by 2008.

Airbus Industries started as a consortium of European aerospace companies in 1970 to compete with the American companies Boeing, Lockheed, and McDonnell Douglas. The Airbus A300 was launched in 1974 with a capacity of about 250 passengers, the A320 in 1981 with a capacity of about 150, and the A330 in 1987 with a capacity of about 295. All of these were twin-engine medium-range aircraft. Boeing launched the twin-engine 767 in 1982 to compete against the A300. The A340, launched in 1993, was Airbus's first four-engine long-range aircraft with a capacity of about 310 passengers. In 1995 Boeing launched the twin-engine 777, with a capacity of about 350. The A330 outsold the 767, but Boeing's 777 outsold the A330 and A340.

The longer history of Boeing notwithstanding, by 2003 Airbus surpassed Boeing in the number of planes delivered to customers and remained ahead in 2008. However, on the measure of orders and value of planes delivered, Boeing surpassed Airbus during the 2006–2008 period.

Next-Generation Products

In 1992 Airbus and Boeing began discussions to cooperate on joint design and manufacture of the next-generation jumbo jet. In 1995 Boeing officials told their Airbus counterparts that the latter's project proposal was too risky. Airbus officials countered that the greater risk was in *not* proceeding with the project. The two companies parted ways because of differing strategic visions of the future. Airbus's vision was that passenger traffic at the world's busiest airports would continue to grow and exceed airport capacity, creating demand for gigantic planes. Boeing's vision was that passengers would be traveling shorter distances point-to-point between smaller airports, avoiding the really congested airport hubs.

Airbus decided to go ahead with developing the A380, a new-generation "Super Jumbo Jet" capable of carrying 550 passengers, at a cost of $12 billion, planned for delivery in 2005. It would be more like a cruise ship than an airplane—with spiral staircases, elevators, onboard medical centers, and zoned seating for families. Airports would need to be redesigned to handle disembarking passengers through two levels of jetways. By May 2004, Airbus had 129 firm orders for the A380 at $280 million per plane. Boeing's more modest vision (the 747-8) was to stretch the 747 fuselage to accommodate about 510 passengers but make it more fuel efficient, at a development cost of $3 billion. The Boeing 747-8 was planned for delivery in 2009. Airbus 380 was designed to be about 20% more fuel efficient than Boeing 747.

Change in Plans

After the events of September 11, 2001, airlines cut back capital expenditures and emphasized fuel efficiency. Boeing designed the 787 Dreamliner to replace the 767 and meet the new priorities of existing customers. The 787 Dreamliner is a twin-engine aircraft capable of carrying about 250 passengers and is highly efficient because it uses lightweight composite materials for most of its construction. The price per aircraft is less than $200 million. To compete with the 787 Dreamliner, Airbus announced the A350, derived from the A330 but with a composite body, to be available in 2013. Since December 2004 the A350 has received 100 firm orders compared to 291 for Boeing's 787 Dreamliner (since May 2004). Analysts feel there is a larger market for the 787 Dreamliner compared to the A380, but profit margins are also lower.

From 2001, when development began on the A380, there have been problems with product design and manufacturing. Initially the aircraft was overweight, which would lead to higher fuel consumption and operating costs. So, Airbus engineers had to redesign the wings and fuselage to be constructed from lighter composite materials instead of aluminum. Manufacturing facilities were located in several countries; the fuselage was made in Germany, the wings in Britain, and the tail section in Spain. These would all be shipped to Toulouse, France, for assembly. Later, the plane would be flown to Hamburg, Germany, to add interior elements such as seats, galleys, and toilets. The coordination of work teams across companies in different countries turned out to be a major challenge for this technically complex manufacturing process. For example, wiring of the 80-yard fuselage required 30,000 cables extending 300 miles hidden behind walls. Wiring snafus led to a six-month delay in initial deliveries. Cultural and political barriers also led to communications problems and further delays. The company experienced two CEO changes in a three-month period. Eventually, the first delivery to Singapore Airlines in October 2007 was two years behind schedule and $6.8 billion over budget.

The Boeing 787 also was plagued by delays. Parts and subsystems for the Dreamliner came from Australia, Canada, France, Japan, India, and the United States, among others. The company experienced technology glitches with the wireless network used to serve in-flight entertainment, as well as a shortage of metal parts fasteners. Sections arrived for final assembly with missing parts. There was a lack of documentation from overseas suppliers and flight guidance software was not ready. Apparently Boeing did not monitor and control its suppliers well enough on this project, which lead to costly delays. Estimates indicate that Boeing may have to pay its airline customers $4 billion in penalties for missed deadlines. On July 8, 2007 (the date coinciding numerically with the model number 787), Boeing had a launch party for the Dreamliner with 15,000 guests hosted by Tom Brokaw. It wanted to have the real aircraft on display but had to make do with a hollow mock-up. By then, it had received about 600 orders. Now the first delivery has been pushed back to late 2009, some 15 months behind schedule.

Discussion Questions

1. Using the framework of Figure 1.1 (see Chapter 1, p. 11), do the Airbus A380 and the Boeing 787 qualify as high-technology products? Justify your position.

2. Draw a road map of the historical evolution of the products from Airbus and Boeing between 1970 and 2005. What insights do you gather from this

road map about the nature of the commercial aircraft industry and the competitive rivalry of the two firms?

3. Both companies experienced severe delays in the development of their next-generation products. What could Airbus and Boeing have done to prevent such delays?

4. Analyzing consumer behavior and other trends in the travel business, which company's vision of the industry's future (Airbus or Boeing) do you think will materialize? Why? How does this affect the firm's positioning strategy?

5. In 1995 Airbus and Boeing decided not to cooperate on the design and manufacture of a next-generation passenger jet. In hindsight, do you think this was the right decision? Justify your position.

Sources: Cole, Jeff, "How Airbus, Boeing Differ on Designing Jumbo Jets," *Wall Street Journal,* November 3, 1999, p. A1; Lunsford, J. Lynn, "Boeing CEO Fights Headwind," *Wall Street Journal,* April 25, 2008, p. B1; Matlack Carol, "The Escalating Woes at Airbus," *Business Week,* March 30, 2006, online at; www.businessweek.com/globalbiz/content/mar2006/gb20060330_075258.htm?chan=search Michaels, Daniel, "For Airbus, Making Huge Jet Requires New Juggling Acts," *Wall Street Journal,* May 27, 2004, p. 1; and Michaels, Daniel, "Airbus, Amid Turmoil, Revives Troubled Plane," *Wall Street Journal,* October 15, 2007, p. 1.

Wikipedia, "Boeing 747," http://en.wikipedia.org/wiki/Boeing_747, accessed May 12, 2008.

Wikipedia, "Airbus 380," http://en.wikipedia.org/wiki/Airbus_a380, accessed May 12, 2008.

Wikipedia, "Airbus," http://en.wikipedia.org/wiki/Airbus, accessed May 12, 2008.

Wikipedia, "Boeing 777," http://en.wikipedia.org/wiki/Boeing_777, accessed May 15, 2008.

Wikipedia. "Boeing 787," http://en.wikipedia.org/wiki/Boeing_787, accessed May 15, 2008.

Wikipedia, "Airbus A350," http://en.wikipedia.org/wiki/Airbus_A350, accessed May 15, 2008.

GOOMZEE MOBILE MARKETING

By Erika Beede, director of marketing, Goomzee Corporation, Missoula, Montana, April 2008.

In March 2008, the market research/analysis company eMarketer reported that of the nearly $2.7 billion spent worldwide on mobile marketing, U.S. spending had reached $878 million; eMarketer predicted that worldwide spending on mobile marketing will grow to $6.5 billion by 2012. Other industry professionals cited U.S. mobile advertising of $1.4 billion in 2006 and forecast that worldwide spending on mobile advertising would range anywhere from $14 billion to $20 billion by 2012.

Although these predictions vary, the indisputable fact is that mobile advertising is growing rapidly; indeed, many of the leading brands are adopting mobile advertising as a new method of communication with consumers. Companies like Coca-Cola, McDonald's, and Google have all implemented mobile advertising in their marketing tactics. A key reason is the explosion in the number of mobile users.

The number of mobile subscribers in the United States reached 250 million in November 2007. Of those subscribers, 58% used text messaging, up from 41% in 2006, and 35% in 2005. Table C.4 shows the spending across various media by U.S. advertisers; Table C.5 shows the percentage of the U.S. population that uses text messaging (SMS).

Given the significant growth in mobile subscribers and consumer demand for instant information, many marketers view the mobile phone as the "third screen" for advertising; the first screen, of

TABLE C.4 Spending in Various Media by U.S. Advertisers			
U.S. Ad Spending*	2006	2007	2011 (forecast)
Overall	$220.00	$283.80	$287.80
Online	$ 16.90	$ 21.40	$ 42.00
Mobile	$ .428	$ .878	$ 5.19
Real estate ad spending	$ 11.00	$ 11.00	$ 11.00

* Billions of dollars

TABLE C.5 Text Message Users as a Percent of U.S. Households	
Age Group	**% Active SMS**
18–24	76%
25–34	58%
35–44	43%
45–54	33%
55–64	22%
65 and older	19%

Source: Forrester Research, December 2006.

course, was the television, and the second screen was the computer, with ads delivered over the Internet. Marketers have experimented using the mobile platform in a variety of ways since the early 2000s, including delivering coupons through text messages, holding contests in which customers use their phones to respond or vote, and providing information over the mobile platform. For example, TV shows such as *American Idol* and *Deal or No Deal* offered voting options through the mobile phone, and people can sign up to receive mobile alerts from ESPN, MySpace, Yahoo! Mail, and MSN Instant Messenger.

Targeting the Real Estate Industry

Although many industries are incorporating text message marketing into their advertising campaigns, the real estate industry has served as a pioneer in this regard. Goomzee is one of a handful of companies that have sprung up to service the growing mobile opportunities in this industry. At its basic level, real estate agents list a text message code on the signs in front of the properties they are selling; possible buyers driving by the property can then enter that text code into their phone and receive an immediate text message back with a description of the property (see Figure C.1). Simultaneously, the listing agent is notified that a prospective buyer has just texted the particular property's code, and he or she can follow up with that potential buyer in a timely fashion. Some competitors in this space include Cell-Signs, House4Cell, DriveBuyTechnologies, OTAir, and RealtyUnwired.

FIGURE C.1 Real Estate Text Message on Mobile Phone

Understanding the Real Estate Market

The real estate industry is composed of brokers and agents who work either individually or out of group offices; they may have joined area Realtor associations and they may participate in multiple listing services (MLSs). In addition to real estate brokers and agents, builders, lenders, homeowner/title insurance agents, and moving companies are also part of the industry fabric. Moreover, real estate properties include land and lots as well as new and used residential, commercial, and condominiums, for sale or lease.

To serve the needs of this industry with a mobile platform, four key questions are salient:

1. What do real estate agents need to do their job well?
2. Who are their consumers?
3. How do they spend their money in terms of marketing?
4. How can mobile marketing help them?

Research shows that nearly three-fourths of consumers are very or extremely interested in being able to get property information while they are standing in front of a home for sale. The average age of the first-time home buyer in the United States in 2007 was 32 years old. Not only do real estate buyers like a large inventory of homes to look at, but they also want to be able to unobtrusively access information independently without having to rely on a real estate agent early in the process (self-service). Buyers demand instant information and, although there may be a flyer box in front of the house, often the box is empty or the flyers are weathered. For the real estate agent, no flyer in front of a home plus increased competition results in a lost potential customer. During a downturn in the market, this is something the agent cannot afford to risk.

Table C.6 shows some useful information about how real estate buyers get their information.

In analyzing a typical real estate transaction and given the economic and regulatory environment in 2008, real estate agents faced a number of challenges in building their sales pipeline and managing cash flows. With the subprime mortgage crisis, "Do Not Call" lists, increased competition from other agents, and online discount brokerages, these challenges continued to stack up. In 2008, the typical sales commission for a property was 6%, which the home seller paid to the listing agent. The seller's agent, or listing agent, often was required to split this commission with the buyer's agent (often a 50-50 split, with 3% going to the buyer's agent). Furthermore, real estate agents paid a percentage to the office they worked for. For each $100,000 of a home's selling price, out of a potential $6,000 commission, $3,000 may go to the buyer's agent and of the remaining $3,000, 5% may go to the broker, leaving the agent $2,850 per $100,000 of a home's selling price. From that, agents must cover expenses such as desk fees, association and MLS dues, marketing expenses, and eventually pay themselves.

Goomzee Connect

Goomzee's mobile advertising and lead generation service, Goomzee Connect, leverages two-way text messaging (SMS) technology. The service enables people to send a text message to a company

TABLE C.6 Sources of Information for Home Buyers, 2006	
Realtor	36%
Internet	24%
Yard sign/house flyer	15%
Friend, neighbor, relative	8%
Home builder	8%
Print newspaper ad	5%
Knew the seller	3%
Home magazine/book	1%

TABLE C.7 Real Estate Ad Revenues for Online/Newspapers/Other Media, 2007 and 2012*		
	2007	**2012**
Newspapers	$ 4,842	$ 3,295
Online	$ 2,582	$ 3,453
All other	$ 4,038	$ 4,458
Total	**$11,462**	**$11,206**

Source: Borrell Associates Inc., 2007.
* in $ hundred thousands; $11,462,000,000.

to express interest in, for example, a house that's for sale. Goomzee calls this "Point of Interest Selling." In this case, the real estate agent listing the house would be able to automatically send information on the house as well as contact information to the person who sent the text inquiry. For every text inquiry, the agent is optionally alerted with buyer contact information and provided lead management tools for follow-up and sale closure.

Goomzee faced several important decisions as it went to market. Product development, including the specific feature set, was a key decision area. These development decisions were also related to pricing considerations. Goomzee considered the following three areas in its pricing decision:

1. Goomzee had to cover its costs of doing business, including the costs of providing the service as well as administrative and overhead costs.

2. Goomzee had to evaluate what competitors were charging for similar mobile marketing (text messaging) services in the real estate industry. In addition to the direct competitors who also offered two-way text message marketing services, traditional advertising venues were a source of indirect competition (flyers, toll-free numbers, AM radio transmitter services,[1] TV and radio ads, newspaper ads, virtual Web-based tours, billboards, and even vehicle wraps). These traditional advertising media could also be considered complementary forms of advertising, with text message marketing used in conjunction with them.

3. Most importantly, Goomzee had to evaluate the real estate agent's budget for marketing. Despite the industry's fluctuations over the past decade, real estate professionals still spent a relatively steady amount overall for advertising, just over $11 billion total. However, where they spend their advertising dollars is expected to change in the next five years, as shown in Table C.7. Newspaper advertising will take a $1.5 billion fall, and a majority of those dollars, about $1 billion, will go into additional online advertising spending, following the trends of many other industries throughout the United States. The rest of their advertising expenditures will fall into the "All other" category, which includes everything from sign riders and flyers to television and radio ads.

During a survey performed by Goomzee, some real estate agents indicated that they would pay between $25 and $100 for a single qualified lead. As Table C.8 shows, Goomzee's pricing structure ultimately was based on the number of text message codes, the period of service, and, if applicable, the number of users (for multiuser plans). It included a three-month trial period in which the agent purchased three codes for $99 for the three-month period. Goomzee also offered a 12-month rate and an extended contract.

Goomzee's own marketing process was based on a familiar sales cycle, categorizing potential agents first as "suspects" (agents who had not yet heard of the Goomzee service), then as "prospects" (who were aware of the service and indicated some interest), and finally, converting them into actual customers and managing the customer relationship over time.

The goal of marketing to suspects, who may be potential customers of text messaging services, is to create awareness of Goomzee's service. Marketing had to identify these people, determine

---

[1] A real estate agent would record a message with information about the house, and then put a radio transmitter inside the house; the yard sign would say "tune your radio to [this station] to hear more about this house." The transmission quality depended upon the distance the potential buyer's car traveled from the house; the farther the distance, the fuzzier the transmission.

TABLE C.8 Goomzee Pricing Structure			

Individual Pricing*

Codes	3 months	12 months**	Per Code/Per Month
3	99.00	249.00	11.00/6.92
5	165.00	349.00	11.00/5.82
10	299.00	499.00	9.97/4.16
20	499.00	799.00	8.31/3.33

* FREE Sign Riders, Setup
** SAVE from 40–60%

Office (Multi-User) Pricing***

Users	Codes	Monthly	Per User
10	25	199.00	19.90
20	50	349.00	17.45
30	100	449.00	14.97

***Includes setup, training, support
***Contact Goomzee for customized plan

their location, and direct them to the Goomzee website for efficient and simple demonstrations, information, and online ordering if they wanted to sign up. Goomzee used the following tools to create awareness:

Press Releases/Publicity: Goomzee sends press releases about its service on a regular basis. Some of the targets for these press relationships include industry-specific contacts in real estate, mobile marketing experts, and local and national media. Print publications, radio, TV, and online news sources are all critical. Appendix C.A at the end of this case includes two examples of such press announcements and the resulting media/press they generated. To effectively cultivate publicity, Goomzee's marketing manager focused on networking with key editors in various media; a "spam" approach to sending out press releases is simply ineffective. The outcome of Goomzee's efforts to cultivate relationships with the press resulted in a number of features in local, regional, and national media.

New Media: The company has a blog that can be seen at http://blog.goomzee.com. It also participates in social networks such as LinkedIn, as well as real estate industry blogs including Active Rain and Inman News. Participation in the wide variety of mobile industry blogs is also helpful, as well as commenting on newsletters such as that sent by Fierce Wireless, the wireless industry's daily monitor.

Trade Shows: Goomzee participates in real estate trade shows and sponsors industry events, gathering information on possible "suspects" at these events. For example, Goomzee exhibited at the Austin Board of Realtors' annual convention in Austin, Texas (the Realty Roundup 2007), and in April 2008 was sponsor of another Austin-based real estate event (Spotlight Austin).

Brand Reputation: Goomzee's efforts to create awareness also focus on creating brand reputation. For example, in order to cultivate an image of corporate social responsibility, Goomzee has donated services to its local Big Brother, Big Sister organization and has sponsored a local author.

Moving customers from "suspect" to "prospect" requires follow-up on leads generated from trade shows, events, and referrals from other agents. In addition, customers who have gone through the online product demonstration also require follow-up. In these follow-up communications, distinguishing prospects' levels of interest is key: Are the agents just not interested now, or are they

FIGURE C.2 Screen Shot of Goomzee Website

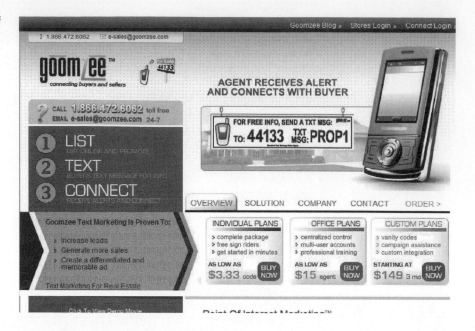

never going to be interested? For all interested prospects, Goomzee keeps some regular communication going, typically through an e-mail newsletter (also known as *e-mail drip marketing*). These e-mail newsletters vary between generic marketing blasts and personalized (rather than mass) messages. Appendix C.B provides samples of these types of communication.

Driving interested prospects to the website is also important, both in terms of effectiveness and efficiency. Figure C.2 shows the company's home page. A series of customer testimonials helps build credibility and interest. Agent testimonials included:

> I've been using your service for over 10 months now. I have over 200 leads, averaging almost a lead per day and no callback issues. Since December '07, I have received more text requests than phone calls from my signs.

> I really appreciate your keeping me informed [of product updates]. It's clear that you want to make your product the best. I think you guys are awesome!

A home seller testimonial stated:

> I wanted to write and tell you all the positive feedback I have been receiving regarding the text-messaging on my house. So many times the flyers on For Sale homes are gone and you just want to know the "411" [basic information] on the house and this is an ideal solution.

For more interested prospects, these follow-up communications must overcome any objections or concerns the real estate agent has about the service. Some agents may require a series of touch points as they navigate the transition from prospect to customer.

Once an agent has signed up for the service, customer relationship marketing strategies are key. These include keeping the agents connected to Goomzee, as well as ongoing advertising to communicate a strong brand reputation. It is also important to ask current clients for introductions and referrals.

Impeccable customer service is crucial to keeping agents connected. This includes providing them regular feedback on features that Goomzee mobile marketing offers. For example, the company's research shows that to get the best results from text messaging, agents need to fully utilize all the service's capabilities. Another key part of effective customer relationship marketing is always asking customers how to improve, and listening genuinely to their responses. For example, clients asked for the following features:

1. They thought it would be valuable to have the ability to "alert" leads with new property information for those properties that they previously texted—for example, price changes or open house dates. So, Goomzee upgraded its technology to include a campaign update notification capability, allowing agents to send a limited number of updates to those customers who had previously requested information about a particular listing.

2. They wanted more interactive administration for managing their Goomzee accounts. New "pop-ups" were implemented, verifying what steps were being taken as they navigated throughout the system. For example, agents who were using a text-code on a property that sold needed to expire the code so it was no longer live. When they choose to expire a code, a pop-up window appears, verifying that they want to expire the existing campaign. This ensures they are not expiring the wrong code. These "pop-ups" appear for many of the actions that are taken to create or edit campaigns.

3. They wanted to simplify management of their campaigns, so Goomzee added searching, filtering, and paging. This feature allows clients to navigate through more complex accounts, containing multiple users and several hundred text codes. For example, clients who manage the multiuser accounts are able to quickly search or filter data much more quickly than if they were to sort through individual records one at a time.

4. They expressed interest in delivering sales lead alerts to more than just themselves. Some agents work on teams, so they wanted other team members to receive the alerts as well as follow up on sales leads. Goomzee now allows multiple e-mail addresses to receive the same sales lead alerts.

5. They wanted to provide photos to consumers texting their properties. After more than six months of careful research and development, Goomzee designed Goomzee Mobile Virtual Tours (patent pending). This feature took months of careful planning and technological design as Goomzee looked to clients, homebuyers, and the future of real estate marketing for information on how to best design this type of service.

As part of listening to and interacting with customers, a company should have a rich, sophisticated view of its customers and the industry in which its customers operate—but it should never presume to know—or act as if it actually knows—more than the customers do!

Goomzee is planning ahead, knowing there will be constant changes in mobile technology, as well as social changes. For example, Goomzee must comply with rules and standards set forth by industry regulators, such as governmental regulations set to protect consumers' privacy. Moreover, specifications are created and enforced by each of the individual mobile carriers before they approve companies, like Goomzee, to provide mobile marketing and advertising to their customers.

Discussion Questions

1. Does it make sense that Goomzee focused on the real estate industry? Why or why not? If not, what target market would you have recommended and why?

2. Would you characterize Goomzee as market oriented? Why or why not? If not, what should Goomzee be doing differently in this regard?

3. Critique Goomzee's pricing strategy.

4. Critique Goomzee's advertising and promotion strategy.

5. Do you think Goomzee will continue to be successful in the future? Why? What additional steps should it take to ensure its ongoing success?

Sources: "Mobile Advertising Spending to Surpass $6.5 Billion in 2012," Marketing Charts, April 2, 2008, online at www.marketingcharts.com/direct/mobile-advertising-spending-to-surpass-65-billion-in-2012-4097; Zeumer, Alexandra, "Worldwide Mobile Ad Spending to Hit $19.1 Billion by 2012," Shaping the Future of the Newspaper, SFN blog, posted March 27, 2008, at www.sfnblog.com/index.php/2008/03/27/1455-emarketer-worldwide-mobile-ad-spending-to-hit-191-billion-by-2012; and Ask, Julie, "U.S. Mobile Marketing Forecast, 2006 to 2011," Jupiter Research, October 20, 2006, online at www.jupiterresearch.com/bin/item.pl/research:vision/1259/id=98111.

APPENDIX C.A   GOOMZEE PRESS RELEASES

PRESS RELEASE #1

FOR IMMEDIATE RELEASE
Goomzee Corporation
Contact: Erika Beede
E-mail: erika@goomzee.com
Cellular: 406-531-3904
Website: www.goomzee.com

Montana-Based Goomzee Offers Realtors and Auto Dealers Innovative Text-Message Marketing Service

MISSOULA, MT February 07, 2007—Connecting buyers and sellers has never been easy, until now. Goomzee Corporation, a software solutions provider, identified and extensively researched two problematic factors for real estate and automotive sales; service differentiation and lead generation. Goomzee Connect has since been designed as an affordable, interactive mobile text-message marketing system. Problem solved. This is the first marketing tool of its kind to be presented to the automotive industry.

Similar to text message voting as seen on today's popular television programs, Goomzee Connect instead delivers instant information about a product or service direct to the consumer's mobile phone. Sellers can monitor and manage inquiries via a user-friendly Web-based campaign management tool.

Mobile marketing researchers predict an explosion for mobile advertising in 2007, given it is relatively inexpensive and it allows businesses, of any size, to reach out to highly targeted consumers. It's a buyer's world, especially for the saturated real estate and automotive industries. As the number of real estate professionals increases, a need for differentiation among one another is presented. Automotive dealers struggle to capture leads during evening and weekend hours and efficiently identify what consumers are looking at. Buyers within these two industries are looking for a no-hassle way to inquire about product features and eliminate pressure from salespeople. Goomzee Connect offers a platform for sellers to synchronize with buyers via an opt-in, text-for-info service 24/7.

Primary pilot customers have been selected in several national markets to round-out the initial system development, identify pricing models and allow for enhanced market adoption once sales begin mid-February.

"Even in the short run we've been trying it, it's worked out really well for me," stated Jeremy Williams, of Windermere Real Estate, in a *Missoulian* interview by reporter Tyler Christensen. Christensen also interviewed another pilot auto dealer customer, Karl Tyler Chevrolet.

"If there's any way to increase car sales in a different marketing light, I'm open to it," declares Dave Peppenger, Internet sales manager for Karl Tyler Chevrolet. "When we first started with eBay, we had no idea what to expect." Before long, Karl Tyler Chevrolet was leading the state in Internet sales.

For more information about Goomzee, visit www.goomzee.com.

Press Coverage: Front page *Missoulian,* Clear Channel Public Radio broadcast with Pete Denault, Montana Associated Technology Roundtable (MATR.net)

PRESS RELEASE #2

FOR IMMEDIATE RELEASE
Goomzee Corporation
Contact: Erika Beede
E-mail: erika@goomzee.com
Office: 406-542-9955
Website: www.goomzee.com

Iron Horse Properties
Contact: Melissa Meuter
E-mail: mmeuter@ironhorsemt.com
Office: 406-863-3060
Website: www.ironhorseproperties.com

Flathead and Lake County Boaters Can Request Property Information from the Water

WHITEFISH, MT July 11, 2007—Expect to see text-for-info signs popping up throughout the valley in the near future. Soon boaters will be able to receive property information to their cell phones, right from the water. Iron Horse Properties Sotheby's International Realty, the Exclusive Affiliate of Sotheby's International Realty for Flathead and Lake Counties, agreed last week to use Goomzee Connect for their Bigfork and Whitefish offices.

"Last weekend, while boating on Flathead with my parents, we noticed a lot of for sale properties. The only problem was that we didn't want to hassle with getting out of the boat to grab a flyer, if there was even one there. I'm pleased that Iron Horse Properties Sotheby's International Realty has recognized the value in Goomzee's service and it's helping us deliver the latest technology to homebuyers," notes Goomzee's Director of Marketing, Erika Beede. Pat Donovan, Broker Owner of Iron Horse Properties Sotheby's International Realty, adds "We are excited to offer this new technology, making information as convenient and accessible for our clients as possible."

Goomzee Connect delivers convenient, 24-hour access to information using cell phone technology. The service is currently being used by real estate professionals in seven states across America and now Goomzee is helping connect buyers and sellers on land and water.

About Goomzee   Based in the Northern Rocky Mountains, Goomzee Corporation has been connecting buyers and sellers through innovative technology solutions since 2003. Goomzee provides superior customer service to clients worldwide through its software products and professional consulting services. For more information about Goomzee Connect, visit www.goomzee.com.

About Iron Horse Properties Sotheby's International Realty

For more information about Iron Horse Properties Sotheby's International Realty, visit www.ironhorseproperties.com.

Press Coverage: *Flathead Business Journal,* Montana Associated Technology Roundtable (MATR.net)

PRESS RELEASE #3

FOR IMMEDIATE RELEASE
Goomzee Corporation
Contact: Erika Beede
E-mail: erika@goomzee.com
Office: 406-542-9955
Cellular: 406-531-3904
Website: www.goomzee.com

Goomzee Positions to Meet Demands of Mobile Expansion

MISSOULA, MT April 29, 2008—In their commitment to meet increased consumer demand for mobile access to information, real estate mobile marketing company Goomzee announced today the release of its 2nd generation text message marketing platform, Goomzee Connect 2.0. The company is also expanding into a new office and adding team members to support increased demand for their services.

Living in an always on, always connected world, consumers expect to have a plethora of information at their fingertips at the click of a button. To accommodate those expectations, Goomzee enhanced its software solution. Over nine months and countless man hours went into a complete re-engineering of their web based campaign management and two-way text messaging system, geared to meet current and future needs of buyers and sellers. "We pulled out all the stops. We gave our engineers in the U.S. and Europe free rein to architect and build a best-of-breed software system to support client needs, from individuals to the largest global enterprises," states Goomzee CEO and Founder, Mike Sparr. "I am pleased to say that all expectations were exceeded, and our clients recognize that value."

Most of the 2.0 enhancements improve efficiency for sellers. Users can easily navigate the system to manage their campaigns, even for the most technologically challenged individuals. To facilitate enterprise needs, Goomzee enhanced the user ability to manage large amounts of data with live search, filtering, and pagination updates. The system also accommodates multi-user accounts; teams, offices, and associations for maximum flexibility.

Facilitating the live exchange between buyer and seller has been a major focus of development of Goomzee Connect 2.0. Agents can send updates or communicate with leads via text message, e-mail and Skype, all via the web admin. Agents can now receive alert notifications on multiple e-mail addresses, supporting sales teams.

The system is built upon the same frameworks as *Fortune* 500 companies' enterprise software systems and many features boast more advanced technology. Goomzee envisions global opportunities for its innovative marketing platform and has invested heavily in building a scalable, enterprise solution. The software can be translated into any language for worldwide deployment.

"We work closely with customers, observe how they use our technology, and develop enhancements based upon their feedback and our own extensive research. Our strong ties in the mobile industry, both in the U.S. and abroad, give our product team valuable insight. This allows us to offer innovative solutions to clients and keep them ahead of the curve. Our hosted software is updated regularly and new features are added monthly, so clients benefit from continual enhancements at no extra cost," adds Sparr.

Goomzee plans to add 3 new employees in their new office to handle the addition of new clients, features, and service with room for expanding their team as the business continues to grow.

About Goomzee Connect   Goomzee Connect is a mobile marketing service that provides instant delivery of product and service information to mobile phones via text (SMS) messaging. With today's dynamic and on-the-go lifestyles, Goomzee provides the latest technology, allowing sellers to meet these growing consumer demands. For more information about Goomzee, visit www.goomzee.com.

Press Coverage: RISMedia, Montana Associated Technology Roundtable (MATR.net)

APPENDIX C.B GOOMZEE E-MAIL MARKETING

E-Mail #1: Personalized

(e-mail sent out to those who registered to view the product demo on Goomzee website)

Hi contact_first_name,

I wanted to take a moment to check in with you, after taking some time to view the Demo of Goomzee Connect at our website www.goomzee.com. I hope you found the information about our text marketing service, which we provided in the short video, helpful as well as answered some of the questions you may have had. If any additional questions have risen since, please don't hesitate to contact me at any time; I would be more than happy to address them.

Goomzee Connect is a great way to create interactive marketing through our text message platform and provide the immediate feedback that some of your consumers are looking for. By connecting you to their mobile number, you can follow up with them to answer any additional questions they may have and begin a dialog. We can help you attract consumers so you may create a customer.

If you are interested in learning more, I would be happy to provide you with information. If you are interested in getting started with a plan, I can help you find a package that best fits your business needs.

We appreciate your interest and hope to hear from you soon.

Sincerely,

E-Mail #2: Drip

(e-mail sent to all leads in our pipeline)

Elevate Real Estate in 2008!

As 2008 continues to fly by, so do your potential leads. Create an *interactive* relationship with your customers; when they drive by your properties, provide them with the ability to get connected with you—instantly!

With Goomzee Connect, you can list your properties with the latest mobile phone technology available to consumers. Goomzee was designed specifically for real estate professionals to fulfill their clients' growing need for instant information. Provide property details as well as your contact information in the form of a text message and your customers walk away with your information—stored in their phone! Receive alerts containing their mobile phone number so you can follow-up with them when the time is most convenient to you.

Sign up before March 15th, 2008, for any of our annual plans, and receive a FREE upgrade to our Mobile Virtual Tour service—coming soon!

Forrester Research reported the following breakdown for text message use in the U.S.: Do any of these age groups fall into your customer base?

Active SMS Users as % of U.S. Households

Age Group	Active SMS User %
18–24	76%
25–34	58%
35–44	43%
45–54	33%
55–64	22%
65 and over	19%

Visit **Goomzee.com** today and you'll find agents just like you, experiencing great results from Goomzee Connect. **Get leads, gain customers, and sell properties with Goomzee Connect.**

Questions? Please feel free to contact me directly. We look forward to assisting you in growing your real estate business!

"Nothing capitalizes on the capabilities of mobile like retail. Shoppers are mobile. The shopping process is often spontaneous. The mobile phone is always on and always available. The mobile phone is an ideal platform for retailers to capture the moment and provide utility, value, and even fun, while driving foot traffic and drive sales."—Mike Baker, VP and head of Nokia Ad Business

---

SELCO–INDIA: LIGHTING THE BASE OF THE PYRAMID

There are more than 2 billion people in the world who currently lack reliable access to electricity. About 46% of households in India do not have access to the power grid, and for many others the supply is unreliable. For those people, smoky, dangerous oil or kerosene lanterns are the primary light source. However, tremendous advances in photovoltaic technology have made provision of electricity in a distributed fashion possible—one household and business at a time. Quite simply, solar voltaic devices, such as the SELCO panel shown in the photo, capture and convert the energy in sunlight into electricity.

Photo reprinted with permission of SELCO Solar Light (P) Ltd., Bangalore, India.

The Solar Electric Lighting Company (SELCO) was founded in 1995 by Harish Hande, the current managing director, and Neville Williams to sell and service solar electric home lighting systems in areas of India lacking access to reliable electricity. SELCO's customers range from poor daily-wage laborers to institutions like schools and seminaries. All buy solar panels at the same rate: about $450 for a 35-watt system that can light several 7-watt fluorescent bulbs for four hours between charges. Emphasis is placed on designing solar home systems (SHS) to meet customers' needs and budgets, whether for home or business. Depending on the customer's requirements, the system can be used for inexpensive lighting, water pumping, water heating, communications, computing, and entertainment. These solar systems reduce dependence on other forms of energy and are more reliable, affordable, and safer to use than the other energy sources used in these rural locations.

The SELCO Business Model

SELCO has refined and articulated a business model that delivers an energy service to its ultimate customer while, at every step along the way, providing for a high-quality, highly recognizable,

Photo reprinted with permission of SELCO Solar Light
(P) Ltd., Bangalore, India.

need-based product. SELCO's personnel understand that they must provide not only needed solar energy services to their customers, but also the information, installation, training, financing, and other products and services necessary to develop a sustainable sales and service infrastructure.

SELCO assembles solar home "kits" using components produced exclusively for them by Indian manufacturers. A typical kit consists of 35-watt panel, four 7-watt compact fluorescent lights and a lead-acid battery that can store sufficient energy for the system to work both day and night throughout the year. The batteries are designed to withstand significant discharge each day without rapid deterioration. Cheaper car batteries cannot withstand this, and would become unusable within about 6 months. An electronic charge controller protects the battery from charging or discharging too much, and enables the battery to be used for at least 5 years.

SELCO sets stringent quality specifications, which are passed on to the customer as performance guarantees. The solar photovoltaic modules and batteries are purchased from external suppliers. SELCO initially had problems with the quality of the compact fluorescent lights (CFLs), so it set up a sister business to manufacture these, as well as the charge controllers.

SELCO created a sales and service infrastructure that provides a one-stop energy shop for poor families and businesses seeking reliable energy services. The company owns and operates a decentralized network of 25 sales and service centers. SELCO manages its service centers from the headquarters in Bangalore, India. Strong local management teams can quickly and efficiently respond to changing customer needs, market conditions, local partners, and all other aspects of day-to-day operations. The strong local presence and understanding of customers allows SELCO to effectively tailor systems and services to best meet their needs. These local agents are the initial point of contact for up to 70% of customers; they receive a percentage commission for each SHS sale they initiate. The sales agents are responsible for promoting the business, visiting potential customers, designing systems, and taking payment.

The installation of the system is then carried out by SELCO technicians who are themselves part of the value proposition, coming up with their own innovative ideas—for example, placing a light in the corner of one room and removing bricks to let the light into other rooms, so that a single light provides background illumination in three rooms. Another feature that allows flexibility at low cost is having lights that can be moved from one place to another.

SELCO helps its customers finance their purchases by partnering with rural banks, leasing companies, and microlending organizations to provide the necessary credit.

With its unique combination of product, service, and finance, SELCO is able to offer superior lighting and electricity at a monthly price comparable to using traditional, less effective sources.

Photo reprinted with permission of SELCO Solar Light (P) Ltd., Bangalore, India.

Combining these three elements creates a virtuous cycle. Using high-quality products reduces the cost of ongoing service and maintenance.

In 2007, SELCO operated in the Indian states of Karnataka, Kerala, and Andhra Pradesh. SELCO says it will move to a new region only if it can obtain good contacts for disseminating information and providing finance.

Success Stories

People's lives and livelihoods have been transformed through SELCO solar lights. The immediate benefit to users is the provision of clean, good-quality light, and power for small appliances. Good light improves morale and opportunities in ways that are difficult to quantify. Children are able to study (or, as one homeowner said, "They have no excuse for not studying!"), domestic tasks are done more safely and easily, and there are increased opportunities for income generation. For example, the photo above shows a man using the solar lights to grow silk worms. For vendors, produce from stalls is displayed better, and they do not have to work with the smell and heat from kerosene lamps. Through its solar electric lighting systems, SELCO provides families with the power to change their own lives.

For example, the rose pickers of one village outside Bangalore, India, typically got up before the sun, grabbed a basket with one hand and a lamp with the other, and hurried to the fields so they could bring their wares to market in time for the dawn crowds. These rose pickers were prime candidates for solar-powered headlamps. SELCO partnered with local banks to help the workers get loans to buy them. Wearing the charged lamps in the predawn darkness, the pickers could work with both hands; they doubled their productivity, boosted their take-home pay, and generated enough income to start paying down the headlamp loans.

Mallika, a mother of three, decided to capitalize on the location of the family's house (by the side of a main road) to sell fruits from a small kiosk. Started in 2002, the business grew with time and became a popular tea spot among regular travelers like bus drivers and small delivery vans. This favorable situation led Mallika to expand into coconuts, short eats, tea, sweet meats, and other items, which made the kiosk even more popular.

Mallika opened the kiosk at daybreak but had to close at dusk because it was difficult to stay open later with only the dim light of the kerosene lamp. Mallika decided to find a solution and was introduced to solar lighting by a SELCO representative. Mallika had a solar home system (SHS) installed, purchased under a credit arrangement by SELCO Lanka, and her business started to flourish.

The youngest daughter of the family, Sanidi Ruwanthika, has this to say about the present situation: "After we built our new house, we installed a SELCO SHS. Now, unlike the past, we keep the kiosk open until late night. Our tea kiosk became a popular halting place for lorries and vans plying the road at night. In addition, vans going on long distance trips also started to stop to buy fruits. Since there are no other fruit stalls nearby open in the night, our business grew day by day."

Many of the benefits of SELCO's products are particularly significant for women: They often spend more time in the home and therefore appreciate the improved light and income-generation opportunities. Many women take the responsibility of paying for the SHS and, through this, gain confidence in financial management.

To date, SELCO has:

- Installed more than 75,000 solar systems in rural areas
- Provided lighting and electricity to more than 300,000 people
- Achieved profitability on annual sales of $3 million

To maintain this success, SELCO continues with promotional activities including:

- Running awareness campaigns in rural areas to demonstrate solar home systems
- Identifying potential beneficiaries
- Training local technicians, installers, and service personnel
- Educating users

However, there are some challenges on SELCO's horizon. A worldwide increase in demand for solar gear has led to cost increases. This, coupled with an inability to raise prices, threatens SELCO's business model. SELCO would also like to triple its number of installations by 2010.

Discussion Questions

1. Explain the innovation adoption process for SELCO systems in terms of the six factors that affect customers's technology adoption decisions (Chapter 7).

2. Of the various models of social entrepreneurship in Chapter 13, which model does SELCO follow? Which of the challenges and success factors for base-of-the-pyramid strategies affect SELCO? Be specific in your elaboration.

3. What role has the marketing function played in SELCO's success? What suggestions do you have to make their marketing more effective?

4. SELCO has been approached by people who would like a franchise to sell SELCO's systems. Should SELCO sell franchises? Why or why not?

5. Is SELCO's business model sustainable long-term? What are the threats to the business model and how can SELCO prepare for these?

6. Is the SELCO business model sufficiently scalable to support its growth goals? If not, what should it do?

7. What proactive actions can help SELCO cross the chasm in the Indian market?

Sources: "About Us," SELCO website, at www.selco-india.com/about_us.html; "SELCO India Links Energy Services to Better Quality of Life," Center for Business as an Agent of World Benefit, Case Western Reserve University, online at http://worldbenefit.case.edu/innovation/bankInnovation View.cfm?idArchive=411; Sen, Snigdha, "Lighting Up Rural India," *Business 2.0,* December 1, 2006, online at http://money.cnn.com/magazines/business2/business2_archive/2006/12/01/ 8394996/index.htm; and "Making a Business from Solar Home Systems," The Ashden Awards for Sustainable Energy, 2005, online at www.ashdenawards.org/files/reports/SELCO%202005%20Technical%20report.pdf.

AUTHOR INDEX

SUBJECT INDEX